B

Foto: H. Gericke, Freiburg

H. KNESER (left) and W. SÜSS (right) discussing organisation matters in 1957. Legend has it that in the 1950s SÜSS brazenly exploited an opportunity to approach KONRAD ADENAUER, first Chancellor of the Federal Republic of Germany, on one of his walks during a vacation at Bühler Höhe (Black Forest) and asked him bluntly for financial support. As a matter of fact, there exist bills in the files of the Institute showing that support was received from the office of the Federal Chancellor (Bundeskanzleramt).

Perspectives
in Mathematics

Anniversary of Oberwolfach 1984

Edited by

W. Jäger
J. Moser
R. Remmert

Birkhäuser Verlag
Basel · Boston · Stuttgart

Library of Congress Cataloging in Publication Data

Main entry under title:

Perspectives in mathematics.

   1. Mathematics –– Addresses, essays, lectures.
2. Mathematisches Forschungsinstitut Oberwolfach ––
Adresses, essays, lectures.   I. Jäger, W. (Willi),
1940 —   .   II. Moser, Jürgen, 1928 —   .
III. Remmert, Reinhold.
QA7. P47 1984   510   84—16922
ISBN 3-7643-1624-1

CIP-Kurztitelaufnahme der Deutschen Bibliothek

Perspectives in mathematics : anniversary of Oberwol-
fach 1984 / ed. by W. Jäger ... – Basel ; Boston ;
Stuttgart : Birkhäuser, 1984.
   ISBN 3-7643-1624-1

NE: Jäger, Willi [Hrsg.]

© 1984 Birkhäuser Verlag Basel
Printed in Germany
ISBN 3-7643-1624-1

Grußwort

Das Jubiläum des Mathematischen Forschungsinstituts Oberwolfach gibt uns Anlaß zu einem herzlichen Glückwunsch an die „Genossenschaft der Mathematiker", die diese wissenschaftliche Forschungsstätte trägt und mit Leben erfüllt.

Mehr als zweitausend Mathematiker aus fast allen Ländern der Erde kommen in jedem Jahr zu wissenschaftlicher Begegnung mit Kollegen der Fachwissenschaft und der Nachbardisziplinen in dieses Tusculum der Mathematik, das in einem der noch stillen Täler des schönen Schwarzwaldes liegt. Dieser Jubiläumsband berichtet von der geschichtlichen Entwicklung von Oberwolfach zu einer internationalen Forschungsstätte und gibt Zeugnis von der Breite und Tiefe der wissenschaftlichen Arbeit, die hier geleistet wird. Die Erträge des wissenschaftlichen Gesprächs bei den Tagungen und der stillen Arbeit in der gepflegten Fachbibliothek in Oberwolfach wirken in aller Welt, wo Mathematiker in der Wissenschaft tätig sind.

Vier Jahrzehnte sind jetzt schon vergangen, seit – in schwerer Kriegszeit – die Mathematik auf dem Lorenzenhof in Oberwolfach eine Heimat gefunden hat. „Wir sehen in die Zukunft wie gegen eine schwarze Wand", wird als Ausspruch eines der Gründer des Instituts im Herbst 1944 überliefert. Aber „man arbeitete an etwas, das den Krieg überdauern sollte, man baute an einer Festung des Geistes, die den erwarteten Stürmen standhalten sollte", so hat es IRMGARD SÜSS, die Frau des um die Gründung des Instituts verdienten Freiburger Mathematik-Ordinarius WILHELM SÜSS, für die Annalen von Oberwolfach festgehalten.

Zu einer „Festung des Geistes" ist Oberwolfach in der Zwischenzeit wahrhaft geworden. Im Jahre 1959 ist die Gesellschaft für mathematische Forschung e. V. als Rechtsträger des Instituts eingetreten. Die glückliche Organisationsstruktur dieser Rechtsträgerschaft und das hohe Engagement der Wissenschaftler, die hier Verantwortung tragen, haben die wissenschaftlichen Erfolge des Forschungsinstituts ermöglicht. Das Land Baden-Württemberg ist stolz darauf, dieser mathematischen Forschungsstätte von internationalem Rang die Heimat bieten zu können.

Die Landesregierung von Baden-Württemberg ist sich auch der Verantwortung für das Mathematische Forschungsinstitut Oberwolfach bewußt und wird wie in der Vergangenheit, so auch in der Zukunft, die wirtschaftlichen Grundlagen für die hier geleistete wissenschaftliche Arbeit gewährleisten.

Ad multos annos!

Stuttgart, im Juli 1984

Professor Dr. HELMUT ENGLER
Minister für Wissenschaft und Kunst
des Landes Baden-Württemberg

Zum Geleit

Das Mathematische Forschungsinstitut in Oberwolfach feiert Geburtstag. Mathematiker aus aller Welt senden Glückwünsche. Mit diesem Band reihen wir uns ein in den Kreis der Gratulanten. Wir haben versucht, Autoren zu gewinnen, die an ausgewählten Problemen über den heutigen Stand der Mathematik berichten und Perspektiven für künftige Entwicklungen geben. Es ist auch ein Band der beglückten Rückschau.

Die Auswahl der Gebiete konnte nicht vollständig sein; wir hoffen indessen, daß die behandelten Fragen ein breites Gesamtbild geben. Unser Dank gilt zunächst denen, die ihre Beiträge termingerecht einreichten, und darüber hinaus allen, die uns ermutigten und bei den Vorarbeiten halfen. In Gesprächen mit Herrn Klaus Peters entstand vor Jahren die Idee zu diesem Buch; beim Birkhäuser Verlag, dem Lorenzenhof von Anbeginn verbunden, lag die Herstellung in bewährten Händen.

Das Institut in Oberwolfach hat in den letzten Jahrzehnten die Entwicklung unserer Wissenschaft im In- und Ausland ganz wesentlich mitbestimmt. Im Krieg gegründet, war es zunächst ein Refugium der Mathematik. In den fünfziger Jahren prägte der erste Hausherr, W. Süss, die einmalige Atmosphäre des Hauses, die heute weltweit gerühmt wird. H. Kneser und Th. Schneider setzten seine Arbeit fort; unter M. Barner, der das Institut seit 1963 leitet, wurde Oberwolfach für Mathematiker aus der ganzen Welt, aus Ost und West, aus Nord und Süd, zu einem Begriff. Hier werden Ideen ausgetauscht, hier werden Freundschaften geschlossen, hier entstehen mathematische Arbeiten, die in den besten Fachzeitschriften veröffentlicht werden. Die unbürokratische und lautlose Art, in der Herr Barner selbstlos die Geschicke Oberwolfachs in guten und schweren Zeiten lenkt, ist beispielhaft. Ihm, seinen Mitarbeitern und Helfern im Hause zu danken, ist uns ein Herzensanliegen.

Wir Mathematiker schätzen uns glücklich, im Lorenzenhof eine Stätte der Begegnung zu besitzen. Mathematiker können auf vieles verzichten, auf Oberwolfach wohl kaum. Möge es gelingen, diese Insel mit ihrer wissenschaftlichen Lebendigkeit und ihrer harmonischen Geselligkeit zu erhalten. Wir hoffen, daß die äußeren Bedingungen, unter denen das Institut weiterhin gedeihen kann, auch in Zukunft bestehen werden und daß es der jungen Generation möglich sein wird, den Geist von Oberwolfach zu wahren.

Oberwolfach, im Mai 1984

W. Jäger   J. Moser   R. Remmert

Preface

The Mathematical Research Institute in Oberwolfach is celebrating a birthday. Mathematicians from all over the world are sending congratulations. With this volume, we join the circle of well-wishers. We have tried to obtain authors who, in selected problems, report on the state of mathematics today and give directions for future development. This is also a volume of affectionate reminiscence.

The choice of topics could not be complete; we hope, however, that the questions treated provide a broad overall picture. Our thanks go first of all to those who encouraged us and helped us with the preliminary work. The idea for this book was born years ago in conversations with KLAUS PETERS. At Birkhäuser Verlag, connected with Oberwolfach from the very beginning, its production was in trustworthy hands.

The Institute in Oberwolfach has, in recent decades, exercised a fundamental influence on the development of our science both at home and abroad. Founded during the War, it was primarily a refuge for mathematics. During the 1950s, the first director, W. SÜSS, set the unique atmosphere of the house which is today renowned around the world. H. KNESER and TH. SCHNEIDER carried his work on under M. BARNER, who has directed the Institute since 1963, Oberwolfach has come to be a concept for mathematicians from all over the world; from East and West, North and South. It is here that ideas are exchanged, friendships are made, and mathematical works are initiated to be published in the leading journals. The unbureaucratic and unobtrusive manner in which Professor BARNER guides the destinies of Oberwolfach in good and bad times is exemplary. Our thanks to him, to his associates and assistants in the house come from the heart.

We mathematicians consider ourselves fortunate to have the Lorenzenhof as a meeting place where important international contacts are established. Mathematicians can go without many things, but not without Oberwolfach. May it prove possible to maintain this island of intellectual vivacity and harmonious fellowship. We hope that the external conditions under which the Institute can continue to flourish will also exist in the future and that the young generation will be able to preserve the spirit of Oberwolfach.

Oberwolfach, May 1984

W. JÄGER   J. MOSER   R. REMMERT

Foto: Irmgard Süss, Freiburg

WILHELM SÜSS, born March 7, 1895, died May 21, 1958;
Founder and first Director of the Institute.

HELLMUTH KNESER, born April 16, 1898, died August 23, 1973;
Director of the Institute 1958–1959.

Foto: Archiv Oberwolfach

THEODOR SCHNEIDER, born May 7, 1911; Director of the Institute 1959–1963.

Foto: G. Fischer, Düsseldorf

MARTIN BARNER, born April 19, 1921; Director of the Institute since 1963.

Foto: Archiv Oberwolfach

The old house, first home of the Institute.

Charles Ehresmann   Besuch auf dem Lorenzenhof 25-27 April 1946

Heinz Hopf   10. – 18. August 1946.

Abschiedsabend Heinz Hopf 17. August 1946:

Hubert Cremer   Hellmuth Kneser   Jürgen Rueger
Helmut Hasse   Ernst Witzel   Emmanuel
Henry Görtler   Hans-Heinrich Ostmann
Hermann Boerner.   Günnig Grüder
W. Threlfall   Hans Heubach
                 Gerrit Bol

Hildegard Süss   Wilhelm Süss.
Klaus Dederich.   Use-Hertha Löllner.
Walter Stakowski   Maria Bertling

        Henri   Cartan,   1-2 Nov. 1946

"Toute âme, en puissance, est divine.
Notre but est de manifester le Divin qui est en nous, en maîtrisant la nature
extérieure et intérieure.
Parvenons-y par le travail, par l'adoration, par la maîtrise de l'esprit ou par
la philosophie, par l'une ou plusieurs de ces voies ou par toutes, et soyons libre.
C'est là toute la religion. Les doctrines, les dogmes, les rites, les livres, les temples
et les formes ne sont que des détails secondaires."
        "L'Evangile Universel de Vivekananda", Romain Rolland
Ne cherche-t'on pas à vivre ce "Rajayoga" à Lorenzenhof?
        Christian & Christiane Pauc, 7-13 August 194.

    Eduard Stiefel 23-28.9.47.
    H.T. act igh 1.-3.10. 47
J. Dieudonné   7 – 13/8/1949

First page of the first guest book of the Institute.

Zwei kurze Kapuzinerpredigten.

1.

Stolpernd über Stock und Stein
drang ich in den Schwarzwald ein.
Aber da stieß ich auf ein Gebiet,
Wo man vor Variablen die Bäume nicht sieht.
Schon wollt' ich verzagen,
Doch mußt' ich mir sagen,
Was hilft schon das Klagen:
Im Schwarzwald ist's schwarz und im Münster
ist's finster:
Drum fasse Mut und fürwahr – dich – mich !
Schon schimmert's Helsinki – Zürcherisch,
Und ich gewahrte zu meinem Vergnügen
(Und müßte lügen, wollt' ich es nügen)
Daß doch noch Pfleger* das Funktionenfeld
pflegen,
Die sich mit dem Einspann, wenigstens quasi,
begnügen.

[*) Pfleger = Plural von Pfleger = Pfleger
+ Kinji + ...]

2.

Summoned by Siessen's cable,
we came through sunshine and Nebel,
gathered under Lorenzhof's gable,
we made it a tower of Babel,
a napkin-ring was our label,
with food was laden the table
as plentiful as unser Schnäbel
of soul, body and mind were able
to swallow. The blackboards were stable
enough for truth and for fable
put on it by talk and by chalk.
Now grateful home we walk.

Mit herzlichem Dank für genossene Gastfreund-
schaft
          25. Okt. 1951   Hermann Weyl
                                 Joachim Weyl
                                 Martha B Weyl

"Kapuzinerpredigten" by Hermann Weyl, pages 25–26 of first guest book.

Wir wandelten mit viel Pläsier
im R_{2n}, meist im R_4.
Gar kühn durchfliegend diesen Raum
bemerkten wir den Schwarzwald kaum;
denn dieser liegt so nebenbei
bescheiden gänzlich im R_3.
Doch stolz bekennt der Philosoph,
daß einzig der Lorenzenhof
für die gelehrte Ritterschar
die fundamentale Basis war.
Mit herzlichem Dank!
27.X.51 Hubert Cremer

Page 27 of first guest book.

The old house, with the guest building under construction.

Foto: G. Fischer, Düsseldorf

Present buildings of the Institute.

Foto: G. Fischer, Düsseldorf

The guest building when completed.

Foto: G. Fischer, Düsseldorf

Bird's eye view of the Institute and its grounds.

Contents

Perspectives in Mathematics
Anniversary of Oberwolfach 1984
© Birkhäuser Verlag, Basel

Das Mathematische Forschungsinstitut Oberwolfach

H. GERICKE

Sonnenbergstraße 31, D-78 Freiburg (i. Br.) (FRG)

Nach einem Bericht von W. Süss aus dem Jahre 1953,
zusammengestellt und ergänzt von H. Gericke (Freiburg i. Br.)

Wilhelm Süss, der das Mathematische Forschungsinstitut in Oberwolfach 1944 gegründet und bis zu seinem Tod 1958 geleitet hat, verfaßte 1953 einen
umfassenden Bericht, der wahrscheinlich für die Ministerien und Instanzen
gedacht war, von denen das Institut Hilfe und Förderung erhalten konnte.
Stücke aus diesem Bericht werden mit „Süss" und der Seitenzahl (in römischen
Ziffern wie in der Vorlage) zitiert; der vollständige Bericht kann in der Bibliothek des Instituts eingesehen werden. Eigene Anmerkungen stehen in spitzen
Klammern ⟨ ⟩.

„I. Vorgeschichte des Instituts

Nur wenige Fächer besitzen in Deutschland zentrale Forschungsinstitu-
te. Seit der Gründung der KWI ⟨Kaiser-Wilhelm-Institute, jetzt: Max Planck-
Institute⟩ zur Förderung der Wissenschaften sind hauptsächlich die Gebiete der
Physik, Chemie, Biologie sowie Teilgebiete der Medizin mit solchen Instituten
ausgestattet. Für gewisse staatliche und allgemeine Aufgaben waren ferner die
PHYSIKALISCH-TECHNISCHE REICHSANSTALT und die CHE-
MISCH-TECHNISCHE REICHSANSTALT errichtet worden. Der Vorstand
der DEUTSCHEN MATHEMATIKER-VEREINIGUNG hatte deshalb
schon in den Jahren vor dem Krieg den Wunsch, auch für die Mathematik in
Deutschland ein Institut zu gewinnen, das wesentlich nur in der Forschung tätig
sein sollte und für die Mathematik diejenigen Aufgaben zu übernehmen hätte,
die von den Instituten der einzelnen Universitäten und Technischen Hochschulen üblicher Weise nicht durchgeführt werden können." (Süss, S. I.)

1944 war eine Situation eingetreten, die die Gründung eines solchen
Instituts begünstigte. Süss schreibt (S. II):

„Durch das Kriegsgeschehen waren zahlreiche Hochschulinstitute nicht
mehr arbeitsfähig, so daß die dort befindlichen Fachkräfte brach lagen. Zugleich

zeigte sich, daß viele Aufgaben in gemeinsamer Arbeit rascher und besser bewältigt werden konnten, als bei den räumlich weiten Entfernungen zwischen den verschiedenen Mitarbeitern an derselben Fragestellung.

Günstige Umstände boten schließlich die Möglichkeit, eine erste notdürftige Unterkunft für den Keim des gedachten Instituts in dem verhältnismäßig ruhig und sicher gelegenen Lorenzenhof in Oberwolfach zu finden."

Vom damaligen Reichsforschungsrat wurde W. Süss ⟨1944⟩ beauftragt, ein „Mathematisches Reichsinstitut" in der genannten Unterkunft zu errichten. Einzelheiten hat Frau Irmgard Süss in einer Schrift „Entstehung des Mathematischen Forschungsinstituts" berichtet*).

Süss, S. II: „Als hauptsächliche Arbeitsmittel dienten Teile gefährdeter Bibliotheken, sowie einige Rechenmaschinen und andere Instrumente von Universitätsinstituten, die nach Oberwolfach verlagert wurden. Einige Zeitschriften-Serien konnte das Institut damals noch kaufen. Als Mitarbeiter wurden eine Reihe von Professoren mit ihren jüngeren Mitarbeitern aus bombenbeschädigten Instituten gewonnen. Bei Kriegsende waren die Mitarbeiter des Leiters die folgenden:

	damals	*heute* ⟨1953⟩
Behnke, Prof. Dr. H.	Universität Münster i. W.	Münster i. W.
Boerner, Prof. Dr. H.	Universität München	Gießen
Bol, Prof. Dr. G.	Universität Greifswald	Freiburg i. Br.
Görtler, Prof. Dr. H.	Universität Freiburg	Freiburg i. Br.
ter Hell, Assistent Dr. H.	Universität Freiburg	Schwand/Baden
Maak, Dozent Dr. W.	Universität Hamburg	München (Prof.)
Pisot, Dr. Ch.	Universität Straßburg	Bordeaux (Prof.)
Roger, Prof. Dr. F.	Universität Bordeaux	
– aus Kriegsgefangen-		
schaft beurlaubt –		
Schneider, Dozent Dr. Th.	Universität Göttingen	Göttingen (Prof.)
Schubart, Assistent Dr. H.	Universität Freiburg	Karlsruhe
Seifert, Prof. Dr. H.	Universität Heidelberg	Heidelberg
Sperner, Prof. Dr. E.	Universität Straßburg	Bonn
Threlfall, Prof. Dr. W.	Universität Frankfurt a. M.	verstorben

und 5 wissenschaftliche Hilfskräfte.

Im Institut selbst wurde damals ganz allein mathematische Grundlagenforschung getrieben, während die gesamte mathematische Zweckforschung an

---

*) 1967, Zur Einweihung des Neubaues. Auch abgedruckt in „General Inequalities 2, the Proceedings of the Second International Conference on General Inequalities held at Oberwolfach in 1978". Dieser Band enthält auch mehrere Bilder von Oberwolfach.

entsprechende Spezialisten in arbeitsfähigen Hochschulinstituten vermittelt wurde."

Aus der Reihe der 33 (von W. Süss vollständig aufgeführten) vom Institut organisierten Forschungsaufträge, aus denen auch einige Buch- und Zeitschriften-Veröffentlichungen der Nachkriegszeit hervorgegangen sind, seien hier nur die folgenden genannt:

Erhard Schmidt, Berlin, und G. Feigl, Breslau: Entwicklungen nach reellen Funktionen.

H. Behnke und A. Kratzer, Münster: Funktionentheoretische Grundlagen der modernen Analysis nebst Anwendungen.

R. König, Jena, und K. H. Weise, Kiel: Mathematische Grundlagen der höheren Geodäsie.

L. Collatz, Karlsruhe: Praxis der Eigenwertprobleme.

H. Seifert und W. Threlfall, Braunschweig: Hypergeometrische Differentialgleichungen.

W. Magnus, Berlin: Formeln und Sätze für die speziellen Funktionen der mathematischen Physik.

E. Kamke, Tübingen: Partielle Differentialgleichungen.

H. Kneser, Tübingen: Die mathematische Behandlung des Flatterverhaltens von Flugzeugen.

F. Rellich, Dresden: Die singulären Eigenwertprobleme der mathematischen Physik (unter Beschränkung auf Differentialgleichungen).

H. Görtler, Freiburg: Grenzschichttheorie.

H. Cremer, Breslau: Aufstellung von rechnerisch leicht zu handhabenden Kriterien dafür, daß sämtliche Nullstellen eines gegebenen Polynoms mit reellen Koeffizienten einen negativen Realteil besitzen.

G. Bol, Freiburg: Die Lösungen von Differentialgleichungen in der Umgebung einer singulären Stelle.

E. Hopf, München: Berechnung der Luftströmung hinter einer Luftschraube.

O. Volk, Würzburg: Auflösung von Systemen von Gleichungen, die mit dem Zünderproblem zusammenhängen.

Süss, S. V.:

„II.   Entwicklung seit Kriegsende

Mit Kriegsende war die reiche Finanzierung des Instituts durch den Reichsforschungsrat hinfällig geworden. Während der Reichsforschungsrat für das Institut einen Etat von jährlich 186 000,– RM in Aussicht genommen hatte (4 Professoren, 10 wissenschaftliche Assistenten, einige Rechenkräfte, Bibliothek und Instrumente etc.), konnte das finanziell schwache Land Baden in den ersten Nachkriegsjahren nur mit Beträgen, die von 18 000,– auf 10 000,– RM sanken, die *Existenz* eines Institut-*Keimes* retten. Von den vielen Aufgaben, die sich das Institut ursprünglich gestellt hatte, (wovon in der nächsten Nummer berichtet wird) mußten die meisten vorläufig ganz vernachlässigt oder aufgegeben werden. Andererseits schien bei der neuen Situation der Nachkriegszeit besonders wichtig zu sein, die deutschen Gelehrten wieder mit der Wissenschaft des Auslandes in Verbindung zu bringen. Durch das fördernde Entgegenkommen wohlwollender Kollegen in der französischen Militärregierung gelang es schon kurze Zeit nach der Besetzung, in immer stärkerem Maß den Kontakt insbesondere mit französischen, englischen und Schweizer Mathematikern zu gewinnen. Die Organisation solcher Besuche gehörte zu den beschränkten Möglichkeiten einer Aktivität des Instituts. Aus ihr entwickelte sich dann die systematische Einrichtung von speziellen Vortragsveranstaltungen (Symposia) auf solchen mathematischen Gebieten, für die wir besonders ausgezeichnete Spezialisten für Vorträge und Diskussionen als Gäste gewinnen konnten.

III.   Beschreibung des MATHEMATISCHEN FORSCHUNGSINSTITUTs in Oberwolfach

Die gegenwärtige Unterkunft des Instituts ist noch immer das Anwesen Lorenzenhof in dem 6 km von der Bahnstation Wolfach entfernten Oberwolfach, das die Regierung für das Institut gemietet hat. Das Institut liegt in einem sehr ruhigen Seitental der Kinzig im schönsten Teil des mittleren Schwarzwaldes. Es besitzt 19 Räume verschiedenster Größe, von denen im unteren Stockwerk der geräumigste als Bibliothek, ein anderer für Vortragszwecke, ein dritter als Eßzimmer und Gemeinschaftsraum benutzt wird. Die übrigen Räume dienen zur Unterbringung der Mitarbeiter und Gäste, wofür insgesamt 32 Betten zur Verfügung stehen. Bei Tagungen, zu denen mehr Besucher erscheinen, konnte stets die zusätzliche Bettenzahl in den in der Nähe befindlichen Gasthöfen gewonnen werden. Alle Mitarbeiter und die jeweiligen Gäste werden im Institut selbst verpflegt.

Bei den beschränkten Mitteln hat sich das Institut allein den Aufbau einer Bibliothek nach dem Kriege zum Ziel gesetzt, während für Instrumente und Maschinen Geld nicht zur Verfügung stand. Durch Kauf, durch zahlreiche Geschenke und durch Tausch gegen Veröffentlichungen des Instituts ist eine verhältnismäßig reichhaltige Bibliothek im Laufe der Jahre zustande gekommen. Die gut organisierte Verbindung mit der Universitätsbibliothek und dem Mathematischen Institut in Freiburg bringt nach Bedarf weitere notwendige Literatur rasch zur Benutzung nach Oberwolfach zur Ergänzung der dortigen Bestände, so daß bisher noch immer bei allen Arbeiten Literaturschwierigkeiten nicht aufgetreten sind.

Nebenher darf erwähnt werden, daß zur gelegentlichen Geselligkeit ein Flügel vorhanden ist, auf dem eine auffallend große Zahl bekannter Mathematiker sich auch als ausgezeichnete Pianisten erwiesen haben.

IV.   Ursprüngliche und gegenwärtige Ziele des Instituts

Bei der Gründung des Instituts waren folgende Aufgaben zur Durchführung des Instituts gedacht, das ja eine für die deutsche mathematische Forschung zentrale Stellung erhalten sollte:

a) Förderung der mathematischen Wissenschaft im weitesten Sinne, also sowohl der sog. reinen wie der angewandten Mathematik, wobei die Anwendungen ohne Beschränkung gedacht waren, soweit sich irgendwelche Möglichkeiten zur praktischen Durchführung zeigten.

b) Im Mittelpunkt sollten Arbeitsbesprechungen, Kolloquia (Symposia) stehen, bei welchen Spezialisten über die neuesten Forschungsergebnisse, über ungelöste Fragen und die besonderen Schwierigkeiten vortragen; die Diskussionen führen dann zu Plänen, auf welche Weise versucht werden kann, die Schwierigkeiten zu überwinden. Entsprechende Forschungsaufträge waren als Ergebnis solcher Untersuchungen gedacht.

c) Das Institut sollte geistiger und organisatorischer Mittelpunkt wichtiger Forschungskomplexe sein und insbesondere auch eine lebende Verbindung zwischen reiner und angewandter mathematischer Forschung unterstützen.

d) Durch Heranziehung zahlreicher jüngerer Mathematiker zur Mitarbeit am Institut, mindestens in der Ferienzeit der Hochschulen, soll die Kopplung zwischen reiner und angewandter Forschung auch der Nachwuchsgeneration weitergegeben werden. Dabei ist auch daran gedacht gewesen, den Geist der Zusammenarbeit mehrerer Forscher an einer größeren Aufgabe, der sich an vielen Stellen erfolgreich gezeigt hat, auch für die Zukunft lebendig zu erhalten.

e) Am Institut selbst sollten durch die Berufung hervorragender Spezialisten einige Hauptgebiete der modernen mathematischen Forschung bearbeitet werden. Für das Institut waren dabei Beschränkungen auf die Theorie gedacht, doch in solcher Form, daß die Weitergabe für die praktische Verwertung und jedenfalls die Verbindung mit den Interessen der Anwendungsgebiete und die Bereitschaft zur Zusammenarbeit mit ihnen nicht außer acht gelassen werden sollte.

f) Einrichtung einer ausführlichen Mathematiker-Kartei zur Beratung bei der Vermittlung von Mitarbeitern.

g) Schaffung einer Zentralstelle für mathematisches Berichtswesen, welche der Literaturbeschaffung, der Herstellung von Photokopien und Mikrofilmen, dem Referate-Wesen in jeder Weise, auch gelegentlich durch Übersetzungen dienen soll. Eine zentrale Auskunfts- und Prüfstelle für mathematische Fragen, Benennung von Spezialisten, die zu deren Beantwortung geeignet erscheinen. Schaffung und Herausgabe erforderlicher mathematischer Literatur

(Mitarbeit bei Zeitschriften und Monographien). Vermittlung bei der Herstellung notwendig erscheinender Tabellierung von Funktionen und dergleichen.

h) Mitarbeit an Ausbildungsfragen im gesamten mathematischen Unterrichtswesen, insbesondere bei Hochschulen und Höheren Schulen, bei letzteren sowohl den Unterricht in den Schulen wie auch die Heranbildung der Lehrer betreffend."

Süss, S. IX:
„V.   Veröffentlichungen des Instituts

In den ersten Jahren nach dem Krieg war keine Möglichkeit zur Publikation deutscher Zeitschriften und Bücher mathematischen Inhalts vorhanden. Das Institut beschloß deshalb, mit seinem großen Kreis von Freunden und Mitarbeitern geeignete Publikationsmöglichkeiten zu schaffen. Ein erster Weg, die Ergebnisse deutscher mathematischer Forschung in der Welt bekannt zu machen, war allerdings durch das Angebot der Militärregierung gegeben, für die

A)   FIAT REVIEW OF GERMAN SCIENCE
     ⟨ = Field Information Agency Technical (British)⟩
die beiden Bände „REINE MATHEMATIK" (Band I und II der FIAT RE-VIEW) zu schreiben. Das Institut übernahm die vollständige Herstellung des Manuskripts dieser FIAT REVIEW mit seinen Mitarbeitern im Institut selbst und an verschiedenen Universitäten, die innerhalb von 10 Monaten einen vollständigen Bericht über die gesamte deutsche mathematische Literatur aus den Jahren 1939 bis 1946 verfaßten. In 30 Kapiteln wird über alle Spezialgebiete der reinen Mathematik einschließlich der Geschichte, Grundlagen, Algebra, Zahlen- und Gruppentheorie, gesamten Analysis, Geometrie und Topologie die gesamte deutschsprachige mathematische Literatur der genannten Jahre im Zusammenhang von 33 Autoren berichtet. Die Bände der FIAT REVIEW sind heute im Buchhandel unter dem Titel ‚Naturforschung und Medizin in Deutschland 1939–1946‘ erhältlich.

B)   Der zweite Publikationsplan des Instituts betraf eine Reihe von Monographien und Lehrbüchern unter dem Titel STUDIA MATHEMATICA.
     ⟨Süss nennt nun die damals erschienenen und geplanten Bände. Hier seien ohne diese Unterscheidung nur ein paar Beispiele herausgegriffen:⟩
     E. Sperner: Einführung in die Analytische Geometrie und Algebra.
     G. Bol: Elemente der Analytischen Geometrie.
     G. Bol: Projektive Differentialgeometrie.
     G. Pickert: Einführung in die höhere Algebra.

W. Maak: Differential- und Integralrechnung.
H. Kneser: Funktionentheorie.

C)   Als erste wissenschaftliche Zeitschrift der Mathematik konnte das Institut mit einer internationalen Redaktion, deren Mitglieder zu dem engeren Freundeskreis des Instituts gehören, die neue mathematische Zeitschrift ARCHIV DER MATHEMATIK herausbringen. Sie erscheint seit 1952 im Verlag BIRKHÄUSER, Basel, alle zwei Monate mit einem Heft. Als Tauschorgan für fremde Zeitschriften konnte sie dem Institut Kenntnis der hauptsächlichsten mathematischen Literatur der ganzen Welt vermitteln.

D)   Seit den Zeiten von Felix Klein hat man in Deutschland immer wieder versucht, die Kluft zwischen dem wissenschaftlichen Niveau der Mathematiker an den Höheren Schulen und an den Universitäten zu überbrücken. Da das Institut sich auch an diesen Bemühungen beteiligen will, gibt es zusammen mit Prof. Behnke in Münster und Prof. Lietzmann in Göttingen die MATHEMATISCH-PHYSIKALISCHEN SEMESTERBERICHTE heraus (Verlag Vandenhoeck u. Ruprecht, Göttingen), die in geeigneten Berichten über die Fortschritte der Wissenschaft und Originalaufsätzen über spezielle Probleme, welche für den Schulmann verständlich sind, die wissenschaftliche Verbindung zwischen Schule und Hochschule in der Mathematik herstellen sollen.

VI.   Allgemeiner Tätigkeitsbericht

Eine Reihe deutscher Mathematiker hat sich im Laufe der Jahre am Institut mit der Fertigstellung mathematischer Arbeiten beschäftigt. So sind z. B. einige *Dissertationen* von jungen Mathematikern in Oberwolfach geschrieben worden.
Da aber mangels genügender Mittel Forschungsarbeiten vom Institut nicht finanziert werden konnten, mußte die gesamte Forschung auf der Basis freiwilliger Mitarbeit und Zusammenarbeit aufgebaut werden." (Süss, S. XI.)
〈Als die deutschen Universitäten ihre Arbeit wieder aufnahmen, gingen die meisten Mitarbeiter des Instituts an ihre früheren Arbeitsstätten zurück oder erhielten Rufe auf Lehrstühle. Das Institut hatte – wie gesagt – keine Mittel zur Finanzierung von Forschungsarbeiten und hat sie auch heute noch nicht. Die Mitarbeiter waren an den nächstgelegenen Universitäten tätig, nur ein oder zwei Stipendiaten der Deutschen Forschungsgemeinschaft arbeiteten im Institut selbst. Aber viele Mathematiker des In- und Auslandes besuchten das Institut, manchmal für ein paar Tage, manchmal auch für ein paar Wochen, hielten Vorträge und besprachen mit Kollegen ihre Arbeiten. Allmählich entwickelten sich daraus Tagungen über begrenzte Spezialgebiete. Dabei wurde von vornherein an intensive Arbeit in kleinen Kreisen gedacht, auch sollten nicht unbedingt fertige Ergebnisse vorgetragen, sondern auch offene Fragen besprochen

werden.⟩ Süss schreibt (S. XI ff.): „Die stärksten Impulse glaubte das Institut der mathematischen Forschung durch *Vorträge und Diskussionen* über aktuelle Probleme der Mathematik geben zu können. Schon während des letzten Kriegsjahres veranstaltete das Institut fast täglich einen Vortrag oder ein Referat, woran sich jedesmal eine Diskussion der Mitarbeiter anschloß.

Später gingen wir zur *systematischen Behandlung von Spezialfragen in Kolloquien* über.

A)   Das erste größere Kolloquium dieser Art war der *Topologie* gewidmet. Hierzu waren 37 Mathematiker aus der Schweiz, Frankreich, Belgien, Österreich und Deutschland erschienen. Darunter H. Hopf, B. Eckmann und E. Stiefel (Zürich), H. Cartan (Paris), Ch. Ehresmann (Strasbourg), G. Hirsch (Brüssel), L. Vietoris (Innsbruck), H. Kneser (Tübingen), G. Köthe (Mainz), G. Nöbeling (Erlangen), H. Seifert und W. Threlfall (Heidelberg), E. Sperner (Bonn). Das Kolloquium zeigte uns Deutschen deutlich, daß Deutschland, das früher einmal der hauptsächliche Ursprung der Topologie gewesen ist, in der topologischen Forschung weitgehend in den Hintergrund getreten ist. Die ausländischen Besucher dieses Kolloquiums waren im wesentlichen die Gebenden. Das Kolloquium hatte aber durch die zahlreichen Diskussionen zwischen den Vertretern der verschiedenen topologischen Schulen für alle Teilnehmer eine große Bedeutung, wie uns aus vielen Briefen bekannt geworden ist. Im Mittelpunkt dieses Kolloquiums stand die Abbildungstheorie von H. Hopf sowie die Theorie von de Rham nebst der Theorie der gefaserten und geblätterten Räume.

B)   Ein zweites größeres Kolloquium sollte den Kreis französischer Mathematiker, der sich um den Namen N. *Bourbaki* schart und viele zukunftsvolle junge französische Mathematiker umfaßt, mit jungen deutschen Mathematikern (Universitätsdozenten) bekanntmachen. Mit Unterstützung der Militärregierung konnte der Besuch für je 15 französische und deutsche Mathematiker finanziert werden, welche während 18 Tagen mit täglich 3 Vorträgen ihre Probleme und Methoden gegenseitig sich bekanntmachten. Die Hauptvertreter der französischen Gruppe waren J. Dieudonné (Nancy) und Ch. Ehresmann (Strasbourg), auf deutscher Seite G. Bol (Freiburg), H. Kneser (Tübingen), G. Köthe (Mainz), H.-H. Ostmann (Berlin) und der Unterzeichnete. Wir hoffen, daß die Bekanntschaft der kommenden Mathematiker-Generation in Frankreich und Deutschland wissenschaftlich, menschlich und auch politisch von besonderem Wert sein wird. Für uns Deutsche war es von großer Bedeutung, nicht nur einen fast vollständigen Überblick über die Bestrebungen des Bourbaki-Kreises zu erhalten, der ja in den letzten Jahren bereits 12 Bücher publiziert hat, sondern auch eine große Zahl der Mitarbeiter in ihrer Arbeit an speziellen Problemen nach den Bourbaki-Methoden genauer kennenzulernen. An den Universitäten Berlin, Mainz und Tübingen ist der direkte Einfluß dieser Methoden bereits auch stark zu erkennen. Er geht im wesentlichen auf jenes Kolloquium in Oberwolfach zurück.

C)   Eine Sonderveranstaltung galt der Bekanntschaft der *Mathematiker* benachbarter Universitäten *beiderseits des Rheins*. Unter den 40 Teilnehmern befanden sich 10 aus Frankreich, 4 aus der Schweiz. (Von den Franzosen nenne ich G. Bouligand, H. Cartan, Chabauty, Deny, unter den Schweizern Eckmann, Ostrowski, von den Deutschen Bol, Pickert, Rohrbach, Wittich.) Das Kolloquium hat seine Aufgabe insofern gut gelöst, als die Bekanntschaft der Mathematiker benachbarter Universitäten zu dem Entschluß geführt hat, die Veranstaltung regelmäßig an den Universitäten Basel, Strasbourg und Freiburg jährlich zu wiederholen. Mit gutem Erfolg fand eine derartige Wiederholung bereits in Basel statt, eine solche in Strasbourg steht bevor.

D)   Ein ganz besonders intensives Kolloquium veranstalteten einige Spezialisten der *Logistik und der mathematischen Grundlagenforschung* unter der geistigen Leitung von P. Bernays (Zürich). Hier standen im Mittelpunkt Widerspruchsfreiheitsbeweise der Zahlentheorie und Analysis und Entscheidbarkeitsprobleme sowie Probleme und Arbeiten von Tarski. Die deutschen Teilnehmer, Arnold Schmidt (Marburg) und hauptsächlich Schüler von H. Scholz (Münster) bezeichneten diese Zusammenkunft für sie als außerordentlich wertvoll, da ihnen bei dieser Gelegenheit bisher ungedruckte Ergebnisse aus dem Arbeitskreis von Bernays und aus USA bekannt geworden sind.

E)   Die *moderne Algebra und Zahlentheorie*, deren ursprüngliche Heimat im wesentlichen Deutschland gewesen ist, während sie heute mit außerordentlichem Erfolg an vielen Orten besonders in USA gepflegt wird, hat zweimal in der Berichtszeit eine größere Anzahl von Interessenten in Oberwolfach zusammengeführt. Von den teilnehmenden Ausländern seien genannt R. Baer (Illinois), B. H. Neumann (Manchester) und H. Neumann (Hull). Die bedeutendsten deutschen Vertreter der Algebra waren dabei H. Hasse und E. Witt (Hamburg) mit einer größeren Zahl sehr fähiger jüngerer Mitarbeiter, darunter M. Kneser und P. Roquette. Hier standen im Mittelpunkt die modernen Erweiterungen der Klassenkörpertheorie, Funktionenkörper sowie Probleme der modernen Gruppentheorie. Wir hatten den Eindruck, daß die neueren Ergebnisse der algebraischen Schule von Hasse das besondere Interesse der Teilnehmer – insbesondere auch der Ausländer – gefunden hat, während neuere Entwicklungen aus der Gruppentheorie amerikanischer Prägung den deutschen Teilnehmern manche Anregung brachte, besonders solche, die für die Grundlagen der Geometrie von Bedeutung sind, die in Deutschland immer wieder eine Reihe von Bearbeitern finden.

F)   Die geschlossenste Veranstaltung, die wohl am weitesten sich ausgewirkt hat, war eine Tagung über *Fragen der komplexen Funktionentheorie*, zu der über 40 Teilnehmer sich eingefunden haben. Wir hatten die besondere Freude des Besuchs von H. Weyl (Princeton-Zürich) und R. Nevanlinna (Helsinki-Zürich), die auch selbst sich durch Vorträge an der Veranstaltung beteiligten. Führende

Vorträge hielten außerdem A. Pfluger (Zürich), H. Kneser und sein Mitarbeiter Stoll (Tübingen), H. Behnke und K. Stein (Münster), E. Peschl (Bonn), H. Petersson (Hamburg), H. Wittich (Karlsruhe) u. a. An 11 Tagen wurden jeweils entweder vormittags oder nachmittags insgesamt 21 Vorträge gehalten, an die sich ausführliche Diskussionen anschlossen, die wesentliche Teile der zweiten Hälfte des Tages ausfüllten. Nevanlinna, Pfluger, Ullrich und Wittich behandelten die moderne Funktionentheorie auf offenen Riemannschen Flächen, deren Hauptvertreter seit 20 Jahren Nevanlinna selbst ist. Mit Ausnahme von Pfluger betrafen die Fragestellungen und Methoden kaum die heute in USA und Rußland in der Funktionentheorie bevorzugten Methoden der Funktionalanalysis und der Variationsprinzipien. Sie gaben einen guten Überblick über die Weiterarbeit an klassischen Problemen in Mitteleuropa, die auch heute noch von großer Bedeutung für wichtige Fragen sind. Für Funktionen mehrerer komplexer Veränderlicher ist seit längerer Zeit die Schule von Behnke tätig. Einen ausgezeichneten Überblick über die Leistungen dieser Schule gaben die Vorträge von Behnke und Stein, die neuerdings im Anschluß an H. Cartan und H. Hopf topologische Methoden in ihre Untersuchungen hineinziehen, während der Vortrag von Stoll einen großen Fortschritt auf dem gleichen Gebiet bei meromorphen Funktionen mehrerer Veränderlicher und die Wertverteilungstheorie von H. Kneser bei Funktionen mehrerer Veränderlicher brachte. Petersson sprach über multiplikative Funktionen für gegebene Kongruenzgruppen und Peschl über reguläre Automorphismen des $2n$-dimensionalen euklidischen Raumes.

G)   In jedem Jahr fanden bisher mindestens einmal, gelegentlich auch zweimal Kolloquia der *Geometrie* statt. Hierbei stand die Differentialgeometrie ganz im Vordergrund entsprechend den Spezialgebieten von G. Bol und dem Institutsleiter Süss. Die bedeutendsten ausländischen Teilnehmer waren E. Bompiani, F. Conforto (Rom), H. Hopf (Zürich), H. Hadwiger (Bern), Vincensini (Marseille), Ancochea (Madrid), E. T. Davies (Southampton). Von deutscher Seite waren u. a. vertreten W. Blaschke (Hamburg), F. Löbell (München), K. Strubecker (Karlsruhe), K. H. Weise (Kiel). Hierbei nahm den größten Teil die Differentialgeometrie der großen geometrischen Gruppen ein, besonders der projektiven, die heute im Mittelpunkt der Bolschen Untersuchungen steht und berufen erscheint, eine größere Auswirkung in der internationalen Literatur zu finden.

Ferner spielten Untersuchungen zur Geometrie des Raumes von Finsler und E. Cartan eine größere Rolle, zu denen Davies (Southampton) und H. Rund (Kapstadt-Freiburg, jetzt Bonn) sowie einige Freiburger wichtigere Beiträge geliefert haben. Ferner Fragen der Verbiegung aus der klassischen Differentialgeometrie im großen.

H)   Für das Jahr 1953 sind bis jetzt folgende Kolloquia geplant:
1) Ein *Hecke-Kolloquium* vom 14. – 19. April, das die Aufgabe haben

soll, die ungelöst gebliebenen Probleme der Schule von E. Hecke gemeinsam zu erörtern. Bisher sind 11 Vorträge angemeldet, z. B. von Conforto (Rom), H. Braun, Petersson (Hamburg), Eichler (Münster), Maass (Heidelberg). Zugesagt haben ihre Teilnahme auch Kloosterman, van der Blij und Springer aus Holland.

2) In der Woche nach Ostern ein philosophisches Kolloquium über die Frage, was die Mathematik zur Verbindung von Naturwissenschaften und Geisteswissenschaften beitragen kann.

3) Ein Kolloquium im August über Fragen der *Geometrie der Zahlen und diophantische Approximationen*.

4) Ende August ein Kolloquium über *reelle Analysis*, die bisher nur gelegentlich durch Vorträge z. B. von Favard, Haupt und Köthe vertreten war.

5) Schließlich ein Kolloquium zur *Differentialgeometrie* im Oktober.

6) Geplant ist auch eine Veranstaltung aus einem Spezialgebiet der ANGEWANDTEN MATHEMATIK, worüber noch keine endgültigen Entscheidungen gefallen sind." (Süss, S. XVI.)

Die auf Seite 391 bis 394 genannten Tagungen fanden zu den folgenden Terminen statt:

A)   Topologie: 2. – 4. April 1949.
B)   Französisch-deutsche Arbeitsgemeinschaft: 9. – 25. August 1949.
C)   Treffen der Mathematiker beiderseits des Rheins: 24. – 26. Nov. 1950. – Nach dem Tod von Süss (1958) und vor allem, als gegenseitige Besuche allgemein leicht geworden waren, hörten diese Zusammenkünfte auf.
D)   Logik und Grundlagenforschung: im Sept. 1949 und im Juni 1952.
E)   Algebra und Zahlentheorie: im August 1951 und im Sept. 1952.
F)   Komplexe Funktionentheorie: im Okt. 1951.
G)   Geometrie: im August 1951 und im März 1952.

Daß eigentliche Tagungen erst 1949 begannen, hängt damit zusammen, daß damals die durch die Lebensmittelkarten und durch die Zonengrenzen bedingten Schwierigkeiten aufhörten.

Außer dem kurzen Wochenendtreffen C) fanden die Tagungen in den Semesterferien statt; im Semester waren die Kollegen nicht abkömmlich. So war das Haus während eines großen Teiles des Jahres, besonders im Winter, nur von wenigen Insassen benutzt, was die Gebefreudigkeit des zuständigen Ministeriums durchaus negativ beeinflußte. Es wurde auch gelegentlich davon gesprochen, das Institut als ausgelagertes Freiburger Universitätsinstitut anzusehen und es, nachdem in Freiburg seit 1952 wieder angemessene Räume zur Verfügung standen, sozusagen nach Freiburg zurückzuholen, d. h. aufzulösen. W. Süss hat solche Überlegungen, die freilich meines Wissens höchstens gesprächsweise, aber nie in offizieller Form an ihn herangetragen wurden, nie ernsthaft in Erwägung gezogen, und er konnte 1952 schon auf die Anerkennung hinweisen, die das Institut auch im Ausland gefunden hatte.

In dieser Situation ist also der Bericht entstanden. Dadurch ist auch motiviert, daß die Ziele des Instituts darin sehr weit gesteckt erscheinen. Es kam eben darauf an, den zuständigen Stellen darzulegen, daß das Institut nicht etwa eine Art Sommerfrische der Freiburger Mathematiker sei, sondern daß wichtige Aufgaben vorlagen, die das Institut eigentlich erfüllen sollte und auch erfüllen könnte, wenn die nötigen Mittel zur Verfügung gestellt würden.

Es ist mir nicht bekannt, daß das Institut nunmehr erhebliche zusätzliche Mittel erhalten hätte; aber es war schon ein Erfolg, wenn die bisherigen Mittel nicht gekürzt wurden.

In den folgenden Jahren war man darum bemüht, die Möglichkeiten des Hauses auszuschöpfen; es wurde auch nicht-mathematischen Arbeitsgruppen zur Verfügung gestellt, z. B. auch zu Wochenendtagungen von Seminaren. Als Beispiel seien die Veranstaltungen im Jahr 1958, dem Todesjahr von W. Süss, angeführt:

Veranstaltungen in Oberwolfach im Jahre 1958

01.03. – 08.03.	Kolloquium über Statistik.
24.03. – 28.03.	Probleme der Nichtlinearen Mechanik.
07.04. – 12.04.	Arbeitsgemeinschaft über Stellenringe und Schnittmultiplizitäten.
23.04. – 01.05.	Arbeitstagung über Fragen der Theoretischen Physik.
15.05. – 18.05.	Treffen von Stipendiaten der Deutschen Forschungsgemeinschaft (Psychologen).
26.05. – 31.05.	Gruppen in der Geometrie.
05.06. – 08.06.	und 13.06. – 17.06. Zusammenkunft von Stipendiaten der Studienstiftung des Deutschen Volkes.
02.07. – 09.07.	Wissenschaft und Schule: Maß und Inhalt (Fortbildungskurs für Studienräte).
19.07. – 20.07.	Völkerrechtliches Seminar, Prof. Kaiser.
23.09. – 29.09.	Ferienseminar über Hochenergiephysik.
03.10. – 08.10.	Arbeitsgemeinschaft über algebraische Gruppen.
14.10. – 19.10.	Mathematikgeschichtliches Kolloquium.
19.10. – 28.10.	Geometrie.
29.10. – 02.11.	Funktionalanalysis.

Ein Höhepunkt im Leben des Instituts war die Feier des 60. Geburtstags von W. Süss am 7. März 1955. H. Behnke hat sie ausführlich geschildert[*]. Für das Jahr 1958 war Süss, der schon 1940 – 1945 Rektor der Universität Freiburg gewesen war, wieder zum Rektor gewählt worden; darin lag auch eine Anerken-

---

[*] Abschied vom Schloß in Oberwolfach. Eine Ansprache am 23. Juni 1972. Jber. Deutsch. Math.-Verein. **75** (1973), 51–61. Dort ist auch ein Bild des „Schlosses" und des Bibliotheksraumes beigegeben.

nung für seine Haltung in jener schwierigen Zeit. Eine schwere Krankheit hinderte ihn daran, das Rektorat zu übernehmen; ihr ist er am 21. Mai 1958 erlegen. Die Leitung des Instituts hatte er vorher an Hellmuth Kneser übergeben, mit dem er seit seiner Greifswalder Zeit (1928) eng befreundet war, und der als Ordinarius in Tübingen seit der Gründung des Instituts an dessen wissenschaftlichen Tätigkeiten stets tatkräftig und wirksam mitgearbeitet hat.

In der Art von Süss konnte das Institut nicht weitergeführt werden. Süss war allein und persönlich für das Institut verantwortlich gewesen; er war selbst erschrocken gewesen, als ihm das ein paar Jahre vorher vom Ministerium eindeutig klargemacht wurde. Er hatte viele Schwierigkeiten durch seine Improvisationsgabe überwunden. Dabei war die Verfassung des Instituts, wie Hellmuth Kneser es einmal ausdrückte, „eine Anarchie mit einem tüchtigen Oberanarchen an der Spitze". Überflüssig, zu sagen, daß der Institutsbetrieb bei dieser „Anarchie", lies: Freiheit des Einzelnen und freiwillige Mitarbeit jedes Einzelnen, wo es gerade erforderlich war, bestens funktionierte. Es stellten sich jetzt die folgenden Aufgaben:

1) Es mußte ein Rechtsträger für das Institut gefunden werden.

2) Die Finanzierung des Instituts mußte dauerhaft gesichert werden.

3) Am Haus waren Reparaturen und Renovierungsarbeiten erforderlich geworden, auch benötigte das Institut dringend mehr Raum.

Zu 1). Auf Grund verschiedener Überlegungen und Besprechungen mit den zuständigen Ministerien erschien die Gründung eines „eingetragenen Vereins" als Rechtsträger des Instituts als die zweckmäßigste Lösung. Nach Zusammenkünften interessierter Kollegen am 11. – 13. März 1959 in Oberwolfach und am 6. April 1959 in Heidelberg wurde am 17. Juni 1959 in Oberwolfach die „Gesellschaft für mathematische Forschung" gegründet und am 14. Juli 1959 unter dem Aktenzeichen VR XIII/1 in das Vereinsregister in Freiburg eingetragen.

Anmerkung: Die Gründungsmitglieder mußten persönlich ihre Unterschrift beim Notariat Freiburg leisten. Deshalb wurden nur wenige (es mußten mindestens sieben sein) und hauptsächlich Freiburger dazu ausgewählt.

Die Präambel der Satzung lautet:

„Die Gesellschaft für mathematische Forschung setzt sich das Ziel, der Mathematik auch in Deutschland Möglichkeiten zur Entwicklung zu verschaffen, wie sie durch verschiedene Institutionen in anderen Ländern bereits gegeben sind. Dies soll in Anknüpfung an die durch Wilhelm Süss in Oberwolfach begründete Tradition der wissenschaftlichen Aussprache begonnen werden. Die Gesellschaft will einen Mittelpunkt intensiver wissenschaftlicher Zusammenarbeit der verschiedenen Generationen bilden und den Gedankenaustausch mit ausländischen Forschern erleichtern."

Protokoll der Gründungsversammlung
der Gesellschaft für mathematische Forschung (e.V.).

Am 12./13. März 1959 traten in Oberwolfach die Herren
Baer, Behnke, Bol, Gericke, Görtler, Hirzebruch, Hellmuth Kneser,
Köthe, Maak, Claus Müller, Roquette, Sperner, Stein und Weise
zu einer Konferenz über die Zukunft der mathematischen Forschung
in Deutschland und den Bestand und Ausbau des Mathematischen
Forschungsinstituts Oberwolfach zusammen. Auf Grund von Besprechungen mit Vertretern des Bundesministeriums des Innern und des Kul-
tusministeriums des Landes Baden-Württemberg wurde die Gründung
einer Gesellschaft für mathematische Forschung als angemessen ange-
sehen. Nach einer weiteren Zusammenkunft in Heidelberg am 26.4.1959
sind die Unterzeichneten im Auftrag der oben genannten Kollegen am
17.6.1959 um 16^{15} Uhr im Mathematischen Institut in Freiburg
zusammengetreten und haben die Gründung der Gesellschaft für
mathematische Forschung mit der beiliegenden Satzung beschlossen.

Der Wissenschaftliche Beirat (s.Satzung § 8, Absatz 1) besteht
aus den Herren

Baer, Prof.Dr.R.,	Falkenstein im Taunus,	Gartenstr. 11
Behnke, Prof.Dr.H.,	Münster i.Westf.,	Rottendorffweg 17
Bol, Prof.Dr.G.,	Stegen/Dreisamtal über	Freiburg i.Br.
Gericke, Prof.Dr.H.,	Freiburg-Littenweiler,	Sonnenbergstr. 31
Görtler, Prof.Dr.H.,	Freiburg i.Br.,	Sonnhalde 9o
Hirzebruch, Prof.Dr.F.,	Bonn a.Rh.,	Endenicher Allee 7
Kneser, Prof.Dr.H.,	Tübingen,	Brunsstr. 31
Köthe, Prof.Dr.G.,	Heidelberg,	Mönchhofstr. 26
Maak, Prof.Dr.W.,	Göttingen,	Ewaldstr. 69
Müller, Prof.Dr.Claus,	Richtrich bei Aachen,	Hauptstr. 7 b
Roquette, Prof.Dr.P.,	Saarbrücken,	Math. Institut
Schneider, Prof.Dr.Th.,	Erlangen,	Am Röthelheim 56
Sperner, Prof.Dr.E.,	Ahrensburg/Holstein,	Hagener Allee 41
Stein, Prof.Dr.K.,	München 9,	Ulmenstr. 14
Weise, Prof.Dr.K.H.,	Kiel,	Clausewitzstr. 14

Als „Zweck des Vereins" wird in § 2 der Satzung angegeben:
„1. Intensivierung der mathematischen Forschung,
 2. Verstärkung der wissenschaftlichen Zusammenarbeit,
 3. Fortbildung in der Mathematik und ihren Grenzgebieten.
Diese Ziele werden in internationalem Rahmen angestrebt."

§ 5 nennt als „Organe des Vereins" den Vorstand, den Wissenschaftlichen Beirat, den Verwaltungsrat und die Mitgliederversammlung.

Der Wissenschaftliche Beirat besteht aus 10 – 20 Mitgliedern; er konstituierte sich erstmals auf der Gründungsversammlung. Seither werden seine Mitglieder auf Vorschlag des Wissenschaftlichen Beirats vom Vorstand ernannt. Die Mitgliedschaft dauert vier Jahre; Wiederwahl ist zulässig.

Der Wissenschaftliche Beirat ist für die grundsätzliche Gestaltung des wissenschaftlichen Programms des Forschungsinstituts verantwortlich. Er bestellt den Direktor des Forschungsinstituts. Er wählt seinen Vorsitzenden und den Schatzmeister. Diese drei Personen bilden den Vorstand der Gesellschaft.

Der Verwaltungsrat beschließt über den Haushalt der Gesellschaft. Er besteht aus einem Vertreter des Ministeriums für Wissenschaft und Kunst von Baden-Württemberg, (früher auch einem Vertreter des Bundeswissenschaftsministeriums), einem Vertreter der Universität Freiburg, dem Schatzmeister und einem weiteren Vertreter des Wissenschaftlichen Beirats und dem Direktor des Forschungsinstituts.

(Der Notar, der die Eintragung in das Vereinsregister vollzog, sagte mir gelegentlich, eine so komplizierte Vereinssatzung sei ihm noch nie vorgekommen; er glaube nicht, daß sie lange funktionieren werde.)

Aus juristischen und finanziellen Gründen wäre eine Aufnahme des Instituts in die Max-Planck-Gesellschaft wünschenswert gewesen. Man hat sich auch darum bemüht. Die Max-Planck-Gesellschaft lehnte aber nach sorgfältiger Prüfung die Aufnahme ab, weil das Institut seinen Aufgaben und seiner Struktur nach nicht in die Satzung der Max-Planck-Gesellschaft paßte. (Die Max-Planck-Gesellschaft betreibt Institute, die ausschließlich reine Forschungsaufgaben haben und auf die Persönlichkeit des Leiters zugeschnitten sind.)

Zu 2). Das Land Baden-Württemberg konnte die Kosten für ein Institut, das den Mathematikern im ganzen Bundesgebiet diente, nicht allein tragen. W. Süss hatte schon vor einigen Jahren einen Zuschuß vom Bundeskanzleramt erwirkt, der inzwischen vom Bundesinnenministerium übernommen worden war. Diese Mittel reichten aber auf die Dauer nicht aus, um das Institut personell, sachlich und räumlich so auszustatten, wie es zu Erfüllung seiner Aufgaben wünschenswert und notwendig war. In den Jahren 1961 – 1964 stellte die Thyssen-Stiftung namhafte Mittel zur Verfügung, die insbesondere die Ausführung dringender Reparaturen und Verbesserungen am Haus ermöglichten. Diese Mittel waren als Anfangsförderung gedacht, und es gelang auch, seit 1965 eine ausreichende Finanzierung je zur Hälfte durch den Bund und das Land Baden-Württemberg zu erhalten. Auf Grund eines allgemeinen Vertrages zwischen dem Bund und den Ländern wird das Institut seit 1977 vom Land Baden-Württemberg allein finanziert.

Zu 3). Die Unterbringung der Tagungsgäste im „Schloß" war zwar in den Nachkriegsjahren annehmbar gewesen, allmählich waren die Verhältnisse aber kaum mehr zumutbar. H. Behnke hat den Kummer eines Bewohners eines Vier-Bett-Zimmers recht farbig geschildert.

Als die Stiftung Volkswagenwerk beschloß, allgemein den Bau von Gästehäusern zu fördern, konnte das Institut die Mittel für den Bau eines Gästehauses mit den erforderlichen Wirtschaftsräumen erhalten.

Der Architekt hat sich bemüht, den Bau durch terrassenförmige Anlage dem Gelände anzupassen. Bei der inneren Gliederung und Einrichtung wurde dafür gesorgt, daß alles das vorhanden ist, was notwendig ist, damit der Büro- und Wirtschaftsbetrieb reibungslos vonstatten geht und die Gäste sich wohlfühlen und in guter Atmosphäre arbeiten können. Am 16. Oktober 1967 wurde das Haus in einer Feierstunde mit Vertretern der Stiftung Volkswagenwerk und der Ministerien seiner Bestimmung übergeben.

Schon vor der Fertigstellung hatte die Stiftung Volkswagenwerk den Grund und Boden angekauft und die Gebäude in Besitz genommen. Diese wurden dem Mathematischen Forschungsinstitut abgabenfrei zur Verfügung gestellt; das Institut hat nur die laufenden Unterhaltskosten zu tragen.

Nunmehr konnte auch, zunächst vorsichtig, an eine durchgreifende Sanierung des Altbaus, des „Schlosses", gedacht werden. Es stellte sich aber heraus, daß der Bau so erhebliche Schäden aufwies, daß eine Sanierung nicht möglich war; jedenfalls wären die Kosten dafür im Voraus gar nicht abschätzbar gewesen. So mußte man sich zum Abbruch und einem Neubau entschließen. Die Mittel dafür stellte die Stiftung Volkswagenwerk, der das Gebäude ja gehört, zur Verfügung. Es wurde aber ganz nach den Wünschen des Instituts gebaut und eingerichtet. Der Neubau enthält neben Vortrags- und Diskussionsräumen ein Musikzimmer und die umfangreiche Bibliothek.

Unmittelbar bevor der Abbruch begann, hielten Freunde des Hauses am 23. Juni 1972 eine Abschiedsfeier, auf der H. Behnke die auf S. 34 (Fußnote) zitierte Ansprache hielt. Sicher war das alte Haus vielen ans Herz gewachsen, aber nach nunmehr neun Jahren – das Vortrags- und Bibliotheksgebäude wurde am 13./14. Juni 1975 eingeweiht – ist den Mathematikern auch das neue Haus vertraut und lieb geworden.

Hellmuth Kneser leitete das Institut bis zur Gründung der Gesellschaft für Mathematische Forschung 1959. Dann wurde er Vorsitzender des Wissenschaftlichen Beirats, Direktor des Mathematischen Forschungsinstituts wurde Theodor Schneider, der dieses Amt 1963 an Martin Barner übergab.

Es ist nicht meine Absicht, hier die Verdienste der Direktoren zu würdigen. Zum Funktionieren des Instituts gehört aber auch, daß Verwaltung und Hauswirtschaft von Männern und Frauen besorgt werden, die das Institut in gewissem Sinne als *ihr* Institut ansehen, die sich persönlich dafür verantwortlich

fühlen, daß die Institutsgäste sich wohlfühlen und daß die äußeren Voraussetzungen für einen guten wissenschaftlichen Betrieb gewährleistet sind; die Tag
für Tag, und manchmal zu ungewöhnlichen Zeiten, Arbeiten erledigen, die man
nur dann bemerken würde, wenn sie *nicht* pünktlich und sorgfältig ausgeführt
werden. Auch sie dürfen es sich als Verdienst anrechnen, daß das Institut sich so
großer Beliebtheit erfreut.

Perspectives in Mathematics

Perspectives in Mathematics
Anniversary of Oberwolfach 1984
© Birkhäuser Verlag, Basel

Differential Geometry and Computer Graphics

THOMAS BANCHOFF

Brown University, Department of Mathematics
Providence, RI 02912 (USA)

About one hundred years ago there was a thriving industry dedicated to the production of mathematical models. No mathematics department which aspired to any stature could afford to be without its cases of models in plaster and wire and cardboard and string, illustrating theorems from differential and algebraic geometry, topology, real and complex analysis and mathematical physics. Catalogues would describe in some detail the additions to their offerings, referring to recent research papers or theses [14], [15]. The models represented a resource of research and teaching complementary to a library, as well as illustrations for theses, e. g. those of Kummer [18] and Schilling [19].

Somehow in the present century these model collections often have fallen into disuse. Unlabelled or mislabelled models gather dust in display cases in mathematical institutions, generally ignored by teachers as well as students, only occasionally attracting some curious stare such as one might give to an artifact in an archaeological museum — somehow we suspect that the object must have had some ritual significance to some earlier society but that meaning has been lost. In spite of the fact that in each generation there are mathematicians who depend on three-dimensional models to help them visualize mathematical relationships, by and large our teaching and research have relied on two-dimensional drawings. Although such representations are quite adeqaute for many topics, they fall short when it comes to investigating some of the subjects of most interest over the past hundred years and especially today, namely objects undergoing change particularly in the neighborhodd of critical "catastrophic" transition points, and objects which exist principally in four dimensions or higher.

Fortunately it is not necessary to revive the studios which turned out their effective three-dimensional products a century ago. The medium which takes their place in our present day is geometric graphics, a fact-growing area of mathematics and computer science research which is destined to change the way we deal with many of the mathematical objects most familiar to us and to suggest entirely new areas for investigation.

In this paper we will present some examples of research carried on over the last fifteen years at Brown University, in collaboration with a number of computer scientists and student apprentices. A volume dedicated to the Mathematisches Forschungsinstitut at Oberwolfach seems to be a particularly appropriate place to present such a collection of examples. First of all Oberwolfach is close to the Darmstadt studios of Alexander von Brill who worked with so many eminent German mathematicians in fashioning the model collections still to be seen in many mathematics institutes in Germany and other parts of the world. Secondly, on a personal note, I gave my first talk in Europe 17 years ago at Oberwolfach — at a Geometrietagung in which all my illustrations were drawings or cardboard models. In suceeding years I have returned several times to conferences in topology, differential geometry and general inequalities, each time bringing more and more of the new films and slides that express the growth of our computer graphics project. Some of my most fruitful collaborations have begun at Oberwolfach and the friendships I have made in these several visits have been very important to me. For all these reasons I am happy to have the opportunity to present some examples of our work in this volume in honor of the Mathematisches Forschungsinstitut in Oberwolfach.

The images in this article were generated using the computer facilities at Brown University in collaboration with David Laidlaw, who did the computer programming and David Margolis who is responsible for the photography. The research was partially supported by the following contracts:

NSF MCS83-02180, ONR N00014-83-K-0146,
ONR N00041-83-K-0148, and DARPA order number 4786.

These images were produced on a Lexidata Solidview Graphics System and a Matrix OCR camera powered by a VAX 11/780 running Berkeley UNIX.

Deformation of the Swallowtail Catastrophe Surface

Among the most interesting objects of study in geometry are singular phenomena which can be deformed into essentially different configurations by arbitrarily small perturbations. The nature of the resulting deformations depends on the sorts of perturbations which are allowed, and a given configuration can represent quite different phenomena in different contexts.

An example of such a situation is found in the swallowtail catastrophe, which arises naturally as the catastrophe surface of the normal mapping for a convex plane curve. For a given curve $X(t)$ with well-defined Frenet frame $T(t)$, $N(t)$, $B(t)$ at every point, we may consider the parallel curve $Y^u(t)$ at distance u:

$$Y^u(t) = X(t) + u N(t).$$

For a given plane curve we may also consider the "sandpile surface"

$$Z(t, u) = X(t) + u N(t) + u B(t)$$

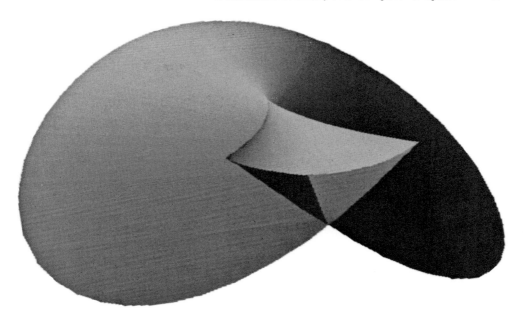

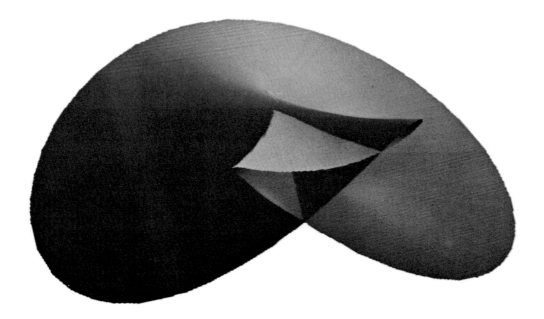

which stacks the various parallel curves in planes perpendicular to the fixed vector $B(t)$. To see where this is singular we form

$$Z_t = \frac{ds}{dt}((1 - u\varkappa(t) \, T(t) - u\tau(t) \, N(t) + u\tau(t) \, B(t))$$

$$Z_u = \qquad\qquad N(t) + B(t)$$

$$Z_t \times Z_u = \frac{ds}{dt}(-2u\tau(t) \, T(t) + (1 - u\varkappa(t)) \, (-N(t) + B(t))).$$

Then if $\tau(t) \equiv 0$, the surface $Z(t,u)$ will be singular exactly when $Z_t \times Z_u = 0$, i. e., when $u = \dfrac{1}{\varkappa(t)}$. The singularity set then is a curve lying over the evolute curve $E(t) = X(t) + \dfrac{1}{\varkappa(t)} N(t)$.

The singular points of $E(t)$, where

$$0 = \frac{ds}{dt} T(t) - \frac{ds}{dt} T(t) + \left(\frac{1}{\varkappa(t)}\right)' N(t),$$

occur precisely over the cusps of the evolute and these correspond to swallowtail points on the surface.

The illustration shows the swallowtail which arises from the singularity of the normal mapping of a parabola. A close-up view of the swallowtail point itself clearly shows two arcs of cuspidal points along with a double curve, all ending at the singular point. Topologically this image in a neighborhood of the swallowtail is a cone over a figure eight, equivalent by an ambient homeomorphism of Euclidean 3-space to a Whitney pinch point. This fact is used in an essential way in the proof of the triple tangency theorem for space curves with isolated torsion zeros [11]. Compare also the case of a polyhedral catastrophe surface of swallowtail type [2].

However if we perturb the original curve $X(t)$ so that it now has non-zero torsion, it follows that $Z_t \times Z_u$ is never zero so there are no singular points whatever in this neighborhood. Naturally there must be double points under small perturbations since there are transversal crossings in the original surface. If an arc of double points ends at a point then that point is necessarily a singularity of the surface, so it must happen that the double point arc of the perturbed surface leaves the neighborhood by means of a new arc of double points created in a neighborhood of one of the two cuspidal edges.

Computer generated images confirm this analysis. We deform the parabola to a twisted cubic $(t, t^2, \epsilon t^3)$ and observe that when ϵ is positive, one cuspidal edge becomes a double curve while the other cuspidal edge becomes smooth, and for ϵ negative the situation is reversed.

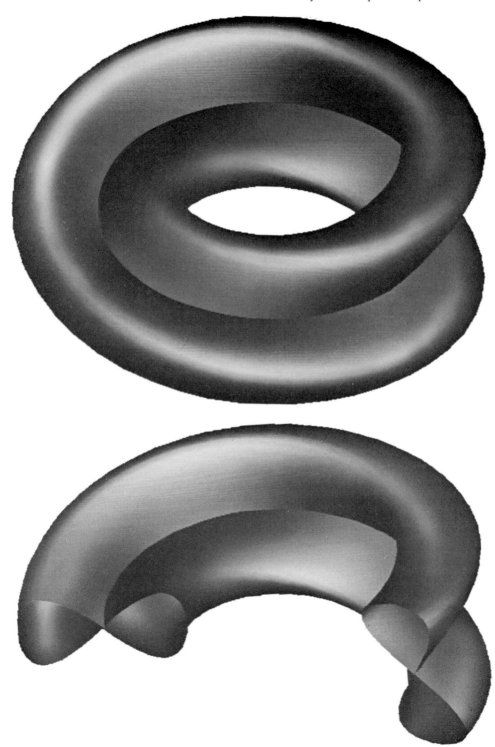

Note that $Z_{uu} = 0$ and

$$Z_{ut} = \frac{ds}{dt}\left(-\varkappa(t)\,T(t) - \tau(t)\,N(t) + \tau(t)\,B(t)\right)$$

so the Gaussian curvature of the surface Z is always nonpositive. We have $K = 0$ if and only if

$$0 = -Z_{ut} \cdot Z_t \times Z_u = -2\left(\frac{ds}{dt}\right)^2 \tau(t)$$

so that the curvature of the perturbed surface is everywhere negative.

The Figure Eight Klein Bottle

In spite of the fact that the "standard" immersion of the Klein bottle is very familiar to mathematicians, it is quite difficult to find a parametric representation of this object. A different immersion of the Klein bottle is however quite easy to model as a "twisted surface of revolution". This immersion was first encountered by the author in conjunction with the construction of minimal surface in a 3-sphere (the bicylinder boundary) [1]: It was constructed earlier by Anthony Phillips in another context.

We begin with an immersion of the circle as a figure eight curve $\varphi \to (\sqrt{2} + \cos\varphi, 0, \cos\varphi\,\sin\varphi)$ and we then rotate this around the z-axis while giving a twist in the vertical plane about the point $(\sqrt{2} + \cos\theta, \sin\theta, 0)$:

$$f(\theta, \varphi) = \left(\left(\cos\left(\frac{n\theta}{2}\right)(\sqrt{2} + \cos\varphi) + \sin\left(\frac{n\theta}{2}\right)(\cos\varphi\,\sin\varphi)\right)\cos\theta,\right.$$
$$\left(\cos\left(\frac{n\theta}{2}\right)(\sqrt{2} + \cos\varphi) + \sin\left(\frac{n\theta}{2}\right)(\cos\varphi\,\sin\varphi)\right)\sin\theta,$$
$$\left.-\sin\left(\frac{n\theta}{2}\right)(\sqrt{2} + \cos\varphi) + \cos\left(\frac{n\theta}{2}\right)(\cos\varphi\,\sin\varphi)\right).$$

If n is even this gives an immersion of the torus, and if n is odd, an immersion of the Klein bottle. In the latter case the preimage of the circle $(\sqrt{2}\cos\theta, \sqrt{2}\sin\theta, 0)$ of double points consists of two curves each with a Mobius band neighborhood while in the familiar "glass-blown" model of the Klein bottle the preimage of the double point set consists of two curves with orientable neighborhoods.

A nice feature of the computer graphics presentation is the ability to "open up" the surface to display its interior structure more effectively.

This example is illustrated in color in [9], which describes computer programs for the generation of the images and the scripting of scenarios for animation sequences.

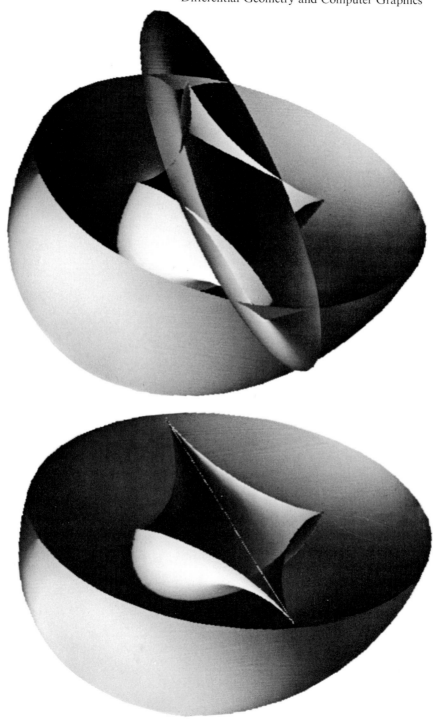

Focal Surfaces of the Ellipsoid

In 1873, Cayley carried out a detailed analysis of the centro-surface of the ellipsoid, the locus of centers of principal curvature [12]. By a particularly laborious calculation he was able to present a single diagram to illustrate the character of these two focal surfaces in the generic case of three unequal axes. The great advantage of high-speed interactive computer graphics is that we can display an entire sequence of pictures as we hold two of the axes fixed and vary the third. We consider the one-parameter family

$$\frac{X^2}{(.5)^2} + \frac{y^2}{b^2} + \frac{z^2}{(.7)^2} = 1$$

where b runs through the values from .4 to .8 and we show the configurations we obtain for .4, .5, .6, .7, and .8.

In order to obtain a better view of the relationship between the original ellipsoid and its two focal surfaces, we show only the lower half, corresponding to values $z \leq 0$, together with the two half focal surfaces which are determined by this half-surface. When b is less than .5, both half focal surfaces lies below the plane containing the boundary ellipse where $z = 0$. If b is between .5 and .7 one lies above and one below, and if b is greater than .7, both lie above. The cases $b = .5$ and $b = .7$ are surfaces of revolution with two equal axes.

It is instructive to consider the intersection of the ellipsoid and the two focal surfaces with the three coordinate planes. In the cases where two axes are equal, we obtain an astroid and a doubly covered segment for two planes and a circle together with a point for the third.

If we now vary the parameter b, the single point will break into a small astroid while the doubly covered segment will become a thin ellipse. We illustrate the case when $b = .6$. In each of the coordinate planes one piece of the focal surface intersection will be given by the evolute of focal curve of the ellipse of intersection of the ellipsoid and the coordinate plane. The other part of the focal surface will come from an ellipse in the coordinate plane. In the case of the plane containing both the major and the minor axis of the ellipsoid, the astroid and the ellipse are tangent at four points which are the umbilics of the ellipsoid, where the two principal curvatures are identical. In this coordinate plane each intersection with the coordinate plane of a focal surface consists of two arcs of the ellipse plus two arcs of the astroid, fitting together to yield a curve with everywhere defined tangent line and four inflection points where the arcs match up.

In order to determine the intersection of the focal surfaces of the ellipsoid with a coordinate plane we observe that, by symmetry, the normals to an ellipse in a coordinate plane all lie in the plane so the focal set includes the evolute astroid of that ellipse. If the ellipse is parametrized by $(a \cos(t), 0, c \sin(t))$ then

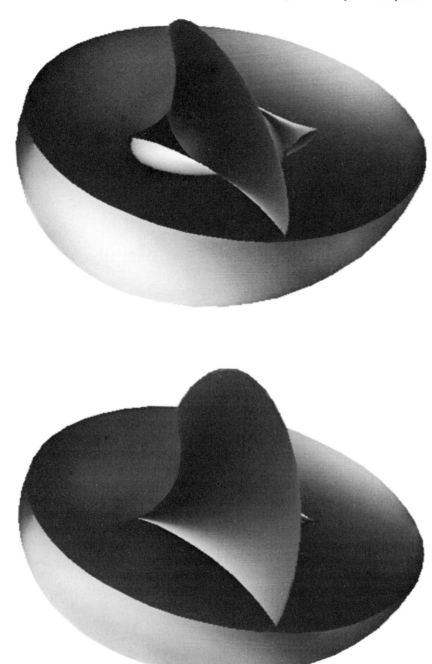

this astroid is $\left(\left(\dfrac{b^2 - c^2}{a}\right)\cos^3(t), 0, \left(\dfrac{c^2 - a^2}{c}\right)\sin^3(t)\right)$ corresponding to the

principal curvature $ac/(a^2\sin^2 t + c^2\cos^2 t)^{\frac{3}{2}}$.

In the orthogonal direction at the point $(a\cos(t_0), 0, c\sin(t_0))$ we find the

curvature of the ellipse $\dfrac{y^2}{(b\sin(t_0))^2} + \dfrac{z^2}{(c\sin(t_0))^2} = 1$ and use Meusnier's

theorem to obtain the principal curvature $ac/(b^2(a^2\sin^2(t_0) + c^2\sin^2(t_0))^{\frac{1}{2}}$

The locus of centers of curvature in this case is

$$\left(\left(\dfrac{a^2 - b^2}{a}\right)\cos t, 0, \left(\dfrac{c^2 - b^2}{c}\right)\sin(t)\right),$$

an ellipse.

The two principal curvatures will be equal at an umbilic point, and this occurs on a coordinate plane if $ac/b^2\sqrt{a^2\sin^2 t + c^2\cos^2 t} = ac/(a^2\sin^2 t + c^2\cos^2 t)^{\frac{3}{2}}$, i. e., if $b^2 = a^2\sin^2 t + c^2\cos^2 t$ for some value of t. When b^2 is between a^2 and c^2, this occurs exactly four times. The same algebraic condition guarantees that the ellipse and the astroid are tangent at these umbilic points.

For a modern treatment of this topic and an application to the bouyancy of ships, see the work of Zeeman [22].

The Veronese Surface

A number of different principles arise in the course of examining projections of geometric objects from 4-space into 3-space and eventually into planes and lines. Quite naturally singularities of composed functions come up, especially when there is a constant background motion to help maintain an impression of 3-dimensionality of the image. Some of the principles are quite obvious once you think of them. But it is the images which make you think of them. Most of the basic ideas already appear in the study of one of our favorite surfaces in 4-space, the *Veronese Surface*.

Recall that the embedding of the real projective plane into $\mathbb{R}^6$ described by Veronese uses quadratic functions taking the same value at antipodal points on the unit sphere $x^2 + y^2 + z^2 = 1$, namely

$$V(x, y, z) = (x^2, y^2, z^2, \sqrt{2}\,xy, \sqrt{2}\,yz, \sqrt{2}\,zx).$$

Note that the image of V is contained in the hyperplane $u_1 + u_2 + u_3 = 1$ as well as the sphere $\Sigma(u_i)^2 = 1$ so the image lies in a small 4-sphere.

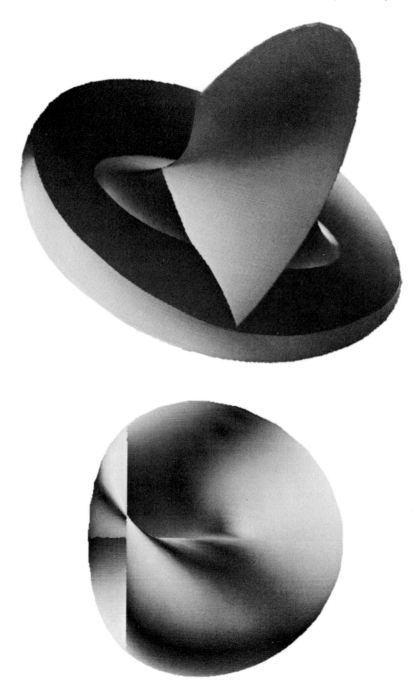

We can suppress the third coordinate and combine the first two by projecting to 4-space to get

$$(x, y, z) \rightarrow \left(\sqrt{2}\,xz, \sqrt{2}\,yz, \frac{1}{\sqrt{2}}(z^2 - x^2), \sqrt{2}\,xy \right)$$

a smooth embedding which we still call the *Veronese surface*.

1)   For almost all orthogonal projections $\pi: \mathbb{R}^4 \rightarrow \mathbb{R}^3$ to 3-space, the only singularities of the composed mapping $\pi \circ f: M^2 \rightarrow \mathbb{R}^3$ will be Whitney pinch points, either *elliptic*, with normal form $(u, v) \rightarrow (u, 2uv, u^2 - v^2)$ or *hyperbolic*, with normal form $(u, 2uv, u^2 + v^2)$. In each case the unique tangent line at the origin is given by the first coordinate axis and if we project this axis, we have a singular point at the origin where the rank of composed mapping is zero. In the elliptic case a small perturbation of the ramification point at the origin

$$(u, v) \rightarrow (2uv - 2 \in v, u^2 - v^2 + 2 \in u)$$

yields a singularity set with image a hypocycloid of three cusps. In the hyperbolic case, a small perturbation of the corner point at the origin

$$(u, v) \rightarrow (2uv - \in v, u^2 + v^2 + \in u)$$

yields a singularity set with image a convex curve together with a convex curve with a single cusp.

    Both of these unfoldings can be observed in the case of the Veronese surface as it is projected into 3-space and then rotated about an axis, as seen in the film *The Veronese Surface*. The deformations given above appear in another context in the monograph *Cusps of Gauss Mappings* [6].

2)   For almost all projections $\varrho: \mathbb{R}^3 \rightarrow \mathbb{R}^2$ the singularity set of the composed mapping $\varrho \circ \pi \circ f: M^2 \rightarrow \mathbb{R}^2$ will have rank at least one at all points and the pinch points for the map $\pi \circ f$ lie on all of these singularity curves. We can identify pinch points as those points which lie on the singularity curves of the projections into the three coordinate planes for almost any choice of basis in $\mathbb{R}^3$. This fact appears, in an essential way in the chain level version of the Whitney duality theorem established by C. McCrory and the author in [3].

3)   For almost all projections $\xi: \mathbb{R}^3 \rightarrow \mathbb{R}^1$, the level sets of the composed mapping $\xi \circ \pi \circ f: \mathbb{R}^4 \rightarrow \mathbb{R}^1$ will be a union of simple closed curves in M^2 except for critical values of this composed mapping. For almost all ξ, the level sets corresponding to critical values will contain just one critical point, either an isolated point corresponding to a local maximum or minimum, or a transversal crossing corresponding to a saddle point. If the critical point happens to be a pinch point of $\pi \circ f$, then the image of the level set under $\pi \circ f$ will either have an isolated point or a point when two curves have a common tangent. In the case of

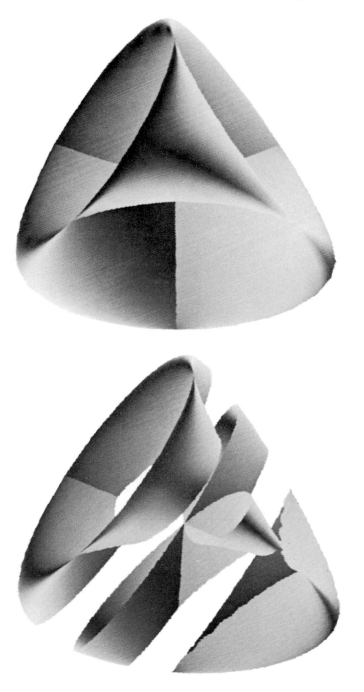

the Veronese surface almost any plane containing the unique tangent line at a pinch point image will intersect the image either in a single point or in a pair of tangent ellipses.

4)   If a pinch point p of $\pi \circ f$ is not a critical point of $\xi \circ \pi \circ f$, then the image under $\pi \circ f$ of the level set containing p has a cusp at $\pi \circ f(p)$. Images of level sets on one side of this set will have a transversal crossing and on the other side they will have a pair of inflection points. This fact is used in the refinement of the triple point theorem for stable mappings of surfaces into 3-space using an idea of R. Goldstein and E. Turner in the case of an immersed surface.
    A sequence of color pictures of level sets of the Veronese surface can be seen in the expository article [20].

5)   For almost all projections $\varrho : \mathbb{R}^3 \to \mathbb{R}^2$ and $\xi : \mathbb{R}^3 \to \mathbb{R}^1$, the envelope of the images under ϱ of the level curves of $\xi \circ \pi \circ f$ will be the image of the singularity curve of $\varrho \circ \varrho \circ f$. This fact, apparent in almost any view of the family of circles of latitude on the sphere, is more complicated in the case where the image of $\varrho \circ \pi \circ f$ has a cusp at p, in which case the image of the level curve of $\xi \circ \pi \circ f$ through p will have higher order of contact at $\varrho \circ \pi \circ f(p)$. (The same behavior was observed in the study of deformations of the Gauss mapping of the perturbation of the isolated parabolic umbilic given by the monkey saddle.) [5], [6].

6)   The original Veronese surface is a tight C^∞ mapping of a surface into 5-dimensional Euclidean space with image contained in no affine hyperplane, and Kuiper has demonstrated that this is the only possible image of a tight smooth mapping up to affine transformations. (See Kuiper's article in this volume for an up-to-date survey of this subject, and also [17]). L. Coghlan in his thesis has characterized the images of tight C^∞ locally stable mappings of the real-projective plane into $\mathbb{R}^3$, and he shows that the intersection of the images with the boundary of its convex hull is a convex surface with either two or four open convex discs removed, any two of them having boundary curves which meet at the images of a Whitney pinch point [13]. Examples of such mappings are the cross-cap and Steiner's Roman surface, obtained in projections of the Veronese surface from $\mathbb{R}^4$ to two different 3-dimensional subspaces. Both models are exhibited in Hilbert and Cohn-Vossen's *Anschauliche Geometrie* [16]. Since for a tight immersion the composition with an orthogonal projection is also a tight mapping, there must be a non-stable smooth tight mapping of the real projective plane into $\mathbb{R}^3$ in any 1-parameter family of projections joining the cross-cap to the Roman surface. One such deformation is given by

$$\left(\sqrt{2}\, xy, \sqrt{2}\, yz, \cos\alpha \left(\frac{1}{\sqrt{2}} \right) (z^2 - x^2) + \sin\alpha \sqrt{2}\, xy \right)$$

or in geographical coordinates by

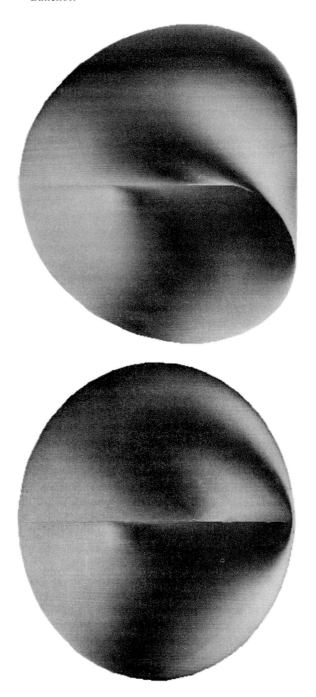

$$(\theta, \varphi) \rightarrow \frac{1}{\sqrt{2}} \left(\cos \theta \sin 2 \varphi, \sin \theta \sin 2 \varphi, \right.$$

$$\left. \cos \alpha \left(\frac{1}{\sqrt{2}} \right) (\sin^2 \varphi - \cos^2 \theta \cos^2 \varphi) + \sin \alpha \sin 2 \theta \right).$$

This will possess a degenerate singularity when $\tan \alpha = \sqrt{2}$. At this singular position there is a segment on the boundary of the convex hull which is covered four times. Arbitrarily near to this value of α there are stable mappings with six pinch points and others with two pinch points. In the former case there are three double curves intersecting at a triple point and in the latter there is one double curve and no triple point. One of the projections of the non-stable transition is an equilateral triangle. (We would like to thank Louis Michel for first pointing out the relationship of this example to Kostant's convexity theorem). We illustrate some of the images in a neighborhood of this singular position.

For additional treatments of the Veronese Surface including computer graphics illustrations, see [4], [8], and [10].

Bibliography

[1]   Banchoff, T. *Minimal Submanifolds of the Bicylinder Boundary*, Boletim da Sociedade Brasileira de Matemática **7** (1976), 37—57.

[2]   Banchoff, T. *Polyhedral Catastrophe Theory I: Maps of the line to the Line*, Dynamical Systems, Academic Press (1973), 7—22.

[3]   Banchoff, T. *Whitney Duality and Singularities of Projections* (with Clint McCrory), Proceedings of Escuela Latino-americana de Mathematica, Rio de Janeiro, Springer-Verlag Lecture Notes in Mathematics **597** (1977), 68—81.

[4]   Banchoff, T. *Computer Animation and the Geometry of Surfaces in 3- and 4-Space*, Proceedings of the International Congress of Mathematicians, Helsinki (1978), (Invited 45 minute address), 1005—1013.

[5]   Banchoff, T. *Sur les points paraboliques des surfaces:* erratum et complements (with R. Thom), C. R. Acad. Sc. Paris, t. **291** (27 Octobre 1980), 503—505.

[6]   Banchoff, T. *Cusps of Gauss Mappings* (with T. Gaffney and C. McCrory), Pitman Advanced Publishing Program, **55** (1982), London, pp. 1—88.

[7]   Banchoff, T. *Circular and Countercircular Images of Plane Curves* (with E. Beckenbach), General Inequalities **3** (1983), ISNM 64, Birkhäuser-Verlag, 321—337.

[8]   Banchoff, T. *The Nine-Vertex Complex Projective Plane* (with W. Kühnel), Mathematical Intelligencer **5** (1983), 11—22.

[9]   Banchoff, T. *DIAL: A Diagrammatic Animation Language* (with S. Feiner and D. Salesin), IEEE Computer Graphics and Applications, Vol. 2, No. 7 (1982), 43—54.

[10]   Banchoff, T. *Computer Graphics in Geometric Research*, Recent Trends in Mathematics, Teubner-Texte **50** (1983), 316—327.

[11]   Banchoff, T. *Counting Tritangent Plane of Space Curves* (with T. Gaffney and C. McCrory), (to appear in *Topology*).

[12]   Cayley, A. *On the Centro-Surface of an Ellipsoid*, Trans. Cambridge Philos. Soc. **12** (1873) 319—365.

[13]   Coghlan, L. *Tight Stable Mappings of Surfaces*, Thesis, Brown University, 1984.

[14]   Dyck, W. *Katalog mathematischer und mathematisch-physikalischer Modelle, Apparate und Instrumente*, München, 1982.

[15]   Dyck, W. *Spezial-Katalog der Mathematischen Ausstellung*, Berlin, 1893.

[16]   Hilbert, D. and Cohn-Vossen, S. *Anschauliche Geometrie*, Göttingen, 1932.

[17]   Kuiper, N. *Stable Surfaces in Euclidean 3-Space* Math. Scand. **36** (1975) 83—96.

[18]   Kummer, R. *Die Flächen mit unendlichvielen Erzeugungen durch Translationen von Kurven*. Dissertation Leipzig, 1894.

[19]   Schilling, C. *Die Minimalflächen fünfter Klasse mit dem Stereoscop-Bild eines Modells derselben*. Dissertation Göttingen, 1880.

[20]   Segal, G. *The Fourth Dimension*. Science Digest Vol. **92** (1984) 68—71.

[21]   Veronese, G. *Grundzüge der Geometrie von mehreren Dimensionen und mehreren Arten gradliniger Einheiten in elementarer Form entwickelt*. (1894) Leipzig.

[22]   Zeeman, E. C. *Catastrophe Theory:* Selected Papers (1972—1977) Addison-Wesley Reading, Mass. 1977.

Perspectives in Mathematics
Anniversary of Oberwolfach 1984
© Birkhäuser Verlag, Basel

The Role of Mathematics in Theoretical Statistics

O. E. BARNDORFF-NIELSEN and D. R. COX

Afdeling for teoretisk Statistik, Matematisk Institut,
Aarhus Universitet, DK-8000 Aarhus C (Denmark)
and Department of Mathematics,
Imperial College of Science & Technology, London (U. K.)

1   Introduction

Statistics is a fascinating subject for a number of compelling reasons, among which we mention the following three. First, there is the challenge of formulating general principles and concepts for the collection, analysis and interpretation of data. Secondly, by contrast with the extreme specialization of so much scientific research, statistical notions find genuinely fruitful application in fields of activity ranging from high energy physics to social anthropology and from historical studies to aspects of current public affairs. Successful application is, moreover, often far from routine; applied statistical work can be demanding and intellectually rewarding. Thirdly, the conversion of general concepts into a mathematical framework powerful enough to handle a wide range of applications opens a rich mine of mathematical problems of varied character.

Applications of classical ideas in the theory of functions of a real variable and in linear algebra are commonplace in statistical discussions even at a quite elementary level and in the present paper we have chosen to emphasize a few rather more general topics calling for mathematical treatment. The choice of topics is, of course, a reflection of our own interests.

The mathematical basis of most statistical theory is the theory of probability, a most vigorous branch of mathematics in its own right, and one which we do not attempt to discuss here. Some aspects of the theory of statistics are, however, not especially probabilistic. For example, the theory of experimental design leads to

(a) combinatorial problems leading to applications of finite geometries, graph theory, and so on;

(b) optimization problems connected with topics in convex analysis;

(c) aspects of the theory of permutation groups, in connection with the randomization of experimental designs.

We shall, however, here concentrate on theoretical problems connected with the analysis of data.

As in other areas of application of mathematics, the level of generality and abstractness appropriate can be difficult to judge. There is strong evidence that undue preoccupation with mathematical considerations in their own right, with wide separation from the motivating scientific problems, is stultifying. Further, many statistical investigations are concerned with rather special situations arising in specific fields and undue generality in their discussion would be inhibiting. It would be very wrong to undervalue such special investigations. Nevertheless the very breadth and unpredictable variety of applications call for discussions of considerable generality and the deployment of powerful mathematics, leading hopefully to ultimate simplification and unification. We believe there to be scope for a range of mathematical ideas, along with a need to handle special problems by more elementary and informal methods. Fruitful evolution calls for diversity of approach.

We illustrate our discussion with one or two examples. Very simple examples, while not without practical importance, are primarily useful to illustrate ideas in their most primitive and striking form. While it is natural in a paper like this to concentrate on such simple illustrations, we have mentioned also rather less simple ones in order to give some idea of the flavour of applications.

A final general comment concerns the role of an axiomatic route to the development of statistical theory. It might seem an attractive approach to statistical inference to formulate clearly acceptable axioms about the nature of statistical reasoning and to deduce from these axioms procedures for the analysis of data. Unfortunately, while attempts along these lines have been made, they have served mainly to isolate the limitations of the initial axioms. Controversy over such 'philosophical' issues has continued in one form or another for 150—200 years and, while of great interest, has had relatively little impact on applications, in our judgment at least.

2   Statistical analysis: Some key ideas

2.1   Formulation

While no single formulation covers conveniently all the types of data that may arise, it is very frequently useful to consider the data as an $n \times q$ matrix, the so-called data matrix. The n rows label n distinct individuals under study, the individuals being runs of an experiment, people, animals, firms, and so on. The columns represent q types of measurement made on each individual.

Example 2.1. Suppose that on each of n widely separated days the wind direction at noon in a certain spot is recorded as an angle in $[0, 2\pi)$. If these are the only data, we have an $n \times 1$ data matrix. If wind speed also is recorded, the data matrix is $n \times 2$, and so on.

Example 2.2. Suppose that on each of n patients suffering from a certain disease we record length of time from starting treatment either to death or to the end of the study, whether dead (1) or alive (0) at the end of the study, one or more indicators of 'quality of life', descriptors of the treatment applied and measures of the patient's medical history before entry, and laboratory and other prognostic measurements taken on entry. A rather small study might have $n \cong 100$ and q, the number of important measures per patient, about 10; a large study might have $n \cong 10^4$ and $q \cong 10^3$ and such quantities of data are by no means unusual in many fields.

The columns of the matrix can be classified in various ways, but here we mention only the division into

(a) response variables whose distribution we wish to study,

(b) explanatory variables, to be used to 'explain' systematic features in the behaviour of the response.

In Example 2.1, there are no explanatory variables.

In Example 2.2, response variables are survival time and quality of life and the object is to study how these are influenced by the treatment applied and the initial state of the patient.

Finally let $\mathscr{S}_j$ denote the set of logically conceivable values for the entries in the jth column of the data matrix. In concrete applications $\mathscr{S}_j$ may be just two points (0,1), a finite or countably infinite set or more generally any subset of R^1 (digitized for computation). Thus each row of the data matrix takes values in $\mathscr{S} = \mathscr{S}_1 \times \cdots \times \mathscr{S}_q \subset R^q$. Conceptually it may be advantageous to allow more general possibilities, such as that each entry is multidimensional or even takes values in some function space, as when, for example, the first entry represents a velocity component as a function of time in some study of turbulent flow.

Of course in any particular phase of analysis only some of the columns in the data matrix may be judged relevant, and in the discussion below we suppose at all points that the data matrix is not necessarily the full matrix as originally recorded, but that appropriate.

2.2   Initial analysis and formulation

Especially in complex problems, it is important to begin by simple analysis of a largely informal kind in which columns of the data matrix (variables) are examined one or at most two at a time, the main features of their distribution assessed and possibly anomalous values detected. It would be good to have more theory to guide this stage of analysis. For large data matrices, formidable computational problems can be involved in managing the data in an accessible form; discussion of these issues belongs to theoretical computer science.

Most theoretical discussions in statistics stem, however, from the following very major steps.

I.   For the ith individual let y_i and x_i denote respectively the response and explanatory variables, these being in general vectors. Then we regard y_i as the observed value of a random variable Y_i having a distribution depending on x_i.

II.   We regard the probability distribution, or at least certain aspects of it, as the thing of real interest, the data being of concern only in so far as they give information about the probability distribution.

III.   We normally restrict the distribution of Y_i and the dependence on x_i to some particular family of distributions, called the *statistical model* for the problem. Such models may be modified during the course of analysis.

We now comment briefly on these steps in the context of the two examples of Section 2.1.

Example 2.1 (cont'd). There being no explanatory variable here, the simplest model would be that $Y_1, \ldots, Y_n$ are independent and identically distributed random variables with support $[0, 2\pi)$. Of course possible seasonal variation and correlation between adjacent values are here being provisionally assumed unimportant. Further it would be common to assume the distribution absolutely continuous and hence determined by a probability density $p(.)$, say; note that there is some idealization here in that the recorded data are inevitably rounded, i.e. discrete. There are now two types of model for consideration:
(a) nonparametric, in which $p(.)$ is assumed only to lie in some fairly rich space of densities, being restricted to be continuous, to be symmetric around some angle (in an obvious sense), to have at most k maxima, or in some such way;
(b) parametric, in which $p(.)$ is provisionally assumed to have a known functional form described by an unknown vector parameter ω taking values in some region Ω of d dimensional Euclidean space, called the parameter space; of course more abstract parameter spaces can be contemplated. The most important example is to take $p(.)$ as having the von Mises form

$$p(y; \alpha, \varkappa) = \{I_0(\varkappa)\}^{-1} \exp\{\varkappa \cos(y - \alpha)\}. \tag{2.1}$$

Here $\omega = (\alpha, \varkappa)$, the parameter space is $[0, 2\pi) \times R^+$ and $I_0(.)$ is a Bessel function. For $\varkappa > 0$, the density has a maximum at $y = \alpha$, the concentration of the density around α increasing with the value of $\varkappa$. An even more restricted form, idealized but useful for discussion, is to suppose that the parameter $\varkappa$ is known, equal to $\varkappa_0$, say, so that the unknown parameter is the angle α and the parameter space $[0, 2\pi)$.

Example 2.2 (cont'd). Suppose for simplicity that all individuals in Example 2.2 are observed until death and that the time to death is taken as the single response variable. If there were no explanatory variables, about the simplest parametric model that might be contemplated would have $Y_1, \ldots, Y_n$ independently and identically distributed on the positive real line with the exponential density

$$\varrho \, e^{-\varrho y}$$

with parameter ϱ and mean $1/\varrho$. To represent the effect of explanatory variables, we might have $Y_1, \ldots, Y_n$ independently exponentially distributed with parameters $\varrho_1, \ldots, \varrho_n$, where

$$\varrho_i = \varrho_0 \, e^{\beta^T x_i}. \tag{2.2}$$

Here x_i is the vector of explanatory variables for the ith individual, β is an unknown parameter vector and ϱ_0 is the baseline parameter corresponding to an individual with $x = 0$. The parameter vector is thus $\omega = (\beta, \varrho_0)$.

Such a model allows the effect of explanatory variables on survival time to be represented in a reasonably flexible way. We shall see in Section 5 how the strong link with the exponential distribution can be removed.

To summarize, a parametric statistical model for a given data matrix consists of

(a) the space $\mathscr{S}$ of possible values,

(b) a division of the columns into response and explanatory variables,

(c) a family of probability distributions for the responses, indexed by an unknown parameter vector ω,

(d) the parameter space of possible values for ω.

The physical idea behind the model is that for given explanatory variables it represents what would happen if a large number of individuals were taken under the conditions of the investigation. As such the parameters describe the behaviour of a large collection or population of individuals and provide also a base for predicting the response of a new individual.

Statistical models are chosen on the basis of previous experience of similar applications, of analysis of the data itself, and sometimes on the basis of a theoretical analysis of the system under study. Of course models leading to simple methods of analysis and elegant theory are particularly appealing. Beyond stressing the critical importance of model choice, we shall in the rest of the discussion suppose that a parametric family of models is specified.

Quite often the parameter ω is partitioned as $\omega = (\psi, \chi)$, interest being focused on the value of ψ, the value of χ being of no immediate concern. Then ψ is called the parameter of interest and χ is called the nuisance parameter. Thus in Example 2.1, with $\omega = (\alpha, \varkappa)$, it might happen that interest focused on the parameter $\varkappa$ and not on the angle α, which would then be a nuisance parameter.

3   More detailed development

3.1   Some key ideas

The mathematical basis for parametric statistical inference is thus the *model function* $p(y; \omega)$ which for any given value of the parameter ω specifies the probability or probability density of any of the possible observational outcomes y of the response variable. Thus $p(y; \omega)$ establishes the connection between the observational or sample space $\mathcal{Y}$ and the parameter space Ω, and we think of a parametric statistical model as a triplet $(\mathcal{Y}, \Omega, p)$. The term *probability function* is used for $p(y; \omega)$ considered as a function of y for a fixed ω, while for a fixed observational y the function L of ω given by $L(\omega) = p(y; \omega)$ is called the *likelihood function* for ω.

The concept of likelihood, introduced by R. A. Fisher in Fisher (1912) and embodied in the likelihood function, is the most important unifying concept in theoretical statistics. From this flows generally applicable methods of (i) estimating parameters and assessing the uncertainty of the estimates; (ii) specifying regions of 'plausible' or 'likely' values of parameters according to a standard scale of plausibility; (iii) testing hypotheses, specified as subsets of the full parameter space Ω.

There are further concepts closely related to likelihood, such as *sufficiency* and *ancillarity*. These also originated with Fisher and have been developed by many others.

Through the application and development of these concepts and methods, in connection with a great variety of problems, it gradually became clear that many model functions share mathematical properties of relevance for statistical inference. This has led to the delineation and study of various main classes of statistical models, notably the *transformation models* and the *exponential models*. In the investigation of these classes very diverse and advanced mathematical results are now being employed. In particular, the theories of (quasi and relatively) invariant measures, of differential and integral geometry, and of Lie algebras and Lie groups have in recent years been shown to provide powerful tools for studying the structures and properties of the model classes.

In the following we discuss in a little more detail a number of these ideas and developments.

3.2   Large sample, intermediate and exact results

Broadly speaking the results of statistical theory can be classified as exact, as large-sample asymptotic or as what we shall term intermediate asymptotic. Exactness means two rather different things. The first is that the functions of the data used in analysis are in some sense uniquely appropriate or optimal. Secondly any distributional calculations are done without mathematical approximation. The derivation and application of the Student t distribution is an instance. Of course all this refers to a given probability model: in any application such models are at best very good approximations, so that it is important not to overinterpret the word 'exact'.

Exact results are available for important classes of problems of both theoretical interest and of practical concern. Nevertheless there are many situations, especially in relatively complex studies of dependency, where approximations are necessary and these are based on large sample theory, in which the dimension, n, of the observed random variable is supposed large. The limit laws of probability theory, especially the central limit theorem, are used to suggest approximations, such as that maximum likelihood estimators have normal distributions. The notion of allowing $n \to \infty$ is a mathematical device to generate approximations and does not have (or hardly ever has) a direct physical interpretation.

From the viewpoint of asymptotic expansions the ordinary large sample theory has the character of the simplest form of first order asymptotics, and asymptotic analysis offers the possibility of refining the first order results, thereby helping to bridge the gap between the realm of exact results and the large sample theory, leading to intermediate asymptotics. In fact, the use of expansions for this purpose is almost as old as the subject of mathematical statistics itself, dating back to the turn of the century. One of the oldest, and the most versatile, type of expansion is the Edgeworth expansion, introduced in Edgeworth (1905).

It may be recalled that the central limit theorem is proved by showing that the characteristic function (Fourier-Stieltjes transform) $E(e^{itZ_n})$ of the distribution of the standardized sum Z_n of n independent and identically distributed random variables tends to $e^{-\frac{1}{2}t^2}$ as $n \to \infty$. With the aid of suitable general theorems it follows that Z_n has asymptotically the standard normal distribution (whose characteristic function is $e^{-\frac{1}{2}t^2}$). The function $E(e^{itZ_n})$ can, under suitable restrictions, be expanded in powers of $1/\sqrt{n}$ and formal inversion suggests that the density of Z_n is asymptotically

$$\varphi(z)\left\{1+\frac{1}{6\sqrt{n}}\varrho_3 H_3(z)+\frac{1}{72n}[3\varrho_4 H_4(z)+\varrho_3^2 H_6(z)]+\cdots\right\}, \qquad (3.1)$$

where the $H_r(.)$ are Hermite polynomials, orthogonal with respect to the density $\varphi(z)=(2\pi)^{-\frac{1}{2}}e^{-\frac{1}{2}z^2}$ and γ_3 and γ_4 are the standardized third and fourth cumulants of the originating random variables. The expression (3.1) gives the first few terms of the Edgeworth expansion. Cramér (1928) was the first to prove that (3.1) is indeed an asymptotic expansion. When the random variables are vectors there is a generalization.

For a modern account of Edgeworth approximations and Edgeworth expansions, using tensor products and the advanced theory of differentiability, we refer to Skovgaard (1981); see also Chambers (1967) and Bhattacharya and Rao (1976).

The early interest in Edgeworth series and related expansions was rooted in the use of these to describe observed non-normal populations and to obtain numerical approximations to characteristics of theoretically defined distributions. With the development of techniques for obtaining exact results and of calculators the importance of the expansions decreased. However, within the last five years there has been a strong resurgence of interest in these expansions and in the related saddlepoint expansions. This is because it has emerged that these expansions are capable of yielding important insights of a kind that do not seem accessible by other means. In fact, there is now the prospect of a mathematically beautiful and readily applicable general theory providing a substantial improvement on the ordinary large sample theory, and at the same time encompassing a great part of the exact theory. Below we shall mention some instances of the kind of result we have in mind.

3.3   Likelihood and probability

The primitive interpretation of the likelihood function $L(\omega) = L(\omega; y)$ is that for any two parameter values ω' and ω the likelihood ratio $L(\omega')/L(\omega)$ measures how plausible or likely ω' is compared to ω as the actual parameter value, in the light of the data y. According to this, likelihood provides only a relative measure of plausibility and any factors in $L(\omega; y)$ that depend on y only are therefore immaterial. Any function on Ω obtained from the original likelihood function by deleting or adding such a factor is consequently considered to be equivalent to the original and is also denoted by $L(\omega)$. However, it is often useful to employ a particular representative of this class of equivalent versions of the likelihood function, namely the *normed likelihood function* $\bar{L}(\omega)$ defined by

$$\bar{L}(\omega) = L(\omega)/L(\hat{\omega}), \tag{3.2}$$

where $\hat{\omega}$ denotes the most likely value of ω, i.e. the value $\hat{\omega}$ such that $L(\hat{\omega})$ $= \sup_{\omega \in \Omega} L(\omega)$. (We shall disregard here the possibility that there is no unique maximum point of $L(\omega)$.)

Considered as a mapping from $\mathcal{Y}$ to Ω, $\hat{\omega}$ is called the *maximum likelihood estimator* of ω, while the actually observed value of the mapping is termed the *maximum likelihood estimate*.

It is often more natural and convenient to work with the log likelihood function $l = \log L$ rather than with the likelihood function L itself. We write $\bar{l}$ for $\log \bar{L}$. The maximal value of $\bar{l}$ is 0 and for any $c \geq 0$ one speaks of the set

$$D_c = \{\omega : \bar{l}(\omega) \geq -c\} \tag{3.3}$$

as the *likelihood region* corresponding to c log likelihood units. Any point in D_c has higher likelihood than any point outside. By describing D_c, for instance in diagrammatic form, for a few selected values of c one may summarize how the model $(\mathcal{Y}, \Omega, p)$ and the data together indicate the degree to which the various parameter values are adoptable as the actual or 'true' value.

However, the likelihood concept *per se* does not give definite guidance to how the c values should be selected. In other words, likelihood does not provide an operational interpretation of the likelihood regions D_c. To obtain such an interpretation, in probabilistic terms, one phrases the question: what is the probability that D_c contains the actual parameter value ω, i.e. what is $P_\omega\{\omega \in D_c\}$? It turns out that in very considerable generality this probability approximately depends on ω and on the particular model only through the dimension of the parameter ω. Denoting this dimension by d we have, in particular, that for d not too large ($d \leq 5$) the probability is approximately .95, .99 and .999 according as $c = d+1$, $d+3$ and $d+5$.

If ω is partitioned into a parameter of interest ψ and a nuisance parameter χ one may for many purposes proceed by introducing the *profile likelihood function* $\tilde{L}$ for ψ,

$$\tilde{L}(\psi) = \sup_{\omega|\psi} \tilde{L}(\omega), \tag{3.4}$$

and then acting as if $\tilde{L}(\psi)$ was a genuine likelihood function. For instance, if $\tilde{l} = \log \tilde{L}$ and if d now denotes the dimension of ψ we have that under broad conditions the profile likelihood region

$$\tilde{D}_c = \{\psi : \tilde{l}(\psi) \geq -c\}$$

contains the actual value of ψ with a probability of approximately .99 if $c = d+3$ and $d \leq 5$.

These results follow from a general theorem which states that the likelihood ratio statistic for testing the null hypothesis $\psi = \psi_0$,

$$w = -2\tilde{l}(\psi_0), \tag{3.5}$$

is when $\psi = \psi_0$ asymptotically distributed in a χ^2-distribution with d degrees of freedom, the probability density function of which is

$$\{\Gamma(\tfrac{1}{2}d)\, 2^{\frac{1}{2}d}\}^{-1} x^{\frac{1}{2}d-1} e^{-\frac{1}{2}x}. \tag{3.6}$$

Closely related to this and of importance in itself is another general theorem according to which the maximum likelihood estimator $\hat{\omega}$ approximately follows a d-dimensional normal distribution with mean ω and variance matrix $\hat{j}^{-1}$, where $\hat{j} = j(\hat{\omega})$ and

$$j(\omega) = -\frac{\partial^2 l(\omega)}{\partial \omega \, \partial \omega^{\mathsf{T}}}$$

is minus the matrix of second order derivatives of the log likelihood function. This conclusion also applies to profile likelihoods. The matrix j is sometimes called the observed information matrix.

These asymptotic or approximate propositions are large sample results, derived from the central limit theorem, and much statistical literature is concerned with establishing these results under ever wider conditions. Of much current interest are cases where the single observations of which y consists are stochastically dependent in some way or where the observational sampling scheme involves censoring or some form of optional stopping rule other than simply sampling for a fixed length of time or till a fixed number of observations have been taken. Especially noteworthy here is the way in which the deep and highly developed probabilistic theory of martingales (Shiryayev, 1981; Helland, 1982), though developed for its abstract interest, has found application in rigorous and general discussions of methodology for survival analysis, cf. for instance Andersen and Gill (1982).

The most essential requirement for the propositions to hold is that the amount of information in the data y concerning the parameter of interest should be large. The information is expressed in the likelihood function as a whole but may often be adequately summarized by the matrix $\hat{j}$, or its profile analogue.

Refinements and relevant modifications of these large sample results, including exact theory, are also part of the mainstream of statistics, and here many major areas of mathematics, such as differential and integral geometry, measure theory (decomposition, factorization and invariance of measures), abstract algebra and convex analysis come into the picture. We shall exemplify this after we have discussed the inferential and model structural background for such improvements.

3.4   Ancillarity and sufficiency

Any function, real or vector valued, of the data y is termed a statistic. A statistic t for which the normed likelihood function $\bar{L}$ depends on y through t only is said to be *sufficient* and the coarsest such function is the *minimal sufficient* statistic. It can be shown that the conditional distribution of y given the minimal sufficient statistic t does not depend on the parameter ω. This together with the defining property of minimal sufficiency shows that in a definitive sense t contains all the information on ω provided by the data y, under the given model.

Obviously, the maximum likelihood estimator $\hat{\omega}$ depends on y through the minimal sufficient statistic t. In certain important cases $\hat{\omega}$ is actually a one-to-one function of t. When this is not so one may seek an auxiliary statistic a such that $\hat{\omega}$ and a together are in one-to-one correspondence with t. As will be discussed later, it is, in fact, possible in considerable generality to construct a so that a is either exactly or approximately *distribution constant*, i. e. its distribution does not depend on ω and hence does not yield any information on ω. In this case a is said to be *ancillary* and the so-called *conditionality principle*, due essentially to R. A. Fisher, prescribes that inference on ω should proceed via the conditional distribution of $\hat{\omega}$ given a, thus substituting the original model with a derived model.

While sufficiency is a readily understandable concept and was quickly and generally accepted and investigated by statisticians, the idea embodied in the conditionality principle has taken a long time to gain a strong foothold. This is partly because the theoretical study of this idea has uncovered certain paradoxical aspects. However, to some extent these aspects seem to derive from a slightly inappropriate or incomplete formalization of the idea and, anyway, in many concrete situations the conditionality approach is found to be essential and sound. Despite the conceptual difficulties this approach is therefore subject to considerable interest and development; see Cox & Hinkley (1974, Chapter 2), Efron and Hinkley (1978), Cox (1980), Hinkley (1980), Barndorff-Nielsen (1980, 1983, 1984) and McCullagh (1984).

It is a useful property of the conditionality approach that there exists a general and simple expression for the model function of the conditional model. In fact, for arbitrary values of ω and of the ancillary statistic a, let us define a conditional distribution for $\hat{\omega}$ by specifying its probability function as

$$p^*(\hat{\omega}; \omega | a) = c|\hat{j}|^{\frac{1}{2}} \bar{L}. \tag{3.7}$$

Here $p^*(.; \omega | a)$ is the density (Radon-Nikodym derivative) of the defined distribution with respect to geometric measure λ on the range of $\hat{\omega}$ (thus λ is Lebesgue measure when this range is an open subset of some Euclidean space). Furthermore, $|\hat{j}|$ is the determinant of the observed information and c is a norming constant determined so as to make the integral of $c|\hat{j}|^{\frac{1}{2}} \bar{L}$ with respect to λ equal to 1. In general, c is a function of both ω and a, though in important cases c does not depend on ω; and in many applications c can be well approximated by $(2\pi)^{-\frac{1}{2}d}$, where d is the dimension of ω. Now, $p^*(\hat{\omega}; \omega | a)$ is generally either exactly or to a high degree of approximation equal to the actual conditional probability function of $\hat{\omega}$ given a. (No similar result exists without the conditionality approach.) We make a few further comments on this relation at a number of places in the following; for a detailed discussion of the derivation, properties and applications of (3.7) we refer to Barndorff-Nielsen (1983) (see also McCullagh (1984)). As a particular application it is possible from (3.7) to show the existence of a so-called Bartlett adjustment factor b which has the property that if the log likelihood ratio statistic (3.5) is modified to $w' = w/b$ then w' is often considerably closer to following the χ^2-distribution (3.6) than is w, another intermediate asymptotic result. Moreover b is simply expressible in terms of the norming constant c of (3.7), cf. Barndorff-Nielsen and Cox (1984).

Attempts have been made to extend the above concepts of sufficiency and ancillarity in order to deal similarly with cases where only a component ψ of ω is of interest, and some of these attempts have led to generally accepted procedures. (A fairly comprehensive survey is given in Barndorff-Nielsen (1978).) For instance, one may ask whether there exists a statistic t whose distribution depends on the interest parameter ψ only and which can be said to contain all the available information on ψ. Such a statistic could then be said to be *sufficient with respect to* ψ. Two definitions to this effect and dealing with substantially different cases are due to Fraser (1956) and Barnard (1963) respectively. The second of these definitions is based on considerations of (maximal) invariance under groups of transformations of the sample space $\mathcal{Y}$. Quite recently, Remon (1984) has combined these, seemingly disparate, ideas into a single definition which thus provides a broader solution. According to this a statistic t is said to be sufficient with respect to the parameter of interest ψ if the profile likelihood function (3.4) for ψ depends on y through t only and if the distribution of t depends on ψ only.

4   Two general families of models

4.1   Transformation models

Suppose there is a group G acting on the sample space $\mathcal{Y}$, i. e. there exists a mapping γ from G into the space of one-to-one transformations of $\mathcal{Y}$ such that for any g, g' in G, $\gamma(g'g) = \gamma(g') \circ \gamma(g)$, where $\circ$ denotes composition of mappings. For conciseness we write gy for $\gamma(g)y$ and $g\pi$ for the lifting by $\gamma(g)$ of a measure π on $\mathcal{Y}$. The set $Gy = \{gy : g \in G\}$ is called the *orbit* of the point $y \in \mathcal{Y}$. The orbits constitute a partition of $\mathcal{Y}$ and any function on $\mathcal{Y}$ which generates the same partition is said to be *maximal invariant*. The above action induces an action of G on the space of all probability measures on $\mathcal{Y}$ by the prescription $(g, P) \to gP$. If the class $\mathcal{P}$ of probability measures given by a statistical model is invariant under this action and if the action is transitive, i. e. if $\mathcal{P}$ consists of a single orbit, then the model is called a *pure transformation model* or, simply, a *transformation model*; while if the action is not transitive we speak of a *composite transformation model*. Since $\mathcal{P}$ is in one-to-one correspondence with the parameter space Ω we may, of course, equally well think of G as acting on Ω. The *index parameter* of a composite transformation model is a subparameter λ such that λ is a maximal invariant function of ω.

Example 4.1. Location-scale model. This is a prototype of the (pure) transformation models. Let $y_1, \ldots, y_n$ be independent and identically distributed observations, each following the probability function

$$\sigma^{-1} f\left(\frac{y-\mu}{\sigma}\right),$$

where f is a known probability density function on the real line. For instance, f may be the normal density $f(z) = (2\pi)^{-\frac{1}{2}} \exp(-z^2/2)$, or the Cauchy density $\pi^{-1}(1+z^2)^{-1}$, etc. The location parameter μ is an arbitrary real number while the scale parameter σ is positive. The model function for the data $y = (y_1, \ldots, y_n)$ is then

$$p(y; \mu, \sigma) = \sigma^{-n} \prod_{i=1}^{n} f\left(\frac{y_i - \mu}{\sigma}\right). \tag{4.1}$$

To describe this as a transformation model we let G be the group of elements $[\mu, \sigma]$ with the group operation defined by

$$[\mu', \sigma'][\mu, \sigma] = [\mu' + \sigma'\mu, \sigma'\sigma], \tag{4.2}$$

i.e. G is the positive affine group. The action of G on the sample space $\mathcal{Y} = R^n$ is determined by

$$([\mu, \sigma], y) \rightarrow (\mu + \sigma y_1, \ldots, \mu + \sigma y_n). \tag{4.3}$$

By means of the maximum likelihood estimator $(\hat{\mu}, \hat{\sigma})$ one may construct a maximal invariant statistic u as

$$u = (u_1, \ldots, u_n) = [\hat{\mu}, \hat{\sigma}]^{-1} y = \left(\frac{y_1 - \hat{\mu}}{\hat{\sigma}}, \ldots, \frac{y_n - \hat{\mu}}{\hat{\sigma}}\right). \tag{4.4}$$

This statistic is ancillary and according to the conditionality principle inference on (μ, σ) is to be performed under the conditional model for $(\hat{\mu}, \hat{\sigma})$ given u. This conditional probability function for $(\hat{\mu}, \hat{\sigma})$ turns out to be of the form

$$p(\hat{\mu}, \hat{\sigma}; \mu, \sigma | u) = c_0(u) \hat{\sigma}^{n-2} \sigma^{-n} \prod_{i=1}^{n} f\left(\frac{\hat{\mu} - \mu}{\sigma} + \frac{\hat{\sigma}}{\sigma} u_i\right). \tag{4.5}$$

Fisher (1934) used this example to illustrate his ideas of ancillarity and conditional inference and he termed the ancillary u the *configuration* of the sample $y_1, \ldots, y_n$. Note how the ancillary can be considered to determine the structure of the distribution (4.5) of the maximum likelihood estimator.

Fisher never formulated this or other related examples in group theoretic terms, and the pioneering effort in establishing that framework is due mainly to Fraser; see Fraser (1979) and references therein. Many of the most important

models in statistics are transformation models, the acting groups being subgroups of the general affine group $GA(n)$. Note that for these inference proceeds
via conditioning on the maximal invariant and ancillary statistic.

The class of transformation models includes virtually all models for
normal or Gaussian variates that allow a complete and explicit analysis. For
these latter models the action of G on $\mathcal{Y}$ is transitive so that the maximal
invariant is trivial and no conditioning is involved. (A more detailed discussion
is available in Barndorff-Nielsen (1982).) A special case, of central practical
importance, is the so-called normal theory linear model, whose theory is best
treated via vector space methods. There are a great many developments
stemming from that theory. Mathematically and statistically interesting treatments of models of this kind, using advanced algebraic methods, have been given
by Andersson (1975), Pukhal'sky (1981), Andersson, Brøns and Jensen (1983)
and Jensen (1975). These authors, however, do not view the models from the
standpoint of transformation models but utilize other types of invariance
properties that seem particular to the multivariate normal distribution. (Recent
comprehensive expositions, along more traditional lines, of the statistical
analysis of multivariate normal models are available in Muirhead (1982) and
Anderson (1984).)

In Fraser's theory of transformation models the action of G on $\mathcal{Y}$ is
assumed to be free, i.e. $g'y = gy$ implies $g' = g$, whatever y, g and g'. This
assumption can, however, be lifted. This is of interest *inter alia* in connection
with (3.7). In fact, it can be shown that for arbitrary transformation models p^* is
exactly equal to the conditional model function for the maximum likelihood
estimator, cf. Barndorff-Nielsen (1983). The proof of this comprise a lemma on
decomposition of measures and a demonstration that $|\hat{j}|^{\frac{1}{2}} d\hat{\omega}$ determines an
invariant measure on the parameter space, viewed as a homogeneous space
relative to the group action.

Example 4.1. Location-scale model (cont'd). Applying (3.7) to the model
(4.1), with $\omega = (\mu, \sigma)$ and a equal to the configuration u as given by (4.4), we find
$|\hat{j}| = D(u)\hat{\sigma}^{-4}$, where

$$D(u) = \{\Sigma g''(u_i)\}\{n + \Sigma u_i^2 g''(u_i)\} - \{\Sigma u_i g''(u_i)\}^2$$

and where $g = -\log f$. Furthermore, with $L(\mu, \sigma)$ given from (4.1), we have
$L(\hat{\mu}, \hat{\sigma}) = p(u; 0, 1)\hat{\sigma}^{-n}$, and hence

$$p^*(\hat{\mu}, \hat{\sigma}; \mu, \sigma | u) = c D(u)^{\frac{1}{2}} p(u; 0, 1)^{-1} \hat{\sigma}^{n-2} \sigma^{-n} \prod_{i=1}^{n} f\left(\frac{y_i - \mu}{\sigma}\right).$$

By the definition of c as a norming constant we find, on comparing with (4.5), that we must have $c = c_0(u) \, D(u)^{-\frac{1}{2}} p(u; 0, 1)$. Thus

$$p^*(\hat{\mu}, \hat{\sigma}; \mu, \sigma | u) = p(\hat{\mu}, \hat{\sigma}; \mu, \sigma | u).$$

It may be noted that from (4.5) one can, by integration, determine the distribution of the quantity $(\hat{\mu} - \mu)/\hat{\sigma}$. This distribution is fully known, i.e. it does not depend on the unknown parameters μ and σ, and it is available for separate inference on μ. We have here a generalization of Student's distribution for $(\bar{y} - \mu)/s$ from a normal sample. Quantities, like $(\hat{\mu} - \mu)/\sigma$, which depend on both the data y and the parameter ω and which have fully known distribution are called *pivots*.

Passing to the more general class of composite transformation models, we still have that the acting groups G are, typically, subgroups of the general affine group $GA(n)$. The von Mises model for two-dimensional unit vectors, mentioned earlier, is a simple instance of such a model, the precision $\varkappa$ being the index parameter and G being the group of rotations of the unit circle, i.e. $SO(2)$. There are important extensions of the von Mises model to unit vectors of dimension three (the Fisher model) and higher, with acting groups $SO(k)$, $k = 3, 4, \ldots$. These, then, are models for observations on the unit spheres S^{k-1}. Furthermore, there are related models for observations on other classical manifolds, such as the unit hyperboloids H^{k-1} (acting group $SO^{\uparrow}(1, k-1)$), the Stiefel manifolds $S(r, m)$ (acting group $O(r)$), and the Grassmann manifolds $G(r, m)$ (acting group $O(r)$). For some details and leads to the literature, see Barndorff-Nielsen, Blæsild, Jensen and Jørgensen (1982).

4.2   Exponential models

Exponential models are characterized by having model function of the form

$$p(y; \omega) = a(\omega) \, b(y) \exp\{\theta^{\mathrm{T}}(\omega) \, t(y)\}, \tag{4.6}$$

where $\theta(\omega)$ and $t(y)$ are $k \times 1$ vectors. We assume that the representation (4.6) is minimal, i.e. it is not possible to represent $p(y; \omega)$ in this way with a dimension lower than k. Then $t = t(y)$ is minimal sufficient and k is said to be the order of the model. Let $\Theta_0 = \theta(\Omega)$ and let

$$\Theta = \{\theta \in R^k : \int b(y) \exp\{\theta^{\mathrm{T}} t(y)\} \, d\mu(y) < \infty\}. \tag{4.7}$$

The set Θ is a convex subset of R^k.

If $\Theta_0 = \Theta$ then the exponential model is said to be *full*. The normal model with parameter (μ, σ) and the von Mises model with parameter $(\alpha, \varkappa)$ are examples of full exponential models, of order 2. For a full model we may take θ

as the parameter and, letting $K(\theta) = -\log a(\theta)$, we have the important property that $K(\theta)$ is the socalled cumulant transform of the distribution of $t = t(y)$. Thus the mean, variance matrix, and higher order cumulants of t are simply determined as derivatives of $K(\theta)$. Furthermore, $K(\theta)$ is a closed convex function on Θ and this and the fact that the log likelihood function is

$$l(\theta) = \theta^{\mathrm{T}} t - K(\theta)$$

is the basis for applying convex duality analysis in the study of the statistical properties of exponential models, cf. Barndorff-Nielsen (1978), Chentsov (1982), and Barndorff-Nielsen and Blæsild (1983a).

Under a full exponential model there are no ancillary statistics and formula (3.7) yields an approximation to the unconditional distribution of $\hat{\theta}$. This approximation is the leading term of the saddlepoint expansion for the distribution of $\hat{\theta}$; such expansions and their statistical applications are discussed in Barndorff-Nielsen and Cox (1979).

In statistical physics models of the full exponential type occur often, both in classical statistical mechanics where they are known as Boltzmann-Gibbs laws and in the study of critical phenomena, the Ising model being a prime case. Saddlepoint approximations are of importance in this area too. However, the exponential model properties studied in statistical physics are mostly of a kind different from those considered in theoretical statistics.

When the exponential model (4.6) is not full but Θ_0 is a d-dimensional submanifold of Θ one speaks of a *curved exponential model* or a *(k, d) exponential model*. The manifold Θ_0 becomes a Riemannian manifold when endowed with the so-called Fisher metric or expected information metric $i = i(\omega)$, defined as the mean value $i(\omega) = E_\omega\{j(\omega)\}$ of the observed information $j(\omega)$ defined in Section 3.3. In addition to the metric tensor i there is a one parameter family of connections $\{\overset{\alpha}{\nabla}: -1 \leqq \alpha \leqq 1\}$, defined in terms of i and of the mean values of the products

$$\frac{\partial l}{\partial \omega_r} \frac{\partial l}{\partial \omega_s} \frac{\partial l}{\partial \omega_t} \quad \text{and} \quad \frac{\partial^2 l}{\partial \omega_r \partial \omega_s} \frac{\partial l}{\partial \omega_t} \quad (r, s, t = 1, 2, \ldots, d)$$

of partial derivatives of the log likelihood function l. The asymptotic properties of the curved exponential family are largely determined by the geometric properties of the manifold as embodied in the Fisher metric i and the connections $\overset{\alpha}{\nabla}$, see Amari (1982a, b) and Amari and Kumon (1983). Of particular interest are those curved exponential models which are also transformation models. All the composite transformation models referred to at the end of the section on transformation models provide examples of this, by fixation of the index parameter. In this context, see also Eriksen (1984).

As mentioned previously, (3.7) gives exactly the conditional distribution of the maximum likelihood estimator when the model is a transformation model. However, there are interesting examples of exponential models that are not of the transformation type and for which (3.7) is still exact, cf. Barndorff-Nielsen (1983) and Barndorff-Nielsen and Blæsild (1983b), and it is an open problem of some interest to describe all exactness cases of (3.7).

5   Survival analysis and point processes

We conclude with some comments on an important special type of application, the analysis of survival data. The objective is to give some quick appreciation of the range of special models and problems that arise in just one area. A simple special case has already been outlined in Example 2.2.

Example 2.2 (cont'd). The model (2.2) for n individuals based on independent exponential distributions has as likelihood function

$$\prod_{i=1}^{n} \varrho_0 \, e^{\beta^T x_i} \exp\left(-\varrho_0 \, e^{\beta^T x_i} y_i\right), \tag{5.1}$$

each individual contributing a factor. The log likelihood function is thus

$$l(\varrho_0, \beta) = n \log \varrho_0 + \sum_{i=1}^{n} \beta^T x_i - \varrho_0 \sum_{i=1}^{n} e^{\beta^T x_i} y_i. \tag{5.2}$$

Although (5.2) has a quite simple form, reduction by sufficiency is not possible when all the vectors x_i are distinct and the exact arguments outlined in Section 4 are not available. Recourse is necessary to large-sample or intermediate asymptotics; details of the latter have not yet been fully worked out. Numerical maximization of (5.2) is usually the starting point in applications.

There are many generalizations of Example 2.2 of practical importance and theoretical interest. First, some individuals may be censored, i.e. after a period of observation y_i it may be that failure (death) of the ith individual has not occurred. Then, provided that the censoring is 'uninformative', such individuals contribute to the likelihood instead of the density only the survival probability

$$\exp\left(-\varrho_0 \, e^{\beta^T x_i} y_i\right).$$

Next it is possible that the vector of explanatory variables is a function of time, $x_i(t)$, say, as, for example, when a time-varying treatment is applied. Then it can be shown that (5.2) is replaced by

$$l(\varrho_0, \beta) = n \log \varrho_0 + \sum_{i=1}^{n} \beta^T x_i(y_i) - \varrho_0 \sum_{i=1}^{n} \int_0^{y_i} e^{\beta^T x_i(u)} \, du. \tag{5.3}$$

All these results centre on the exponential distribution in that an individual with $x_i \equiv 0$ has exponentially distributed failure-time with density

$\varrho_0 e^{-\varrho_0 y}$. The most vivid interpretation of the parameter ϱ_0 is as the hazard or age-specific failure rate

$$\lim_{\Delta \to 0+} \frac{P(y < Y < y + \Delta | y < Y)}{\Delta} \tag{5.4}$$

which is constant if and only if the random variable Y is exponentially distributed.

Non-exponential distributions can be recovered by taking (5.4) equal to an appropriate function of $y, \varrho_0(y)$, say. When explanatory variables are present, (2.2) can then be generalized by taking (5.4) for the ith individual in the form

$$\varrho_0(y) e^{\beta^T x_i}. \tag{5.5}$$

This is the natural adaptation to the present case of the idea of the complete intensity function of a stochastic point process and indeed most of the present discussion can be extended to point processes.

In some contexts $\varrho_0(y)$ is taken as a function of known form but with unknown parameters, $\varrho_0(y) = \varrho_0 y^\gamma$ being one special case with $\gamma = 0$ recovering (2.2).

Another possibility is to allow $\varrho_0(y)$ to be an arbitrary unknown function. Then, however, the likelihood function contains an unknown function as well as unknown parameters, and a modification of the simple notion of likelihood is necessary before ideas corresponding to those of Section 4 can be applied to extract information about the parameter vector β, which would often be the aspect of most concern.

This modified or partial likelihood function is produced as follows. Order the failure-times, assumed distinct, $y_{(1)} < y_{(2)} < \cdots < y_{(r)}$. Let $x_{(j)}$ refer to the individual failing at $y_{(j)}$ and let $\mathcal{R}_{(j)}$ be the set of individuals at risk of failure at $y_{(j)} - 0$, i.e. the set of individuals who have not already failed or been censored. Then, given a failure at $y_{(j)}$ the probability that the failure was on the individual in fact observed to fail is

$$\exp(\beta^T x_{(j)}) / \sum_{k \in \mathcal{R}_{(j)}} \exp(\beta^T x_k) \tag{5.6}$$

independently of the function $\varrho_0(y)$. Multiplication of the factors (5.6) from each failure leads to the function

$$\sum_j \beta^T x_{(j)} - \sum_j \log \left\{ \sum_{k \in \mathcal{R}_{(j)}} \exp(\beta^T x_k) \right\}. \tag{5.7}$$

Formula (5.7) is an example of a partial log likelihood function (Cox, 1975).

To prove rigorously in any generality that maximum likelihood estimates and associated statistics derived from (5.7) have the same large-sample asymptotic properties as 'ordinary' maximum likelihood statistics raises a formidable probability problem resolved by an appeal to the theory of martingales. Essentially one considers the gradient of (5.7) as a random step function evolving in time as failures occur. With that process can be associated a martingale (Brémaud, 1981) and application of a martingale central limit theorem yields the desired results (Andersen and Gill, 1982; Davis, 1983).

Finally it is worth stressing the wide-ranging role of statistical considerations in such studies, embracing,

(i) the design of observational and experimental studies aiming to ensure clear interpretation;
(ii) the monitoring of data quality;
(iii) techniques for graphical and other descriptive analysis of large sets of data;
(iv) the formulation of flexible and interpretable models as a base for more searching analysis;
(v) the development of techniques for model checking, as well as for analysis in the light of a given model;
(vi) the examination of computational and numerical analytical aspects especially associated with relatively complicated models with parameters of high dimensionality;
(vii) the critical interpretation of conclusions.

For general reading on the topics outlined in this paper, we suggest Barndorff-Nielsen (1978) on exponential families, Barndorff-Nielsen (1982) and Barndorff-Nielsen, Blæsild, Jensen & Jørgensen (1982) on transformation models and exponential transformation models, Cox & Hinkley (1974) on the general ideas of statistical inference, and Cox & Oakes (1984) and Andersen & Gill (1982) on the analysis of survival data.

We have not discussed the various areas of applied probability though many parts of these are of importance in constructing appropriate parametric models for statistical analysis. Stereology is one such area, which is currently in fruitful interaction with statistics, cf. Baddeley, Gundersen, Jensen and Sundberg (1984). In the context of the present paper this field is of some particular interest because of its significant applications and developments of results of integral geometry.

Support from a NATO Research Grant (RG 011.80) is acknowledged with thanks, as is the support of the second author by the Science and Engineering Research Council.

References

Amari, S.-I., (1982a). Differential geometry of curved exponential families—curvatures and information loss. *Ann. Statist.* **10**, 357—385.

Amari, S.-I., (1982b). Geometrical theory of asymptotic ancillarity and conditional inference. *Biometrika* **69**, 1—17.

Amari, S.-I. and M. Kumon, (1983). Differential geometry of Edgeworth expansion in curved exponential family. *Ann. Inst. Statist. Math.* **35**, 1—24.

Andersen, P. K. and R. D. Gill, (1982). Cox's regression model for counting processes: a large sample study. *Ann. Statist.* **10**, 1100—1120.

Anderson, T. W., (1984). *Introduction to Multivariate Statistical Analysis.* 2nd edition. Wiley, New York.

Andersson, S. A., (1975). Invariant normal models. *Ann. Statist.* **3**, 132—154.

Andersson, S. A., H. Brøns and S. T. Jensen, (1983). Distribution of eigenvalues in multivariate statistical analysis. *Ann. Statist.* **11**, 392—415.

Baddeley, A., H. J. G. Gundersen, E. B. Jensen and R. Sundberg, (1984). Recent trends in stereology. Research Report 98, Dept. Theor. Statistics, Aarhus Univ.

Barnard, G. A., (1963). Some logical aspects of the fiducial argument. *J. R. Statist. Soc.* **B 25**, 111—114.

Barndorff-Nielsen, O. E., (1978). *Information and Exponential Families.* Wiley, Chichester.

Barndorff-Nielsen, O. E., (1980). Conditionality resolutions. *Biometrika* **67**, 293—310.

Barndorff-Nielsen, O. E., (1982). Parametric statistical models and inference: some aspects. Research Report 81, Dept. Theor. Statistics, Aarhus Univ.

Barndorff-Nielsen, O. E., (1983). On a formula for the distribution of the maximum likelihood estimator. *Biometrika* **70**, 343—365.

Barndorff-Nielsen, O. E., (1984). On conditionality resolution and the likelihood ratio ancillary for curved exponential models. *Scand. J. Statist.* **11**. (To appear).

Barndorff-Nielsen, O. E. and P. Blæsild, (1983a). Exponential models with affine dual foliations *Ann. Statist.* **11**, 753—769.

Barndorff-Nielsen, O. E. and P. Blæsild, (1983b). Reproductive exponential families. *Ann. Statist.* **11**, 770—782.

Barndorff-Nielsen, O. E., P. Blæsild, J. L. Jensen and B. Jørgensen, (1982). Exponential transformation models. *Proc. R. Soc.* **A 379**, 41—65.

Barndorff-Nielsen, O. E. and D. R. Cox, (1979). Edgeworth and saddle-point approximations with statistical applications. (With discussion.) *J. R. Statist. Soc.* **B 41**, 279—312.

Barndorff-Nielsen, O. E. and D. R. Cox, (1984). Bartlett adjustments to the likelihood ratio statistic and the distribution of the maximum likelihood estimator. *J. R. Statist. Soc.* **B 46**. (To appear).

Bhattacharya, R. N. and R. R. Rao, (1976). *Normal Approximations and Asymptotic Expansions.* Wiley, New York.

Brémaud, P., (1981). *Point Processes and Queues; Martingale Dynamics.* Springer-Verlag, New York.

Chambers, J. M., (1967). On methods of asymptotic approximation for multivariate distributions. *Biometrika* **54**, 367—384.

Chentsov, N. N., (1982). *Statistical Decision Rules and Optimal Conclusions. Transl. Math. Mono.* **53**, Amer. Math. Soc., Providence, R. I.

Cox, D. R., (1975). Partial likelihood. *Biometrika* **62**, 269—276.

Cox, D. R., (1980). Local ancillarity. *Biometrika* **67**, 273—278.

Cox, D. R. and D. V. Hinkley, (1974). *Theoretical Statistics.* Chapman & Hall, London.

Cox, D. R. and D. Oakes, (1984). *Analysis of Survival Data.* Chapman & Hall, London.

Cramér, H., (1928). On the composition of elementary errors. *Skand. Aktuarietidskr.* **11**, 13—74, 141—180.

Davis, M. H. A., (1983). The martingale theory of point processes and its application to the analysis of failure-time data. In *Electronic Systems Effectiveness and Life Cycle Costing.* Berlin: Springer-Verlag. Ed. J. K. Skwirzynski, 135—155.

Edgeworth, F. Y., (1905). The law of error. *Cambridge Philos. Trans.* **20**, 113—141.

Efron, B. and D. V. Hinkley, (1978). Assessing the accuracy of the maximum likelihood estimator: observed versus expected Fisher information. (With discussion.) *Biometrika* **65**, 457—487.

Eriksen, P. S., (1984). (k, l) exponential transformation models. *Scand. J. Statist.* **11.** (To appear).

Fisher, R. A., (1912). On an absolute criterion for fitting frequency curves. *Messenger of Mathematics* **41**, 155—160.

Fisher, R. A., (1934). Two new properties of mathematical likelihood. *Proc. R. Soc.* **A 144**, 285—307.

Fraser, D. A. S., (1956). Sufficient statistics with nuisance parameters. *Ann. Math. Statist.* **27**, 838—842.

Fraser, D. A. S., (1979). *Inference and Linear Models.* McGraw-Hill, Toronto.

Helland, I., (1982). Central limit theorems for martingales with discrete or continuous time. *Scand. J. Statist.* **9**, 79—94.

Hinkley, D. V., (1980). Likelihood as approximate pivotal distribution. *Biometrika* **67**, 287—292.

Jensen, S. T., (1975). Covariance hypotheses which are linear in both the covariance and the inverse covariance. Preprint 1, Inst. Math. Statist., Univ. Copenhagen. (To appear in *Ann. Statist.*)

McCullagh, P., (1984). Local sufficiency. *Biometrika* **71.** (To appear.)

Muirhead, R. J., (1982). *Aspects of Multivariate Statistical Theory.* Wiley, New York.

Pukhal'sky, E. A., (1981). Minimal sufficient statistics for normal models with algebraic structure. *Theory Prob & Its Appl.* **26**, 564—572.

Remon, M., (1984). On a concept of partial sufficiency: *L*-sufficiency. *Int. Statist. Review* **52.** (To appear).

Shiryayev, A. N., (1981). Martingales: recent developments, results and applications. *Int. Statist. Rev.* **49**, 199—234.

Skovgaard, I. M., (1981). Edgeworth expansions in statistics. Preprint 12, Inst. Math. Statist., Univ. Copenhagen.

Perspectives in Mathematics
Anniversary of Oberwolfach 1984
© Birkhäuser Verlag, Basel

Rigid Analytic Geometry

SIEGFRIED BOSCH

Universität Münster, Mathematisches Institut,
Einsteinstrasse 62, D-4400 Münster (FRG)

When K. Hensel invented the p-adic numbers in 1905, he was probably not aware of the fact that his fields $\mathbb{Q}_p$ would once be of considerable interest from a function theoretic viewpoint. Moreover, first attempts to develop a function theory over such fields were not spectacular enough to draw much attention among the mathematical community. Due to the well-known topological difficulties (see section 1), it was necessary to restrict oneself to analytic functions on *simple domains* such as discs or annuli. Here, of course, a function is called analytic if it has a globally convergent power series or Laurent series expansion.

A serious attempt towards the discussion of analytic functions on *general domains* was made by M. Krasner in the forties. His theory of analytic continuation, which applies to functions of one variable, is still widely used today. However, a generalization to several variables is far from being obvious. There are profound obstacles which could not be overcome till now.

Another attempt towards a theory of analytic functions on *general domains* was made by J. Tate in 1961, when he gave a seminar at Harvard entitled "Rigid analytic spaces". Tate had discovered that, in the non-Archimedean case, elliptic curves with non-integral j-invariant can be viewed as analytic tori. In principal, this result can be understood in terms of function theory on *simple domains*. However a satisfying treatment of the subject needs a more sophisticated theory of analytic functions which allows the notion of a Riemann surface or of a global analytic space. Thus motivated, Tate tried to work out the details in his seminar. Thereby he laid the foundations of modern rigid analytic geometry, the theory of rigid analytic functions of several variables [19].

Rigid analysis was further promoted by H. Grauert and R. Remmert in the sixties. Just as Tate was influenced from algebraic geometry, Grauert and Remmert were thinking in terms of complex analysis. Their ideas revealed new aspects of the theory and were picked up by many other mathematicians. Subsequently, a completion and substantial simplification of the theory was achieved. In the present article, we take the opportunity to give an account of the basic concepts. Also we will discuss some applications of rigid analysis. For details concerning the general theory, the reader is referred to the monography [2].

Today, rigid analytic geometry is a well-established disciplin in mathematics which has interesting applications. To a certain extent, this is due to the conferences which have been organized on the subject. They helped to communicate new ideas and methods, and to integrate various developments in non-Archimedean analysis. In Germany, there were the facilities of the "Mathematisches Forschungsinstitut Oberwolfach" which could be used as a convenient and reliable meeting place having a warm atmosphere. It ist one of the merits of Oberwolfach that an international group of experts in non-Archimedean analysis could grow together, which now meets regularly in France, Germany and the Netherlands.

1   Rigid analysis on bounded domains in k^n

Let us fix an algebraically closed field k with a complete non-Archimedean valuation (supposed to be non-trivial). For any prime p, the field $\mathbb{C}_p$, completion of the algebraic closure of $\mathbb{Q}_p$, is a good example for k. We want to use k as a ground field and define analytic functions over k. In order to get a feeling for the subject, let us look at some elementary results of classical complex analysis whose non-Archimedean analogues one would like to prove:

(a) *Let f be analytic on a disc $D \subset \mathbb{C}$. Then f has a power series expansion which converges and represents f on D.*

(b) *Let f be analytic on an open set $U \subset \mathbb{C}$. Then, in terms of compact convergence, f can be approximated by rational functions having poles outside U.*

(c) *Let f be analytic on a compact Riemann surface. Then f is constant.*

The listed results are all of the following type: A local assumption on the function f, namely analyticity, and a topological assumption on the domain where f is defined yield a global property for f. Due to the total disconnectedness of non-Archimedean fields it is obvious that results of this type cannot be true over k if we use a *local* definition of analyticity. Thus a stronger notion is necessary. There are several possibilities. For example, one can take a classical result on analytical functions, translate it to the non-Archimedean case and use it as an axiom defining analyticity. If we do this with result (a), we end up with Tate's rigid analysis whereas Runge's theorem (b) is the point of departure for Krasner's theory of analytical elements.

Now let us describe a little bit more seriously the basics of Tate's theory. First one considers model spaces, so-called *affinoid varieties*, on which analytic functions — we say *affinoid functions* — can be defined easily. If V is an affinoid variety, the set of affinoid functions on V is denoted by $\mathcal{O}(V)$. The definitions read as follows:

(i) *The unit ball* $\mathbb{B}^n := \{(x_1, \ldots, x_n) \in k^n \,|\, |x_i| \leq 1, \, i = 1, \ldots, n\} \subset k^n$ *is an affinoid variety. The set* $\mathcal{O}(\mathbb{B}^n)$ *consists of all power series functions in n variables with coefficients in k which converge on* $\mathbb{B}^n$.

(ii) *Let* $f_1, \ldots, f_r \in \mathcal{O}(\mathbb{B}^n)$. *Then* $V := Z_{\mathbb{B}^n}(f_1, \ldots, f_r)$, *the common zero set of* $f_1, \ldots, f_r$, *is an affinoid variety;* $\mathcal{O}(V)$ *consists of all restrictions of functions in* $\mathcal{O}(\mathbb{B}^n)$ *to* V.

(iii) *Let $V \subset \mathbb{B}^m$, $W \subset \mathbb{B}^n$ be affinoid varieties as in* (ii). *An affinoid morphism $\varphi : V \to W$ is a map which, coordinatewise, is defined by affinoid functions on V.*

One verifies that each affinoid morphism $\varphi : V \to W$ induces a k-algebra homomorphism $\varphi^* : \mathcal{O}(W) \to \mathcal{O}(V)$ (just compose affinoid functions on W with φ), and that the affinoid varieties and morphisms form a category which has "nice" properties. The rings $\mathcal{O}(V)$ are accessible by methods of functional analysis and commutative algebra. They share several properties with finitely generated k-algebras. For example, they are Noetherian, and there is even a version of the Noether normalization lemma (where the role of free polynomial rings is played by the algebras $\mathcal{O}(\mathbb{B}^d)$, $d \geq 0$). The name *affinoid* has been chosen for this reason.

So far the unit ball $\mathbb{B}^n$ is the only open subset $V \subset k^n$ on which analytic functions have been defined. The crucial question is how to define $\mathcal{O}(U)$ if U is an open subset of an affinoid variety V. There is an easy answer if U is not too general. Namely, let us assume that

$$U = \{x \in V \mid |f_1(x)|, \ldots, |f_r(x)| \leq |g(x)|\}$$

where $f_1, \ldots, f_r, g \in \mathcal{O}(V)$ have no common zero on V. (The set U is called a *rational subdomain* of V). Let $X_1, \ldots, X_r$ be the coordinate functions on $\mathbb{B}^r$ and consider the zero set

$$U' := Z_{V \times \mathbb{B}^r}(f_1 - g X_1, \ldots, f_r - g X_r).$$

Then U' is an affinoid subvariety of $V \times \mathbb{B}^r$, and the projection $V \times \mathbb{B}^r \to V$ induces an injective affinoid morphism $\varphi : U' \to V$ which identifies U' with U. Now set $\mathcal{O}(U) := \mathcal{O}(U')$ and interpret the algebra homomorphism $\varphi^* : \mathcal{O}(V) \to \mathcal{O}(U')$ as the map which restricts affinoid functions on V to U. Thereby we have equipped U with the structure of an affinoid variety. In fact, we have defined the *presheaf $\mathcal{O}$ of affinoid functions on V* which associates to each rational subdomain $U \subset V$ its algebra of affinoid functions $\mathcal{O}(U)$. It goes without saying that our definition provides what is expected in "simple" situations where U is a disc or an annulus, etc.

Although general open subsets of V are far from being rational subdomains, it is nevertheless clear that these domains form a basis of the topology of V. Moreover, the above considerations can be carried out for more general open subsets of V, so-called *affinoid subdomains*, which decompose into finitely many rational subdomains of V.

It might be interesting to know on which open subsets $U \subset k^1$ we can now consider analytic functions. Since any "closed" disc in k^1 is easily transformed onto $\mathbb{B}^1$, one can show that these are precisely the finite unions of standard sets. A *standard set* in k^1 is a "closed" disc from which a finite number of "open" discs has been removed. However, using the methods of the next section, the definition of analytic functions can be extended to *all* open subsets of k^1.

2   Tate's acyclicity theorem and its consequences

Due to the total disconnectedness of the ground field k, we cannot expect that the presheaf $\mathcal{O}$ of affinoid functions is a sheaf on an affinoid variety V. For example, set $V := \mathbb{B}^1$ and consider an open covering of $\mathbb{B}^1$ by disjoint "closed" discs U_i of radius < 1. Then each U_i is an affinoid subdomain of $\mathbb{B}^1$. The characteristic function χ of a particular disc U_{i_0} defines a function on $\mathbb{B}^1$ which is affinoid on each U_i (namely a constant); however, χ is not affinoid itself since it cannot be represented by a convergent power series on $\mathbb{B}^1$. Thus, in accordance with the remarks made at the beginning of section 1, affinoidness is not a local property.

Nevertheless, in a modified sense, there does exist a local aspect of affinoidness. This fact is one of the miracles in rigid analysis; it is the essence of Tate's acyclicity theorem which we present here in a simplified version.

Theorem. *Let V be an affinoid variety, and let $V = \bigcup\limits_{i=1}^{n} U_i$ be a finite covering of V by affinoid subdomains U_i. Let $f : V \to k$ be a function on V. Then f is affinoid if and only if each restriction $f|_{U_i}$ is affinoid on U_i, $i = 1, \ldots, n$.*

Let us call an open subset of V *admissible* if it is an affinoid subdomain of V, and let us call an open covering $U = \bigcup\limits_{i \in I} U_i$ of an affinoid subdomain $U \subset V$ *admissible* if the index set I is finite and if all U_i are admissible open. Then Tate's theorem says that, within the framework of admissible open sets and coverings, the presheaf $\mathcal{O}$ of affinoid functions is a *sheaf* on V. This fundamental fact is basic for the construction of global spaces in rigid analysis. Namely, one uses sheaf-theoretic methods which are similar to the well-established methods from algebraic geometry or complex analysis.

Of course, it is necessary to carry out all constructions in terms of admissible open sets and admissible open coverings. The concept of *Grothendieck topologies* provides a useful formalism for this procedure. Giving a simplified version, we say that a Grothendieck topology (or simply G-topology) on a set X consists of

(i) *a system S of subsets of X (called admissible open sets), and*
(ii) *for each $U \in S$, a system Cov U of coverings of U by sets of S (called admissible open coverings of U).*

One requires some compatibility properties, just a minimal amount to allow the definition of sheaves and their Čech cohomology; see [2]. Then it is easily seen that most standard terms of topology such as continuity, connectedness, quasi-compactness and so on have their counterparts in terms of G-topologies. Furthermore, each topology on X may be viewed as a G-topology in a trivial way.

For an affinoid variety V, we have already considered its so-called *weak G-topology* which consists of all affinoid subdomains of V, and of all finite coverings of such sets by such sets. The presheaf of affinoid functions is a sheaf with respect to this G-topology. There is a rather technical procedure which refines weak G-topologies on affinoid varieties and simultaneously extends their sheaves of affinoid functions (by means of projective limits). This procedure can be carried out in a unique best possible way, the method is straightforward. The resulting G-topology is called the *strong G-topology* on V. It is considerably finer than the weak one.

From now on we will always view an affinoid variety V as a locally ringed space $(V, \mathcal{O})$, where V carries the strong G-topology and where $\mathcal{O}$ is the (extended) sheaf of affinoid functions on V. Using these spaces as model spaces, the definition of global spaces is without problems.

Definition. *A (global) rigid analytic variety is a locally ringed space* $(X, \mathcal{O})$ *(with respect to a G-topology on X) which admits an admissible open covering* $\{X_i\}_{i \in I}$ *such that* $(X_i, \mathcal{O}|_{X_i})$ *is affinoid for all* $i \in I$.

The sheaf $\mathcal{O}$ is called the sheaf of *rigid analytic functions* on X; the adjective "rigid" is to point out that analyticity is meant in terms of G-topologies and not of ordinary topologies. *Morphisms of rigid analytic varieties* are defined in the usual way; for affinoid varieties, they correspond to the affinoid morphisms discussed in section 1. Furthermore, one defines *analytic subsets* and *coherent modules* on rigid analytic varieties. The procedure is well-known from algebraic geometry or complex analysis. In order to keep our presentation simple, nothing has been said so far about non-algebraically closed ground fields k and non-reduced structure sheaves $\mathcal{O}$. In fact, the above definition can be generalized to these cases in a canonical way, and we will assume this tacidly from now on.

We conclude this section by discussing some examples of rigid analytic varieties. Since any such variety can be recovered from a given admissible open covering, the usual pasting procedures may be used for construction. Thus the affine n-space k^n becomes a rigid analytic variety if one uses an increasing sequence of balls as admissible open affinoid covering. A similar procedure applies to Zariski-closed subsets of k^n. Hence each affine scheme X of finite type over k may be analytified and thus becomes a rigid analytic variety X^{an}. Since the strong G-topology on affinoid varieties is finer than the Zariski topology (a property which is not enjoyed by the weak G-topology), one concludes that each Zariski-open subset of X gives rise to an admissible open subset of X^{an}. Therefore the analytification process can be extended to schemes of locally finite type over k. In particular, interesting algebraic varieties such as complete curves, surfaces, or abelian varieties can be viewed as rigid analytic varieties and can be investigated by rigid analytic methods. On the other hand, there is a list of rigid analytic varieties which originate from complex analysis. An important class of such varieties is formed by the analytic tori, which are defined as quotients of a multiplicative group $(k^*)^n$ by a discrete subgroup Γ of rank n. By "discrete" we mean that any admissible open affinoid subset of $(k^*)^n$ contains only finitely many points of Γ.

We could start now with the discussion of the different branches of rigid analysis. For example, we could take complex analysis as a yardstick and talk about the extent to which rigid analysis has been established. However, we cannot do this seriously here for reasons of space; also it might be not so interesting for non-specialists. Instead we have chosen to demonstrate the power of rigid analysis by presenting some of its applications to algebraic geometry. The remainder of the article is devoted to this purpose.

3   Rigid analysis and algebraic geometry

For all applications of rigid analysis to algebraic geometry, the theory of proper maps is fundamental. The notion of properness is based on the classical notion of relative compactness. Notice that, classically, a separated space X is compact if $X \Subset X$ or, more generally, if there exist two finite open coverings $\{U_j\}_{j \in J}$ and $\{U'_j\}_{j \in J}$ of X such that $U'_j \Subset U_j$ for all $j \in J$.

In rigid analysis, one uses a relative version of relative compactness. Let $\sigma : U \to V$ be a morphism of affinoid varieties, and consider an admissible open affinoid subvariety $U' \subset U$. Then U' is said to be *relatively compact in U over V* (in signs $U' \Subset_V U$) if, for some $n \geq 0$, there is a closed immersion $U \to \mathbb{B}^n \times V$ extending σ such that the image of U' is contained in a product $\mathbb{B}^n_\varepsilon \times V$, where $\mathbb{B}^n_\varepsilon$ is a ball in $\mathbb{B}^n$ of radius $\varepsilon < 1$. Furthermore, a morphism of rigid analytic varieties $\varphi : X \to Y$ is called *separated* if the diagonal morphism $X \to X \times_Y X$ is a closed immersion. Using these notions, proper morphisms are introduced as follows.

Definition (Kiehl). *A morphism of rigid analytic varieties $\varphi : X \to Y$ is called proper if*

(i) *φ is separated, and*

(ii) *there exists an admissible open affinoid covering $\{V_i\}_{i \in I}$ of Y, and for each $i \in I$, there exist two admissible open affinoid coverings $\{U_{i1}, \ldots, U_{in_i}\}$ and $\{U'_{i1}, \ldots, U'_{in_i}\}$ of $\varphi^{-1}(V_i)$ such that $U'_{ij} \Subset_{V_i} U_{ij}$ for all i and j.*

If $Y = \operatorname{Sp} k$ is the one-point-space, X is called proper (over k). Proper rigid analytic varieties are the analogues of compact complex analytic spaces or complete algebraic varieties, whereas proper morphisms of rigid analytic varieties correspond to proper morphisms in complex analysis or algebraic geometry. As a matter of fact, we will see that, for proper morphisms, there is a perfect interplay between rigid analysis and algebraic geometry.

As in complex analysis or algebraic geometry, the key result for proper morphisms is the direct image theorem which states that all higher direct images of coherent modules are coherent again. This result (as well as the related theorem on formal functions) is due to Kiehl [10]; it has far-reaching consequences. For example, the following corollaries are derived in the usual way.

Proper Mapping Theorem. *Let* $\varphi : X \to Y$ *be a proper morphism of rigid analytic varieties, and let* A *be a (closed) analytic subset of* X. *Then* $\varphi(A)$ *is an analytic subset of* Y.

Stein Factorization. *Each proper morphism of rigid analytic varieties* $\varphi : X \to Y$ *factors uniquely into a proper morphism* $\varphi' : X \to Y'$ *and a finite morphism* $\pi : Y' \to Y$ *such that* φ' *is surjective, has connected fibres and satisfies* $\varphi'_*(\mathcal{O}_X) = \mathcal{O}_{Y'}$.

Of course, connectedness is meant here in terms of Grothendieck topologies (or in terms of the Zariski topology, which amounts to the same). Furthermore, one shows that Serre's fundamental article [18] describing the relationship between analytic and algebraic geometry can be carried over almost verbatim; the results remain unchanged (see [11]). In particular, one obtains

Chow's Theorem. *Each analytic subset of a complete algebraic variety is algebraic.*

Using the proper mapping theorem, Chow's theorem and the fact that analytifications of projective varieties are proper, one easily verifies that a morphism $X \to Y$ of schemes of finite type over k is proper (in the sense of algebraic geometry) if and only if its analytification $X^{an} \to Y^{an}$ is proper (in the sense of rigid analytic geometry). From all this one may conclude that, for non-Archimedean ground fields, rigid analysis is a veritable replacement of complex analysis as a transcendental method in algebraic geometry.

Finally, we want to refer briefly to the problem of algebraization of rigid analytic varieties. First, in dimension 1, all proper rigid analytic varieties are algebraic. This result fails to be true in higher dimensions. For example,

Theorem [5], [6]. *Let* $T = (k^*)^n/\Gamma$ *be an analytic torus of dimension* n. *Then the following conditons are equivalent:*

(i) T *is algebraic.*

(ii) *The field of meromorphic functions on* T *has transcendence degree* n *over* k.

(iii) *Let* H *denote the group of analytic homomorphisms* $(k^*)^n \to k^*$. *There is a homomorphism* $\sigma : \Gamma \to H$ *such that* $\Phi(\gamma, \gamma') := (\sigma(\gamma))(\gamma')$, $\gamma, \gamma' \in \Gamma$, *is symmetric and such that the bilinear form* $-\log|\Phi|$ *is positive definit.*

Since for any irreducible proper rigid analytic variety X, the field M of meromorphic functions is an algebraic function field of transcendence degree $\leq \dim X$ (see [1]), and since M equals the rational function field if X is algebraic, the definition of *Moishezon varieties* makes sense. As in complex analysis, there are Moishezon varieties which are not algebraic. However, by what we have stated above, analytic tori are algebraic as soon as they are Moishezon. The theory of Moishezon varieties is still at its early stages. A thorough investigation might be worth-while.

4   Rigid analytic uniformization of algebraic curves and of abelian varieties

As we have mentioned before, the invention of rigid analysis was closely related to uniformization problems in algebraic geometry. In the early seventies, Tate's uniformization of elliptic curves with non-integral j-invariant [2,17] was generalized in two ways, namely, by Mumford [12] to curves of higher genus having split degenerate reduction and by Raynaud [16] to abelian varieties of general dimension. Mumford's curves were thoroughly investigated from the rigid analytic viewpoint by Gerritzen [7], [8]. Thereby a substantial simplification and extension of Mumford's work could be achieved. In the case of abelian varieties, the same was done by Lütkebohmert and the author of this article in [3]. From today's point of view, the central fact in uniformization theory is the following:

Stable Reduction Theorem. *Let C be a complete non-singular curve of genus $g \geq 2$ over a complete non-Archimedean field k. Then there is a finite algebraic extension k' of k such that $C_{k'}$ has stable reduction over the valuation ring $\overset{\circ}{k'}$ of k' (in the sense of Deligne-Mumford [4]).*

This result has been obtained by Deligne and Mumford [4] in the case where the valuation on k is discrete. Their proof uses deep results from algebraic geometry such as Grothendieck's semi-abelian reduction of abelian varieties [9] and, of course, the existence of Néron models [13], [15]. Namely, they showed that the semi-abelian reduction of the Jacobian of C implies stable reduction for C. It is a surprizing fact that there is a direct proof in terms of rigid analysis which uses algebraic geometry only on an elementary level and which works for arbitrary ground fields k. This was first realized by van der Put in [14]; a complete proof of the Stable Reduction Theorem is contained in [3].

The curve C has stable reduction over k if, among other things, C can be extended to a scheme C^0, proper and flat over the valuation ring $\overset{\circ}{k}$ of k, such that the associated reduction $\tilde{C} := C^0 \times_{\overset{\circ}{k}} \tilde{k}$ (where $\tilde{k}$ is the residue field of k) is reduced and non-singular, up to a finite number of ordinary double points. Then the analytification C^{an} of C may be interpreted as the formal completion of C^0, and thereby C^{an} is equipped with an analytic reduction map $\pi : C^{an} \rightarrow \tilde{C}$ which enjoys the following properties:

(i) *if $\tilde{x} \in \tilde{C}$ is non-singular, $\pi^{-1}(\tilde{x})$ is analytically isomorphic to an "open" disc;*

(ii) *if $\tilde{x} \in \tilde{C}$ is an ordinary double point, $\pi^{-1}(\tilde{x})$ is analytically isomorphic to an "open" annulus.*

The main problem in the analytic proof of the Stable Reduction Theorem is to find an analytic reduction map $C^{an} \rightarrow \tilde{C}$ of the above type. It is then easy to descend from C^{an} to C^0. Furthermore, the reduction map $\pi : C^{an} \rightarrow \tilde{C}$ is the key tool for attacking the uniformization of C. Namely, one has the following description of C^{an}: Consider the π-inverse of the non-singular locus $\tilde{C} - S(\tilde{C})$ of

$\tilde{C}$. This is a disjoint union of components which are smooth over $\overset{\circ}{k}$. Then C^{an} is obtained by connecting these components by means of the annuli $\pi^{-1}(\tilde{x})$, $\tilde{x} \in S(\tilde{C})$, and the universal covering $\hat{C}$ of C^{an} is constructed by resolving all loops which are generated by this process. If $\tilde{C}$ has only rational components (this is the case of Mumford's split degenerate reduction), $\hat{C}$ can be viewed as an open analytic subspace of $\mathbb{P}^1$, the automorphisms of $\hat{C}$ over C^{an} being fractional linear transformations. More precisely, if g is the genus, the curve C^{an} has an analytic uniformization as follows: There are $2g$ disjoint "open" discs $D_1, \ldots,$

$D_g, D'_1, \ldots, D'_g \subset \mathbb{P}^1$ such that C^{an} is obtained from $\mathbb{P}^1 - \overset{g}{\underset{i=1}{\cup}} D_i \cup D'_i$ by identify-

ing the circumference of D_i with the circumference of D'_i for each i. Thereby one gets an explicit description of C^{an} which allows the computation of several invariants of C in an easy way. If $\tilde{C}$ has non-rational components, the uniformization of C goes along the same lines, although $\hat{C}$ cannot be viewed as a subspace of $\mathbb{P}^1$ in this case.

Now let us indicate how the uniformization of abelian varieties is related to the stable reduction of curves. Let J be the Jacobian of a complete non-singular curve C. The Stable Reduction Theorem provides deep insight into the analytic structure of C. Thereby it is possible to construct line bundles with prescribed properties and to carry out explicit computations. Namely, one constructs an open analytic subgroup $\bar{J}$ of J, the group of *normalized* line bundles on C; it has a canonical reduction $\tilde{J}$ which is isomorphic to the Jacobian of $\tilde{C}$. Hence $\tilde{J}$ is an extension of an abelian variety by a multiplicative group. If the valuation of the ground field k is discrete, $\bar{J}$ may be interpreted as the identity component of the formal completion of the Néron model N of J. Thus we see that N has semi-abelian reduction (which is the assertion of the Monodromy Theorem [9]). In fact, even the existence of N can be deduced along these lines.

In order to construct the universal covering of J, one looks at the analytic cohomology group $H^1(C, \mathbb{Z})$. It is free of rank $r \leq g$ ($=$ genus of C); in fact, the rank r reflects the number of loops in C as discussed above. Using Picard functors, one interprets $H^1(C, \mathbb{Z})$ as the $\mathbb{Z}$-module of analytic group homomorphisms $\mathbb{G}_m \to J$ or $\mathbb{G}_m \to \bar{J}$, where $\mathbb{G}_m$ denotes the multiplicative group over k and where $\bar{\mathbb{G}}_m$ is its subgroup of "units". Thus there is a closed subgroup $\bar{\mathbb{G}}_m^r \hookrightarrow \bar{J}$ which reduces to the multiplicative part of $\tilde{J}$ and which may be extended to an analytic homomorphism $\mathbb{G}_m^r \to J$. Then $\hat{J} := \mathbb{G}_m^r \times \bar{J}/(\text{diagonal})$ is the universal covering of J. The projection map $\hat{J} \to J$ has a discrete kernel Γ which is free of rank r, so that $J = \hat{J}/\Gamma$. In particular, the following assertions are equivalent:

    (i) *C is a Mumford curve*,
    (ii) rank $H^1(C, \mathbb{Z}) = g$,
    (iii) *J is an analytic torus*.

Since, up to isogeny, any abelian variety A can be embedded into a product of Jacobian varieties, the above uniformization of Jacobians implies the uniformization of A. Namely, one constructs the analogues $\bar{A}$ and $\hat{A}$ of the groups $\bar{J}$ and $\hat{J}$, and shows $A = \hat{A}/\Gamma$. In this case, the rank of Γ has to be interpreted as the rank of the cohomology group $H^1(A', \mathbb{Z})$, where A' is the dual abelian variety of A.

The thorough understanding of uniformization problems of the above type seems to be a major breakthrough in rigid analysis, which leads to interesting further applications. In particular, it shows that rigid analysis can provide an interpretation in down-to-earth terms of complicated matters in formal algebraic geometry.

Bibliography

[1]   Bosch, S.: Meromorphic functions on proper rigid analytic varieties. Séminaire de Théorie des Nombres, Bordeaux (1983)

[2]   Bosch, S., Güntzer, U., and Remmert, R.: Non-Archimedean analysis. A systematic approach to rigid analytic geometry. Grundlehren der Mathematischen Wissenschaften, Bd. 261, Springer-Verlag (1984)

[3]   Bosch, S., and Lütkebohmert, W.: Stable reduction and rigid analytic uniformization of abelian varieties, to appear

[4]   Deligne, P., and Mumford, D.: The irreducibility of the space of curves of given genus. Publ. Math. IHES **36** (1969)

[5]   Fresnel, J., and van der Put, M.: Géométrie analytique rigide et applications. Birkhäuser, Boston—Basel—Stuttgart (1981)

[6]   Gerritzen, L.: On non-archimedean representations of abelian varieties. Math. Ann. **196**, 323—346 (1972)

[7]   Gerritzen, L.: Zur nichtarchimedischen Uniformisierung von Kurven. Math. Ann. **210**, 321—337 (1974)

[8]   Gerritzen, L.: Unbeschränkte Steinsche Gebiete von $\mathbb{P}_1$ und nichtarchimedische automorphe Formen. J. Reine Angew. Math. **297**, 21—34 (1978)

[9]   Grothendieck, A.: Groupes de monodromie en géométrie algébrique (SGA 7 I). Lecture Notes in Mathematics 288. Springer, Berlin—Heidelberg—New York (1972)

[10] Kiehl, R.: Der Endlichkeitssatz für eigentliche Abbildungen in der nichtarchimedischen Funktionentheorie. Inventiones math. **2**, 191—214 (1967)

[11] Köpf, U.: Über eigentliche Familien algebraischer Varietäten über affinoiden Räumen. Schriftenreihe Math. Inst. Univ. Münster, 2. Serie. Heft **7** (1974)

[12] Mumford, D.: An analytic construction of degenerating curves over complete local fields. Compositio Math. **24**, 129—174 (1972)

[13] Néron, A.: Modèles minimaux des variétés abéliennes sur les corps locaux et globaux. Publ. Math. IHES **21** (1964)

[14] van der Put, M.: Stable reductions of algebraic curves. Report Rijksuniversiteit Groningen (1981)

[15] Raynaud, M.: Modèles de Néron. C. R. Acad. Sc. Paris, t. **262** (7 février 1966), Série A, 345—347

[16] Raynaud, M.: Variétés abéliennes et géométrie rigide. Actes, Congrès intern. math., t. **1**, 473—477 (1970)

[17] Roquette, P.: Analytic theory of elliptic functions over local fields. Hamburger Math. Einzelschriften. Neue Folge, Heft **1** (1970)

[18] Serre, J.-P.: Géométrie algébrique et géométrie analytique. Ann. de l'Inst. Fourier **6**, 1—42 (1956)

[19] Tate, J.: Rigid analytic spaces. Private Notes (1962). Reprinted in Inventiones math. **12**, 257—289 (1971)

Perspectives in Mathematics
Anniversary of Oberwolfach 1984
© Birkhäuser Verlag, Basel

The Yang-Mills Equations on Euclidean Space

S. K. DONALDSON

The Institute for Advanced Study
Princeton, NJ 08540 (USA)

I will use this article to describe a number of recent developments and current problems related to the Yang-Mills equations. It is probably now well-known that these equations arose in elementary particle physics and can be seen as non-linear versions of the Maxwell field equations. We will discuss them purely from the mathematical point of view as interesting (and perhaps compa-ratively simple) examples of non-linear partial differential equations related to geometry. Even so, this account does not pretend to give a complete or balanced picture and is necessarily confined by the limits of the writer's interests and knowledge. However I hope to be able to illustrate the range of different techniques that are being brought to bear on these problems. In many ways this article is an amplification of a lecture given by Prof. M. F. Atiyah at the 1983 Arbeitstagung in Bonn, and I have learnt these ideas from conversations with him and with Dr. C. H. Taubes.

First we set down the two equations that we shall discuss.

The Yang-Mills equations on $\mathbb{R}^4$. (References [BL], [P])

The geometrical setting for these requires a compact Lie group G (which we will suppose to be simple), a G-bundle P over $\mathbb{R}^4$ and a *connection A* on P. By definition this connection is a solution to the *Yang-Mills equations* if

$$d_A^* F_A = 0$$

where F_A is the curvature of A and d_A^* is the natural operator extending the usual d^* using the connection A. Thus the equation in long hand with respect to a trivialisation of the bundle takes the rather gross form:

$$\left[\left(\frac{\partial}{\partial x^i} + A_i \right) \varepsilon_{jk}^{lm} \left(\frac{\partial A_l}{\partial x^m} - \frac{\partial A_m}{\partial x^l} + \frac{1}{2} [A_l, A_m] \right) \right]_{\text{skew}} = 0$$

where "skew" means making the expression totally anti-symmetric in i, j, k. Thus it is clearly a nonlinear partial differential equation for A and equally clearly one is eager to avoid writing down this explicit form.

The Yang-Mills equations are formally the Euler-Lagrange equation for the Lagrangian functional:

$$E_1(A) = \int\limits_{\mathbb{R}^4} |F_A|^2 \, d\mu.$$

Both the equation and the Lagrangian are conformally invariant and it is usual to use this to supply "boundary conditions" for the problem by demanding that the bundle and connection extend to $S^4 = \mathbb{R}^4 \cup \{\infty\}$. Moreover it will be technically convenient in this article to suppose that the fibre over ∞ has a fixed trivialisation $P_\infty \cong G$ and to work throughout with these "framed connections".

The Yang-Mills-Higgs Equations on $\mathbb{R}^3$. (The primary mathematical reference is [J.T.].)

Here we require a connection on a G-bundle over $\mathbb{R}^3$ and a *Higgs field Φ*, a section of the associated bundle of Lie algebras adP. The Yang-Mills-Higgs equations relating the pair are

$$\begin{cases} d_A^* F_A = [d_A\Phi, \Phi] \\ d_A^* d_A\Phi = 0 \end{cases}$$

which are the Euler-Lagrange equations for:

$$E_2(A, \Phi) = \int\limits_{\mathbb{R}^3} |F_A|^2 + |d_A\Phi|^2 \, d\mu.$$

Again we have boundary conditions "at infinity"; we should require, say, that $E_2(A, \Phi)$ is finite and that the Higgs field Φ approaches a fixed adjoint orbit in the Lie algebra of G at large distances from the origin. For example, if $G = SU(2)$ the adjoint orbits in $\mathfrak{su}(2) \cong \mathbb{R}^3$ are 2-spheres and we sould suppose that for fixed $L > 0$:

$$|\Phi_x| \to L \quad \text{as} \quad x \to \infty.$$

This time the equations are not conformally invariant but they transform naturally under *constant* scale changes of the metric on $\mathbb{R}^3$, and the Higgs field Φ transforms with weight -1. Thus we can by a scale change suppose that (in the $SU(2)$ case) $|\Phi| \to 1$ at infinity. Again it will be technically convenient to use a fixed frame at infinity compatible with the Higgs field. This is slightly awkward since compactifying to S^3 is not appropriate; rather we have to choose a fixed point on the "2-sphere at infinity" and a trivialisation of the fibre there.

Similarities between the equations

There are a number of features that these two equations share in common. First, locally, the Yang-Mills-Higgs equations can be obtained from the pure Yang-Mills by "dimension reduction". Thus we consider a bundle over $\mathbb{R}^4$ with a lifting of the translations in one preferred direction to the bundle and a connection invariant under this translation. Taking a transversal slice gives a connection over $\mathbb{R}^3$ and the remaining component of the connection over $\mathbb{R}^4$ becomes naturally a section of the bundle of Lie algebras adP, the Higgs field. For such invariant connections the Yang-Mills equations reduce to the Yang-Mills-Higgs and moreover the two Lagrangian integrands correspond.

The boundary conditions in the two problems are very different and no non-trivial translation invariant connection over $\mathbb{R}^4$ can extend to S^4. There is, however, a qualitative correspondence in that in each case the solutions have a global *topological* invariant. For the Yang-Mills equations this is the standard topological classification of principal G-bundles P over S^4. The classification goes by the group:

$$\pi_3(G) \cong \mathbb{Z}$$

canonically (for simple Lie groups G) and the resulting "Pontrayagin index" is denoted by k. Changing the orientation if necessary we can suppose $k \geq 0$.

For the Yang-Mills-Higgs equations things are slightly more complicated. Now the behaviour of the field Φ restricted to a large 2-sphere "at infinity" defines an element of:

$$\pi_2(G/J)$$

where G/J is the adjoint orbit fixed by the boundary conditions. For example, in the simplest case $G = SU(2)$, $J = S^1$ and the classification is by an integer "topological charge" $k \in \pi_2(S^2) \cong \mathbb{Z}$ which we can again suppose positive.

In each case very little is known about solutions to the general equations and most attention has been given to certain first order equations which imply the second order. These first order equations have the distinctive feature that their solutions—"*instantons*" and "*monopoles*" respectively, are parametrised by manifolds or *moduli spaces*.

The instantons have, by definition, self-dual curvature;

$$F_A = * F_A$$

and the monopoles satisfy the Bogomolny equation:

$$d_A \Phi = * F_A$$

It is fairly simple to show that not only do these equations imply the second order ones, but that their solutions are *absolute minima* of the functionals E_1, E_2 within the fixed topological type. Again the two equations correspond under dimension reduction.

Next we have in each case explicit "fundamental solutions" to the equations for group $G = SU(2)$. In the Yang-Mills case this is the 1-instanton depending upon a *centre* in $\mathbb{R}^4$ and a *scale* $\lambda > 0$ (a manifestation of conformal invariance). Taking the centre at 0 the explicit formula (not used below) is:

$$A_\lambda(x) = \left(\frac{1}{\lambda^2 + |x|^2} \right) (\mathcal{O}_1 i + \mathcal{O}_2 j + \mathcal{O}_3 k)$$

$$\mathcal{O}_1 = x_1 \, dx_2 - x_2 \, dx_1 + x_3 \, dx_4 - x_4 \, dx_3$$

$$\mathcal{O}_2 = x_1 \, dx_3 - x_3 \, dx_1 + x_4 \, dx_2 - x_2 \, dx_4$$

$$\mathcal{O}_3 = x_1 \, dx_4 - x_4 \, dx_1 + x_2 \, dx_3 - x_3 \, dx_2.$$

In the Yang-Mills-Higgs case the fundamental solution is the *Prasad-Sommerfield monopole* which, like the 1-instanton, has rotational symmetry about a centre. This time the scale is fixed by the boundary condition $|\Phi| \to 1$. The explicit formula is:

$$A(x) = \left(\frac{1}{\sinh r} - \frac{1}{r} \right) (\alpha_1 i + \alpha_2 j + \alpha_3 k) \quad \alpha_1 = x_2 \, dx_3 - x_3 \, dx_2$$
$$\text{and cyclic permutations.}$$

$$\Phi(x) = \left(\frac{1}{\tanh r} - \frac{1}{r} \right) \left(\frac{x_1}{r} i + \frac{x_2}{r} j + \frac{x_3}{r} k \right)$$

Finally one has a lowest order intuition that perhaps the general solution is made up of combinations of these fundamental solutions of particle-like solutions with some kind of non-linear interaction between them. For example, Taubes proved that if k points p_i are taken in $\mathbb{R}^3$, sufficiently far apart, then there is a solution to the Bogomolny equations with charge k, modelled on k copies of the Prasad-Sommerfield monopoles with centres at p_i ([J. T.] Chapter 3). The corresponding fact for the Yang-Mills equations is that there is a k-instanton modelled on k copies of the basic 1-instanton each with sufficiently small scale size λ. One aim of the subject is to understand how far this intuitive idea can be generalized and made precise.

Section 1. Variational aspects and homotopy Approximation

We begin with a classical example. If M is some complete Riemannian manifold the infinite dimensional space ΩM of (based) loops in M is endowed with the "energy" functional:

$$E(\gamma) = \int_0^1 |\dot{\gamma}|^2 \, ds.$$

The critical points of E are the geodesic loops and with suitable restrictions the ideas of Morse Theory can be applied to E to mediate between information about the homotopy type of ΩM and information about the geodesic loops.

Now consider the space $\Omega^2 M$ of based maps of the 2-sphere to M. There is an analogous energy functional:

$$E_0(f) = \int_{S^2} |df|^2 \, d\mu$$

whose critical points are the *harmonic maps* ([E. L.]). It is well-known that a naive generalisation of the results of the one-dimensional theory to the functional E_0 is not possible. For example, consider the case $M = S^2$, with the usual metric; then the *holomorphic maps* from S^2 to S^2 (regarding S^2 as a Riemann surface in the usual way) are harmonic and it can be shown ([E. L.]) that within the component $\Omega_k^2(S^2)$ of the function space representing maps of degree $k \geq 0$ these give the absolute minima of E_0 and that there are no higher critical points.

Hence if a gradient flow for E_0 with suitable good properties could be defined the infinite dimensional space $\Omega_k^2(S^2)$ would retract onto the finite dimensional manifold R_k of (based) holomorphic (thus *rational*) maps, but it is easy to see that these spaces have different homotopy types.

Thus it is a very important problem to understand precisely why such Morse Theory methods fail and whether any vestige of them persists. Evidence that something might be possible is given by a theorem of Segal:

Theorem [S1]. *If $k \geq d$ then the inclusion $R_k \hookrightarrow \Omega_k^2(S^2)$ induces an isomorphism of homotopy groups in dimensions less than $2d$.*

Segal worked with explicit representations of the rational maps as quotients of coprime polynomials or as configurations of zeros and poles in $\mathbb{C}$. Analogous results comparing the rational maps into other homogeneous spaces G/J with the function space $\Omega^2(G/J)$ have been proved by M. A. Guest and by F. C. Kirwan.

Now turn back to the two functionals $E_1(A)$, $E_2(A, \Phi)$ of the Introduction whose critical points are solutions of the Yang-Mills and Yang-Mills-Higgs equations respectively.

It has long been realised that the Yang-Mills equations in four dimensions behave in a broadly similar fashion to the harmonic mapping equation for two-dimensional domains. For example, both are conformally invariant (see [B] for an account from this point of view). The relevant infinite dimensional "function space" is now $\Omega^3(G)$ (see [AJ])-taking based maps since we are using framed connections. The component $\Omega_k^3(G)$ corresponds up to homotopy to the space $\mathscr{C}_k$ of all connections on a bundle with Pontrayagin index k. Inside $\mathscr{C}_k$ the finite dimensional moduli space $M_k(G)$ of self-dual connections (analogous to holomorphic maps) or instantons gives the absolute minimum of the energy functional E_1. One would like to be able to have ways of passing information between:

(a) *The homotopy type of* $\Omega^3_k(G)$.

(b) *The homotopy type of the absolute minimum* M_k.

(c) *The existence and Morse indices of higher critical points.*

Likewise for the Yang-Mills-Higgs equations we again have an infinite dimensional space on which the functional $E_2(A, \Phi)$ is defined. It has the homotopy type, determined by the "Higgs field at infinity", of the relevant component of $\Omega^2(G/J)$ ([T2]) and we have again the inclusion of the finite dimensional moduli manifold of monopoles, solutions to the Bogomolny equations, as the absolute minimum of E_2.

So each of these two variational problems fits into the same formal pattern and each is analogous to the problem of the harmonic maps of the two dimensional sphere. In neither case is there a standard theory that can be applied or, more technically the "Condition C" of Palais and Smale is not satisfied (see [U1] for an account). We will sketch the reasons why this is so.

In the introduction we saw that the explicit fundamental solution to the Yang-Mills equations on $\mathbb{R}^4$ contained a scale parameter $\lambda > 0$. Letting λ tend to 0 we obtain a family of instantons on the bundle with $k = 1$ all with the same (absolute minimum) value of the energy E_1. This family converges over $\mathbb{R}^4 \setminus \{0\}$ to the flat trivial connection but does not converge over the origin. Similarly if we were to follow the gradient flow of the Yang-Mills functional E_1 the resulting path might develop a number of point-like singularities over which topological information is lost.

In the Yang-Mills-Higgs case the scale size of the fundamental solution is fixed by the boundary conditions. Rather we would get bad behaviour if the gradient flow of the Yang-Mills-Higgs functional E_2 produced a family of (A, Φ) which resembled a number of monopoles whose centres separate indefinitely far apart with time.

In a recent series of papers Taubes has attacked these severe analytical difficulties. The starting point for this attack was work of Uhlenbeck which implied, roughly speaking, that the only bad behaviour in these two variational problems was produced in the manner sketched above ([U2], [U3], [S2]). The analogous theory in the harmonic mapping problem was developed earlier ([SU]). In the first article of this series Taubes proved:

Theorem ([T1]). *There exists a solution to the* SU(2) *Yang-Mills-Higgs equations which is not a solution to the first order Bogomolny equations.*

This was proved by applying the ideas of Morse Theory to the Hopf element in:

$$\pi_1(\Omega^2(\mathrm{SU}(2)/S^1)) \cong \pi_3(S^2)$$

and a deep argument which ruled out the indefinite separations of the energy density under a minimisation procedure.

This theorem settled an outstanding problem on the existence of non-minimal critical points for the functional $E_2(A, \Phi)$. For the Yang-Mills functional E_1 the corresponding problem is unsolved and appears to be very hard. Taubes developed [T3] similar techniques in this case which although not yet able to solve the problem shed some light on it.

Moreover, for groups $G = SU(2)$, $SU(3)$ Taubes showed that the Morse indices of non-minimal critical points of either the Yang-Mills or Yang-Mills-Higgs functionals *increased* as the topological invariant k became large ([T2]). Thus for large k these higher critical points (if they exist at all) would plausibly not affect the topological discussion in low dimensions, in line with the situation for the harmonic mapping problem. In this direction Taubes proved, for $G = SU(2)$, $SU(3)$:

Theorem ([T3]). *The moduli space $M_k(G)$ of instantons on $\mathbb{R}^4$ is connected.*

One would hope that these methods of Taubes yield further information in the future.

Before this work of Taubes, and advancing from a different direction, Atiyah and Jones studied the homology of the moduli spaces of instantons, absolute minima of the Yang-Mills functional E_1. Led in part by the analogy with Segal's results on the spaces of rational maps they conjectured that the inclusion:

$$M_k(G) \underset{i}{\hookrightarrow} \mathscr{C}_k \cong \Omega_k^3(G)$$

induces isomorphisms of homology in a given dimension once the Pontrayagin index k is taken sufficiently large. In fact they worked only with the group $SU(2)$ and they were able to prove in this case that:

Theorem ([AJ])

$$i_* : H_d\big(M_k(SU(2))\big) \to H_d\big(\Omega_k^3(SU(2))\big)$$

is surjective for $d \ll k$.

Atiyah and Jones used an explicitly known family of instantons which are, in a certain precise way, superpositions of k copies of the fundamental solution, and related the homology of these spaces to that of the *configuration space* of k distinct points in $\mathbb{R}^4$. Similar arguments indicated that the same approximation in homology might occur for the moduli spaces of monopoles, solutions to the Bogomolny equations.

It turns out that we can in some ways *compare* these three variational problems, none of which satisfies the Palais-Smale condition C. First consider the $SU(2)$ monopoles. These are the absolute minima of E_2 on a space having the same homotopy type as the based maps from S^2 to S^2. This latter space carries the energy functional E_0. Moreover, as we shall see in Section 3 the *absolute minima* of the two functions are also precisely related, as was conjectured first by Atiyah and Murray.

Theorem ([D 3]) *Choose an isomorphism* $\mathbb{R}^3 \cong \mathbb{C} \times \mathbb{R}$. *There is a natural one-to-one correspondence between the moduli space of* $SU(2)$ *monopoles of topological charge* k *and the (based) rational maps from* S^2 *to* S^2.
(There are similar conjectures for other groups.)

So, using the Theorem of Segal on the topology of the spaces of rational maps we find that the homotopy approximation property—analogue of the conjecture of Atiyah and Jones—holds for the monopole problem. Moreover, this is at least consistent with the result of Taubes on the growth of the Morse indices of higher critical points.

It would be more interesting, however, if the homotopy approximation could be proved by direct arguments on the lines of those of Taubes. Conjecturally the higher critical points of the Yang-Mills-Higgs functional $E_2(A, \Phi)$ precisely account for the difference in homotopy between the spaces of rational maps R_k and the infinite dimensional function space $\Omega_k^2(S^2)$. The theorem of Segal would then be, in a sense, explained by the increase of the Morse indices of E_2 as the topological charge k becomes large. Similarly it would be interesting to understand why the two functionals $E_0(f)$, $E_2(A, \Phi)$, defined on spaces of the same homotopy type and with the same absolute minima, have higher critical points in one case but not the other.

What is more our other variational problem—the Yang-Mills equations on $\mathbb{R}^4$ may also be cast in the same form. In this case the infinite dimensional space on which the Yang-Mills functional E_1 is defined is, homotopically:

$$\Omega^3(G) \cong \mathrm{Maps}(S^2, \Omega G)$$

The space of loops in a Lie group is in a natural way an infinite dimensional Kähler manifold [P. S.] and has many properties in common with finite dimensional homogeneous spaces such as complex Grassmannians and flag manifolds.

Again as we shall see in Section 3 we can identify the rational curves in ΩG (i. e., based holomorphic maps $S^2 \to \Omega G$) with the instantons once we pick a complex structure $\mathbb{C}^2 \cong \mathbb{R}^4$. So this finally puts our three problems on the same footing; as the study of functionals defined on spaces of the homotopy type:

$$\Omega^2(X)$$

for complex Kähler X, whose absolute minima are the holomorphic maps. Whether this unification is of any help in understanding the higher critical points remains to be seen.

Section 2. The construction of solutions

Instantons

In this Section and the next we shall be concerned entirely with the first order equations; the self-duality and Bogomolny equations whose solutions are instantons and monopoles respectively.

The first part of the story is now classical. (The best reference [A] may be hard to obtain.) We conformally compactify $\mathbb{R}^4$ to S^4, which is to be thought of as the quaternionic projective line $\mathbb{H}\mathbb{P}^1$, and introduce the Penrose *twistor space*.

$$\mathbb{C}\mathbb{P}^3 = \mathbb{H}^2 \setminus \{0\}/\mathbb{C}^*$$

This has a "real structure" σ given by multiplication in $\mathbb{H}^2$ by the quaternion j, and a fibration:

$$\pi : \mathbb{C}\mathbb{P}^3 \to \mathbb{H}\mathbb{P}^1$$

Ward used this twistor map to reduce the self-duality equations locally to holomorphic geometry or globally to (real) algebraic geometry. A self-dual connection over S^4 lifts to an algebraic bundle E over $\mathbb{C}\mathbb{P}^3$ with a real structure:

$$\overline{\sigma^*(E)} \cong E^*$$

The process can be inverted and gives the famous Ward correspondence between such self-dual connections and algebraic bundles with real structure, trivial on the fibres of π (a 4-dimensional family of projective lines in $\mathbb{C}\mathbb{P}^3$).

The algebraic bundles arising thus were classified by Atiyah, Hitchin, Drinfeld and Manin and this gave the explicit "ADHM representation" of instantons (for the classical groups). Restricting for simplicity to the group $SU(2) \cong Sp(1)$ this is:

Theorem ([ADHM]). *There is a natural 1-to-1 correspondence between:*

(a) *Equivalence classes of $Sp(1)$ instantons over S^4, with Pontrayagin index k.*

(b) *Equivalence classes under $O(k, \mathbb{R})$ of Pairs (B, b) where B is a $(k \times k)$ symmetric quaternionic matrix, b is a $(k \times 1)$ quaternionic vector such that:*

$$B^* B + b^* b$$

is a real matrix, and for each $x \in \mathbb{H}$: $\begin{pmatrix} B - x \, 1 \\ b \end{pmatrix} : \mathbb{H}^{k+1} \to \mathbb{H}^k$ *is surjective.*

Given such a pair (B, b) the ADHM construction produces a self-dual connection over S^4 by pulling back the natural connection on the "Hopf bundle" over $\mathbb{H}\mathbb{P}^k$ via a certain map

$$\mathbb{H}\mathbb{P}^1 \to \mathbb{H}\mathbb{P}^k.$$

More explicitly, working over the finite part $\mathbb{H} \subset \mathbb{H}\mathbb{P}^1$ we take the bundle whose fibre over $x \in \mathbb{H}$ is:

$$\mathrm{Ker} \begin{pmatrix} B - x\,1 \\ b \end{pmatrix} : \mathbb{H}^{k+1} \to \mathbb{H}^k$$

with the natural connection induced by orthogonal projection (in fact the construction produces most naturally *anti*-self-dual connections).

This process easily gives a good understanding of the fundamental solution of the Introduction and shows in a precise way how to generalize it to larger values of k.

Monopoles

As we mentioned in the Introduction the Bogomolny equations are, locally, a special form of the self-duality equations. Thus the same twistor methods apply to this equation. From the geometric point of view this has been developed by Hitchin [H1], [H2] in an especially attractive version using the space $\mathbb{T}$ of oriented straight lines in $\mathbb{R}^3$.

A line in $\mathbb{R}^3$ has an associated direction vector in S^2 and for each direction the corresponding family of lines is parametrised by a real affine plane. Thus $\mathbb{T}$ fibres over the Riemann sphere and has in fact naturally the structure of a complex surface. Then the version of the Ward correspondence appropriate to monopoles produces holomorphic bundles over $\mathbb{T}$. Hitchin found however (working with the group $SU(2)$) that this bundle, and in turn the monopole, was determined entirely by an algebraic curve in T; the *spectral curve S*.

Given a solution to the $SU(2)$ Bogomolny equations and a line in $\mathbb{R}^3$, with coordinate t along the line, Hitchin considers the linear ordinary differential operator:

$$(\nabla_t + i\Phi)$$

acting on sections of the associated $\mathbb{C}^2$ vector bundle E over the line. (Recall that by definition the Higgs field Φ is a section of the bundle $ad\,P \subset \mathrm{End}\,E$.) He found that for generic directions in S^2 there are k lines in $\mathbb{R}^3$ for which this operator has a solution decaying at both ends of the line. Moreover these lines, considered as points of $\mathbb{T}$, vary holomorphically with the direction vector and give rise to a k-fold branched cover:

$$\mathbb{T} \supset S \rightarrow S^2 \cong \mathbb{C}\mathbb{P}^1$$

of an algebraic curve S which is by definition the spectral curve. Conversely given this curve the solution to the Bogomolny equations could be recovered, explicitly in principle.

This work was extended by Murray [M] to other Lie groups G. He found that in general one should consider a configuration of curves in $\mathbb{T}$; one curve for each vertex of the Dynkin diagram of G. If two vertices are joined by an edge in the diagram the intersections of the corresponding curves satisfy a certain constraint.

Meanwhile W. Nahm developed a way to produce monopoles by an extension of the ADHM construction. As in the ADHM construction of instantons Nahm considers a family of linear maps between quaternionic vector spaces, but now the spaces are infinite dimensional and the maps are given by ordinary differential operators, rather than matrices. Likewise the constraints on the operators are no longer real algebraic equations but a non-linear system of ordinary differential equations.

More precisely Nahm [N1] considers three $(k \times k)$ matrices $T_i(z)$, depending on one real variable z and satisfying the equations (now called "*Nahm's equations*"):

$$\frac{d T_i}{dz} = [T_j, T_k] \quad ((ijk)\ cyclic\ permutation\ of\ (123))$$

(It is not immediately obvious but easy to verify that this system has the same symmetry—the three dimensional affine euclidean group—as the original problem.) These matrices are required to satisfy certain extra conditions; most importantly (for the group $SU(2)$) they are to be defined and regular over the interval $(0,2)$ with prescribed poles at the endpoints $z = 0,2$. Then Nahm defines a vector bundle over $\mathbb{R}^3$ whose fibre at (x_1, x_2, x_3) is the kernel of the operator:

$$\delta - x = \frac{d}{dz} + i(T_1 - x_1) + j(T_2 - x_2) + k(T_3 - x_3)$$

acting on quaternion valued functions over $(0,2)$, relative to suitable boundary conditions. On this bundle he found a way to define a connection and Higgs field giving a solution to the Bogomolny equations; in a precise formal analogue of the ADHM construction.

These methods of Hitchin and Nahm are intimately related, as was shown by Hitchin [H2]. To give the flavour of this consider the matrix valued function:

$$T_2 + i T_3$$

corresponding to a solution of Nahm's equations. These equations imply that:

$$\frac{d}{dz}(T_2 + iT_3) = i[T_1, T_2 + iT_3]$$

so the conjugacy class of $T_2 + iT_3$ is independent of $z \in (0,2)$ and in particular the eigenvalues are constant and give, generically, k points in $\mathbb{C}$. But the system has $SO(3)$ symmetry so there are a family of matrices on the same footing as $T_2 + iT_3$ and considering all their eigenvalues we get a subset of $SO(3) \times_{S^1} \mathbb{C} = \mathbb{T}$ presented as a k-fold branded cover of S^2. Although not at all obvious this subset is the *same* as Hitchin's spectral curve S and combining the two methods Hitchin found precise conditions for a curve $S \subset \mathbb{T}$ to correspond to a monopole, and deduced that all monopoles could be constructed by Nahm's method.

"Strange Duality"

This rather mysterious, but already fruitful, way of viewing the ADHM construction of instantons and Nahm's construction of monopoles was noted by E. Corrigan and Nahm. A connection over $\mathbb{R}^4$ for one of the classical groups can be defined by the covariant derivatives ∇_i acting on sections of a vector bundle E associated to the fundamental representation. We form the operator:

$$D = \nabla_0 + i\nabla_1 + j\nabla_2 + k\nabla_3$$

acting on sections of $E \otimes_{\mathbb{C}} \mathbb{H}$ (this is the Dirac operator associated to the connection and it can be defined invariantly) together with the adjoint:

$$D^* = -\nabla_0 + i\nabla_1 + j\nabla_2 + k\nabla_3$$

We have the Weitzenboch formula:

$$D^*D = -(\nabla_0^2 + \nabla_1^2 + \nabla_2^2 + \nabla_3^2)$$
$$+ i([\nabla_2, \nabla_3] + [\nabla_0, \nabla_1]) + j([\nabla_3, \nabla_1] + [\nabla_0, \nabla_2])$$
$$+ k([\nabla_1, \nabla_2] + [\nabla_0, \nabla_3])$$

By definition the commutators $[\nabla_i, \nabla_j]$ are algebraic operators given by components F_{ij} of the curvature and one sees straight from the formula that the connection is (anti) self-dual if and only if:

$$D^*D = -(\nabla_0^2 + \nabla_1^2 + \nabla_2^2 + \nabla_3^2) = \Delta$$

that is if and only if D^*D is a *real* operator.

Now recall that the quadratic constraint appearing in the ADHM construction is that the map:

$$\beta = \begin{pmatrix} B \\ b \end{pmatrix}$$

be such that again $\beta^* \beta = B^* B + b^* b$ is a *real* operator. So, formally, the original operator equation $\{D^* D \text{ real}\}$ is transformed by the ADHM construction into the matrix equation $\{\beta^* \beta \text{ real}\}$. This condition also appears in Nahm's construction of monopoles. In that case we have Nahm's operator:

$$\delta = \frac{d}{dz} + i T_1 + j T_2 + k T_3$$

and again Nahm's equations for the matrices $T_i(z)$ are equivalent to the condition that $\delta^* \delta$ be a real operator.

We cannot attempt to explain this apparent duality here, but it will appear again in the next Section.

Section 3. Relations with complex variables

Now I shall discuss some complex analytic aspects of the first order Yang-Mills and Yang-Mills-Higgs equations. These are related to the twistor methods of Section 2, but are really logically independent. Specifically, we have the identifications stated in Section 1 of the moduli spaces of:

instantons with holomorphic maps $S^2 \to \Omega G$ once a complex structure is fixed on $\mathbb{R}^4$.

$SU(2)$-*monopoles* with holomorphic maps $S^2 \to S^2$, once an isomorphism $\mathbb{R}^3 \cong \mathbb{C} \times \mathbb{R}$ is fixed.

It turns out that both of these indentifications can be obtained as applications of a general principle relating real and complex moduli problems. We can illustrate this principle by the following simple example:

Suppose one wishes to classify the $(n \times n)$ complex matrices M satisfying the equation:

$$[M^*, M] = 0.$$

The natural symmetry group of the problem is $U(n)$ acting by conjugation. A matrix which commutes with its adjoint can be diagonalised by a unitary transformation and conversely any matrix with an orthonormal base of eigenvectors commutes with its adjoint matrix. Thus the classification of the solutions to this real equation modulo the real group $U(n)$ is effected by the coefficients of the characteristic polynomial of M and the resulting moduli space is the *complex space* $\mathbb{C}^n$.

While this example is very familiar the correct generalisation is less well know (see [KN]). Given any representation of a compact Lie group K:

$$\varrho : K \to SU(m)$$

with associated representation of the complexified group:

$$\varrho : K^{\mathbb{C}} \to SL(m, \mathbb{C})$$

There is a map $\mu : \mathbb{C}^m \to K$ such that the quotient $\mu^{-1}(0)/K$ can be identified, up to a "precisely known" error, with the complex quotient:

$$\mathbb{C}^m/K^{\mathbb{C}}$$

In the example above we have the adjoint representation of $U(n)$. The map μ is:

$$\mu(M) = i[M, M^*]$$

and the "precisely known" error comes from the standard fact that the invariants given by the coefficients of the characteristic polynomial do not detect matrices with non-diagonalisable Jordan Normal Form. In the general case the map μ has variational description—a necessary and sufficient condition that $\mu(p) = 0$ is that the norm of p be minimal among all points in the complex orbit $K^{\mathbb{C}} \cdot p$. These ideas can be given a natural and exact form within the subject of "Geometric Invariant Theory". The same basic principle was noted by Atiyah and Bott [AB] in the context of the Yang-Mills equations over Riemann surfaces. In that case they found an infinite dimensional example of the phenomenon, with the complex Lie group being the group of all complex linear automorphisms of a vector bundle. Later it was found that these ideas could be successfully applied to the first order Yang-Mills equations over compact complex surfaces [D2].

Now we turn to the first of our problems: the instantons on $\mathbb{R}^4$. Fixing a complex structure we first identify these with certain *holomorphic bundles* on the projective plane. Recall that the (anti) self-duality equation is equivalent to the operator equation:

$$D^* D = \varDelta$$

for $D = \nabla_0 + i\nabla_1 + j\nabla_2 + k\nabla_3$.

Fixing a complex structure $z = x_0 + ix_1$, $w = x_2 + ix_3$ we can naturally introduce the complex differential operators:

$$\bar{\partial}_1 = \nabla_0 + i\nabla_1$$
$$\bar{\partial}_2 = \nabla_2 + i\nabla_3$$

and the anti self-duality equation becomes equivalent to the pair

$$[\bar{\partial}_1, \bar{\partial}_2] = 0$$
$$[\bar{\partial}_1, \bar{\partial}_1^*] + [\bar{\partial}_2, \bar{\partial}_2^*] = 0.$$

The first equation is complex and transforms naturally under the complex automorphisms of the bundle, moreover modulo this complex group and relative to suitable boundary conditions the solutions are in one to one correspondence with the holomorphic bundles over $\mathbb{C}\mathbb{P}^2$ trivial on the line at infinity. The second equation is in a precise sense the condition $\mu = 0$ appropriate to this complex group action [D1]. Thus the finite dimensional theory would lead one to hope that the instantons on $\mathbb{R}^4$ are in one-to-one correspondence with the holomorphic bundles on the projective plane trivial (and trivialised) on the line at infinity.

This is in fact true (at least for classical Lie groups) but the simplest proof does not use this infinite dimensional formalism; rather it uses the ADHM construction to *transform* to a finite dimensional problem. Recall that the matrix equation appearing in the ADHM description was: $\{B^* B + b^* b \text{ real}\}$, for quaternionic matrices (B, b). For simplicity we make the assumption that $b = 0$ (in fact this never happens for matrices satisfying the non-degeneracy condition of the ADHM construction). Then the condition on the quaternionic matrix $B \{B^* B \text{ real}\}$ can be rewritten in terms of a pair of *complex* matrices B_1, B_2 as:

$$[B_1, B_2] = 0$$
$$[B_1, B_1^*] + [B_2, B_2^*] = 0$$

the second equation is of the type $\mu = 0$ to which the finite dimensional theory applies; and this remains true when the b term is carried along. The first matrix equation is entirely complex and turns out to be that required in an explicit construction of algebraic bundles due to Horrocks and Barth. Notice that this is precisely the "strange dual" of the operator equations above.

The other step in the argument, identifying holomorphic bundles on the plane trivial on a line with rational maps:

$$S^2 \rightarrow \Omega G$$

is due to Atiyah and is again an application of a general principle relating "unitary" and complex structures.

For the best known example of this principle, take the manifold of full flags in $\mathbb{C}^n$:

$$\mathbb{F} = \{0 \subset W_1 \subset \cdots \subset W_{n-1} \subset \mathbb{C}^n\}$$

This is a homogeneous space for the compact group $U(n)$ hence it is obviously compact with a natural metric:

$$\mathbb{F} = U(n)/T.$$

On the other hand it is a homogeneous space for the complex group $GL(n, \mathbb{C})$ and hence is obviously complex algebraic:

$$\mathbb{F} = GL(n, \mathbb{C})/B.$$

Now consider the loop space of a compact group G, and recall that we are working with based loops; this is to be thought of as the unitary description. Similarly we may form the space of loops in the associated complex group $G^{\mathbb{C}}$ and consider the subgroup P of loops which extend holomorphically over the unit disc in $\mathbb{C}$. Then it can be shown that [Pr]:

$$\Omega G^{\mathbb{C}}/P \cong \Omega G$$

analogous to the description of flag manifolds.

Then it turns out that writing down the definition by transition functions of a holomorphic bundle on $\mathbb{CP}^2$, trivial on a given line, gives a direct identification of such bundles with holomorphic maps: $\mathbb{CP}^1 \to \Omega G^{\mathbb{C}}/P$. (As always we are using based maps and similarly bundles with a fixed trivialisation at infinity.)

For the monopoles the story, in so far as it is known yet, runs on similar lines. The main difference is that even the "Dual" description using Nahm's construction leads to an infinite dimensional problem and new technical features arise [D3].

References

[A]   Atiyah, M. F.,*The Geometry of Yang-Mills fields*, Fermi Lectures, Scuola Normale Superiore, Pisa (1979).

[AB]   Atiyah, M. F. and R. Bott, *The Yang-Mills Equations over Riemann surfaces*, Phil. Trans. Roy. Soc. London (1983).

[ADHM] Atiyah, M. F., V. G. Drinfeld, N. J. Hitchin and Yu. I. Manin, *Construction of instantons*, Phys. Lett. **65A** (1978).

[AJ]   Atiyah, M. F. and J. D. S. Jones, *Topological aspects of the Yang-Mills equations*, Commun. Math. Phys. **61** (1978).

[B]   Bourguignon, J. P., *Harmonic curvature for gravitational and Yang-Mills fields*, In: Springer Lecture Notes **949** (1982).

[BL]   Bourguignon, J. P. and H. B. Lawson, *Yang-Mills Theory: its physical origins and differential geometric aspects*, in, Annals. of Math. Studies **102** Princeton (1982).

[D1]   Donaldson, S. K., *Anti self-dual Yang-Mills connections over complex algebraic surfaces and stable vector bundles*, Submitted to the Proc. Lond. Math. Soc.

[D2]   Donaldson, S. K., *Instantons and Geometric Invariant Theory*, To Appear in Commun. Math. Phys.

[D3]   Donaldson, S. K., *Nahm's equations and the classification of monopoles*, Submitted to the Commun. Math. Phys.

[EL]   Eells, J. and L. Lemaire, *Report on harmonic maps*, Bull. Lond. Math. Soc. **10** (1978).

[H1]   Hitchin, N. J., *Monopoles and geodesics*, Commun. Math. Phys. **83** (1982).

[H2]   Hitchin, N. J., *On the construction of monopoles*, Commun. Math. Phys. **89** (1983).

[JT]   Jaffe, A. and C. Taubes, *Vortices and monopoles*, Birkhäuser (1980).

[KN]   Kempf, G. and L. Ness, *Lengths of vectors in representation spaces*, in, Springer Lecture Notes **732** (1978).

[M]   Murray, M. K., *Monopoles and spectral curves for arbitrary Lie Groups*, Commun. Math. Phys. **90** (1983).

[N1]   Nahm, W., *All multimonopoles for arbitrary gauge groups*, (preprint) TH 3172—CERN (1982).

[N2]   Nahm, W., *The algebraic geometry of multimonopoles*, University of Bonn, Preprint.

[P]   Parker, T., *Gauge Theories on four dimensional Riemannian manifolds*, Commun. Math. Phys. **85** (1982).

[Pr]   Pressley, A., *Decompositions of the space of loops on a Lie Group*, Topology **19** (1980).

[P.S.] Pressley, A. and G. B. Segal, Oxford U. P. to appear.

[S1]   Sedlacek, S., *A direct method for minimizing the Yang-Mills functional*, Comm. Math. Phys. **86** (1982).

[S2]   Segal, G. B., *The topology of spaces of rational functions*, Acta Math. **143** (1979).

[T1]   Taubes, C. H., *The existence of a non-minimal solution to the SU(2) Yang-Mills-Higgs equations on* $\mathbb{R}^3$, Comm. Math. Phys. **86** (1982).

[T2]   Taubes, C. H., *Stability in Yang-Mills theories*, Comm. Math. Phys. to appear.

[T3]   Taubes, C. H., *Path connected Yang-Mills moduli spaces*, U. C. Berkeley preprint.

[U1]   Uhlenbeck, K. K., *Variational problems for gauge fields*, In Annals of Math. Studies **102** (1982).

[U2]   Uhlenbeck, K. K., *Connections with* L^p *bounds on curvature*, Commun. Math. Phys. **83** (1982).

[U3]   Uhlenbeck, K. K., *Removable singularities in Yang-Mills fields*, Commun. Math. Phys. **83** (1982).

[SU]   Sacks, J. and K. K. Uhlenbeck, *The existence of minimal two spheres*, Ann. of Math. **113** (1981).

Perspectives in Mathematics
Anniversary of Oberwolfach 1984
© Birkhäuser Verlag, Basel

Gauss Maps of Surfaces

JAMES EELLS

University of Warwick, Mathematical Institute,
Coventry, Warwickshire CVA 7AL (G. B.)

1   Introduction*

In this essay we trace the recent evolution of the theory of minimal branched immersions—equivalently, conformal harmonic maps—of a Riemann surface M into a Riemannian manifold N. Our point of departure is a study of the Gauss map of an immersion into Euclidean space—there being two key results especially relevant to our development:

Let $\varphi : M \to \mathbb{R}^n$ be a conformal immersion and $\gamma_\varphi : M \to Q_{n-2}$ its Gauss map into the complex quadric (§ 3). Then
(3.7) φ *is harmonic iff* γ_φ *is anti-holomorphic* (Theorem of Chern [13]);
(3.13) φ *has constant mean curvature iff* γ_φ *is harmonic* (Theorem of Ruh-Vilms [66]).

We proceed by allowing increasingly general ambient manifolds N, and examining the properties of the associated Gauss maps[1]. Thus we consider Gauss maps of conformal immersions into space forms (§ 3); and Gauss lifts into associated Grassmann bundles (§ 4). With (3.7) and (3.13) as guidelines we finally obtain in (5.3)—a result found in collaboration with S. Salomon [23]— *a parametrization of the conformal harmonic maps of M into N in terms of (partially) holomorphic maps of M into a suitable twistor space over N.*

That is a theorem involving first order Gauss maps—motivated by a fundamental theorem of Calabi [9], [10] parametrizing isotropic harmonic maps into real projective spaces, and involving higher order Gauss maps. (See (5.12) and (5.11)).

Twistor theory in 4 dimensional Riemannian geometry [3] has been used to parametrize self-dual connexions in terms of holomorphic vector bundles over the appropriate twistor space. In case dim $N = 4$ our twistor space has the same underlying real structure, but a different almost complex structure. It is worth noting that self-dual connexions are the instantons of gauge theory

---

* Lecture given at the Oberwolfach Tagung on *Differentialgeometrie im Großen* in May 1983.
[1]) Of course, the Gauss map is an indispensible tool for the study of the geometry of submanifolds. It had its origins in Gauss's description of curvature of a surface M in $\mathbb{R}^3$, relating the intrinsic geometry of M to the extrinsic geometry with respect to $\mathbb{R}^3$. See [20], [56], [58].

(4-dimensional base), whereas our conformal harmonic maps correspond to the chiral models (2-dimensional domain).

We consider only 2-dimensional domains M, for they display many special features. For instance,

a) the qualitative properties of Gauss maps depend only on the conformal structure of M; thus (3.7) is a result very special to 2-dimensional domains;

b) our conformal immersions give rise to various holomorphic k-adic differentials on M. Such a quadratic differential was found by Hopf [43] and that led to an extensive generalization via the notion of isotropy (Calabi [9], [10]; see (6.4)).

Our results are established by representation—theoretic methods (only sketched here); in particular, type decompositions of complex tensor fields over M, and of vector bundles. For instance,

a) conformality of an immersion $\varphi : M \to N$ is equivalent to the vanishing of the (2,0)-part $(\varphi^* h)^{2,0}$ of the first fundamental form $\varphi^* h$, h being the Riemannian metric of N. And harmonicity of φ insures that $(\varphi^* h)^{2,0}$ is a holomorphic quadratic differential (2.4);

b) umbilicity of φ is described in terms of the (2,0)-part of its second fundamental form β_φ;

c) the basic relation $d\gamma_\varphi = \beta_\varphi$ between the Gauss map γ_φ and the second fundamental form is derived from the tensor decomposition $TQ_{n-2} = K^* \otimes K^\perp$ of the tangent bundle of the quadric.

Our exposition has shamefully ignored large areas of major interest and immediate proximity: In particular, the quantitative aspects of Gauss maps related to curvature, as well as value distribution theory. In our final section (§ 6) we make a few remarks on those topics, and pose various related problems.

During the preparation of this paper the author was on a Mission Scientifique du Fonds National Belge de la Recherche Scientifique à l'Université Libre de Bruxelles (1983). He has received valuable advice from L. Lemaire, J. Rawnsley, S. Salamon, and J. C. Wood. He records here his thanks to all.

2   Conformal harmonic maps

General background reference [21, § 10].

Let M be a Riemann surface; i.e., an oriented 2-dimensional smooth surface equipped with a conformal equivalence class of Riemannian metrics. Let N be an n-dimensional smooth Riemannian manifold with metric h, sometimes denoted by brackets $<, >$. Let $<, >^{\mathbb{C}}$ be its complex bilinear extension to the complexified tangent bundle $T^{\mathbb{C}}N = TN \otimes_{\mathbb{R}} \mathbb{C}$.

A map $\varphi : M \to N$ is *harmonic* if

$$\nabla_{\bar{z}} \partial_z \varphi = 0; \quad \text{equivalently,} \quad \nabla_z \partial_{\bar{z}} \varphi = 0; \tag{2.1}$$

here $\partial_z \varphi = \varphi_z = \dfrac{1}{2}\left(\dfrac{\partial \varphi}{\partial x} - i\,\dfrac{\partial \varphi}{\partial y}\right)$

$\partial_{\bar z} \varphi = \varphi_{\bar z} = \dfrac{1}{2}\left(\dfrac{\partial \varphi}{\partial x} + i\,\dfrac{\partial \varphi}{\partial y}\right)$

in an isothermal chart of M, and ∇ is the covariant differential of the induced bundle $\varphi^{-1}T^{\mathbb{C}}N$. If $(\Gamma^\gamma_{\alpha\beta})$ is the system of Christoffel symbols associated to the Levi-Civita connexion of h, then (2.1) has the explicit representation

$$\varphi^\gamma_{z\bar z} + \Gamma^\gamma_{\alpha\beta}\varphi^\alpha_z \varphi^\beta_{\bar z} = 0 \quad (1 \leqslant \gamma \leqslant n). \tag{2.2}$$

The left member of (2.1) or (2.2) is called the *tension field* of φ; and is also denoted by τ_φ.

(2.3) We shall say that a map $\varphi : M \to N$ *is conformal* if its first fundamental form $\varphi^* h$ has type decomposition with vanishing (2,0)-part:

$$(\varphi^* h)^{2,0} = <\varphi_z,\, \varphi_z>^{\mathbb{C}} dz^2 \equiv 0$$

indeed

$$<\varphi_z,\, \varphi_z>^{\mathbb{C}} = \frac{1}{4}\left(|\varphi_x|^2 - |\varphi_y|^2 - 2i<\varphi_x,\, \varphi_y>\right)$$

(2.4) If $\varphi : M \to N$ *is harmonic, then* $(\varphi^* h)^{2,0}$ *is a holomorphic quadratic differential.*

For $\partial_{\bar z}<\varphi_z,\, \varphi_z>^{\mathbb{C}} = 2<\nabla_{\bar z}\varphi_z,\, \varphi_z>^{\mathbb{C}} \equiv 0$

by (2.1).

(2.5) *A non constant conformal harmonic map* $\varphi : M \to N$ *is a minimal branched immersion*; i.e., it is a conformal immersion except at isolated points where the differential $d\varphi = 0$, and around each of these there are normal charts in which φ has the form [36]

$$\varphi^1(z) = c\,\mathrm{Re}(z^k) + o(|z|^k)$$
$$\varphi^2(z) = c\,\mathrm{Im}(z^k) + o(|z|^k)$$
$$\varphi^\alpha(z) = o(|z|^k) \quad \text{for} \quad 3 \leqslant \alpha \leqslant n.$$

Thus in treating conformal harmonie immersions φ, we shall admit a discrete set of points at which $d\varphi = 0$.

(2.6)   If φ maps M into a Kähler manifold N then the complexification of its differential decomposes: $d^{\mathbb{C}}\varphi = d'\varphi + d''\varphi$; see [21, § 9].
And $\varphi: M \to N$ is *holomorphic* iff $d''\varphi \equiv 0$; otherwise said, iff the differential $d\varphi$ is $\mathbb{C}$-linear. It follows easily from (2.1) that *every holomorphic map is harmonic* [24].

(2.7)   We note also that a holomorphic map $\varphi: M \to N$ *is conformal*, using (2.3) and the fact that φ preserves types.

3   Gauss maps into Grassmannians

(3.1)   Let $N(c)$ be an n-dimensional Riemannian space form; i.e., a connected complete Riemannian manifold of constant sectional curvature c. Its universal cover is therefore the Euclidean sphere S^m (of radius $1/\sqrt{c}$) if $c > 0$; the Euclidean space $\mathbb{R}^n$ if $c = 0$; or the hyperbolic n-sphere (represented as the open disc of radius $1/\sqrt{-c}$) if $c < 0$.

(3.2)   For any conformal immersion $\varphi: M \to N(c)$ its *Gauss map* γ_φ assigns to each point $x \in M$ the 2-dimensional oriented totally geodesic subspace (a space form of the same curvature) determined by the image of the tangent map $d\varphi(x): T_x M \to T_{\varphi(x)} N(c)$.

The case $N(o) = \mathbb{R}^n$

(3.3)   Let $\Lambda^2 \mathbb{R}^n$ denote the Euclidean vector space of 2-vectors of $\mathbb{R}^n$. Denote by $G_2(\mathbb{R}^n)$ the Grassmannian of unit 2-vectors; or equivalently, of oriented 2-dimensional linear subspaces of $\mathbb{R}^n$.

Each $\alpha \in G_2(\mathbb{R}^n)$ determines a complex 1-dimensional subspace of the complexification $\mathbb{C}^n = \mathbb{R}^n \otimes_{\mathbb{R}} \mathbb{C}$ of $\mathbb{R}^n$. Thus we have the natural embedding (as defined below)

$$i: G_2(\mathbb{R}^n) \to \mathbb{C}P^{n-1} \tag{3.4}$$

into the complex projective $(n-1)$-space; that is easily seen to be a smooth embedding, if $G_2(\mathbb{R}^n)$ is endowed with its structure of the homogeneous space $SO_n/SO_2 \times SO_{n-2}$. In fact, if α is represented by the exterior product $v \wedge w$ of orthonormal vectors, then $i(\alpha) = [z]$, the complex line determined by $z = v + \sqrt{-1}\, w \in \mathbb{C}^n$; and have the *isotropy condition*

$$<z, z>^{\mathbb{C}} = 2\sqrt{-1} <v, w> = 0, \tag{3.5}$$

$<, >^{\mathbb{C}}$ denoting the complex bilinear extension of the inner product $<, >$ of

$\mathbb{R}^n$. Thus the image of i is the complex quadric hypersurface Q_{n-2} in $\mathbb{C} P^{n-1}$ (a Kähler manifold), whose equation in homogeneous coordinates $z_1, \ldots, z_n$ is

$$\sum_{k=1}^{n} z_k^2 = 0.$$

Otherwise said, *we have a canonical identification i of $G_2(\mathbb{R}^n)$ with the quadric Q_{n-2}.*

(3.6)   The *Gauss map* $\gamma_\varphi : M \to Q_{n-2}$ of the conformal immersion $\varphi : M \to \mathbb{R}^n$ assigns to each point $x \in M$ (where $d\varphi(x) \neq 0$) the oriented 2-space parallel to $d\varphi(x) \, T_x M$ through the origin in $\mathbb{R}^n$. From (3.5) we obtain the following result of Chern [13]:

(3.7)   *The map i in (3.4) identifies γ_φ with $[\partial_{\bar{z}} \varphi]$. A conformal immersion $\varphi : M \to \mathbb{R}^n$ is harmonic iff its Gauss map $\gamma_\varphi : M \to Q_{n-2}$ is anti-holomorphic.*

(3.8)   The *second fundamental form* β_φ of a conformal immersion $\varphi : M \to \mathbb{R}^n$ can be canonically identified with the differential of the Gauss map:

$$\beta_\varphi = d\gamma_\varphi.$$

That property requires a detailled analysis of the geometry of the tangent bundle

$$TQ_{n-2} = K^* \otimes K^\perp,$$

where K is the bundle whose fibre over $L \in Q_{n-2}$ is L itself. Thus $d\gamma_\varphi : TM \to T^*M \otimes V(\mathbb{R}^n, M)$, where $V(\mathbb{R}^n, M)$ denotes the normal bundle of M in $\mathbb{R}^n$.

(3.9)   As a complement to (3.7) we have the following result [40]:
    A conformal immersion $\varphi : M \to \mathbb{R}^n$ has holomorphic Gauss map iff φ is totally umbilic; i.e., $(\beta_\varphi)^{2,0} \equiv 0$.
    Indeed, the complexification $\beta_\varphi^c = \beta_\varphi' + \beta_\varphi''$, where $\beta_\varphi' = \nabla' \partial' \varphi$, $\beta_\varphi'' = \nabla'' \partial'' \varphi$ (because, in the notation of [21, §9], $\nabla' \partial'' \varphi + \nabla'' \partial' \varphi = 0$). Then from (3.8) we find $\beta_\varphi^c = \partial' \gamma_\varphi + \partial'' \gamma_\varphi$, which has type (1,1) iff $\partial'' \gamma_\varphi = 0$.

(3.10)   The *third fundamental form of* $\varphi : M \to \mathbb{R}^n$ is γ_φ^{*k}, where k is the Kähler metric of Q_{n-2}. A direct calculation gives [57]

$$\gamma_\varphi^{*k} = \langle \beta_\varphi, \tau_\varphi \rangle - K^M g \tag{3.11}$$

where g is a conformal metric on M and K^M is its Gaussian curvature.

(3.12)   The *mean curvature* of $\varphi : M \to \mathbb{R}^n$ is the normal field $\tau_\varphi/2$. We let

$D(\tau_\varphi/2)$ denote its normal covariant differential; and say that φ has *constant mean curvature* if $D(\tau_\varphi/2) \equiv 0$.

An important theorem of Ruh-Vilms [66] takes the form:

(3.13) *A conformal immersion $\varphi : M \to \mathbb{R}^n$ has constant mean curvature iff its Gauss map $\gamma_\varphi : M \to Q_{n-2}$ is harmonic.*

That follows easily by differentiating equation (3.8), and making use of Codazzi's equation (see (4.5) below).

Those immersions with totally geodesic Gauss map ($\nabla d\gamma_\varphi \equiv 0$) have been classified [12].

(3.14) The next result is due to Obata [57]:

Let $\varphi : M \to \mathbb{R}^n$ be a conformal immersion. Its Gauss map $\gamma_\varphi : M \to Q_{n-2}$ is conformal iff φ is pseudo-umbilic; i.e., there is a smooth function $\lambda : M \to \mathbb{R}$ such that

$$< \beta_\varphi, \tau_\varphi > \ = \lambda g.$$

Furthermore, φ maps M harmonically into a Euclidean hypersphere S^{n-1} of $\mathbb{R}^n$ (with $j : S^{n-1} \to \mathbb{R}^n$ the inclusion map) iff $\Phi = j \circ \varphi$ is pseudo-umbilic with constant mean curvature $\neq 0$.

The first assertion is an immediate consequence of (3.11):

$$(\gamma_\varphi^* k)^{2,0} = \ < \beta_\varphi^{2,0}, \tau_\varphi > ,$$

since a conformal metric on M has type $(1,1)$; and that is 0 iff $< \beta_\varphi, \tau_\varphi >$ has type $(1,1)$—and consequently is a function multiple of g. The second assertion is a direct calculation, starting from the formula

$$\tau_\Phi = (di)\,(\tau_\varphi) - 2v,$$

where v is the unit field of S^{n-1} in $\mathbb{R}^n$.

Combining various preceding results, we obtain

(3.15) *Let $\varphi : M \to \mathbb{R}^n$ be a conformal immersion. Then $\gamma_\varphi : M \to Q_{n-2}$ is conformal and harmonic iff φ maps M harmonically into $\mathbb{R}^n$ or into some hypersphere.*

(3.16) **Example [65].** Let $\varphi : S \to \mathbb{R}^n$ be a conformal immersion of the sphere S, with constant mean curvature. Then φ maps S minimally into some hypersphere. Indeed, γ_φ is harmonic by (3.13); and conformal by (2.4), because every holomorphic quadratic differential on S is identically 0. Ruh [65] also shows that if $\varphi : S \to S^4$ has trivial normal bundle, then φ maps S onto a great 2-sphere in S^4.

The case $N(1) = S^n$

(3.17) Now let $\varphi : M \to S^n$ be a conformal immersion (which we are *careful not to confuse* with the induced conformal immersion $\Phi = j \circ \varphi\ M \to \mathbb{R}^{n+1}$, where $j : S^n \to \mathbb{R}^{n+1}$ denotes the standard inclusion map); and consider its Gauss map $\gamma_\varphi : M \to G_3(\mathbb{R}^{n+1})$ as in (3.2); here $G_3(\mathbb{R}^{n+1})$ is the Grassmannian of oriented 3-spaces in $\mathbb{R}^{n+1}$. Then ([11], [44] for the first assertion; and [57] for the second):

φ *is harmonic iff* γ_φ *is harmonic. And* φ *is pseudo-umbilic iff* γ_φ *is conformal.*
The second statement follows as in (3.14), now using [57]

$$\gamma_\varphi^* k = <\beta_\varphi, \tau_\varphi> - K^M g + g. \tag{3.18}$$

Also

$$\gamma_\varphi = \Phi \wedge \gamma_\Phi \tag{3.19}$$

Remark. There are analogous results for conformal immersions $\varphi : M \to N(-1)$, hyperbolic n-space.

4   Gauss maps into Grassmann bundles

Gauss lifts

(4.1)   For any oriented Riemannian n-manifold N let $\pi : Q(N) \to N$ denote the associated bundle whose fibre model is Q_{n-2}; Thus

$$Q(N) = SO(N) \times_{SO_n} Q_{n-2}$$

With any conformal immersion $\varphi : M \to N$ we have its *Gauss lift* $\tilde{\varphi}$:

$$\begin{array}{ccc} & Q(N) & \\ {}^{\tilde{\varphi}}\nearrow & & \downarrow{\pi} \\ M & \xrightarrow{\varphi} & N \end{array}$$

which assigns to each point $x \in M$ the oriented 2-space $d\varphi(x)\, T_x M$ in $T_{\varphi(x)} N$.

(4.2)   **Example.** Take $N = S^n$. Then the elements of $Q(S^n)$ can be considered as orthogonal pairs (V, W) of oriented subspaces in $\mathbb{R}^{n+1}$ dimensions $2, n-2$ resp. We have the homogeneous representations

$$SO_{n+1}/SO_2 \times SO_{n-2} = Q(S^n) \to S^n = SO_{n+1}/SO_n;$$

and the following diagrams of Riemannian fibrations:

$$\begin{array}{ccc} & Q(S^n) & \\ \pi_1 \swarrow & \downarrow \pi = \pi_2 & \searrow \pi_3 \\ Q_{n-1} & S^n & G_{n-2}(\mathbb{R}^{n+1}) \end{array}$$

Here $\pi_1(V, W) = V$, $\pi_3(V, W) = W$; and

$$\pi(V, W) = (V \oplus W)^{\perp}.$$

Given a conformal immersion $\varphi : M \to S^n$ and writing $j \circ \varphi = \Phi : M \to \mathbb{R}^{n+1}$ once again, we observe [44]:

(4.3) $\gamma_1 = \pi_1 \circ \tilde{\varphi} = \gamma_\Phi$, the *Gauss map of* $\Phi : M \to \mathbb{R}^{n+1}$;
     $\gamma_3 = \pi_3 \circ \tilde{\varphi} = (\varphi \wedge \gamma_1)^{\perp} = \gamma_\varphi^{\perp}$, *the normal Gauss map of* $\varphi : M \to S^n$;
     $\pi \circ \tilde{\varphi} = \varphi = (\gamma_1 \oplus \gamma_3)^{\perp}$.
Furthermore, $\tilde{\varphi}$ *is* π_1-*horizontal* [44]. *Therefore,* $\varphi : M \to S^n$ *has constant mean curvature iff* $\tilde{\varphi} : M \to G_2(S^n)$ *in harmonic.*

(4.4)   **Example.** Thinking of a complex line in $\mathbb{C}^n$ as an oriented plane in $\mathbb{R}^{2n}$ defines a totally geodesic embedding of $\mathbb{C}P^{n-1}$ in Q_{2n-2}. If $\varphi : M \to N$ is a holomorphic immersion of M into a Kähler manifold N, then the Gauss lift factors:

$$\begin{array}{ccc} & \mathbb{G}_1(N) \subset Q(N) & \\ \tilde{\varphi} \nearrow & & \downarrow \\ M & \xrightarrow[\varphi]{} & N \end{array}$$

In particular, taking $N = \mathbb{C}P^n$, we have an associated diagram as in (4.2); and Gauss maps related as in (4.3); [55], [45].

Gauss sections

(4.5)   Although (4.3) interrelates various Gauss maps into space forms $N(c)$ with $c \neq 0$, the case of $N(o) = \mathbb{R}^n$ is exceptional in that context. (Indeed, in that case the immersion φ itself is not built into $\tilde{\varphi}$). For a proper unification, we proceed as follows, starting as in (4.1):

The vertical subbundle of $TQ(N)$ has the decomposition (analogous to that in (3.8)):

$$T^V Q(N) = K^*(N) \otimes K^{\perp}(N),$$

where
$$K(N) = \{(W, w) \in Q(N) \times TN : w \in W\}$$
$$K^{\perp}(N) = \{(W, w) \in Q(N) \times TN : w \perp W\}.$$

With that identification,

(4.6) $\nabla d\varphi = (d\tilde{\varphi})^{V}$.

Now Codazzi's equation [49 II, p. 25] for $\varphi : M \to N$ gives the normal component of the curvature operator of N in the form

(4.7) $D_{Y}(\nabla d\varphi)(X, Z) - D_{X}(\nabla d\varphi)(Y, Z) = (R^{N}(d\varphi(X), d\varphi(Y)) d\varphi(Z))^{\perp}$

for all vector fields X, Y, Z on M. A direct calculation yields the following result of C. M. Wood [72]:

Let $^{V}\nabla$ denote the covariant differential in the vertical bundle $T^{V}Q(N)$. Treat $\tilde{\varphi}$ as a section of $\varphi^{-1}Q(N)$, and call

$$\tau^{V}(\tilde{\varphi}) = \text{Trace } {}^{V}\nabla(d\tilde{\varphi})^{V} \qquad (4.8)$$

its *tension field* (as a section—not as a map (as in (2.1))). Say that $\tilde{\varphi}$ *is a harmonic section* if $\tau^{V}(\tilde{\varphi}) \equiv 0$. Then

$$\tau^{V}(\tilde{\varphi}) = \nabla\tau(\varphi) - (Ric_{\varphi})^{\perp} d\varphi, \qquad (4.9)$$

where $Ric_{\varphi} = \text{Trace } R^{N}(d\varphi, -) d\varphi$;

$$(Ric_{\varphi})^{\perp} d\varphi(X) = \sum_{i=1}^{2} (R^{N}(d\varphi(e_{i}), d\varphi(X))^{\perp} d\varphi(e_{i}),$$

in terms of a local orthonormal frame field e_{1}, e_{2}.

As a consequence, *if* $N(c)$ *is a* space form, then $(Ric_{\varphi})^{\perp} d\varphi \equiv 0$, and $\varphi : M \to N(c)$ *has constant mean curvature iff* $\tilde{\varphi}$ *is a harmonic section.*

If k is the induced Riemannian metric on the bundle space $Q(N)$ and k^{V} its vertical component, then the *third fundamental form of* $\varphi : M \to N$ is

$$\tilde{\varphi}^{*}k^{V}(X, Y) = <\nabla_{X}d\varphi, \nabla_{Y}d\varphi> \qquad (4.10)$$

Again, a direct calculation yields [72]

$$Ric_{\varphi} d\varphi = K^{M}g - <\tau_{\varphi}^{V}, \nabla d\varphi> + \tilde{\varphi}^{*}k^{V}. \qquad (4.11)$$

Remark. A formula analogous to (4.9) has been found in [47] in the context of harmonic foliations.

5   Gauss maps into twistor bundles

First order constructions

(5.1)   Let N be an oriented Riemannian n-manifold. We define the vector bundle $\Pi \subset TQ(N)$ as follows: Each $q \in Q(N)$ represents an oriented Euclidean 2-space in $T_{\pi(q)}N$, and therefore a complex line L_q. The fibre Π_q is the subspace of $T_qQ(N)$ spanned by the lift of L_q to the horizontal subspace $T_q^H Q(N)$ and the vertical subspace $T_q^V Q(N)$. These components have complex structures J_q^H and J_q^V, respectively. And thereby we have two complex structures J_1, J_2 defined on the bundle Π, as follows:

$$J_1 = \left. \begin{matrix} J^H \text{ on } T^H Q(N) \cap \Pi \\ J^V \text{ on } T^V Q(N) \end{matrix} \right\} ; \quad J_2 = \left. \begin{matrix} J^H \text{ on } T^H Q(N) \cap \Pi \\ -J^V \text{ on } T^V Q(N) \end{matrix} \right\} .$$

It must be emphasized that these structures reflect very different geometric aspects of N, as the following example shows:

(5.2)   **Example.** Let dim $N = 4$. Then as oriented planes in $T_{\pi(q)}N$, both q and its orthogonal complement determine complex lines, and thus a complex structure on $T_{\pi(q)}N$. Consequently J_1 and J_2 are almost complex structures on the manifold $Q(N)$. Its fibre $Q_2 = S_+ \times S_-$; we have corresponding bundles $Q_{\pm}(N)$ with fibres $S_{\pm}$.

$(Q_+(N), J_1)$ is the standard twistor space of N. The structure J_1 is integrable iff N is anti-self dual (i.e., the component W_+ of the Weyl curvature W is 0); see [3].

By way of contrast, the structure J_2 is never integrable [23]. Nonetheless, the twistor space $(Q_+(N), J_2)$ is of fundamental importance to us. For instance:

Suppose that N is anti-self dual and Einstein. Then every J_2-holomorphic map $\psi : M \to Q_+(N)$ is harmonic (with respect to the natural fibred metric on $Q_+(N)$). Indeed, the Kähler form ω of $(Q_+(N), J_2)$ satisfies $(d\omega)^{1,2} \equiv 0$; and in the present circumstances that is just the required condition [53].

(5.3)   The following result is due to Eells-Salamon [23]:

Let $\varphi : M \to N$ be a conformal immersion of a Riemann surface M into an oriented Riemannian n-manifold N; let $\tilde{\varphi} : M \to Q(N)$ be its Gauss lift, as in (4.1). Holomorphicity of $\tilde{\varphi}$ is of course defined in terms of Π. The assignment $\varphi \rightsquigarrow \tilde{\varphi}$ is a bijection between conformal harmonic maps $\varphi : M \to N$ and J_2-holomorphic maps $\tilde{\varphi} : M \to Q(N)$.

(In that statement we exclude constant maps φ and vertical maps $\tilde{\varphi}$.) Again, the proof involves essentially analysis of types: We decompose the induced complex tangent bundle $\varphi^{-1} T^{\mathbb{C}} N = T' \oplus T''$. Then the covariant differential $\nabla \tilde{\varphi}$ is a section of $T^* M \otimes \varphi^{-1} \Lambda^2 TN$; its (2,0)-component $\eta = (\nabla \tilde{\varphi})^{2,0}$ itself decomposes: $\eta = \eta_1 + \eta_2$, where η_1 and η_2 are sections of $T' M \otimes \Lambda^{2,0}$ and $T'' \otimes \Lambda^{2,0}$, respectively.

Then $\eta_1 \equiv 0$ iff $\tilde{\varphi}$ is J_1-holomorphic;
$\eta_2 \equiv 0$ iff $\tilde{\varphi}$ is J_2-holomorphic
iff φ is harmonic.

(5.4) **Example.** Take $N = \mathbb{R}^n$. Then associated with a conformal immersion $\varphi : M \to \mathbb{R}^n$ we have the diagram

$$Q(\mathbb{R}^n) = \mathbb{R}^n \times Q_{n-2} \xrightarrow{\varrho} Q_{n-2}$$

$$M \xrightarrow{\varphi} \mathbb{R}^n$$

where ϱ denotes the indicated projection. It is anti-holomorphic as a map $(Q(\mathbb{R}^n), J_2) \to Q_{n-2}$, so the Gauss map $\gamma_\varphi = \varrho \circ \tilde{\varphi}$ is anti-holomorphic; and conversely, as we have seen in (3.7).

(5.5) **Example.** Take $N = S^n$. Then the Gauss lift $\tilde{\varphi}$ of (4.2) associated with a conformal harmonic immersion $\varphi : M \to S^n$ is J_2-holomorphic. In case $n = 3$ the examples of Lawson [51] all have J_2-holomorphic Gauss lifts. Ditto for the examples of Bryant [7] in case $n = 4$.

(5.6) **Example.** Let N be a Kähler manifold. Then with the identification made in (4.4), Theorem (5.3) can be sharpened to a bijection between conformal harmonic maps $\varphi : M \to N$ and J_2-holomorphic maps $\psi : M \to \mathbb{G}_1(N)$, using the natural analogue of the definitions of J_2-structure on $\mathbb{G}_1(N)$.

(5.7) *A map $\psi : M \to Q(N)$ is horizontal iff it is both J_1- and J_2-holomorphic* [23]. In that case it is certainly conformal; as is the composition $\varphi = \pi \circ \psi : M \to N$. *If ψ is horizontal and harmonic, then $\varphi = \pi \circ \psi$ is harmonic.*

Higher order constructions

(5.8) A map $\varphi : M \to N$ is said to be *(real) isotropic* if

$$< \partial_z^\alpha \varphi, \partial_z^\beta \varphi >^{\mathbb{C}} \equiv 0 \text{ for all } \alpha, \beta \geq 1.$$

From (2.3) we see that isotropic maps are conformal. A detailed study — in considerably greater generality — of isotropy via horizontality has been made by Rawnsley [63].

(5.9) **Examples.** Every harmonic map $\varphi : S^2 \to \mathbb{R} P^n$ is isotropic [9], [10]. Every harmonic map $\varphi : S^2 \to \mathbb{C} P^N$ is complex isotropic [18], [32], [8]. Every harmonic map (from a torus) $\varphi : T^2 \to \mathbb{C} P^n$ of degree $\neq 0$ is (complex) isotropic [26], [27].

(5.10) For any decomposition $n = r + s$ we let $\mathbf{G}_r$ denote the Grassmannian of complex r-spaces in $\mathbb{C}^{n+1}$:

$$\mathbf{G}_r = \mathbf{G}_r(\mathbb{C}^{n+1}) = U_{n+1}/U_r \times U_{s+1};$$

and

$$\mathbf{G}_r(\mathbb{C} P^n) = U_{n+1}/U_r \times U_s \times U_1$$

the corresponding Grassmann bundle of $\mathbb{C} P^n$. Thus $\mathbf{G}_r(\mathbb{C} P^n) = \{(V, W) : V \in \mathbf{G}_r, W \in \mathbf{G}_s, V \perp W\}$. We have the following diagram analogous to that in (4.2):

$$\begin{array}{ccc}
 & \mathbf{G}_r(\mathbb{C} P^n) & \\
\pi_1 \swarrow & \downarrow \pi & \searrow \pi_3 \\
\mathbf{G}_r & \mathbb{C} P^n & \mathbf{G}_s
\end{array}$$

where $\pi_1(V, W) = V$, $\pi_3(V, W) = W$; and

$$\pi(V, W) = (V \oplus W)^\perp$$

Now it is convenient to change slightly the representation of $\mathbf{G}_r(\mathbb{C} P^n)$, replacing the pairs (V, W) by (V, X), where $V, X (= W^\perp)$ are complex subspaces of dimensions $r, r+1$, with $V \subset X$. Then $\pi(V, X) = V^\perp \cap X$.

(5.11) If $f : M \to \mathbb{C} P^n$ is a holomorphic map, which we shall also suppose is *full* (i.e., $f(M)$ lies in no proper projective subspace), then its rth *associated map* $f_r : M \to \mathbf{G}_{r+1}$, is defined by

$$f_r(x) = \mathrm{Span} \ \{f(x), \partial_z f(x), \ldots, \partial_z^r f(x)\};$$

see [73] for details.

Then $\psi = (f_{r-1}, f_r) : M \to \mathbf{G}_r(\mathbb{C} P^n)$; and the composition $\varphi = \pi \circ \psi : M \to \mathbb{C} P^n$; in general, φ is neither holomorphic nor anti-holomorphic. We define its *order* (more precisely, its ∇_z-order) to be

$$\max_{x \in M} \dim \mathrm{Span} \ \{\nabla_z^\alpha \varphi(x) : 1 \leqq \alpha\}$$

The next result is a reformulation of the parametrization theorem of [27]:

The assignment $\psi \rightsquigarrow \pi \circ \psi = \varphi$ is a bijection between the full horizontal holomorphic maps $\psi : M \to \mathbf{G}_r (\mathbb{C} P^n)$ and the full complex isotropic harmonic maps $\varphi : M \to \mathbb{C} P^n$ of order r.

(5.12) Now take $n = 2r$. Consider the subbundle $\mathscr{H}_r = SO_{2r+1}/U_r$ of $\mathbf{G}_r(\mathbb{C}P^{2r})$ over the real projective space $\mathbb{R}P^{2r} \subset \mathbb{C}P^{2r}$. Thus $\mathscr{H}_r = \{Vr\text{-dimensional:} \ V \perp \bar{V}\}$, and the projection $\pi : \mathscr{H}_r \to \mathbb{R}P^{2r}$ is given by $\pi(V) = (V \oplus \bar{V})^{\perp}$.

As a corollary of (5.11)—and our chief motivation and inspiration—we recover the theorem of Calabi [9], [10]:

There is a bijection between full horizontal holomorphic maps $\psi : M \to \mathscr{H}_r$ *and full isotropic harmonic maps* $\varphi : M \to \mathbb{R}P^{2r}$.

(5.13) Generalizing (5.10), we write $n = r + s + t$ and form the Riemannian fibrations

$$\mathscr{H}_{r,s,t} = U_{r+s+t}/U_r \times U_s \times U_t$$

$$\begin{array}{ccc} & \pi_r \swarrow & \downarrow \pi & \searrow \pi_t \\ \mathbf{G}_r(\mathbb{C}^n) & \mathbf{G}_s(\mathbb{C}^n) & \mathbf{G}_t(\mathbb{C}^n) \end{array}$$

Say that a map $\varphi : M \to \mathbf{G}_s(\mathbb{C}^n)$ is *full* if the image spaces $\varphi(x)$ do not all lie in a proper subspace of $\mathbb{C}^n$. Say that a map $\psi : M \to \mathscr{H}_{r,s,t}$ is *full* if $\pi_r \circ \psi$ and $\pi_t \circ \psi$ are.

Then [30]:

The assignment $\psi \rightsquigarrow \pi \circ \psi = \varphi$ *is a bijection between full horizontal holomorphic maps* $\psi : M \to \mathscr{H}_{r,s,t}$ *and full strongly isotropic maps* $\varphi : M \to \mathbf{G}_s(\mathbb{C}^n)$ *of order* r.

We refer to [28], [29], [33] for similar parametrization theorems for (real and complex) isotropic harmonic maps of Riemann surfaces into other symmetric spaces—possibly furnished with indefinite metrics.

(5.14) **Remark.** It is important to emphasize that the full horizontal holomorphic maps appearing in (5.11)—(5.13) are obtained explicitly from bundle constructions.

6   Problems and prospects

Existence

(6.1) **Problem.** *Let M be a compact Riemann surface and N a complete oriented Riemannian n-manifold. When is there a nonconstant conformal harmonic map* $\varphi : M \to N$?

a) In case $n = 2$ we are dealing with branched coverings (2.6); see [21, § 11], [22].

b) If the universal cover of N carries a smooth strictly convex function, then any harmonic map $\varphi : M \to N$ which is null homotopic is constant; in particular, any harmonic map $\varphi : S^2 = N$ is constant.

By way of contrast, every compact N^3 has an embedded closed minimal surface [61]. If N^3 is a 3-sphere with any Riemannian metric h, then N^3 has an embedded minimal 2-sphere [71]. If $\pi_2(N^3) \neq 0$, there is an essential minimally embedded S^2 in N^3 [54].

c) If N is compact and has strictly negative sectional curvature, there are only finitely many nonconstant conformal harmonic maps $M \to N$ [52], [1].

(6.2)   **Problems.** *Let $\mathscr{H}$ be a homotopy class of maps $M \to N$. When does it contain a conformal harmonic map?*

In particular, suppose that N is a compact simply connected 4-manifold.

a) Find an example of a homotopy class $\mathscr{H} \in \pi_2(N)$ not represented by a (conformal) harmonic map $S^2 \to N$. It is known that there is a system of generators of $\pi_2(N)$ which can be so represented [67].

b) Which integral homology classes of dimension 2 are represented by conformal harmonic maps of compact Riemann surfaces? What is the minimum genus of such a representation?

c) For any compact N, if M is a compact oriented surface of genus $M \geqq 1$ and $\mathscr{H}$ a homotopy class of maps $M \to N$ whose elements induce an injection on the fundamental groups, then there is a conformal structure μ on M and a conformal harmonic map $(M, \mu) \to (N, h)$ inducing the same action on π_1 ([69], [68]).

d) For some further results in case dim $N = 3$, see [31], [37].

Methods

(6.3)   To attack those problems there are two very different methods available, to supplement classical variational theory:

a) The first uses the results of § 5 to transfer the problem to its corresponding holomorphic version. And then appeals to our extensive knowledge of holomorphic function theory of Riemann surfaces; e. g., to construct harmonic maps of prescribed degree from a torus $T^2 \to \mathbb{C}P^n$ [25], [27]; to construct minimal immersions of M in S^4 and S^6 [7], [6]. That method requires much further development; especially, the differential systems and their integral manifolds arising in Theorem (5.3) above.

In a related context, the Teichmüller theory of M was used in [69], [68] to prove c) of (6.2). There are fragments indicating the presence of a decent moduli space for isotropic harmonic maps into Hermitian symmetric spaces N. And perhaps for the J_2-holomorphic maps $M \to Q(N)$ of (5.3).

b) The second belongs to geometric measure theory. Striking applications have been made by Pitts [61] and F. Smith [71], producing the results mentioned in (6.1). There have been recent fundamental advances in regularity theory for energy-minimizing harmonic maps, by Giaquinta-Giusti and Schoen-Uhlenbeck. Many further such incursions can be expected in the near future.

Gauss maps

(6.4)   **Problem.** *Characterize those maps* $\gamma : M \to Q_{n-2}$ *which are the Gauss maps of conformal immersions* $\gamma : M \to \mathbb{R}^n$ *with constant mean curvature* $(n \geq 4)$. Certainly γ must be harmonic by (3.13). On the other hand, not every harmonic map γ arises in this manner, even for $n = 4$. A substantial amount of work has been done on this problem by Hoffman-Osserman [42], [39]. For the case $n = 3$ see Kenmotsu [48].

(6.5)   **Problem.** *Suppose that M is compact. Characterize those homotopy classes of maps* $M \to Q_{n-2}$ *which contain the Gauss map of a conformal immersion of* $M \to \mathbb{R}^n$ (*or* T^n) *with constant mean curvature.*
    Not every homotopy class contains a Gauss map in case $n = 4$.

Isotropy

(6.6)   **Problem.** *Let M be a compact Riemann surface and N an Hermitian symmetric space. Under what circumstances is a conformal harmonic map* $\varphi : M \to N$ *isotropic?* Some examples were mentioned in (5.9), and others can be found in [30], [62]. However, we do not yet have sufficiently good conditions for isotropy of a surface of genus p mapped into $\mathbb{C}P^n$ $(n \geq 3)$; or of a sphere S^2 mapped into $\mathbb{G}_2(C^n)$ or Q_n.
    Isotropy conditions are described
        a) via holomorphic k-adic differentials on M. See [10] and [27, § 7], following the original idea of Hopf [43]; and also [14], [17], where those computations were rederived using the method of moving frames.
        b) They can also be interpreted as conservation laws of the associated variational principle defining harmonic maps [4], [62].
        Much more can be expected from both interpretations.

Twistor constructions

(6.7)   Section 5 calls for a higher order version of Theorem (5.3), presumably involving some form of isotropy. Some first steps have already been taken by Rawnsley [64], in case N is a Kähler manifold; his twistor bundle is $\mathbb{G}_r(N) \to N$, as in (5.6)—which admits natural almost complex structures J_1, J_2 as in (5.1). However, an effective bijective correspondence is lacking, due to complications involving fullness of the maps (a difficulty which does not arise in (5.3)).
        Of course, such a program will involve higher order fundamental forms of maps and higher order Gauss maps (see [4], [14], [34]).

Variational aspects

(6.8)   Harmonic maps are the extrema of the energy functional

$$E(\varphi) = \tfrac{1}{2} \int |d\varphi|^2.$$

They are closely related to those of the area functional

$$A(\varphi) = \int |\Lambda^2 \, d\varphi|.$$

In fact, *for any map φ we have $A(\varphi) \leq E(\varphi)$, with equality iff φ is conformal* [24]. Thus if φ is conformal and minimizes A, then it minimizes E. Also [67], *if (φ, μ) is an extremal of E with respect to variations of both φ and the conformal structure μ of M, then $\varphi : (M, \mu) \to (N, h)$ is a conformal harmonic map.*

(6.9)   Experience over the past twenty years has taught us that only in exceptional circumstances can we except a homotopy class of maps to contain a map minimizing E; and that not every homotopy class will contain a harmonic map [25]. In particular, we cannot expect to have a systematic variational method producing (weak) extrema of E. Regularity is in somewhat better shape—yet another simplifying aspect of 2-dimensional domains [35].

Thus we have been forced to turn elsewhere: To twistor representations, thereby exploiting the interrelationships between harmonicity and holomorphicity—such as is found in the Weierstrass representation formula [40], [38], [7].

(6.10)   However, we certainly must not lose track of those variational origins. In particular, there is need to

a) develop further the stability properties of conformal harmonic maps $\varphi : M \to N$, for both functionals E and A;

b) calculate/estimate the Morse index of such maps—for they are often not minima of E. And find a form of critical point theory for $E : C^\infty(M, N) \to \mathbb{R}$, sufficient at least to give information on the energy spectrum [26];

c) analyze the effect on conformal harmonic maps by a change of metric h on N. In particular, how does the complex structure J_2 in (5.3) vary with h?

d) *For $\varphi : M \to \mathbb{R}^n$ conformal and harmonic we have*

$$A(\gamma_\varphi) = \int |K^M|,$$

K^M denoting the curvature function of M [41]. Such quantitative results are of course intimately related to integral geometry, and thus to value distribution theory; a key paper here is that of Chern-Osserman [16]. That theory needs to be extended, especially to isotropic harmonic maps of Riemann surfaces into Riemannian and Hermitian symmetric spaces.

References

[1] T. Adachi and T. Sunada, *Energy spectrum of certain harmonic mappings.*

[2] F. J. Almgren, *Some interior regularity theorems for minimal surfaces and an extension of Bernstein's theorem.* Ann. of Math. 84 (1966), 277—292.

[3] M. F. Atiyah, N. J. Hitchin and I. M. Singer, *Self-duality in four-dimensional geometry.* Proc. Roy. Soc. London A362 (1978), 425—461.

[4] P. Baird and J. Eells, *A conservation law for harmonic maps.* Geo. Symp. Utrecht (1980). Springer Notes 894, 1—25.

[5] W. Blaschke, *Sulla geometria differenziale delle superficie S_2 nello spazio euclideo S_4.* Ann. di Mat. 28 (1949), 205—209.

[6] R. Bryant, *Submanifolds and special structures on the octonians.* J. Diff. Geo. 17 (1982), 185—232.

[7] R. Bryant, *Conformal and minimal immersions of compact surfaces into the 4-sphere.* J. Diff. Geo. 17 (1982), 455—473.

[8] D. Burns, *Harmonic maps from $\mathbb{C}P^1$ to $\mathbb{C}P^n$.* Harmonic maps Symp. New Orleans (1980). Springer Notes 949 (1982), 48—55.

[9] E. Calabi, *Quelques applications de l'analyse complexe aux surfaces d'aire minima.* Topics in Complex Manifolds, Univ. Montréal (1967), 59—81.

[10] E. Calabi, *Minimal immersions of surfaces in Euclidean spheres.* J. Diff. Geo. 1 (1967), 111—125.

[11] X. P. Chen, *Harmonic mapping and Gauss mapping.*

[12] B. Y. Chen and S. Yamaguchi, *Classification of surfaces with totally geodesic Gauss image.* Ind. Math. J. 32 (1983), 143—154.

[13] S. S. Chern, *Minimal surfaces in an Euclidean space of N dimensions.* Diff. and Comb. Topology. Princeton (1965), 187—198.

[14] S. S. Chern, *On the minimal immersions of the two-sphere in a space of constant curvature.* Prob. in Analysis. Princeton (1970), 27—40.

[15] S. S. Chern, *On minimal spheres in the four-sphere.* Studies and essays presented to Y. W. Chen. Taiwan (1970), 137—150.

[16] S. S. Chern and R. Osserman, *Complete minimal surfaces in Euclidean n-space.* J. Analyse Math. 19 (1967), 15—34.

[17] S. S. Chern and J. Wolfson, *Minimal surfaces by moving frames.* Amer. J. Math. 105 (1983), 59—83.

[18] A. M. Din and W. J. Zakrzewski, *General classical solutions in the $\mathbb{C}P^{n-1}$ model.* Nucl. Phys. B 174 (1980), 397—406.

[19] A. M. Din and W. J. Zakrzewski, *Properties of the general classical $\mathbb{C}P^{n-1}$ model.* Phys. Letters 95 B (1980), 419—422.

[20] P. Dombrowski, 150 years after Gauss's *"Disquistiones generales circa superficies curvas"* Astérique 62 (1979).

[21] J. Eells and L. Lemaire, *A report on harmonic maps.* Bull. London Math. Soc. 10 (1978), 1—68.

[22] J. Eells and L. Lemaire, *On the construction of harmonic and holomorphic maps between surfaces.* Math. Ann. 252 (1980), 27—52.

[23] J. Eells and S. Salamon, *Constructions twistorielles des applications harmoniques.* C. R. Paris 296 (1983), 685—687.

[24] J. Eells and J. H. Sampson, *Harmonic mappings of Riemannian manifolds.* Amer. J. Math. 86 (1964), 109—160.

[25] J. Eells and J. C. Wood, *Restrictions harmonic maps of surfaces.* Topology 15 (1976), 263—266.

[26] J. Eells and J. C. Wood, *The existence and construction of certain harmonic maps.* Symp. Math. Ist. Naz. Alta Mat. Roma 26 (1982), 123—138.

[27] J. Eells and J. C. Wood, *Harmonic maps from surfaces to complex projective spaces.* Advances in Math. 49 (1983), 217—263.

[28] S. Erdem, Thesis Univ. Leeds (1983).

[29] S. Erdem and J. F. Glazebrook, *Harmonic maps of Riemann surfaces to indefinite complex hyperbolic and projective spaces.* Proc. London Math. Soc. 47 (1983), 547—562.

[30] S. Erdem and J. C. Wood, *On the construction of harmonic maps into a Grassmannian.* J. London Math. Soc. 28 (1983), 161—174.

[31] M. Freedman, J. Hass, and P. Scott, *Least area incompressible surfaces in 3-manifolds.* Inv. Math. 71 (1983), 601—642.

[32] V. Glaser and R. Stora, *Regular solutions of the $\mathbb{C}P^n$ models and further generalizations.* CERN Preprint (1980).

[33] J. F. Glazebrook, *Isotropic harmonic maps to Kähler manifolds, and related topics.* Thesis. Univ. of Warwick (1983).

[34] P. Griffiths and J. Harris, *Algebraic geometry and local differential geometry.* Ann. E. N. S. 12 (1979), 355—432.

[35] M. Grüter, *Regularity of weak H-surfaces.* J. Reine Angew. Math. 329 (1981), 1—15.

[36] R. D. Gulliver, R. Osserman, and H. Royden, *A theory of branched immersions of surfaces.* Amer. J. Math. 95 (1973), 750—812.

[37] J. Hass, *Embedded minimal surfaces in three and four dimensional manifolds.* Thesis. Univ. California Berkeley (1981).

[38] N. J. Hitchin, *Monopoles and geodesics.* Comm. Math. Phys. 83 (1982), 579—602.

[39] D. A. Hoffman, *When is a map a Gauss map?*

[40] D. A. Hoffman and R. Osserman, *The geometry of the generalized Gauss map.* Mem. Amer. Math. Soc. 236 (1980).

[41] D. A. Hoffman and R. Osserman, *The area of the generalized Gaussian image and the stability of minimal surfaces in S^n and in $\mathbb{R}^n$.* Math. Ann. 260 (1982), 437—452.

[42] D. A. Hoffman and R. Osserman, *The Gauss map of surfaces in $\mathbb{R}^n$.*

[43] H. Hopf, *Uber Flächen mit einer Relation zwischen den Hauptkrümmungen.* Math. Nachr. 4 (1950/1), 232—249.

[44] T. Ishihara, *The harmonic Gauss maps in a generalized sense.* J. London Math. Soc. 26 (1982), 104—112.

[45] T. Ishihara, *The Gauss map and non-holomorphic harmonic maps.* Amer. J. Math.

[46] T. Ishihara, *The Gauss map of Kähler immersions into complex hyperbolic spaces.* Tokushima J. Math.

[47] F. W. Kamber and P. Tondeur, *Curvature properties of harmonic foliations.* Ill. J. Math. (1983).

[48] K. Kenmotsu, *Weierstrass formula for surfaces of prescribed mean curvature.* Math. Ann. 245 (1979), 89—99.

[49] S. Kobayashi and K. Nomizu, *Foundations of differential geometry I, II.* Interscience (1963, 1969).

[50] H. B. Lawson, *Lectures on minimal submanifolds.* IMPA (1970). Math. Lecture Series 9 (1980). Publish or Perish.

[51] H. B. Lawson, *Complete minimal surfaces in S^3.* Ann. of Math. 92 (1970), 335—374.

[52] L. Lemaire, *Harmonic mappings of uniformly bounded dilatation.* Topology 16 (1977), 199—201.

[53] A. Lichnerowicz, *Applications harmoniques et variétés kählériennes.* Symp. Math. Ist. Naz. Alta Mat. 3 (1970), 341—402.

[54] W. H. Meeks and S. T. Yau, *Topology of three-dimensional manifolds and the embedding problems in minimal surface theory.* Ann. of Math. 112 (1980), 441—484.

[55] S. Nishikawa, *The Gauss map of Kähler immersions.* Tohoku Math. J. 27 (1975), 453—460.

[56] J. C. C. Nitsche, *Vorlesungen über Minimalflächen.* Grundlehren Band 199. Springer (1975).

[57] M. Obata, *The Gauss map of immersions of Riemannian manifolds in spaces of constant curvature*. J. Diff. Geo. 2 (1968), 217—223.

[58] R. Osserman, *A survey of minimal surfaces*. Van Nostrand Math. Studies 25 (1969).

[59] R. Osserman, *Minimal surfaces, Gauss maps, total curvature, eigenvalue estimates and stability*. The Chern Symp. (1979). Springer (1980), 199—227.

[60] M. Pinl. *B-Kugelbilder reeller Minimalflächen in R_4*. Math. Z. 59 (1953), 290—295.

[61] J. T. Pitts, *Existence and regularity of minimal surfaces on Riemannian manifolds*. Princeton Notes Series 27 (1981).

[62] J. Ramanathan, *Harmonic maps from S^2 to $G_{2,4}$*.

[63] J. H. Rawnsley, *On the rank of horizontal maps*. Math. Proc. Camb. Phil. Soc. 92 (1982), 485 — 488.

[64] J. H. Rawnsley, *Twistor spaces and isotropic harmonic maps of Riemann surfaces*. (Warwick Preprint (1983)).

[65] E. A. Ruh, *Minimal immersions of 2-spheres in S^4*. Proc. Amer. Math. Soc. 28 (1971), 219—222.

[66] E. A. Ruh and J. Vilms, *The tension field of the Gauss map*. Trans. Amer. Math. Soc. 149 (1970), 569—573.

[67] J. Sacks and K. Uhlenbeck, *The existence of minimal immersions of 2-sphere*. Ann. of Math. 113 (1981), 1—24.

[68] J. Sacks and K. Uhlenbeck, *Minimal immersions of closed Riemann surfaces*. Trans. Amer. Math. Soc. 271 (1982), 639—652.

[69] R. Schoen and S. T. Yau, *Existence of incompressible minimal surfaces and the topology of three dimensional manifolds with non-negative scalar curvature*. Ann. of Math. 10 (1979), 127—142.

[70] Y. L. Shen, *On submanifolds in Riemannian manifolds of constant curvature*.

[71] F. R. Smith, *On the existence of embedded minimal 2-spheres in the 3-sphere, endowed with an arbitrary metric*. Thesis. Univ. of Melbourne (1982).

[72] C. M. Wood, *Some energy-related functionals and their vertical variational theory*. Thesis. Univ. of Warwick (1983).

[73] H. H. Wu, *The equidistribution theory of holomorphic curves*. Ann. of Math. Studies 64 (1970).

[74] J. Eells and S. Salamon, *Twistorial construction of harmonic maps of surfaces into four-manifolds*.

[75] J. H. Rawnsley, *f-structures, f-twistor spaces and harmonic maps*.

[76] S. Salamon, *Harmonic and holomorphic maps*.

Note Added in Proof

Concerning (6.7), F. E. Burstall and J. H. Rawnsley have established higher order (isotropy) versions of (5.3) and (5.6) for Riemannian and Kähler manifolds—without fullness considerations.

[75] supercedes [64].

[74] and [76] provide details of the announcement [23].

Perspectives in Mathematics
Anniversary of Oberwolfach 1984
© Birkhäuser Verlag, Basel

Neuere Fortschritte in der diophantischen Geometrie

GERD FALTINGS

Gesamthochschule Wuppertal, Fachbereich 7 – Mathematik,
Gausstraße 20, D-5600 Wuppertal 1 (FRG)

§ 1 Vorwort

Im Folgenden soll versucht werden, zu erläutern, wie kürzlich die Vermutungen von Tate, Shafarevich und Mordell bewiesen worden sind. Es ist aber nicht die Absicht des Verfassers, einen vollständigen Überblick über das Gebiet zu geben.

Worum geht es also? Bei der Tate-Vermutung (genauer gesagt, dem hier gezeigten Teil dieser Vermutung) betrachtet man eine abelsche Varietät A über einem Zahlkörper K. A ist also eine glatte zusammenhängende projektive K-Varietät, welche gleichzeitig die Struktur einer kommutativen algebraischen Gruppe besitzt. Wenn $\bar{K}$ den algebraischen Abschluß von K bezeichnet, so betrachtet man die $\bar{K}$-rationalen Punkte $A(\bar{K})$ von A. Diese bilden dann eine abelsche Gruppe, und man weiß, daß diese Gruppe divisibel ist. Weiter ist bekannt, daß die Torsion von $A(\bar{K})$ von der Form $(\mathbb{Q}/\mathbb{Z})^{2g}$ ist, wobei g die Dimension von A sein soll.

Wenn $\pi = \mathrm{Gal}(\bar{K}/K)$ die absolute Galois-Gruppe von K bezeichnet, so operiert π auf $A(\bar{K})$, und man kann nach der Struktur dieses π-Moduls fragen. Diese Frage ist noch weitgehend ungeklärt, und sie berührt viele wichtige offene Probleme. Bekannt ist jedoch einiges über die Operationen von π auf der Torsion in $A(\bar{K})$: Jeder K-Endomorphismus der algebraischen Gruppe A induziert einen π-linearen Endomorphismus auf $A(\bar{K})$ und auch auf der Torsion $\mathrm{Tor}(A(\bar{K}))$. Wenn

$$\hat{\mathbb{Z}} = \varprojlim_{n} \mathbb{Z}/n\mathbb{Z}$$

die profinite Komplettierung von $\mathbb{Z}$ bezeichnet, ergibt sich somit eine Abbildung

$$\mathrm{End}_K(A) \otimes_{\mathbb{Z}} \hat{\mathbb{Z}} \to \mathrm{End}_\pi(\mathrm{Tor}(A(\bar{K}))).$$

Man weiß nun, daß diese Abbildung stets ein Isomorphismus ist. Dies folgt aus der Tate-Vermutung, welche folgendes besagt: Man wähle eine Primzahl l und bilde den Tate-Modul

$$T_l(A) = \mathrm{Hom}(\mathbb{Q}_l/\mathbb{Z}_l, A(\bar{K})).$$

($\mathbb{Q}_l = l$-adische Komplettierung von $\mathbb{Q}$, usw.). $T_l(A)$ ist als abelsche Gruppe isomorph zu $\mathbb{Z}_l^{2g}$, und π operiert stetig darauf.

Es gilt nun (Tate-Vermutung):

a) $T_l(A) \otimes_{\mathbb{Z}_l} \mathbb{Q}_l$ ist halbeinfacher π-Modul

b) Die Abbildung $\mathrm{End}_K(A) \otimes_{\mathbb{Z}} \mathbb{Z}_l \longrightarrow \mathrm{End}_\pi(T_l(A))$ ist ein Isomorphismus.

Wenn man eine abelsche Varietät A über einem Zahlkörper K vorgibt, so kann man sie zu einer Varietät über dem Ring R der ganzen Zahlen in K ausdehnen. Diese Ausdehnung ist nicht eindeutig bestimmt, doch gibt es ein kanonisches Modell von A über R, das Néron-Modell. Man kann es folgendermaßen grob charakterisieren: Wenn man eine Ausdehnung von A über R vorgibt, so betrachtet man für die Primideale $\mathfrak{p} \subset R$ die Reduktion modulo $\mathfrak{p}$. Dies sind für fast alle $\mathfrak{p}$ wieder glatte abelsche Varietäten. Die endlich vielen $\mathfrak{p}$, für die das nicht gilt, heißen Stellen schlechter Reduktion. Das minimale Modell hat die Eigenschaft, daß die Anzahl dieser Stellen so klein wie möglich wird. Es gilt nun die Shafarevich-Vermutung: Es gibt nur endlich viele Isomorphieklassen abelscher Varietäten A über K der Dimension g, welche gute Reduktion außerhalb einer vorgegebenen endlichen Menge S von Primstellen von K haben.

Daraus folgen andere Varianten. Man kann zum Beispiel auch Isomorphieklassen prinzipal polarisierter abelscher Varietäten betrachten, oder Isomorphieklassen von Kurven vom Geschlecht $g > 1$ (benutze Torelli). Die letzte Aussage ist die ursprüngliche Vermutung von I. R. Shafarevich. Man weiß nun, daß aus der Shafarevich-Vermutung die Vermutung von Mordell folgt, also daß es auf einer Kurve X über K vom Geschlecht $g > 1$ nur endlich viele rationale Punkte gibt.

Die angewandten Beweismethoden stammen aus der algebraischen Geometrie. Dort wurden sie entwickelt, um die entsprechenden Sätze für Funktionenkörper zu beweisen. Allerdings gelang es lange nicht, sie auf Zahlkörper zu übertragen, obwohl es dazu eine generelle Methode gab. Dies gelang zuerst dem Verfasser, wobei er noch die Theorie der Kompaktifizierungen von Modulräumen benutzte. So konnte er zunächst die Tate-Vermutung beweisen, worüber er im Mai 1983 auf einem Vortrag in Oberwolfach berichtete. Anschließend zeigte sich zu seiner großen Überraschung, daß die angewandten Methoden auch zum Beweis der Shafarevich-Vermutung ausreichten.

Alles in allem kann man dies als einen weiteren Erfolg der von A. Grothendieck eingeleiteten Neuformulierung der algebraischen Geometrie ansehen.

§ 2 Historischer Abriß

Die Suche nach ganzzahligen oder rationalen Lösungen von Gleichungen ist schon sehr alt. Bekannt ist zum Beispiel das Fermat-Problem, also die

Suche nach Paaren (x, y) rationaler Zahlen mit

$$x^n + y^n = 1.$$

Es gibt stets die trivialen Lösungen $(0,1)$ und $(1,0)$, und es wird vermutet, daß dies alle sind, falls n größer als zwei ist. Dies ist bekannt für $n \leq 125\,000$, aber ein allgemeiner Beweis wurde bis jetzt noch nicht gefunden.

Die Fermat-Gleichung definiert eine algebraische Kurve vom Geschlecht

$$g = \frac{(n-1)\,(n-2)}{2}.$$

Wir wollen uns im folgenden auf algebraische Kurven beschränken, da der Fall höher dimensionaler Varietäten noch ganz ungeklärt ist. Außerdem betrachten wir nur Lösungen in $\mathbb{Q}$, doch bleiben alle Aussagen gültig für beliebige Zahlkörper, und sogar für alle endlich erzeugten Erweiterungen von $\mathbb{Q}$.

Eine grobe Klassifikation der algebraischen Kurven erfolgt durch ihr Geschlecht. Dies ist eine ganze nicht negative Zahl, und ganz allgemein verhalten sich Kurven recht verschieden, je nachdem ob ihr Geschlecht 0,1 oder größer als 1 ist. Als Beispiel mögen die folgenden Kurven dienen:

a) $y \ = x^2$
b) $y^2 = x^3 - x$
c) $y^2 = x^5 - x$

Diese haben Geschlecht 0,1 und 2, so daß wir von jeder Klasse einen Vertreter aufgelistet haben.

Wir interessieren uns für rationale Punkte auf Kurven, daß heißt, für rationale Lösungen der sie definierenden Gleichungen.

Für Kurven vom Geschlecht 0 sind diese recht leicht zu finden, wie man auch am Beispiel a) sieht. Falls es überhaupt einen rationalen Punkt gibt, so kann man alle solche Punkte als rationale Funktion eines Parameters erhalten. Für Kurven vom Geschlecht 1 wird die Sache schon schwieriger. Bekanntlich tragen diese eine Gruppenstruktur, wobei man als neutrales Element einen beliebigen rationalen Punkt wählen kann. Die rationalen Punkte bilden dann eine abelsche Gruppe, und man kann im Allgemeinen nicht mehr sagen als der folgende, von L. J. Mordell 1922 ([17]) bewiesene Satz aussagt:

Satz: Sei X eine Kurve vom Geschlecht 1 über $\mathbb{Q}$. Dann ist die abelsche Gruppe $X(\mathbb{Q})$ endlich erzeugt.

Es kann also unendlich viele Punkte geben, wobei man als Maß für ihre Anzahl den Rang der abelschen Gruppe $X(\mathbb{Q})$ nehmen kann. Es wird vermutet, daß dieser Rang beliebig groß werden kann.

Es bleiben nun noch die Kurven vom Geschlecht größer als 1 übrig. In seiner Arbeit stellte Mordell die Vermutung auf, daß diese stets nur endlich viele rationale Punkte besitzen. Er konnte dies jedoch nicht zeigen, und das Problem schien lange unangreifbar zu sein.

Den ersten Fortschritt erzielte 1928 A. Weil in seiner Promotionsschrift ([35]), indem er den Satz von Mordell auf abelsche Varietäten höherer Dimensionen verallgemeinerte. Das Resultat heißt heute „Satz von Mordell-Weil", und man kann Beweise zum Beispiel in [13] oder [20] finden. Was besagt nun dieser Satz in unserem Fall?

Jede Kurve X kann man in ihre Jacobi'sche $J = J(X)$ einbetten. Dann ist $J(\mathbb{Q})$ eine endlich erzeugte abelsche Gruppe, und die Mordell-Vermutung besagt, daß diese abelsche Gruppe einen endlichen Durchschnitt mit X besitzt, falls X verschieden von J ist (das heißt, falls das Geschlecht von X größer als 1 ist).

Man hat versucht, die Mordell-Vermutung auf diese Weise zu zeigen, doch ist dies meines Wissens bis heute noch niemandem gelungen. Man hätte sie zum Beispiel folgern können, wenn man gezeigt hätte, daß jede endlich erzeugte Untergruppe von J endlichen Durchschnitt mit X hat. Dies ist zwar heute bewiesen, doch braucht man dazu schon die Mordell-Vermutung.

Den nächsten Fortschritt erzielte 1929 C. L. Siegel ([28]). Er zeigte, daß auf einer affinen Kurve vom Geschlecht größer als Null nur endlich viele ganze Punkte liegen. Das heißt in unseren Beispielen, daß die Gleichungen b) und c) nur endlich viele ganzzahlige Lösungen (x, y) besitzen. K. Mahler verallgemeinerte dies später ([14]) auf den Fall, daß x und y rationale Zahlen sind, in deren Nenner nur endlich viele vorgegebene Primzahlen aufgehen dürfen. Man kann diese Ergebnisse zum Beispiel in [13] nachlesen. Bis in jüngste Zeit war dies für Zahlkörper der „Stand der Technik".

Fortschritte gab es zunächst wieder für Funktionenkörper, als Yu. I. Manin ([15]) und H. Grauert ([11]) 1963 bzw. 1965 die Mordell-Vermutung für Funktionenkörper bewiesen. Dabei benutzten sie die Differentiation im Grundkörper, und außerdem ergaben sich Komplikationen durch konstante Kurven. Dies führte dazu, daß eine Übertragung auf Zahlkörper nicht möglich war.

Doch wurden dadurch weitere Forschungen angeregt, und A. N. Parshin bewies 1968 ([24]), daß die Mordell-Vermutung aus der Shafarevich-Vermutung folgt. Außerdem bewies er für Funktionenkörper der Charakteristik 0 einen Teil der Shafarevich-Vermutung, welcher zum Beweis von Mordell ausreichte. Die ganze Vermutung zeigte dann in Charakteristik null 1971 S. Arakelov ([11]), und in positiver Charakteristik 1978 L. Szpiro ([29]). Wir erläutern kurz für den Grundkörper $\mathbb{Q}$ das Argument von Parshin:

Wenn X eine Kurve vom Geschlecht größer Eins über $\mathbb{Q}$ ist, und $x \in X(\mathbb{Q})$ ein rationaler Punkt, so konstruiert man eine Überlagerung $Y(x)$ von X, welche genau über x verzweigt ist. Man hat so viel Kontrolle über die Stellen schlechter Reduktion von $Y(x)$, daß man die Shafarevich-Vermutung anwenden kann, und er ergibt sich, daß man auf diese Weise nur endlich viele Isomorphie-

Klassen von Kurven $Y(x)$ erhält. Daraus folgert man, daß es auch nur endlich viele Punkte $x \in X(\mathbb{Q})$ gibt.

Man hat natürlich versucht, alles auf Zahlkörper zu übertragen. Arakelov ([2]) hat auch eine Methode dazu entwickelt. Sie besteht kurz gesagt darin, daß man metrisierte Objekte über $\mathrm{Spek}(\mathbb{Z})$ betrachten muß, wobei die Metrik der Struktur an der unendlichen Stelle von $\mathbb{Q}$ entspricht. Eine direkte Übertragung mit Hilfe dieser Methode scheiterte jedoch, da einige Hilfsmittel über Funktionenkörpern (Hodge-Theorie, Kodaira-Spencer Klasse, Frobenius) kein Analogon für Zahlkörper haben. Der Erfolg stellte sich erst ein, als man diese Überlegungen auf die Tate-Vermutung anwandte.

Dies war schon 1966 von J. Tate ([33]) für endliche Körper bewiesen worden, und J. G. Zarhin ([37], [38]) verallgemeinerte dies auf Funktionenkörper über endlichen Körpern.

Die Beweismethode läßt sich etwa wie folgt charakterisieren:

Sei $A/\mathbb{Q}$ eine abelsche Varietät, $T_l(A)$ ihr Tate-Modul, $\pi = \mathrm{Gal}(\bar{\mathbb{Q}}/\mathbb{Q})$ die absolute Galois-Gruppe. Man zeigt, daß jeder π-invariante Unterraum

$$W \subseteq T_l(A) \otimes_{\mathbb{Z}_l} \mathbb{Q}_l$$

Bild eines Idempotents aus $\mathrm{End}_K(A) \otimes_{\mathbb{Z}} \mathbb{Q}_l$ ist. Daraus lassen sich die zu zeigenden Behauptungen leicht folgern. Man reduziert auf den Fall, daß A prinzipal polarisiert und W ein maximal isotroper Teilraum von $T_l(A) \otimes_{\mathbb{Z}_l} \mathbb{Q}_l$ ist. Dann gehört zu W eine l-divisible Untergruppe

$$G_\infty = \bigcup_n G_n$$

von $A[l^\infty]$, und die abelschen Varietäten $A_n = A/G_n$ sind wieder prinzipal polarisiert. Es reicht nun, wenn unendlich viele dieser A_n zueinander isomorph sind, und dazu zeigt man, daß die A_n nur endlich viele verschiedene $\mathbb{Q}$-rationale Punkte im Modulraum $\underline{A}_g$ der prinzipal polarisierten abelschen Varietäten liefern.

Für endliche Grundkörper ist damit der Beweis beendet, denn $\underline{A}_g$ besitzt dann überhaupt nur endlich viele rationale Punkte. Für Zahl- oder Funktionenkörper zeigt man noch, daß die Höhe der A_n entsprechenden Modulpunkte beschränkt ist. Dazu folgt nun ein Exkurs über Höhen.

§3 Höhen

Wir beschränken uns wieder auf den Grundkörper $\mathbb{Q}$. Klassisch wurde die Höhe eines Punktes $x \in \mathbb{P}^n(\mathbb{Q})$ folgendermaßen definiert: Schreibe:

$$x = (x_0 : \ldots \ldots : x_n)$$

mit teilerfremden ganzen Zahlen $x_0, \ldots, x_n$. Dann sei

$$H(x) = \sqrt{x_0^2 + \cdots + x_n^2}, \quad \text{(Höhe)}$$
$$h(x) = \log(H(x)). \qquad \text{(logarithmische Höhe)}$$

Aus der Definition folgt sofort, daß für $c \in \mathbb{R}$ die Menge der Punkte $x \in \mathbb{P}^n(\mathbb{Q})$ mit $h(x) \leqq c$ endlich ist.

Wir haben schon erwähnt, daß S. Arakelov eine Theorie zur Übersetzung von algebraischer Geometrie in die Zahlentheorie entwickelt hat. Dabei entspricht Spek $(\mathbb{Z})$ einer affinen algebraischen Kurve. Bekanntlich erhält man aber eine voll befriedigende Theorie nur für komplette algebraische Kurven, so daß man zu Spek $(\mathbb{Z})$ noch einen unendlichen Punkt hinzunehmen muß, entsprechend der archimedischen Bewertung von $\mathbb{Q}$.

Man muß also die algebraisch geometrischen Objekte auf Spek $(\mathbb{Z})$ mit einer Struktur im Unendlichen versehen, und die Erfahrung zeigt, daß diese meist in einer Metrik besteht. Zum Beispiel ist ein Geradenbündel auf Spek $(\mathbb{Z})$ nichts anderes als ein projektiver $\mathbb{Z}$-Modul P vom Rang 1. Dieser ist notwendig frei, also von der Form

$$P = \mathbb{Z} \cdot m$$

mit einem Erzeugenden m, und alle solche P's sind isomorph. Dies ändert sich, wenn man noch eine Struktur im Unendlichen einführt, in diesem Falle eine hermite'sche Metrik $\| \ \|$ auf $P \otimes_{\mathbb{Z}} \mathbb{C}$. Diese wird charakterisiert durch $\|m\|$, und man kann dem Paar $(P, \| \ \|)$ eine Invariante zuordnen, nämlich den Grad:

$$\mathrm{grad}(P, \| \ \|) = -\log(\|m\|).$$

Man nennt ein solches Paar $(P, \| \ \|)$ ein metrisiertes Geradenbündel auf Spek $(\mathbb{Z})$, und die Gruppe der metrisierten Geradenbündel entspricht der Picard-Gruppe einer kompletten algebraischen Kurve.

Man kann nun die Definition der Höhe umformulieren: Auf $\mathbb{P}^n_{\mathbb{Z}}$ existiert das universelle Geradenbündel $\mathcal{O}(1)$, welches man wie folgt mit einer hermite'schen Metrik versieht: Es gibt eine Surjektion

$$\mathcal{O}^{n+1} \to \mathcal{O}(1).$$

Nach Basiserweiterung zu $\mathbb{C}$ trägt $\mathcal{O}^{n+1}$ die kanonische konstante Metrik, und damit wird eine Metrik auf $\mathcal{O}(1)$ induziert.

Ein Punkt $x \in \mathbb{P}^n(\mathbb{Q})$ definiert eine Abbildung

$$\varrho \colon \mathrm{Spek}(\mathbb{Z}) \to \mathbb{P}^n_{\mathbb{Z}}.$$

Das Pullback $\varrho^*(\mathcal{O}(1))$ ist dann in natürlicher Weise ein metrisiertes Geraden-bündel, und

$$h(x) = \mathrm{grad}\left(\varrho^*(\mathcal{O}(1))\right).$$

Etwas allgemeiner sei X ein eigentliches $\mathbb{Z}$-Schema, $\underline{L}$ ein Geradenbündel auf X, und $\underline{L} \otimes_{\mathbb{Z}} \mathbb{C}$ sei mit einer hermite'schen Metrik versehen. Jeder Punkt $x \in X(\mathbb{Q})$ definiert wieder eine Abbildung

$$\varrho : \mathrm{Spek}(\mathbb{Z}) \to X,$$

und man setzt

$$h_L(x) = \mathrm{grad}\left(\varrho^*(\underline{L})\right).$$

Dies hängt im Wesentlichen nur von $\underline{L} \otimes_{\mathbb{Z}} \mathbb{Q}$ ab, was heißen soll, daß bei Änderung der hermite'schen Metrik oder der Ausdehnung von $\underline{L} \otimes_{\mathbb{Z}} \mathbb{Q}$ von $X \otimes_{\mathbb{Z}} \mathbb{Q}$ auf X sich die Funktion h_L nur um eine auf $X(\mathbb{Q})$ beschränkte Funktion ändert.

h_L ist linear in $\underline{L}$, und es ist klar, daß für ein amples Geradenbündel $\underline{L}$ die Menge der $x \in X(\mathbb{Q})$ mit $h_L(x) \leq c$ wieder stets endlich ist.

Wir wollen diese Überlegungen anwenden auf $\underline{A}_g$, den groben Modulraum der prinzipal polarisierten abelschen Varietäten der Dimension g. Dieser ist leider nicht komplett, so daß man eine Komplettierung wählen muß, zum Beispiel die von W. L. Baily und A. Borel konstruierte, oder die von A. Ash, D. Mumford, M. Rapoport und Y. Tai ([4]). Dies führt zu einigen technischen Schwierigkeiten, und wir geben nur das Endresultat der Überlegungen an:

Sei $A/\mathbb{Q}$ eine prinzipal polarisierte abelsche Varietät über $\mathbb{Q}$. Wir nehmen an, daß A semistabile Reduktion besitzt, also sich zu einer semiabelschen Varietät

$$p : A \to \mathrm{Spek}(\mathbb{Z})$$

über $\mathrm{Spek}(\mathbb{Z})$ ausdehnt. (Die Fasern von p sind Erweiterungen abelscher Varietäten mit Tori.) Dann ist

$$\omega_{A/\mathbb{Z}} = p_*(\Omega^g_{A/\mathbb{Z}})$$

ein Geradenbündel auf $\mathrm{Spek}(\mathbb{Z})$, welches an der unendlichen Stelle eine natürliche hermite'sche Metrik besitzt. Denn

$$\omega_{A/\mathbb{Z}} \otimes_{\mathbb{Z}} \mathbb{C} \cong \Gamma\left(A(\mathbb{C}), \Omega^g_A\right)$$

ist der Raum der holomorphen g-Formen auf dem komplexen Torus $A(\mathbb{C})$, und man definiert darauf eine Norm durch

$$\|\alpha\|^2 = (-1)^{\frac{g(g-1)}{2}} \left(\frac{i}{2}\right)^g \int\limits_{A(\mathbb{C})} \alpha \wedge \bar{\alpha} \quad \text{für} \quad \alpha \in \Gamma\left(A(\mathbb{C}), \Omega_A^g\right).$$

Dann setze man $h(A) = \text{grad}(\omega_{A/\mathbb{Z}}, \| \, \|)$, und die Menge der Isomorphieklassen der A mit $h(A) \leq c$ ist endlich. Natürlich überträgt sich alles auf beliebige Zahlkörper.

§ 4 Die Tate-Vermutung

Sei A eine abelsche Varietät über einem Zahlkörper K, $T_l(A)$ ihr Tate-Modul, $\pi = \text{Gal}(\bar{K}/K)$. Wir wollen zeigen:

a) $T_l(A) \otimes_{\mathbb{Z}_l} \mathbb{Q}_l$ ist halbeinfacher π-Modul.
b) Die Abbildung $\text{End}_K(A) \otimes_{\mathbb{Z}} \mathbb{Z}_l \to \text{End}_\pi(T_l(A))$ ist bijektiv.

Wir beschränken uns auf den Fall, daß A semistabile Reduktion hat, und daß der Grundkörper K gleich $\mathbb{Q}$ ist. Natürlich überträgt sich alles auf den allgemeinen Fall.
Wir haben schon dargelegt, daß man die folgende Aussage zeigen muß:
    Sei $G_\infty \subseteq A[l^\infty]$ eine l-divisible Untergruppe, $A_n = A/G_n$. Dann ist $h(A_n) = h(A)$.
    Zuerst betrachtet man dazu ganz allgemein eine Isogenie von abelschen Varietäten:

$$\varphi: A_1 \longrightarrow A_2 = A_1/G$$
$$p_1 \searrow \quad \swarrow p_2$$
$$\text{Spek}(\mathbb{Z})$$

Diese liefert eine Abbildung

$$\varphi^*: p_{2*}(\Omega_{A_2/\mathbb{Z}}^g) \to p_{1*}(\Omega_{A_1/\mathbb{Z}}^g)$$

Mit Hilfe dieser Abbildung kann man $h(A_1)$ und $h(A_2)$ vergleichen:
    Da φ^* die Metriken um einen Faktor $\sqrt{\text{grad}(\varphi)}$ verändert, ist ($\#$ = Ordnung)

$$h(A_2) - h(A_1) = \tfrac{1}{2}\log\left(\text{grad}(\varphi)\right) - \log\left(\# \,\text{Coker}\, \varphi^*\right)$$

Wenn $s: \text{Spek}(\mathbb{Z}) \to G$ den Nullschnitt bezeichnet, so ist

$$\# \,\text{Coker}(\varphi^*) = \# s^*(\Omega_{G/\mathbb{Z}}^1),$$

und somit

$$h(A_2) - h(A_1) = \tfrac{1}{2} \log \left(\operatorname{grad}(\varphi) \right) - \log \left(\# s^*(\Omega^1_{G/\mathbb{Z}}) \right).$$

Wir wenden dies auf die Stufen G_n einer l-divisiblen Untergruppe

$$G_\infty = \bigcup_n G_n \subseteq A[l^\infty]$$

an. Die dabei auftretenden Invarianten von G_n haben eine einfache Form, nämlich

$$\operatorname{grad}(\varphi_n) = \operatorname{ord}(G_n) = l^{nh}, \quad \# s^*(\Omega^1_{G_n/\mathbb{Z}}) = l^{nd}$$

Dabei ist h die Höhe und d die Dimension der l-divisiblen Gruppe G_∞. Es ist also

$$h(A/G_n) - h(A) = n \cdot \log(l) \left(\frac{h}{2} - d \right),$$

und wir wollen zeigen, daß

$$h = 2d.$$

Dazu sei $W = T_l(G) \subseteq V = T_l(A)$ das zu G gehörige Untergitter. Dann ist h der Rang von W, und aus der Theorie der l-divisiblen Untergruppen ([26], [32]) erhält man einen Zusammenhang zwischen d und der Galois-Aktion von π auf W:

Sei $\chi = \det_W : \pi \to \mathbb{Z}_l^*$

der Charakter, mit dem π auf $\Lambda^h W$ operiert. Wenn $I \subseteq \pi$ die Trägheitsgruppe zu l bezeichnet, so ist

$$\chi | I = \chi_0^d | I,$$

wobei χ_0 der zyklotomische Charakter sei, über den π auf den Einheitswurzeln von l-Potenz-Ordnung operiert.

Da χ unverzweigt außerhalb l ist (semistabile Reduktion), und da $\mathbb{Q}$ keine unverzweigten Erweiterungen besitzt, ist auch auf ganz π

$$\chi = \chi_0^d.$$

Nunmehr ist aber nach dem bereits von A. Weil bewiesenen Teil der Weil-Vermutungen $\Lambda^h W$ „rein vom Gewicht $h/2$", das heißt, für fast alle Primzahlen p

hat (F_p = Frobenius)

$$\chi(F_p) = \chi_0(F_p)^d = p^d$$

absoluten Betrag $p^{\frac{h}{2}}$. Somit ist notwendigerweise

$$h = 2d,$$

und die Tate-Vermutung bewiesen.

Als Korollar folgt sofort:
Zwei abelsche Varietäten A_1 und A_2 sind genau dann isogen, falls ein π-Isomorphismus

$$T_l(A) \otimes_{\mathbb{Z}_l} \mathbb{Q}_l \cong T_l(A_2) \otimes_{\mathbb{Z}_l} \mathbb{Q}_l$$

existiert.

Es gibt zwischen ihnen genau dann eine Isogenie vom Grad prim zu l, falls sogar

$$T_l(A_1) \cong T_l(A_2).$$

§ 5 Die Shafarevich-Vermutung

Sei S eine endliche Menge von Stellen des Zahlkörpers K. Wir wollen zeigen, daß es nur endlich viele prinzipal polarisierte abelsche Varietäten der Dimension g über K gibt, welche gute Reduktion außerhalb S haben. Wir beschränken uns wieder auf semistabile A's, und nehmen an, daß $K = \mathbb{Q}$. Wie immer funktioniert der Beweis auch im allgemeinen Fall. Wir gehen in zwei Schritten vor, indem wir zunächst nur die Endlichkeit bis auf Isogenie beweisen und dann zeigen, daß in jeder Isogenie-Klasse nur endlich viele Isomorphie-Klassen liegen.

Für den ersten Schritt reicht es nach der Tate-Vermutung, die π-Darstellungen $T_l(A) \otimes_{\mathbb{Z}_l} \mathbb{Q}_l$ zu betrachten. Diese sind halbeinfach und unverzweigt außerhalb S und l, und damit charakterisiert durch die Spuren der Frobenius-Elemente F_p auf ihnen. Es zeigt sich sogar, daß man endlich viele Elemente $F_{p_1}, \ldots, F_{p_r}$ finden kann, deren Spuren schon die Isomorphie-Klasse von $T_l(A) \otimes_{\mathbb{Z}_l} \mathbb{Q}_l$ bestimmen.

Diese Spuren sind ganze Zahlen, und nach den Weil-Vermutungen (bzw. Satz von Weil) sind ihre absoluten Beträge $\leq 2g\sqrt{p_i}$. Sie können also jeweils nur endlich viele Werte annehmen und es folgt die Endlichkeit der Isogenie-Klassen.

Wir können uns also bei der Betrachtung der Isomorphie-Klassen auf abelsche Varietäten B beschränken, welche isogen zu einem festen A sind. Aus der Formel für die Änderung von $h(\ )$ bei Isogenien folgt dann sofort, daß

$$\exp\left(2 \cdot h(B) - 2 \cdot h(A)\right) \in \mathbb{Q}^*$$

eine positive rationale Zahl ist, und daß eine Primzahl l in den so für B_1 und B_2 gebildeten Zahlen mit derselben Potenz eingeht, falls zwischen B_1 und B_2 eine Isogenie vom Grad prim zu l existiert, also falls die π-Moduln $T_l(B_1)$ und $T_l(B_2)$ isomorph sind.

Da es bei festem l nur endlich viele Möglichkeiten für das Gitter

$$T_l(B) \subseteq T_l(B) \otimes_{\mathbb{Z}_l} \mathbb{Q}_l \cong T_l(A) \otimes_{\mathbb{Z}_l} \mathbb{Q}_l$$

gibt (Jordan-Zassenhaus), ist die l-Potenz in $\exp\left(2h(B) - 2h(A)\right)$ gleichmäßig beschränkt. Wenn wir nun noch zeigen, daß genügend große Primzahlen l darin gar nicht vorkommen können, so erhalten wir eine obere Schranke für $h(B)$ und damit die Endlichkeit der Isomorphie-Klassen der betrachteten B's. Dazu reicht es, wenn für „große" l bei jeder Isogenie

$$\varphi: B_1 \rightarrow B_2 = B_1/G$$

von l-Potenz-Ordnung $h(B_1)$ und $h(B_2)$ übereinstimmen. Dazu darf man voraussetzen, daß G von l annuliert wird, und dann ist wieder

$$h(B_2) - h(B_1) = \log(l)\left(\tfrac{1}{2}h - d\right),$$

wobei h die Dimension des $\mathbb{F}_l$-Vektorraums

$$W = G(\bar{\mathbb{Q}}) \subseteq B_1[l](\bar{\mathbb{Q}}) \cong T_l(B_1)/l \cdot T_l(B_1)$$

bezeichnet, und d definiert ist durch

$$l^d = \#\left(s^*(\Omega^1_{G/\mathbb{Z}})\right).$$

Wir wollen zeigen, daß $h = 2d$. Dazu bezeichnen wir wieder mit

$$\chi = \det{}_W : \pi \rightarrow \mathbb{F}_l^*$$

den Charakter, mit dem π auf $\Lambda^h W$ operiert. χ ist unverzweigt außerhalb l, und wie vorher ergibt sich (unter Benutzung von [26]), daß

$$\chi = \chi_0^d \quad (\chi_0 = \text{zyklotomischer Charakter})$$

Man wählt nun eine feste Primzahl $p \notin S$. Dann ist

$$p^d = \chi_0^d (F_p)$$

ein Eigenwert von F_p auf $\Lambda^h(T_l(B_1)/l \cdot T_l(B_1))$, und somit ist l ein Teiler von $P_h(p^d)$, wobei P_h das charakteristische Polynom von F_p auf $\Lambda^h T_l(B_1)$ oder auch $\Lambda^h T_l(A)$ bezeichnet.

Nun ist $P_h(T)$ unabhängig von l und hat Nullstellen vom Betrag $p^{\frac{h}{2}}$. (Weil-Vermutungen). Wenn man also l so groß wählt, daß es keine der von Null verschiedenen ganzen Zahlen

$$P_h(p^d), \quad \begin{array}{c} 0 \le h \le 2g \\ 0 \le d \le g \\ h \ne 2d \end{array}$$

teilt, so muß für l die Behauptung gelten, und wir sind fertig.

§ 6 Offene Probleme

Es werden einige Fragen aufgezeigt, deren Beantwortung mir von Interesse zu sein scheint. Sie sind teilweise vage und teilweise recht präzise, und natürlich ist es klar, daß dies keine vollständige Liste relevanter Probleme sein kann.

1   Effektivität

Man sucht eine obere Schranke für die Höhe eines rationalen Punktes auf einer Kurve (vom Geschlecht größer Eins). Der Beweis der Mordell-Vermutung ist nicht effektiv, hauptsächlich weil man bei der Shafarevich-Vermutung aus jeder Isogenie-Klasse einen Vertreter wählt, über den man a priori wenig weiß. Ähnliches gilt übrigens auch für den Satz von Mordell-Weil oder den Satz von Siegel über ganze Punkte.

2   Höherdimensionale Varietäten

Wir haben nur rationale Punkte auf Kurven untersucht. Dort gibt es entweder einfache Parametrisierungen der rationalen Punkte (bei Geschlecht 0 und 1) oder Endlichkeit. Vielleicht kann man Ähnliches auch bei Flächen erreichen, wobei die Extremfälle einmal rationale Flächen sind, zum anderen Flächen vom allgemeinen Typ. Man kann sich auch fragen, ob auf algebraischen

Mannigfaltigkeiten mit amplen kanonischen Geradenbündeln stets nur endlich viele rationale Punkte liegen.

3   Verständnis rationaler Punkte

Man will wissen, wann und warum eine Kurve einen rationalen Punkt besitzt. Als Beispiel mögen die Resultate von B. Mazur dienen, der die $\mathbb{Q}$-rationalen Punkte auf gewissen Modulkurven [16] bestimmt hat. Die dabei verwandten Methoden sind leider sehr speziell und nicht allgemein zu verwenden. A. Grothendieck vermutet, daß man die rationalen Punkte aus der algebraischen Fundamentalgruppe der Kurve ablesen kann.

4   Galois-Operationen

Wenn A eine abelsche Varietät der Dimension g über einem Zahlkörper ist, und $K \hookrightarrow \mathbb{C}$ eine komplexe Einbettung von K, so setze man

$$T(A) = H_1(A(C), \mathbb{Z}).$$

Dann ist $T(A)$ eine freie abelsche Gruppe vom Rang $2g$, und für jede Primzahl l ist

$$T(A) \otimes_{\mathbb{Z}} \mathbb{Z}_l \cong T_l(A).$$

Gibt es eine algebraische Untergruppe

$$G(A) \subseteq GL(T(A) \otimes_{\mathbb{Z}} \mathbb{Q}),$$

mit Lie-Algebra

$$\mathfrak{g}(A) \subseteq \mathfrak{gl}(T(A) \otimes_{\mathbb{Z}} \mathbb{Q}),$$

so daß für jedes l $\mathfrak{g}(A) \otimes_{\mathbb{Q}} \mathbb{Q}_l$ die Lie-Algebra des Bildes der absoluten Galoisgruppe $\pi = \mathrm{Gal}(\bar{K}/K)$ ist? Aus der Shafarevich-Vermutung folgt, daß es eine Unteralgebra

$$M \subseteq \mathrm{End}(T(A))$$

gibt, so daß für jedes l $M \otimes_{\mathbb{Z}} \mathbb{Z}_l$ die vom Bild von π erzeugte Unteralgebra ist. Ein natürlicher Kandidat für $G(A)$ ist die sogenannte Mumford-Tate-Gruppe ([7]).

5   Die Gruppe ⅢⅢ

Für eine abelsche Varietät A über einem Zahlkörper K definiert man bekanntlich die Tate-Shafarevich-Gruppe $ⅢⅢ_{A/K}$ als Kern der Abbildung der Galoiskohomologie

$$H^1(K, A(\bar{K})) \to \prod_v H^1(K_v, A(\bar{K}_v))$$

(Produkt über alle Stellen)

$ⅢⅢ_{A/K}$ ist eine Torsionsgruppe, und man vermutet, daß sie endlich ist. Für jede Primzahl l ist die l-Torsion $ⅢⅢ_{A/K}[l^\infty]$ eine direkte Summe aus einer endlichen Gruppe und von Summanden $\mathbb{Q}_l/\mathbb{Z}_l$. Deren Anzahl läßt sich wie folgt bestimmen: Wenn R den Ring der ganzen Zahlen in K bezeichnet, und $H^1_{fl}(R, A[l^\infty])$ die flache Kohomologie der l-Torsion von A (die sogenannte Selmer-Gruppe), so gibt es eine exakte Injektion

$$A(K) \otimes_\mathbb{Z} \mathbb{Q}_l/\mathbb{Z}_l \hookrightarrow H^1_{fl}(R, A[l^\infty]),$$

deren Kokern bis auf eine endliche Gruppe $ⅢⅢ_{A/K}[l^\infty]$ ist. Man kann sich zunächst fragen, ob der Rang des l-divisiblen Teils von $H^1_{fl}(R, A[l^\infty])$ unabhängig von l ist (das ist richtig für Funktionenkörper über endlichen Körpern), und ihn dann mit dem Rang der Mordell-Weil-Gruppe $A(K)$ vergleichen. Das letztere führt dazu, Nicht-Torsionspunkte auf abelschen Varietäten zu suchen. Dazu ist bis jetzt keine befriedigende Methode bekannt.

Schließlich noch eine Warnung an den Leser: Es kann durchaus sein, daß $ⅢⅢ_{A/K}$ nicht immer endlich ist. Literatur: [34]

6   Die Vermutung von Birch-Swinnerton-Dyer

Zu jeder abelschen Varietät A/K definiert man in bekannter Weise eine L-Reihe

$$L(A, s) = \prod_v L_v(A, s).$$

Die obige Vermutung fordert nun einmal die analytische Fortsetzung von $L(A, s)$, zum anderen macht sie Aussagen über spezielle Werte dieser L-Reihe. Über den zweiten Teil ist so gut wie gar nichts bekannt (zum Beispiel taucht in der vermuteten Formel die Ordnung von $ⅢⅢ_{A/K}$ auf), während man den ersten Teil so formulieren kann, daß zu der abelschen Varietät eine automorphe Form gehören soll. Auch hier fehlt noch eine allgemeine Idee. Es sei nur bemerkt, daß nach der Tate-Vermutung die automorphe Form schon die Isogenie-Klasse von A bestimmt. Literatur: [34], [36]

Literaturverzeichnis

[1]   Arakelov, S., *Families of curves with fixed degeneracies*, Math. USSR Izv. **5** (1979), 1277—1302.

[2]   Arakelov, S., *An intersection theory for divisors on an arithmetic surface*, Math. USSR Izv. **8** (1974), 1167—1180.

[3]   Artin, M., *Algebraisation of formal moduli I, in Global Analysis*, Univ. Tokyo Press, Tokio 1969. pg. 21—71.

[4]   Ash, A., D. Mumford, M. Rapoport, Y. Tai, *Smooth compactification of locally symmetric varieties*, Math. Sci. Press, Brookline 1975.

[5]   Baily, W. L., A. Borel, *Compactification of arithmetic quotients of bounded symmetric domains*, Ann. of Math. **84** (1966), 442—528.

[6]   Deligne, P., D. Mumford, *The irreducibility of the space of curves of a given genus*, Publ. math. IHES **36** (1968), 75—110.

[7]   Deligne, P., J. S. Milne, A. Ogus, K. Shih, *Hodge Cycles, Motives, ans Shimura Varieties*, Springer Lecture Notes 900, Springer, Berlin 1982.

[8]   Faltings, G., *Calculus on arithmetic surfaces*, erscheint in Ann. of Math.

[9]   Faltings, G., *Arakelov's theorem for abelian varieties*, Invent. math. **73** (1983), 337—347.

[10]   Faltings, G., *Endlichkeitssätze für abelsche Varietäten über Zahlkörpern*, Invent. math. **73** (1983), 349—366.

[11]   Grauert, H., *Mordells Vermutung über rationale Punkte auf algebraischen Kurven und Funktionenkörper*, Publ. math. IHES **25** (1965), 131—149.

[12]   Hartshorne, R., *Algebraic geometry*, Springer Verlag, New York 1977.

[13]   Lang, S., *Diophantine geometry*, Interscience, New York 1962.

[14]   Mahler, K., *Über die rationalen Punkte auf Kurven vom Geschlecht 1*, J. reine und angew. Math. **170** (1934), 168—178.

[15]   Manin, Yu. I., *Rational points on an algebraic curve over function fields*, Translations AMS **50** (1966), 189—234.

[16]   Mazur, B., *Modular Curves and the Eisenstein Ideal*, Publ. math. IHES **47** (1977), 33—186.

[17]   Mordell, L. J., *On the rational solutions of the indeterminate equation of third and fourth degrees*, Proc. Cambridge Phil. Soc. **21** (1922), 179—192.

[18]   Moret-Bailly, L., *Variétés abéliennes polarisées sur les corps de fonctions*, C. R. Acad. Paris **296** (1983). 267—270.

[19]   Mumford, D., *Geometric invariant theory*, Springer Verlag, Berlin 1965.

[20]   Mumford, D., *Abelian varieties*, Oxford Univ. Press, Oxford 1974.

[21]   Mumford, D., *Stability of projective varieties*, l'Ens. Math. **23** (1977), 39—100.

[22]   Mumford, D., *Hirzebruch's proportionality theorem in the non-compact case*, Invent. math. **42** (1977), 239—272.

[23]   Namikawa, Y., *Toroidal compactification of Siegel spaces*, Springer Lecture Notes **182** (1980).

[24]   Parshin, A. N., *Algebraic curves over function fields I*, Math. USSR Izv. **2** (1968), 1145—1170.

[25]   Parashin, A. N., *Quelques conjectures de finitude en géometrie diophantienne*, Actes congres intern. math. Nizza (1970), I, 467—471.

[26]   Raynaud, M., *Schémas en groupes de type $(p,\ldots,p)$*, Bull. Soc. Math. France **102** (1974), 241—280.

[27]   Shafarevich, I. R., *Algebraic number fields*, Translations AMS **31** (1963), 25—39.

[28]   Siegel, C. L., *Über einige Anwendungen diophantischer Approximationen*, Abh. Preuss. Akad. Wiss. Phys.-Math. Kl. 1929, Nr. 1.

[29]   Szpiro, L., *Sur le théoreme de rigidité de Parsin et Arakelov*, Asterique **64** (1974), 169—202.

[30]   Szpiro, L., *Séminaire sur les pinceaux de courbes de genre au moins deux*, Asterique **86** (1981).

[31]   Tate, J., *Algebraic cycles and poles of zeta functions*, in *Arithmetical algebraic geometry*, 93—110. Harper and Row, New York 1965.

[32]   Tate, J., *p-divisible groups*, in *Proc. Conference on local fields*, 158—183, Springer-Verlag, Berlin 1967.

[33]   Tate, J., *Endomorphisms of abelian varieties over fields of finite characteristics*, Invent. math. **2** (1966), 281—315.

[34]   Tate, J., *On the conjectures of Birch and Swinnerton-Dyer and a geometric analogue*, Sem. Bourbaki **306** (1965/66).

[35]   Weil, A., *L'arithmétique sur les courbes algébriques*, Acta Math. **52** (1928), 281—315.

[36]   Weil, A., *Über die Bestimmung Dirichletscher Reihen durch Funktionalgleichungen*, Math. Ann. **168** (1967), 149—156.

[37]   Zarhin, J. G., *Isogenies of abelian varieties over fields of finite characteristics*, Math. USSR Sbornik **24** (1974), 451—461.

[38]   Zarhin, J. G., *A remark on endomorphisms of abelian varieties over function fields of finite characteristics*, Math. USSR Izv. **8** (1974), 477—480.

Perspectives in Mathematics
Anniversary of Oberwolfach 1984
© Birkhäuser Verlag, Basel

Foundational Ways[1])

SOLOMON FEFERMAN

Stanford University, Department of Mathematics,
Stanford, CA 94305 (USA)

1   The foundational enterprise: carrying on in mathematics and logic

In this century mathematicians have become conscious of the unity and structure of their subject to an unprecedented extent. Early on, logic offered grand claims to explain this and to rescue mathematics from its "crisis". which if it threatened one, seemed to threaten all. Global views about the nature of mathematics were propounded. These then took hold of foundational discussions to such an extent that the logical foundations of mathematics is currently thought of only in terms of such grand positions: logicism, formalism, platonism and constructivism. They are all rather tired-looking now, if not suffering from senescence and still more basic ills. Long since, most mathematicians have given up worrying about the crises and gone about their daily affairs, attending to logical hygiene only as needed. If asked, they will say they are really formalists, or that Zermelo-Fraenkel meets their needs, or whatever[2]).

Among those mathematicians who still take foundational concerns seriously, there ist a growing band who aim to recapture this territory from logic in favor of mathematics. Some offer up a new global scheme, but now one that is purely mathematical such as category theory[3]). Others propose a more phenomenal view: mathematics is as mathematics does. These critics of logical foundations are united in the view that mathematics is more reliable than any of the foundational schemes which had been propounded by the logicians to "secure" it. They further complain that the logical analysis of mathematics bears little or no relation to actual practice[4]).

Given that the "big" logical positions were extremely overstated and defective in substantial ways, this kind of reaction is natural. However, I believe it has gone too far in the opposite direction. The aim here is to restore the

---

[1]) This is a "slimmed-down" version of a paper entitled "Working foundations" (Feferman 1984a) which is to be part of the proceedings of a symposium on the foundations of mathematics held in Florence in June 1981.

[2]) For a typical expression of such views see Dieudonné 1982.

[3]) I have argued particularly against the scheme of categorical foundations in Feferman 1977.

[4]) See for example Davis & Hersh 1981, Hersh 1979, Lakatos 1976, 1978 and MacLane 1982. Logicians who reached similar conclusions are Goodman 1979, Kreisel 1967, 1976, 1977 and Wang 1974.

position of the logical or metamathematical approach to the foundations of mathematics but at a more every-day "local" level which is not preoccupied with the grand schemes. On my view, this is a direct continuation of work that mathematicians themselves have carried on from the very beginning of our subject up to the present. The distinctive role of logic lies in its more conscious, systematic approach and its different ways of slicing up the subject.

If one analyzes foundational activity (whether carried on by mathematicians as a matter-of-course or more consciously by logicians), one finds that it falls into one of five or six characteristic modes. Each of the following sections is devoted to one of these *foundational ways*, and their presentation is the same throughout. Each section begins with a general explanation of that kind of foundational activity. This is then illustrated by some familiar examples from mathematics. It then goes on to give some examples from metamathematics, some which will be familiar but probably far from all[5]). References will be given only for the (presumably) unfamiliar work.

As usual, a disclaimer: this is not a survey of mathematical logic; it only has to do with that small part of the subject (these days) which has foundational concerns. Even that is hardly surveyed; the illustrative examples are meant to be typical, but evidently depend on my own knowledge and interests. I do believe there is much to interest mathematicians in the rest of logic, but that is another story which is more often told[6]).

2   Conceptual clarification

The first mode of foundational activity we consider (also called *conceptual analysis* or *explication*) usually takes place at an advanced stage in the organization of a subject (cf. §6 below on axiomatics). Typically, one has a settled fund of well-understood basic concepts, and there is the the question of giving precise definitions of other frequently used informal concepts in terms of these. There are usually some specified requirements to be met, and in some cases it may be shown that only one such definition is possible. In other cases, the requirements only partially determine the exact definiton, and the criteria for acceptance are subtler and often require the test of practice for complete acceptance.

Some classical examples from mathematical analysis are the definitions of *limit, continuous function, smooth function* and *area*. More recent examples are the concepts of *dimension* in topology and of *natural mapping* and *universal construction* in algebra and topology.

---

[5]) Many more such examples are given in the paper "Working foundations" from which this paper is derived (cf. ftn. 1).

[6]) A good source of material to begin with is the *Handbook of Mathematical Logic* edited by Barwise.

The best known examples from logic are the definitions of *satisfaction* and *truth* (by Tarski) and of *mechanically computable function* (by Turing and Post).

Tarski's semantical concepts are all referred to suitably specified *formal languages L*. These were used in the 50's to explain the informal idea of a *transfer principle* in algebra, the most famous example of which is *Lefschetz's principle* in algebraic geometry. One has a transfer of results stated in L from one structure M to another structure M' if $M \equiv_L M'$, which means that M and M' satisfy the same statements from L. The strongest present formulation of Lefschetz's principle in these terms has been given by Eklof 1973 (following Feferman 1972). A variety of further interesting transfer principles in algebra may be found in Cherlin 1976.

The notions of effective computability mentioned above operate on finitely presented data, e.g. the natural numbers or finite strings of symbols. However, they have been the basis for, or suggested, precise definitions of computability in other situations. For example, the concept of *finite effective construction* as applied in geometry and algebra has been plausibly explained relative to arbitrary mathematical structures as domains in Friedman 1971. Various notions of *infinitary construction* have also been proposed, but with less sureness about the conceptual analysis; even so, the definitions taken have turned out to be exceptionally useful. For one approach and relevant literature see Fenstad 1980.

There are two important concepts in logical work which have not yet been satisfactorily defined. The first is that of *identity of proofs*. Obviously, the significance of this goes far beyond logic; mathematicians constantly compare proofs and judge whether they are (essentially) the same or different. One would think that a *theory of proofs* would establish this as a basic notion (comparable to isomorphism in algebra or homeomorphism in topology). In fact there is a highly developed theory of proofs (inaugurated by Hilbert) but no convincing notion of identity has yet been produced within it. The paper Prawitz 1971 presents interesting relevant notions and results.

A second informal concept which has not yet been defined precisely is that of *natural well-ordering*. The most familiar example of such is the ordering of "figures" $\omega^{x_0} \cdot k_0 + \omega^{x_1} \cdot k_1 + \cdots + \omega^{x_n} \cdot k_n$ in Cantor normal form for all ordinals up to the ordinal ε_0 (the least solution of $\omega^x = \omega$). This was used by Gentzen (in 1936) to give a semi-finitary consistency proof of arithmetic. Since then technical proof theory has been dominated by the use of larger and larger naturally presented effective well-orderings. Consistency proofs on cooked-up but non-natural effective well-orderings can always be given, so it is important to know just what makes an ordering natural. This question has been approached through category theory, but without any satisfactory answer thus far; cf. e.g. Feferman 1968 and Girard 1981.

3   Dealing with problematic concepts and principles (Part I)

At each stage in the development of mathematics there is a body of ideas and methods generally regarded as well-understood, and with which mathematicians in the mainstream confidently carry on their work. It may happen subsequently that the place of such ideas in our understanding undergoes considerable revision (as happened e. g. with the conception of the geometric line). The concern here, though, is with the problems raised *at that stage* by concepts and principles which have a certain plausibility or utility, but about which there is less certainty or security than for the accepted core. In this section, we deal with one foundational way of dealing with such situations, called the method of *interpretation* or *models;* in the next we shall deal with another principal way.

Familiar examples of problematic notions at various stages in the development of mathematics are *zero, negative numbers, imaginary* and *complex numbers*, and *points at infinity*. Examples of troublesome principles are the *parallel postulate* and, more recently, the *well-ordering principle*. A typical example of interpretation is that of the complex numbers as pairs of real numbers. Implicit in the sucess of this interpretation is that a series of conditions are met about complex numbers, which we now describe as being a field generated from the reals by adjunction of a root of $x^2 + 1 = 0$.

More generally in present day terms, one has a set of axiomatic requirements containing the problematic notion or principle, and the method of interpretation is to yield a model for these axioms. In the case of a problematic principle, providing a model in addition for its negation shows this to be a non-trivial exercise (otherwise the principle in question is a consequence of the remaining axioms). But there might further be grounds for considering the negated principle to have its own plausibility, e. g. as with the parallel postulate. Put in still more logical terms, denote the basic system of axioms (or *axiomatic theory*) S and denote by P the principle in question and $\neg P$ its negation. The adjunction of $P(\neg P)$ to S is denoted by $S + P(S + \neg P)$. Providing a model for $S + P$ establishes the *consistency* of P with S, while providing one for $S + \neg P$ establishes *independence* of P from S.

Most mathematicians are familiar with the consistency and independance results in set theory due to Gödel and Cohen. Here the axiomatic theory S taken as basic is the system ZF of Zermelo-Fraenkel set theory. The first principle P to consider is AC, the *axiom of choice*, which Zermelo had proved to be equivalent to the well-ordering principle. AC ist needed to establish a good theory of infinite cardinal numbers and to order them linearly in the transfinite sequence $\aleph_\alpha$. The first question to answer is that of where the cardinality of the continuum $2^{\aleph_0}$ falls in this sequence. Certainly $2^{\aleph_0} \geq \aleph_1$ and the Cantor continuum hypothesis CH proposes $2^{\aleph_0} = \aleph_1$. An obvious generalization, denoted GCH, is that $2^{\aleph_\alpha} = \aleph_{\alpha+1}$ for all ordinals α. With all efforts to prove or disprove them unsuccessful, CH and GCH are problematic principles relative to

$ZF + AC$. As is well known, Gödel established that (i) $ZF + AC + GCH$ has a model, while Cohen established that (ii) $ZF + \neg AC$ has a model and (iii) $ZF + AC + \neg CH$ has a model. In each case, the interpretation is given back in terms of a model M for ZF: Gödel uses M to form a *submodel* M_0 called the *constructible sets*, while Cohen uses it to form an *extension model* $M' = M_0[G]$ by adjunction of certain *generic sets* G (analogous to transcendental extensions in algebra). At any rate, in each case what is achieved can be formulated as a *relative consistency result*, i.e. if ZF is consistent then so also is $ZF + AC + GCH$, etc.

In logic it has been observed that one can usefully sharpen these relative consistency results as follows. Suppose S, T are two axiomatic theories, not necessarily with the same basic language (L_S = language of S, L_T = language of T) but where $L_S \subseteq L_T$. Let Σ be a class of statements in L_S. Suppose further that every axiom of S is a theorem of T (i.e. is provable from the axioms of T). Then T is said to be a *conservative extension of S for the class Σ* if whenever $A \in \Sigma$ and A is a theorem of T then A is already a theorem of S. When this holds, T gives no new results about statements in Σ not already given by S. Note that as long as Σ contains statements like $0 \neq 0$, each conservation result implies relative consistency. There are usually two kinds of cases of interest, first where $L_S = L_T$ but Σ is properly contained in L_S, and second where $\Sigma = L_S$ and L_S is properly contained in L_T. In the latter case we simply say that T is a conservative extension of S.

Kreisel observed that $ZF + AC + GCH$ is a conservative extension of ZF for the class Σ of purely number-theoretic statements (as expressed in L_{ZF}). The reason is that Gödel's constructible sets model for $ZF + AC + GCH$, obtained from a model for ZF, is standard for the natural numbers — in other words, though the meaning of "set" changes in the interpretation, the meaning of "natural number" does not. This conservation result proved to be significant in an unexpected way. In 1965 Ax and Kochen proved, using set-theoretical arguments, that a certain effectively given set F_p of first-order axioms for the p-adic numbers $\mathbb{Q}_p$ is complete, i.e. every statement or its negation in L_{F_p} is provable from F_p. It follows that there is (in principle) a decision method for validity in $\mathbb{Q}_p$; given a statement A in L_{F_p} just run effectively through the theorems of F_p until we hit A or $\neg A$. The statement that (theoremhood in) F_p is decidable is itself equivalent to a statement of elementary number theory, call it D_p. It happens that Ax and Kochen used the unusual hypothesis CH in their argument (of course, with AC). In other words they showed $ZF + AC + CH$ proves D_p, and observed as a consequence of Kreisel's conservation result that already ZF proves D_p. Later, Cohen 1969 produced an explicit decision method for $\mathbb{Q}_p$ by different arguments which could be readily formalized in ZF, thus confirming the prediction in principle by conservation.

Here are some less familiar examples of interpretations in logic. First are the models of the *λ-calculus* produced by Scott 1972 and (following Plotkin) Scott 1976. The λ-calculus is a formalism for defining functions, and its basic operation is the binary one of *application*, written fx (or $f(x)$). Formally, this allows any element f of the universe (over which the variables range) to be applied to any other element x, in other words each member f of the universe acts as a *total function*. It is fairly obvious what axiomatic requirements are to be met under this interpretation; the main one asserts that any expression $\tau[x]$ of the language containing a free variable x determines a function f whose values are given for all x by $fx = \tau[x]$; this function is usually denoted $f = \lambda x \cdot \tau[x]$ (thus explaining the use of "λ-calculus" as designation for this formal system). The problematic feature of the λ-calculus is that it allows *self-application*, i.e. formation of ff. This possiblity does not exist in familiar universes of functions, where the domain of arguments of a function is always prior in some sense to the function itself. It is by no means obvious how to construct models of the λ-calculus, which is what Scott achieved.

As with the previous example there is further utility to the interpretation, in this case applying to theoretical computer science. Programs involving recursion determine functions as minimal fixed points of functional equations $f = \tau[f]$. It is not hard to derive explicit fixed-points $\varphi = \tau[\varphi]$ in the formalism of the λ-calculus. Scott's models for the λ-calculus give definite meaning to each such expression φ, even when the recursion is not independently analyzed. More generally, Scott has used this to provide what is called a *denotational semantics for programming languages*, in which each element of a program is given meaning as a mathematical object.

Of course the problems raised by *self-reference* (in natural language) and *self-membership* (in the theory of classes) are old, and were at the source of the "foundational crisis" at the beginning of this century. It seems that an essential ingredient of the paradoxes is the combination of negation with self-application. This suggests that use only of *positive* instances of self-application need not lead to inconsistencies. Examples of such are the statement, *this statement is true*, and the notions, the *class of all classes* or the *category of all categories*. Axiomatic theories permitting these and other instances of positive self-application have been developed and shown to be conservative extensions of known consistent theories by means of suitable interpretations. For a survey of work in this direction see Feferman 1984. An alternative (and earlier) treatment of the problem of self-membership in category theory is described in the next section.

A number of further examples dealing with problematic notions and principles by the method of interpretation are given in Section 3 of the paper Feferman 1984a. One of the most interesting has to do with Brouwer's concept of *(free) choice sequence*, which is central to his intuitionistic redevelopment of mathematical analysis. These are supposed to be infinite sequences $\alpha = (a_0, a_1, \ldots, a_n, \ldots)$ generated without any known law, e.g. randomly or by a sequence of arbitrary choices, and of which only a finite amount of information is available at any given time. According to Brouwer, this informal understanding leads one to accept certain *continuity principles* concerning statements involving choice sequences. These principles in turn have consequences such as that every function of real numbers is continous, which *prima facie* contradict classical statements, but only when one uses the classical law of excluded middle (*L. E. M.*). Kleene and Vesley 1965 formulated Brouwer's ideas about choice sequences in an axiomatic theory *C S* with intuitionistic logic (no use of *L. E. M.*) and showed it to be consistent by means of a complicated *realizability interpretation* (a method of interpretation for intuitionistic systems originally developed by Kleene in 1945).

4   Dealing with problematic concepts and principles (Part II)

Here we consider a second foundational way of dealing with the problem situation described at the beginning of the previous section. Instead of interpreting the problematic concepts or principles in some sort of direct way, one tries to *replace* or find *substitutes* for them which do the same work, or even to *eliminate* them entirely. In the latter case one wants to preserve the useful consequences, i. e. establish a conservation result. In logical work the process of elimination is frequently established by *syntactic transformations*.

The classical example from mathematics of a concept which was eliminated while saving its applications is that of *infinitesimal;* this idea could be dispensed with once the notions of *limit, derivative*, etc. were given $'\varepsilon, \delta'$ definitions. (As is well known, Abraham Robinson restored infinitesimals as respectable objects of mathematical study; but that is another story, more properly falling under Section 5 below.) The replacement of *multi-valued function* in complex analysis by *single-valued function on a Riemann surface* is an example of a problematic concept dealt with by substitution.

Some examples from logic of these foundational tacks are the treatments of the notions, the *class of all classes* and the *category of all categories* in the formal framework of the *BG* (Bernays-Gödel) theory of *sets and classes*. Here the problematic notions are replaced by the *class of all sets* and the *category of all small categories* (respectively) where, following the terminology of MacLane 1961, one uses the distinction between set and class in *BG* to distinguish between *small* and *large categories*. Another solution to this problem situation had been given in the framework of *ZF* (a theory of sets only) by the use of *Grothendieck universes*. On the face of it, that required assuming the existence of many inaccessible cardinals. It was shown in Feferman 1969 how to use the *reflection principle in ZF* (reflecting properties of the universe of all sets down to subuniverses which are themselves sets) in order to avoid such unusual assumptions.

There are a number of frequently used syntactic transformations which serve general procedures of elimination, going back to Hilbert, Gödel, Herbrand and Gentzen in the late 20's and early 30's (see Feferman 1984a, Section 4 for more detail and references). An interesting recent example of a different kind of syntactic transformation was that introduced by Kreisel and Troelstra 1970 to *eliminate choice sequences*. This yields conservation of the theory *C S* over a constructive theory without choice sequences, permitting certain strengthenings of the Kleene-Vesley result for CS mentioned in the preceding section.

5   Dealing with problematic methods and results

In the preceding two sections the emphasis was on troublesome concepts or principles which are relatively basic in character and which usually emerge early in the development of a subject. In this section we concentrate on mathematically more advanced methods or results which, typically, raise questions about their validity at a later stage of development. The problem here is to 'justify' the methods or results, by providing them with a 'rigorous foundation'. This means giving precise *sufficient* conditions under which these methods can be applied or for which the results hold. It is rare that such conditions are *necessary*, and there is then room for improvement. That is generally a process which takes place over a period of time and involves successive steps of sharpening, extending and generalizing.

There are many examples from mathematics of this mode of foundational activity, of which the following are just a few: *Descartes-Euler theorem, Fourier series, calculus of variations, Dirichlet's principle, Stoke's theorem, Jordan curve theorem, generalized functions, probability theory* and *algebraic geometry*. The foundations of the mathematically isolable parts of other subjects follows the same pattern, e.g. with *quantum mechanics*.

One finds many fewer examples of this kind from logic. The most well known is for *the theory of transfinite ordinals and cardinals*, initially developed informally by Cantor and currently given precise formulation in axiomatic set theories such as $ZF + AC$ or $BG + AC$. Of more interest is the work by Robinson 1966 providing foundations for *infinitesimal analysis*. (A good introduction to this is provided by the article Stroyan 1977.) For a few further (less far-reaching) examples, cf. 1984a, Section 5.

6   Organizational foundations and axiomatization

This is one of the most familiar kinds of foundational work, and as a result its significance tends to be over-emphasized. It usually takes place at a relatively advanced stage in the development of a subject, when it is ripe for *organization* and *systematic exposition*. What is required is a *choice and ordering of concepts and results*. Basic results which cannot be justified by further appeal within the given framework are taken as *axioms*. Usually, *typical patterns of proof* emerge in mathematical organization and some attention may be given there to *proof principles*, but the explicit systematization of *methods of proof* is taken up only in logical work. There is frequently some overlap in axiomatic development with the foundational moves described in the preceding Sections 2—5.

Euclidean geometry provides of course the classical example of axiomatization. But while long considered a model for rational thought, it was relatively isolated of its kind until the 19th century. At that time the growing professionalization of mathematics and the expansion of organized higher mathematical education for engineers and scientists led mathematicians to devote more and more effort to systematic organization and to a renewed involvement in axiomatization. In geometry this was revived with *non-Euclidean geometry, projective geometry* and Hilbert's *re-axiomatization of Euclidean geometry*. In algebra these efforts gave rise in the 19th and 20th century to the axioms for *groups, rings, fields, vector spaces, algebras*, etc. Axiomatization took hold in analysis in the 20th century with *metric spaces, topological spaces, Banach spaces, Hilbert spaces, measure spaces*, etc. Evidently axiomatics is pandemic within present-day mathematics.

The work in the 19th century on the axiomatic *characterization of the basic number systems* $\mathbb{N}$, $\mathbb{Q}$, $\mathbb{R}$ and $\mathbb{C}$[7]) forms a bridge to the logical approach to axiomatic foundations in the 20th century. This involves a shift away from axiomatics within traditional subject areas (such as geometry, algebra, analysis) to the study of principles concerning ideas, structures and principles which may be involved in *all* areas. The most extensively developed example of that in the 20th century is of course *axiomatic set theory*[8]). This is the current expression of the platonistic philosophy of mathematics, and there is considerable agreement about its core from those who accept that position. However, there is also a frontier of active research on the so-called *large cardinal axioms* whose status is far from settled (the article of Kanamori and Magidor 1978 gives a good relatively recent survey).

Axiomatization becomes more exciting at the outer fringes of a subject, where one must consider whether propositions that can't be settled by previously accepted axioms should be regarded as new axioms. It is also exciting when there are genuinely competing over-all schemes for organization and axiomatization. One of the main sources of this kind of stimulus for logic has been in the *axiomatic foundations of constructive mathematics*. Initially constructivity in this century was dominated by Brouwer's ideas, which he propounded in startling and sometimes mystifying ways. However, these have since been captured and made widely accessible by the axiomatic analysis which began with Heyting's work in the 30's and has stretched to the theories of choice sequences treated more recently by Kleene/Vesley and Kreisel/Troelstra (mentioned in Sections 3, 4 above). Other approaches within the constructivist framework were more understandable than Brouwer's to begin with, but raised questions which axiomatization helped to resolve. Two such particularly to be mentioned are

---

[7]) By Méray, Weierstrass, Dedekind, Cantor, Peano, Frege and others.
[8]) At the hands of Zermelo, Skolem, Fraenkel, von Neumann, Bernays, Gödel and others.

those due to Markov and Bishop. Less well-known but also distinctive and coherent is the approach due to Martin-Löf. In many respects these schools or approaches account for the same body of mathematics and do so in closely related ways; in other respects they are divergent and in some sense competitors. Not only that, in each case competitive axiomatizations have been proposed which must be considered for adoption. The book Beeson 1984 contains a detailed comparative study of most of the systems which have been developed for these different approaches (outside of Brouwer's theory of choice sequences).

Once axiomatic foundations of a given subject are fairly well settled, it is common to undertake a *fine analysis* of the role of different axioms or groups of axioms in different parts of practice. A number of examples and references are given in Feferman 1984a, Section 6.

7   Reflective expansion

One further foundational way was taken up in the paper from which the present one is derived, and which will only be indicated rather briefly here. This is the idea of *reflective expansion of concepts and principles.* In the historical development of mathematics new concepts emerge and then become clarified and established as basic ingredients of our thought. Thus, at different stages, one obtained the concepts of *whole number*, of *point, line* and *plane*, of *ordered pair*, of *function* and of *set*. Other concepts (and associated principles) are then derived by reflection on these. In a way that process is a form of generalization, but of the following particular character: at a certain point one reflects on what has led one to accept and work with given concepts, and recognizes that much more is *implicit* in doing so. For example, Euclidean geometry of the plane and space were eventually explained in terms of $\mathbb{R}^2$ ($= \mathbb{R} \times \mathbb{R}$) and $\mathbb{R}^3$ ($= \mathbb{R} \times \mathbb{R} \times \mathbb{R}$) using the real number system $\mathbb{R}$ and pairing and tripling operations. Reflection on that led to the concepts of *n-tuple* and *n-dimensional space* $\mathbb{R}^n$. (The concept of n-tuple is itself obtained by reflection on the notion of ordered pair.) Reflection on the ordinal enumeration of the whole numbers, led to the concept of *ordinal number*. Reflection on the formation of derivative and integral as operators on functions led to the idea of *function operators* or *functionals*.

The process of reflective expansion of concepts and principles has been taken up in a theoretical way in logic, starting with Kreisel 1970. A case study on which I have concentrated is that in which one is led from the natural number system ($\mathbb{N}, 0, Sc$) together with the basic ideas of inductive proof and definition on $\mathbb{N}$, to its reflective expansion into so-called *predicative mathematics;* cf. Feferman 1978. Since then I have formulated and studied more generally a concept of the *reflective closure of a theory*, in work which is as yet unpublished; this is an area of continuing research, which should cover ground all the way from finitist mathematics to the generation of extraordinarily large cardinals.

This concludes my round-up of modes of foundational activity, in logic as in mathematics. While each way has a characteristic problem situation and typical examples, the importance over-all of foundational work lies in its essential, continuing role in the refinement and improvement of mathematical understanding. I have tried to show that this is carried on almost as a matter of course, independent of any grand philosphical position as to the nature of mathematics. But it is also carried on as a matter of necessity, in a way integral to what makes mathematics such a distinctive body of thought. Any philosophy of mathematics worthy of its name must address itself to the more difficult question of what *that* amounts to.

Bibliography

Ax, J. and S. Kochen
1965   Diophantine problems over local fields I, II. *Amer. J. Math.* 87, 605—630.
Barwise, J. (ed.)
1977   *Handbook of Mathematical Logic* (North-Holland)
Beeson, M.
1984   *Foundations of Constructive Mathematics: Metamathematical Studies* (Springer) (to appear).
Cherlin, G.
1976   Model theoretic algebra: selected topics, *Lecture Notes in Maths.* 521.
Cohen, P. J.
1969   Decision procedure for real and p-adic fields, *Comm. Symp. Pure Appl. Math.* 22, 131—151.
Davis, P. J. and R. Hersh
1981   *The Mathematical Experience* (Birkhäuser).
Dieudonné, J.
1982   Mathématiques vides et mathématiques significatives, in *Penser les mathématiques* (Editions du Seuil), 15—38.
Eklof, P.
1973   Lefschetz' principle and local functors, *Proc. A.M.S.* 37, 333—339.
Feferman, S.
1968   Systems of predicative analysis II. Representations of ordinals, *J. Symbolic Logic* 33, 193—220.
1969   Set-theoretical foundations of category theory (with an Appendix by G. Kreisel), in *Lecture Notes in Maths.* 106, 201—247.
1972   Infinitary properties, local functors, and systems of ordinal functions, in *Lecture Notes in Maths.* 255, 63—97.
1977   Categorical foundations and foundations of category theory, in *Logic, Foundations of Mathematics and Computability Theory*. (Reidel), 149—169.
1978   A more perspicuous formal system for predicativity in *Konstruktionen versus Positionen* I, (Walter de Gruyter), 87—139.
1984   Toward useful type-free theories I, *J. Symbolic Logic* (to appear).
1984a   Working foundations (to appear in *Synthèse*).
Fenstad, J. E.
1980   *General Recursion Theory. An axiomatic approach* (Springer).
Friedman, H.
1971   Algorithmic procedures, generalized Turing algorithms and elementary recursion theories, in *Logic Colloqium* 1969 (North-Holland) 361—390.

Girard, J.-Y.
1981 $\prod_2^1$-logic, Part I: Dilators, *Annals of Math. Logic* 21, 75—219.
Goodman, N. D.
1979   Mathematics as an objective science, *Amer. Math. Monthly* 86, 540—551.
Hersh, R.
1979   Some proposals for reviving the philosophy of mathematics, *Advances in Mathematics* 31, 31—50.
Kanamori, A. and M. Magidor
1978   The evolution of large cardinal axioms in set theory, in *Lecture Notes in Maths.* 669, 99—275.
Kleene, S. C. and R. E. Vesley
1965   *The Foundations of Intuitionistic Mathematics, especially in relation to recursive functions* (North-Holland).
Kreisel, G.
1967   Mathematical logic: what has it done for the philosophy of mathematics, in *Bertrand Russell, Philosopher of the Century* (Allen and Unwin), 201—272.
1976   What have we learned from Hilbert's second problem? *Proc. Sympos. Pure Math.* XXVIII, (AMS) 93—130.
1977   Review of *Brouwer* 1975, Bull. AMS 83, 86—93.
Kreisel, G. and A. S. Troelstra
1970   Formal systems for some branches of intuitionistic analysis, *Annals Math. Logic* 1, 229—387.
Lakatos, I.
1976   *Proofs and Refutations: the Logic of Mathematical Discovery* (Cambridge University Press).
1978   *Mathematics, Science and Epistemology: Philosophical Papers vol. 2* (Cambridge Univ. Press).
MacLane, S.
1961   Locally small categories and the foundations of mathematics, in *Infinitistic Methods* (Pergamon) 25—43.
1981   Mathematical models: a sketch for the philosophy of mathematics, *Amer. Math. Monthly* 88, 462—72.
Prawitz, D.
1971   Ideas and results in proof theory, in *Proc. Second Scandinavian Logic Symposium* (North-Holland) 235—307.
Robinson, A.
1966   *Non-Standard Analysis* (North-Holland; rev. ed. 1974).
Scott, D.
1972   Continuous lattices, in *Lecture Notes in Maths.* 274, 97—136.
1976   Data Types as lattices, *SIAM J. Comput.* 5, 522—587.
Stroyan, K. D.
1977   Infinitesimal analysis of curves and surfaces, in *Barwise* 1977, 197—231.
Wang, H.
1974   *From Mathematics to Philosophy* (Routledge & Kegan Paul).

Perspectives in Mathematics
Anniversary of Oberwolfach 1984
© Birkhäuser Verlag, Basel

Von der Brownschen Bewegung zum Brownschen Blatt: Einige neuere Richtungen in der Theorie der stochastischen Prozesse

HANS FÖLLMER

Mathematikdepartment, ETH-Zentrum,
CH-8092 Zürich (Switzerland)

1   Einleitung

Für $x \in R^1$ und $t, \tau > 0$ ist

$$p_t(x, y) = \frac{1}{\sqrt{2\pi t \tau}}\, e^{-\frac{(x-y)^2}{2t\tau}} \quad (y \in R^1) \tag{1.1}$$

die Dichte der Normalverteilung mit Mittelwert x und Varianz $t\tau$, also eine Gaußsche Glockenkurve. Wenn man nun dies Bild in Bewegung setzt, also t als Zeitparameter auffaßt und $p_t(x, y)$ als Übergangsdichte für die zufällige Bewegung eines Teilchens auf der reellen Achse, so kommt man zur Brownschen Bewegung mit Diffusionskonstante τ. Die Entwicklung der Wahrscheinlichkeitstheorie in den letzten 40 Jahren hat gezeigt, wie reich die mathematische Struktur ist, die sich dabei entfaltet. Zugleich ist klar geworden, daß die Brownsche Bewegung nicht irgendein Beispiel ist, sondern daß sie in der Wahrscheinlichkeitstheorie eine fundamentale Rolle spielt. Man kann das in verschiedener Weise präzisieren. Zum Beispiel ist die Brownsche Bewegung das natürliche Bezugsmodell für die klassischen Grenzwertsätze der Wahrscheinlichkeitstheorie, im Sinne von funktionalen Varianten des zentralen Grenzwertsatzes und von starken Approximationen; vgl. [17]. Zugleich ist die Brownsche Bewegung die Basis der stochastischen Analysis, insbesondere der Theorie der stetigen Martingale, des Itô-Kalküls und der stochastischen Differentialgleichungen; einige dieser Punkte werden wir im nächsten Abschnitt kurz in Erinnerung rufen.

Die Brownsche Bewegung mit Diffusionskonstante τ induziert eine Wahrscheinlichkeitsverteilung auf dem Raum der stetigen Pfade; für $\tau = 1$ ist es das Wienermaß. Wenn man nun seinerseits das Wienermaß in Bewegung setzt, stößt man auf eine Fülle neuer Fragestellungen. Wenn man zum Beispiel die Diffusionskonstante τ der Brownschen Bewegung als Zeitparameter auffaßt, so führt das auf eine Einbettung des Wienermaßes in das Brownsche Blatt. Das kann man in verschiedener Weise interpretieren, zum Beispiel als Übergang

— von einem eindimensionalen zu einem unendlich-dimensionalen Diffusionsprozeß,
— vom eindimensionalen zum unendlich-dimensionalen Itô-Kalkül,
— von einem einparametrigen Martingal (X_t) zu einem zweiparametrigen Martingal $(X_{t,\tau})$,
— von einem zeitlich indizierten Markoffschen Prozeß zu einem räumlich indizierten Markoffschen Feld.

Jedes dieser Stichworte deutet aktuelle Forschungsrichtungen der letzten Jahre an. Dabei geht es einerseits um völlig neue Probleme, in einigen Punkten ergibt sich aber auch ein tieferes Verständnis des Wienermaßes selbst. Ähnlich wie die Anstöße zur Entwicklung einer mathematischen Theorie der Brownschen Bewegung von den Anwendungen her kamen, so hängen auch einige dieser Richtungen mit neueren Entwicklungen in der mathematischen Physik zusammen; zum Teil sind sie von daher motiviert. Aber auch nach rein mathematischen Kriterien ist klar, daß es sich um Fragestellungen handelt, die von grundlegender Bedeutung sind, und die wohl auch in den kommenden Jahren die Entwicklung der Theorie der stochastischen Prozesse stark beeinflussen werden.

Im folgenden versuche ich, zu einigen dieser Richtungen eine kurze und möglichst elementare Einführung zu geben. Es ist natürlich nur eine Auswahl. Auch der Titel ist nicht allzu wörtlich gemeint. Er soll nur eine allgemeine Richtung im Sinne der obigen Stichworte andeuten; nicht alle der im folgenden angeschnittenen Fragen hängen mit der Brownschen Bewegung oder dem Brownschen Blatt direkt zusammen. Umgekehrt gehen wir auf einen zentralen Aspekt des Brownschen Blattes überhaupt nicht ein, nämlich auf seine Rolle in funktionalen Varianten des zentralen Grenzwertsatzes für räumlich indizierte Familien von Zufallsvariablen und auf die wichtige Querverbindung zur mathematischen Statistik; vgl. [17].

M. Nagasawa und J. D. Deuschel möchte ich für eine Reihe von Bemerkungen und Korrekturen herzlich danken.

2   Einige Eigenschaften der Brownschen Bewegung

In diesem Abschnitt erinnern wir zunächst an einige Aspekte der Brownschen Bewegung, auf die wir uns im folgenden beziehen wollen, und zwar an das Wienermaß und einige seiner Trägereigenschaften, an die Rolle der Brownschen Bewegung als Martingal und als stochastischer Integrator, und an die Verbindung zur Potentialtheorie, die sich zur Zeit in einer unendlich-dimensionalen Variante neu aktiviert. Wir setzen also ein im Jahre 1923, machen aber dann im Teil (e) noch eine Anmerkung zur Frühgeschichte.

a) Das Wienermaß

1923 konstruierte N. Wiener ein exaktes Modell für die Brownsche Bewegung, also eine Wahrscheinlichkeitsverteilung P auf dem Raum $C[0, 1]$ der stetigen Pfade, für die der stochastische Prozeß der Koordinatenabbildungen $W_t(w) = w(t)$ ein zentrierter Gaußscher Prozeß mit Kovarianzfunktion $E[W_s W_t] = \min(s, t)$ ist. Den Raum $C[0, 1]$, versehen mit dem *Wienermaß P*, bezeichnet man auch als *Wienerraum*.

Schon durch Wiener selbst, und dann vor allem durch die tiefliegenden Untersuchungen von P. Lévy zur Feinstruktur der Brownschen Pfade [64] wurden die Trägereigenschaften des Wienermaßes genauer eingegrenzt; vgl. [52]. Zum Beispiel ist P auf die Menge derjenigen stetigen Funktionen konzentriert, die auf jedem Intervall $[0, t]$ die quadratische Variation

$$\lim_n \sum_{k \leq t2^n} (W_{(k+1)2^{-n}} - W_{k2^{-n}})^2 = t \tag{2.1}$$

haben. Für eine solche Funktion gilt aber die *Itô-Formel*, d.h. für jede C^2-Funktion F ist

$$F(W_t) - F(W_o) = \int_0^t F'(W_s)\, dW_s + \frac{1}{2} \int_0^t F''(W_s)\, ds, \tag{2.2}$$

wenn man das erste Integral als Limes der Summen

$$\sum_{k \leq t2^n} F'(W_{k2^{-n}})\, (W_{(k+1)2^{-n}} - W_{k2^{-n}}) \tag{2.3}$$

definiert. Auf dieser Itô-Formel beruht die spezielle Struktur des stochastischen Kalküls über der Brownschen Bewegung. Ein weiteres Beispiel für eine Trägereigenschaft des Wienermaßes ist Lévy's Stetigkeitsmodul: das Maß P ist auf die Menge der stetigen Funktionen mit

$$\lim_{h \downarrow 0} \sup_{s \leq t \leq s+h} \frac{|W_t - W_s|}{\sqrt{2h \log h^{-1}}} = 1 \tag{2.4}$$

konzentriert. Die Untersuchung der Feinstruktur der Brownschen Pfade ist keineswegs abgeschlossen, vgl. zum Beispiel Yor [99]. In Abschnitt 3 werden wir sehen, wie sich sogar klassische Trägereigenschaften wie (2.1) und (2.4) durch Einbettung des Wienermaßes in eine Dynamik weiter verschärfen lassen.

b) Die Brownsche Bewegung als Martingal

Die Brownsche Bewegung ist ein stetiges Martingal, und in der Klasse der stetigen Martingale spielt sie eine fundamentale Rolle. Um das genauer zu formulieren, bezeichnen wir mit $\mathscr{F}_t$ die von $(W_s)_{0 \le s \le t}$ erzeugte σ-Algebra. Ein stochastischer Prozeß $(M_t)_{0 \le t \le 1}$ auf dem Wienerraum heißt *adaptiert* bezüglich der *Filtrierung* $(\mathscr{F}_t)_{0 \le t \le 1}$, wenn M_t für jedes t meßbar ist bezüglich $\mathscr{F}_t$. Ein adaptierter Prozeß (M_t) ist ein *Martingal*, wenn die bedingten Erwartungen der Inkremente bezüglich der jeweiligen Vorgeschichte verschwinden, wenn also für $s \le t$

$$E[M_t - M_s | \mathscr{F}_s] = 0 \qquad (2.5)$$

gilt. Da die Inkremente des Wienermaßes (W_t) unabhängig von der Vorgeschichte und zentriert sind, ist (W_t) ein Martingal. Außerdem folgt aus der speziellen Form der Approximation (2.3), daß die in der Itô-Formel (2.2) auftretenden stochastischen Integrale (lokale) Martingale sind. Speziell ist

$$W_t^2 - t = 2 \int_0^t W_s \, dW_s \qquad (2.6)$$

ein Martingal, und durch diese Eigenschaft ist die Brownsche Bewegung in der Klasse der stetigen Martingale charakterisiert. Aus dieser Charakterisierung von Lévy folgt, daß sich jedes stetige Martingal – jetzt bezüglich einer allgemeinen Filtrierung auf irgendeinem Wahrscheinlichkeitsraum – durch eine stochastische Zeittransformation auf eine Brownsche Bewegung reduzieren läßt, und daß jedes Martingal über dem Wienerraum sich als stochastisches Integral der Brownschen Bewegung im Sinne von c) darstellen läßt; vgl. [50].

c) Stochastische Integration

Aus (2.1) folgt, daß ein typischer Pfad des Wienermaßes auf keinem Intervall von beschränkter Variation ist. Für einen adaptierten Prozeß (H_t) auf dem Wienerraum kann man also das Integral $\int H_t \, dW_t$ nicht im Sinne der klassischen Integrationstheorie definieren. Für die speziellen Integranden in der Itô-Formel kommt man mit der pfadweisen Konstruktion (2.3) direkt zum Ziel. Für allgemeinere Integranden braucht man aber die L^2-Konstruktion von Itô [51]. Dazu setzt man

$$\int_0^1 H \, dW = \sum_i h_i (W_{t_{i+1}} - W_{t_i}) \qquad (2.7)$$

für elementare Integranden der Form $H_t(w) = \sum_i h_i(w) \, I_{(t_i, t_{i+1})}(t)$, aus den Martingaleigenschaften der Brownschen Bewegung ergibt sich die Isometrie

$$E\left[\left(\int_0^1 H \, dW\right)^2\right] = E\left[\int_0^1 H_t^2 \, dt\right],$$

und über diese Isometrie wird das stochastische Integral auf eine allgemeine Klasse von Integranden fortgesetzt; vgl. [50].

Mit dem enormen Aufschwung der allgemeinen Martingaltheorie in den 70er Jahren, der sich an den sukzessiven Veröffentlichungen des „Séminaire de Probabilités" der französischen Schule besonders eindrücklich ablesen läßt, hat sich insbesondere die Theorie der stochastischen Integration zu einem vielseitigen und technisch subtilen Instrumentarium entwickelt, das sich gerade auch in den Anwendungen als äußerst effizient erwiesen hat. An die Stelle der Brownschen Bewegung ist dabei die Klasse der Semimartingale

$$X = M + A \tag{2.8}$$

bezüglich einer allgemeinen Filtrierung getreten. Dabei bezeichnet $M = (M_t)$ ein lokales Martingal und $A = (A_t)$ einen previsiblen Prozeß mit Pfaden von beschränkter Variation; vgl. Dellacherie-Meyer [19]. Es hat sich dabei übrigens gezeigt, daß man mit der naiven Idee der *pfadweisen Konstruktion* weiter kommt, als es zunächst den Anschein hatte. Hat zum Beispiel der Integrand (H_s) stetige Pfade, so gilt fast sicher

$$\int_0^t H_s \, dX_s = \lim_n \sum_{T_k^n \leq t} H_{T_{k+1}^n}(X_{T_{k+1}^n} - X_{T_k^n}), \tag{2.9}$$

wenn man die Diskretisierung an die zufällige Entwicklung des Integranden anpaßt und die Stopzeiten $T_k^n = \inf\{t > T_{k-1}^n \,\|\, |H_t - H_{T_{k-1}^n}| > 2^{-n}\}$ benutzt; vgl. Bichteler [7]. In diesem Sinn kann man auch eine große Klasse von stochastischen Differentialgleichungen pfadweise lösen [7]. Ein wichtiger Vorteil der pfadweisen Konstruktion ist, das man dabei nicht über die Doob-Meyer-Zerlegung (2.8) des Semimartingals gehen muß, die ja von der jeweiligen Struktur des zugrundeliegenden Maßes und der Filtrierung abhängt. Für stetige Semimartingale erhält man speziell die pfadweise Existenz der quadratischen Variation

$$<X>_t = X_t^2 - 2 \int_0^t X_s \, dX_s = \lim_n \sum_{T_k^n \leq t} (X_{T_{k+1}^n} - X_{T_k^n})^2,$$

und daraus ergibt sich, wiederum pfadweise in Analogie zu (2.2), die *allgemeine Itô-Formel* für stetige d-dimensionale Semimartingale $X = (X^1, \ldots X^d)$:

$$F(X_t) - F(X_o) =$$
$$= \sum_{i=1}^d \int_0^t \frac{\partial}{\partial x_i} F(X_s) \, dX_s^i + \frac{1}{2} \sum_{i,j=1}^d \int_0^t \frac{\partial^2}{\partial x_i \, \partial x_j} F(X_s) \, d<X^i, X^j>_s \tag{2.10}$$

wobei

$$<X^i, X^j> = \frac{1}{2}(<X^i + X^j> - <X^i> - <X^j>) \tag{2.11}$$

den *Kovarianzprozeß* von X^i und X^j bezeichnet.

Nun kann man ja von dieser historischen Entwicklung der stochastischen Integration, die von der Brownschen Bewegung zur Klasse der Semimartingale geführt hat, einmal ganz absehen und systematisch ansetzen, nämlich beim Begriff des stochastischen L^p-Integrators. Damit ist ein stochastischer Prozeß $X = (X_t)$ gemeint, für den man eine L^p-wertige Integrationstheorie mit previsib-len Integranden entwickeln kann. Die Minimalforderung dafür ist, daß die analog zu (2.7) definierten elementaren Integrale $\int H^n \, dX$ in L^p gegen 0 konver-gieren, wenn die elementaren Integranden H^n gleichmäßig gegen 0 gehen; für $p = 0$ ist dabei die stochastische Konvergenz gemeint. Nun besagt aber ein tiefliegen-der Satz von Dellacherie [19] und Bichteler [7], daß die Klasse der L^0-Integratoren gerade mit der Klasse der Semimartingale übereinstimmt; vgl. auch [87]. In diesem Sinn hat also die historische Entwicklung zum richtigen Ergebnis geführt.

Es gibt aber im Umkreis der Itô-Formel Phänomene, die aus dem „kano-nischen" Rahmen der Semimartingale doch wieder hinausführen. Ein solches Beispiel ist Fukushima's Erweiterung der Itô-Formel auf Funktionen im Dirich-letraum eines reversiblen Markoff-Prozesses [37]. Im Falle der Brownschen Bewegung sieht das aus wie folgt. Sei F eine absolutstetige Funktion mit quadra-tisch integrierbarer Ableitung F'. Dann ist $F(W)$ i. a. kein Semimartingal mehr, wohl aber ein *Dirichlet-Prozeß*, d. h. die Summe eines lokalen Martingals und eines Prozesses mit stetigen Pfaden, die zwar i. a. nicht mehr von beschränkter Variation sind, wohl aber von quadratischer Variation 0. Nach Yor und Bouleau [99] hat man die explizite Formel

$$F(W_t) - F(W_o) = \int_0^t F'(W_s) \, dW_s - \int_{-\infty}^{\infty} F'(a) \, L_t^{da}, \qquad (2.12)$$

wobei L_t^a die Lokalzeit der Brownschen Bewegung im Punkte a bezeichnet [99]. Perkins [78] hat gezeigt, daß L_t^a in der Variablen a ein Semimartingal ist, der letzte Term ist also als stochastisches Integral zu verstehen. Vom Standpunkt der allgemeinen Theorie der Semimartingale aus ist die Rolle der Dirichletprozesse noch nicht voll geklärt; sie treten aber in letzter Zeit in verschiedenen Zusam-menhängen auf, vgl. zum Beispiel (4.6).

d) Die Querverbindung zur Potentialtheorie

Sei P_x auf $C([0, \mathscr{F}], R^d)$ die Verteilung der d-dimensionalen Brown-schen Bewegung mit Start in x. Für den Koordinatenprozeß $W = (W^1, \ldots, W^d)$ gilt dann $< W^i, W^j >_t = \delta_{ij} t$, die Itô-Formel (2.10) nimmt also hier die Form

$$F(W_t) - F(W_o) = \sum_{i=1}^d \int_0^t \frac{\partial}{\partial x_i} F(W_s) \, dW_s^i + \frac{1}{2} \int_0^t \Delta F(W_s) \, ds \qquad (2.13)$$

an. Ist F harmonisch auf dem Gebiet G, so verschwindet der letzte Term, d. h. $F(W)$ verhält sich wie ein lokales Martingal, solange sich die Brownsche Bewegung in G aufhält. Ist G beschränkt und F stetig auf $\bar{G}$, so folgt mit dem Stopsatz für Martingale

$$E_x[F(X_T)] = F(x) \quad (x \in G), \tag{2.14}$$

wobei $T = \inf\{t > 0 \,|\, X_t \notin G\}$ den Zeitpunkt des ersten Austretens aus G bezeichnet. (2.14) läßt sich als *stochastische Lösung des Dirichlet-Problems* auffassen, und das ist nur eine der vielfältigen Beziehungen zwischen der Brownschen Bewegung und der Potentialtheorie des Laplace-Operators; vgl. Doob [26]. Aus diesen Beziehungen entwickelte sich in den Arbeiten von Doob, Hunt, Meyer, Dynkin u. a. die probabilistische Potentialtheorie für allgemeine Klassen von Markoffschen Prozessen [11], [16]. In den 70er Jahren kam die Theorie der Dirichleträume für reversible Markoffsche Prozesse hinzu [37]. Wir werden im Abschnitt 3c sehen, wie sich zur Zeit die Beziehung zwischen Brownscher Bewegung und Potentialtheorie wieder neu aktiviert, und zwar in einer unendlich-dimensionalen Version, die sich nicht mehr in den Rahmen der Huntschen Axiomatik einordnet, bei der aber die Theorie der Dirichleträume eine natürliche Anwendung findet.

Das älteste Anwendungsbeispiel für die Querverbindung zwischen Brownscher Bewegung und klassischer Potentialtheorie ist die auf Kakutani [55] zurückgehende Untersuchung der *Selbstüberschneidungen* der Brownschen Pfade. Durch Berechnung der potentialtheoretischen Kapazität eines Brownschen Pfades folgt, daß die Brownsche Bewegung in Dimension $d \leq 3$ fast sicher unendlich viele Doppelpunkte hat, für $d > 2$ fast sicher keinen Tripelpunkt, für $d > 3$ fast sicher keinen Doppelpunkt; vgl. [29]. Diese Resultate lagen bereits Mitte der 50er Jahre vor. Es zeigt sich aber in letzter Zeit, daß die Selbstüberschneidungen der Brownschen Bewegung im R^d bzw. gewisser Irrfahrten auf Z^d eine ganz entscheidende Rolle für Konstruktionsprobleme in der Quantenfeldtheorie spielen, vgl. die tiefliegenden Untersuchungen von Brydges, Fröhlich, Spencer [12], auch Dynkin [31]. Wir werden in Abschnitt 3c sehen, wie sich durch Einbettung der Brownschen Bewegung in den unendlich-dimensionalen Ornstein-Uhlenbeck-Prozeß neue Präzisierungen dieser klassischen Resultate ergeben.

e) Eine historische Anmerkung

Es ist wohlbekannt, daß A. Einstein im Jahre 1905 die Übergangsdichte (1.1) für die Bewegung eines Teilchens unter dem Einfluß der Kollisionen mit vielen sehr viel kleineren Teilchen ableitet, und zwar aus der Annahme, daß die Inkremente stationär und unabhängig sind; vgl. [74]. Eine Verfeinerung dieser Überlegung führt zur Approximation des Wienerprozesses durch den Ornstein-Uhlenbeck-Prozeß [74]. Das Problem, diesen stochastischen Prozeß nun seinerseits durch einen exakten Grenzübergang aus den Prinzipien der Newtonschen Mechanik abzuleiten, blieb lange Zeit offen. Erst 1971 wurde es von R. Holley für den eindimensionalen Fall gelöst, 1981 von Dürr, Goldstein und Lebowitz auch für höhere Dimensionen: vgl. [28].

Weniger bekannt ist, daß L. Bachelier [5] bereits im Jahre 1900 die Übergangsdichte (1.1) und darüber hinaus weitere Eigenschaften der Brownschen Bewegung erhielt, zum Beispiel die Beziehung zur Wärmeleitungsgleichung („loi du rayonnement de la probabilité") und die Verteilung von $\max_{s \leq t} W_s$. In [5] geht es nicht um die Bewegung eines physikalischen Teilchens, sondern um die mathematische Beschreibung von Kursschwankungen auf der Pariser Börse. Diese ökonomische Interpretation führt übrigens immer noch zu mathematisch interessanten Fragen. Zum Beispiel hängt die Black-Scholes-Formel für die Evaluierung von Optionen mit der Darstellbarkeit gewisser Martingale als stochastische Integrale zusammen; vgl. [44].

Die mathematische Frühgeschichte der Brownschen Bewegung reicht aber noch weiter zurück, mindestens bis zu einer Arbeit von T. Thiele im Jahre 1880. Bei der Diskussion des kumulativen Effekts von Meßfehlern tritt hier nicht nur die Brownsche Bewegung als stochastischer Prozeß mit unabhängigen normalverteilten Inkrementen auf, sondern auch schon eine erste Version des Kalman-Filters; vgl. [62].

3   Dynamisierung des Wienermaßes

Oft kann man Informationen über eine Wahrscheinlichkeitsverteilung dadurch gewinnen, daß man sie in eine geeignete Dynamik einbettet. Wir illustrieren das zunächst, indem wir eine Eigenschaft der zweidimensionalen Normalverteilung durch Einbettung in eine zweidimensionale Brownsche Bewegung ableiten. Danach wenden wir dasselbe Prinzip eine Stufe höher an, indem wir nun ihrerseits die Brownsche Bewegung dynamisieren, und zwar in zwei Varianten, 1) durch die Girsanov-Transformation (nach Bismut), und 2) durch Einbettung des Wienermaßes in das Brownsche Blatt bzw. in den unendlich-dimensionalen Ornstein-Uhlenbeck-Prozeß (nach Malliavin und Stroock). Beide Varianten führen zum Malliavin-Kalkül, d. h. zu einer partiellen Integration für Wiener-Funktionale, mit der man Informationen über die Regularität ihrer Verteilungen und über die Struktur ihrer bedingten Erwartungen gewinnen kann. Variante 1) entspricht einer deterministischen Strömung auf dem Wienerraum und ist insofern die einfachere. Dafür eröffnet die Variante 2) die reiche Struktur der probabilistischen Potentialtheorie auf dem Wienerraum, und damit auch neue Möglichkeiten zur Analyse des Wienermaßes selbst. Zugleich ist das Brownsche Blatt die natürliche Basis für den unendlich-dimensionalen Itô-Kalkül, insbesondere für unendlich-dimensionale stochastische Differentialgleichungen, wie sie zum Beispiel bei der raum-zeitlichen Renormierung von großen stochastischen Interaktionssystemen auftreten.

a) Die Hyperkontraktivität der zweidimensionalen Normalverteilung

Sei (X, Y) ein zentrierter Gaußscher Vektor mit $E[X^2] = E[Y^2] = 1$ und Kovarianz $E[XY] = \varrho$. Die bedingte Erwartung einer Funktion von Y in bezug auf X läßt sich für $p \geqq 1$ als Kontraktion im L^p-Raum über der standardisierten Normalverteilung auffassen. Nelson's Hyperkontraktivität besagt, daß man für

$$p - 1 \geqq \varrho^2 (q - 1) \tag{3.1}$$

sogar eine Kontraktion von L^p auf L^q hat; vgl. [75], [42]. Die zweidimensionale Normalverteilung ist ja nun wirklich ein klassisches Objekt, und es ist erstaunlich, daß dieses Ergebnis erst durch die Querverbindung zur Quantenfeldtheorie zustande kam. Der folgende Beweis durch Einbettung in eine zweidimensionale gekoppelte Brownsche Bewegung (X_t, Y_t) mit Kovarianzprozeß $< X, Y >_t = \varrho t$ stammt von Neveu [76]. Sei $(\mathscr{F}_t)$ bzw. $(\mathscr{G}_t)$ die von (X_t) bzw. (Y_t) erzeugte Filtrierung. Es genügt zu zeigen, daß für beschränkte $\mathscr{F}_1$- bzw. $\mathscr{G}_1$-meßbare Funktionen F und $G > 0$ die Abschätzung

$$E[FG] \leqq \|F\|_q, \|G\|_p \tag{3.2}$$

gilt für $q^{-1} + q'^{-1} = 1$. Für die beiden Martingale $M_t = E[F^{q'}|\mathscr{F}_t]$ und $N_t = E[G^p|\mathscr{G}_t]$ ist (3.2) gleichbedeutend mit

$$E[M_1^{q'^{-1}} N_1^{p^{-1}}] \leqq E[M_o^{q'^{-1}} N_o^{p^{-1}}] . \tag{3.3}$$

Aber unter der Annahme (3.1) folgt aus der Itô-Formel (2.10), daß der Prozeß

$$Z_t = M_t^{q'^{-1}} N_t^{p^{-1}}$$

in ein lokales Martingal und in einen stetigen Prozeß mit fallenden Pfaden zerfällt, und das impliziert (3.3).

b) Dynamisierung des Wienermaßes durch die Girsanov-Transformation

Sei P das Wienermaß $C[0, 1]$, (W_t) der Koordinatenprozeß und $(\mathscr{F}_t)$ die zugehörige Filtrierung. Wir gehen nun — in der Sprechweise von D. Williams [97] — von der römischen Zeit t zur griechischen Zeit τ über und betrachten die deterministische Entwicklung

$$X_\tau = W + \tau U \quad (\tau \geqq 0) \tag{3.4}$$

auf $C[0, 1]$. Dabei sei $U = (U_t)_{0 \leqq t \leqq 1}$ von der Form $U_t = \int_0^t u_s \, ds$ mit einem an $(\mathscr{F}_t)$ adaptierten beschränkten Prozeß (u_t). Sei nun F eine Fréchet-differenzierbare Funktion mit Ableitung $D_w F$ an der Stelle w, aufgefaßt als Maß $D_w F(dt)$ auf $[0, 1]$. Dann ist

$$\frac{d}{d\tau} F(X_\tau(w))|_{\tau=0} = D_w F(U) = \int_0^1 u_s(w) \, D_w F[s, 1] \, ds,$$

also

$$E\left[\frac{d}{d\tau} F(X_\tau)|_{\tau=0}\right] = E\left[\int_0^1 u_s \, D_W F[s, 1] \, ds\right]. \tag{3.5}$$

Man kann aber die Ableitung (3.5) auch anders berechnen, und zwar mit Hilfe der (Cameron-Martin-Maruyama-)Girsanov-Transformation. Ist nämlich P_τ das zum Wienermaß P äquivalente Maß mit Dichte

$$G_\tau = \exp\left[-\tau \int_0^1 u_s \, dW_s - \frac{1}{2} \tau^2 \int_0^1 u_s^2 \, ds\right], \tag{3.6}$$

so ist X_τ unter P_τ eine Brownsche Bewegung [50]. Es gilt also

$$E[F(X_\tau) \, G_\tau] = E[F(W)] = \text{const},$$

durch Differentiation folgt

$$E\left[\frac{d}{d\tau} F(X_\tau)|_{\tau=0}\right] = -E\left[F(W) \frac{d}{d\tau} G_\tau(W)|_{\tau=0}\right] = E\left[F(W) \int_0^1 u_s \, dW_s\right],$$

und durch Identifikation mit (3.5) erhält man die Gleichung

$$E\left[F(W) \int_0^1 u_s \, dW_s\right] = E\left[\int_0^1 u_s \, D_W F(s, 1) \, ds\right]. \tag{3.7}$$

Das ist die *partielle Integration* auf dem Wienerraum, auf der der Malliavin-Kalkül in der Version von Bismut [9] beruht. Nur reicht das Argument in dieser Form noch nicht aus, weil viele der wahrscheinlichkeitstheoretisch interessanten Funktionale auf dem Wienerraum gar nicht Fréchet-differenzierbar sind.

Sei zum Beispiel $X: C[0, 1] \to C[0, 1]$ die fast überall pfadweise konstruierte Lösung einer stochastischen Differentialgleichung

$$dX = \sigma(X) \, dW + b(X) \, dt \tag{3.8}$$

mit glatten Koeffizienten. Für Fréchet-differenzierbares G auf $C[0, 1]$ ist $F = G(X)$ selbst nicht Fréchet-differenzierbar. Trotzdem zeigt Bismut [9], daß

(3.7) gilt, und zwar mit

$$D_W F(s, 1) = \sigma(X_s)\, \varphi_s^{-1} \int\limits_s^1 \varphi_t\, D_X\, G(\mathrm{d}t), \qquad (3.9)$$

wobei (φ_t) eine Lösung der Linearisierung $\mathrm{d}\varphi = \sigma'\varphi\,\mathrm{d}W + b'\varphi\,\mathrm{d}t$ von (3.8) bezeichnet. Hier ist ein direktes Argument von Bichteler. Zunächst genügt es ja für die Differentiation (3.5), wenn

$$\frac{1}{\tau}\,[F(W + \tau U) - F(W)]$$

in $L^2(P)$ gegen ein geeignetes $D_W F(U)$ konvergiert. Die Lösung X einer stochastischen Differentialgleichung mit glatten Koeffizienten ist aber insofern eine glatte Abbildung auf $C[0, 1]$, als $F = G(X)$ in diesem abgeschwächten Sinne differenzierbar ist, wobei $D_W F$ gerade durch (3.9) gegeben ist; vgl. [8]. Es geht also im Grunde darum. einen Kalkül für Funktionale auf dem Wienerraum zu entwickeln, dessen Regularitätsbedingungen an die spezielle Struktur des Wienermaßes angepaßt sind, und der damit für die wahrscheinlichkeitstheoretischen Anwendungen flexibler ist als der übliche Fréchet-Kalkül auf dem Banachraum $C[0, 1]$. Systematische Ansätze zur Präzisierung dieser Regularitätsbedingungen findet man bei Shigekawa bzw. in [50], bei Stroock [89], und natürlich in der fundamentalen Arbeit von Malliavin [67], auf die wir im nächsten Abschnitt eingehen.

Das erste Ziel des Malliavin-Kalküls war es, mit direkten wahrscheinlichkeitstheoretischen Methoden die Glattheit der Übergangswahrscheinlichkeiten für Lösungen von stochastischen Differentialgleichungen unter Hörmander-Bedingungen an die zugehörigen partiellen Differentialgleichungen nachzuweisen. Die Grundidee ist einfach die: das Wienermaß P ist „glatt", wenn also ein Funktional F auf dem Wienerraum „glatt" ist und außerdem „nichtdegeneriert", so muß auch die Verteilung von F glatt sein. Bismut [9] und Bichteler-Jacod [8] zeigen, wie man von der partiellen Integration (3.7) aus in dieser Richtung weitergeht. Hier ist der erste Schritt, mit dem man zur Existenz einer Dichtefunktion kommt. Für eine C^1-Funktion f liefert (3.7), angewandt auf das Funktional $f(F)$, die Abschätzung

$$|E[f'(F(W))\, D_W F(U)]| =$$

$$= \left| E\left[f(F(W)) \int\limits_0^1 u_s\,\mathrm{d}W_s \right] \right| \leq \|f\|_\infty\, E\left[\left| \int\limits_0^1 u_s\,\mathrm{d}W_s \right| \right]. \qquad (3.10)$$

Daraus folgt die Eixistenz einer Dichtefunktion für die Verteilung von F unter dem Maß $\mathrm{d}Q = DF(U)\,\mathrm{d}P$, damit aber auch unter dem Wienermaß P falls F „nichtdegeneriert" ist. Inzwischen gibt es auch eine Reihe von Anwendungen auf Situationen, an die man mit den bekannten analytischen Methoden ohnehin nicht direkt herankommt, zum Beispiel auf bedingte Diffusionen [10], auf Sprungprozesse [8], und auf die lokalen Übergangswahrscheinlichkeiten von unendlichen Teilchensystemen [49].

Der Anwendungsbereich des Malliavin-Kalküls ist aber nicht auf die Frage der Existenz und Regularität von Übergangsdichten beschränkt. Wie Bismut gezeigt hat [9], liefert er auch einen effektiven Zugang zu Problemen der stochastischen Filtertheorie, zum Beispiel zur folgenden Frage. Wir haben bereits den Satz von Itô erwähnt, daß sich jedes quadratisch integrierbare Funktional F auf dem Wienerraum als stochastisches Integral

$$F(W) = \int_0^1 H_s \, dW_s$$

darstellen läßt. Im Prinzip kann man den Integranden (H_s) aus der quadratischen Variation $\langle M \rangle_t = \int_0^t H_s^2 \, ds$ des Martingals $M_t = E[F|\mathscr{F}_t]$ rekonstruieren. Man möchte aber möglichst explizite Formeln haben. Nun ist aber (H_s) auch, $P \times dt$-fast sicher, durch die Gleichung

$$E\left[F(W) \int_0^1 u_s \, dW_s \right] = E\left[\int_0^1 H_s u_s \, ds \right]$$

für beschränkte adaptierte Prozesse (u_s) festgelegt. Also folgt aus (3.7) die Beziehung

$$H_s = E[D_W F[s, 1]|\mathscr{F}_s].$$

Für Fréchet-differenzierbares F ist das die Filterformel von Clark, für die in (3.9) betrachteten Funktionale von Lösungen einer stochastischen Differentialgleichung ist es die Filterformel von Haussmann [45].

Im nächsten Abschnitt ersetzen wir die deterministische Bewegung (3.4) durch einen unendlich-dimensionalen Diffusionsprozeß. Das ist zwar technisch aufwendiger, eröffnet aber eine Reihe von neuen Möglichkeiten.

   c) *Einbettung der Brownschen Bewegung in das Brownsche Blatt*

In den 50er Jahren führten Kitagawa, Chentsov und Yeh das *Brownsche Blatt* ein, definiert als stetiger zentrierter Gauß-Prozeß

$$B = (B_{t,\tau})_{0 \le t \le 1, \tau \ge 0}$$

mit Kovarianzfunktion $E[B_{t,\tau} B_{s,\sigma}] = \min(t, s) \min(\tau, \sigma)$. Für jedes $\tau > 0$ ist

$\tau^{-\frac{1}{2}} B_\tau$ ein Wienerprozeß, d. h. $B_\tau = (B_{t,\tau})_{0 \leq t \leq 1}$ ist eine Brownsche Bewegung mit Diffusionskonstante τ. Der stochastische Prozeß $(B_\tau)_{\tau \geq 0}$ auf $C[0, 1]$ läßt sich als *unendlich-dimensionale Brownsche Bewegung* auffassen. Zur Feinstruktur des Brownschen Blattes vgl. [77], [32], [96], [17]. Zum Beispiel gilt in Analogie zu (2.4) fast sicher

$$\lim_{h \downarrow 0} \sup_{|R| \leq h} \frac{|B(R)|}{\sqrt{2h \log \dfrac{1}{h}}} = 1, \tag{3.11}$$

wenn R die Rechtecke $(s, t) \times (\sigma, \tau)$ durchläuft, $|R|$ die Fläche von R bezeichnet und $B(R) = B_{t,\tau} - B_{t,\sigma} - B_{s,\tau} + B_{s,\sigma}$ die „stochastische Fläche" [77].

Wir nehmen nun an, daß das Brownsche Blatt B und der Wienerprozeß W unabhängig voneinander auf einem gemeinsamen Wahrscheinlichkeitsraum $(\Omega, \mathscr{F}, P)$ definiert sind. Das Wienermaß, also die Verteilung von W unter P auf dem Raum $C[0, 1]$, bezeichnen wir im folgenden mit μ. Durch

$$X_\tau = W + B_\tau \quad (\tau \geq 0) \tag{3.12}$$

erhalten wir einen stochastischen Prozeß mit Zustandsraum $C[0, 1]$, die unendlich-dimensionale Brownsche Bewegung mit Startverteilung μ. Für jedes τ ist X_τ in römischer Zeit eine Brownsche Bewegung mit Diffusionskonstante $1 + \tau$, hat also dieselbe Verteilung wie $\sqrt{1 + \tau}\, W$. Für eine Fréchet-differenzierbare Funktion F auf $C[0, 1]$ folgt $E[F(X_\tau)] = E[F(\sqrt{1 + \tau}\, W)]$ und

$$\frac{d}{d\tau} E[F(X_\tau)]_{\tau = 0} = \frac{1}{2} E[D_W F(W)]. \tag{3.13}$$

Um nun wieder zu einer partiellen Integration auf dem Wienerraum zu kommen, brauchen wir eine direkte Differentiation von F entlang $(X_\tau)_{\tau \geq 0}$, wobei aber jetzt — im Unterschied zu (3.5) — die Itô-Formel für die unendlich-dimensionale Brownsche Bewegung ins Spiel kommt. Sei also F eine C^2-Funktion auf $C[0, 1]$. Dann ist

$$F(X_\tau) - F(X_0) = D_W F(B_\tau) + \frac{1}{2} \int_0^1 \int_0^1 B_\tau(s)\, B_\tau(t)\, D_W^2 F(ds, dt) + o(\|B_\tau\|^2),$$

also

$$E[F(X_\tau) - F(X_0) \mid W] \approx \frac{1}{2} \tau \int_0^1 \int_0^1 \min(s, t)\, D_W^2 F(ds, dt)]. \tag{3.14}$$

Wenn wir für $w \in C[0, 1]$

$$\mathcal{L}F(w) = \frac{1}{2} \int_0^1 \int_0^1 \min(s,\,t)\,D_w^2 F(\mathrm{d}s,\,\mathrm{d}t) - \frac{1}{2} \int_0^1 w(s)\,D_w F(\mathrm{d}s) \qquad (3.15)$$

setzen, erhalten wir also aus (3.13) und (3.14) die Gleichung

$$\int \mathcal{L}F\,\mathrm{d}\mu = 0. \qquad (3.16)$$

Daraus ergibt sich aber eine zu (3.7) analoge partielle Integration. Wegen

$$\mathcal{L}(f(F)) = (f'(F))\,\mathcal{L}F + \frac{1}{2}(f''(F,\,F))\,\langle F,\,F \rangle$$

mit

$$\langle F,\,F \rangle\,(w) = \int_0^1 \int_0^1 \min(s,\,t)\,D_w F(\mathrm{d}s)\,D_w F(\mathrm{d}t)$$

folgt nämlich für $g = f'$ aus (3.15), angewandt auf das Funktional $f(F)$, die Beziehung

$$\int g'(F)\,\langle F,\,F \rangle\,\mathrm{d}\mu = -2 \int g(F)\,\mathcal{L}F\,\mathrm{d}\mu. \qquad (3.17)$$

Insbesondere erhält man wieder, analog zu (3.10) ein Kriterium für die Absolutstetigkeit der Verteilung von F.

Nun sind aber die eigentlich interessanten Wienerfunktionale in der Regel nicht glatt im Sinne des Fréchet-Kalküls; in der Regel sind sie überhaupt nur μ-fast sicher definiert. Die Verteilungen von X_τ ($\tau \geq 0$) sind aber alle singulär zueinander; insbesondere ist die Verteilung von X_τ für jedes $\tau > 0$ singulär zum Wienermaß. Der erste Schritt zum eigentlichen Malliavin-Kalkül besteht nun darin, daß man das Brownsche Blatt in den unendlich-dimensionalen Ornstein-Uhlenbeck-Prozeß

$$Y_\tau = e^{-\frac{\tau}{2}}\,(W + B_{e^\tau - 1}) \quad (\tau \geq 0) \qquad (3.18)$$

transformiert. $(Y_\tau)_{\tau \geq 0}$ ist ein Markoffscher Prozeß auf $C[0,\,1]$. Zu den Eigenschaften der Halbgruppe vgl. Meyer [70]; der infinitesimale Erzeuger ist gerade der Operator $\mathcal{L}$ aus (3.15). Das Wienermaß μ ist invariant unter (Y_τ), daher die Beziehung (3.16). Darüber hinaus ist (Y_τ) in bezug auf μ reversibel, d. h. invariant unter der Zeitumkehr; vgl. Abschnitt 4. Man kann also $\mathcal{L}$ zu einem selbstadjungierten Operator auf $L^2(\mu)$ fortsetzen, und zwar in der Form

$$\mathcal{L} = -\sum_{n=0}^{\infty} \frac{n}{2}\,P_n \qquad (3.19)$$

wobei P_n die Projektion auf $\mathcal{H}_n$ in der Wiener-Itô-Zerlegung $L^2(\mu) = \bigoplus_{n \geq 0} \mathcal{H}_n$ bezeichnet. Stroock entwickelt den Malliavin-Kalkül in axiomatischer Form, ausgehend von den Eigenschaften von $\mathcal{L}$ als „symmetrischer Diffusionsoperator" bzw. von den entsprechenden Eigenschaften der Halbgruppe [89], [90]. Malliavin [67] entwickelt den Kalkül im Sinne der stochastischen Differentiation von Semimartingalen: Ausgangspunkt ist hier die Klasse der Wienerfunktionale F, für die

$$M_\tau^F = F(Y_\tau) - \int_0^\tau \mathcal{L} F(Y_\sigma) \, d\sigma \qquad (3.20)$$

ein Martingal ist mit quadratischer Variation

$$\langle M^F \rangle_\tau = \int_0^\tau \langle F, F \rangle (Y_\sigma) \, d\sigma. \qquad (3.21)$$

Für C^2-Funktionen F folgen diese Beziehungen aus der unendlich-dimensionalen Itô-Formel, für allgemeinere F sind $\mathcal{L} F$ und $\langle F, F \rangle$ durch (3.20) und (3.21) *definiert*. Unter zusätzlichen Regularitätsannahmen an $\mathcal{L} F$ und $\langle F, F \rangle$ ergeben sich dann Aussagen über die Regularität der Verteilung von F; vgl. [67], [88].

Mit dem unendlich-dimensionalen Ornstein-Uhlenbeck-Prozeß kommt zugleich die Potentialtheorie reversibler Markoffscher Prozesse ins Spiel, insbesondere die Theorie der Dirichleträume im Sinne von Fukushima [37]. Es sei hier nur die folgende bemerkenswerte Anwendung erwähnt. Eine Eigenschaft von Funktionen in $C[0, 1]$ gilt *quasi-überall*, wenn das Komplement der entsprechenden Menge $A \subseteq C[0, 1]$ die Kapazität 0 hat, wobei die potentialtheoretische Kapazität über die zum Operator $\mathcal{L}$ gehörende Dirichletform definiert ist [38]. Das ist aber gleichgedeutend mit

$$P[Y_\tau \in A \quad \text{für jedes} \quad \tau \geq 0] = 1. \qquad (3.22)$$

Insbesondere ist dann $\mu(A) = 1$ für das Wienermaß μ. Die Umkehrung gilt im allgemeinen nicht. Fukushima zeigt aber mit direkten Kapazitätsabschätzungen, daß sich eine ganze Reihe der bekannten klassischen Trägereigenschaften des Wienermaßes zu einer „quasi-überall"-Eigenschaft verschärfen lassen [38]. Zum Beispiel gilt das für die Eigenschaften (2.1) und (2.4); für (2.1) wurde das bereits von D. Williams mit Hilfe der starken Markoff-Eigenschaft von (Y_τ) gezeigt. Einiges folgt natürlich aus bekannten Resultaten zur Feinstruktur des Brownschen Blattes. Zum Beispiel impliziert (3.11), übersetzt mit (3.18), daß (2.4) mit „$\leq$" auch quasi-überall gilt. Für das nächste Beispiel erweitern wir zunächst das Brownsche Blatt auf den Parameterbereich $(0, \mathcal{F}) \times (0, \mathcal{F})$ und auf den Wertebereich R^d. Nach Orey-Pruitt [77] hat die Menge

$$\{B_{t,\tau} = a \quad \text{für ein} \quad (t,\tau) \in (0,\mathscr{F}) \times (0,\mathscr{F})\}$$

die Wahrscheinlichkeit 1 oder 0 in Dimension $d < 4$ bzw. $d \geq 4$. Aus der Darstellung

$$Y_\tau = e^{-\frac{\tau}{2}} B_{e^\tau} \tag{3.23}$$

folgt damit für $d = 2, 3$, daß die Menge

$$A = \{w \in C((0,\mathscr{F}), R^d): w(t) = 0 \quad \text{für ein} \quad t > 0\}$$

positive Kapazität hat, obwohl ja $\mu(A) = 0$ gilt. Für $d \geq 4$ ist die Kapazität $= 0$.

Zu der in Abschnitt 2c erwähnten Frage der Selbstüberschneidungen zeigt Fukushima [38], daß die Eigenschaft der Brownschen Pfade, keine Doppelpunkte zu haben, quasi-überall gilt in Dimension $d \geq 7$. Auf der anderen Seite hat Kono gezeigt, daß die Brownsche Bewegung in Dimension $d \leq 5$ mit positiver Kapazität Doppelpunkte hat — obwohl das ja für $d = 4,5$ bezüglich des Wienermaßes fast sicher nicht passiert. Der Fall $d = 6$ ist offen.

d) Stochastische Differentialgleichungen über dem Brownschen Blatt

Vor 40 Jahren hatte K. Itô gerade die Grundlagen zur Theorie der stochastischen Differentialgleichungen über der Brownschen Bewegung gelegt [51]. Das ursprüngliche Motiv war rein mathematischer Natur: Itô wollte Kolmogorov's analytische Beschreibung des infinitesimalen Verhaltens eines stetigen Markoffschen Prozesses stochastisch interpretieren, und zwar durch eine direkte Konstruktion des Prozesses aus den Pfaden der Brownschen Bewegung als Lösung der Gleichung

$$dX = \sigma(X,t)\,dW + b(X,t)\,dt. \tag{3.24}$$

Dabei war insbesondere das Problem der stochastischen Integration zu lösen; vgl. 2b). Seitdem hat die Theorie der stochastischen Differentialgleichungen eine enorme Bedeutung gewonnen, mit einer Fülle von Anwendungen auch außerhalb der Mathematik, die ihrerseits auf die Entwicklung der mathematischen Theorie stark zurückwirkten.

In den letzten Jahren haben sich gute Gründe ergeben, auch hier eine Stufe höher zu gehen und unendlich-dimensionale stochastische Differentialgleichungen zu betrachten, insbesondere stochastische Differentialgleichungen über dem Brownschen Blatt. Zum Beispiel führt die Untersuchung des makroskopischen Verhaltens großer Systeme von stochastischen Differentialgleichungen in der Regel auf raum-zeitlich indizierte Gaußsche Felder. Wenn man das nun als zeitliche Entwicklung eines räumlichen Feldes interpretieren will, stellt sich die Frage nach dem Bewegungsgesetz dieser Zeitentwicklung, und das läßt sich in der Regel als unendlich-dimensionale stochastische Differentialgleichung präzisieren. Das folgende elementare aber grundlegende Beispiel stammt von K. Itô [53].

Sei $(W_t^i)_{i=1,2,\ldots}$ eine abzählbare Kollektion von unabhängigen Brownschen Bewegungen. Zunächst besagt das Gesetz der großen Zahlen, daß die empirischen Verteilungen

$$\mu_t^n \equiv \frac{1}{n} \sum_{i=1}^{n} \delta_{W_t^i} \quad (t \geq 0)$$

für $n \uparrow \infty$ fast sicher gegen eine deterministische Bewegung $\mu_t (t \geq 0)$ im Raum der Wahrscheinlichkeitsverteilungen auf dem R^1 konvergieren, und zwar ist μ_t die Normalverteilung mit Dichte $p_t(0, \cdot)$. Die makroskopische Fluktuation ergibt sich aus dem zentralen Grenzwertsatz:

$$Z_t^n \equiv \sqrt{n} \, (\mu_t^n - \mu_t) \quad (t \geq 0),$$

aufgefaßt als stochastischer Prozeß mit Werten im Raum $\mathscr{D}'$ der Distributionen auf dem R^1, konvergiert gegen einen $\mathscr{D}'$-wertigen zentrierten Gauß-Prozeß $(Z_t)_{t \geq 0}$ mit Kovarianzfunktion

$$E[Z_s(\varphi) \, Z_t(\psi)] = \mathrm{cov}(\varphi(W_s^1), \psi(W_t^1)) \quad (\varphi, \psi \in \mathscr{D}).$$

Dabei ist die Konvergenz der endlich-dimensionalen Randverteilungen des Systems $\{Z_t^n(\varphi); t \geq 0, \varphi \in \mathscr{D}\}$ gemeint. Bis hierhin erhält man also die makroskopische Beschreibung sofort aus den klassischen Grenzwertsätzen der Wahrscheinlichkeitstheorie. Wenn man sich nun aber für die infinitesimale Struktur des Prozesses (Z_t) interessiert, ist das Problem bereits in diesem elementaren Beispiel sehr viel subtiler. Wie Itô [53] zeigt, wird man auf eine stochastische Differentialgleichung der Form

$$\mathrm{d}Z_t = \sigma(Z_t) \, \mathrm{d}B_t + \frac{1}{2} \partial^2 Z_t \, \mathrm{d}t \tag{3.25}$$

über dem Brownschen Blatt $B = (B_{t,x})_{t \geq 0, x \in R^1}$ geführt. Dabei ist ∂ die Differentiation auf $\mathscr{D}'$ und $\sigma(Z_t) = \partial \circ \sqrt{P_t(0, \cdot)}$.

Die Situation kompliziert sich natürlich sofort, wenn sich die einzelnen Komponenten nicht mehr unabhängig voneinander verhalten. Betrachten wir zum Beispiel die folgende „mean-field"-Interaktion für ein endliches Teilchensystem:

$$\mathrm{d}X_t^{n,i} = \mathrm{d}W_t^i + \frac{1}{n} \sum_{j=1}^{n} b(X_t^{n,i}, X_t^{n,j}) \, \mathrm{d}t \quad (1 \leq i \leq n) \tag{3.26}$$

mit unabhängigen Brownschen Bewegungen (W^i) und einer glatten Funktion $b(x, y)$. McKean [68] hat gezeigt, daß bei unabhängigen und identisch verteilten Startwerten das System für $n\uparrow\infty$ gegen eine Kollektion $(X_t^i)_{i=1,2,\ldots}$ von unabhängigen Diffusionen

$$\mathrm{d}X_t^i = \mathrm{d}W_t^i + [\int b(X_t^i, y)\, \mu_t\,(\mathrm{d}y)]\, \mathrm{d}t \tag{3.27}$$

konvergiert. Dabei ist μ_t, die empirische Verteilung im Zeitpunkt t, eine schwache Lösung der nichtlinearen parabolischen Differentialgleichung

$$\frac{\partial u}{\partial t} = \frac{1}{2}\frac{\partial^2}{\partial x^2}\, u - \frac{\partial}{\partial x}\, [u\int b(x, y)\, u(y)\, \mathrm{d}y], \tag{3.28}$$

die sich als stochastisches Analogon zur Vlasov-Gleichung auffassen läßt. Das Gesetz der großen Zahlen wirkt sich also hier als „*Fortpflanzung des Chaos*" aus: die Unabhängigkeit der Anfangszustände, die im endlichen System zur Zeit $t > 0$ verloren geht, wird im Vlasov-McKean-Limes $n\uparrow\infty$ wieder hergestellt. Schon auf dieser Ebene des Gesetzes der großen Zahlen gibt es übrigens tiefliegende Probleme, wenn die durch $b(x, y)$ beschriebene Interaktion singulär wird. Zum Beispiel wurde erst kürzlich von Kotani und Osada [58] ein erster Fortschritt bei der direkten Analyse des Falles $b(x, y) = \delta(x - y)$ erzielt, der mit der Burger-Gleichung zusammenhängt.

In den letzten Jahren hat man nun auch die Struktur der Fluktuationen untersucht. Insbesondere hat Tanaka [93] eine zu (3.25) analoge stochastische Differentialgleichung über dem Brownschen Blatt abgeleitet. Die hier benötigten Varianten des zentralen Grenzwertsatzes hängen übrigens eng mit der Asymptotik von U-Statistiken zusammen; vgl. [83]. Diese Querverbindung zur Mathematischen Statistik kommt besonders deutlich bei Sznitman [92] zum Ausdruck, der das n-Teilchen-System mit Hilfe einer Girsanov-Transformation über der Limesdynamik (3.27) darstellt. Alle Konvergenzargumente spielen sich jetzt im Exponenten der Girsanov-Formel ab, und hier kommen die Grenzwertsätze für U-Statistiken direkt zum Zuge.

Insgesamt hat sich die raum-zeitliche Renormierung für interaktive Teilchensysteme in den letzten Jahren zu einer sehr aktiven Forschungsrichtung entwickelt. Als weitere Beispiele nennen wir hier nur [18] für kritische Phänomene im Zusammenhang mit der obigen mean-field-Interaktion und [80] für Renormierung im Sinne des hydrodynamischen Limes. Für die Beziehungen zwischen unendlich-dimensionalen stochastischen Differentialgleichungen, Dirichletformen und Quantenfeldtheorie vgl. [3]. Zugleich, und zum Teil in Wechselwirkung mit diesen Anwendungen, entwickelt sich die allgemeine Theorie der unendlich-dimensionalen stochastischen Differentialgleichungen intensiv weiter; auch hier beweist übrigens das Instrumentarium des Martingaltheorie seine Effizienz und Eleganz.

4   Zeitumkehr von Diffusionen in Dimension 1 und ∞

Seit Q die Verteilung eines stochastischen Prozesses $(X_t)_{0 \leq t \leq T}$, und sei $\hat{Q}$ die Verteilung des zeitumgekehrten Prozesses $\hat{X}_t = X_{T-t}$. Die Frage ist dann, wie sich die jeweilige Struktur von Q in Eigenschaften von $\hat{Q}$ übersetzt. So bleibt zum Beispiel die Markoff-Eigenschaft erhalten, weil sie sich symmetrisch als bedingte Unabhängigkeit von Vergangenheit und Zukunft formulieren läßt. Dagegen geht die Semimartingal-Eigenschaft bei Zeitumkehr im allgemeinen verloren. Ist zum Beispiel X eine Bijektion vom Wienerraum auf das Einheitsintervall, so ist das erzeugte Martingal $X_t = E[X|\mathscr{F}_t]$ nach Zeitumkehr kein Semimartingal mehr [95].

Seit der klassischen Arbeit von Kolmogorov „Über die Umkehrbarkeit der statistischen Naturgesetze" [57] tritt die Frage der Zeitumkehr in verschiedenen Zusammenhängen immer wieder auf. Zu einigen ihrer Aspekte in der Theorie der Markoffschen Prozesse vgl. [72], [15], [4], [84]; zum Beispiel ist schon die Invarianz der Markoff-Eigenschaften eine subtile Frage, wenn man die konstante Zeit T durch eine zufallsabhängige Stopzeit ersetzt. Im folgenden illustrieren wir die Zeitumkehr zunächst im einfachen Spezialfall einer Diffusion im R^1. Danach gehen wir zu unendlich-dimensionalen Diffusionen über und stellen die Verbindung zur Theorie der Markoffschen Felder her.

a) Zeitumkehr in der Girsanov-Transformation

Sei also Q auf $C[0, T]$ die Verteilung eines eindimensionalen Diffusionsprozesses mit stochastischer Differentialgleichung

$$dX = dW + b(X) \, dt. \tag{4.1}$$

Die Drift b sei beschränkt, im übrigen aber nur meßbar, die Startverteilung von X_0 sei durch eine strikt positive Dichte $\varrho(x, 0)$ gegeben. Wenn wir mit P_x das Wienermaß zum Startpunkt x bezeichnen, dann ist Q absolutstetig bezüglich des σ-endlichen Maßes $P = \int\limits_{-\infty}^{\infty} P_x \, dx$, und die Dichte ist durch die Girsanov-Transformation gegeben:

$$\frac{dQ}{dP} = \varrho(W_o, 0) \exp\left[\int\limits_0^T b(W_s) \, dW_s - \frac{1}{2} \int\limits_0^T b^2(W_s) \, ds \right] \tag{4.2}$$

vgl. [50]. Insbesondere folgt, daß die Verteilung von X_t durch eine strikt positive Dichte $\varrho(x, t)$ gegeben ist. Sei nun $\hat{Q}$ die Verteilung des zeitumgekehrten Prozesses, also das Bild von Q unter der Abbildung φ, die jedem Pfad $w \in C[0, T]$ den zeitumgekehrten Pfad φw mit $(\varphi w)(t) = w(T - t)$ zuordnet. Aus der Symmetrie der Übergangsdichte (1.1) der Brownschen Bewegung folgt zunächst, daß das

Maß P *reversibel* ist, also invariant unter φ. Also ist auch $\hat{Q}$ absolutstetig zu P mit Dichte

$$\frac{\mathrm{d}\hat{Q}}{\mathrm{d}P} = \frac{\mathrm{d}Q}{\mathrm{d}P} \circ \varphi = \varrho(W_T, 0) \exp\left[\left(\int_0^T b(W_s)\,\mathrm{d}W_s\right) \circ \varphi - \frac{1}{2}\int_0^T b^2(W_s)\,\mathrm{d}s\right]. \tag{4.3}$$

Zugleich folgt aus der Absolutstetigkeit von $\hat{Q}$, daß $\hat{Q}$ wieder durch eine Girsanov-Transformation gegeben sein muß, also

$$\frac{\mathrm{d}\hat{Q}}{\mathrm{d}P} = \varrho(W_o, T) \exp\left[\int_0^T \hat{b}(W_s, s)\,\mathrm{d}W_s - \frac{1}{2}\int_0^T \hat{b}^2(W_s, s)\,\mathrm{d}s\right] \tag{4.4}$$

mit einer noch zu identifizierenden Driftfunktion $\hat{b}(x, t)$; vgl. [66] Ch. 7. Die Identifikation ergibt sich aus dem Vergleich von (4.3) und (4.4). Man muß dabei nur die Zeitumkehr des stochastischen Integrals in (4.3) berechnen. Für glattes b geht das am einfachsten durch Umrechnung des Itô-Integrals auf ein bezüglich der Zeitumkehr symmetrisches Stratonovic-Integral; vgl. [71] p. 201. Im allgemeinen Fall ergibt sich die Lösung aus der in (2.12) bereits erwähnten verallgemeinerten Itô-Formel

$$\int_0^T b(W_s)\,\mathrm{d}W_s = B(W_T) - B(W_o) - \frac{1}{2}\int b(a)\,L_T^{da}, \tag{4.5}$$

wobei B eine Stammfunktion zu b bezeichnet. Wegen der Invarianz von $\int b(a)\,L_T^{da}$ unter φ folgt nämlich aus (4.5)

$$\left(\int_0^T b(W_s)\,\mathrm{d}W_s\right) \circ \varphi = -\int_0^T b(W_s)\,\mathrm{d}W_s - \int b(a)\,L_T^{da}.$$

Durch Identifikation von (4.3) und (4.4) erhält man, daß $V(x, t) = \log \sigma(x, T-t)$ entlang der Pfade der Brownschen Bewegung einen Dirichlet-Prozeß im Sinne der Erläuterung von (2.12) ergibt, mit Martingalkomponente

$$\int_0^T \frac{\partial}{\partial x} V(W_s, s)\,\mathrm{d}W_s = \int_0^T (b(W_s) + \hat{b}(W_s, s))\,\mathrm{d}W_s.$$

Daraus folgt die Beziehung

$$b(x) + \hat{b}(x, t) = \frac{\partial}{\partial x} \log \varrho(x, T-t) \tag{4.6}$$

zwischen Vorwärtsdrift, Rückwärtsdrift und den sukzessiven Dichten.

Im reversiblen Fall ist $\hat{b} = b$, also $2b = \dfrac{\partial}{\partial x} \log \varrho(x)$, und das ist Kolmogorov's klassische Beziehung zwischen Drift und reversibler Gleichgewichtsdichte. Diese explizite Berechnung reversibler Gleichgewichte ist wohl die bekannteste Anwendung der Zeitumkehr. Für nicht-reversible Verteilungen tritt die Dualitätsbeziehung (4.6) bei Nagasawa [72] auf.

Die Frage der Zeitumkehr stellt sich nicht nur in Verbindung mit der Untersuchung von Gleichgewichten. Auch in transienten Situationen kommt sie in natürlicher Weise ins Spiel, und hier bestimmt sie die Struktur der raum-zeit-harmonischen Funktionen des Prozesses. Wenn wir zum Beispiel auf die Brownsche Bewegung mit Startverteilung μ die Zeitumkehr anwenden, so erhalten wir die auf die Endverteilung μ hin konditionierte Brownsche Bewegung. Aus (4.6) mit $b = 0$ folgt, daß deren Drift durch

$$\hat{b}(x, t) = \frac{\partial}{\partial x} \log h(x, t)$$

gegeben ist, wobei

$$h(x, t) = \int p_{T-t}(x, y)\, \mu(\mathrm{d}y)$$

eine raum-zeit-harmonische Funktion der Brownschen Bewegung ist, also Lösung der dualen Wärmeleitungsgleichung $\left(\dfrac{1}{2} \dfrac{\partial^2}{\partial x^2} + \dfrac{\partial}{\partial t} \right) h = 0$. $\hat{Q}$ ist also die Verteilung eines Doobschen h-Prozesses; vgl. [26]. Das ist ein ganz allgemeiner Zusammenhang: wenn man einen Markoffschen Prozeß von seinem Martinrand aus zurücklaufen läßt, und zwar mit verschiedenen Startverteilungen μ auf dem Rand, so entspricht das der Zeitumkehr der verschiedenen h-Prozesse, wobei h die (raum-zeit-)harmonischen Funktionen des Prozesses durchläuft; vgl. [59].

b) Anmerkung zur stochastischen Mechanik

Bereits 1931 erscheint die Frage der Zeitumkehr für die Brownsche Bewegung auf einem Zeitintervall $[0, T]$, und implizit auch die Idee des h-Prozesses, in einer Arbeit von Schrödinger „Über die Umkehrung der Naturgesetze" [82], und zwar mit dem Hinweis auf „merkwürdige Analogien zur Quantenmechanik, die mir sehr des Hindenkens wert erscheinen". Hier kündigt sich also bereits Nelson's „Stochastische Mechanik" an, in der die Schrödinger-Gleichung in eine stochastische Newton-Gleichung für Diffusionsprozesse übersetzt wird, wobei die Definition der stochastischen Beschleunigung auf der Drift b und zugleich auf der Drift $\hat{b}$ der Zeitumkehr beruht; vgl. Nelson [74], Nagasawa [73], Meyer [71]. Man kann diesen Zusammenhang auch als Äquivalenz der Schrödinger-Gleichung mit einem stochastischen Kontrollproblem formulieren, in dessen Lagrangefunktion wieder die Struktur der Zeitumkehr eingeht; vgl. Guerra-Morato [43], Yasue [98]. Schrödinger's Schlußfrage in [82], ob man diese formalen Äquivalenzen „wirklich zu einem besseren Verständnis der Quantenmechanik ausnutzen kann", ist wohl noch offen. Aber Jona-Lasinio u. a. haben gezeigt, daß man aufgrund dieser Beziehung mit Erfolg das technische Instrumentarium für Diffusionsprozesse ins Spiel bringen kann, insbesondere die Theorie der kleinen Störungen von Wentzell und Freidlin [35], um Probleme des semiklassischen Limes für die Schrödingergleichung zu behandeln [54]. Die Äquivalenz ist übrigens auch in umgekehrter Richtung interessant und führt zu neuen Fragen in der Theorie der Diffusionsprozesse; vgl. [73].

Im nächsten Abschnitt gehen wir auf eine andere Querverbindung zwischen der Zeitumkehr von Diffusionen und der mathematischen Physik ein, nämlich auf die „dynamische" Charakterisierung von Markoffschen Feldern im Sinne der statistischen Mechanik.

c) Zeitumkehr und Gibbsmaße

Für unendlich-dimensionale Diffusionsprozesse führt Kolmogorov's klassische Beziehung zwischen Drift und reversiblem Gleichgewicht auf eine Charakterisierung der reversiblen Gleichgewichtsverteilungen als Gibbsmaße im Sinne der statistischen Mechanik. Man kann das in verschiedener Weise präzisieren, zum Beispiel in der folgenden.

Sei I eine abzählbare Indexmenge, zum Beispiel ein d-dimensionales Gitter Z^d. Wir betrachten das unendliche System der stochastischen Differentialgleichungen

$$\mathrm{d}X^i = \mathrm{d}W^i + b^i(X)\,\mathrm{d}t \quad (i \in I), \tag{4.8}$$

wobei $(W^i)_{i \in I}$ eine Kollektion von unabhängigen Brownschen Bewegungen bezeichnet und die Driftfunktionen b^i in dem Sinne interaktiv sind, daß $b^i(x)$ nicht nur von der i-ten Koordinate des Vektors $x = (x_i)_{i \in I}$ abhängt. Zur Existenz solcher Diffusionsprozesse auf geeigneten Zustandsräumen $S \subseteq R^I$ vgl. [85], [27], [63]. Unter einigen Regularitätsannahmen kann man nun zeigen [94], daß die Zeitumkehr $\hat{X}^i_t = X^i_{T-t}$ durch die stochastischen Differentialgleichungen

$$\mathrm{d}\hat{X}^i = \mathrm{d}\hat{W}^i + \hat{b}^i(X, t)\,\mathrm{d}t \quad (i \in I) \tag{4.9}$$

beschrieben sind, wobei die Driftfunktionen

$$\hat{b}^i(x, t) = \frac{d}{\mathrm{d}x_i} \log \varrho^i(x, T-t) - b^i(x) \tag{4.10}$$

von den sukzessiven Verteilungen μ_t des Diffusionsprozesses abhängen, aber nur über die bedingten Dichten

$$\varrho^i(x, t) \equiv \varrho^i\left(x_i | x_j \, (j \neq i), t\right)$$

der i-ten Koordinate, gegeben die Koordinaten $j \neq i$. Nehmen wir nun an, daß wir eine Diffusion mit *lokaler* Interaktion haben, daß also $b^i(x)$ nur von den Koordinaten j in der Nachbarschaft von i abhängt („räumliche Markoff-Eigenschaft"), oder jedenfalls nur „schwach" von weit entfernten Koordinaten j. Dann ist die Lokalität der Zeitumkehr gleichbedeutend mit der Lokalität der sukzessiven bedingten Verteilungen des Prozesses. Für eine reversible Gleichgewichtsverteilung μ gilt $\hat{b} = b$, und daraus folgt, daß die bedingten Dichten von μ nur lokal von den Außenbedingungen abhängen, d.h. μ ist ein *Gibbsmaß* im Sinne der statistischen Mechanik. Ist die Interaktion sogar lokal im Sinne der räumlichen Markoff-Eigenschaft, so sind die reversiblen Gleichgewichtsverteilungen *Markoffsche Felder* im Sinne des nächsten Abschnittes.

Diese „dynamische" Charakterisierung von Gibbsmaßen ist in den letzten Jahren in einer ganzen Reihe von Situationen geklärt worden, neben den Gittermodellen vom obigen Typ [27], [36], [49] zum Beispiel auch für interaktiv diffundierende unendliche Teilchensysteme im R^d [61]. Daneben gibt es eine umfangreiche Literatur für interaktive Markoff-Prozesse in stetiger und diskreter Zeit auf Konfigurationsräumen der Form $S = \{0, 1\}^I$ oder $S = Z^I$, mit entsprechenden Charakterisierungen der reversiblen Gleichgewichtsverteilungen; vgl. z. B. [65], [86], [21].

Wie im Abschnitt 3, so kann man auch in diesem Zusammenhang mit Hilfe der Dynamik oft weitere Informationen über die Struktur der Gleichgewichtsverteilungen gewinnen, obwohl man an der Dynamik per se vielleicht gar nicht interessiert ist. Ein Beispiel ist die „dynamische" Analyse des Ising-Modells von Holley-Stroock [48]. Die in a) erwähnte Anwendung der Zeitumkehr auf transiente Situationen steht für unendlich-dimensionale Diffusionen erst am Anfang.

Im folgenden Abschnitt gehen wir auf Markoffsche Gittermodelle vom Typ des Ising-Modells näher ein. Vorher wollen wir aber noch eine ganz andere Querverbindung zwischen Prozessen und Feldern erwähnen, die sich in letzter Zeit als fruchtbar erweist. Für eine große Klasse von Gaußschen Feldern kann man nämlich die Kovarianzfunktion des Feldes als Green-Funktion eines symmetrischen Markoffschen Prozesses interpretieren, und es stellt sich dann die Frage, wie sich Eigenschaften des Prozesses, insbesondere der Feinstruktur der Pfade, in Eigenschaften des Feldes übersetzen. Zum Beispiel hat Dynkin [30] gezeigt, wie sich räumliche Markoff-Eigenschaften des Feldes auf das Pfadverhalten des Prozesses zurückführen lassen. Auch die bereits in Abschnitt 2c erwähnte Beziehung zwischen den Selbstüberschneidungen der Brownschen Pfade und Eigenschaften von euklidischen Feldern gehört in diesen allgemeinen Zusammenhang; vgl. [31].

5   Markoffsche Felder

In Abschnitt 3 haben wir den Übergang von der Brownschen Bewegung zum Brownschen Blatt dynamisch interpretiert, also als Einbettung des Wiener-maßes in einen unendlich-dimensionalen Diffusionsprozeß. Man kann das aber auch als Übergang zu einem räumlich indizierten stochastischen Feld auffassen, speziell als Übergang von einem Markoffschen Prozeß zu einem Markoffschen Feld. Zur genauen Formulierung der räumlichen Markoff-Eigenschaft des Brownschen Blattes vgl. [81], [96]; im Prinzip geht es darum, daß bei gegebener Situation auf dem Äußeren eines Gebietes die Prognose für das, was im Inneren passiert, nur von der Situation „auf dem Rand" abhängt.

Die Idee des Markoffschen Feldes hat in den 70er Jahren eine erhebliche Brisanz entwickelt, und zwar gleichzeitig in der Wahrscheinlichkeitstheorie und in der mathematischen Physik. In der Quantenfeldtheorie kam die räumliche Markoff-Eigenschaft durch Nelson's Axiomatisierung der euklidischen Felder ins Spiel und wurde dann von Osterwalder und Schrader zur Reflektionspositi-vität erweitert [75], [40]. In der Statistischen Mechanik wurden Markoffsche Gitterfelder zur mathematisch exakten Analyse von kooperativen Effekten benutzt, insbesondere von kritischen Phänomenen vom Typ der Phasenüber-gänge. In den Arbeiten von Dobrushin u. a. wurde die wahrscheinlichkeitstheo-retische Bedeutung der Phasenübergänge freigelegt und die allgemeine Theorie der Markoffschen Felder mit vorgegebenen lokalen bedingten Verteilungen entwickelt [79]. Spitzer u.a. klärten die Rolle der Markoffschen Felder als reversible Gleichgewichte für stochastische Teilchenbewegungen mit lokaler Interaktion; vgl. 4c. Insgesamt hat die neubelebte Wechselwirkung mit Frage-stellungen und Methoden der mathematischen Physik eine ganze Reihe neuer wichtiger Entwicklungen in der Wahrscheinlichkeitstheorie ausgelöst; weitere Beispiele sind die Theorie selbstähnlicher Prozesse [22] und der nicht-zentralen Grenzwertsätze [23]. Auch auf klassische Fragen der Wahrscheinlichkeitstheo-rie hat es Rückwirkungen gegeben; vgl. zum Beispiel H. Kesten's Lösung des Perkolationsproblems [56].

Wir gehen im folgenden nur auf einen einzigen Aspekt dieser sehr vielfäl-tigen Entwicklung ein, nämlich auf Dobrushin's Kontraktionsmethode, die es erlaubt, für eine große Klasse von Markoffschen Feldern die kritischen Phäno-mene gerade auszuschließen. Wir illustrieren das kurz für die globale Markoff-Eigenschaft und wenden dann die Methode auf einen ganz anderen Problem-kreis in der Theorie der räumlich indizierten stochastischen Prozesse an, näm-lich auf die Konvergenz von zweiparametrigen Martingalen.

a) Markoffsche Felder auf Z^d

Sei $\Omega = S^I$ der kompakte Raum der Konfigurationen $\omega : I \to S$, wobei S einen endlichen Zustandsraum und I das d-dimensionale Gitter Z^d bezeichnet. Ein *Markoffsches Feld* ist eine Wahrscheinlichkeitsverteilung P auf Ω, deren bedingte Verteilungen

$$\pi_i(x|\eta) = P\left[\omega(i) = x | \omega(j) = \eta(j)\ (j \neq i)\right] \tag{5.1}$$

nur von den Werten der Konfiguration η in der jeweiligen Nachbarschaft $N(i) \equiv \{j \in I : \|j - i\| = 1\}$ abhängen. Es gilt also die *räumliche Markoff-Eigenschaft*

$$\pi_i(\cdot|\eta) = \pi_i(\cdot|\zeta) \quad \text{falls} \quad \eta = \zeta \quad \text{auf} \quad N(i). \tag{5.2}$$

So wie man sich in der Theorie der Markoffschen Prozesse die Halbgruppe bzw. deren infinitesimale Charakteristika vorgibt und von daher die Eigenschaften des Prozesses untersucht, so ist jetzt die Kollektion $(\pi_i)_{i \in I}$ der bedingten Verteilungen vorgegeben. Das Problem ist zunächst einmal die Bestimmung der Menge $\mathscr{G}(\pi)$ derjenigen Markoffschen Felder, die mit dieser Kollektion im Sinne von (5.1) verträglich sind; vgl. [79].

Die Frage der *Existenz* führt auf Konsistenzbedingungen an die Familie (π_i). Spitzer und Averintsev zeigten, daß die bedingten Verteilungen genau dann konsistent sind, wenn sie im Sinne der Statistischen Mechanik durch ein Interaktionspotential beschreibbar sind. Ein Beispiel ist das ferromagnetische Ising-Modell

$$\pi_i(x|\eta) = \frac{1}{Z_i(\eta)} \exp\left[\beta \sum_{j \in N(i)} x\eta(j)\right] \tag{5.3}$$

mit $S = \{+1, -1\}$ und $\beta > 0$; an jedem Gitterpunkt wird also der Wert ± 1 angenommen, und die Wahrscheinlichkeit für „$+$" steigt mit der Magnetisierung $\sum_{j \in N(i)} \eta(j)$ der Umgebung.

Die Frage der *Eindeutigkeit* hängt damit zusammen, wie stark die Interaktion ist, wie sehr also die bedingten Verteilungen (5.1) von den Randbedingungen η abhängen. Im allgemeinen Fall läuft die Bestimmung der konvexen Menge $\mathscr{G}(\pi)$ im Sinne einer Choquet'schen Integraldarstellung auf die Bestimmung der Extremalpunkte hinaus. Gibt es mehr als einen, ist also das Markoffsche Feld durch seine bedingten Verteilungen nicht eindeutig festgelegt, so entspricht das in der physikalischen Interpretation einem „kritischen Phänomen". Im Ising-Modell z. B. ist das Feld in Dimension $d = 1$ eindeutig bestimmt, aber in Dimension $d \geqq 2$ kommt es für genügend großes β zu einer „spontanen Magnetisierung", d. h. es gibt zwei verschiedene translationsinvariante Extremalpunkte P^+ und P^- mit Magnetisierung

$$\int \omega(i)\, P^+(d\omega) = -\int \omega(i)\, P^-(d\omega) > 0, \tag{5.4}$$

vgl. [79]. Aizenman [1] und Higuchi [46] haben gezeigt, daß dies in Dimension

$d = 2$ auch die einzigen Extremalpunkte sind, d.h. es gilt

$$P = \alpha P^+ + (1 - \alpha) P^- \tag{5.5}$$

für jedes $P \in \mathscr{G}(\pi)$.

b) Dobrushin's Kontraktionsmethode

Dobrushin hat eine allgemeine Kontraktionsmethode entwickelt, die es ermöglicht, für eine große Klasse von Interaktionen die Eindeutigkeit und darüber hinaus eine Reihe von Regularitätseigenschaften des zugehörigen Markoffschen Feldes nachzuweisen. Sie beruht auf der Annahme, daß die Interaktion nicht zu stark ist, und zwar im folgenden Sinn:

$$c \equiv \sup_k \sum_i C_{ik} < 1 \tag{5.6}$$

wobei

$$C_{ik} \equiv \sup \left\{ \frac{1}{2} \| \pi_k(\cdot | \eta) - \pi_k(\cdot | \zeta) \| : \eta = \zeta \quad \text{außer in } i \right\}$$

den Einfluß von i auf k bemißt: $\| \cdot \|$ bezeichnet die totale Variation. Sei nun P ein Markoffsches Feld mit bedingten Verteilungen (π_i), und sei $\tilde{P}$ irgendeine Wahrscheinlichkeitsverteilung auf Ω mit bedingten Verteilungen $(\tilde{\pi}_i)$. Wir setzen

$$b_i \equiv \frac{1}{2} \int \| \pi_i(\cdot | \eta) - \tilde{\pi}_i(\cdot | \eta) \| \, \tilde{P}(\mathrm{d}\eta).$$

Dann erhält man Dobrushin's fundamentalen *Vergleichssatz* [20] in der folgenden Variante:

$$| \int f \, \mathrm{d}P - \int f \, \mathrm{d}\tilde{P} | \leq \sum_k (bD)_k \, \delta_k(f) \tag{5.7}$$

für jede stetige Funktion f auf Ω, wobei $D = \sum_{n \geq 0} C^n$ die Summe der Potenzen der Matrix $C = (C_{ik})$ bezeichnet und

$$\delta_k(f) \equiv \sup \{ | f(\omega) - f(\eta) | : \omega = \eta \quad \text{außer in } k \}$$

die maximale Schwankung von f in der k-ten Koordinate.

Aus der Bedingung (5.6) ergibt sich insbesondere die *Eindeutigkeit*: hat $\tilde{P}$ dieselben bedingten Verteilungen (π_i) wie P, so folgt aus (5.7) mit $b \equiv 0$, daß $\tilde{P}$ mit P übereinstimmt. Die Bedingung (5.6) liefert aber mehr als nur die Eindeutigkeit. Zum Beispiel ergeben sich aus (5.7) Abschätzungen für das exponentielle Abklingen der Korrelationen, aus denen insbesondere ein zentraler Grenzwertsatz folgt; vgl. z. B. [60]. Ein anderes Beispiel ist die Verstärkung der Markoffschen Eigenschaft, auf die wir im nächsten Abschnitt eingehen. Wir haben die Kontraktionsmethode hier übrigens nur in ihrer einfachsten Version skizziert. Für flexiblere Varianten, insbesondere bei nicht-kompaktem Zustandsraum S, vgl. [24], [25]. Auch die Markoff-Eigenschaft (5.2) kann dabei erheblich abgeschwächt werden.

c) Die globale Markoff-Eigenschaft

Das Brownsche Blatt ist ein Markoffsches Feld, aber wir haben in Abschnitt 3 b) gesehen, daß wir es auch als unendlich-dimensionalen Markoffschen Prozeß interpretieren können, wenn man die eine Koordinate als Zeitrichtung auffaßt. Analog stellt sich für die hier betrachteten Markoffschen Felder auf Z^d die Frage, ob man sie als Markoffsche Ketten mit Zustandsraum $S^{Z^{d-1}}$ auffassen kann, und das ist gerade die Frage nach der globalen Markoff-Eigenschaft.

Im allgemeinen folgt aus (5.2) nur die *lokale* Markoff-Eigenschaft, d. h. für *endliche* Teilmengen $V \subseteq Z^d$ gilt

$$E[f \,|\, \mathscr{F}_{V^c}] = E[f \,|\, \mathscr{F}_{\partial V}] \tag{5.8}$$

für $\mathscr{F}_V$-meßbare Funktionen f. Für eine Funktion, die nur von den Koordinaten in V abhängt, hängt also die bedingte Erwartung bezüglich der von den Koordinaten $j \notin V$ erzeugten σ-Algebra nur von den Koordinaten im Rande ∂V von V ab. Aus Dobrushin's Bedingung (5.6) folgt nun aber auch die *globale* Markoff-Eigenschaft, d. h. (5.8) gilt für alle Teilmengen V von Z^d, insbesondere für jeden Halbraum. Das liegt daran, daß die bedingte Verteilung $P[\cdot \,|\, \mathscr{F}_{V^c}] \,(\eta)$ sich als ein Markoffsches Feld auf S^V auffassen läßt, das auch wieder die Bedingung (5.6) erfüllt; also ist es eindeutig bestimmt durch seine bedingten Verteilungen, in diese gehen aber nur die Werte der Konfiguration η auf dem Rande von V ein; vgl. [2], [33].

Für attraktive Interaktionen wie z. B. im ferromagnetischen Ising-Modell hat man neben der Kontraktionsmethode auch sehr effiziente Monotonie-Argumente zur Verfügung. Damit kann man auch für die Extremalpunkte P^+ und P^- in (5.4) die globale Markoff-Eigenschaft nachweisen [33], [41], [6]. Daraus folgt aber, daß auch jede Mischung $P = \alpha P^+ + (1-\alpha)P^-$ die globale Markoff-Eigenschaft für Halbräume V hat, denn es gilt

$$P[\cdot \,|\, \mathscr{F}_{V^c}] \,(\eta) = P^+ [\cdot \,|\, \mathscr{F}_{V^c}] \, I_{\{\bar{\eta} > 0\}} + P^- [\cdot \,|\, \mathscr{F}_{V^c}] \, I_{\{\bar{\eta} < 0\}},$$

wenn $\bar{\eta}$ den mittleren Wert von η auf dem Rande ∂V im Sinne des punktweisen Ergodensatzes bezeichnet (eine der vielen mathematischen Bemerkungen, die bei einem Nachmittagstee in Oberwolfach zustandekommen; in diesem Fall hatte H. v. Weizsäcker die auslösende Idee). Higuchi [47] zeigt mit einem analogen Argument, daß man zu beliebigen Teilmengen V von Z^d übergehen kann. Für das zweidimensionale Ising-Modell ist damit die Frage der Markoffschen Eigenschaft vollständig geklärt. Ein Beispiel von R. Israel zeigt, daß man die globale Markoff-Eigenschaft nicht allgemein für Extremalpunkte erwarten kann.

In stetigen Feldern eröffnet die globale Markoff-Eigenschaft die Querverbindung zur Theorie der unendlich-dimensionalen stochastischen Differentialgleichungen und zur zugehörigen unendlich-dimensionalen Potentialtheorie, speziell zur Theorie der Dirichletformen; vgl. Albeverio, Høegh-Krohn [3].

d) Martingale mit mehrdimensionalem Parameter

Man kann das Brownsche Blatt als ein Martingal mit zweidimensionalem Parameter auffassen und über diesem Martingal einen zweiparametrigen stochastischen Kalkül entwickeln; vgl. Cairoli-Walsh [14]. Auch die allgemeine Theorie der zweiparametrigen Martingale ist insofern stark an der Struktur des Brownschen Blattes orientiert, als ihre wesentlichen Sätze auf der folgenden Hypothese über die zugrundeliegende Filtrierung $(\mathscr{F}_t)_{t \in R^2_+}$ beruhen:

Für $t = (t_1, t_2)$ sind die σ-Algebren $\mathscr{F}^1_{t_1} = \bigvee_{t_2 \geqq 0} \mathscr{F}_{(t_1,t_2)}$ und

$$\mathscr{F}^2_{t_2} = \bigvee_{t_1 \geqq 0} \mathscr{F}_{(t_1,t_2)} \text{ bedingt unabhängig bezüglich } \mathscr{F}_t; \qquad (5.9)$$

vgl. [69]. Diese Annahme ist für die Filtrierung $\mathscr{F}_{(t_1,t_2)} = \sigma(B_{s_1,s_2} : s_i \leqq t_i)$ des Brownschen Blattes erfüllt. Aber vom Standpunkt der Markoffschen Felder aus ist sie sehr restriktiv. Andererseits zeigen Gegenbeispiele, daß man nur mit Vorsicht über die Annahme (5.9) hinausgehen kann; zu starke räumliche Interaktion führt auch in der Martingaltheorie zu „kritischen Phänomenen". Wir wollen das anhand der Frage der *fast sicheren Konvergenz* von Martingalen illustrieren. Dabei beschränken wir uns auf diskrete Parameter $t \in Z^2_+$; mit „$\leqq$" bezeichnen wir die komponentenweise Halbordnung.

Sei P eine Wahrscheinlichkeitsverteilung auf $\Omega = S^I$ mit $I = Z^2$, und sei

$$X_t = E[X|\mathscr{F}_t] \quad (t \in Z^2_+) \qquad (5.10)$$

das von einer beschränkten Zufallsvariablen X erzeugte Martingal, wobei $\mathscr{F}_t$ die von den Koordinaten $0 \leqq i \leqq t$ erzeugte σ-Algebra bezeichnet. Unter der Annahme (5.9) konvergiert X_t für $t \uparrow \mathscr{F}$ fast sicher; vgl. Cairoli [13]. Andererseits haben Dubins und Pitman gezeigt, daß ohne (5.9) ein beschränktes Martingal fast sicher divergieren kann. Man muß also die über (5.9) hinausgehende diagonale Interaktion sorgfältig kontrollieren. Wir zeigen nun, wie Dobrushin's Kontraktionsmethode auch hier zu Ergebnissen führt. Zunächst folgt aus einer auf Blackwell-Dubins und Hunt zurückgehende Modifikation des eindimensionalen Konvergenzsatzes für Martingale, daß der Prozeß

$$X_t^{(1)} = E[X \mid \mathscr{F}_{t_1}^1 \mid \mathscr{F}_{t_2}^2] \quad (t \in Z_+^2)$$

P-fast sicher konvergiert; unter der Annahme (5.9) wäre das bereits gleichbedeutend mit der Konvergenz von (X_t); vgl. [91]. Durch Iteration erhält man die Konvergenz von

$$X_t^{(n)} = E[X_t^{(n-1)} \mid \mathscr{F}_{t_1}^1 \mid \mathscr{F}_{t_2}^2]$$

für jedes n. Wir nehmen nun an, daß P ein Markoffsches Feld ist, und daß die Kontraktionsbedingung (5.6) erfüllt ist. Dann kann man mit Hilfe des Vergleichssatzes (5.7) zeigen, daß $X_t^{(n)}$ gegen X_t konvergiert, und zwar gleichmäßig in t; vgl. [34]. Daraus ergibt sich die gewünschte fast sichere Konvergenz des Martingals (X_t). Auch in diesem Zusammenhang zeigt sich also, daß die Kontraktionsbedingung (5.6) nicht nur die Eindeutigkeit garantiert, sondern darüber hinaus auch weitere kritische Phänomene ausschließt.

Literatur

[1]   Aizenman, M., *Translation invariance and instability of phase coexistence in the two-dimensional Ising model*. Comm. Math. Phys. **73**, 83—94 (1980).

[2]   Albeverio, S., R. Høegh-Krohn, *Uniqueness and global Markov property for Euclidean fields and lattice systems*. In: Quantum fields-algebras, processes (ed. L. Streit), 303—330, Springer (1980).

[3]   Albeverio, S., R. Høegh-Krohn, *Diffusion fields, Quantum fields, Fields with values in Lie groups*. In: Advances in Probability, Stochastic differential equations, Ed. M. Pinsky, Dekker (1982).

[4]   Azéma, J., *Théorie générale des processus et retournement du temps*. Ann. Ecole Norm. Sup. **6**, Ser. 4, 459—519 (1973).

[5]   Bachelier, L., *Théorie de la spéculation*. Ann. Sci. Ec. Norm. Sup. III-17, 21—86 (1900).

[6]   Belissard, J., R. Høegh-Krohn, *Compactness and the maximal Gibbs state for random Gibbsian fields on the lattice*. Comm. Math. Phys. **84**, 297—327 (1982).

[7]   Bichteler, K., *Stochastic Integration and L^p-Theory of semimartingales*. Ann. Prob. **9**, 49—89 (1981).

[8]   Bichteler, K., J. Jacod, *Calcul de Malliavin pour les diffusions avec sauts*. Sem. Probabilités XVII. Lect. Notes in Math. **986**, Springer (1983).

[9]   Bismut, J. M., *Martingales, the Malliavin Calcules and Hypoellipticity under general Hörmander's conditions*. Z. Wahrscheinlichkeitstheorie verw. Geb. **56**, 469—505 (1981).

[10]   Bismut, J. M., D. Michel, *Diffusions conditionelles*. J. Funct. Anal **45**, 274—292 (1982).

[11]   Blumenthal, R. M., R. Getoor, *Markoff Processes and Potential Theory*. Academic Press, New York (1968).

[12]   Brydges, D., J. Fröhlich, T. Spencer, *The random walk representation of classical spin systems and correlation inequalities*. Comm. Math. Phys. **83**, 123—150 (1982).

[13]   Cairoli, R., *Une inégalité pour martingales à indices multiples et ses applications*. Sém. Probabilités IV, Springer Lecture Notes in Math. **124**, 1—27 (1970).

[14] Cairoli, R., J. B. Walsh, *Stochastic integrals in the plane*. Acta Math. **134**, 111—183 (1975).

[15] Chung, K. L., J. B. Walsh, *To reverse a Markov process*. Acta Math. **123**, 225—251 (1969).

[16] Chung, K. L., *Lectures from Markov processes to Brownian motion*. Springer (1982).

[17] Csörgö, M., P. Révész, *Strong Approximations in Probability and Statistics*. Academic Press, New York (1983).

[18] Dawson, D. A., *Critical dynamics and fluctuations for a mean field model of cooperative behaviour*. J. Stat. Physics **31**, 29—85 (1983).

[19] Dellacherie, C., P. A. Meyer, *Probabilités et Potentiel*. Théorie des Martingales. Hermann, Paris (1980).

[20] Dobrushin, R. L., *Prescribing a system of random variables by conditional distributions*. Theor. Prob. Appl. **15**, 458—486 (1970).

[21] Dobrushin, R. L. et al. (Ed), *Locally interacting systems and their application in Biology*. Springer Lecture Notes in Math. **653** (1978).

[22] Dobrushin, R. L., *Gaussian and their subordinated self-similar random generalized fields*. Ann. Probability **7**, 1—28 (1979).

[23] Dobrushin, R. L., P. Major, *Non-central limit theorems for non-linear functionals of Gaussian fields*. Z. Warscheinlichkeitstheorie **50**, 27—52 (1979).

[24] Dobrushin, R. L., E. A. Pecherski, *Uniqueness conditions for finitely dependent random fields*. In: Coll. Math. Soc. J. Bolyai **27**, Random Fields, 223—261, North-Holland (1981).

[25] Dobrushin, R. L., E. A. Pecherski, *A criterion of the uniqueness of Gibbsian fields in the non-compact case*. To appear (1983).

[26] Doob, J. L., *Classical Potential theory and its classical counterpart*. Springer (1983).

[27] Doss, H., G. Royer, *Processus de diffusion associés aux mesures de Gibbs*. Z. Wahrscheinlichkeitstheorie **46**, 125—158 (1979).

[28] Dürr, D., S. Goldstein, J. L. Lebowitz, *A mechanical model for the Brownian motion of a convex massive particle*. Z. Wahrscheinlichkeitstheorie verw. Geb. **62**, 427—448 (1983).

[29] Dvoretsky, A., P. Erdös, S. Kakutani, *Double points of Brownian motion in n-space*. Acta Sci. Math. Szeged **12**, 75—81 (1950).

[30] Dynkin, E. B., *Markov processes and random fields*. Bull. Amer. Math. Soc. **3**, 975—999 (1980).

[31] Dynkin, E. B., *Markov Processes as a Tool in Field Theory*. J. Funct. Analysis **50**, 167—187 (1983).

[32] Ehm, W., *Sample function properties of multi-parameter stable processes*. Z. Wahrscheinlichkeitstheorie **56**, 195—228 (1981).

[33] Föllmer, H., *On the global Markov property*. In: Quantum fields-algebras, processes (ed. L. Streit), 293—302 (1980).

[34] Föllmer, H., *Almost sure convergence of multiparameter martingales for Markov random fields*. Ann. Probability **12** (1984).

[35] Freidlin, M. I., A. D. Wentzell, *Random perturbations of dynamical systems*. Springer (1984).

[36] Fritz, J., *Stationary measures of stochastic gradient systems, Infinite lattice models*. Z. Wahrscheinlichkeitstheorie **59**, 479—490 (1982).

[37] Fukushima, M., *Dirichlet forms and Markov processes*. North Holland (1980).

[38] Fukushima, M., *Basic Properties of Brownian motion and a capacity on the Wiener space*. J. Math. Soc. Japan **36**, 161—175 (1984).

[39] Georgii, H.-O., *Canonical Gibbs measures*. Springer Lecture Notes in Math. **760** (1979).

[40] Glimm, J., A. Jaffe, *Quantum Physics. A Functional Point of View*. Springer (1981).

[41] Goldstein, S., *Remarks on the global Markov property*. Comm. Math. Phys. **74**, 223—234 (1980).

[42] Gross, L., *Logarithmic Sobolev inequalitics*. Amer. J. Math. **97**, 1061—1083 (1976).

[43] Guerra, F., L. M. Morato, *Quantization of dynamical systems and stochastic control theory*. Phys. Review D **27**, 1774—1786 (1983).

[44] Harrison, J. M., S. R. Pliska, *Martingales and stochastic integrals in the theory of continuous trading*. Stoch. Processes and their appl. **11**, 215—260 (1981).

[45] Haussmann, U. G., *On the integral representation of functionals of Itô processes*. Stochastics **3**, 17—27 (1979).

[46] Higuchi, Y., *On the absence of the non-translation invariant Gibbs states for the two-dimensional Ising model*. In: Coll. Math. Soc. J. Bolyai **27**, Random Fields, 517—534, North-Holland (1981).

[47] Higuchi, Y., *A remark on the global Markov property for the d-dimensional Ising model*. Preprint (1983).

[48] Holley, R., D. Stroock, *Applications of the stochastic Ising model to the Gibbs states*. Comm. math. Phys. **48**, 249—265 (1976).

[49] Holley, R., D. Stroock, *Diffusions on an infinite dimensional torus*. J. Funct. Anal. **42**, 29—63 (1981).

[50] Ikeda, N., S. Watanabe, *Stochastic differential equations and diffusion processes*. North Holland (1981).

[51] Itô, K., *Differential equations determining Markov processes*. Zenkoku Shijo Sugaku Danwakai **244**, 1352—1400 (1942).

[52] Itô, K., H. P. McKean, Jr., *Diffusion processes and their sample paths*. Springer (1965).

[53] Itô, K, *Distribution-valued processes arising from independent Brownian motions*. Math. Zeitschrift (erscheint demnächst).

[54] Jona-Lasinio, G., F. Martinelli, E. Scoppola, *The semiclassical limit of Quantum Mechanics: a qualitative theory via stochastic mechanics*. Physics Reports **77**, p. 313 (1981).

[55] Kakutani, S., *On Brownian motion in n-space*. Proc. Acad. Japan **20**, 648—652 (1944).

[56] Kesten, H., *Percolation theory for mathematicians*. Birkhäuser (1982).

[57] Kolmogorov, A. N.: *Zur Umkehrbarkeit der statistischen Naturgesetze*. Math. Ann. **113**, 766—772 (1937).

[58] Kotani, S., H. Osada, *Propagation of chaos for Burger's equation*. Preprint (1983).

[59] Kunita, H., T. Watanabe, *On certain reversed processes and their application to Potential theory and boundary theory*. J. Math. Mechanics **15**, 393—434 (1966).

[60] Künsch, H., *Decay of correlation under Dobrushin's uniqueness condition and its application*. Comm. Math. Physics. **84**, 207—222 (1982).

[61] Lang, R., *Unendlich-dimensionale Wienerprozesse mit Wechselwirkung*. Z. Wahrscheinlichkeitstheorie verw. Geb. **39**, 277—299 (1977).

[62] Lauritzen, S. L., *Time Series Analysis in 1880: a discussion of contributions made by T. N. Thiele*. Int. Statistical Rev. **49**, 319—331 (1981).

[63] Leha, G., G. Ritter, *On interacting diffusion processes in finite and infinite dimensions*. Preprint Erlangen Univ. (1983).

[64] Lévy, P., *Processus stochastiques et mouvement Brownien*. Gauthier-Villars, Paris (1948).

[65] Liggett, T. M., *The stochastic evolution of infinite systems of interacting particles*. In: Ecole de Saint Flour VI, Springer Lecture Notes in Math. **598**, 187—248 (1977).

[66] Liptser, R. S., A. N. Shiryaev, *Statistics of Random processes I*. Springer (1977).

[67] Malliavin, P., *Stochastic calcules of variations and hypoelliptic operator*. Conf. Stoch. Diff. Equations Kyoto, 195—263. Wiley (1978).

[68] McKean, H. P., *Propagation of chaos for a class of non-linear parabolic equations*. Lecture Series in Diff. equ. **7**, Cath. Univ., 41—57 (1967).

[69] Meyer, P., *Théorie élementaire des processus à deux indices*. Springer Lecture Notes in Math. **863**, 1—39 (1981).

[70] Meyer, P., *Note sur les processus d'Ornstein-Uhlenbeck*. Sém. Probabilités XVI, Springer Lecture Notes Math. **920**, 95—132 (1982).

[71] Meyer, P., *Géometrie différentielle stochastique (bis)*. Sém Probabilités XVI. Springer Lecture Notes in Math. **921**, 165—207 (1982).

[72]   Nagasawa, M., *Time reversion of Markov processes*. Nagoya Math. J. **24**, 117—204 (1964).

[73]   Nagasawa, M., *Segregation of a population in an environment. J. Math. Biology* **9**, 213—235 (1980).

[74]   Nelson, E., *Dynamical theories of Brownian motion*. Princeton University Press (1967).

[75]   Nelson, E., *The free Markov field*. J. Funct. Analysis **12**, 211—227 (1973).

[76]   Neveu, J., *Sur l'espérance conditionelle par rapport à un mouvement brownien*. Ann. Inst. H. Poincaré XII, 105—109 (1976).

[77]   Orey, S., W. Pruitt, *Sample functions of the N-parameter Wiener process*. Ann. Probability **1**, 138—163 (1973).

[78]   Perkins, E., *Local time is a semimartingale*. Z. Wahrscheinlichkeitstheorie **60**, 79—118 (1982).

[79]   Preston, C., *Random Fields*. Lecture Notes in Mathematics. Springer (1976).

[80]   Rost, H., *Hydrodynamik gekoppelter Diffusionen: Fluktuationen im Gleichgewicht*. In: Dynamics and Processes. Springer Lecture Notes Math. 1031 (1983).

[81]   Rozanov. Y. A., *Markov random fields*. Springer (1982).

[82]   Schrödinger, E., *Über die Umkehrung der Naturgesetze*. Sitzungsber. Preuss. Akad. Wiss. Berlin, Phys. Math. 144—153 (1931).

[83]   Serfling, R. J., *Approximation theorems of Mathematical Statistics*. Wiley (1980).

[84]   Sharpe, M., *Some transformations of diffusions by time reversal*. Ann. Probability **8**, 1157—1162 (1980).

[85]   Shiga, T., A. Shimizu, *Infinite dimensional stochastic differential equations and their applications*. J. Math. Kyoto Univ. **20**, 395—416 (1980).

[86]   Spitzer, F., *Infinite Systems with locally interacting components*. Ann. Probability **9**, 349—364 (1981).

[87]   Stricker, C., *Caracterisation des semimartingales*. Sém. Probabilités XVII. Springer Lecture Notes Math. (1984).

[88]   Stroock, D. W., *The Malliavin calculus and its applications to second order parabolic differential equations*. Math. Systems Theory **14**, 25—65, 141—171 (1981).

[89]   Stroock, D. W., *The Malliavin Calculus and its applications*. In: Stochastic Integrals, 384—432, Springer Lecture Notes in Math. **851** (1981).

[90]   Stroock, D. W., *Some applications of stochastic calcules to partial differential equations*. Ecole d'Eté Probabilités. Springer Lecture Notes in Mathematics **976**, 267—382 (1983).

[91]   Sucheston, L., *On one-parameter proofs of almost sure convergence of multiparameter processes*. Z. Wahrscheinlichkeitstheorie verw. Geb. **63**, 43—50 (1983).

[92]   Sznitman, A. S., *An example of non-linear diffusion process with normal reflecting boundary conditions and some related limit theorems*. Thèse, Uni. Paris VI (1983).

[93]   Tanaka, H., *Limit theorems for certain diffusion processes with interaction*. In: Proc. of Taniguchi Int. Symp. on Stochastic Analysis, Katata and Kyoto. Ed. by K. Itô (Kinokuniya, 1984).

[94]   Wakolbinger, A., H. Föllmer, *Time reversal of infinite-dimensional diffusions*. Preprint (1984).

[95]   Walsh, J., *A non reversible semi-martingale*. Sém. Prob. XVI, Springer Lecture Notes in Math. **920**, 212 (1982).

[96]   Walsh, J., *Propagation of singularities in the Brownian sheet*. Ann. Probability **10**, 279—288 (1982).

[97]   Williams, D., „*To begin at the beginning ...*". In: Stochastic Integrals, Springer Lecture Notes in Math. **851** (1981).

[98]   Yasue, K., *Stochastic calculus of variations*. J. Funct. Anal. **41**, 327—340 (1981).

[99]   Yor, M., *Sur la transformée de Hilbert des temps locaux Browniens et une extension de la formule d'Itô*. Sém. Probabilités XVI, Springer Lecture Notes in Math. **920**, 238—247 (1982).

Perspectives in Mathematics
Anniversary of Oberwolfach 1984
© Birkhäuser Verlag, Basel

Entwicklungen in der komplexen Analysis mehrerer Veränderlichen

OTTO FORSTER und KARL STEIN

Mathematisches Institut der Universität,
Theresienstraße 39, D-8000 München 2 (FRG)

In diesem Artikel soll über einige Themen aus der komplexen Analysis mehrerer Veränderlichen berichtet werden, die in den vergangenen vier Jahrzehnten bei Tagungen im Mathematischen Forschungsinstitut Oberwolfach zur Sprache kamen. Die dort gehaltenen Vorträge und vor allem auch die daran anschließenden Diskussionen haben oft neue Untersuchungen angeregt, die zu weiteren Fortschritten in der komplexen Analysis geführt haben.

Angesichts der Fülle des Stoffs mußten die Referenten eine sehr einschränkende Auswahl treffen; auf viele wichtige Entwicklungen konnte leider nicht eingegangen werden. Insbesondere fehlt hier ein Bericht über das sog. $\bar{\partial}$-Problem (dabei handelt es sich um eine gewisse Verallgemeinerung der Cauchy-Riemannschen Differentialgleichungen), das einen Zusammenhang zwischen der Theorie der partiellen Differentialgleichungen und der komplexen Analysis herstellt. Diesem sehr aktiven Gebiet werden seit einigen Jahren auch eigene Tagungen gewidmet.

So werden hier nur kurze Übersichten (bei denen Vollständigkeit nicht angestrebt ist) zu den folgenden Fragenkreisen gegeben:

1. Komplexe Räume, Garben- und Cohomologietheorie;
2. Holomorph-vollständige Räume;
3. Modifikationen komplexer Räume;
4. Modulprobleme, Deformationen;
5. Äquivalenzrelationen in komplexen Räumen.

Es ist den Referenten ein Anliegen, an den bedeutenden Einfluß zu erinnern, den der Ergebnisbericht von H. Behnke und P. Thullen „Theorie der Funktionen mehrerer komplexer Veränderlichen" aus dem Jahre 1934 (Springer-Verlag, Berlin-Heidelberg) auf die nachfolgende Entwicklung der komplexen Analysis mehrerer Veränderlichen ausgeübt hat. Im Jahr 1970 ist eine zweite Auflage dieses Berichts — im folgenden als „B.-Th.-Bericht II" zitiert — erschienen, die durch Beiträge verschiedener Autoren zum damaligen Stand der Theorie ergänzt ist.

1   Komplexe Räume, Garben- und Cohomologietheorie

Eine der wesentlichen Veränderungen, die sich in der Funktionentheorie mehrerer Veränderlichen nach dem Kriege vollzogen hat, ist das Eindringen der Garben- und Cohomologietheorie.

Dies spiegelt sich schon in der Entwicklung des Begriffs des komplexen Raumes wieder. Klassischerweise betrieb man die mehrdimensionale Funktionentheorie hauptsächlich in Gebieten des $\mathbb{C}^n$ oder (unverzweigten) Riemannschen Gebieten über dem $\mathbb{C}^n$ bzw. über dem projektiven Raum $\mathbb{P}_n(\mathbb{C})$, wie etwa im Ergebnis-Bericht von Behnke-Thullen. Eine andere beliebte Art der Abschließung des $\mathbb{C}^n$ war der sog. Osgoodsche Raum $\mathbb{P}_1(\mathbb{C})^n$, das n-fache Produkt der Riemannschen Zahlensphäre. (Welche Art der Abschließung zu bevorzugen sei, war damals fast ein Glaubenskrieg; die systematische Untersuchung des Sachverhalts war einer der Ursprünge der Theorie der Modifikationen). Die abstrakten komplexen Mannigfaltigkeiten der komplexen Dimension > 1 tauchen anfangs nur vereinzelt in der Funktionentheorie auf (Carathéodory [6], Teichmüller [35]) und wurden erst nach dem Kriege dort allmählich heimisch (u.a. durch den Einfluß von H. Hopf, vgl. Abschnitt 3). Ein anderer Ursprung für die Beschäftigung mit komplexen Mannigfaltigkeiten war die Theorie der Kähler-Mannigfaltigkeiten (Kähler [17], Hodge [16]), die vor allem durch die Arbeiten von Kodaira [21] zu einem mächtigen Hilfsmittel in der Algebraischen Geometrie wurde.

Im Gegensatz zur Funktionentheorie einer Veränderlichen gibt es bei mehreren Veränderlichen nicht-uniformisierbare Verzweigungen (z. B. für die Funktion $\sqrt{z_1 z_2}$); man kann sich deshalb nicht auf komplexe Mannigfaltigkeiten beschränken. Dies führte Behnke-Stein [3] zur Definition von komplexen Räumen, die lokal analytisch verzweigte Überlagerungen von Gebieten des $\mathbb{C}^n$ sind. Cartan [8] definierte komplexe Räume, die lokal Normalisierungen von analytischen Hyperflächen in Gebieten des $\mathbb{C}^n$ sind. J. P. Serre [30] stellte schließlich die Definition der komplexen Räume auf eine garbentheoretische Grundlage. Danach wird eine komplexe Struktur auf einem (separierten) topologischen Raum X dadurch fixiert, daß man für jede offene Menge $U \subset X$ den Ring $\mathcal{O}_X(U)$ der auf U holomorphen Funktionen angibt, d. h. eine Garbe von Ringen $\mathcal{O}_X$ auf X auszeichnet und so einen „geringten Raum" $(X, \mathcal{O}_X)$ erhält. Sei speziell Y eine analytische Teilmenge einer offenen Menge $G \subset \mathbb{C}^n$. Dann definiert man die Garbe $\mathcal{O}_Y$ als Quotienten $\mathcal{O}_Y = (\mathcal{O}_G/\mathcal{J}_Y)|\,Y$, wobei $\mathcal{O}_G$ die Garbe der im gewöhnlichen Sinn holomorphen Funktionen auf offenen Mengen $U \subset G \subset \mathbb{C}^n$ bezeichnet und $\mathcal{J}_Y \subset \mathcal{O}_G$ die Idealgarbe aller auf Y verschwindenden holomorphen Funktionen ist. Ein komplexer Raum im Sinne von Serre ist ein geringter Raum $(X, \mathcal{O}_X)$, der lokal zu einem Raum $(Y, \mathcal{O}_Y)$ isomorph ist, der wie oben mit Hilfe einer analytischen Menge Y definiert wird. In einer wichtigen Arbeit von Grauert-Remmert [14] über komplexe Räume wird bewiesen, daß die komplexen Räume im Sinne von Behnke-Stein eine Teilklasse der komplexen Räume im Sinne von Serre bilden (die sog. normalen komplexen Räume, auf denen der 1. Riemannsche Hebbarkeitssatz gilt) und mit den von Cartan definierten komplexen Räumen übereinstimmen.

Eine weitere Verallgemeinerung, nämlich die komplexen Räume mit „nilpotenten Elementen", wurden von Grauert [13], in Analogie zum Vorgehen von Grothendieck in der Algebraischen Geometrie, eingeführt. Die lokalen Modelle sind hier geringte Räume der Gestalt $(Y, (\mathcal{O}_G/\mathcal{J})|Y)$, wobei Y eine analytische Menge in einer offenen Menge $G \subset \mathbb{C}^n$ und $\mathcal{J} \subset \mathcal{O}_G$ irgend eine kohärente Idealgarbe mit Nullstellengebilde Y ist (i. a. ist $\mathcal{J}$ echt kleiner als die Garbe $\mathcal{J}_Y$ aller auf Y verschwindenden holomorphen Funktionen). Die komplexen Räume im Sinne von Serre werden dann zur Unterscheidung als *reduziert* bezeichnet.

Es sei noch auf die unendlichdimensionalen Banach-analytischen Räume von Douady (vgl. Abschnitt 4) hingewiesen, sowie auf die rigid-analytischen Räume (siehe Bosch-Güntzer-Remmert [5]), bei denen statt des Körpers der komplexen Zahlen ein nicht-archimedisch bewerteter Körper zugrunde gelegt wird.

Auf einem komplexen Raum $(X, \mathcal{O}_X)$ (dieser Begriff sei im vorliegenden Abschnitt fortan im Sinne von Grauert verstanden) betrachtet man neben der Strukturgabe $\mathcal{O}_X$ sog. analytische Garben, das sind Garben von Moduln über der Garbe von Ringen $\mathcal{O}_X$. Beispiele hierfür sind etwa die Idealgarbe $\mathcal{J}_Y \subset \mathcal{O}_X$ einer analytischen Menge $Y \subset X$ oder die Garbe der holomorphen Schnitte eines holomorphen Vektorbündels über X. Ist $\mathcal{F}$ eine beliebige $\mathcal{O}_X$-Modulgarbe und sind $f_1, \ldots, f_r \in \mathcal{F}(W)$ endlich viele Schnitte von $\mathcal{F}$ über einer offenen Menge $W \subset X$, so ist über W die „Relationengarbe" $\mathcal{R} = \mathcal{R}(f_1, \ldots, f_r) \subset \mathcal{O}_X^r|W$ wie folgt definiert: Für eine offene Menge $U \subset W$ besteht $\mathcal{R}(U)$ aus allen r-tupeln $(\varphi_1, \ldots, \varphi_r) \in \mathcal{O}_X(U)^r$, so daß $\varphi_1 f_1 + \cdots + \varphi_r f_r = 0$. Eine $\mathcal{O}_X$-Modulgarbe $\mathcal{F}$ heißt *kohärent*, wenn sie lokal endlich erzeugt über $\mathcal{O}_X$ ist und die Relationengarbe zwischen je endlich vielen Schnitten von $\mathcal{F}$ ebenfalls lokal endlich erzeugt ist. Grundlegend für die Theorie der kohärenten analytischen Garben sind die Sätze von Oka [22] über die Kohärenz der Strukturgarbe des $\mathbb{C}^n$ und von Cartan [7] über die Kohärenz der Idealgarbe einer analytischen Menge im $\mathbb{C}^n$. (Daraus folgen dann die entsprechenden Sätze für beliebige komplexe Räume.)

Man hat nun den mächtigen Apparat der Cohomologietheorie von Garben zur Verfügung, der von J. Leray stammt und der vor allem von H. Cartan in seinem Seminar in Paris seit 1948 ausgebaut und für die komplexe Analysis nutzbar gemacht wurde. (Eine lehrbuchmäßige Darstellung der Cohomologietheorie von Garben findet sich bei Godement [12].)

Einer der Fundamentalsätze der Garben- und Cohomologietheorie in der Komplexen Analysis ist das sog. Theorem B von Cartan-Serre, das besagt, daß für jede kohärente analytische Garbe $\mathcal{F}$ auf einem Steinschen Raum X alle Cohomologiegruppen $H^q(X, \mathcal{F})$ für $q \geqq 1$ verschwinden (vgl. Abschnitt 2). Für kompakte komplexe Räume X (die im Vergleich zu den Steinschen Räumen in gewisser Weise den anderen Extremfall komplexer Räume darstellen) gilt, daß alle Cohomologiegruppen $H^q(X, \mathcal{F})$, $q \geqq 0$, endlich-dimensionale Vektorräume sind. Dies wurde zunächst von Kodaira [20] auf kompakten komplexen Mannigfaltigkeiten X für lokalfreie Garben $\mathscr{F}$ unter Anwendung von Sätzen aus der globalen Theorie der elliptischen Differentialoperatoren bewiesen; später gaben dann Cartan-Serre [9] einen Beweis, der für beliebige kompakte komplexe Räume X und kohärente analytische Garben $\mathscr{F}$ gilt und der sich auf einen funktional-analytischen Satz von L. Schwartz stützt.

Sei X eine n-dimensionale kompakte komplexe Mannigfaltigkeit und $\mathscr{F}$ die Garbe der holomorphen Schnitte eines holomorphen Vektorraumbündels F über X. Aufgrund des Endlichkeitssatzes kann man die sog. *Euler-Poincaré-Charakteristik*

$$\chi(X, F) = \sum_{q \geqq 0} (-1)^q \dim H^q(X, \mathscr{F})$$

definieren. (Es gilt $H^q(X, \mathscr{F}) = 0$ für $q > n$.) Es ist der Inhalt des berühmten Satzes von Riemann-Roch-Hirzebruch, daß man $\chi(X, F)$ allein aus topologischen Invarianten des Tangentialbündels T_X und des Bündels F berechnen kann. Genauer gilt $\chi(X, F) = \mathrm{td}(T_X) \, \mathrm{ch}(F) [X]$. Dabei sind der Chern-Charakter ch und die Todd-Klasse td gewisse Invarianten, die man mit Hilfe der Chernklassen für ein beliebiges Vektorraumbündel E über X als Elemente des Cohomologierings

$$H^{**}(X, \mathbb{Q}) = \sum_{q=1}^{n} H^{2q}(X, \mathbb{Q})$$

definieren kann; $\mathrm{td}(T_X) \, \mathrm{ch}(F) [X]$ bedeutet den Anteil des Cup-Produkts $td(T_X) ch(F)$, der in $H^{2n}(X, \mathbb{Q}) = \mathbb{Q}$ liegt. Dieser Satz wurde von Hirzebruch [15] 1956 unter Benutzung schwieriger topologischer Methoden, u.a. der Thomschen Cobordismus-Theorie, für projektiv-algebraische Mannigfaltigkeiten X bewiesen. Später fand Grothendieck (s. Borel-Serre [4]) einen rein algebraischen Beweis dafür und verallgemeinerte den Satz gleichzeitig auf eigentliche Abbildungen $X \to S$. Atiyah-Singer [1], [2] bewiesen 1963 einen Index-Satz für elliptische Differentialoperatoren auf kompakten Mannigfaltigkeiten. Dieser Atiyah-Singersche Indexsatz liefert, angewandt auf den $\bar{\partial}$-Operator, den Satz von Riemann-Roch für beliebige (nicht notwendig projektiv-algebraische) kompakte komplexe Mannigfaltigkeiten. (Dies war insbesondere wichtig für die Kodairaschen Untersuchungen kompakter komplexer Flächen.)

Bei der Grothendieckschen Verallgemeinerung des Riemann-Rochschen Satzes auf eigentliche Abbildungen $\pi: X \to S$ spielt der Begriff der Bildgarben eine wichtige Rolle. Sei $\mathscr{F}$ eine Garbe auf X. Ordnet man jeder offenen Teilmenge $U \subset S$ die Cohomologiegruppe $H^q(\pi^{-1}(U), \mathscr{F})$ zu, so erhält man eine Prägarbe, deren zugeordnete Garbe die q-te *Bildgarbe* von $\mathscr{F}$ heißt und mit $R^q \pi_* \mathscr{F}$ bezeichnet wird. Grothendieck bewies im algebraischen Fall, daß die Bildgarben kohärenter Garben bei eigentlichen Abbildungen stets wieder kohärent sind.

Der analytische Fall ist viel schwieriger zu behandeln und wurde von Grauert [13] gelöst: Ist $\pi\colon X \to S$ eine eigentliche holomorphe Abbildung komplexer Räume und $\mathscr{F}$ eine kohärente analytische Garbe auf X, so sind alle Bildgarben $R^q\, \pi_*\, \mathscr{F}$ kohärent. (Einfachere Beweise wurden von Forster-Knorr [10] und Kiehl-Verdier [19] gegeben; für Verallgemeinerungen auf „relativ-analytische Räume" siehe [11], [18].) Wendet man den Bildgarbensatz auf den Fall an, daß S ein Punkt ist, erhält man wieder den Endlichkeitssatz von Cartan-Serre. Von Remmert [23] wurde folgender wichtige Abbildungssatz bewiesen: Sei $\pi\colon X \to S$ eine eigentliche holomorphe Abbildung. Dann ist das Bild $\pi(Y)$ jeder analytischen Teilmenge $Y \subset X$ analytisch in S. Dieser Satz ist ebenfalls ein Spezialfall des Grauertschen Bildgarbensatzes: Da die Garbe $R^0\, \pi_*\, \mathscr{O}_Y$ kohärent ist, ist $\mathrm{Supp}\,(R^0\, \pi_*\, \mathscr{O}_Y) = \pi(Y)$ analytisch.

Im Zusammenhang mit dem Bildgarbensatz sind auch die *Halbstetigkeitssätze* zu erwähnen, die ebenfalls zuerst von Grothendieck in der algebraischen Geometrie bewiesen wurden und dann auf den analytischen Fall übertragen worden sind. Sei $\pi\colon X \to S$ eine eigentliche holomorphe Abbildung und $\mathscr{F}$ eine kohärente analytische Garbe auf X, die platt über S liegt. Für einen Punkt $s \in S$ bezeichne $X_s = \pi^{-1}(s)$ die Faser über s und $\mathscr{F}(s)$ die analytische Beschränkung von $\mathscr{F}$ auf X_s. Dann gilt:

Für jedes $q \geqq 0$ ist die Funktion $s \mapsto \dim H^q(X_s, \mathscr{F}(s))$ nach oben halbstetig auf S (sogar bzgl. der Zariski-Topologie) und die Euler-Poincaré-Charakteristik $\chi(X_s, \mathscr{F}(s))$ ist lokal-konstant. Diese Aussagen spielen für das Studium von Familien komplexer Räume eine wichtige Rolle; für einen Beweis siehe Riemenschneider [24]. Schneider [28], [29] verallgemeinerte die Halbstetigkeitssätze auf relativ-analytische Räume. Weitere Verallgemeinerungen der Bildgarbensätze auf nicht-eigentliche Abbildungen finden sich bei Riemenschneider [25], Siu [32], [33], Siegfried [31].

Resultate ganz anderer Art aus der Cohomologietheorie analytischer Garben wurden von Scheja [26], [27] bewiesen. Z. B. gilt: Sei $\mathscr{F}$ eine kohärente analytische Garbe auf dem komplexen Raum X und $A \subset X$ eine analytische Teilmenge. Falls für jeden Punkt $x \in A$ gilt $\dim_x A \leqq \mathrm{codh}\,\mathscr{F}_x - p - 2$, so ist die Beschränkungsabbildung $H^p(X, \mathscr{F}) \to H^p(X \setminus A, \mathscr{F})$ ein Isomorphismus. Dabei ist $\mathrm{codh}\,\mathscr{F}_x$ die homologische Codimension des $\mathscr{O}_{X,x}$-Moduls $\mathscr{F}_x$. Für eine freie $\mathscr{O}_X$-Modulgarbe auf einer n-dimensionalen komplexen Mannigfaltigkeit ist die homologische Codimension gleich n, so daß man im Spezialfall $\mathscr{F} = \mathscr{O}_X$ und $p = 0$ insbesondere den sog. zweiten Riemannschen Hebbarkeitssatz erhält, daß man holomorphe Funktionen über 2-codimensionale Ausnahmemengen im $\mathbb{C}^n$ holomorph fortsetzen kann. Im Zusammenhang mit Fortsetzbarkeitsproblemen sei noch auf die Thimmschen Lückengarben (Thimm [36], Siu-Trautmann [34]) hingewiesen.

Literatur

[1]   Atiyah, M. F. and I. M. Singer, *The index of elliptic operators and compact manifolds*. Bull. Amer. Math. Soc. **69** (1963), 422—433.

[2]   Atiyah, M. F. and I. M. Singer, *The index of elliptic operators I*. Ann. of Math. **87** (1968), 484—530.

[3]   Behnke, H. and K. Stein, *Modifikationen komplexer Mannigfaltigkeiten und Riemannscher Gebiete*. Math. Ann. **124** (1951), 1—16.

[4]   Borel, A. et J. P. Serre, *Le théorème de Riemann-Roch (d'après Grothendieck)*. Bull. Soc. Math. France **86** (1958), 97—136.

[5]   Bosch, S., U. Güntzer and R. Remmert, Non-Archimedean Analysis, Springer-Verlag 1984.

[6]   Carathéodory, C., *Über die analytischen Abbildungen von mehrdimensionalen Räumen*, pp. 93—101 in: Verhandlungen Int. Math. Congr. Zürich 1932, Bd. 1.

[7]   Cartan, H., *Idéaux et modules de fonctions analytiques de variables complexes*. Bull. Soc. Math. France **78** (1950), 28—64.

[8]   Cartan, H., *La notion d'espace analytique générale et fonction holomorphe sur un tel espace*. Séminaire Cartan, ENS Paris 1951/52, Exposé 13.

[9]   Cartan, H. et J. P. Serre, *Un théorème de finitude concernant les variétés analytiques compactes*. C. R. Acad. Sci. Paris **237** (1953), 138—130.

[10]   Forster, O. und K. Knorr, *Ein Beweis des Grauertschen Bildgarbensatzes nach Ideen von B. Malgrange*. Manuscr. Math. **5** (1971), 19—44.

[11]   Forster, O. und K. Knorr, *Relativ-analytische Räume und die Kohärenz von Bildgarben*. Invent. Math. **16** (1972), 113—160.

[12]   Godement, R., *Théorie des faisceaux*. Hermann, Paris 1964.

[13]   Grauert, H., *Ein Theorem der analytischen Garbentheorie und die Modulräume komplexer Strukturen*. Publ. IHES, Vol. 5, Paris 1960.

[14]   Grauert, H. und R. Remmert, *Komplexe Räume*. Math. Ann. **136** (1958), 245—318.

[15]   Hirzebruch, F., *Neue topologische Methoden in der algebraischen Geometrie*. Springer 1956. Dritte erweiterte Auflage unter dem Titel: Topological methods in Algebraic Geometry. Springer 1978.

[16]   Hodge, W. V. D., *The theory and applications of harmonic integrals*. Cambridge 1941.

[17]   Kähler, E., *Über eine bemerkenswerte Hermitesche Metrik*. Abh. Math. Sem. Univ. Hamburg **9** (1933), 173—186.

[18]   Kiehl, R., *Relativ analytische Räume*. Invent. Math. **16** (1972), 40—112.

[19]   Kiehl, R. und L. Verdier, *Ein einfacher Beweis des Kohärenzsatzes von Grauert*. Math. Ann. **195** (1971), 24—50.

[20]   Kodaira, K., *On cohomology groups of compact analytic varieties with coefficients in some analytic faisceaux*. Proc. Nat. Acad. Sci. USA **39** (1953), 865—868.

[21]   Kodaira, K., *Collected Works*. Vol. I-III. Iwanami Shoten, Princeton University Press 1975.

[22]   Oka, K., *Sur les fonctions analytiques de plusieurs variables*. VII. Sur quelques notions arithmétiques. Bull. Soc. Math. France **78** (1950), 1—27.

[23]   Remmert, R., *Holomorphe und meromorphe Abbildungen komplexer Räume*. Math. Ann. **133** (1957), 328—370.

[24]   Riemenschneider, O., *Über die Anwendung algebraischer Methoden in der Deformationstheorie komplexer Räume*. Math. Ann. **187** (1970), 40—55.

[25]   Riemenschneider, O., *Halbstetigkeitssätze für 1-konvexe holomorphe Abbildungen*. Math. Ann. **192** (1971), 216—226.

[26]   Scheja, G., *Riemannsche Hebbarkeitssätze für Cohomologieklassen*. Math. Ann. **144** (1961), 345—360.

[27]   Scheja, G., *Fortsetzungssätze der komplex-analytischen Cohomologie und ihre algebraische Charakterisierung*. Math. Ann. **157** (1964), 75—94.

[28] Schneider, M., *Halbstetigkeitssätze für relativ analytische Räume*. Inven. Math. **16** (1972), 161—176.

[29] Schneider, M., *Bildgarben und Fasercohomologie für relativ analytische Räume*. Manuscr. Math. **7** (1972), 67—82.

[30] Serre, J. P., *Géométrie algébrique et géométrie analytique*. Ann. Inst. Fourier **6** (1955/56), 1—42.

[31] Siegfried, P., *Un théorème de finitude pour les morphismes q-convexes*. Comm. Math. Helv. **49** (1974), 417—459.

[32] Siu, Y. T., *A pseudoconcave generalization of Grauert's direct image theorem I, II*. Ann. Scuola Norm, Sup. Pisa **24** (1970), 278—330, 439—489.

[33] Siu, Y. T., *A pseudoconvex-pseudoconcave generalization of Grauert's direct image theorem*. Ann. Scuola Norm. Sup. Pisa **26** (1972), 647—664.

[34] Siu, Y. T. and G. Trautmann, *Gap-sheaves and extension of coherent analytic subsheaves*. Lect. Notes in Math. **172**, Springer-Verlag 1971.

[35] Teichmüller, O., *Veränderliche Riemannsche Flächen*. Dtsch. Math. **7** (1944), 344—359.

[36] Thimm, W., *Lückengarben von kohärenten analytischen Modulgarben*. Math. Ann. **148** (1962), 372—394.

2   Holomorph-vollständige Räume

Einen der wesentlichen Anstöße für die Entwicklung der Theorie der analytischen Funktionen mehrerer Veränderlichen gab die Entdeckung der Tatsache, daß es im $\mathbb{C}^n$ für $n \geqq 2$ im Gegensatz zu $n = 1$ Gebiete G gibt, so daß sich jede in G holomorphe Funktion in ein größeres Gebiet $G' \supset G$ holomorph fortsetzen läßt. Eine Reihe solcher Fortsetzungssätze wurde von Hartogs [19] 1906 bewiesen. Z. B. gilt: Ist $G \subset \mathbb{C}^n$, $n \geqq 2$, ein Gebiet und $K \subset G$ eine kompakte Teilmenge, so daß $G \setminus K$ zusammenhängt, so läßt sich jede in $G \setminus K$ holomorphe Funktion nach G fortsetzen. Dies führt zum Begriff des Holomorphiegebiets und der Holomorphiehülle. Ein Gebiet $G \subset \mathbb{C}^n$ heißt *Holomorphiegebiet*, wenn es das genaue Existenzgebiet einer holomorphen Funktion ist. Von Cartan-Thullen [6] stammt folgende wichtige Charakterisierung: Ein Gebiet $G \subset \mathbb{C}^n$ ist genau dann Holomorphiegebiet, wenn es holomorph-konvex ist, d. h. wenn für jede kompakte Menge $K \subset G$ die Menge

$$\hat{K} := \{x \in G : |f(x)| \leqq \sup |f(K)| \quad \text{für alle } f \in \mathcal{O}(G)\}$$

wieder kompakt ist. Dabei bezeichnet $\mathcal{O}(G)$ den Ring aller holomorphen Funktionen $f : G \to \mathbb{C}$. Die *Holomorphiehülle* eines Gebiets G im $\mathbb{C}^n$ (d. h. das größte Gebiet, in das man alle holomorphen Funktionen in G simultan fortsetzen kann) existiert i. a. nicht als Gebiet im $\mathbb{C}^n$, sondern nur als unverzweigtes Riemannsches Gebiet über dem $\mathbb{C}^n$. Deshalb muß man auch unverzweigte Holomorphiegebiete über dem $\mathbb{C}^n$ betrachten. Daß auch diese holomorph-konvex sind, ist viel schwieriger zu zeigen als für schlichte Holomorphiegebiete und wurde erst 1953 von Oka [27] im Zusammenhang mit der Lösung des Levi-Problems, auf das wir noch zurückkommen werden, bewiesen.

P. Cousin [7] hatte 1895 die Sätze von Mittag-Leffler und Weierstraß über die Existenz meromorpher Funktionen mit vorgegebenen Hauptteilen bzw. mit vorgegebenen Null- und Polstellenflächen auf den $\mathbb{C}^n$ übertragen. Man versuchte diese Sätze nun auch für Holomorphiegebiete zu beweisen; dies waren aber zur Zeit der 1. Auflage des Behnke-Thullen-Berichts 1934 noch offene Probleme (die Frage nach der Gültigkeit dieser Sätze nennt man das Cousin-I bzw. Cousin-II-Problem). In einer Reihe von Arbeiten machte Oka [25] die entscheidenden Fortschritte zur Lösung dieser Probleme. So bewies er 1937, daß in einem Holomorphiegebiet des $\mathbb{C}^n$ jedes Cousin-I-Problem lösbar ist und zeigte 1939, daß in einem Holomorphiegebiet die Hindernisse gegen die Lösbarkeit des Cousin-II-Problems rein topologischer Natur sind. Stein [29] gab dann in der Sprache der algebraischen Topologie explizit Bedingungen für die Lösbarkeit des Cousin-II-Problems in Holomorphiegebieten an. In einer weiterführenden Arbeit [30] führte Stein den Begriff der holomorph-vollständigen Mannigfaltigkeit ein (in [30] „R-konvex" genannt). Eine komplexe Mannigfaltigkeit X heißt *holomorph-vollständig*, wenn folgende Axiome erfüllt sind:

1) X ist holomorph-konvex (siehe oben),
2) X ist holomorph-separabel (d.h. die holomorphen Funktionen auf X trennen die Punkte von X).

(In [30] wurden noch zwei weitere Axiome gefordert, aber Grauert [12] konnte zeigen, daß sie aus den übrigen folgen). Viele der Aussagen, die für Holomorphiegebiete gelten, können auf holomorph-vollständige Mannigfaltigkeiten übertragen werden. Cartan-Serre [5] bewiesen für die holomorph-vollständigen Mannigfaltigkeiten (nun Steinsche Mannigfaltigkeiten genannt) die berühmten Theoreme A und B (die dann später auch auf Steinsche Räume, d.h. holomorph-vollständige komplexe Räume im Sinne von Grauert, übertragen worden sind). Sei $\mathscr{F}$ eine kohärente analytische Garbe auf der Steinschen Mannigfaltigkeit X. Dann gilt: a) Die globalen Schnitte aus $H^0(X, \mathscr{F})$ erzeugen jede Faser $\mathscr{F}_x$ über dem lokalen Ring $\mathcal{O}_{X,x}$ (Theorem A), b) $H^q(X, \mathscr{F}) = 0$ für alle $q \geqq 1$ (Theorem B). Diese Theoreme haben viele Anwendungen. Z.B. folgt, daß auf einer Steinschen Mannigfaltigkeit X jedes Cousin-I-Problem lösbar ist und genau dann jedes Cousin-II-Problem lösbar ist, falls $H^2(X, \mathbb{Z}) = 0$.

Schon E.E. Levi [21] stellte fest, daß der Rand eines Holomorphiegebiets im $\mathbb{C}^n$ (von Levi wurde der Fall $n = 2$ betrachtet) einer gewissen Konvexitätsbedingung genügen muß. Sei $G \subset \mathbb{C}^n$ ein relativ-kompaktes Gebiet mit glattem Rand. G heißt (streng) *pseudokonvex*, wenn es zu jedem Randpunkt von G eine offene Umgebung U und eine zweimal stetig differenzierbare Funktion φ in

U gibt, so daß $U \cap G = \{z \in U : \varphi(z) < 0\}$ und die hermitesche Form $\left(\dfrac{\partial^2 \varphi}{\partial z_i \, \partial \bar{z}_j} \right)$

positiv semi-definit (bzw. positiv definit) ist. Levi stellte fest, daß ein Holomorphiegebiet mit glattem Rand notwendig pseudokonvex ist und für streng pseudokonvexe Gebiete lokal auch die Umkehrung gilt. Daß streng pseudokonvexe Gebiete im $\mathbb{C}^n$ auch global Holomorphiegebiete sind, wurde zuerst 1942 von

Oka [26] für $n = 2$ bewiesen, dann im allgemeinen Fall 1953/54 von Oka [27], Bremermann [4] und Norguet [24].

Der Begriff des streng pseudokonvexen Gebietes läßt sich auch auf komplexe Mannigfaltigkeiten und komplexe Räume übertragen. Grauert [15] bewies, daß jedes streng pseudokonvexe Gebiet $G \subset \subset X$ in einem komplexen Raum holomorph-konvex ist und für jede kohärente analytische Garbe $\mathscr{F}$ auf X die Cohomologiegruppen $H^q(G, \mathscr{F})$ für $q \geq 1$ endlich-dimensional sind. Dieser Endlichkeitssatz wurde von Grauert [16] zum Beweis von Verschwindungssätzen auf kompakten komplexen Räumen angewandt: Ein Geradenbündel E über einem kompakten komplexen Raum X heißt nach Grauert negativ, wenn der Nullschnitt eine streng pseudokonvexe Umgebung besitzt; das duale Bündel heißt dann positiv (über komplexen Mannigfaltigkeiten stimmen diese Begriffe mit den üblichen in der Kählertheorie definierten Begriffen überein). Es gilt nun: Ist $\mathscr{F}$ eine kohärente analytische Garbe über dem kompakten komplexen Raum X und L ein positives Geradenbündel auf X, so verschwinden die Cohomologiegruppen $H^q(X, \mathscr{F} \otimes L^k)$ für alle $q \geq 1$ und genügend große k. Daraus leitet Grauert einen neuen Beweis und gleichzeitig eine Verallgemeinerung des Kodairaschen Satzes ab, daß kompakte Kählermannigfaltigkeiten mit Hodgescher Metrik projektiv-algebraisch sind.

Durch Ausschöpfungsmethoden gewinnt man aus den Sätzen über streng pseudokonvexe Gebiete folgende Charakterisierung Steinscher Räume (Narasimhan [22]): Ein komplexer Raum X ist genau dann Steinsch, wenn es eine streng plurisubharmonische Funktion $\varphi : X \to R$ gibt, so daß die Mengen $\{x \in X : \varphi(x) < c\}$ für alle $c \in \mathbb{R}$ relativ-kompakt in X liegen.

Das Studium der streng pseudokonvexen Gebiete im $\mathbb{C}^n$ hat wieder einen enormen Auftrieb erhalten nach dem Beweis folgenden Satzes von Fefferman [9]: Seien G, G' zwei beschränkte streng pseudokonvexe Gebiete im $\mathbb{C}^n$ mit $\mathscr{C}^\infty$-glattem Rand. Dann läßt sich jede biholomorphe Abbildung $f : G \to G'$ zu einer $\mathscr{C}^\infty$-Abbildung $\bar{G} \to \bar{G}'$ fortsetzen. Ein vereinfachter Beweis wurde von Bell-Ligocka [3] gegeben; siehe dazu auch Diederich-Lieb [8].

Wir haben schon den Satz von Oka erwähnt, daß die Hindernisse gegen die Lösbarkeit des Cousin-II-Problems in Holomorphiegebieten rein topologischer Natur sind. Allgemeiner bezeichnet man als *Okasches Prinzip* die Gültigkeit von Sätzen, die es erlauben, analytische Probleme auf Steinschen Räumen auf topologische Probleme zurückzuführen. Grauert [13], [14] bewies ein Okasches Prinzip für Faserbündel. U.a. gilt: Sei G eine komplexe Liegruppe und X ein Steinscher Raum. Dann besteht eine bijektive Beziehung zwischen topologischen und analytischen Isomorphieklassen von G-Prinzipalfaserbündeln auf X. Die Sätze von Grauert wurden von Forster-Ramspott [10], [11] verallgemeinert. Als Anwendung davon erhält man Aussagen über vollständige Durchschnitte, z.B.: In einer n-dimensionalen Steinschen Mannigfaltigkeit X sei Y eine Untermannigfaltigkeit der Dimension $< n/2$. Dann ist Y idealtheoretischer vollständiger Durchschnitt in X genau dann, wenn das Normalenbündel von Y trivial ist. Sätze über mengentheoretische vollständige Durchschnitte in

Steinschen Mannigfaltigkeiten wurden von Banica-Forster [2] und Schneider [28] bewiesen.

Bei der Anwendung des Okaschen Prinzips benützt man, daß Steinsche Mannigfaltigkeiten und Räume besondere topologische Eigenschaften haben. Nach Andreotti-Frankel [1] hat eine (komplex) n-dimensionale Steinsche Mannigfaltigkeit X den Homotopietyp eines (reell) n-dimensionalen CW-Komplexes, insbesondere verschwinden die Gruppen $H^q(X, \mathbb{Z})$ für $q > n$ und $H_n(X, \mathbb{Z})$ ist frei. Von Narasimhan [23], L. Kaup [20] und Hamm [18] wurden diese Aussagen schließlich auf Steinsche Räume mit Singularitäten verallgemeinert.

Literatur

[1]   Andreotti A. and T. Frankel, *The Lefschetz theorem on hyperplane sections*. Ann. of Math. **69** (1959), 713—717.

[2]   Banica, C und O. Forster, *Complete intersections in Stein manifolds*. Manuscr. Math. **37** (1982), 343—356.

[3]   Bell, S. and E. Ligocka, *A simplification and extension of Fefferman's theorem on biholomorphic mappings*. Invent. Math. **57** (1980), 283—289.

[4]   Bremermann, H. J., *Über die Äquivalenz der pseudokonvexen Gebiete und der Holomorphiegebiete im Raum von n komplexen Veränderlichen*. Math. Ann. **128** (1954), 63—91.

[5]   Cartan, H., *Faisceaux analytiques sur les variétés de Stein: démonstration des théorèmes fondamentaux*. Séminaire H. Cartan 1951/52, exposé 19.

[6]   Cartan, H. und P. Thullen, *Zur Theorie der Singularitäten der Funktionen mehrerer komplexer Veränderlicher, Regularitäts- und Konvergenzbereiche*. Math. Ann. **106** (1932), 617—647.

[7]   Cousin, P., *Sur les fonctions de n variables complexes*. Acta Math. **19** (1895), 1—62.

[8]   Diederich, K. und I. Lieb, *Konvexität in der Komplexen Analysis*. DMV-Seminar **2**, Birkhäuser 1981.

[9]   Fefferman, C., *The Bergman kernel and biholomorphic mappings of pseudoconvex domains*. Invent. Math. **26** (1974), 1—65.

[10]   Forster, O. und K. J. Ramspott, *Okasche Paare von Garben nicht-abelscher Gruppen*. Invent. Math. **1** (1966), 260—286.

[11]   Forster, O. und K. J. Ramspott, *Analytische Modulgarben und Endromisbündel*. Invent. Math. **2** (1966), 145—170.

[12]   Grauert, H., *Charakterisierung der holomorph-vollständigen Räume*. Math. Ann. **129** (1955), 233—259.

[13]   Grauert, H., *Holomorphe Funktionen mit Werten in komplexen Lieschen Gruppen*. Math. Ann. **133** (1957), 450—472.

[14]   Grauert, H., *Analytische Faserungen über holomorph-vollständigen Räumen*. Math. Ann. **135** (1958), 263—273.

[15]   Grauert, H., *On Levi's problem and the imbedding real-analytic manifolds*. Ann. Math. **68** (1958), 460—472.

[16]   Grauert, H., *Die Bedeutung des Levischen Problems für die analytische und algebraische Geometrie*. Proc. Int. Congr. Math. Stockholm 1962, pp. 86—101.

[17]   Grauert, H. und R. Remmert, *Theorie der Steinschen Räume*. Springer 1977.

[18]   Hamm, H., *Zum Homotopietyp Steinscher Räume*. Journ. f. d. r. u. a. Math. **338** (1983), 121—135.

[19]   Hartogs, F., *Einige Folgerungen aus der Cauchyschen Integralformel bei Funktionen mehrerer Veränderlichen*. Sb. Bayer. Akad. Wiss. München **36** (1906), 223—242.

[20]   Kaup, L., *Eine topologische Eigenschaft Steinscher Räume*. Nachr. Akad. Wiss. Göttingen, Math.-Nat. Kl. Jg. 1966, pp. 213—224.

[21]   Levi, E. E., *Studii sui punti singolari essenziali delle funzioni analitiche di due o più variabili complessi*. Annali Mat. pur. appl. **17** (1910), 61—87.

[22]   Narasimhan, R., *The Levi problem for complex spaces I, II*. Math. Ann. **142** (1961), 355—365; **146** (1962), 195—216.

[23]   Narasimhan, R., *On the homology of Stein spaces*. Invent. Math. **2** (1967), 377—385.

[24]   Norguet, F., *Sur les domains d'holomorphie des fonctions univalentes de plusieurs variables complexes*. Bull. Soc. Math. France **82** (1954), 137—159.

[25]   Oka, K., *Sur les fonctions analytiques de plusieurs variables*. Iwanami Shoten Tokyo 1961, 1983. Sammlung der funktionentheoretischen Arbeiten Okas, darunter:

[26]   VI. *Domaines pseudoconvexes*. Tohoku Math. J. **49** (1942), 15—52.

[27]   IX. *Domaines finis sans point critique intérieur*. Japan. Journ. of Math. **23** (1953), 97—155.

[28]   Schneider, M., *Vollständige, fastvollständige und mengentheoretisch-vollständige Durchschnitte in Steinschen Mannigfaltigkeiten*. Math. Ann. **260** (1982), 151—174.

[29]   Stein, K., *Topologische Bedingungen für die Existenz analytischer Funktionen komplexer Veränderlichen zu vorgegebenen Nullstellenflächen*. Math. Ann. **117** (1941), 727—757.

[30]   Stein, K., *Analytische Funktionen mehrerer komplexer Veränderlichen zu vorgegebenen Periodizitätsmoduln und das zweite Cousinsche Problem*. Math. Ann. **123** (1951), 201—222.

3   Modifikationen komplexer Räume

Im April 1949 hielt Heinz Hopf in Oberwolfach einen Vortrag mit dem Titel „Komplex-analytische Mannigfaltigkeiten", der bei den Zuhörern, zu denen auch der Ältere der beiden Referenten gehörte, einen sehr starken Eindruck hinterließ. Hopf beschrieb in seinem Vortrag unter anderem ein Verfahren, von ihm σ-*Prozeß* genannt, bei welchem ein Punkt x einer zweidimensionalen komplexen Mannigfaltigkeit X durch eine Riemannsche Zahlensphäre σ so ersetzt wird, daß wieder eine komplexe Manjnfaltigkeit entsteht; dabei wurde σ durch das Büschel der komplexen Linienelemente in x repräsentiert. Zwar handelt es sich hier um eine lokale Version einer in der algebraischen Geometrie seit langem (spätestens seit einer Arbeit von Castelnuovo [4] aus dem Jahre 1893) bekannten Operation. Neu und wichtig war aber, daß der σ-Prozeß in der Funktionentheorie mit Nutzen verwendet werden kann, worauf von Hopf nachdrücklich hingewiesen wurde. In einer Note in den Rendiconti di Matematica e delle sue applicazioni [19] (1951) hat Hopf, nach einem Vortrag in Rom, den σ-Prozeß nochmals ausführlich dargestellt.

Der Hopfsche Vortrag und die Note [19] stehen am Anfang einer Entwicklung in der komplexen Analysis, an der zahlreiche Autoren mitgewirkt haben.

Angeregt durch den Vortrag von Hopf wurde von Behnke-Stein [5] — im Zusammenhang mit Überlegungen zur Kompaktifizierung komplexer Mannigfaltigkeiten — der Begriff des σ-Prozesses erweitert zum Begriff der *Modifikation*. Seien X, Y komplexe Räume und M, N nirgendsdichte abgeschlossene Teilmengen von X bzw. Y. Es sei $f\colon Y \setminus N \to X \setminus M$ eine biholomorphe Abbildung mit folgender Eigenschaft: Strebt ein Punkt $y \in Y \setminus N$ gegen N, so strebt

$f(y)$ gegen M (dies bedeutet: Zu jeder Umgebung $U(M)$ in X und zu jedem Punkt $y \in N$ gibt es eine Umgebung $V(y)$ in Y, derart, daß $V(y) \setminus (V(y) \cap N)$ vermöge f in $U(M) \setminus M$ abgebildet wird). Dann heißt das Quintupel $\mathfrak{M} = (Y, N, f, X, M)$ eine *Modifikation von X in M*.*)

Im folgenden werden der Einfachheit halber die vorkommenden komplexen Räume immer als reduziert vorausgesetzt, wenn nicht ausdrücklich anderes vermerkt ist.

Die Modifikation $\mathfrak{M} = (Y, N, f, X, M)$ heißt *stetig* bzw. *eigentlich* bzw. *holomorph* bzw. *meromorph*, wenn f zu einer stetigen bzw. eigentlichen bzw. holomorphen bzw. meromorphen Abbildung $\varphi : Y \to X$ fortsetzbar ist; φ heißt dann *Modifikationsabbildung*. Der Begriff der meromorphen Abbildung kann sowohl im Sinne einer Definition von Stoll [32, II] (S-Meromorphie) wie auch im Sinne einer solchen von Remmert [31] (R-Meromorphie) genommen werden (die beiden Definitionen sind nicht äquivalent — auch nicht für den Fall kompakter Bildräume — wie Hirschowitz [11] gezeigt hat). Zu den genannten Arten von Modifikationen siehe Stoll [32, I], Kreyszig [25], Grauert-Remmert [9], (vgl. auch Remmert [30]), Aeppli [1]. In [1] werden insbesondere topologische Aspekte zum Begriff der Modifikation behandelt.

Von Grauert-Remmert wurde in [9] u. a. bewiesen: Seien X eine komplexe Mannigfaltigkeit, Y ein normaler komplexer Raum, M und N analytische Mengen in X bzw. Y, $\mathfrak{M} = (Y, N, f, X, M)$ eine stetige Modifikation. $\mathfrak{M}$ sei (in einem naheliegenden Sinne) „wesentlich". Dann ist N entweder leer oder rein 1-codimensional in Y. — Weitere Beweise dieses Satzes sind von Kerner [23] und Kuhlmann [28] angegeben worden. Der Beweis von Kuhlmann ist algebraischer Natur: Es wird ein Satz über vollständige lokale Ringe hergeleitet, aus dem der Satz von Grauert-Remmert folgt.

Eigentliche holomorphe Modifikationen haben die wichtige Eigenschaft, daß durch sie die Ringe der meromorphen Funktionen auf den beteiligten komplexen Räumen jeweils isomorph aufeinander bezogen werden.

Im Zusammenhang mit R-meromorphen Abbildungen spielen eigentliche holomorphe Modifikationen eine besondere Rolle. Eine R-meromorphe Abbildung F des komplexen Raumes Z_1 in den komplexen Raum Z_2 ist eine mengenwertige Abbildung (Korrespondenz) von Z_1 in Z_2 mit folgenden Eigenschaften: 1) Der Graph $G_F \subset Z_1 \times Z_2$ von F ist eine analytische Menge in $Z_1 \times Z_2$; 2) die Projektion $p_1 : G_F \to Z_1$ des komplexen Unterraumes G_F von $Z_1 \times Z_2$ in Z_1 ist eine eigentliche holomorphe Modifikationsabbildung. G_F entsteht also aus Z_1 durch eine eigentliche holomorphe Modifikation.

---

*) Die Bezeichnung „Modifikation" wurde den Autoren bei einer Diskussion über einen passenden Namen für den Übergang von X zu Y von Frau Elisabeth Behnke vorgeschlagen. — Es sei betont, daß die obige Definition auch für nichtreduzierte komplexe Räume sinnvoll ist. In [5] wurden die komplexen Räume als sogenannte Riemannsche Gebiete (d.h. in heutiger Terminologie als normale komplexe Räume) vorausgesetzt, und die Definition war etwas spezieller als oben gefaßt.

Sei N eine nichtleere kompakte analytische Menge in dem komplexen Raum Y, die in Y nirgends dicht und nirgends diskret ist. N heißt *exzeptionell in* Y (vgl. [7, 8, 22]), wenn es eine eigentliche holomorphe Modifikation $\mathfrak{M} = (Y, N, f, X, M)$ mit der Modifikationsabbildung $\varphi: Y \to X$ gibt, derart, daß M diskret und $\varphi^{-1}(M) = N$ ist und daß außerdem gilt: Zu jeder Umgebung U von M in X und jeder in $V := \varphi^{-1}(U)$ holomorphen Funktion g gibt es eine in U holomorphe Funktion f, so daß $g = f \circ \varphi$ ist. Von Grauert [7] wurden Bedingungen für die Exzeptionalität von N angegeben: Notwendig und hinreichend hierfür ist, daß es eine offene und relativ kompakte Umgebung W von N in Y gibt, die streng pseudokonvex ist, derart, daß N m a x i m a l e kompakte analytische Teilmenge von W ist (dies bedeutet, daß jede nirgendsdiskrete kompakte analytische Teilmenge von W in N enthalten ist). Daraus folgt das (allerdings nur hinreichende) Kriterium: N ist sicher dann exzeptionell, wenn es eine kohärente analytische Garbe $\mathfrak{m}$ von Keimen holomorpher Funktionen mit N als Nullstellengebilde gibt, derart, daß das $\mathfrak{m}$ zugeordnete verallgemeinerte Normalenbündel „schwach negativ" ist. Eine weitere Folgerung ergibt sich für den Fall, daß Y eine zweidimensionale komplexe Mannigfaltigkeit ist: Seien $N_1, \ldots, N_k$ die irreduziblen Komponenten von N; sei jeweils $c_{\nu\mu} = N_\nu \cdot N_\mu$ die Schnittzahl von N_ν und $N_\mu (\nu, \mu = 1, \ldots, k)$. Dann ist notwendig und hinreichend für die Exzeptionalität von N, daß die Matrix $(c_{\nu\mu})$ negativ definit ist (insbesondere sind unter dieser Bedingung alle Selbstschnittzahlen $c_{\nu\nu}$ der N_ν negativ). Erfüllt N diese Bedingung und ist (Y, N, f, X, M) eine zugehörige eigentliche Modifikation mit der Modifikationsabbildung $\varphi: Y \to X$, so ist X ein normaler komplexer Raum. Die Punkte von M sind dann nicht notwendig gewöhnliche Punkte von X; dies ist jedoch nach Mumford [29] (siehe auch Hirzebruch [18]) der Fall, wenn die irreduziblen Komponenten von N in bestimmter Weise singularitätenfrei in Y eingebettete Riemannsche Zahlensphären sind.

Die Frage, wann eine analytische Menge N in einem komplexen Raum Y vermöge einer eigentlichen holomorphen Modifikationsabbildung in eine niederdimensionale analytische Menge, die nicht notwendig diskret ist, übergeführt werden kann, wurde u. a. von Knorr-Schneider behandelt. Siehe hierzu [24]; dort wird die oben wiedergegebene notwendige und hinreichende Bedingung von Grauert auf „relativexzeptionelle" analytische Mengen verallgemeinert.

Spezielle eigentliche holomorphe Modifikationen sind die σ-*Modifikationen*. Seien wieder $\mathfrak{M} = (Y, N, f, X, M)$ eine eigentliche holomorphe Modifikation und $\varphi: Y \to X$ die zugehörige Modifikationsabbildung; M sei eine nichtleere analytische Menge in X, es gelte $\varphi^{-1}(M) = N$. φ heißt *monoidale Transformation von X mit dem Zentrum M*, wenn die folgenden beiden Bedingungen erfüllt sind: 1) Ist $\mathscr{J}$ die Idealgarbe von M, so ist die analytische Urbildgabe $\varphi^* \mathscr{J}$ inversibel, d.h. $\varphi^* \mathscr{J}$ wird in Y lokal überall durch jeweils einen holomorphen Funktionskeim, der kein Nullteiler ist, erzeugt (das Nullstellengebilde von $\varphi^* \mathscr{J}$ ist dann N, und es gilt $\dim_y N = \dim_y Y - 1$ für alle $y \in N$); 2) ist Z ein komplexer Raum, der nicht notwendig reduziert ist, und ist $\psi: Z \to X$ eine holomorphe Abbildung, derart, daß die Garbe $\psi^* \mathscr{J}$ inversibel ist, so gibt es stets genau eine holomorphe Abbildung $\Psi: Z \to Y$ mit $\psi = \varphi \circ \psi$ (siehe hierzu Hironaka-Rossi [16], vgl. auch Ancona-Tomassini [3]). Ist φ eine solche monoidale Transformation, so wird $\mathfrak{M}$ auch als σ-*Modifikation (Éclatement)* von X mit dem Zentrum M und der Übergang von M zu $\varphi^{-1}(M)$ als *Aufblasen (Blowing up)* von M bezeichnet; ist $M = \{x_0\}$ einpunktig und x_0 ein gewöhnlicher Punkt von X, so heißt φ auch eine *quadratische Transformation von X*.

Die Namen „quadratische Transformation" und „monoidale Transformation" stammen aus der klassischen algebraischen Geometrie, wo sie bei speziellen birationalen Abbildungen auftraten.

1) Man betrachte die quadratische Cremona-Transformation $(z_0 : z_1 : z_2) \mapsto (z_1 z_2 : z_0 z_2 : z_0 z_1)$ der projektiven Ebene $\mathbb{P}_2$. Diesen birationalen Automorphismus von $\mathbb{P}_2$ kann man wie folgt zerlegen: Zunächst wird in den drei Punkten $(1;0:0)$, $(0:1:0)$ und $(0:0:1)$ der σ-Prozeß durchgeführt; anschließend werden die strikt Transformierten der drei Geraden $z_0 = 0$, $z_1 = 0$ und $z_2 = 0$ mit der Umkehrung des σ-Prozesses jeweils wieder zu einem Punkt zusammengeblasen.

2) In der klassischen algebraischen Geometrie verstand man unter einer monoidalen Fläche (oder kurz Monoid) eine Fläche $F \subset \mathbb{P}_3$ vom Grad d, die einen singulären Punkt $P \in F$ der Multiplizität $d-1$ besitzt. Eine allgemeine Gerade durch P schneidet dann F noch in einem einzigen weiteren Punkt. Deshalb liefert die Projektion mit Zentrum P auf eine nicht durch P laufende Ebene eine birationale Abbildung $F \to \mathbb{P}_2$. Bei dieser wird der Punkt P zu einer Kurve vom Grad $d-1$ aufgeblasen; außerdem werden $d(d-1)$ Geraden auf F (mit Vielfachheit zu zählen) jeweils zu einem Punkt zusammengeblasen.

Es gilt folgende allgemeine Existenzaussage für σ-Modifikationen ([16], vgl. auch Kuhlmann [26]): Ist M eine nirgends dichte analytische Menge im komplexen Raum X, so existiert eine σ-Modifikation $\mathfrak{M} = (Y, N, f, X, M)$, und diese ist bis auf Äquivalenz eindeutig bestimmt. Für den Fall, daß X eine zweidimensionale komplexe Mannigfaltigkeit und M einpunktig ist, ergibt sich (bis auf Äquivalenz) der Hopfsche σ-Prozeß.

In manchen Fällen lassen sich eigentliche holomorphe Modifikationen in σ-Modifikationen „zerlegen", d. h. sie können durch Hintereinanderschalten geeigneter σ-Modifikationen gewonnen werden. So gilt nach Hopf [19, 20]: Seien X, Y zweidimensionale komplexe Mannigfaltigkeiten, sei $\mathfrak{M} = (Y, N, f, X, \{x_0\})$ eine eigentliche holomorphe Modifikation mit der Modifikationsabbildung $\varphi: Y \to X$; die analytische Menge $N = \varphi^{-1}(x_0)$ sei eindimensional. Dann ist $\mathfrak{M}$ bis auf Äquivalenz auf genau eine Art in σ-Prozesse zerlegbar. In der algebraischen Geometrie entspricht dieser Sachverhalt einer Aussage von Zariski [33] (dort Abschnitt 24, „Lemma").

σ-Modifikationen stellen ein wesentliches Hilfsmittel zur Auflösung der Singularitäten komplexer Räume dar. Im Falle normaler komplexer Räume der Dimension 2 wurde die Auflösung der Singularitäten von Hirzebruch [17] — im

Anschluß an eine Arbeit von Jung [21] aus dem Jahre 1908 über die lokale Darstellung einer algebraischen Funktion von zwei Veränderlichen — bewerkstelligt. In der algebraischen Geometrie ist die Möglichkeit der Singularitätenauflösung für komplexe zweidimensionale projektiv algebraische Varietäten seit langem bekannt. Für algebraische Varietäten beliebiger Dimension über Körpern der Charakteristik Null wurde das Desingularisationsproblem von Hironaka in seiner berühmten Arbeit [12] im positiven Sinne gelöst; die Auflösbarkeit der Singularitäten von im Unendlichen abzählbaren komplexen Räumen wurde ebenfalls von Hironaka [14] (vgl. auch [13, 15]) nachgewiesen. Für den Fall komplexer Räume der Dimension 3 siehe auch Kuhlmann [27]. Für dreidimensionale algebraische Varietäten war die Auflösung der Singularitäten schon von Zariski [33] (1944) geleistet worden.

Zu neueren Entwicklungen in der Theorie der Modifikationen siehe [3].

Literatur

[1]   Aeppli, A, *Modifikationen von reellen und komplexen Mannigfaltigkeiten*. Comm. Math. Helv. **31** (1956), 219—301.

[2]   Aeppli, A., *Reguläre Modifikation komplexer Mannigfaltigkeiten, regulär verzweigte Überlagerungen*. Comm. Math. Helv. **33** (1959), 1—22.

[3]   Ancona, V. and G. Tomassini, *Modifications analytiques*. Lect. Notes in Math. **943,** Springer-Verlag 1982.

[4]   Castelnuovo, G., *Sui multipli di una serie lineare di gruppi di punti appartenente ad una curva algebraica*. Rend. Circ. Palermo **7** (1893).

[5]   Behnke, H. und K. Stein, *Modifikationen komplexer Mannigfaltigkeiten und Riemannscher Gebiete*. Math. Ann. **124** (1951), 1—16.

[6]   Fischer, G., *Complex analytic geometry*. Lect. Notes in Math. **538,** Springer-Verlag 1976.

[7]   Grauert, H., *Über Modifikationen und exzeptionelle analytische Mengen*. Math. Ann. **146** (1962), 331—368.

[8]   Grauert, H., *Die Bedeutung des Levischen Problems für die analytische und algebraische Geometrie*. Proc. of the Int. Congr. of Math., Stockholm 1962, 86—101.

[9]   Grauert, H. und R. Remmert, *Zur Theorie der Modifikationen I*. Stetige und eigentliche Modifikationen komplexer Räume. Math. Ann. **129** (1955), 274—296.

[10] Hartshorne, R., *Algebraic geometry*. Grad. Texts in Math. **52,** New York–Heidelberg–Berlin 1977.

[11] Hirschowitz, A., *Les deux types de méromorphie différent*. Journ. f. d. r. u. ang. Math. **313** (1980), 157—160.

[12] Hironaka, H., *Resolution of singularities of an algebraic variety over a field of characteristic zero*. Ann. of Math. **79** (1964), I: 109—203, II: 205—326.

[13] Hironaka, H., *On resolution of singularities (Characteristic zero)*. Proc. of the Int. Congr. of Math., Stockholm 1962, 507—521.

[14] Hironaka, H., *Bimeromorphic smoothing of a complex-analytic space*. 1971 (hektographiert) 111 S.

[15] Hironaka, H., *Introduction to the theory of infinitely near singular points*. Memorias de Matematica del Instituto „Jorge Juan" **28,** Madrid 1974.

[16] Hironaka, H. and H. Rossi, *On the equivalence of imbeddings of exceptional complex spaces*. Math. Ann. **156** (1964), 313—333.

[17]   Hirzebruch, F., *Über vierdimensionale Riemannsche Flächen mehrdeutiger analytischer Funktionen von zwei komplexen Veränderlichen*. Math. Ann. **126** (1953), 1—22.

[18]   Hirzebruch, F., *The topology of normal singularities of an algebraic surface (d'après un article de D. Mumford)*. Sém. Bourbaki **15** (1962/63), N° 250.

[19]   Hopf, H., *Über komplex-analytische Mannigfaltigkeiten*. Rend. di Mat. e delle sue appl. Ser. V, **10** (1951), 169—182.

[20]   Hopf, H., *Schlichte Abbildungen und lokale Modifikationen vierdimensionaler komplexer Mannigfaltigkeiten*. Comm. Math. Helv. **29** (1955), 132—156.

[21]   Jung, H. W. E., *Darstellung der Funktionen eines algebraischen Körpers zweier unabhängiger Veränderlichen x, y in der Umgebung einer Stelle x = a, y = b*. Journ. f. d. r. u. ang. Math. **133** (1908), 289—314.

[22]   Kaup, B., *Über Kokerne und Pushouts in der Kategorie der komplex-analytischen Räume*. Math. Ann. **189** (1970), 60—76.

[23]   Kerner, H., *Bemerkung zu einem Satz von H. Grauert und R. Remmert*. Math. Ann. **157** (1964), 206—209.

[24]   Knorr, K. und M. Schneider, *Relativ-exzeptionelle analytische Mengen*. Math. Ann. **193** (1971), 238—254.

[25]   Kreyszig, E., *Stetige Modifikationen komplexer Mannigfaltigkeiten*. Math. Ann. **128** (1955), 479—492.

[26]   Kuhlmann, N., *Projektive Modifikationen komplexer Räume*. Math. Ann. **139** (1960), 217—238.

[27]   Kuhlmann, N., *Über die Auflösung der Singularitäten dreidimensionaler komplexer Räume*. I: Math. Ann. **151** (1963), II: Math. Ann. **154** (1964), 387—405.

[28]   Kuhlmann, N., *Über die Reinheit von Entartungs- und Verzweigungsmengen*. Math. Ann. **178** (1968), 24—43.

[29]   Mumford, D., *The topology of normal singularities of an algebraic surface and a criterion for simplicity*. Inst. des Hautes Etudes Scientifiques, Publ. math. **9** (1961), 229—246.

[30]   Remmert, R., *Über stetige und eigentliche Modifikationen komplexer Räume*. Coll. de Topologie de Strasburg 1954, 1—17.

[31]   Remmert, R., *Holomorphe und meromorphe Abbildungen komplexer Räume*. Math. Ann. **133** (1956), 328—370.

[32]   Stoll, W., *Über meromorphe Modifikationen*. I, II: Math. Z. **61** (1954), 206—324 und 467—488. III: Math. Z. **62** (1955), 189—210. IV, V: Math. Ann. **130** (1955), 147—182 und 272—316.

[33]   Zariski, O., *Reduction of singularities of algebraic threedimensional varieties*. Ann. of Math. **45** (1944), 472—542.

4   Modulprobleme, Deformationen

Schon Riemann fragte sich, von wieviel Moduln (= Parametern) eine Riemannsche Fläche vom Geschlecht g abhängt, und zeigte, daß diese Anzahl für $g \geqq 2$ gleich $3g - 3$ ist. Diese Aussage wurde später durch Arbeiten von Teichmüller (Abschn. 1, Lit. [35]), Ahlfors, Bers, Rauch, Grothendieck präzisiert, und man gelangte schließlich (s. [12]) zur Konstruktion einer Familie $\pi : \mathfrak{X} \to S$ aller Riemannscher Flächen vom Geschlecht g. Dabei sind $\mathfrak{X}$ und S komplexe Mannigfaltigkeiten, π eine reguläre holomorphe Abbildung und die Fasern $\pi^{-1}(s)$, $s \in S$, durchlaufen alle kompakten Riemannschen Flächen vom Geschlecht g mit ausgezeichneter Basis der 1. ganzzahligen Homologiegruppe. Die Dimension von S ist gleich $3g - 3$.

Allgemeiner ist beim Modulproblem eine gewisse Klasse von analytischen Objekten gegeben, die in (zu präzisierender) natürlicher Weise mit der
Struktur eines komplexen Raums versehen werden soll. Ein wichtiges Beispiel ist
das folgende: Gegeben ist ein komplexer Raum X; es soll die Menge aller
kompakten analytischen Teilräume von X mit der Struktur eines komplexen
Raumes versehen werden. In der algebraischen Geometrie existiert das analoge,
u.a. von Hilbert und Chow behandelte Problem, die algebraischen Teilmengen
des $\mathbb{P}_n$ zu parametrisieren. In verallgemeinerter, moderner Form wurde dies
Problem von Grothendieck [13] gelöst.

Das Modulproblem für die kompakten analytischen Unterräume eines
komplexen Raums wurde von Douady [4] gelöst, der darüber in Oberwolfach
1965 vortrug. Um zu präzisieren, was die natürliche komplexe Struktur auf der
Menge der Unterräume ist, wird das Problem analog zum Vorgehen von Grothendieck als universelles Problem formuliert: Zu dem vorgegebenen komplexen Raum X sucht man einen komplexen Raum H und einen analytischen
Unterraum $Z \hookrightarrow X \times H$, der eigentlich und platt über H bzgl. der Projektion
$X \times H \to H$ liegt, so daß folgendes gilt: Ist S ein komplexer Raum und $Y \subset X \times S$
ein analytischer Unterraum, der eigentlich und platt über S liegt, so gibt es genau
eine holomorphe Abbildung $\alpha: S \to H$, so daß Y das Urbild von Z bei der
Abbildung $id_X \times \alpha : X \times S \to X \times H$ ist. Man überlegt sich leicht, daß die Lösung
$Z \hookrightarrow X \times H$ des Problems im Falle der Existenz eindeutig bestimmt ist. Die
Punkte von H stehen in bijektiver Beziehung zu den algebraischen Unterräumen
von X; dem Punkt $h \in H$ entspricht dabei der Unterraum $Z_h := Z \cap (X$
$\times \{h\}) \subset X \times \{h\} \cong X$. Zur Lösung seines Problems entwickelte Douady die
Theorie der Banach-analytischen Räume; obwohl der am Ende konstruierte
Raum H ein lokal endlich dimensionaler komplexer Raum im klassischen Sinn
ist, braucht man im Verlaufe des Beweises notwendig unendlich-dimensionale
Räume. (Die Arbeit von Douady gab auch den Anstoß zu einem systematischen
Studium der unendlich-dimensionalen Funktionentheorie).

Verwandt mit dem Modulproblem ist die von Kodaira-Spencer [15]
initiierte Deformationstheorie komplexer Strukturen. Ein Problem ist dabei
folgendes: Man geht aus von einer kompakten komplexen Mannigfaltigkeit X
und möchte alle komplexen Strukturen auf der X unterliegenden differenzierbaren
Mannigfaltigkeit parametrisieren, die in der „Nähe" der gegebenen komplexen
Struktur liegen. Unter Heranziehung der Theorie der fast-komplexen Strukturen und Ergebnissen aus der Theorie elliptischer Differentialoperatoren auf
kompakten Mannigfaltigkeiten bewiesen Kodaira-Nirenberg-Spencer [16] folgendes Resultat: Ist $H^2(X, \Theta) = 0$, wobei Θ die Garbe der holomorphen Vek-
torfelder auf X bezeichnet, so existiert eine holomorphe Familie $\pi: \mathfrak{X} \to S$ von
kompakten komplexen Mannigfaltigkeiten $X_s = \pi^{-1}(s)$ über einer Nullumgebung S im $\mathbb{C}^m$, $m = \dim H^1(X, \Theta)$, so daß X_0 isomorph zur gegebenen Mannigfaltigkeit X ist und die übrigen X_s alle (in einem zu präzisierenden Sinn) benach-
barten komplexen Strukturen durchlaufen, d.h. die Deformation $\mathfrak{X} \to S$ vollständig ist. Kuranishi [17] verallgemeinerte das Resultat auf den Fall, daß

$H^2(X, \Theta) \neq 0$. In diesem Fall ist die Basis S i.a. nur eine analytische Menge in einer Nullumgebung des $\mathbb{C}^m$.

Die Methoden aus der Theorie der fastkomplexen Strukturen und Potentialtheorie auf Mannigfaltigkeiten sind nicht mehr anwendbar, wenn man zu komplexen Räumen mit Singularitäten übergeht. Die ersten theoretischen Grundlagen dafür wurden von Grauert-Kerner [11] gelegt. Verselle (= vollständige und effektive) Deformationen von Keimen komplexer Räume mit isolierten Singularitäten wurden von Tjurina [21] und Grauert [9] konstruiert. Die Konstruktion verseller Deformationen kompakter komplexer Räume ist noch komplizierter und wurde mit verschiedenen Methoden von Douady [5], Grauert [10], Palamodov [20], Forster-Knorr [8] gelöst.

Ein anderes Deformationsproblem komplexer Strukturen, das auch schon von Kodaira-Spencer betrachtet wurde, ist das der holomorphen Vektorbündel auf einer vorgegebenen kompakten komplexen Mannigfaltigkeit (oder kompakten komplexen Raum). Nach Vorarbeiten von W. Fischer [6] wurde von Forster-Knorr [7] die Existenz der versellen Deformation eines holomorphen Vektorbündels E auf einem kompakten komplexen Raum X bewiesen. Beim Beweis wird die Tatsache benutzt, daß man die Vollständigkeit einer Deformation nur im formalen Sinne nachweisen muß (Schuster [22]). Für dieses Deformationsproblem ist $H^1(X, \operatorname{End} E)$ der Zariski-Tangentialraum der Basis S der versellen Deformation; falls $H^2(X, \operatorname{End} E) = 0$, ist S glatt.

Bei der versellen Deformation eines Vektorbündels kann folgendes Phänomen auftreten: Sei E etwa das (topologisch triviale) Bündel $\mathcal{O}(k) \oplus \mathcal{O}(-k)$, $(k \geq 1)$, über $\mathbb{P}_1$. Dann ist die Basis der versellen Deformation von E eine Nullumgebung des $\mathbb{C}^{2k-1}$ (d.h. genauer der Keim des $\mathbb{C}^{2k-1}$ im Nullpunkt), trotzdem treten in der versellen Deformation nur endlich viele Isomorphieklassen von Bündeln auf, nämlich die Bündel $\mathcal{O}(k') \oplus \mathcal{O}(-k')$, $0 \leq k' \leq k$. Will man deshalb die Menge aller Isomorphieklassen von analytischen Strukturen auf einem vorgegebenen topologischen Vektorbündel in natürlicher Weise mit der Struktur eines komplexen Raumes versehen, so muß man sich auf sog. *stabile* Bündel beschränken, bei denen solche Erscheinungen nicht auftreten können. Dieser Begriff ist auf projektiv-algebraischen Varietäten definiert, auf denen ein ampler Divisor ausgezeichnet ist. (Nach einem Satz von Serre fallen auf einer projektiv-algebraischen Varietät über dem Körper $\mathbb{C}$ die Begriffe holomorphes Vektorbündel und algebraische Vektorbündel zusammen.) Maruyama [18] bewies die Existenz des Modulraums der Vektorbündel zu vorgegebenem Hilbert-Polynom auf einer projektiv-algebraischen Varietät. Der allgemeine Existenzsatz sagt jedoch nichts darüber aus, wie der Modulraum im einzelnen aussieht, ja nicht einmal, ob er nicht-leer ist. Dies erfordert jedesmal eingehendere Untersuchungen. So bewies Barth [2], daß der Modulraum der stabilen Vektorbündel vom Rang 2 auf dem $\mathbb{P}_2$ mit vorgegebenen Chernklassen $c_1 \equiv 0(2)$ und c_2 eine irreduzible quasi-projektive rationale Mannigfaltigkeit der Dimension $4c_2 - c_1^2 - 3$ ist. (Ist diese Zahl negativ, so ist der Modulraum leer.) Hulek [14] bewies das analoge Resultat für ungerade 1. Chernklasse.

Die Theorie der Vektorbündel auf dem $\mathbb{P}_n$ wurde in den letzten zehn Jahren intensiv erforscht (siehe dazu Okonek-Schneider-Spindler [19]); es sind aber noch längst nicht alle Probleme gelöst. Eines der ungelösten Probleme ist die Frage, ob es auf dem $\mathbb{P}_n$ für großes n holomorphe Vektorbündel vom Rang 2 gibt, die nicht direkte Summen von Geradenbündeln sind. (Dies hängt mit der Frage zusammen, ob alle 2-codimensionalen Untermannigfaltigkeiten des $\mathbb{P}_n$ vollständige Durchschnitte sind.) Barth-Van de Ven [2] zeigten, daß es zumindest keine unendlichen Serien solcher Bündel gibt, genauer, daß jedes holomorphe 2-Bündel auf $\mathbb{P}_n$, das sich auf alle $\mathbb{P}_m$, $m > n$, fortsetzen läßt, notwendig spaltet. Es sei noch erwähnt, daß einer der Anstöße für die intensive Beschäftigung mit holomorphen Vektorbündeln auf dem $\mathbb{P}_n$ der Zusammenhang zwischen gewissen 2-Bündeln auf dem $\mathbb{P}_3$ und Problemen der theoretischen Physik (Yang-Mills-Felder) war, siehe dazu Atiyah [1].

Literatur

[1]   Atiyah, M. F., *Geometry of Yang-Mills fields*. Lezioni Fermiane, Scuola Norm. Sup. Pisa 1979.

[2]   Barth, W., *Moduli of vector bundles on the projective plane*. Invent. Math. **42** (1977), 63—91.

[3]   Barth, W. and A. Van de Ven, *A decomposability criterion for algebraic 2-bundles on projective spaces*. Invent. Math. **25** (1974), 91—106.

[4]   Douady, A., *Le problème des modules pour les sous-espaces analytiques compacts d'un espace analytique donné*. Ann. Inst. Fourier **16** (1966), 1—95.

[5]   Douady, A., *Le problème des modules locaux pour les espaces C-analytiques compactes*. Ann. Sci. ENS, Série IV, **4** (1974), 569—602.

[6]   Fischer, W., *Zur Deformationstheorie komplex-analytischer Faserbündel*. Schriftenreihe Math. Inst. Univ. Münster, Heft 30 (1964).

[7]   Forster, O. und K. Knorr, *Über die Deformation von Vektorraumbündeln auf kompakten komplexen Räumen*. Math. Ann. **209** (1974), 291—346.

[8]   Forster, O. und K. Knorr, *Konstruktion verseller Familien kompakter komplexer Räume*. Lecture Notes in Math. **705,** Springer 1979.

[9]   Grauert, H., *Über die Deformation isolierter Singularitäten analytischer Mengen*. Invent. Math. **15** (1972), 171—198.

[10]   Grauert, H., *Der Satz von Kuranishi für kompakte komplexe Räume*. Invent. Math. **25** (1974), 107—142.

[11]   Grauert, H. und H. Kerner, *Deformation von Singularitäten komplexer Räume*. Math. Ann. **153** (1964), 236—260.

[12]   Grothendieck, A., *Construction de l'espace de Teichmüller*. Séminaire Cartan ENS 13e année (1960/61), exposé 17.

[13]   Grothendieck, A., *Techniques de construction et théorèmes d'existence en géométrie algébrique, IV: Les schémas de Hilbert*. Séminaire Bourbaki **13** (1960/61), exposé 221.

[14]   Hulek, K., *Stable rank-2 vector bundles on $\mathbb{P}_2$ with c_1 odd*. Math. Ann. **242** (1979), 241—266.

[15]   Kodaira, K. and D. C. Spencer, *On deformations of complex analytic structure I, II*. Ann. of Math. **67** (1958), 328—466.

[16]   Kodaira, K., L. Nirenberg and D. C. Spencer, *On the existence of deformations of complex analytic structures*. Ann. of Math. **68** (1958), 450—459.

[17]   Kuranishi, M., *On the locally complete families of complex analytic structures.* Ann. of Math. **75** (1962), 536—577.

[18]   Maruyama, M., *Moduli of stable sheaves I, II.* Math. J. Kyoto Univ. **17** (1977), 91—126, **18** (1978), 557—614.

[19]   Okonek, C., M. Schneider und H. Spindler, *Vector bundles on complex projective spaces.* Progress in Math. **3**, Birkhäuser 1980.

[20]   Palamodov, V. P., *Deformations of complex spaces.* Russian Math. Surveys **31** (1976), 129—197.

[21]   Tjurina, G. N., *Locally semiuniversal flat deformations of isolated singularities of complex spaces.* Mathematics of the USSR — Izvestija **3** (1969), 967—999.

[22]   Schuster, H. W., *Formale Deformationstheorien.* Habilitationsschrift, München 1971.

5   Äquivalenzrelationen in komplexen Räumen

Die in diesem Abschnitt auftretenden komplexen Räume werden stets als reduziert vorausgesetzt.

Sei X eine Menge, $R \subset X \times X$ eine Äquivalenzrelation in X und X/R die aus den Äquivalenzklassen von R bestehende Quotientenmenge. Man hat die Quotientenabbildung $q : X \to X/R$, die jedem Element von X seine Äquivalenzklasse zuordnet. Ist M eine Teilmenge von X, so wird unter der R-saturierten Hülle von M die Menge $M_R := q^{-1}(q(M))$ verstanden; M heißt R-saturiert, wenn $M = M_R$ ist.

Sei nun X insbesondere ein topologischer Raum. X/R wird mit der Quotiententopologie versehen, die als die feinste Topologie in X/R, so daß die Quotientenabbildung q stetig wird, definiert ist. R heißt offen bzw. diskret bzw. eigentlich bzw. semieigentlich, wenn q offen bzw. diskret bzw. eigentlich bzw. semieigentlich ist (das letzte bedeutet: Zu jedem Punkt $y \in X/R$ gibt es eine offene Umgebung $V(y)$ in X/R und eine quasikompakte Menge K_y in X, derart, daß alle Fasern von q über den Punkten von $V(y)$ die Menge K_y treffen; (vgl.[20, 23]). Sind alle Fasern von q zusammenhängend, so heißt R einfach. Zu jeder Äquivalenzrelation R in X gehört eine einfache Äquivalenzrelation N_R, deren Elemente die Zusammenhangskomponenten der Äquivalenzklassen von R sind.

Sei weiterhin $X = (X, \mathcal{O})$ ein (reduzierter) komplexer Raum. Auf dem Quotientenraum X/R ist wie folgt eine Garbe von lokalen $\mathbb{C}$-Algebren, die mit $\mathcal{O}/R$ bezeichnet sei, definiert; für jede offene Menge $U \subset X/R$ besteht die $\mathbb{C}$-Algebra $\Gamma(U, \mathcal{O}/R)$ der Schnitte von $\mathcal{O}/R$ über U aus denjenigen Funktionen $\varphi : U \to \mathbb{C}$, für welche $\varphi \circ q$ zu $\Gamma(q^{-1}(U), \mathcal{O})$ gehört. Das Paar $(X/R, \mathcal{O}/R)$ ist dann ein geringter Raum (er wird wieder mit X/R bezeichnet), und die Quotientenabbildung q wird zu einem Morphismus geringter Räume.

Es stellt sich die Frage, wann $X/R = (X/R, \mathcal{O}/R)$ ein komplexer Raum ist. Notwendig hierfür ist sicher jede der beiden folgenden Bedingungen:

1)   Die Topologie von X/R ist Hausdorffsch;

2) R ist eine analytische Menge in $X \times X$. — Hat R diese Eigenschaft, so wird R als analytische Äquivalenzrelation bezeichnet [11]*).

Die Frage nach hinreichenden Bedingungen ist in der Literatur oft behandelt worden. Wir geben im folgenden einige Resultate dazu wieder.

Sei G eine Gruppe holomorpher Automorphismen von X; G erzeugt die Äquivalenzrelation $R_G := \{(x, g(x)) : x \in X, g \in G\} \subset X \times X$ in X. Dann gilt:

a)   Operiert G eigentlich diskontinuierlich auf X, so ist $X/R_G = (X/R_G, \mathcal{O}/R_G)$ ein komplexer Raum (siehe Baily [2] und Cartan [3, 4]).

b)   Ist G eine komplexe Liesche Transformationsgruppe von X und ist die komplexe Struktur von X maximal (siehe [16], § 72), so ist X/R_G genau dann ein komplexer Raum, wenn die Topologie von X/R_G Hausdorffsch ist; trifft dies zu, so ist die komplexe Struktur von X/R_G ebenfalls maximal (siehe Holmann [11]). — In dieser Situation ist also die obige notwendige Bedingung 1) auch hinreichend.

An weiteren hinreichenden Bedingungen seien genannt:

c) R ist eigentlich, und X/R ist lokal-$\mathcal{O}/R$-separabel (dies bedeutet: Jeder Punkt von X/R hat eine offene Umgebung V, derart, daß die Schnitte von $\mathcal{O}/R$ über V die Punkte von V trennen). — Siehe hierzu Cartan [5] (vgl. auch Wiegmann [32]).

d)   Die komplexe Struktur von X ist maximal , X/R ist lokal kompakt und lokal-$\mathcal{O}/R$-separabel. — X/R besitzt dann ebenfalls maximale komplexe Struktur. Siehe Wiegmann [33] und Furushima [7, 8], vgl. auch Andreotti-Stoll [1].

e)   Die komplexe Struktur von X ist maximal, R ist offen und analytisch. — Auch X/R besitzt dann maximale Struktur. Siehe Kaup [14]. Von Holmann [11] war schon früher die Bedingung unter der zusätzlichen Voraussetzung, daß R diskret ist, als hinreichend nachgewiesen worden.

f) X ist ein normaler komplexer Raum mit abzählbarer Topologie, R ist semieigentlich und „holomorph". — Siehe Grauert [9]. (Die Bedingung „R ist holomorph" besagt, daß R analytisch ist und daß, grob gesprochen, es in R-saturierten Umgebungen der Fasern von $q : X \to X/R$ genügend viele faserkonstante holomorphe Funktionen gibt.) Die obigen von R zu erfüllenden Bedingungen sind auch notwendig, damit X/R ein komplexer Raum ist.

Wir zitieren zwei Anwendungen:

(1)   Sei X holomorph konvex, es sei R die Menge aller Paare $(x, x') \in X \times X$, derart, daß $f(x) = f(x')$ ist für jede in X holomorphe Funktion f. R ist eine eigentliche Äquivalenzrelation in X und X/R ist lokal-$\mathcal{O}/R$-separabel. R erfüllt also die Bedingung in (c); daher ist X/R ein komplexer Raum, und dieser ist

---

*) Analytische Äquivalenzrelationen werden in der Literatur auch anders definiert. In [18] (vgl. auch [6]), wird dieser Begriff weiter, in [29] enger als oben gefaßt. Im folgenden hat die Bezeichnung „analytische Äquivalenzrelation" stets die oben angegebene Bedeutung. An Stelle des Terms „analytisch" sind auch die Terme „mengentheoretisch komplex" oder einfacher „komplex" in Gebrauch (vgl. [9]).

holomorph-vollständig (Steinsch). — Die Möglichkeit, durch Quotientenbildung von einem holomorph-konvexen zu einem holomorph-vollständigen Raum überzugehen, wurde von Remmert [24, 25] entdeckt; der Übergang wird als Remmert-Reduktion und der Quotientenraum als Remmert-Quotient bezeichnet. Zur Ausdehnung dieser Begriffe auf sog. relative Quotienten komplexer Räume siehe Knorr-Schneider [17] und Retter [27].

(2)   Sei $f: X \to Y$ eine holomorphe Abbildung, es seien $R = R(f)$ und $N = N(f)$ die Äquivalenzrelationen in X, deren Äquivalenzklassen die Fasern bzw. die Niveaumengen, d. h. die Zusammenhangskomponenten der Fasern von f, sind. Es werde vorausgesetzt, daß die Niveaumengen von f kompakt sind. Dann ist N eigentlich, und es läßt sich zeigen, daß X/N lokal-$\mathcal{O}/N$-separabel ist. Damit erfüllt N die Bedingung in (c), X/N ist also ein komplexer Raum. — Siehe hierzu Cartan [5]; der Spezialfall, daß X als komplexe Mannigfaltigkeit vorausgesetzt ist, war von Stein [29] behandelt worden.

An Konsequenzen von (2) seien genannt:

i)   Die zur Remmert-Reduktion eines holomorph-konvexen Raums gehörende Äquivalenzrelation ist einfach.

ii)   Sind die Niveaumengen der holomorphen Abbildung $f: X \to Y$ sämtlich kompakt, so gestattet f die Faktorisierung $X \xrightarrow{q} X/N(f) \xrightarrow{g} Y$, wobei die Quotientenabbildung q holomorph, eigentlich und einfach, und die Abbildung g diskret ist. Es ist klar, daß jede Faktorisierung von f in eine eigentliche einfache und in eine diskrete holomorphe Abbildung zu der angegebenen in einem naheliegenden Sinne äquivalent ist. Ist f insbesondere eigentlich, so erfüllt auch $R(f)$ die obige Bedingung in (c), und $X/R(f)$ ist ein komplexer Raum. Man hat die Faktorisierung $X \xrightarrow{q} X/N(f) \xrightarrow{g_1} X/R(f) \xrightarrow{g_2} Y$ von f mit $g = g_2 \circ g_1$, dabei ist g_1 endlich und g_2 injektiv. (Vgl. hierzu [26, 6, 16] sowie B.-Th.-Bericht II, Anhang zu Kap. VII von W. Kaup.)

Sei $\varphi: X \to Y$ eine holomorphe Abbildung. φ heißt *maximal*, wenn gilt: Jede holomorphe Abbildung $\psi: X \to Z$, für welche $N(\varphi) \subset N(\psi)$ ist (ψ heißt dann von φ strikt abhängig), wird von φ auf genau eine Art majorisiert, d. h. es gibt zu ψ genau eine holomorphe Abbildung $\alpha(\psi): Y \to Z$ mit $\psi = \alpha(\psi) \circ \varphi$. Ist $h: X \to W$ eine holomorphe Abbildung mit $N(\varphi) = N(h)$ (φ und h heißen dann strikt verwandt) und ist φ maximal, so wird das Paar (φ, Y) eine *komplexe Basis zu h* genannt. In der oben in (ii) betrachteten Situation ist die Quotientenabbildung $q: X \to X/N(f)$ maximal und das Paar $(q, X/N(f))$ eine komplexe Basis zu f. — Eine Bedeutung dieser Begriffsbildungen liegt darin, daß durch eine maximale Abbildung $\varphi: X \to Y$ die Klasse der von φ strikt abhängigen holomorphen Abbildungen auf die Klasse aller holomorphen Abbildungen von Y bijektiv bezogen wird; dies kann bei Untersuchungen über den Zusammenhang zwischen analytischer und algebraischer Abhängigkeit holomorpher Funktionen verwendet werden. Ferner: Existiert zu der holomorphen Abbildung $h: X \to W$ eine komplexe Basis (φ, Y), so gestattet h die Faktorisierung $X \xrightarrow{\varphi} Y \xrightarrow{d} W$ in eine maximale und eine diskrete holomorphe Abbildung. — Es sei angemerkt, daß für meromorphe Abbildungen entsprechende Begriffsbildungen möglich und von Nutzen sind (siehe z. B. [31 II, 34]).

Im allgemeinen gibt es zu einer holomorphen Abbildung $h: X \to W$ keine komplexe Basis, wie einfache Beispiele zeigen. Hinreichende Bedingungen für die Existenz komplexer Basen sind von verschiedenen Autoren angegeben worden, siehe [19, 28, 29, 30, 31 I, 34, 21, 23, 14, 15]. Notwendig ist jedenfalls, daß es unter den analytischen Äquivalenzrelationen in X, die durch holomorphe Abbildungen von X induziert werden und $N(h)$ umfassen, eine kleinste gibt.

Eine allgemeinere Situation wurde von B. Kaup [14, 15] betrachtet: Sei $R \subset X \times X$ eine analytische Äquivalenzrelation und N_R die zugehörige einfache Äquivalenzrelation (die nicht notwendig analytisch ist). Sei weiter $B(R)$ der Durchschnitt aller analytischen Äquivalenzrelationen in X, die N_R umfassen. $B(R)$ ist wieder eine analytische Äquivalenzrelation in X; sie heißt *komplexe Basis von R*. Zu fragen ist: Wann ist $X/B(R)$ ein komplexer Raum? Ist dies der Fall und wird R durch eine holomorphe Abbildung f von X induziert, so liefert die Quotientenabbildung $q: X \to X/B(R)$ eine komplexe Basis zu f. — Nach B. Kaup gilt nun: Sei X ein zusammenhängender normaler komplexer Raum mit abzählbarer Topologie, sei R eine offene analytische Äquivalenzrelation in X; dann ist auch $B(R)$ eine offene Äquivalenzrelation in X. Somit ist $X/B(R)$ ein komplexer Raum (siehe oben (e)), und dieser ist normal.

Literatur

[1]   Andreotti, A. and W. Stoll, *Meromorphic functions on complex spaces*. Notes in Math. 409, Berlin-Heidelberg-New York 1974, 279—309.

[2]   Baily, W. L., *On the quotient of an analytic manifold*. Proc. Nat. Acad. Sci. USA (1954), 804—808.

[3]   Cartan, H., *Quotient d'une variété analytique par un groupe discret d'automorphismes*. Sém. E.N.S. (1953/54), Exposé 12.

[4]   Cartan, H., *Quotient d'un espace analytique par un groupe d'automorphismes*. Algebraic Geometry and Topology. A symposium in honor of S. Lefschetz, Princeton 1957, 90—102.

[5]   Cartan, H., *Quotients of complex analytic spaces*. Contributions to Function Theory, Bombay 1960, 1—16.

[6]   Fischer, G., *Complex analytic geometry*. Lect. Notes in Math. **538,** Springer-Verlag 1976.

[7]   Furushima, M., *On semi-proper equivalence relations on complex spaces*. Mem. Fac. Sci. Kyushu Univ. A 34 (1980), 127—130.

[8]   Furushima, M., *Remarks on semi-proper equivalence relations on complex spaces*. Mem. Fac. Sci. Kyushu Univ. A 34 (1980), 351—355.

[9]   Grauert, H., *Set theoretic complex equivalence relations*. Math. Ann. **265** (1983), 137—148.

[10]   Holmann, H., *Quotienten komplexer Räume*. Math. Ann. **142** (1961), 407—440 (1961).

[11]   Holmann, H., *Komplexe Räume mit komplexen Transformationsgruppen*. Math. Ann. **150** (1963), 327—360.

[12]   Kaup, B., *Über offene analytische Äquivalenzrelationen auf komplexen Räumen*. Math. Ann. **183** (1969), 6—16.

[13]   Kaup, B., *Relationen auf komplexen Räumen*. Comm. Math. Helv. **46** (1971), 48—64.
[14]   Kaup, B., *Offene analytische Äquivalenzrelationen und komplexe Basen*. Hab.-Schrift Freiburg (Schweiz) 1973.
[15]   Kaup, B., *Zur Konstruktion komplexer Basen*. Manuscripta Math. **15** (1975), 385—408.
[16]   Kaup, L. und B. Kaup, *Holomorphic functions of several variables*. Verlag de Gruyter 1983.
[17]   Knorr, K. und M. Schneider, *Relativexzeptionelle analytische Mengen*. Math. Ann. **193** (1971), 238—254.
[18]   Kiehl, R., *Äquivalenzrelationen in analytischen Räumen*. Math. Z. **105** (1968), 1—20.
[19]   Koch, K., *Zur Theorie der Funktionen mehrerer komplexer Veränderlichen. Die analytische Projektion*. Schriftenreihe des Math. Inst. d. Univ. Münster, Heft 6 (1953).
[20]   Kuhlmann, N., *Über holomorphe Abbildungen komplexer Räume*. Arch. Math. **15** (1964), 81—90.
[21]   Kuhlmann, N., *Niveaumengen holomorpher Abbildungen und nullte Bildgarben*. Manuscripta Math. **1** (1960), 147—189.
[22]   Kuhlmann, N., *Über analytisch abhängige holomorphe und meromorphe Abbildungen*. Manuscripta Math. **1** (1969), 339—353.
[23]   Kuhlmann, N., *Komplexe Basen zu quasieigentlichen holomorphen Abbildungen*. Comm. Math. Helv. **48** (1973), 340—353.
[24]   Remmert, R., *Sur les espaces analytiques holomorphiquement séparables et holomorphiquement convexes*. C. R. Acad. Sci. Paris **243** (1956), 118—121.
[25]   Remmert, R., *Reduction of complex spaces*. Princeton Seminars on Analytic Functions, vol. 1, Sem. **1** (1960), 190—205.
[26]   Remmert, R. und K. Stein, *Eigentliche holomorphe Abbildungen*. Math. Z. **73** (1960), 159—189.
[27]   Retter, K., *Relative komplexe Quotienten*. Manuscripta Math. **34** (1981), 279—291.
[28]   Stein, K., *Analytische Projektion komplexer Mannigfaltigkeiten*. Colloque sur les fonctions de plusieurs variables. Bruxelles 1953, 97—107.
[29]   Stein, K., *Analytische Zerlegungen komplexer Räume*. Math. Ann. **132** (1956), 63—93.
[30]   Stein, K., *Die Existenz komplexer Basen zu holomorphen Abbildungen*. Math. Ann. **136** (1958), 1—8.
[31]   Stein, K., *Maximale holomorphe und meromorphe Abbildungen I, II*. Am. Journal of Math. **85** (1963), 298—313 und **86** (1964), 823—868.
[32]   Wiegmann, K. W., *Strukturen auf Quotienten komplexer Räume*. Comm. Math. Helv. **44** (1969), 93—116.
[33]   Wiegmann, K. W., *Some remarks on a quotient theorem by Andreotti and Stoll*. Rev. Roum de Math. pures et appl. **23** (1978), 965—971.
[34]   Wolffhardt, K., *Existenzsätze für maximale holomorphe und meromorphe Abbildungen*. Math. Z. **85** (1964), 328—344.

Perspectives in Mathematics
Anniversary of Oberwolfach 1984
© Birkhäuser Verlag, Basel

Hyperbolic Differential Operators

LARS GÅRDING

University of Lund, Department of Mathematics,
Box 725, S-22007 Lund (Sweden)

1   Introduction

Linear mathematical models of wave propagation lead to hyperbolic operators, notably the wave operator

$$L = c^{-2}\partial_1^2 - \partial_2^2 - \cdots - \partial_n^2, \ \partial_k = \partial/\partial x_k, \ c > 0.$$

Its main feature is perhaps that it has a fundamental solution, i.e. a Schwartz distribution $F(x)$ such that $LF(x) = \delta(x)$, with support in the cone

$$x_1 \geq 0, \ c^2 x_1^2 - x_2^2 - \cdots - x_n^2 \geq 0.$$

Taking x_1 as time t and $y = (x_2, \ldots, x_n)$ as space variables, this means that an initial disturbance at $t = 0$, $y = 0$ propagates at time t to the ball $|y|^2 \leq c^2 t^2$. We shall turn this property into a general definition.

A differential operator $P = P(D)$ with constant coefficients,

$$P(D) = \sum a_\alpha D^\alpha, \ D_k = \partial_k/i,$$

where the a_α are allowed to be square matrices, is said to be hyperbolic with respect to some $N \in R^n \setminus 0$ if it has a fundamental solution $E = E(P, N, x)$ with support in a proper closed cone meeting the hyperplane $xN = \sum x_k N_k = 0$ only at the origin. For this there is a wellknown[1] equivalent algebraic condition: $Q(N) \neq 0$, $\det P(\xi + tN) \neq 0$ for all real ξ when $\operatorname{Im} t$ is large enough negative. Here Q is the principal part of $\det P$. In section 3 below, dealing with hyperbolic constant coefficient differential operators, the fine properties of the fundamental solution E are discussed, in particular its support and its singularities. The emphasis here and in the rest of the paper is on scalar operators.

The fundamental solution E determines virtually all there is to say about Cauchy's problem with boundary data on a non-reflecting hypersurface. Section 3 also treats mixed problems where some boundary data are also given on a

---

[1] See Atiyah-Bott-Gårding [1] I 127—129.

reflecting hyperplane and presents an algebraic criterion for this problem to be hyperbolic in the sense of having suitably defined fundamental solutions with support properties consistent with finite propagation velocity.

The notion of hyperbolicity given here depends on the notion of a distribution. If we allow fundamental solutions to be distributions on smaller classes than that of infinitely differentiable functions with compact supports, e. g. Gevrey classes, the algebraic condition changes accordingly. This kind of improper hyperbolicity will not be considered.

Section 4 is devoted to linear hyperbolic operators with smooth coefficients. Hyperbolicity will again be intrinsic with reference to the existence of fundamental solutions with supports expressing finite propagation velocity. For strongly hyperbolic operators $P(x, D)$, i. e. those whose principal characteristic polynomials have simple zeros, the theory of Cauchy's problem and the mixed problem has a long history but also recent advances. To achieve something for not strongly hyperbolic operators one has to use all the power of microlocal analysis, recapitulated in section 2, and it is remarkable that there is a class of not strongly hyperbolic operators, the regularly hyperbolic ones, which can be characterized both intrinsically and algebraically.

Section 5 deals with non-linear hyperbolic differential equations. While local existence for Cauchy's problem is not problematic, at least not for strong hyperbolicity, global existence is more delicate. Some recent work on this problem is summarized. There is also a presentation of some results on microlocal regularity of solutions of non-linear hyperbolic equations.

Some more or less obvious problems are indicated in their context. Mostly they have to do with singularities, with unexplored non regular boundary problems and with conditions for global existence.

I have chosen to present hyperbolicity as a conceptual notion. This leads to a general theory and there has been no place to go into applications. They are important enough, but in many cases the general theory is not enough and can only serve as a background. The importance of such situations for the general theory remains to be seen.

2   Microlocal analysis

The modern theory of hyperbolic differential equations borrows some of its deepest insights from a relatively recent branch of mathematics, microlocal analysis. In this theory, the singularities of distributions are analyzed locally in terms of Fourier transforms and all calculations proceed modulo smooth, i. e. C^∞ functions. This section reviews some of its main points, relying on Hörmander [4].

Wave front sets. The basic concept of microlocal analysis is the wave front set of a distribution and its singularity function. Let u be a distribution and

$x \in R^n$ a point. Let $f \in C_0^\infty$ be supported near x and let it be 1 close to x. Consider the Fourier transform of fu,

$$\hat{fu}(\eta) = \int e^{-iy\eta} f(y) u(y) \, dy$$

and let $s_u(x, \xi)$ be the least upper bound for supp f tending to x of numbers s for which $|\eta|^s \hat{fu}(\eta)$ is in L^2 at infinity in some cone around $\xi \neq 0$. The function $s_u(x, \xi)$, the singularity function of f, is obviously homogeneous of degree zero in ξ and semi-continuous from below. That $u \in C^\infty$ close to x means of course that s_u is infinite close to x. That $s_u(x, \xi)$ is large negative means that the Fourier transform of fu with supp f close to x is large in the ξ direction. The closure of the set where $s_u(x, \xi) < \infty$ is called the singularity spectrum or wave front set of u and is traditionally denoted by $WF(u)$. In other words, (x, ξ) belongs to $WF(u)$ if and only if $\hat{fu}(\xi)$ is not rapidly decreasing in any cone around ξ when the support of f tends to x. The projection of $WF(u)$ on x-space is the ordinary singular support of u, the complement of the maximal open set where u reduces to a C^∞ function.

The singularity function behaves well under smooth changes of variables, $x \rightarrow y(x)$. It is not difficult to see that

$$s_u(x, \xi) = s_v(y, \eta)$$

when $u(x) = v(y)$ and $\xi \, dx = \eta \, dy$ identically. This means that the singularity function of a distribution u on a manifold X is a function of $T^* X \backslash 0$, the cotangent bundle of X minus its zero section.

Oscillatory integrals. Let X be an open part of x and consider formal oscillatory integrals

$$F(x) = \int e^{is(x, \theta)} a(x, \theta) \, d\theta \tag{1}$$

with x in X and θ in R^N for some N. The phase function $s(x, \theta)$ is supposed to be real, C^∞ for $\theta \neq 0$ and homogeneous of degree 1 in θ. The amplitude function $a(x, \theta)$ shall be infinitely differentiable with an asymptotic expansion

$$a \sim a_m + a_{m-1} + \cdots$$

for large θ where $a_{m-k}(x, \theta)$ is homogeneous of degree $m-k$, $k = 0, 1, \ldots,$ and C^∞ when $\theta \neq 0$. The precise conditions for the expansion are that

$$\partial_x^\alpha \partial_\theta^\beta (a - a_m - \cdots - a_{m-k+1}) = O(|\theta|^{m-k-|\beta|})$$

locally uniformly in x. The class of these functions will be denoted by S^m. The basic fact about oscillatory integrals is that if $s_x(x, \theta) \neq 0$ for $\theta \neq 0$, then (1) defines in a natural way a distribution F whose wave front set is contained in the set of points (x, ξ) for which $\xi = s_x(x, \theta)$ and $s_\theta(x, \theta) = 0$.

Let X and Y be open parts of R^n and consider operators A from compactly supported distributions in Y to distributions in X defined by the oscillatory integral

$$Au(x) = (2\pi)^{-n} \int e^{is(x,\eta)} a(x,\eta) \hat{u}(\eta) \, d\eta \tag{2}$$

where the phase s and the amplitude a depend on n variables η. Note that if $s = x\eta$ and $a(x,\eta)$ is a polynomial in η, then the Fourier inversion formula reduces A to $a(x, D)$, $D = \partial/i\partial x$, a differential operator whose characteristic polynomial is $a(x, \xi)$. When $s = x\eta$ and a is an arbitrary amplitude function, A is said to be a pseudodifferential operator with symbol $a(x, \xi)$. In the general case and when $s_x \neq 0$ for $\xi \neq 0$, A is said to be a Fourier integral operator. It is not difficult to see that

$$WF(Au) \subseteq \chi \circ WF(u)$$

where χ is a relation consisting of pairs $(x, \xi), (y, \eta)$ for which $\xi = s_x, y = s_\eta$. When A is pseudodifferential, this relation is the identity, in the general case it is canonical in the sense that $\xi \, dx - \eta \, dy = 0$. More generally, let X and Y be manifolds of dimension n and let $C \subseteq (T^*(x) \setminus 0) \times (T^*(Y) \setminus 0)$ be a manifold of dimension $2n$ which is canonical in this sense. Corresponding to C, considered as a relation, there are locally defined Fourier integral operators that change wave front sets according to C (Hörmander [4]).

Pseudodifferential operators. Making the Fourier transform explicit in the formula

$$Pu = (2\pi)^{-n} \int e^{ix\xi} p(x, \xi) \hat{u}(\xi) \, d\xi$$

for a pseudodifferential operator we can write

$$Pu = \int p(x, y) u(y) \, dy$$

where

$$p(x, y) = \int e^{i(x-y)\xi} p(x, \xi) \, d\xi$$

is the kernel of P. By the properties of oscillating integrals, p is a C^∞ function when $x \neq y$. P is said to be properly supported when the support of its kernel has compact intersections with all sets of the form $x \in K$ or $y \in K$ with K compact. It is easy to see that every pseudodifferential operator is the sum of a properly supported one and a C^∞ operator, i.e. one with a C^∞ kernel.

The properly supported pseudodifferential operators form an algebra under linear operations and composition. The symbol of a product PQ is given by the asymptotic formula

$$\sum \partial_\xi^\alpha p(x, \xi) D_x^\alpha q(x, \xi)/\alpha!.$$

For differential operators this is simply the fact that the left side of

$$e^{-ix\xi}p(x,D)\, q(x,D)\, e^{ix\xi} = e^{-ix\xi}\, p(x,D)\, e^{ix\xi}\, q(x,\xi) = p(x,D+\xi)\, q(x,\xi)$$

is the characteristic polynomial of the differential operator $p(x,D)\, q(x,D)$. When $p = p_m + p_{m-1} + \cdots$, $\sigma(p) = p_m$ is said to be the principal symbol of P. It follows that $\sigma(PQ) = \sigma(P)\,\sigma(Q)$. When $p_m(x,\xi) \neq 0$ for all x and all $\xi \neq 0$, P is said to be elliptic. In this case there is a Q with principal symbol $p_m(x,\xi)^{-1}$ such that the symbol of PQ is ~ 1.

A pseudodifferential operator has a simple effect on singularity functions,

$$s_{Pu}(x,\xi) \geq s_u(x,\xi) - m_p(x,\xi)$$

where $m_p(x,\xi)$ is the least upper bound of numbers m for which $p(y,\eta) = O(|\eta|^m)$ when y is close to x and η conically close to ξ. When $p \in S^m$ and $p_m(y,\eta) \neq 0$ conically close to (x,ξ), there is equality.

This short description fits only the first generation of pseudodifferential operators. There is for instance a more precise symbolic calculus, the Weyl calculus, which is invariant under linear canonical maps (Hörmander [5]).

Positivity. Let

$$H = \sum h_{\alpha\beta}(x)\, D^\alpha f(x)\, \overline{D^\beta f(x)}, \quad |\alpha| = |\beta| = m$$

be a quadratic form for which

$$\sum h_{\alpha\beta}(x)\, \xi^\alpha \xi^\beta \geq c |\xi|^{2m}$$

for some $c > 0$ and suppose that all $h_{\alpha\beta}(x)$ are uniformly continuous. Then, with

$$(f,f) = \|f\|^2 = \int |f(x)|^2\, dx, \quad \|f\|_k^2 = \sum \|D^\alpha f\|^2, \quad |\alpha| \leq k,$$

one has

$$\mathrm{Re} \int H\, dx \geq (c-\varepsilon)\, \|f\|_m^2 - b_\varepsilon\, \|f\|_0^2$$

for all $f \in C_0^\infty$, $\varepsilon > 0$ and some $b_\varepsilon > 0$. (Gårding's inequality, Gårding [2].) For pseudodifferential operators there is a sharper result: if $\sigma(P) \geq c |\xi|^{2m}$ then

$$(Pf,f) \geq c \|f\|_m^2 - \mathrm{const}\, \|f\|_{m-1/2}^2$$

(Hörmander [6]) but it requires more regularity of the symbol of P (Lax and Nirenberg [1]). (For later developments, see Melin [1] and Fefferman and Phong [1].)

Propagation of singularities. Let $P(x, D)$ be a differential operator or a properly supported pseudodifferential operator of order m and suppose that its principal symbol $p_m(x, \xi)$ is real and that $dp_m(x, \xi) \neq 0$ when $\xi \neq 0$ (P is real and of principal type). For such operators we have Hörmander's propagation of singularities theorem ([4] II p 196): outside of $WF(Pu)$, $WF(u)$ is a union of (null) bicharacteristics of P, solutions $t \rightarrow (x, \xi)$ of the Hamiltonian system

$$x_t = \partial p_m(x, \xi)/\partial \xi, \quad \xi_t = -\partial p_m(x, \xi)/\partial x, \quad p_m(x, \xi) = 0.$$

More precisely, the singularity function is constant along the bicharacteristics. This basic result has many refinements when the above assumptions are not satisfied. See for instance Hörmander [2] and Dencker [1].

Parametrices. Let $P(x, D)$ be a properly supported pseudodifferential operator. A right (left) parametrix of P is a sum A of Fourier integral operators such that the wave front set of $APu - u$, $(PAu - u)$, is empty for u smooth and conveniently supported. When P is elliptic, A may be taken as pseudodifferential. Its symbol is then computed by inverting the formula for the symbol of a product. When P is not elliptic, partial parametrices are sometimes used. They achieve only that the wave front sets of the distributions above have empty intersections with given open subsets of $T^* R^n \setminus 0$.

3   Constant coefficients

Hyperbolic differential equations with constant coefficients have been studied in detail. The account that follows relies mostly on Atiyah-Bott-Gårding [1] where earlier work by Herglotz [1], Petrovsky [2] and Gårding [1] was brought to a conclusion, mixing properties of hyperbolic polynomials with some algebraic topology. The remaining problems in this context will be mentioned. As pointed out before, the emphasis is on scalar operators.

Hyperbolic polynomials. Let $\mathrm{hyp}(N, m)$ be the class of scalar hyperbolic polynomials $P(\xi)$ of degree m in n variables with principal parts P_m which are hyperbolic with respect to N: $P_m(N) \neq 0$, $P(\xi + itN) \neq 0$ for all real ξ when $t \leq c$, some constant. It can be shown that this implies the same for large positive t so that $\mathrm{hyp}(-N, m) = \mathrm{hyp}(N, m)$. When P belongs to $\mathrm{hyp}(N, m)$, its principal part P_m belongs to $\mathrm{Hyp}(N, m)$, the homogeneous polynomials in the class. In fact, when $s \rightarrow \infty$, then

$$s^{-m} P(s(\xi + itN)) \rightarrow P_m(\xi + itN).$$

It is easy to see that a homogeneous polynomial P is hyperbolic with respect to N if and only if $P(N) \neq 0$ and all the zeros of $t \to P(\xi + tN)$ are real for all real ξ. In particular, $P(\xi)/P(N)$ is a real polynomial. Principal parts of hyperbolic polynomials are assumed to be real in the sequel.

When P is strongly hyperbolic with respect to N, i.e. all the zeros of $t \to P_m(\xi + tN)$ are real and different when ξ is not proportional to N or, equivalently, $dP_m(\xi) \neq 0$ for all real $\xi \neq 0$, it is easy to see that $P + Q \in \text{hyp}(N, m)$ for all Q of degree $< m$. In the general case, this happens if and only if, (Svensson [1]),

$$\sup_{\xi} |Q(\xi + itN)/P_m(\xi + itN)| \tag{1}$$

tends to zero as $t \to \infty$. In particular, all such Q form a linear space.

It follows from the definition that polynomials hyperbolic with respect to N constitute a multiplicative class invariant under differentiation in the N direction as long as the result is not zero. Hence, starting from linear forms $P(\xi) = \sum a_j \xi_j$, hyperbolic with respect to N when $P(N) \neq 0$, and wave polynomials $\Delta_a(\xi) = a^2 \xi_1^2 - \xi_2^2 - \cdots - \xi_n^2$, hyperbolic with respect to N when $\Delta_a(N) > 0$, we can build up a whole family of hyperbolic polynomials. Another important example is $P = \det(A_1 \xi_1 + \cdots + A_n \xi_n)$ with $A_1, \ldots$ symmetric matrices, P being hyperbolic with respect to N when $\sum A_k N_k$ is positive or negative definite. Let us also note, as an example of the restriction (1), that $\Delta_a^k(\xi) + Q(\xi)$, $\deg Q < 2k$, is hyperbolic as Δ_a^k if and only if

$$Q = Q_1 \Delta_a^{k-1} + Q_2 \Delta_a^{k-2} + \cdots$$

where the degree of Q_j is at most j.

Nuij [1] has proved that the class $\text{Hyp}(N, m)$ is connected and simply connected, in particular every polynomial in the class is a limit of strongly hyperbolic ones.

Systems. Under this heading we treat matrix-valued polynomials $P(\xi) = \sum A_\alpha \xi^\alpha$, $|\alpha| \leq m$ whose determinants are hyperbolic with respect to some N. We say that P is strongly hyperbolic if $P + Q$ is hyperbolic when $\deg Q < m$. When $m = 1$, a necessary and sufficient condition for this is that $A(N)^{-1} A(\xi)$ with $A(\xi) = \sum A_j \xi_j$ be diagonalizable, $= S(\xi) D(\xi) S^{-1}(\xi)$, to a real diagonal matrix $D(\xi)$, uniformly in ξ, i.e. $|S(\xi)| + |S^{-1}(\xi)|$ is bounded when $|\xi| = 1$ (Yamaguti and Kasahara [1]). John [1] has investigated the characteristic polynomial of the equations of crystal optics, a case when $m = 2$, $n = 4$, P is homogeneous and the matrices are 3 by 3. He found strong hyperbolicity in spite of the fact that the determinant is not strongly hyperbolic. Nor is P the limit of hyperbolic polynomials with a strongly hyperbolic determinant. This example is so far not part of a general theory.

Hyperbolicity cones. When $P \in \mathrm{hyp}\,(N, m)$, the m real zeros of $P_m(\xi + tN)$ $= 0$ generate m real sheets of the hypersurface $P_m(\xi) = 0$, also called the characteristics of the corresponding differential equation (see Fig. 1).

The component of the set $P_m(\xi) \neq 0$ that contains N is a convex open cone, the hyperbolicity cone $\Gamma(P, N) = \Gamma(P_m, N)$ of P. It has the property that $P(\xi + i\eta) \neq 0$ when $\eta \in \pm(cN + \Gamma)$, $c > 0$ large, so that P is hyperbolic with respect to all η in Γ and $-\Gamma$. If $L(P)$ is the lineality of P, i. e. the set of ξ for which $P(\eta + t\xi) = P(\eta)$ for all t and η, then $L(P)$ is of course linear, $L(P) = L(P_m)$ as a consequence of the hyperbolicity, and

$$\Gamma(P, N) + L(P) = \Gamma(P, N) \tag{2}$$

In other words, $\Gamma(P, N)$ is a convex wedge with edge $L(P)$.

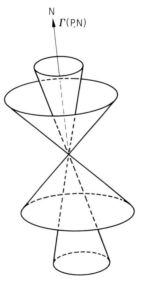

Fig. 1   The hypersurface $P_m(\xi) = 0$ and the hyperbolicity cone $\Gamma(P, N)$

It is clear, that $\Gamma(PQ, N) = \Gamma(P, N) \cap \Gamma(Q, N)$ when P and Q are hyperbolic with respect to N. In the degenerate case $P = \mathrm{const} \neq 0$, $\Gamma(P, N) = R^n$ and if $P = \sum a_j \xi_j$ is real then $P(N) \neq 0$ is the only condition for hyperbolicity and $\Gamma(P, N)$ is the open half-space $P(\xi)\,P(N) > 0$.

Propagation cones. The cone $K = K(P, N)$, defined as the set of x in R^n for which $x\xi \geq 0$ for all ξ in $\Gamma(P, N)$, is a closed proper cone orthogonal to the lineality $L(P)$. It is called the propagation cone of P with respect to N because (see below) $P(D)$ has a fundamental solution with support in K. Some examples: $P = \text{const} \neq 0$ gives $K = (0)$ and if $P = a_1 \xi_1 + \cdots + a_n \xi_n$ is linear, K is the half-line spanned by $P(N)(a_1, \ldots, a_n)$. When P is the wave polynomial $\varDelta_a$, K is given by $\varDelta_{1/a}(x) \geqq 0$, $N_1 x_1 \geqq 0$. It follows from the corresponding property of hyperbolicity cones that $K(PQ, N)$ is the convex hull of $K(P, N)$ and $K(Q, N)$.

Localization. The wave front surface. (Atiyah-Bott-Gårding [1], Hörmander [3]). Consider a rational function $f(\xi) = P(\xi)/Q(\xi)$ of degree $m(f)$ $= m(P) - m(Q)$, $m(P) = \deg P$. For a given ζ and variable ξ we can expand $f(\xi + t\zeta)$ in powers of t^{-1}.

$$f(\xi + t\zeta) = \sum t^{m(f)-k} f_k(\xi, \zeta).$$

The first non-vanishing coefficient, $f_\zeta(\xi)$, obtained for $k = m_\zeta(f)$ is a rational function of ξ called the localization of f at infinity in the direction ζ, briefly the localization at ζ. The number $m_\zeta(f)$ is called its multiplicity,

$$f(\xi + t\zeta) = t^{m(f)-m_\zeta(f)} \left(f_\zeta(\xi) + O(t^{-1}) \right).$$

It follows that $(fg)_\zeta = f_\zeta g_\zeta$ and that $m_\zeta(fg) = m_\zeta(f) + m_\zeta(g)$. Notice that if f is a homogeneous polynomial, then $f_{-\zeta}(\xi) = \pm f_\zeta(\xi) = 0$ is the equation of the tangent cone at ζ of the real hypersurface $f = 0$.

It can be shown that localizations of hyperbolic polynomials preserve hyperbolicity and principal parts and that

$$\varGamma(P_\zeta, N) \supset \varGamma(P, N),$$

strictly when ζ is not in $L(P)$. The cones $\varGamma(P_\zeta, N)$ will be called the local hyperbolicity cones and their duals $K(P_\zeta, N)$ the local propagation cones. They have the property that

$$K(P_\zeta, N) \subset K(P, N)$$

with strict inclusion when ζ is not in $L(P)$. Examples: when $P_m(\zeta) = \text{const} \neq 0$ we have the trivial case $\varGamma(P_\zeta, N) = R^n$, $K(P_\zeta, N) = 0$, when $P_m(\xi) = 0$, $d P_m(\xi) \neq 0$, $\varGamma(P_\zeta, N)$ is a half-space and $K(P_\zeta, N)$ a half-line. When $\dim K(P_\zeta, N) > 1$ we have a case of so-called conical refraction at ζ, named after the corresponding phenomenon is crystal physics. In the general case, the union

$$W(P, N) = \cup K(P_\zeta, N), \quad \zeta \neq 0,$$

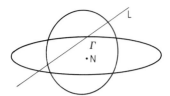

The hyperbolicity cone Γ and the direction N.

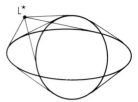

The propagation cone K and its boundary W.

Fig. 2   Projective images of the hypersurface $P_m = 0$, the hyperbolicity cone $\Gamma(P, N)$, the propagation cone $K(P, N)$ and the wave front surface $W(P, N)$ when $m = 5$, $n = 3$ and P_m has a linear factor. The straight parts of W represent the local 2-dimensional propagation cones at the double points of $P_m = 0$. Those on the straight line of $P_m = 0$ meet at a point on W.

is a closed conical set of codimension $\geqq 1$ called the wave front surface of P. Its importance for the propagation of singularities will be seen below. When P is strongly hyperbolic, the pairing $\xi \to -\xi$ of $P_m(\xi) = 0$ reduces the m separate sheets of this hypersurface to $[(m+1)/2]$ sheets and this is also the number of sheets of $W(P, N)$ since, obviously, $K(P_{-\zeta}, N) = K(P_\zeta, N)$ (see Fig. 2. Other instructive figures of this kind in Duff [1], who treats anisotropic elasticity).

Fundamental solutions and their wave front sets. When $P \in \mathrm{hyp}\,(N, m)$, the distribution

$$E = E(P, N, x) = (2\pi)^{-n} \int e^{ix(\xi + i\eta)} \, P(\xi + i\eta)^{-1} \, d(\xi + i\eta)$$

with η fixed in $cN - \Gamma(P, N)$ is a fundamental solution of P, $PE = \delta(x)$. It is independent of η and the Paley-Wiener-Schwartz theorem shows that $E = 0$ outside the propagation cone $K(P, N)$. Putting $P = P_m + Q$, the convergent series

$$P^{-1} = \sum_0^\infty (-1)^k \, Q^k \, P_m^{-k-1}$$

where the argument is $\xi - itN$, t large positive, shows that

$$E(P, N, x) = \sum_0^\infty (-1)^k \, Q(D)^k \, E(P_m^{k+1} N, x)$$

with convergence in the distribution sense.

The singularities of the wave front set can be determined with some precision. The formula for the localization of P at $\zeta \neq 0$,

$$P(\xi + \tau N + t\zeta) = t^{m - m_\zeta} \left(P_\zeta(\xi + \tau N) + O(t^{-1}) \right)$$

shows that, if $g \in C_0^\infty$, the limit of

$$t^{m_\zeta - m} \int e^{-it\zeta x} E(P, N, x) g(x) \, dx =$$
$$= (2\pi)^{-n} \int \int \left(e^{ix(\xi + \tau N)} t^{m_\zeta - m} P(\xi + \tau N + t\zeta)^{-1} \, d\xi \right) g(x) \, dx \qquad (3)$$

as $t \to \infty$ is

$$\int E(P_\zeta, N, x) g(x) \, dx.$$

Hence, if $(\operatorname{supp} g, \zeta)$ is strictly outside the wave front set $WF(E)$ of E, the limit of (3) is zero and this proves that

$$(x, \zeta) \notin WF(E) \Rightarrow E(P_\zeta, N, x) = 0.$$

Hence, we get the left side of the following inclusions of $WF(E)$,

$$\{(x, \zeta);\ x \in \operatorname{supp} E(P_\zeta, N, .)\} \subset WF(E) \subset \{(x, \zeta);\ x \in K(P_\zeta, N)\}. \qquad (4)$$

The other inclusion is more difficult (Hörmander [1] II p 125). Actually, $E(P, N, .)$ is analytic outside $W(P, N)$ (Atiyah-Bott-Gårding [1]) and (4) holds also for the analytic wave front set (Hörmander l.c.).
    The inclusion (4) is an equality when

$$\operatorname{supp} E(P_\zeta, N, x) = K(P_\zeta, N)$$

for all $\zeta \neq 0$. This happens for instance when P is strongly hyperbolic.
    The other possibility, that

$$K(P_\zeta, N) \setminus \operatorname{supp} E(P_\zeta, N, .) \text{ not empty,}$$

is in a way a transient because for large enough k one has

$$K(P_\zeta^k, N) = \operatorname{supp} E(P_\zeta^k, N)$$

for all $\zeta \neq 0$ and hence equality in (4) for P^k (Atiyah-Bott-Gårding [1] p 178). Nevertheless, the precise determination of $WF(E)$ remains a problem. Tsuji [2] has noted that (3) has an asymptotic expansion in t,

$$t^{m - m_\zeta} \sum Q_k(D) E(P^{k+1}, N, x) t^{-k}, \quad Q_0 = 1,$$

and that

$$(x, \zeta) \notin WF(E) \Rightarrow Q_k(D) E(P^{k+1}, N, x) = 0$$

for $k = 0, 1, 2, \ldots$ and he proved equality in (4) when $m_\zeta(P) \leq 2$ for all $\zeta \neq 0$. This is so far the best result.

The Cauchy problem. When $P \in \mathrm{hyp}(N, m)$, the fundamental solution $E = E(P, N, x)$ vanishes of order m at $xN = 0$ in the sense that

$$\int_{yN \geq 0} E(x - y)\, v(y)\, \mathrm{d}y = O((xN)^m_+)$$

when $v \in C_0^\infty$. The same holds true if $yN \geq 0$ is replaced by a region $\phi(y) \geq 0$ where the hypersurface $\phi(y) = 0$ is spacelike in the sense that $\phi_y \in \Gamma(P, N)$ for all y. Hence, if $v, w \in C^\infty$, the function

$$u(x) = \int_{\phi(y) \geq 0} E(x - y)\, (v(y) - Pw(y))\, \mathrm{d}y + w(x)$$

solves Cauchy's problem $Pu = v$ when $\phi > 0$, $u - w = O(\phi^m)$. The solution is unique.

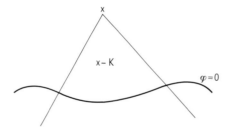

Fig. 3   To Cauchy's problem

Energy inequalities. Associated with $P \in \mathrm{hyp}(N, m)$, $N = (1, 0, \ldots, 0)$, $P(N) > 0$, there are energy forms

$$H(f) = \sum c_{\alpha\beta}\, D^\alpha f(x)\, \overline{D^\beta f(x)}, \quad |\alpha|, |\beta| < m, \tag{5}$$

having the property that, with $x' = (x_2, \ldots, x_n)$,

$$\frac{\mathrm{d}}{\mathrm{d}x_1} \int H(f)\, \mathrm{d}x' =$$

$$- \operatorname{Im} \int P(D) f(x)\, \overline{Q(D) f(x)}\, \mathrm{d}x' + O\left(\int |D^{m-1} f(x)|^2\, \mathrm{d}x'\right) \tag{6}$$

for some Q of degree $m - 1$, the simplest case being $H(f) = |f(x)|^2/2$, $P = D_1$, $Q = 1$. It suffices to construct them for P homogeneous and then the choice

$Q(\xi) = \partial P(\xi)/\partial \xi_1$ and a Fourier transform with respect to x' shows that there is a homogeneous $H(f)$ such that

$$H\left(f(t)\,e^{ix'\xi'}\right) \geq c \sum |f^{(k)}(t)||\xi'|^{m-1-k}|^2, \quad k = 0, \ldots, m-1, \tag{7}$$

provided that P is strongly hyperbolic. Here and in the sequel, c denotes positive constants.

Let us put

$$\|f(t)\|_{r,s}^2 = \sum_{|\alpha| \leq r} \int |(1+|\xi'|)^s \, \xi_2^{\alpha_2} \cdots \xi_n^{\alpha_n} D_1^{\alpha_1} \hat{f}(t,\xi')|^2 \, d\xi' \tag{8}$$

where s is a real number, $r \geq 0$ and integer and $\hat{f}$ is the Fourier transform of f with respect to x'. It is easy to see that (6), (7) gives an energy inequality for P,

$$\|f(t_1)\|_{m-1,s} \leq c\|f(t_2)\|_{m-1,s} + c \int_{t_1}^{t_2} \|Pf(t)\|_{0,s} \, dt \tag{9}$$

where $t_1 > t_2$, f is smooth with compact support in the hyperplanes $x_1 = \text{const}$ and c is allowed to depend on $t_2 - t_1$. When P is not strongly hyperbolic, e. g. when P has multiple characteristics of multiplicity at most r, then (9) holds with $s + r - 1$ instead of s on the right.

When a system $P = D_1 + A_2 D_2 + \cdots + A_n D_n + A_0$ is hyperbolic with respect to $(1, 0, \ldots, 0)$ and uniformly diagonalizable, $A_2 \xi_2 + \cdots + A_n \xi_n = S(\xi') D(\xi') S^{-1}(\xi')$, we can put

$$\int H(f) \, dx' = \int |S(\xi') f(t, \xi')|^2 \, d\xi'$$

and then (9) holds with $m = 1$.

Mixed problems. Let $P \in \text{hyp}\,(N, m)$, $N = (1, 0, \ldots, 0)$, and consider the mixed problem in the half-space $x_n > 0$, the hyperplane $x_n = 0$ not being characteristic or spacelike,

$$x_n > 0 \Rightarrow Pu = v, \quad x_n = 0 \Rightarrow B_1(D)u = w_1, \ldots, B_q(D)u = w_q \tag{10}$$

where $B_1, \ldots, B_q$ are constant coefficient operators and the data $v, w_1, \ldots, w_q$ are required to vanish for large negative x_1. This means that for simplicity we have put the Cauchy data on some hyperplane $x_1 = \text{const}$ equal to zero. We simplify further by taking $v = 0$ and then we are faced with forward radiation in the sense of increasing x_1 from sources in the hyperplane $x_n = 0$. Example: when $P = D_1^2 - D_2^2 - \cdots - D_n^2$ and $B_1(D) = D_n - b_1 D_1 - \cdots - b_{n-1} D_{n-1}$, this is the oblique derivative problem for the wave operator.

The Fourier-Laplace transform

$$\hat{f}(\xi') = \int e^{-ix'\xi'} f(x') \, dx', \quad x' = (x_1, \ldots, x_{n-1}), \quad \xi' = (\xi_1, \ldots, \xi_{n-1})$$

where $-\operatorname{Im}\xi'=cN$, c large positive, reduces the problem to one dimension

$$x_n>0 \Rightarrow P(\xi',D_n)\hat{u}=0, \quad x_n=0 \Rightarrow B_j(\xi',D_n)\hat{u}=\hat{w}_j. \tag{11}$$

Since P is hyperbolic with respect to N, the polynomials $\xi_n \to P(\xi',\xi_n)$ have no real zeros. The zeros ξ_n which give exponential solutions $e^{ix\xi}$ of $Pu=0$ that are small for $x_n>0$ large are those with positive imaginary parts. Let there be m_+ of them and form the polynomial

$$P_+(\xi',\xi_n)=\xi_n^{m_+}+\cdots$$

with these zeros. For (10) to have a unique solution spanned by the corresponding $e^{ix\xi}$ it is necessary and sufficient that the B_j be linearly independent modulo $P_+(\xi',\xi_n)$, a requirement called the Lopatinski condition. For hyperbolic operators it was first formulated by Agmon [1]. For the oblique derivative problem we have

$$P_+=\xi_n(\xi_1^2-\xi_2^2-\cdots-\xi_{n-1}^2)^{1/2}$$

where the square root has positive imaginary part and the condition says that $P_+ =0 \Rightarrow B_1(\xi)\neq 0$. It is satisfied, e. g., when $b_1,\ldots,b_{n-1}$ are real and $b_1>0$ (see Hörmander [1] II, 178—179, for a complete discussion).

In the general case, the Lopatinski condition amounts to the non-vanishing of the Lopatinski determinant

$$L(\xi')=\det L_{ij}(\xi'), \quad i,j=1,\ldots,m_+$$

where

$$L_{ij}(\xi')=(2\pi i)^{-1}\int_\gamma \xi_n^{i-1} B_j(\xi',\xi_n) P_+^{-1}(\xi',\xi_n)\,d\xi_n \tag{12}$$

with γ a loop around the zeros of P_+. Let $\Gamma_n(P,N)$ be the projection of the hyperbolicity cone $\Gamma(P,N)$ onto $\xi_n=0$, $\xi \to (\xi',0)$. It is clear that the functions (12) are analytic when $-\operatorname{Im}\xi'\in cN+\Gamma_n(P,N)$.

When the Lopatinski condition holds, (11) has unique solutions $u =E_k(\xi',x_n)$ defined by

$$x_n=0 \Rightarrow B_j(\xi',D_n) E_k(\xi',x_n)=\delta_{jk}$$

and easy to write down explicitly. Their inverse Fourier transforms $e_j(x)$ then satisfy

$$x_n>0 \Rightarrow P(D) e_j(x)=0, \quad x_n=0 \Rightarrow B_j e_k(x)=\delta_{jk}(x')$$

and vanish for $x_1 < 0$. In Hörmander [1] II, 163—186, there is a proof that such fundamental solutions exist if and only if $q = m_+$ and the Lopatinski condition holds for Im ξ_1 large negative.

The support properties of the distributions $e_k(x)$ depend on the propagation cone of P and the zeros of the Lopatinski determinant. Sakamoto [2] has shown that L has a principal part $L_0(\xi') \neq 0$ defined as the first limit

$$\varepsilon^h L(\xi'/\varepsilon), \quad \varepsilon \to 0, \quad h = 0, 1, 2, \ldots$$

which is finite and that $L_0(N) \neq 0$ is both necessary and sufficient in order for (10) be hyperbolic in the sense that all the e_j are supported in a proper cone. More precisely, let $\Gamma(L, N)$ be the component of $R^{n-1} \setminus (L_0 = 0)$ containing $N = (1, 0, \ldots, 0)$ and let $K_L(P)$ be the dual in the hyperplane $x_n = 0$ of $\Gamma_n(P, N) \cap \Gamma(L, N)$. The supports of the distributions e_k are then contained in $K_L(P) + K(P, N)$. By definition, $K_L(P)$ contains the intersection of $x_n = 0$ with $K(P, N)$ but it may be bigger in which case wave propagation in the boundary may be faster than for Cauchy's problem. This phenomenon is referred to as fast waves. With the uniform Lopatinski condition, meaning that $L_0(\xi') \neq 0$ when Im $\xi_1 \leq 0$ and $\xi_2, \ldots, \xi_m$ are real, fast waves do not occur. The oblique derivative problem, for example, is hyperbolic without fast waves if and only if $(b_1, \ldots, b_{n-1}) = F + iE$, F and E real, satisfies the condition that $1 + (E, E) \geq 0$, F is in the forward lightcone: $(E, F)^2 \leq (F, F)(1 + (E, E))$ and $1 + (E, E) = 0 \Rightarrow F \neq 0$ where the parentheses denote the Lorentz product in $x_n = 0$ (Miyatake [1], Gårding [4]).

Energy inequalities for mixed problems and strongly hyperbolic operators corresponding to the one for Cauchy's problem were first stated by Agmon [1]. They hold under the uniform Lopatinski condition (Sakamoto [1]) but also in other cases. Necessary and sufficient conditions for their validity are known (Agemi and Shirota [1], Sakamoto [4]) but they are rather complicated.

The wave front sets of the fundamental solutions e_k above have so far not got any attention but should be amenable to treatment by the methods used in the free case. In mixed problems with reflection, for instance $P(D)E_1(x, y) = \delta(x - y)$ when $x_n, y_n > 0$ and $B_j E_1(x, y) = 0$ when $x_n = 0$, Tsuji [1] has determined $WF(E_1)$ under a number of restrictive assumptions. See also Wakabayashi [1].

Lacunas and sharp fronts. The wave front set is not the only interesting characteristic of fundamental solutions $E = E(P, N)$ for hyperbolic differential operators. As seen from (4) it is for instance important to know when the support of E coincides with the propagation cone $K(P, N)$.

Examples: When $P = \Delta^k$ is a power of the wave operator $\Delta = D_1^2 - D_2^2 - \cdots - D_n^2$ and $N = (1, 0, \ldots, 0)$ it is a classical fact that supp $E(P, N) = K = K(\Delta, N)$ except when $n > 2$ is even in which case supp $E(P, N) = W = W(\Delta, N) = \partial K$ when $2k < n$ while if $2k \geq n$ $E(P, N)$ is a polynomial of degree $2k - n$ inside

K. If Q has order $< 2k$ and $P+Q$ is hyperbolic with respect to N, all this changes but there remains an essential difference between n even and n odd: $E(P+Q, N, x)$ has respectively has not a C^∞ (actually analytic) extension from $K \setminus W$ to K.

In Atiyah-Bott-Gårding [1] the following definition is proposed. Let Ω be a component of the maximal open set where a distribution u is a C^∞ function and let $x_0 \in \partial \Omega$. Then u is said to have a sharp front at x_0 from Ω if u has a C^∞ extension from Ω to $\Omega \cup M$ where M is some neighborhood of x_0. When u has sharp fronts at all of $\partial \Omega$, Ω is said to be a lacuna for u and when $u = 0$ in Ω, Ω is said to be a strong lacuna. In the examples above $K \setminus W$ is a lacuna or not for $P + Q$ according as n is even or odd. It is a strong lacuna when $P(\xi) + Q(\xi)$ has the form $P(\xi + \xi_0)$ for some ξ_0 and n is even, $2k < n$, but probably only then. There are, however, cases with $k = 1$ and variable coefficient Q when $K \setminus W$ is a strong lacuna (Stellmacher [1]) and wave equations in four variables with variable coefficients not reducible by a change of variables to the constant coefficient case where the same thing happens (and W is curved Günther [1]).

In the three-author-article just quoted, which extends work by Petrovsky [2], the existence and non-existence of lacunas for $E(P, N)$ when P in $\mathrm{hyp}\,(N, m)$ is homogeneous are tied to the topology of the intersection of the complex projective hyperplane X^*: $x\xi = 0$ and the projective hypersurface A^*: $P(\xi) = 0$. The basis is the observation that if $x \in K(P, N) \setminus W(P, N)$ then

$$D^\alpha \left(E(P, N, x) - (-1)^n E(P, -N, x) \right), \quad |\alpha| > m - n$$

is a rational integral

$$\mathrm{const} \quad \int_{\beta^*(x)} (x\xi)^{m-n-|\alpha|} \, \xi^\alpha \, P(\xi)^{-1} \, \omega(\xi)$$

where

$$\omega(\xi) = \sum (-1)^{j-1} \, \xi_j \, d\xi_1 \ldots d\xi_{j-1} \, d\xi_{j+1} \ldots d\xi_n$$

making the integral homogeneous of degree zero and closed and $\beta^*(x)$ is a tube around the projective cycle

$$\alpha^*(x) = \{\xi - i\varepsilon v(\xi)\}, \quad \xi \text{ real } |\xi| = 1, \quad v(\xi) \in \Gamma(P_\xi, N) \cap X^*$$

with v continuous and ε is small enough. When $\alpha^*(x)$ is homologous to zero in $X^* \setminus A^*$, then the integral above vanishes and this condition, due to Petrovsky [2], is both necessary and sufficient for x to be in a lacuna for all powers of P. There is also a corresponding local condition for sharp fronts. Lacunas and sharp front for the fundamental solutions of radiation problems seem not to have been worked out.

4   Variable Coefficients

The first part of this section deals with boundary problems for strongly hyperbolic operators with variable smooth coefficients. Existence and uniqueness for Cauchy's problem is by now classical, the corresponding results for the mixed problem are more recent and require a bit of microlocal analysis. The second part deals with operators which are not strongly hyperbolic. For some time the only substantial results for them required uniform multiplicity of the characteristics. More recently, the situation for operators with characteristics of varying multiplicity has been clarified largely due to the work of V. Ya. Ivrii.

4.1   Strong hyperbolicity

A bit of history. A scalar differential operator $P(x, D)$ with smooth coefficients is said to be strongly hyperbolic if its characteristic polynomial $P(x, \xi)$ has fixed degree m and is strongly hyperbolic at every point. For such operators, Cauchy's problem is locally solvable and the properties of the solution including finite propagation velocity are locally the same as in the case of constant coefficients. The first existence proof, for $m = 2$, is due to Hadamard [1] who constructed a parametrix for the fundamental solution. His method was greatly simplified by Marcel Riesz [1]. The next step was existence proofs by approximation from the analytic case made possible by energy estimates by Schauder [1] for the wave equation and by Petrovsky [1] for systems. This method was so successful that it permitted both of them to pass to non-linear equations. The cases treated by Petrovsky [1] are essentially first order systems whose principal symbols are uniformly diagonalizable to a real diagonal matrix. This class is the true analogue for systems of scalar strongly hyperbolic operators.

In the fifties, the laborious approximation from the analytic case was replaced by a functional analysis argument (Gårding [2]) and the Cauchy problem was solved also for distributions (Leray [1]).

Existence and uniqueness for Cauchy's problem from energy inequalities. The energy inequality (3.9) (of the preceding section) extends to operators

$$P(x, D) = \sum a_\alpha(x) D^\alpha, \quad |\alpha| \leq m$$

with smooth coefficients when $P_m(x, D)$ is hyperbolic with respect to x_1 in some region Ω where $f \in C_0^\infty(\Omega)$ (see fig. 4). In fact, all the steps (3.5), (3.6), (3.7) go through by applying Gårding's inequality and some pseudodifferential theory. Weakening (3.9) a bit we can write

$$\|f\|_{m-1, s}^- \leq c \|Pf\|_{0, s}^- \tag{1}$$

where

$$\|f\|_{r,s} = (\int \|f(t)\|_{r,s}^2 \, dt)^{1/2}, \quad \|f\|_{r,s}^- = (\int_{t \leq 0} \|f(t)\|_{r,s} \, dt)^{1/2}$$

with $\|f(t)\|_{r,s}$ according to (3.8).

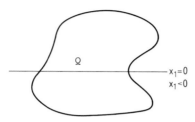

Fig. 4

Let $H^{r,s}$ be the closure of $C_0^\infty (x_1 \leq 0)$ with respect to the norm $\|f\|_{r,s}^-$. By continuity, (1) holds when $f \in H^{m,s}$ has support in a compact part of $\Omega \cap (x_1 \leq 0)$. From this follows a uniqueness result: suppose that, for some $\varepsilon > 0$, $Pu = 0$ in $\Omega \cap (x_1 < \varepsilon)$ and that $\operatorname{supp} u \cap (x_1 \leq \varepsilon)$ is a compact part of Ω (see fig. 4). Then $u = 0$ when $x_1 < 0$. In fact, since $Pu = 0$ when $x_1 < \varepsilon$, u is a smooth function of x_1 whose values are distributions in $(x_2, \ldots, x_n)$. It follows that $u \in H^{m,s}$ for some s so that (1) applies and hence $u = 0$ for $x_1 < 0$.

To produce an existence result we shall follow Hörmander [3] p. 241. Let $\mathring{H}^{-r,-s}$ be the dual of $H^{r,s}$ with respect to an extension of the duality

$$(f,g) = \int f(x) \, g(x) \, dx$$

with $f \in C_0^\infty (x_1 \leq 0)$, $g \in C_0^\infty (x_1 < 0)$. It is not difficult to see that $v \in \mathring{H}^{-r,-s}$ if and only if v can be extended by zero for $x_1 > 0$ to a distribution in all of R^n for which

$$\int (1+|\xi|)^{-2r} (1+|\xi'|)^{-2s} |\hat{v}(\xi)|^2 \, d\xi$$

is finite. Now let $v \in \mathring{H}^{-k,-s}$, $k = m - 1$. We shall see that there exists a $u \in \mathring{H}^{0,-s}$ such that $P'u = v$ in Ω, P' being the adjoint of P. The corresponding result for P then follows with respect to $x_1 > 0$. In fact, by the energy inequality,

$$|(v,f)| \leq c_0 \|f\|_{k,s}^- \leq cc_0 \|Pf\|_{0,s}^-$$

so that, by the Hahn-Banach theorem, there is a $u \in H^{0,-s}$ for which $(v,f) = (u, Pf)$ when $f \in C_0(\Omega)$. If $v \in C_0^\infty$, it follows easily from this that $u \in C^\infty$ in the region of uniqueness.

A less lax treatment of the energy inequality (3.9) will give more precise results about the continuity properties of the inverse of P (see Gårding [5]) but it is remarkable that the precise form of (3.9) is not necessary to produce local uniqueness and existence. It suffices, e. g., that

$$\|f\|_{k,s}^{-} \leqq c \|Pf\|_{0,s+s_0}^{-} \tag{2}$$

for some $k \geqq 0$, some s_0 and all s (Hörmander [3]).

Cauchy's problem with boundary data,

$$Pu = v \quad \text{when} \quad x_1 > 0, \quad u - w = O(x_1^m)$$

can be reduced to the results above when u is required to exist only locally, w is a distribution which is smooth in x_1 and v is a distribution having the same property only for small x_1. Taking $w = 0$ and $v = \delta(x - y)$, $y_1 > 0$ small, we get a fundamental solution $u = E(x, y)$ of P which vanishes when $x_1 < y_1$. Since the solution is unique and P is hyperbolic also with respect to all $x_1 + a_2 x_2 + \cdots$ when $a_2, \ldots$ are small, it follows that, close to y, the support of $x \to E(x, y)$ is contained in a proper cone with its vertex at y. In other words, we have intrinsic hyperbolicity.

By Hörmander's theorem, the singularities of $x \to E(x, y)$ lie on null bicharacteristics

$$dx/dt = \partial P_m(x, \xi)/\partial \xi, \quad d\xi/dt = -\partial P_m(x, \xi)/\partial x$$

issuing from y in the direction of increasing x_1. Just as for constant coefficients, these form $[(m + 1/2)]$ sheets issuing from y, the wave front surface of P and E from y. Outside the outer sheets, $E(x, y)$ vanishes.

Cauchy's problem on a manifold. In order to extend the results above to a manifold X of dimension n where global existence for Cauchy's problem is required, all we have to do is to make some geometric assumptions. It suffices to assume that X has a time function $t(x)$, real and with $t_x \neq 0$, such that $P_m(x, \xi)$ is strongly hyperbolic with respect to t_x for all x and that compact parts of the hypersurfaces $t = $ const remain compact when propagated by the null bicharacteristics of P_m. Existence and uniqueness of a fundamental solution $E(x, y)$ which vanishes for $t(x) < t(y)$ follows. The behavior of $E(x, y)$ near its wave front surface can, at least in principle, be read off from a construction by means of Fourier integral operators of a parametrix for E by Hörmander [4] II (with J. J. Duistermaat). The pairing of the characteristic sheets of $P_m(y, \eta) = 0$ induced by $\eta \to -\eta$ produces some sharp fronts of this parametrix and similar constructions (Gårding [4]).

Mixed problems. Mixed boundary problems for strongly hyperbolic variable coefficient operators have been treated in a satisfactory way when the uniform Lopatinski condition holds. Suppose for instance that $P = D_n^m + \cdots$ has order m and is strongly hyperbolic with respect to $(1, 0, \ldots, 0)$ and that there are m_+ boundary operators $B_k = D_n^{r_k} + \cdots$ of different orders $r_k < m$ such that their principal parts satisfy the uniform Lopatinski condition relative to the principal part of P. Assume also that $P, B_1, \ldots$ have C^∞ coefficients all of whose derivatives are bounded and that their principal parts are constant far away. Then the mixed problem

$$x_n > 0 \Rightarrow Pu = v, \quad x_n = 0 \Rightarrow B_k u = w_k$$

where the C^∞ data $v, w_1, \ldots$ vanish for large negative x_1 has a unique solution with that same property. This holds also when the data are distributions. The emission properties of the solution are given by the principal part of P so that global versions of this result exists. It is due to Sakamoto [1]. Her proof rests on non-trivial energy inequalities and microlocal technique. There is an analogous result for first order systems due to Kreiss [1] (see also Rauch [1]). The proofs have found their way into a book of Chazarin and Piriou [1].

Mixed problems without the uniform Lopatinski condition have been investigated by many authors. The oblique derivative problem for the wave equation without fast waves extends perfectly to variable coefficients (Miyatake [1], Gårding [5]) but when fast waves appear, the situation is not very clear in spite of many interesting papers (Ikawa [1] and earlier references there, Eskin [1] and Tsuji [2]). The general lesson seems to be that good existence theorems without the uniform Lopatanski condition require more regularity of the boundary data for a given regularity of the solution than in the uniform case.

4.2   Weakly hyperbolic operators

Hyperbolicity, weak hyperbolicity. The basic energy inequality (3.9) is characteristic of strong hyperbolicity. In fact, Ivrii and Petkov [1] showed simply that if it holds in a strip $C < t(x) < c$ for an operator $P(x, D)$ with smooth coefficients, then P has order m and is strongly hyperbolic with respect to t_x. They also proved that the analogous inequality for $t(x) = x_1$ and first order systems

$$P(x, D) = A_1(x) D_1 + \cdots + A_n(x) D_n + A_0(x)$$

implies that $A_1(x)$ is invertible and that

$$A_1(x)^{-1} (A_2(x) \xi_2 + \cdots + A_n(x) \xi_n)$$

is diagonalizable to a real matrix, locally uniformly in x and $\xi' = (\xi_2, \ldots, \xi_n)$ when $|\xi'| = 1$.

All this shows that to get further interesting results, one has to abandon strong hyperbolicity. It is then natural to stay with Cauchy's problem and finite propagation velocity and ask for necessary conditions that these still make sense. Just as for constant coefficients, we shall require the existence of fundamental solutions with support properties expressing the finite propagation velocity.

Let $P = P(x, D)$ be a differential operator of order m with C^∞ coefficients in an open subset Ω of R^n. Let C be a proper, closed, convex cone with its vertex at the origin. We shall say that P is (locally) hyperbolic (with respect to C) if every point x_0 in Ω has a neighborhood ω for which there is a fundamental solution $E(x, y)$ of P such that

$$x \notin y + C \Rightarrow E(x, y) = 0. \tag{3}$$

Because of Schwartz's kernel theorem, the following is an equivalent formulation: P has a right linear continuous inverse $E: \mathscr{D}'(\omega) \to \mathscr{D}'(\omega)$ such that

$$\operatorname{supp} u \subset \operatorname{supp} f + C \tag{4}$$

when $u = Ef$ and $f \in C_0^\infty(\omega)$. In fact, E has a kernel, $(Ef)(x) = \int E(x, y) f(y) \, dy$, so that the two conditions above are indeed the same.

As we have seen in the preceding section, strongly hyperbolic scalar operators and first order systems are hyperbolic in the sense above. Conversely, it has been proved that if P is hyperbolic with respect to C, then P is weakly hyperbolic with respect to C in the sense that its principal symbol $P_m(x, \xi)$ (its determinant in case of first order systems) is hyperbolic with respect to all N in the open dual cone of C, the set of ξ in R^n such that $\xi x > 0$ for all x in $C \setminus 0$. This is the Lax-Mizohata theorem, first proved in a simple case by Lax [1] and in general by Mizohata [1]. There is a simple proof in Hörmander [3], who uses ideas from Ivrii and Petkov [1]. The first step is to turn the requirement (4) into an inequality (see fig. 4).

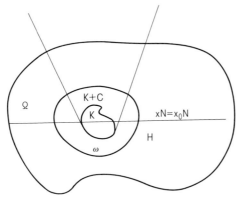

Fig. 4.1 K is supp f.

For any real s and tempered distribution v, let $\|v\|_s$ be the Sobolev norm

$$\|v\|_s^2 = \int (1 + |\xi|^2)^s |\hat{v}(\xi)|^2 \, d\xi$$

where $\hat{v}$ is the Fourier transform of v. When H is a closed half-space of R^n, put

$$\|v\|_{s,H} = \inf \|w\|_s$$

where w runs over all distributions equal to v in H. When $s \geq 0$, then

$$\|v\|_{-s,H} = \sup |\int v(x) w(x) \, dx| / \|w\|_s$$

where $w \in C_0^\infty (H)$. When $s \geq 0$ is an integer, $\|v\|_{s,H}$ is equivalent to

$$\left(\int_H \sum |D^\alpha v(x)| \, dx \right)^{1/2}, \quad |\alpha| \leq s.$$

It follows from (4) that if $N \in \Gamma$ and H is the half-space $xN \leq x_0 N$, then supp $u \cap H$ is compact when ω is small enough. Hence, by the continuity of E, there are numbers p and q, $q \geq 0$ such that

$$\|u\|_{p,H} \leq c \|f\|_{q,H}, \quad u = Ef. \tag{5}$$

Inserting suitable functions u into this inequality, one proves that if P_m is the principal symbol of P and $P_m(x_0, N) \neq 0$, then $P_m(x_0, \xi)$ is hyperbolic with respect to N. Since $P_m(x_0, N) \neq 0$ for some N in Γ, the desired result follows.

The pair (p, q) of (5) is called an index of P. When P is strongly hyperbolic, all (p, q) with $p - q = m - 1$ are indices. When P ist just hyperbolic, the difference may be less.

Linearization of the bicharacteristic system. Microlocally, the difficulties with the hyperbolicity of an operator P occur at the set $\sum$ of points (x, ξ) where $P_m = dP_m = 0$, P_m being the principal symbol. These points are precisely the critical points of the bicharacteristic system

$$dx/dt = P_{m\xi}(x, \xi), \quad d\xi/dt = -P_{mx}(x, \xi)$$

where the indices ξ and x denote the respective gradients. It is therefore natural to consider the linearized system

$$dy/dt = Q_\eta(y, \eta), \quad d\eta/dt = -Q_y(y, \eta)$$

at (x, ξ). Here Q is the Hessian form of P at (x, ξ),

$$P_m(x + y, \xi + \eta) = Q(y, \eta) + \text{higher terms.}$$

Under canonical maps leaving $\sum \xi_i \, dx_i$ invariant, the associated Hamilton map,

$$F: y, \eta \to Q_\eta, \, -Q_y$$

changes by a similarity. When P_m is hyperbolic with respect to some N, so is Q provided it does not vanish, and F has at most one non-zero pair $\pm\lambda$ of real eigenvalues. The others are zero or purely imaginary pairs. When the real pair occurs, P is said to be effectively hyperbolic at (x, ξ). (For a complete analysis of F, see Hörmander [3]).

Regular hyperbolicity. When a differential operator P is hyperbolic regardless of lower order terms which, however, are allowed to influence the index, P is said to be regularly hyperbolic. In this case there is an additional necessary condition besides weak hyperbolicity, namely that P be effectively hyperbolic at all relevant pairs (x, ξ) of the region considered (Ivrii and Petkov [1]). The introduction of the Hamilton map in this connection („the fundamental map") is due to Ivrii. The proof proceeds by the construction of suitable approximative solutions u of $Pu = 0$ contradicting (5). Quite recently, Melrose [1] has announced a converse of this condition: if P is weakly hyperbolic with respect to a time function and P is effectively hyperbolic then P is in fact regularly hyperbolic. (See also Nishitani [1].)

Examples. To produce some interesting examples of weakly hyperbolic operators we shall consider

$$Pu = u_{11} - au_{22} + bu_1 + cu_2 + du \tag{6}$$

where $x = (x_1, x_2)$, the indices denote differentiation with respect to the corresponding variables and the coefficients a, b, c, d are real and smooth when $x_1 \geq 0$ and $a(x) > 0$ except possibly when $x_1 = 0$. The functions u are assumed to be real.

According to these assumptions, P is strongly hyperbolic when $x_1 > 0$ and weakly hyperbolic when $x_1 > 0$. We shall require and construct inequalities (5) of the form

$$\|u\|_0 \leq c \|Pu\|_q, \quad u \in C_0^\infty \, (x_1 \geq 0) \tag{7}$$

where $q > 0$ is an integer and

$$\|v\|_p^2 = \int_0^{t_0} \sum |D^\alpha v(x)|^2 \, dx, \quad |\alpha| \leq p.$$

The notation on the right indicates that the region of integration is $0 < t < t_0$. We keep t_0 fixed. Let us first note that, by integration by parts,

$$\int_0^{t_0} x_1^{-A} |v(x)|^2 \, dx \leq c\|v\|_r^2 \tag{8}$$

when $A > 0$, $r > A + 1$ and v has compact support in $x_1 > 0$.

The integer q in (7) is closely connected with the best estimate $\phi(t)$ in

$$B(x) \leq \phi(x_1) E(x)$$

where E is an energy density of P,

$$E = u_1^2 + a(x) u_2^2 + u^2$$

and

$$B = (1 - 2b)u_1^2 - 2(c + a_2)u_1 u_2 + a_1 u_2^2 + 2(1 - d)uu_1 .$$

In fact, putting $f = Pu$, one has the energy indentity

$$E_1 - 2(au_1 u_2)_2 - B + u_1^2 = 2u_1 f \leq u_1^2 + f^2,$$

so that by an integration,

$$\int E(t, x_2) \, dx_2 \leq \int\!\!\int_0^t \phi(x_1) E(x) \, dx + \int\!\!\int_0^t f^2(x) \, dx.$$

In particular, when $\phi(x_1) = A/x_1$, an easy argument gives

$$\int_0^{t_0} u^2 \, dx = c \int_0^{t_0} (1 + x_1^{-1-A}) f^2(x) \, dx.$$

When $\phi(x_1) = x_1^{-1-\varepsilon}$, $\varepsilon > 0$, the factor of f^2 on the right is replaced by something that grows exponentially as x_1 tends to zero and this rules out an estimate like (7) for any q. When $\phi(x_1) = A/x_1$, we get from (8) an inequality (7) with some $q \leq A + 2$.

How A depends on the coefficients of P can be read off from the inequality $x_1 B \leq AE$ for small x_1, i.e.

$$(1 - 2b)x_1 u_1^2 - 2(c + a_2)x_1 u_1 u_2 + a_1 x_1 u_2^2 + 2(1 - d)x_1 uu_1 \leq A(u_1^2 + au_2^2).$$

Keeping d within fixed bounds and $A \geqq 1$, this implies and is implied by

$$b^* x_1 u_1^2 - 2(c + a_2) x_1 u_1 u_2 + a_1 x_1 u_2^2 \leqq A(u_1^1 + a u_2^2)$$

for some b^* in each case. Requiring this for all u_1, u_2 means that

$$A + Aa \geqq b^* x_1 + a_1 x_1, \ (A - b^* x_1)(Aa - a_1 x_1) \geqq (c + a_2)^2 x_1^2.$$

When $a(x) = x_1 g(x_2)$ vanishes just once for $x_1 = 0$, it suffices for this to take $A > 1$ regardless of the size of the non-principal coefficients of P. When $a(x) = x_1^2 g(x_2)$ vanishes twice, the second equation above leads to the condition $A \geqq 1 + (1 + c^2/g)^{1/2}$ when $x_1 = 0$. In this case A and hence also q is influenced by the non-principal coefficients. But P is still effectively hyperbolic. In fact, when $x_1 = \xi_1 = 0$, $\xi_2 = 1$ we have $Q(y, \eta) = \eta_1^2 - y_1^2 g(x_2)$ and the Hamilton map

$$(y, \eta) \to (2\eta_1, 0, 2 y_1 g(x_2), 0)$$

has the eigenvalues $\pm 2 g(x_2)^{1/2}$. When $a(x)$ vanishes more than twice for $x_1 = 0$, our computations indicate that (7) requires some lower order terms of P to vanish for $x_1 = 0$. More general and more precise results of the nature indicated here, also for the complete Cauchy problem, were given by Oleinik [1] and Menikoff [1].

In the first of the preceding cases, when $a(x) = g(x_2) x_1$, one has $d P_2(x, \xi) \neq 0$ when $P_2(x, \xi) = \xi_1^2 - a(x) \xi_2^2 = 0$. Also in the general case, the analogous condition implies that (5) holds with index (p, q), $q - p = m - 1$ independent of the lower terms of P, and it is also necessary for this (completely regular hyperbolicity, Ivrii and Petkov [1], Ivrii [1]).

A class of weakly hyperbolic, not necessarily effectively hyperbolic, operators P has been treated by Hörmander [3]. Improving on results by Ivrii and Petkov [1], he showed that for (3) to hold when the characteristics of $P_m(x, \xi) = 0$ are at most double it is necessary that

$$|P'_{m-1}(x, \xi)| \leqq \sum \mu_j(x, \xi)$$

at points where $P(x, \xi) = 0$, $d P(x, \xi) = 0$. Here

$$P'_{m-1}(x, \xi) = P_{m-1}(x, \xi) + \frac{i}{2} \sum \partial^2 P_m(x, \xi) / \partial x_j \, \partial \xi_j$$

is the subprincipal part of P and the $\pm i \mu_j$, $\mu_j > 0$, are the purely imaginary non-vanishing eigenvalues of the Hamilton map at (x, ξ). He also gave a sufficient condition for hyperbolicity close to this necessary one and found indices (p, q) with $p - q = m - 2$. Sakamoto [3] treats another case when the operator is hyperbolic and the Hamilton map nilpotent.

Propagation of singularities. Let $P = P(x, D)$ be a hyperbolic differential operator of order m and let u be a solution of $Pu = 0$ in some open set. By Hörmander's theorem, the wave front set $WF(u)$ of u is a union of bicharacteristics outside of the part $\sum$ of $T^* R^n \setminus 0$ where $\xi \to P_m(x, \xi)$ has multiple characteristics or, in other words, $P_m = d\,P_m = 0$. Bicharacteristics of P issuing from $\sum$ reduce to points. When P is effectively hyperbolic, corresponding to the real valued eigenvalues $\neq 0$ of the Hamiltonian map at (x, ξ) in $\sum$, there is one pair of bicharacteristics of its Hessian Q which cross at the origin. Correspondingly, Melrose [1] has announced that if P is effectively hyperbolic, its bicharacteristics have limit points only on $\sum$ and that $WF(u)$ is a union of bicharacteristics joined at these points. This is also the case in our example above when $a(x) = x_1^{2k} g$, $g \neq 0$, where $\sum$: $x_1 = \xi_1 = 0$ (Alinhac [1] and [2], Taniguchi and Tozaki [1]). In other cases, the bicharacteristics do not have limit points on $\sum$ and there is propagation within $\sum$ (Ivrii [2] p. 89). The general picture (Lascar-Sjöstrand [1]) seems to be that the intersection of $WF(u)$ with $\sum$ is contained in the set of points which are arbitrarily close to non-nul bicharacteristics outside $\sum$, i.e. solutions of the bicharacteristic system without the condition that $P_m(x, \xi) = 0$. This formulation also covers conical refraction both for constant and variable coefficients (see Ivrii [3]).

Microlocally, the theory of propagation of singularities is not limited to hyperbolic equations. It exists also for pseudodifferential equations and covers many situations and may not yet be in anything like a final stage. The Springer Lecture Note 856 (1981) by R. Lascar sums up recent development.

5   Non-linear equations

This last section is very short mainly because I chose to deal lightly with the theory of non-linear hyperbolic systems in the plane and the attached theory of shocks. Instead I sum up the local existence results for Cauchy's problem in the general non-linear strongly hyperbolic case and the question of global existence in special cases. There is also a paragraph on the recent theory of microlocal wave propagation for non-linear equations, a subject in its beginnings.

Local existence. The Cauchy problem for a quasilinear equation,

$$P(x, D^{m-1}u, D)u = Q(x, D^{m-1}u), \quad u - w = O(x_1^m)$$

is uniquely solvable close to a point x_0 where $\xi \to P(x, D^{m-1}w(x), \xi)$ is homogeneous of degree m and strongly hyperbolic with respect to x_1. Here $D = \partial/\partial x$, $D^{m-1}u = \{D^\alpha u; |\alpha| < m\}$ and P and Q are supposed to be smooth functions in all their variables. This result for $m = 2$ is due to Schauder [1] and for general

systems to Petrovsky [1] and Leray [1]. The method of proof is that of successive approximations via the map $u \rightarrow v = T(u)$ where v solves

$$P(x, D^{m-1}u, D)v = Q(x, D^{m-1}v), \quad v - w = O(x_1^m).$$

The proof depends on the Sobolev imbedding theorems. One gets regularity
requirements due to the fact that the energy inequality (9) of section 3 requires
only Lipschitz continuity of the principal coefficients of P. More precisely, it
suffices to require that $Q(x, y)$ and the coefficients $P_\alpha(x, y)$ of

$$P(x, y, D) = \sum P_\alpha(x, y) D^\alpha, \quad |\alpha| = m,$$

where $y = \{y^\alpha, |\alpha| < m\}$, have derivatives of order $\leq r > (n+1)/2$ with respect to
x and y which are locally in L^1 with respect to x_1 and locally in L^2 with respect to
$(x_2, \ldots, x_n, y)$ and that the $(x_2, \ldots)$-derivatives of order $\leq r$ of $D^{m-1}w$ are
locally in L^2 when $x_1 = x_{01}$. The corresponding derivatives of the solution u are
then L^2-continuous functions of x_1 (Gårding and Leray, unpublished). For an
ordinary differential equation $du/dt = f(t, u)$, the analogous theory requires
$\partial f(t, u)/\partial u$ to be locally integrable in t locally uniformly with respect to u.

Corresponding results hold for first order systems whose determinant is
strongly hyperbolic (Gårding [7]). For uniformly diagonalizable systems they
are also true but the energy inequality requires a sharp form of Gårding's
inequality (Lax and Nirenberg [1]) and the differentiability requirement is now $r
> (n+4)/2$.

In a known way (Leray [1]) non-linear equations are reduced by dif-
ferentiations to quasilinear ones.

Non-linear mixed problems have been treated in a similar way by
Schauder [2] (with Krzyzanski) and [3]. It should not be too difficult to extend
their results to a general situation when good energy inequalities are available
for the corresponding linear problems.

Global existence. When $u(t, x)$ solves the wave equation

$$u_{tt} - \Delta u = 0, \quad \Delta = \sum D_k^2$$

with smooth compactly supported initial data, it is not difficult to see that

$$u = O\left((1+t)^{-(n-1)/2}\right), \quad t > 0.$$

In other words, the solution is dispersed or evened out for large times. This fact
has been used in a study of the lifespan $T = T(\varepsilon)$ before blowing up of solutions u
of a non-linear equation

$$u_{tt} - \Delta u = F(u, Du, D^2u), \quad F(X) = O(|X|^2) \tag{1}$$

with small initial data

$$u(0, x) = \varepsilon h(x), \ u_t(0, x) = \varepsilon g(x), \ \varepsilon > 0.$$

This study was initiated by John [2], [3] who proved in the quasilinear case and for $n = 3$ that $T(\varepsilon) > \varepsilon^{-N}$ for small ε and all $N > 0$. Later Klainerman and Ponce [1] have shown that $T(\varepsilon) = \infty$ for small ε when $n > 5$. When (1) gets a damping term u on the left it suffices to have $n > 4$. Analogous results for the mixed problem outside a cylinder were obtained by Shibata [1] who has an additional u_t in the equation and gets global existence for small data when $n > 4$.

In connection with models for quantum field theory, global existence for

$$u_{tt} - \Delta u = f(u)$$

with smooth f and smooth Cauchy data for $t = 0$ has attracted much interest. When $f(u) = F'(u)$ with $F(u) > 0$, the energy

$$\int (u_t^2 + u_x^2 + 2F(u)) \, dx$$

is invariant and global existence is easy to get, but if the third term above is not dominated by the sum of the others, uniqueness is an open problem (Strauss [1]).

In this connection let me also mention an important paper by Rabinowitz [1]. It exhibits a striking difference between the linear and non-linear wave equation in one space variable: if the equation $u_{tt} - u_{xx} = 0$, $u = 0$ for $x = 0, \pi$, is modified by a non-linear term $\varepsilon F(x, u)$ where F_u stays away from zero, then the equation has a unique solution of period 2π for small ε.

In one space dimension there is a class of non-linear hyperbolic systems, the conservation laws, which appear in onedimensional gas dynamics and have served as model equations for hydrodynamics. Their theory is reviewed in a recent book, Smoller [1] (see also Lax [2]). The general form of such a conservation law is

$$u_t + f(u)_x = 0, \ u \in R^n.$$

When $n = 1$, u is constant along the characteristics $dx = f'(u) \, dt$. This formula solves Cauchy's problem $u(0, x) = g(x)$ explicitly, but the trouble is that the solution may not be single valued or even defined everywhere also when g is smooth. One remedy is to pass to weak solutions, to restrict f to be convex and to impose an entropy condition restricting the size of the discontinuities (Smoller [1], 266). Under these conditions there will be uniqueness and existence for bounded initial datum and the solution will have a net of jump discontinuities corresponding to the shock waves of gas dynamics. For systems there is a variety of entropy conditions and a remarkable existence proof by a random difference scheme due to Glimm [1] (see Smoller [1] and Harten and Lax [1]). The extension of all this beautiful theory to more than one space variable is a problem of the first oder, unfortunately a very old one, present already in Euler's equations in hydrodynamics.

Propagation of singularities. When a differential operator does not have C^∞ coefficients, the linear theory of propagation of singularities no longer applies. The same holds for non-linear equations. In both cases one must get some information about the wave front sets of products which, in general is much bigger than the wave front sets of the factors. It is therefore a bit of a miracle that something of the linear theory survives in a hostile non-linear environment. One main restriction is that regularity s on the Sobolev scale is manageable only when $s > n/2$, n being the dimension. Only then is the local H^s space a ring.

The first results obtained were for one space variable. Reed [1] proved that if $F \in C^\infty$, a solution u of

$$u_{tt} - u_{xx} = F(u, u_t, u_x)$$

is a C^∞ function at all points (x, t) where the characteristics do not meet the singular supports of the Cauchy data at $t = 0$. This simple state of affairs ceases to hold for analogous higher order equations: the singularities are still carried by characteristics but if two of them meet, the singularity will propagate in the direction away from $t = 0$ on all of them, but it will be weaker outside the original pair. There are analogous results for systems (Rauch and Reed [1], [2]).

For quasilinear equations in several variables the results are less precise, but there are still traces of the linear theory. Say that $u \in H^s(\gamma)$, γ a subset of $R^n \times (R^n \setminus 0)$, if the singularity function s_u of u is $> s$ at every point on γ. A result by Rauch [2] says that if $u \in H^s$, $s > n/2$, and u satisfies the quasilinear wave equation

$$(D_1^2 - D_2^2 - \cdots - D_n^2)\, u = f(u), \quad f \in C^\infty$$

then $u \in H^r(x, \xi)$ implies $u \in H^r(\gamma)$ when $r < 2s + 1 - n/2$ and γ is the bicharacteristic through (x, ξ). On the other hand, in situations like this, the spreading of singularities need not occur along bicharacteristics. Beals [1] has constructed an example: there is a Cauchy problem

$$u_{tt} - u = f(x)u^3, \quad f \in C_0^\infty(R^n), \quad n > 1,$$

such that the singular support of u contains the cut-off cone $t > |x|$, $t < 1$; but the singular supports of the Cauchy data at $t = 0$ reduce to the origin.

The results above by Rauch [2] were extended by Bony [1] to solutions of

$$P(x, \mathscr{D})u = F(\mathscr{D}^{m-1}u)$$

where $F \in C^{\infty}$ and P is strongly hyperbolic of order m. Then the same statement holds with $s > m - 1 + n/2$ and $r < 2s + 1 - m - n/2$. Beals and Reed [1] contains a simplified proof of this and Beals and Reed [2] has analogous results for pseudodifferential equations with smooth coefficients.

A systematic theory of the propagation of singularities for non-linear differential equations, not necessarily hyperbolic has been set up by Bony [1]. His main tools are a modified multiplication operating on the convolution of Fourier transforms and a modified symbolic calculus.

References

Agemi R. and Shirota T. [1], *On necessary and sufficient conditions for the L^2 well-posedness of mixed problems for hyperbolic equations.* J. Fac. Sc. Hokkaido Univ. Series I. 21 (1970) 133—151.

Agmon S. [1], *Problèmes mixtes pour les équations hyperboliques d'ordre supérieur.* Coll. Int. CNRS **117**, Les équations aux derivées partielles. Paris 1962.

Alinhac S. [1], *Parametrix pour un système hyperbolique à multiplicité variable.* Asterisque **34—35** (1976) 3—26.

— [2], *Branching of singularities of a class of hyperbolic operators.* Ind. Univ. Math. J. **27** (1978) 1027—1037.

Atiyah M., Bott R., Gårding L. [1], *Lacunas for hyperbolic differential operators with constant coefficients I, II.* Acta Math. **124** (1970) 10—189 and **131** (1973) 145—206.

Beals M. [1], *Self-spreading and strength of singularities for solutions to semilinear wave equations.* Ann. Math. **118.1** (1983) 187—214.

Beals M. and Reed M. [1], *Propagation of singularities for hyperbolic pseudodifferential operators with non-smooth coefficients.* Comm. Pure Appl. Math. **XXXV** (1982) 169—184.

— [2], *Microlocal regularity theorems for non-smooth pseudodifferential operators and application to non-linear problems.* Preprint 1983.

Bony J.-M. [1], *Calcul symbolique et propagation des singularitées pour les équations aux derivées partielles non linéaires.* Ann. Ec. Norm. Sup. **14** (1981).

Chazarain J. and Piriou A. [1], *Introduction à la théorie des équations aux derivées partielles.* Gauthier Villars. Paris 1981.

Dencker N. [1], *On the propagation of singularities for pseudodifferential operators of principal type.* Arkiv. f. Mat. **20** (1982) 23—60.

Duff, G. F. D. [1], *The Cauchy problem for elastic waves in an anisotropic medium.* Phil. Trans. Roy. Soc. A **252** (1960) 249—273.

Eskin G. [1], *Initial-boundary value problem for second order hyperbolic equations with general boundary conditions.* I. J. d'Anal. Math. **40** (1981) 43—89, II. Preprint 1983.

Fefferman C. and Phong D. H. [1], *On positivity of pseudodifferential operators.* Proc. Nat. Ac. Sc. **75** (1978) 4673—4674.

Gårding L. [1], *Linear hyperbolic differential equations with constant coefficients.* Acta Math. **85** (1950) 1—62.

— [2], *Dirichlet's problem for linear elliptic partial differential equations.* Math. Scand. **1** (1953) 55—72.

— [3], *Solution directe du problème de Cauchy pour les équations hyperboliques.* Coll. Int. CNRS, Nancy 1956, 71—90.

— [4], *Sharp fronts of paired oscillatory integrals.* Publ. RIMS Kyoto U. 12, suppl. (1977) 53—68. Correction ibid. **13** (1977) 821.

— [5], *Le problème de la derivée oblique pour l'équation des ondes.* C. R. Paris **285** (1977) 773—775. Rectification ibid. **285** (1978) 1199.

— [6], *Introduction to hyperbolicity in "Hyperbolicity".* CIME Ligouri Ed. (1976).

— [7], *Problème de Cauchy pour les systèmes quasi-linéaires d'ordre un strictément hyperboliques.* Coll. Int. CNRS **117**, Les équations aux derivées partielles. Paris 1962.

Glimm J. [1], *Solutions in the large for nonlinear hyperbolic systems of equations.* Comm. Pure Appl. Math. **18** (1965) 95—105.

Günther P. [1], *Zur Gültigkeit des Huygensschen Prinzips bei partiellen Differentialgleichungen vom normalen hyperbolischen Typus.* Leipziger Ber. Math. Nat. Kl. **100** (1952) no. 2.

Hadamard J. [1], *Le problème de Cauchy pour les équations aux derivées partielles linéaires hyperboliques.* Paris 1932.

Harten A. and Lax P. D. [1], *A random choice finite difference scheme for hyperbolic conservation laws.* SIAM J. Num. An. **18.2** (1981) 289—365.

Herglotz G. [1], *Über die Integration linearer partieller Differentialgleichungen I. II. III.* Leipziger Ber. Sächs. Ak. d. Wiss. Math. Phys. Kl. **78** (1926) 93—126, **80** (1928) 6—114.

Hörmander L. [1], *The analysis of partial differential operators I, II.* Springer 1983.

— [2], *Propagation of singularities and semi-global existence theorems for pseudodifferential operators of principal type.* Ann. Math. **108** (1978) 569—609.

— [3], *The Cauchy problem for differential equations with double characteristics.* J. d'Anal. Math. **32** (1977) 118—196.

— [4], *Fourier integral operators.* I. Acta Math. **127** (1971) 79—180, II. (with J. J. Duistermaat) Acta Math. **128** (1972) 183—269.

— [5], *The Weyl calculus of pseudodifferential operators.* Comm. Pure Appl. Math. **XXXII** (1979) 359—443.

— [6], *Pseudodifferential operators and non-elliptic boundary problems.* Ann. of Math. **83** (1966) 129—209.

— [7], *On the singularities of solutions of partial differential equations.* Int. Conf. Funct. Anal. and Rel. Topics. Tokyo 1969.

Ikawa M. [1], *Mixed problems for the wave equation IV. The existence and exponential decay of solutions.* J. Math. Kyoto Univ. **19** (1979) 375—411.

Ivrii V. Ja. and Petkov V. M. [1], *Necessary conditions for the correctness of the Cauchy problem for non-strictly hyperbolic equations.* Russian Math. Surveys **29**:**5** (1974) 1—70.

Ivrii V. Ja. [1], *Sufficient conditions for regular and completely regular hyperbolicity.* Trudi Mosk. Mat. Ob. **33** (1976) 3.65. Translated in Trans. Moscow Math. Soc. 1978.

— [2], *Wave front sets of solutions of some hyperbolic pseudodifferential equations.* Trudi Mosk. Mat. Ob. **39** (1979). Translated in Trans. Moscow Mat. Soc. 1981:1.

— [3], *Wave fronts of solutions of some hyperbolic equations and conical refraction.* Soviet. Math. Dokl. **226**:**6** (1976).

— [4], *The propagation of singularities of non-classical boundary value problems for second order hyperbolic equations.* Trans. Moscow Math. Soc. 1983:1 87—99.

John F. [1], *Algebraic conditions for hyperbolicity of systems of partial differential equations.* Comm. Pure Appl. Math. **XXXI** (1978) 89—106. Addendum l.c. 787—793.

— [2], *Blow-up of solutions of non-linear wave equations in three space dimensions.* Manuscripta math. **28** (1979) 235—268.

— [3], *Lower bounds for the life span of solutions of non-linear wave equations in three dimensions.* Comm. Pure Appl. Math. **XXXVI** (1983) 1—35.

Kreiss H. O. [1], *Initial boundary value problems for hyperbolic systems*. Comm. Pure Appl. Math. **XXIII** (1970) 277—298.

Lascar B. and Sjöstrand J. [1], *Equation de Schrödinger et propagation des singularités pour des opérateurs pseudodifferentiels à charactéristiques de multiplicité variable I*. Astérisque **95** (1982) et II, Journées "Equations aux derivées partielles" Sain-Jean-de-Monts Juin 1983.

Lax P. D. [1], *Asymptotic solutions of oscillatory initial value problems*. Duke Math. J. **24** (1957) 627—646.

— [2], *Hyperbolic systems of conservation laws and the mathematical theory of shock waves*. SIAM Reg. conf. series in appl. math. **11** (1973).

Lax P. D. and Nirenberg L. [1], *On stability for difference schemes, a sharp form of Gårding's inequality*. Comm. Pure Appl. Math. **19** (1966) 473—492.

Leray J. [1], *Hyperbolic differential equations*. Inst. Adv. Study, Princeton 1953.

Melin A. [1], *Lower bounds for pseudodifferential operators*. Ark f. Mat. **9** (1971) 117—146.

Melrose B. [1], *The Cauchy problem for effectively hyperbolic operators*. Preprint 1983.

Menikoff A. [1], *The Cauchy problem for weakly hyperbolic equations*. Amer. J. Math. **97.1** (1975) 548—558.

Miyatake S. [1], *A sharp form of the existence theorem for hyperbolic mixed problems of second order*. J. Math. Kyoto U. **17—2** (1977) 199—223.

Mizohata S. [1], *Some remarks on the Cauchy problem*. J. Math. Kyoto U. **1** (1961) 109—127.

Nishitani T. [1], *Local energy integrals of effectively hyperbolic operators I, II*. Preprint 1982.

Nuij W. [1], *A note on hyperbolic polynomials*. Math. Scand. **23** (1968) 69—72.

Oleinik O. A. [1], *On the Cauchy problem for weakly hyperbolic operators*. Comm. Pure Appl. Math. **XXIII** (1970) 569—586.

Petrovsky I. G. [1], *Über das Cauchysche Problem für Systeme von partiellen Differentialgleichungen*. Mat. Sb. **2** (44) (1937) 815—870. Addendum: Some remarks on my papers on the problem of Cauchy. Mat. Sb. **39** (81) (1956) 267—272.

— [2], *On the diffusion of waves and lacunas for hyperbolic equations*. Mat. Sb. **17** (59) (1945) 289—370.

Rabinowitz P. H. [1], *Periodic solutions of non-linear hyperbolic partial differential equations I, II*. Comm. Pure Appl. Math. **XX** (1967) 145—200 and **XXII** (1969) 15—39.

Rauch J. [1], *L_2 is a continuable initial condition for Kriess' mixed problems*. Comm. Pure Appl. Math. **XXV** (1972) 265—275.

— [2], *Singularities of solutions to semilinear wave equations*. J. Math. Pures et Appl. **58** (1979) 299—308.

Rauch J. and Reed M. [1], *Propagation of singularities for semilinear hyperbolic equations in one variable*. Ann. of Math. **11** (1980) 531—552.

— [2], *Non-linear microlocal analysis of semilinear hyperbolic systems in one space dimension*. Duke J. **49.2** (1982) 397—475.

Riesz M. [1], *L'integrale de Riemann-Liouville et le problème de Cauchy*. Acta Math. **81** (1949) 1—223.

Sakamoto R. [1], *Mixed problems for hyperbolic equations I, II*. J. Math. Kyoto U. **10** (1970) 349—373 and 403—417.

— [2], *E-wellposedness for hyperbolic mixed problems with constant coefficients*. J. Math. Kyoto U. **14** (1974) 93—118.

— [3], *Wellposedness of weakly hyperbolic Cauchy problems*. Publ. RIMS **15** (1979) 469—518.

— [4], *L^2-wellposedness for hyperbolic mixed problems*. Publ. RIMS **8** (1972—73) 265—293.

Schauder J. [1], *Das Anfangswertproblem einer quasilinearer Differentialgleichung zweiter Ordnung in beliebiger Anzahl von Veränderlichen*. Fund. Math. **24** (1935) 213—246.

— [2], with Krzyzanski M., *Quasilineare Differentialgleichungen zweiter Ordnung vom hyperbolischen Typus. Gemischte Randwertaufgaben*. Studia Math. **6** (1936) 162—189.

— [3], *Gemischte Randwertaufgaben bei partiellen Differentialgleichungen vom hyperbolischen Typus*. l. c. 190—198.

Shibata Y. [1], *On the global existence of classical solutions of second order fully non-linear hyperbolic equations with first order dissipation in the exterior domain.* Tsukuba J. Math. **7.1** (1983) 1—68.

Smoller J. [1], *Shock waves and reaction-diffusion equations.* Springer 1983.

Stellmacher K. [1], *Eine Klasse hyperbolischer Differentialgleichungen und ihre Integration.* Math. Ann. **130** (1955) 219—233.

Strauss W. [1], *Non-linear wave equations.* Lecture Notes in Physics 73. Invariant wave equations. Springer 1977.

Svensson L. S. [1], *Necessary and sufficient conditions for the hyperbolicity of polynomials with hyperbolic principal part.* Ark. Mat. **8** (1968) 145—162.

Tanigucho K. and Tozaki Y. [1], *A hyperbolic equation with double characteristics which has a solution with branching singularities.* Math. Jap. **25.3** (1980) 279—300.

Tsuji M. [1], *Propagation of singularities for hyperbolic equations with constant coefficients.* Jap. J. Math. **2.2** (1976) 361—410.

— [2], *Characterization of well-posed mixed problems for the wave equation in quarter space.* Proc. Jap. Ac. **50** (1974) 138—142.

— [3], *Singularities of elementary solutions of hyperbolic equations with constant coefficients.* Preprint.

Wakabayashi S. [1], *Propagation of singularities of fundamental solutions of hyperbolic mixed problems.* Publ. RIMS **15** (1979) 553—578.

Yamaguti M. and Kasahara K. [1], *Sur le système hyperbolique à coefficients constants.* Proc. Jap. Acad. **35.9** (1959) 547—555.

Perspectives in Mathematics
Anniversary of Oberwolfach 1984
© Birkhäuser Verlag, Basel

Developments in Combinatorial Optimization

MARTIN GRÖTSCHEL

Universität Augsburg, Mathematisches Institut,
Memminger Straße 6, D-89 Augsburg (FRG)

Abstract

This paper describes developments in combinatorial optimization in the last thirty years and outlines trends of future research. Section 1 introduces a few representative problems of the subject and mentions some applications. Polynomial time solvability and $\mathcal{NP}$-completeness of combinatorial optimization problems are discussed in Section 2. Polyhedral combinatorics, and the theory and practice of cutting planes are surveyed in detail in Section 3. Many of the landmarks of these topics are mentioned, open problems and future developments are outlined. Section 4 describes some of the relations of combinatorial optimization to other branches of mathematics. In particular, some of the major recent breakthroughs that arose from applying the results of other fields to combinatorial optimization (and vice versa) are mentioned. Moreover, lists of promising research areas and concrete open problems are given.

1   Introduction and Applications

The roots of combinatorial optimization lie in easy-looking problems (mostly of economical or technical nature) of the following kind.

(1.1)   Given n cities and distances between these, find a roundtrip through all cities of shortest total length (the TRAVELLING SALESMAN PROBLEM).

(1.2)   Given a road network connecting two cities A and B, find a shortest route (with respect to time or distance) from city A to B (the SHORTEST PATH PROBLEM).

(1.3)   Determine the layout of a printed circuit board so that no two lines (or as few lines as possible) intersect — except in their endpoints (the PLANARITY PROBLEM).

(1.4)   Find a simultaneous permutation of the rows and columns of an (n, n)-matrix such that the sum of the entries above the main diagonal is as large as possible (TRIANGULATION OF INPUT OUTPUT MATRICES).

(1.5)   Given m machines and n jobs which consist of a given sequence of operations on some of the machines, suppose that for each operation a processing time on the associated type of machine is given and that each job has a due date. Find a feasible assignment of operations to machines such that as few of the due dates as possible are violated (a SCHEDU-LING PROBLEM).

(1.6)   Determine the routes of the garbage collection trucks of a city so that all side conditions with respect to working time, capacity of trucks etc. are satisfied and the total distance covered by all trucks is minimized (a ROUTING PROBLEM).

(1.7)   Given, at a university or school, a number of courses, class-rooms and teachers, assign teachers to courses and courses to classrooms so that no two courses are in the same classroom at the same time, no two teachers give the same course, teachers are able to give the course etc. (an ASSIGN-MENT PROBLEM).

(1.8)   Given a pipeline system between a "source" and a "sink", determine the maximal amount of "flow" from the source to the sink through the network subject to capacity constraints etc. (a FLOW PROBLEM).

The problems mentioned above are examples of so-called combinatorial optimization problems. Formally, a *combinatorial optimization problem* can be described by a set of *instances* and a *task*. Each instance is given by a pair (S, c) where S is a finite set and $c: S \to \mathbb{R}$ any function, and for each instance (S, c) the task is to find an element $s \in S$ whose function value $c(s)$ is maximum (or minimum).

The elements of S are called *feasible solutions*, an element of S maximizing (or minimizing) c over S is called *optimal solution* of the instance (S, c). The function c is called *objective function*.

Clearly, every combinatorial optimization problem can be solved by *enumeration*, i. e. by scanning through S, evaluating for each $s \in S$ the objective function $c(s)$, and choosing the element s^* with highest (or lowest) value $c(s^*)$. Thus for a combinatorial optimization problem to be nontrivial we have to assume that the set S and the function c of every instance are structured in some way, i. e. that S and c are describable in much less space than the cardinality of S.

In very many cases S is a set of subsets of a finite set E, and the objective function is specified by giving a value $c(e)$ to each element $e \in E$ and setting $c(T) := \sum_{e \in T} c(e)$ for every set $T \in S$. Problems of this type are called *linear objective combinatorial optimization problems*. These are the best-studied and most important combinatorial optimization problems, and we will restrict our attention to this class of problems in the sequel. Thus we will focus on problems where each instance is given by a finite set E (the *ground set*), a (usually implicitly defined) set $\mathscr{I} \subseteq 2^E$ of feasible solutions and a function $c : E \to \mathbb{R}$ where the task is to find a set $F \in \mathscr{I}$ such that $c(F) := \sum_{e \in F} c(e)$ is maximum or minimum.

For most of the problems (1.1), . . ., (1.8) it is trivial to see how they can be phrased in the way defined above. Consider, for instance, the travelling salesman problem (1.1). With each instance of this problem we associate a complete graph $K_n = (V, E)$ with n nodes (representing the cities) where each pair of distinct nodes is linked by an edge (representing a road connection). The "ground set" is the edge set E. The feasible solutions $\mathcal{I} \subseteq 2^E$ are the "roundtrips" or "tours", which are sets of n edges forming a cycle which passes through every node (hamiltonian cycles). The "value" of each edge $ij \in E$ is the distance c_{ij} between its two endnodes (cities) i and j, and so the "length" of a tour T is $c(T) := \sum_{ij \in T} c_{ij}$. The task is to find a tour T^* such that $c(T^*)$ is minimum.

Most of the problems studied in the early days of this subject came from operations research, industrial management, computer science and military applications. But problems of this kind arise almost everywhere, and therefore combinatorial optimization has found successful applications in fields like archeology, biology, chemistry, geography, linguistics, physics, sociology and others.

This survey is not meant as an overview of the applications of combinatorial optimization. The reader interested in this should consult the appropriate sections of the classified bibliographies Kastning (1976), Hausmann (1978), von Randow (1982), the book Roberts (1978), or the papers Balas & Padberg (1975), Grötschel (1982), Iri (1983) which explain various applications (of special types) of combinatorial optimization problems.

2   Polynomial Time Solvability and $\mathcal{N}\mathcal{P}$-Completeness

The theory of combinatorial optimization—at least in the way I view the subject—aims at a better mathematical understanding of the type of problems introduced in paragraph 1 with the ultimate goal to provide tools for the design of "efficient" algorithms for solving these problems.

The notion of efficiency needs, of course, some clarification. In the early days of combinatorial optimization the efficiency of an algorithm was usually tested empirically by programming the algorithm and running the code on several data sets, measuring time and storage space needed, and fitting these measures to some curves. Such empirical comparisons are still of great value for those who are using implemented algorithms in practice, but from a theoretical viewpoint they are quite unsatisfactory.

A theoretically more appealing concept of efficiency was brought into the field from complexity theory. The notions "solvable in polynomial time" and "NP-complete" introduced in the late sixties have considerably changed the way combinatorial optimizers look at their subject and put new research topics into focus.

In short and informally these concepts can be described as follows. First one has to fix a model of computation. Usually Turing machines or RAM machines are used (for a nonexpert in this field it is sufficient to consider a real world computer). Then one has to decide how the problem instances are to be encoded for the machine and how the "size" of an instance has to be measured.

The standard way to encode numbers (say integers) is to use their binary representation and so the *input length or size* of an integer is the number of digits of this representation. Graphs (a large number of combinatorial optimization problems can be formulated as problems concerning graphs) are usually encoded by means of adjacency or edge lists. The number of nodes and edges of a graph is a convenient measure of the input size of a graph. In this way (other structures can be handled similarly) with every instance of a combinatorial optimization problem an input size, which is an integral number, can be associated.

An algorithm for a combinatorial optimization problem is said to run in *polynomial time* (or to be a "good algorithm") if there is a polynomial $p : \mathbb{N} \to \mathbb{N}$ such that for every instance of the problem of input size at most n the running time of the algorithm is at most $p(n)$. The running time is measured by counting the "steps" the algorithm has to perform until termination, where steps are elementary operations (like comparisons, additions and multiplications) on the machine model considered. By $\mathscr{P}$ we denote the class of problems which can be solved by a polynomial time algorithm. Of course, an algorithm with running time n^{1000} is by no means good in practice, but, for large enough instances, it is still a lot better than an algorithm which performs 2^n steps. On the other hand, it has turned out that for many of the practically relevant problems—once their polynomial time solvability was discovered—good algorithms with low degree polynomial time bounds could be found, say of order at most n^3 or n^4.

The notion of polynomial time solvability was introduced by Edmonds (1965 a) and Cobham (1965). It is customary to call problems which are solvable in polynomial time *easy*.

For the definition of "difficult" problems a few more technicalities are necessary. The success of this so-called "theory of $\mathscr{NP}$-completeness" rests on a fundamental result of Cook (1971) whose far reaching consequences for combinatorial optimization were recognized and popularized by Karp (1972).

To explain the ideas behind this theory let us consider a combinatorial minimization problem Π. We now want to solve the following decision problem. Given an instance (P, c) of Π and an additional number B, decide whether there is a feasible solution, say S, whose value $c(S)$ is at most as large as B. (Clearly, if we could find an optimum solution to (P, c) in polynomial time we could solve this decision problem in polynomial time by calculating the optimum value and comparing it with B.)

We say that problem Π belongs to the class of problems $\mathscr{NP}$ if Π has the following property. There is a polynomial time algorithm which does the following. If for an instance (P, c) and a bound B there is a solution $S \in P$ with $c(S) \leq B$, then the algorithm can verify in polynomial time that $S \in P$ and $c(S) \leq B$.

Note that the algorithm is not required to find S. The only thing the algorithm has to do is, given (P, c), B and S, check whether $S \in P$ and $c(S) \leq B$ in polynomial time. The problems in $\mathscr{NP}$ are called *solvable in nondeterministic polynomial time*. The name stems from an equivalent definition of the class $\mathscr{NP}$

where algorithms are considered which are allowed to make guesses (nondeterministic steps).

For example, consider the travelling salesman problem (1.1). Suppose an instance of this problem and a bound B are given. Now we guess a set T of edges and mark it on the map with a red pencil. We choose a starting point and follow the red edges. If we return to the starting point, have passed all cities exactly once and have encountered all red edges then T is a roundtrip. Now we add up the distances associated with the red edges to obtain $c(T)$ and compare this value with B. This procedure is clearly a polynomial time algorithm of the type we consider, so the travelling salesman problem belongs to $\mathcal{NP}$.

It is obvious that $\mathcal{P} \subseteq \mathcal{NP}$. And it is intuitively convincing that not everything that can be guessed can also be constructed. But despite enormous research efforts in the last decade it could not be decided yet whether $\mathcal{P} \neq \mathcal{NP}$ or not. In my opinion this question is one of the major open problems in mathematics. $\mathcal{P} \neq \mathcal{NP}$ (together with further known results) would imply that there is a host of combinatorial optimization problems relevant for real world applications which are inherently intractable.

The class $\mathcal{NP}$ contains a further important class of problems which are called $\mathcal{NP}$-complete. This class is denoted by $\mathcal{NPC}$. Let us call a problem Π $\mathcal{NP}$-complete (or simply *hard*) if it has the following property: $\Pi \in \mathcal{NP}$, and if Π can be solved in polynomial time, then every problem in $\mathcal{NP}$ can be solved in polynomial time. The $\mathcal{NP}$-complete problems are in a sense the hardest problems in $\mathcal{NP}$, since in order to show that $\mathcal{P} = \mathcal{NP}$ it suffices to prove for just one $\mathcal{NP}$-complete problem that it is in $\mathcal{P}$.

The classification of combinatorial optimization problems into hard and easy ones was one of the main streams of research of this field in the seventies. It turned out that most of the practically relevant problems are in fact $\mathcal{NP}$-complete. This, of course, had significant implications on the directions of further theoretical and algorithmic investigations about which we will report in the subsequent sections.

The results of the studies done in this area of complexity theory are documented in the excellent book Garey & Johnson (1979) where for several hundred generic problems (and some thousand variants) their membership in $\mathcal{P}$ and $\mathcal{NPC}$ is recorded. An ongoing guide of David Johnson in the Journal of Algorithms documents the current progress in this subject.

In the meantime for almost all major problems it has been decided whether they are in $\mathcal{P}$ or in $\mathcal{NPC}$. Today's research in this area mainly concentrates on exploring the borderlines between $\mathcal{P}$ and $\mathcal{NPC}$, to unify results and get a deeper understanding of "difficulty".

Almost no significant progress with respect to the $\mathcal{P} \neq \mathcal{NP}$ problem has been made. There are two major (and usually powerful) techniques available to attack this problem. With "simulation" one could try to prove $\mathcal{P} = \mathcal{NP}$, while "diagonalization" might be a good tool to show that $\mathcal{P} \neq \mathcal{NP}$. Baker, Gill & Solovay (1975), however, showed in a beautiful paper (using oracle techniques which we do not want to describe here) that these two methods do not suffice to solve this important question. It is even conceivable that this problem cannot be solved within the framework of formal set theory, see Hartmanis & Hopcroft (1976). Yet, almost all researchers in combinatorial optimization assume $\mathscr{P} \neq \mathscr{NP}$ as a working hypothesis.

The theory of $\mathscr{NP}$-completeness has provided a rough but very useful classification scheme for combinatorial optimization problems. When studying a problem, nowadays a standard first step is to check whether it belongs to $\mathscr{P}$ or is $\mathscr{NP}$-complete.

Graph theory and combinatorial optimization were two of the main areas of application of the theory of $\mathscr{NP}$-completeness. In the meantime complexity theory has "invaded" other mathematical fields as well, in particular disciplines like number theory or algebra where "sizes of numbers" or "lengths of proofs" are considered.

I think that today complexity theory plays two important roles. It provides a language to distinguish between hard and easy problems and its concepts are convenient tools for the analysis of algorithms.

3   Cutting Planes and Polyhedral Combinatorics

A simple idea but one of the most fruitful approaches in combinatorial optimization is to formulate combinatorial optimization problems as integer linear programs or even as linear programs. Theoretically this is rather easy, but to make this idea an algorithmic success quite a number of new mathematical concepts and algorithmic design techniques had to be developed. For more detailed surveys of this subject the reader should consult the excellent papers Pulleyblank (1983) und Schrijver (1983). Moreover, the book Schrijver (1984a) will appear soon which treats the whole area in depth and contains almost all of the results known to date.

3.1   *Polyhedra Associated with Combinatorial Optimization Problems*

We shall now describe a method with which a polyhedron can be associated with (almost) every combinatorial optimization problem. Suppose we have a minimization problem where each instance is given by a ground set E, a set $\mathscr{I} \subseteq 2^E$ of feasible solutions, and an objective function $c : E \to \mathbb{R}$. Let $\mathbb{R}^E$ denote the real $|E|$-dimensional vector space where for every vector $x \in \mathbb{R}^E$ its components are indexed by the elements of E, i.e. $x = (x_e)_{e \in E}$. For every subset $F \subseteq E$ we define its *incidence* (or characteristic) *vector* $\chi^F = (\chi_e^F)_{e \in E}$ by

$$\chi_e^F = 1 \text{ if } e \in F \quad \text{and} \quad \chi_e^F = 0 \text{ if } e \notin F.$$

With the set $\mathscr{I}$ of feasible solutions we associate the convex hull of the incidence vectors of the elements of $\mathscr{I}$, i.e. a polytope $P_{\mathscr{I}} \subseteq \mathbb{R}^E$ defined by

(3.1) $P_{\mathscr{I}} := \operatorname{conv} \{ \chi^F \in \mathbb{R}^E | F \in \mathscr{I} \}$.

The combinatorial optimization problem

(3.2) $\min \{ c(F) | F \in \mathscr{I} \}$

can now be written (considering c as a vector in $\mathbb{R}^E$) as

(3.3) $\min \{ c^T x | x \in P_{\mathscr{I}} \}$.

Every feasible solution $F \in \mathscr{I}$ corresponds to a vertex of the polytope $P_{\mathscr{I}}$ and vice versa. As $P_{\mathscr{I}}$ is a polytope, problem (3.3) is a linear program, and it is well-known that, for every objective function, the program (3.3) has an optimum solution which is a vertex of $P_{\mathscr{I}}$. Thus, by solving (3.3) one can obtain an optimum incidence vector of a set $F \in \mathscr{I}$ which in turn is an optimum solution of (3.2).

Problem (3.3) can of course be solved in finite time by generating all the vertices of $P_{\mathscr{I}}$ (these are implicitly given through $\mathscr{I}$) and selecting the one with the best value. But this way nothing has been gained by the new representation of the combinatorial optimization problem.

Linear programs are usually given by a linear objective function and a system of linear equations and inequalities, and all (nontrivial) algorithms solving linear programming problems require such systems as input. Thus, in order to use the powerful tools and methods of linear programming it is necessary to find a linear system describing $P_{\mathscr{I}}$. Theoretically, the theorem of Weyl guarantees that for every polyhedron P there are a matrix A and a vector b such that $P = \{ x | A x \leq b \}$. There are even constructive proofs of this theorem, but they are not effective in the sense that one can "easily read" from $\mathscr{I}$ an inequality system describing $P_{\mathscr{I}}$.

So in order to transform a combinatorial optimization problem Π into a linear programming problem one has to solve the following problem:

(3.4)   *For every instance $(\mathscr{I}, c)$ of Π find a system of linear equations and inequalities describing the associated polytope $P_{\mathscr{I}}$.*

One of the (at present) most flourishing branches of combinatorial optimization, called *polyhedral combinatorics*, considers this task as one of its central research topics. We shall now survey some of the ideas developed in this area and some of the successes of this approach. And we shall mention interesting open research problems.

3.2   *Some Examples*

A *graph* $G = (V, E)$ consists of a finite nonempty set V of *nodes* and a set E of *edges* which are two-element subsets of V. (For ease of exposition we do not consider loops and multiple edges here.) The two elements of V, say i and j, forming an edge $e \in E$ are called the endnodes of e, and we write $e = ij$ instead of $e = \{i, j\}$. A *digraph* $D = (V, A)$ consists of a finite nonempty set V of nodes and a set $A \subseteq V \times V$ called *arcs*. If $a = (i, j) \in A$ then i is called the *tail* of arc a and j is called the *head* of a, i and j are also called the *endnodes* of a.

If $G = (V, E)$ $(D = (V, A))$ is a graph (digraph) and $W \subseteq V$ a set of nodes then $E(W)$ $(A(W))$ denotes the set of edges (arcs) with both endnodes in W; $\delta(W) \subseteq E$ denotes the set of edges with one endnode in W and the other in $V \setminus W$; $\delta^+(W) \subseteq A$ (resp. $\delta^-(W) \subseteq A$) denotes the set of arcs with tail (resp. head) in W and head (resp. tail) in $V \setminus W$. We write $\delta(v)$ instead of $\delta(\{v\})$ for $v \in V$.

Matchings

One of the first positive results in this area concerns matchings in biparti-te graphs. It is generally attributed to Birkhoff (1946) and von Neumann (1953). A graph $G = (V, E)$ is called *bipartite* if V can be partitioned into two nonempty, disjoint subsets V_1, V_2 such that each edge has one of its endnodes in V_1 and the other in V_2, i.e. if $E = \delta(V_1) = \delta(V_2)$. A *matching* in G is a set $M \subseteq E$ of edges such that no two edges of M have a common endnode. The Birkhoff-von Neumann theorem characterizes the polytope associated with the matchings in a bipartite graph.

(3.5)   **Theorem.** *Let $G = (V, E)$ be a bipartite graph. Then the convex hull of the incidence vectors of the matchings of G (the* **matching polytope** *of G) is the polytope defined by the following system of inequalities:*

(1) $x_e \geq 0$ *for all $e \in E$,*

(2) $\sum_{e \in \delta(v)} x_e \leq 1$ *for all $v \in V$.* $\square$

This theorem has some interesting consequences in graph theory. It implies, for instance, via LP-duality König's matching theorem

„*The maximum cardinality of a matching in a bipartite graph is equal to the minimum cardinality of a set of nodes that meets all edges.*"

or Hall's famous marriage theorem.

It is easy to see that the system (1), (2) of (3.5) is not sufficient for the description of the matching polytope of nonbipartite graphs. For instance, if G is a cycle of length three then the vector $\frac{1}{2}(1, 1, 1)$ is a vertex of the polytope defined by (1), (2). A major breakthrough was obtained by Edmonds, cf. Edmonds (1965 a), (1965 b), who found a complete system for the matching polytope in general.

(3.6)   **Theorem.** *Let* $G = (V, E)$ *be a graph. Then the convex hull of the incidence vectors of the matchings of* G *is the polytope defined by*

(1) $x_e \geqq 0$ *for all* $e \in E$

(2) $\displaystyle\sum_{e \in \delta(v)} x_e \leqq 1$ *for all* $e \in E$

(3) $\displaystyle\sum_{e \in E(W)} x_e \leqq (|W| - 1)/2$ *for all* $W \subseteq V, |W|$ *odd.* $\square$

Theorem (3.6) is a deep result with many consequences in graph theory and combinatorial optimization. It has meanwhile found generalizations in many directions. The reader interested in this should consult section 6 of Schrijver (1983).

Theorem (3.6) looks—at first sight—quite useless from a linear programming point of view since the number of constraints (3) is exponential in the input size of G. So it already takes exponential time to input the inequalities (1), (2), (3), which means that no algorithm requiring the full system (1), (2), (3) of (3.6) can run in time polynomial in the input size of G. However, and this was the second major achievement, Edmonds was able to devise an algorithm for the solution of linear programs with constraint system (1), (2), (3) which runs in time polynomial in $|E|$. Edmonds exploited LP-duality theory, the complementary slackness theorem, and the fact that the system (1), (2), (3) has a "nice" implicit description which made it possible to avoid the use of all inequalities at once and to generate inequalities whenever necessary. We shall come back to this point in Section 3.4.

Matroids and Generalizations

Another class of polyhedra which is "well understood" is a class of polytopes associated with matroids. A matroid M on E is a pair $(E, \mathscr{I})$ where $\mathscr{I}$ is a subset of the set of all subsets of E satisfying

(3.7) $\emptyset \in \mathscr{I}$,

(3.8) $J \subseteq I \in \mathscr{I} \Rightarrow J \in \mathscr{I}$,

(3.9) $I, J \in \mathscr{I}, |I| < |J| \Rightarrow \exists e \in J \setminus I$ with $I \cup \{e\} \in \mathscr{I}$.

The elements of $\mathscr{I}$ are called the *independent sets* of M.

For instance, if $G = (V, E)$ is a graph, then the set $(E, \mathscr{I})$ where $\mathscr{I} := \{F \subseteq E \mid F \text{ is a forest (i.e. } F \text{ contains no cycle)}\}$ is a matroid on the edge set E of G (the so-called *forest matroid* of a graph). If A is a matrix with entries from some field K and with column index set E, then call a set $I \subseteq E$ independent if the column vectors of A corresponding to the indices of I are linearly independent. The pair $(E, \mathscr{I})$ defined this way is a matroid, the so-called *matric matroid*.

We recommend Welsh (1976) for more information about matroids. The survey paper Iri (1983) and the forthcoming book Recski (1984) particularly focus on applications of matroid theory.

Given a matroid $M = (E, \mathscr{I})$ and a function $c : E \to \mathbb{R}$, then the matroid optimization problem is to find an independent set I such that $c(I)$ is as large as possible. (A special case of this, e.g., is the problem to find a maximum (or minimum) forest in a graph.) The matroid polytope P_M is defined as follows

$$P_M := \operatorname{conv} \{\chi^I \in \mathbb{R}^E \mid I \in \mathscr{I}\}.$$

It is very easy to solve matroid optimization problems with the famous *greedy algorithm*. (Set $I := \emptyset$. Choose an element $e \in E$ such that c_e is as large as possible. If $c_e \geq 0$ and $I \cup \{e\} \in \mathscr{I}$ set $I := I \cup \{e\}$. Remove e from E and continue until E is empty.) Edmonds (1971) interpreted the greedy algorithm as a linear programming algorithm and derived from this the following characterization of the matroid polytope.

(3.10) **Theorem.** *Let $M = (E, \mathscr{I})$ be a matroid and P_M be the associated polytope, then*

$$P_M = \{x \in \mathbb{R}^E \mid \quad (1) \quad x_e \geq 0 \qquad \text{for all } e \in E$$
$$(2) \quad \sum_{e \in F} x_e \leq r(F) \quad \text{for all } F \subseteq E\}. \quad \square$$

Above, for every set $F \subseteq E$, $r(F)$ denotes the largest cardinality of an independent set contained in F. The number $r(F)$ is called the *rank* of F. Edmonds (1970) was able to extend Theorem (3.10) to the intersection of two matroids as follows.

Let $M_1 = (E, \mathscr{I}_1)$ and $M_2 = (E, \mathscr{I}_2)$ be two matroids on the same ground set E. The pair $(E, \mathscr{I})$ with

$$\mathscr{I} := \mathscr{I}_1 \cap \mathscr{I}_2$$

is called the intersection of the matroids M_1 and M_2, and

$$P_{\mathscr{I}} := \operatorname{conv} \{\chi^I \in \mathbb{R}^E \mid I \in \mathscr{I}\}$$

is called the *2-matroid intersection polytope*.

(3.11) **Theorem.**

$P_{\mathscr{I}} = P_{M_1} \cap P_{M_2}$, *in particular*

$P_{\mathscr{I}} = \{x \in \mathbb{R}^E \mid$ *(1)* $x_e \geqq 0$ *for all $e \in E$*

                    (2) $\sum\limits_{e \in F} x_e \leqq min\{r_1(F), r_2(F)\}$ *for all $F \subseteq E\}$.* $\square$

In (2) above r_i denotes the rank function of M_i, $i = 1,2$. This theorem has many interesting special cases. For instance, it gives a complete linear description of the branching polytope of a digraph (see Edmonds (1967)), but the Birkhoff-von Neumann Theorem (3.5), König's matching theorem, Hall's marriage theorem, and Fulkerson's branching theorem can also be derived from (3.11). Optimizing over the intersection of three or more matroids is $\mathscr{N}\mathscr{P}$-complete. This suggests that a similar result extending (3.11) to the intersection of three or more matroid polytopes is unlikely.

The results of Edmonds were the starting point of a new branch of combinatorial optimization, the theory of submodular functions. For a finite set E a function $f: 2^E \to \mathbb{R}$ is called *submodular* if

(3.12) $f(S \cup T) + f(S \cap T) \leqq f(S) + f(T)$ for all $S, T \subseteq E$.

The connection with matroids is the fact that matroid rank functions (see (3.10)) are special submodular functions satisfying in addition $r(\emptyset) = 0$, $S \subseteq T \subseteq E \Rightarrow r(S) \leqq r(T)$, and $r(S) \leqq |S|$ for all $S \subseteq E$. It was observed that most of the properties of matroids resp. matroid polyhedra are in fact consequences of the submodularity of the rank function.

With every submodular function $f: 2^E \to \mathbb{R}$ (without loss of generality we may assume $f(\emptyset) = 0$) one can associate the polyhedron

$$P_f := \{x \in \mathbb{R}^E \mid \sum\limits_{e \in F} x_e \leqq f(F)   \text{ for all } F \subseteq E\}.$$

It is possible to solve $max\{c^T x \mid x \in P_f\}$ by means of an appropriately modified greedy algorithm, and one can show that the vertices of P_f are integral valued if f is integral valued. Moreover, the matroid intersection theorem (3.11) also extends to the intersection of two such polyhedra, see Edmonds & Giles (1977).

In the recent few years there has been an inflation of frameworks based on submodular functions combined with various graph theory concepts (in particular network flow theory). They aim at a unification of those parts of combinatorial optimization for which polyhedral results (like the one described above) or min-max theorems (like König's matching theorem) exist. Concepts like polymatroids (Edmonds (1970)), submodular flows (Edmonds & Giles (1977)), lattice polyhedra (Hoffman & Schwartz (1978)), generalized polymatroids (Frank (1984)), kernel systems (Frank (1979)), base polyhedra of submodular systems (Fujishige (1983)), polymatroid network flows (Hassin (1978)), (Lawler & Martel (1982)) and others are competing for the attraction of further investigators. Schrijver (1984 b), (1984 c) has surveyed, compared and extended these efforts.

It turns out that most of these frameworks are in a certain sense equivalent and that they indeed unify large parts of the existing theory. But it seems to me that the final word is not said yet and that it may take some more time to find a "best possible" setting of the theory of submodular functions (combined with graphs and digraphs) within combinatorial optimization.

All the structure results quoted so far are—of course—closely related to the algorithmic aspects of the optimization problems mentioned. A truly positive result of this theory is that most of the optimization problems associated with matroids, intersections of two matroids (polymatroids etc.) can be solved in polynomial time provided that one can check in polynomial time whether a set is independent in a matroid (an element of a polymatroid etc.) or not.

The most general result in this respect is due to Grötschel, Lovász & Schrijver (1981). It is based on the ellipsoid method (see Section 3.4) and shows that submodular functions can be minimized in polynomial time. More precisely,

(3.12) **Theorem.** *Let $f: 2^E \to \mathbb{Z}$ be a submodular function. Suppose that a positive integer B is known with $|f(S)| \leq B$ for all $S \subseteq E$. Then there exists an algorithm which finds a set $S^* \subseteq E$ such that $f(S^*) \leq f(S)$ for all $S \subseteq E$ and which runs in time polynomial in $|E|$, $\lceil \log B \rceil$ and the time necessary to evaluate the function f.* $\square$

So in particular, if there is an algorithm to evaluate the function f in time polynomial in $|E|$ and $\lceil \log B \rceil$ (this is the case for all practically relevant problems), then the algorithm (3.12) is polynomial in the sense described in Section 2.

A further interesting aspect of submodular functions is that it is $\mathcal{NP}$-complete to maximize them. The same phenomenon is also known for convex functions $f: \mathbb{R}^n \to \mathbb{R}$ which are easy to minimize but difficult to maximize. It turns out that there is a close connection linking submodular and convex functions. In a sense submodular functions can be viewed as the discrete analogues of convex functions. An interesting survey illuminating this aspect and describing more results about and applications of submodular functions is the paper by Lovász (1983).

Matching theory and matroid theory are in several respects well-understood, thus there have been continuing attempts to unify these theories. It seems that the best setting of such a generalization is the following. Suppose $G = (V, E)$ is a graph and $M = (V, \mathcal{I})$ is a matroid on the node set V. If $c: E \to \mathbb{R}$ is an objective function then the *matroid matching problem* is to find an *independent matching* $F \subseteq E$ of maximum total weight $c(F)$. Here an independent matching is a set $F \subseteq E$ which is a matching of G, for which $V(F) := \{i \in V | i$ is contained in some edge of $F\}$ is independent in M. (It is easy to see that the graph matching problem and the matroid intersection problem are special cases of the matroid matching problem.) However, it turned out, that the matroid matching problem is $\mathcal{NP}$-complete, see Lovász (1981). But Lovász, on the positive side, found a polynomial time algorithm to obtain a maximum cardinality matroid matching in case the given matroid is matric. This algorithm is one of the most involved algorithms known in this field. It is based on a geometric representation of the matroid matching problem and uses a number of ingenious new algorithms for problems in affine and projective geometry.

The weighted version of the matric matroid case is not solved yet. It seems plausible that this problem can also be solved in polynomial time, but what — probably — is lacking here, is a better understanding of the convex hull of the incidence vectors of the independent matchings, i.e. a description of this polytope by means of equations and inequalities.

Another outgrowth of matroid theory is the theory of greedoids. Greedoids have been defined in an attempt to obtain a better insight into the combinatorial structures for which the greedy method works.

A greedoid is a pair $(E, \mathcal{I})$ where E is a finite set and $\mathcal{I}$ is a subset of 2^E which satisfies (3.7) and (3.9). So a matroid is a greedoid satisfying (3.8) in addition. The main advocates of greedoids are B. Korte and L. Lovász who are currently working on a research program which aims at characterizing those combinatorial optimization problems which are greedoids, and obtaining richer substructures within the class of greedoids which give rise to certain min-max relations or polynomial time algorithms. For more information see Korte & Lovász (1983).

Shortest Paths, Cuts, and Flows

Up to now we have only considered combinatorial optimization problems which are solvable in polynomial time (with a few exceptions mentioned in side remarks). We shall now turn to a problem which is solvable in polynomial time only if the objective function is restricted in some way, and we shall indicate here some of the subtleties coming up in polyhedral combinatorics.

An instance of the shortest path problem can be described as follows.

Given a directed graph $D = (V, A)$, a function $c: A \to \mathbb{R}$, and two different nodes $s, t \in V$. An (s, t)-*path* P in D is a set of arcs $\{a_1, a_2, \ldots, a_k\}$ such that the tail of a_1 is s, the head of a_k is t, and the head of arc a_i is the tail of arc $a_{i+1}, i = 1, \ldots, k-1$. Moreover, we require that no node appears in P more than twice as an endnode of an arc $a_i, i = 1, \ldots, k$. (If P is an (s, t)-path and $(t, s) \in A$ then $P \cup \{(t, s)\}$ is a *directed cycle*.) The task is to find an (s, t)-path P such that $c(P)$ is as small as possible.

To treat the shortest path problem from a polyhedral point of view the first idea — of course — is to consider the polyhedron

$$P_{SP}(D) := \text{conv } \{\chi^P \in \mathbb{R}^A | P \subseteq A \text{ is an } (s, t)\text{-path}\}$$

and solve min $c^T x$, $x \in P_{SP}(D)$. However, there is no explicit linear description of $P_{SP}(D)$ known to date. And it is very likely that we will never be able to obtain such a characterization since the shortest path problem (in this general formulation) is $\mathcal{NP}$-complete.

On the other hand this problem can be solved in polynomial time if the objective function is nonnegative. (In fact, one can do a little better, namely the problem is solvable in polynomial time if D contains no directed cycle such that the sum of the arc weights of the directed cycle is negative, see for instance Lawler (1976). We do not want to go into these details here.) And in this case a polyhedral result is available. The polytope one should consider is

$$P_S(D) := \text{conv } \{\chi^P \in \mathbb{R}^A | P \subseteq A \quad \text{contains an } (s, t)\text{-path}\}.$$

It is clear that for every objective function $c : A \rightarrow \mathbb{R}$ with $c(a) \geq 0$, for all $a \in A$ there is always an optimum solution of min $\{c^T x | x \in P_S(D)\}$ which is the incidence vector of an (s, t)-path. So, in this case the shortest path problem can also be solved via the linear program min $\{c^T x | x \in P_S(D)\}$. The polytope $P_S(D)$ can be described as follows.

(3.13) **Theorem.** *Let $D = (V, A)$ be a digraph, and let s, t be two different nodes of V. Then*

$$P_S(D) = \{x \in \mathbb{R}^A | (1) \quad 0 \leq x_a \leq 1 \text{ for all } a \in A,$$
$$(2) \quad \sum_{a \in \delta^+(W)} x_a \geq 1 \text{ for all } W \subseteq V \text{ with } s \in W, t \notin W\}. \quad \square$$

A set $\delta^+(W)$ of arcs with $s \in W$ and $t \notin W$ is called an (s, t)-*cut*. Another interesting combinatorial optimization problem is, given a digraph $D = (V, A)$ and arc capacities $c(a) \geq 0$ for all $a \in A$, to find a minimum capacity (s, t)-cut. A polyhedral characterization of (s, t)-cuts is the following.

(3.14) **Theorem.** *Let $D = (V, A)$ be a digraph, and s, t be different nodes of V. Let*

$$P_{CT}(D) := \text{conv } \{\chi^B \in \mathbb{R}^A | B \subseteq A \text{ contains an } (s, t)\text{-cut}\}, \text{ then}$$

$$P_{CT}(D) = \{x \in \mathbb{R}^A | (1) \ 0 \leq x_a \leq 1 \text{ for all } a \in A,$$
$$(2) \quad \sum_{a \in P} x_a \geq 1 \text{ for all } (s, t)\text{-paths } P \subseteq A\}. \quad \square$$

Analogously to shortest paths there is no characterization of the convex hull of the incidence vectors of (s, t)-cuts of a diagraph known (the general cut problem, i.e. c not restricted to nonnegative vectors, is also $\mathcal{NP}$-complete).

The striking similarity of Theorems (3.14) and (3.13) is not just a coincidence. The (s, t)-paths and the (s, t)-cuts form what is called a *blocking pair*. This is a polarity relation, introduced by Fulkerson, which has found many nice applications in combinatorial optimization. The blocking theory has revealed many interesting connections between problems which were formerly considered quite unrelated. We recommend Fulkerson (1971) for a survey of this theory and the related theory of antiblocking polyhedra.

The (s, t)-cuts are important because of their relation to flows. An (s, t)-*flow* in a digraph $D = (V, A)$ is a vector $x \in \mathbb{R}^A$ satisfying

$$x_a \geq 0 \qquad \text{for all } a \in A$$

$$\sum_{a \in \delta^-(v)} x_a = \sum_{a \in \delta^+(v)} x_a \text{ for all } v \in V \setminus \{s, t\}.$$

The value of an (s, t)-flow x is the net amount of flow leaving s, i. e. this value is equal to

$$\sum_{a \in \delta^+(s)} x_a - \sum_{a \in \delta^-(s)} x_a$$

which is clearly equal to $\displaystyle\sum_{a \in \delta^-(t)} x_a - \sum_{a \in \delta^+(t)} x_a$. We say that an (s, t)-flow is subject to capacity $c : A \to \mathbb{R}_+$ if $x_a \leq c_a$ for all $a \in A$. One of the most celebrated theorems in combinatorial optimization is the following one due to Ford & Fulkerson (1956) and Elias, Feinstein & Shannon (1956).

(3.15) **Max-Flow Min-Cut Theorem.** *Let $D = (V, A)$ be a directed graph, let $s, t \in V, s \neq t$, and let $c : A \to \mathbb{R}_+$ be a capacity function. Then the maximum value of an (s, t)-flow subject to the capacity c is equal to the minimum capacity of an (s, t)-cut. If all capacities $c_a, a \in A$, are integer then there exists an integer optimum flow.* $\square$

Theorem (3.15) in particular implies that for a digraph $D = (V, A)$, two nodes $s, t \in V, s \neq t$, and capacities $c_a, a \in A$ the linear program (called *network flow problem*).

$$\max \sum_{a \in \delta^+(s)} x_a - \sum_{a \in \delta^-(s)} x_a$$

$$\sum_{a \in \delta^+(v)} x_a - \sum_{a \in \delta^-(v)} x_a = 0 \text{ for all } v \in V \setminus \{s, t\}$$

$$0 \leq x_a \leq c_a \text{ for all } a \in A$$

has an integral optimum solution whenever all the capacities c_a are integral.

Ford & Fulkerson (1956) gave an algorithm to compute a maximum (s, t)-flow and a minimum (s, t)-cut which is based on the min-max relation of

(3.15). Edmonds & Karp (1972) showed that this algorithm, with some modifications, is polynomial. This algorithm is one of the combinatorial algorithms which is most frequently used in practice. The reason is that quite a large number of real world problems can be formulated as network-flow problems.

Hard Combinatorial Optimization Problems

We now turn our attention to combinatorial optimization poblems which are $\mathcal{NP}$-complete. With each of these problems we can associate a class of polyhedra as described in section 3.1. For instance, consider the travelling salesman problem (TSP) (1.1). With each instance of (1.1) we associate the complete graph $K_n = (V, E)$ with $V = \{1, 2, \ldots, n\}$. Each tour is a hamiltonian cycle of K_n. Thus the vertices of the (symmetric) *travelling salesman polytope*

$$Q_T^n := \text{conv}\,\{\chi^T \in \mathbb{R}^E \mid T \subseteq E \text{ is the edge set of a hamiltonian cycle}\}$$

correspond to the feasible solutions of the n-city problem, and each instance of the TSP can be solved — in principle — via the LP

$$\min\ c^T x,\ x \in Q_T^n$$

where $c \in \mathbb{R}^E$ is a vector describing the distances between the cities.

Enormous research efforts have gone into describing the polytopes associated with hard problems. Up to date no single example of a hard combinatorial optimization problem could be found for which an explicit linear characterization of the associated class of polytopes could be determined. In retrospect, this is no surprise since complexity theory provides good reasons to believe that for $\mathcal{NP}$-complete problems no descriptions of the type discussed in the foregoing subsections for easy problems can ever be obtained, see Karp & Papadimitriou (1982) und Papadimitriou (1984) for precise versions of this statement.

The research effort spent on this type of investigations, however, was not in vain. For many of the practically relevant problems it was possible to determine large classes of facets. These classes of facets could be incorporated into cutting plane algorithms (to be described in Section 3.5) which for quite a number of problems seem to be the practically most efficient methods available at present.

Moreover, it was also possible to use this polyhedral information to determine further special cases of hard problems for which polynomial time algorithms exist. The number of results in this area is so vast that it is impossible to survey the main results and give proper credit. To give at least some examples I would like to mention two of my favourite problems.

The travelling salesman polytope Q_T^n is one of the best studied polytopes. Grötschel & Padberg (1984) have collected all the known results on Q_T^n and described in Padberg & Grötschel (1984) their algorithmic uses. Parts of these results can be summarized in the following.

(3.16) **Theorem.** *Let $n \geq 6$. Then the travelling salesman polytope Q_T^n is contained in the polytope defined by the following system of equations and inequalities:*

(1) $0 \leq x_e \leq 1$ *for all $e \in E$*

(2) $\displaystyle\sum_{e \in \delta(v)} x_e = 2$ *for all $v \in V$*

(3)   *subtour elimination constraints*

$$\sum_{e \in E(W)} x_e \leq |W| - 1 \quad \text{for all } W \subseteq V,\ 3 \leq |W| \leq n - 3 \text{ and } 1 \in W.$$

(4)   *comb constraints*

$$\sum_{e \in E(H)} x_e + \sum_{i=1}^{k} \sum_{e \in E(T_i)} x_e \leq |H| + \sum_{i=1}^{k} (|T_i| - 1) - \frac{k+1}{2}$$

(5)   *clique tree inequalities*

$$\sum_{j=1}^{s} \sum_{E(H_j)} x_e + \sum_{i=1}^{t} \sum_{e \in E(T_i)} \leq \sum_{j=1}^{s} |H_i| + \sum_{i=1}^{t} (|T_i| - t_i) - \frac{t+1}{2}.$$

(In (4) a comb consists of a node set H (called handle) and node sets $T_1, \ldots, T_k$ (called teeth) such that

i) $|H \cap T_i| \geq 1$ for $i = 1, \ldots, k$,

ii) $|T_i \setminus H| \geq 1$ for $i = 1, \ldots, k$,

iii) $T_i \cap T_j = \emptyset$ $1 \leq i < j \leq k$,

iv) $k \geq 3$ and k odd,

v) $1 \in H$.

In (5) a clique tree consists of a set of node sets $H_1, \ldots, H_s \subseteq V$ (called handles) and a set of node sets $T_1, \ldots, T_t$ (called teeth) satisfying

i) $T_i \cap T_j = \emptyset$ $1 \leq i < j \leq t$,

ii) $H_i \cap H_j = \emptyset$ $1 \leq i < j \leq s$, and $s \geq 2$,

iii)   for each $i \in \{1, 2, \ldots, t\}$, $2 \leq |T_i| \leq n - 2$ and some $v \in T_i$ belongs to no H_j for $j = 1, \ldots, s$,

iv)   for each $j \in \{1, 2, \ldots, s\}$ the number of T_i having nonempty intersection with H_j is odd and at least three,

v)   for $i \in \{1, 2, \ldots, t\}$ and $j \in \{1, 2, \ldots, s\}$, if $H_j \cap T_i \neq \emptyset$, then $H_j \cap T_i$ is an articulation set of the subgraph C of K_n with node set $\bigcup_{j=1}^{s} H_j \cup \bigcup_{i=1}^{t} T_i$ and edge set $\bigcup_{j=1}^{s} E(H_j) \cup \bigcup_{i=1}^{t} E(T_i)$, moreover C is connected.

And where t_i is the number of handles tooth T_i intersects.

Moreover, each of the inequalities of the system (1), (3), (4) and (5) defines a facet of Q_T^n, and no two of these inequalities are equivalent with respect to Q_T^n. The dimension of Q_T^n is $|E| - n$, and (2) is a minimal equation system for the affine hull of Q_T^n. □

It is obvious that the number of facets of Q_T^n described in Theorem (3.16) is incredibly large. Nevertheless these are by far not all of the facets of Q_T^n.

The second example I would like to mention is the ($\mathcal{NP}$-complete) *acyclic subdigraph problem*. An instance of this problem can be described as follows. Given a digraph $D = (V, A)$ with arc weights $c_a \in \mathbb{R}$ for all $a \in A$. Find an arc set $B \subseteq A$ which contains no directed cycle (i. e. B is an *acyclic arc set*) such that $c(B)$ is as large as possible. The associated polyhedron is

$$P_{AC}(D) := \mathrm{conv}\,\{\chi^B \in \mathbb{R}^A \,|\, B \subseteq A \text{ is acyclic}\}.$$

The following has been proved in Grötschel, Jünger & Reinelt (1982).

(3.17) Theorem. *Let $D = (V, A)$ be a digraph. Then each of the following inequalities defines a facet of $P_{AC}(D)$. No two of these inequalities are equivalent with respect to $P_{AC}(D)$.*

(1) $x_{ij} \geqq 0$ *for all $(i,j) \in A$,*

(2) $x_{ij} \leqq 1$ *for all $(i,j) \in A$ with $(j,i) \notin A$,*

(3) $\displaystyle\sum_{(i,j)\,\in\,C} x_{ij} \leqq |C| - 1$ *for all directed cycles $C \subseteq A$,*

(4)   *k-fence inequalities*

   $\displaystyle\sum_{e\,\in\,F} x_e \leqq |F| - k + 1$ *for all k-fences $F \subseteq A$,*

(5)   *Möbius ladder inequalities*

   $\displaystyle\sum_{e\,\in\,M} x_e \leqq |M| - \frac{k+1}{2}$ *for all Möbius ladders $M \subseteq A$,*

*In (4) a subdigraph $(V(F), F)$ of D is a **simple k-fence** if $V(F)$ consists of two disjoint node sets $U = \{u_1, \ldots, u_k\}$, $W = \{w_1, \ldots, w_k\}$ of cardinality k and F consists of all arcs (u_i, w_i), $i = 1, \ldots, k$ and all arcs (w_j, u_i) $i,j = 1, \ldots, k, i \neq j$, and where a k-fence is a digraph which can be obtained from a simple k-fence by repeated subdivision of arcs.*

In (5) a Möbius ladder is defined as follows. Let $C_1, C_2, \ldots, C_k$ be a sequence of different dicycles in a digraph $D = (V, A)$ such that the following holds:

(i) $k \geqq 3$ *and k is odd.*

(ii) C_i *and* C_{i+1} *(*$i = 1, \ldots, k-1$*) have a directed path* P_i *in common,* C_1 *and* C_k *have a dipath* P_k *in common.*

(iii)   *Given any dicycle* $C_j, j \in \{1, \ldots, k\}$, *set*
$$J = \{1, \ldots, k\} \cap (\{j-2, j-4, j-6, \ldots\} \cup \{j+1, j+3, j+5, \ldots\}).$$
Then every set $\left(\bigcup_{i=1}^{k} C_i \right) \setminus \{e_i | i \in J\}$ *contains exactly one dicycle (namely* C_j), *where* $e_i, i \in J$, *is any are contained in the dipath* P_i.

(iv)   *The largest acyclic arc set in* $\bigcup_{i=1}^{k} C_i$ *has cardinality* $\left| \bigcup_{i=1}^{k} C_i \right| - \dfrac{k+1}{2}$.

Then we call the arc set $M = \bigcup_{i=1}^{k} C_i$ *a* **Möbius-ladder.**

Let us set

$$P_C(D) := \{x \in \mathbb{R}^A | x \text{ satisfies } (1), (2), (3)\}.$$

The class of digraphs D with $P_C(D) = P_{AC}(D)$ is called *weakly acyclic*. In Section 3.4 we will show that programs over $P_C(D)$ can be solved in polynomial time. Thus, since for this class of digraphs $\max \{c^T x | x \in P_C(D)\}$ equals $\max \{c^T x | x \in P_{AC}(D)\}$ $(c \geqq 0)$, for weakly acyclic digraphs the acyclic subdigraph problem can be solved in polynomial time. This class, for instance, contains the planar digraphs as an (important) subclass. This indicates — and we shall formulate this in more detail later — that polyhedral results can be used to obtain good algorithms, in particular, for special cases of hard problems.

3.3   Integer Linear Programming and Cutting Planes

One of the main approaches in the fifties and sixties was to consider combinatorial optimization problems as integer linear programs and use the simplex method together with so-called cutting planes to solve these problems. The idea of this technique is the following.

Given an instance of a combinatorial optimization problem, find a matrix A and a vector b such that

(3.18) $\{x \in \mathbb{R}^n | A x \leqq b, x \text{ integer}\}$

corresponds to the feasible solutions of the instance; i.e., in the cases we consider, the set defined above should consist of all incidence vectors of feasible solutions.

For instance, if $G = (V, E)$ is a graph with edge weights $c_e \in \mathbb{R}$ for all $e \in E$ then the feasible solutions of the following integer linear program (cf. (3.5), (3.6)).

$$\max \ c^T x$$

$$\sum_{e \in \delta(v)} x_e \leqq 1 \qquad \text{for all } v \in V$$

(3.19) $\qquad x_e \geqq 0 \qquad \text{for all } e \in E$

$$x_e \text{ integer} \qquad \text{for all } e \in E$$

are exactly the incidence vectors of the matchings of G. Similarly, if $K_n = (V, E)$ is a complete graph and $c_e \in \mathbb{R}$ for all $e \in E$, then (cf. (3.16)) the feasible solutions of

$$\min \ c^T x$$

$$\sum_{e \in \delta(v)} x_e = 2 \qquad \text{for all } v \in V$$

(3.20) $\quad \sum_{e \in E(W)} x_e \leqq |W| - 1 \ \text{ for all } \ W \subseteq V, \ 3 \leqq |W| \leqq n - 3$

$$0 \leqq x_e \leqq 1 \qquad \text{for all } e \in E$$

$$x_e \text{ integer} \qquad \text{for all } e \in E$$

are the incidence vectors of the hamiltonian cycles in K_n.

In general, it is not too difficult to find such an integer linear programming formulation for any combinatorial optimization problem.

Gomory's Algorithm

The algorithmic aspect behind such formulations is the following. Let us remove the integrality stipulations from (3.18) and solve the linear program $\max \{c^T x \mid A x \leqq b\}$ (with the simplex method). (We call this LP the *linear programming relaxation* of the combinatorial optimization problem.) In case the optimum solution x^* of the LP is integral, then x^* is an optimum solution of the integer linear program and thus an optimum solution of the combinatorial problem is found. If x^* is not integral one would like to cut off x^* from $\{x \in \mathbb{R}^n \mid A x \leqq b\}$ by adding a further inequality (called *cutting plane*), say $a^T x \leqq a_0$, in such a way that

(3.21) $x^* \notin \{x \in \mathbb{R}^n \mid A x \leqq b, \ a^T x \leqq a_0\}$ and

$$\{x \in \mathbb{Z}^n \mid A x \leqq b\} = \{x \in \mathbb{Z}^n \mid A x \leqq b, \ a^T x \leqq a_0\}.$$

Gomory (1958), (1960) has devised a very simple method with which such an inequality can be read from the simplex tableau corresponding to the optimum solution x^*. And moreover, Gomory proved that by adding a finite number of his type of cutting planes an optimum solution of the corresponding integer program can be found.

However, practical computational experience revealed that Gomory's algorithm is very inefficient, and to my knowledge, there is no commercial code for integer programming problems which uses these cutting planes. There have been a number of attempts in the late sixties to devise new types of cutting planes different from the ones Gomory proposed, but there have been no computationally significant improvements.

This failure was one of the reasons that led to the developments discussed in Section 3.2. Consider for instance Edmond's characterization of the matching polytope (3.6) and the integer LP (3.19). Edmonds proved that the only inequalities that have to be added to (3.19) to obtain the matching polytope are the inequalities (3) of (3.6). Gomory's result shows that after a finite number of applications of his procedure all these inequalities can be obtained, but — and that turned out to be the case in practice — zillions of redundant inequalities may have to be added as well.

This observation suggests that Gomory's approach is too general. It is probably more effective to concentrate on particular problems and to characterize the cutting planes necessary for those problems (i. e. to describe the facets of the convex hull of the incidence vectors of the feasible solutions) in order to use this special type of problem specific cutting planes in LP-based algorithms. This idea will be further explored in Sections 3.4 and 3.5.

The Closure of $\{x \mid A x \leqq b\}$

Gomory's idea has been systematized and been brought into the form of a nice theorem by Chvátal (1973) and — in more generality — by Schrijver (1980). To describe this, let us define for a given rational (m, n)-matrix A and a rational vector b

(3.22) $P := \{x \in \mathbb{R}^n \mid A x \leqq b\}$ and $P_I := \text{conv} \{x \in \mathbb{Z}^n \mid A x \leqq b\}$.

Let a_i, $i = 1, \ldots, n$ denote the column vectors of A. If $\lambda \in \mathbb{R}^m$, $\lambda \geqq 0$ then the inequality

$$\sum_{i=1}^{n} (\lambda^T a_i) \, x_i \leqq \lambda^T b$$

is satisfied by all points in P, and clearly, every integral vector contained in P satisfies

(3.23) $\sum_{i=1}^{n} [\lambda^T a_i] \, x_i \leqq [\lambda^T b]$

where $[\alpha]$ denotes the largest integer not larger than α. Thus, (3.23) is a valid inequality for P_I. Let us denote by $cl(P)$ (closure of P) the set of vectors x in $\mathbb{R}^n$ satisfying $A x \leqq b$ and all inequalities of type (3.23) derived from $A x \leqq b$ in the way described above. To shorten notation we write $cl^r(P)$ for $cl(cl(\ldots cl(P)\ldots))$ where the closure operation cl is applied r times iteratively.

(3.24) **Theorem.** *Let P and P_I be as defined above. Then $cl(P)$ is a polyhedron containing P_I. Moreover, there exists a number $r \in \mathbb{N}$ such that*

$$P_I = cl^r(P). \quad \square$$

The smallest integer r with $P_I = cl^r(P)$ is called the *Chvátal rank* of P.

Total Unimodularity

It has been observed in the fifties that some matrices A associated with combinatorial optimization problems have the property that $\{x \in \mathbb{R}^n \mid Ax \leq b\}$ is a polyhedron with only integral vertices if and only if b is integral. This property is of particular importance, since it implies that if $b \in \mathbb{Z}^m$ the polyhedra P and P_I defined in (3.22) coincide, i. e. no cutting planes have to be added to obtain P_I from P.

Hoffman & Kruskal (1956) introduced the following class of matrices. A matrix A is called *totally unimodular* if the determinants of all its square submatrices are 0, 1 or -1, and they showed:

(3.25) **Theorem.** *Let A be an integral matrix. Then A is totally unimodular if and only if for every integral vector b the polyhedron $\{x \in \mathbb{R}^n \mid Ax \leq b, x \geq 0\}$ has integral vertices only.* $\quad \square$

This in particular implies that for a totally unimodular matrix A and integral vectors c and b the linear program $\{\max c^T x \mid Ax \geq b, x \geq 0\}$ as well as its dual linear program $\min \{b^T y \mid A^T y \geq c, y \geq 0\}$ have integral optimum solutions, provided feasible solutions exist.

Prime examples of totally unimodular matrices are the node-edge incidence matrices of bipartite graphs (these are the matrices defined by the left hand sides of the inequalities (2) of (3.5) and the node-arc incidence matrices of digraphs. (The matrices of network flow problems, cf. (3.15), are submatrices of such matrices.) Total unimodularity of these matrices is easy to prove by induction. Thus, in particular, the important Theorems (3.5) and (3.15) can be considered as consequences of Theorem (3.25).

There are quite a number of other characterizations of totally unimodular matrices, see Schrijver (1984b) for a survey. But none of these is a good characterization in the sense that it allows to check in polynomial time whether a matrix is totally unimodular or not. This problem has recently been solved by Seymour (1980) in the following way.

Tutte (1965) proved that a matrix A is totally unimodular if and only if the matric matroid defined by A over the reals is regular (see Welsh (1976) for a definition). Seymour then showed that a matroid is regular if and only if it can be constructed via three (rather simple) types of compositions starting from three types of matroids: forest matroids, duals of forest matroids and a particular 10-element matroid. Edmonds pointed out that this characterization can be used to decompose in polynomial time a given matric matroid into the three types of matroids or to show that the matroid is not regular, i.e. that the matrix is not totally unimodular.

Another interesting line of research building on the proof techniques developed for the treatment of the concepts defined above is the investigation of so-called *totally dual integral systems* (TDI-systems). These generalize totally unimodular systems in various ways. We do not go into details here and refer to the survey papers Edmonds & Giles (1984), Pulleyblank (1983) and the forthcoming book Schrijver (1984a).

3.4   *Separation Problems and the Ellipsoid Method*

We have already seen that cutting off points by hyperplanes is an old idea in combinatorial optimization which has led (among others) to the developments described in the foregoing sections. A new impetus came into the field through the ellipsoid method.

The ellipsoid method is an algorithm developed by Shor (1970), (1977) which has been considered for some time in nonlinear (in particular nondifferentiable) optimization. Khachiyan (1979) observed that the ellipsoid method can be modified (using some observations from linear algebra, number theory and complexity theory) in such a way that it yields a polynomial time method for the solution of linear programming problems. This result caused great and well-deserved excitement in the world of mathematical programming.

It turned out that the ellipsoid method has even more potential and that it is particularly suited for the design of polynomial time algorithms for combinatorial optimization problems. These observations have been made by Karp & Papadimitriou (1982), Padberg & Rao (1984) and Grötschel, Lovász & Schrijver (1981). We do not want to explain the ellipsoid method (this algorithm is only used as a proof technique and can be replaced by other algorithms like the new "simplex method" of Yamnitsky & Levin (1982)), but we would like to mention the most important consequences which have led to new fields of research.

To present the results correctly we would have to introduce quite a technical machinery. We want to avoid this and state the results in a slightly imprecise form, making, however, the essence of them clear. Let us introduce the following three problems:

(3.26) **Optimization Problem.** *Given a polytope $P \subseteq \mathbb{R}^n$ and a vector $c \in \mathbb{Q}^n$. Find a vector $x^* \in P$ maximizing $c^T x$ over P or prove that P is empty.*

(3.27) **Separation Problem.** *Given a polytope $P \subseteq \mathbb{R}^n$ and a vector $y \in \mathbb{Q}^n$. Decide whether $y \in P$, and if $y \notin P$, find a vector $d \in \mathbb{Q}^n$ such that $d^T y > d^T x$ for all $x \in P$ (i.e. find a hyperplane separating y from P):*

(3.28) **Membership Problem.** *Given a polytope* $P \subseteq \mathbb{R}^n$ *and a vector* $y \in \mathbb{Q}^n$. *Decide whether* y *belongs to* P *or not.*

Clearly, if one can solve the separation problem for P then one can solve the membership problem for P.

In order to speak about polynomial time algorithms for these problems we have to specify the input lengths. For the vectors c in (3.26) resp. y in (3.27), (3.28) this is just the length of their binary encoding (rationals r are encoded by encoding the numerator p and the denominator q of a coprime representation $r = \frac{p}{q}$.) An important consequence of the ellipsoid method is that we do not need to know all the facet defining inequalities or all the vertices of P explicitly to define the input length of P. It is sufficient to consider an upper bound on the maximum input length of a vertex or a facet defining inequality of P and to use this number as the input length. Thus, for the case of 0/1-polytopes $P \subseteq \mathbb{R}^n$, which we are particularly interested in, we can simply use the natural number n as input length of P, despite the fact that P may have a number of vertices and facets which is exponential in n (like the matching polytope (3.6) and the travelling salesman polytope (3.16)).

So, if we speak of a polynomial time algorithm to solve problem (3.26), (3.27) or (3.28) we mean that this algorithm runs in time polynomial in the input length of P (as defined above) and the input length of c resp. y. The following theorem is one of the most useful consequences of the ellipsoid method.

(3.29) **Theorem.** *(a) Let* $P \subseteq \mathbb{R}^n$ *be a polytope with rational vertices, then the optimization problem (3.26) for P can be solved in polynomial time if and only if the separation problem (3.27) for P can be solved in polynomial time.*

(b) Let $P \subseteq \mathbb{R}^n$ *be a full-dimensional polytope with rational vertices for which an interior point is known in advance. Then the optimization problem (3.26) for P can be solved in polynomial time if and only if the membership problem (3.28) for P can be solved in polynomial time.* □

Note that the condition necessary for the validity of (3.29) (b) is almost always fulfilled in the case of the 0/1-polytopes we consider. Most of these polytopes are full-dimensional and contain the zero vector and all unit vectors, so $\frac{1}{n+1}(1,\dots,1)$ is an interior point of these polytopes. If the polytopes are not full-dimensional (like the travelling salesman polytope) then usually an equality representation of the affine hull is known and the polytope can be projected to make it full-dimensional (in a lower dimensional space).

Theorem (3.29) thus states that in order to be able to optimize over P in polynomial time it suffices to be able to decide in polynomial time whether a given point belongs to P or not. The latter problem looks much easier but — as (3.29) states — has the same degree of difficulty as the former.

We want to describe two applications of this theorem which also show how good algorithms for some combinatorial optimization problems can be used to design good algorithms for others.

Consider the problem of finding a minimum capacity (s, t)-cut in a digraph $D = (V, A)$ with capacities $c_a \geq 0$ for all $a \in A$. We know that the shortest (s, t)-path problem can be solved in polynomial time, for instance with the method of Dijkstra (1959). By Theorem (3.14) we can find a minimum (s, t)-cut by solving the linear program

$$\min c^T x$$

(3.30) (1) $0 \leq x_a \leq 1$ for all $a \in A$,

  (2) $\sum_{a \in P} x_a \geq 1$ for all (s, t)-paths $P \subseteq A$.

To be able to solve (3.30) in polynomial time it suffices by Theorem (3.29) to solve the separation problem for the $0/1$-polytope $P_{CT}(D)$ defined by the constraints (1), (2) of (3.30) in polynomial time.

So, given a vector $y \in \mathbb{Q}^n$ we have to check in polynomial time whether y satisfies these two systems of inequalities. To check (1) is trivial. If one of the components of y is smaller than 0 or larger than 1 we obtain a separating inequality $x_a \geq 0$ or $x_a \leq 1$. In order to check the inequality system (2) of (3.30) we may therefore assume that y satisfies (1). Now we consider the components y_a of y as "lengths" of the arcs $a \in A$, and we calculate a shortest (s, t)-path P^* in D with respect to the length vector $y = (y_a)$ (by Dijkstra's method in polynomial time). Now if $\sum_{a \in P^*} y_a \geq 1$ then, since P^* is a shortest path, y satisfies all inequalities (2), otherwise the inequality $\sum_{a \in P^*} x_a \geq 1$ separates y from the polytope $P_{CT}(D)$.

This shows how a polynomial time minimum capacity cut algorithm can be derived via Theorem (3.29) from a polynomial time shortest path algorithm.

In the second example we consider the acyclic subdigraph problem. By Theorem (3.17) we know that the convex hull $P_{AC}(D)$ of the incidence vectors of acyclic subdigraphs of a digraph $D = (V, A)$ satisfies

$$P_{AC}(D) \subseteq P_C(D) := \{x \in \mathbb{R}^n | \text{ (1) } 0 \leq x_a \leq 1 \qquad \text{for all } a \in A$$

$$\text{(2) } \sum_{a \in C} x_a \leq |C| - 1 \qquad \begin{array}{l} \text{for all directed} \\ \text{cycles } C \subseteq A \}. \end{array}$$

Given a vector $y \in \mathbb{Q}^n$ we can easily check inequalities (1), and to check (2) we may assume that y satisfies (1). We now define new "lengths"

$$w_a := 1 - y_a \quad \text{for all } a \in A.$$

For each arc $(i,j) \in A$ we calculate a shortest (j,i)-path P_{ji} in D with respect to the length vector w. Clearly, for each arc $(i,j) \in A$, $C_{ij} := P_{ji} \cup \{(i,j)\}$ is the shortest directed cycle (with respect to w) in D containing (i,j). Let C^* be a shortest of these cycles C_{ij}, $(i,j) \in A$. Suppose $\sum_{a \in C^*} w_a \geqq 1$ then $\sum_{a \in C^*} y_a \leqq |C^*| - 1$ and thus — by our construction — y satisfies all directed cycle inequalities (2). If $\sum_{a \in C^*} w_a < 1$ then $\sum_{a \in C^*} y_a > |C^*| - 1$ and the inequality $\sum_{a \in C^*} x_a \leqq |C^*| - 1$ separates y from $P_C(D)$. Hence by using $|A|$ times a shortest path algorithm we can design a polynomial time separation algorithm for $P_C(D)$.

From Theorem (3.29) we can conclude that we can optimize over $P_C(D)$ in polynomial time. This shows that for the class of weakly acyclic digraphs the acyclic subdigraph problem can be solved in polynomial time.

Note that in both applications described above the number of facets and the number of vertices of the polytopes is exponential in the input size of D. Still one can optimize in polynomial time.

In various applications it is important to test membership or solve the separation problem. The design of such algorithms has been neglected for a long time. Now that the ellipsoid method has shown the polynomial time equivalence of these problems to the optimization problem (3.26), a considerable amount of research is spent on inventing truly good combinatorial separation algorithms which do not suffer from the numerical disadvantages of the ellipsoid method.

For instance, Padberg & Rao (1982) have shown how to solve the separation problem for the matching polytope (3.6), Cunningham (1984) has designed a good combinatorial algorithm to solve this problem for matroid polytopes (3.10). But there is still a lot to do. A major achievement would be the invention of a good combinatorial algorithm for the polymatroid separation problem. This would imply a good combinatorial algorithm for the minimization of submodular functions (3.12). The only known polynomial time method for this problem described in Grötschel, Lovász & Schrijver (1981) utilizes the ellipsoid method.

Generalizations of the ellipsoid method and further applications of this method to problems in number theory, geometry, and combinatorial optimization are described in the forthcoming book Grötschel, Lovász & Schrijver (1985).

We want to mention a further example. In an outstanding paper H. W. Lenstra jun. (1983) has shown that for fixed $n \in \mathbb{N}$, given a rational (m,n)-matrix A and a vector $b \in \mathbb{Q}^m$, one can check in polynomial time whether or not there is an integral vector x satisfying $A x \leqq b$. This result implies that integer linear programming problems in fixed dimension can be solved in polynomial time. This result follows quite easily from a newly developed method to find a reduced basis of a lattice (see Lenstra, Lenstra & Lovász (1982)) and the ellipsoid method.

More generally, in fixed dimension one can even minimize a convex function over the integral points of a convex body in polynomial time (cf. Grötschel, Lovász & Schrijver (1985)).

3.5   Cutting Plane Algorithms in Practice

In the foregoing sections we have mainly concentrated on theoretical issues. Now we turn to computer implementations of the ideas described before. We have outlined Gomory's cutting plane algorithm and mentioned its practical failure. The ellipsoid method together with the investigations of special combinatorial polytopes has shed a new light on this subject and seems to indicate that good cutting plane algorithms might exist provided that the cutting planes are selected carefully. We shall describe now what this means in practice.

The approach works in the same way for $\mathcal{NP}$-complete problems and problems in $\mathcal{P}$. We also want to point out that this approach dates back to the fifties where Dantzig, Fulkerson & Johnson (1954) have used it for the solution of a large Travelling Salesman Problem. The methods described in this paper have been neglected for a long time and found a revival only recently. We outline the basic issues of this technique by means of the $\mathcal{NP}$-complete acyclic subdigraph problem.

Each instance of this problem is given by a digraph $D = (V, A)$ and arc weights c_a, $a \in A$. As described in Section 3.1 resp. 3.2 we associate with each instance of this problem a polytope $P_{AC}(D)$ and want to solve

$$(3.31) \quad \max \{c^T x \mid x \in P_{AC}(D)\}.$$

By Theorem (3.17) we know that $P_{AC}(D)$ is contained in the polytope defined by

$$(3.32) \quad
\begin{array}{lll}
(1) & 0 \leq x_a \leq 1 & \text{for all } a \in A \\[2mm]
(2) & \sum_{a \in C} x_a \leq |C| - 1 & \text{for all directed cycles } C \subseteq A \\[2mm]
(3) & \sum_{a \in F} x_a \leq |F| - k + 1 & \text{for all } k\text{-fences } F \subseteq A \\[2mm]
(4) & \sum_{a \in M} x_a \leq |M| - \dfrac{k+1}{2} & \text{for all Möbius ladders } M \subseteq A
\end{array}$$

and that (almost) all of these inequalities define facets of $P_{AC}(D)$. Instead of solving (3.31) we try to optimize $c^T x$ over (3.32). Using the ellipsoid method we know that we can optimize over the polytope $P_C(D)$ defined by (1) and (2) of (3.32) in polynomial time. But the ellipsoid method is very inefficient in practice, so we replace it by (the nonpolynomial) simplex algorithm. Practical experience has shown that the simplex algorithm works extremely fast on the average. (Recent work of Borgwardt (1982) has established a theoretical explanation of this.) Moreover, numerical experiments indicate that one can indeed optimize over $P_C(D)$ in this way for quite large digraphs.

For this we have to use the separation algorithm for $P_C(D)$ described in Section 3.4. By choosing data structures carefully one can implement this separation algorithm (based on shortest path techniques) so that it runs in $0(|V(D)|^3)$ time. For large digraphs this is quite a lot, so one often does a preprocessing by running fast problem specific heuristics that try to find violated inequalities. The design of such heuristics is guided by a careful analysis of fractional solutions that come up during practical experience with such an algorithm. It turned out empirically that such heuristics can significantly speed up the actual performance.

Now it may happen that a solution x^* satisfies all inequalities (1), (2), i. e. x^* optimizes $c^T x$ over $P_C(D)$, but that it is not integral. Then we try to cut off x^* using the inequalities (3) and (4) of (3.32). These inequalities are best possible cutting planes, since they define facets of $P_{AC}(D)$, but we do not know any (nontrivial) algorithm that checks whether x^* satisfies the inequalities (3), (4) or not. (This situation usually occurs in $\mathcal{NP}$-hard problems. There are "recognizable classes" of facets and some, which are not "well-behaved".) In such a case, we again design heuristics, hoping that they will find some violated inequalities of type (3) or (4). It might happen (and in practice it usually does) that by iterating this cutting plane recognition procedure (using separation heuristics combined with exact separation algorithms) one ends up with an integral solution. In this case we have solved the acyclic subdigraph problem.

If, however, the last optimum solution x^* is not integral and no inequality of type (1),...,(4) can be found that is violated by x^*, then we go to our last resort: branch & bound, cf. (4.6). Practical experience shows that in most cases only a few branching steps are necessary to get to the optimum integral solution.

There are quite a number of further tactical issues involved. One has to decide how many cutting planes to add in each step in order to keep the LP small, whether inequalities which are nonbinding in the present optimum solution should be removed etc. Our practical experience with this type of algorithms shows that there is no general answer. Each problem has to be studied individually. But there is a good chance that such investigations result in practically quite efficient methods.

Computational experience with cutting plane methods for hard problems as described above is reported for instance in Crowder & Padberg (1980), Barahona & Maccioni (1982), Grötschel, Jünger & Reinelt (1983), Crowder, Johnson & Padberg (1983). Such an algorithm has been implemented for the polynomially solvable matching problem as well. A surprising outcome of the computational experiments reported in Grötschel & Holland (1984) is the empirically observed fact that this (theoretically nonpolynomial) algorithm is as fast as the best combinatorial matching algorithms. Thus, it seems that the type of cutting plane algorithms described above deserves further attention, even in the case of polynomially solvable problems.

4.   **Future Developments**

It is impossible to survey — given bounded space — all flourishing branches of combinatorial optimization and to discuss all significant recent results. In Section 3 I have made an attempt to outline the developments in polyhedral combinatorics, the subject closest to my own research interest, and I have already pointed out various directions of future research in this area.

I will now present a (nonsystematic and probably unbalanced) collection of further topics and problems which I think have future potential from a theoretical or practical viewpoint and which are worth studying. I will give only few comments and quote only very few references in order to keep within my page limits. I am sure that my opinion is biased but I hope that some readers may find something of interest.

4.1   *Relations to other Branches of Mathematics*

In the first two sections (4.1) and (4.2) I would like to point out some of my views about possible general future developments in the relations of combinatorial optimization to other mathematical fields, and I will give a few remarkable examples.

(4.1)   **Integer Programming and Number Theory**

It seems that integer programming and number theory (in particular the geometry of numbers) study the same objects, but from very different viewpoints and using quite unrelated methods. There should be a way to make the deep results collected in number theory in the last centuries profitable for integer programming. And vice versa, number theory might benefit from some of the concepts and algorithms developed in integer programming.

For instance, ways to compute the Smith or Hermite normal form of a matrix are known for decades, but only recently Kannan & Bachem (1979) found an algorithm — using some nice combinatorial "tricks" — that calculates these normal forms in polynomial time. On the other hand, H. W. Lenstra jun. (1983), as mentioned in Section 3.4, proved — using number theoretic arguments — that integer programming problems are solvable in polynomial time in fixed dimension. In Grötschel, Lovász & Schrijver (1984) a (still minor) attempt is made to build a bridge between these two disciplines. In particular, a number of algorithmic versions of various results known in the geometry of numbers are proved.

It seems to me that the algorithm of Lenstra, Lenstra & Lovász (1982) to find a reduced basis in a lattice might be a first good tie between these two areas (and algebra in addition). The algorithm was developed to derive certain results in polyhedral theory (in particular about polytopes associated with combinatorial optimization problems) from the ellipsoid method. It turned out that it can also be utilized, cf. Lenstra et al. (1982), to factor polynomials over the rationals in polynomial time. Kannan and Lovász observed that one can derive from the basis reduction algorithm a method which, given an algebraic number and a bound on the degree of its minimal polynomial, computes the minimal polynomial in polynomial time. The basis reduction algorithm has recently been applied by H. te Riele and A. Odlyzko, see te Riele (1983), to disprove the long standing Mertens' Conjecture. Moreover, the basis reduction algorithm is currently being used by several people to break cryptosystems, see e. g. Adleman (1983) and Lagarias & Odlyzko (1983).

Cryptography, anyway, seems to be one of the reasons that has attracted number theorists to study complexity questions. The exciting developments in prime testing and factoring of integers clearly show that nontrivial number theory is able to contribute substantially to a better understanding of very down-to-earth complexity or integer programming problems.

(4.2)   Relations of Combinatorial Optimization to Other Mathematical Disciplines

I have already mentioned before various contacts and fruitful cooperations of the theory of combinatorial optimization with other branches of mathematics. There is no way to be complete, but I would like to mention a few more of these which I think are worth investigating.

Polyhedral combinatorics should be able to benefit significantly from the developments in *convex geometry*, in particular the (general) *theory of polyhedra*. Although the objects of study in these two areas are more or less the same there have been very few applications of the "general theory" (as for instance described in the book Grünbaum (1967)) to the study of concrete polyhedra (as described in Sections 3.1 and 3.2). Mainly due to the work of V. Klee and his collaborators the contacts get closer and more of the geometers get interested in "real-world polyhedra" like matching polytopes and travelling salesman polytopes. Maybe the new techniques developed in the theory of polyhedra will lead to a proof of the Hirsch conjecture, or at least to proofs of this conjecture for more classes of interesting combinatorial polyhedra (see for instance Klee & Kleinschmidt (1984)). These lines of research may also produce a polynomial time version of the simplex method, something of real practical interest.

A striking example of a fertile application of *commutative algebra* are the results of Stanley concerning the enumeration of faces of various dimensions of polytopes. Stanley extended and applied the theory of graded algebras and their Hilbert functions, and Cohen-Macauly rings to give (among others) tight upper bounds on the number of faces in each dimension in terms of the number of vertices. Subsequently, he combined these methods with some recent results in *algebraic geometry* to complete the proof of McMullen's conjectured characterization of the face-counting vectors of simplicial polytopes. An excellent account of these developments is given in Billera (1983). Billera also describes the first attempts of an application of these methods to the study of integer solutions to systems of linear inequalities. I am sure that the power of these methods has not fully been recognized yet, and that we may expect further interesting results from this approach.

I also believe that the methods of *algebra* have not been exploited enough yet. Of course, algebra is so vast a field that in the concrete situation of a combinatorial optimization problem it is almost impossible to guess which of the algebraic techniques might fruitfully apply. Algebraic concepts certainly have influenced the field. There are matching polynomials, chromatic polynomials, chain groups and the like. But as Stanley's approach shows, in certain special cases the use of more sophisticated groups, rings etc. associated with a combinatorial object may result in deep new insights. This area of research is almost untouched.

In general, the question we address here is "What is a 'good' representation of a combinatorial optimization problem?". The survey in Section 3 shows how combinatorial optimization problems can be represented by means of polyhedra, and it also proves that this method has lead to theoretically and practically exciting new developments. But there is a host of further possibilities.

The paper Lovász (1982), for instance, describes two interesting further approaches. First Lovász discusses a method introduced by W. W. and S. R. Li to associate a certain polynomial with a graph. The ideas of this and the proof techniques come from *algebraic geometry*. Using Hilbert's Nullstellensatz it is possible to determine the degrees of these polynomials and to obtain estimations from these degrees for the chromatic number and the stability number of a graph. In the second approach *topology* is used. With each graph G a neighbourhood complex $\mathcal{N}(G)$ is associated whose connectivity gives a lower bound on the chromatic number. Both applications use nontrivial theory to obtain quite surprising connections between algebraical resp. topological invariants and combinatorial parameters. Applications of this kind, however, are quite sporadic. This is probably due to the fact that there are only very few people who know enough from either area to see nontrivial connections.

The relations of combinatorial optimization to other branches of mathematical programming are manifold. Nonlinear programming, stochastic programming, and integer programming borrow from and stimulate each other. Two significant contributions of nonlinear programming to combinatorial optimization are the development of the ellipsoid method (see Section 3.4) and its extension to a powerful tool in combinatorial optimization, and moreover, the technique of Lagrangean relaxation of (hard) combinatorial optimization problems together with the design of subgradient algorithms (incorporated in branch & bound schemes) for the solution of these Lagrangean relaxations (cf. Fisher (1981) and Geoffrion (1974) for surveys).

4.2   *General Areas of Future Research*

The next problem areas concern developments in combinatorial optimization which, I think, need further study and which may contribute significantly to a better theoretical understanding of combinatorial optimization or to a better use of the theory in practice.

(4.3)   **Generalizations of Min-Max Results**

The presently best survey of min-max results in combinatorial optimization theory is Schrijver (1983). Schrijver has collected all results known to date, classified them with respect to area and generality and described their relation. I have already mentioned in Section 3.2 (for the special case of submodular functions) that major research projects are carried out to unify these results and find a general setting which allows a better understanding of the fact that certain combinatorial objects stand in a min-max relation to others (cf. König's Theorem in Section 3.2) while other quite similar ones do not.

Major contributions to this area are, for instance, the results of Mader (1978 a) (1978 b) on edge resp. vertex disjoint S-paths and Seymour's results on flows in matroids, cf. Seymour (1977) (1981).

Further progress in this min-max theory is highly desirable not only from a theoretical point of view. Min-max relations usually are good optimality criteria and therefore often form the backbone of polynomial time algorithms.

(4.4)   **Speed-Up and Lower Bounds for Easy Problems**

Algorithmic research with respect to problems solvable in polynomial time mainly concentrates on the two subjects mentioned in the heading. There has been significant progress with respect to the first in the recent years, while almost no nontrivial results can be reported about the second.

There are two ways to get better algorithms for easy problems. Either one finds a new method with better time or space complexity or one modifies one of the existing algorithms in some way. The first case — of course — is rather rare. The research efforts on the second in the recent years were quite successful and have not only brought up a list of new "tricks" with which such speed-ups can be obtained, but also a general theory of algorithmic techniques and data handling procedures which provides a powerful tool for algorithm improvements. A very nice account of this theory can be found in Tarjan (1978). Gabow (1983) reports about the success of the scaling technique.

For example, the speed-up techniques were particularly successful with respect to calculating maximum flows in networks and shortest paths in digraphs. The original (nonpolynomial) Ford-Fulkerson algorithm for network flows has been modified and remodified in various ways and quite substantial running time improvements were obtained, see Tarjan (1983a) for a survey of this. There are a number of competing shortest path algorithms each of which has been subject to various modifications. Recent improvements on shortest paths methods in planar graphs are reported in Frederickson (1983). A new variant of Dantzig's algorithm with good expected running time is presented in Bloniarz (1983).

A surprising example of speed improvement by a new method was the $0(n^{\log_2 7})$ matrix multiplication algorithm of Strassen (1969). Stimulated by this, there has been further progress in the meantime, see for instance Coppersmith & Winograd (1982) where a matrix multiplication algorithm with running time less than $0(n^{2.495548})$ is described. These authors give a speed-up theorem and by this they also show that there is no best matrix multiplication algorithm. A similarly striking case is the planarity algorithm of Hopcropft & Tarjan (1972) which proved that for a graph $G = (V, E)$ planarity can be tested in $0(|V|)$ time.

Whenever such improvements are obtained a question that arises is whether or not this new algorithm is best possible with respect to time (or space) complexity. This leads to the task of finding lower bounds for the computational complexity of a combinatorial optimization problem (with respect to some machine model, like RAM or Turing machines). So the question is, can one prove that for every instance of a combinatorial optimization problem of input length n at least $p(n)$ steps are necessary for its solution, where $p: \mathbb{N} \rightarrow \mathbb{N}$ is some function (e. g. a polynomial).

There are some trivial bounds. For instance if two (n, n)-matrices have to be multiplied then each entry of the two matrices has to be touched at least once. Thus, at least $2n^2$ steps are necessary to compute the product of two (n, n)-matrices. Similarly, for the determination of certain graph parameters all nodes or all edges have to be examined at least once. So one gets lower bounds $|V|$ or $|E|$ for the number of steps necessary to calculate this parameter, see Rivest & Vuillemin (1978).

It is somewhat astonishing that such lower bounds are often the only ones available, and for most easy combinatorial optimization problems the gap between the complexity of the best known algorithm and the best lower bound for its solution is considerable. A discussion of methods to establish lower bounds can be found in Weide (1977).

(4.5) Heuristics for Hard Problems

Most of the combinatorial optimization problems that come up in the real world are $\mathcal{NP}$-complete. But solutions have to be found, e. g. for the routing of garbage collection trucks or the layout of a computer chip, and mathematicians cannot hide behind an intractability proof and leave the problem to the imagination of economists or engineers. Practice demands the design of heuristic algorithms which produce a "good" solution of the problem.

Heuristic methods for hard problems are widely used by practitioners with (more or less) satisfactory success. For many years the judgement of the quality and effectiveness of heuristic methods was largely based on empirical computational experience. That is, some test runs on "representative" real world and some "representative" randomly generated problem instances were performed and the method which yielded the "best result on the average" was chosen to be used.

The recent years have seen an increasing mathematical interest in the performance analysis of heuristics. In particular, two new tools — "worst-case-analysis" and "average case analysis" — have been developed and provide reasonable means to judge the quality of a heuristic algorithm.

In *worst-case analysis* one tries to prove a performance guarantee for a heuristic, i. e. to show that for every instance of an optimization problem a certain algorithm produces a solution whose value differs from the optimum value by no more than, say, p per cent. Of course, one would like to have a very fast algorithm with best possible performance guarantee.

Results of this type are surveyed in Fisher (1980), Garey & Johnson (1979), Grötschel (1982) and Korte (1979). It turns out that hard problems differ very much with respect to their "approximability".

For instance, no performance guarantee for polynomial time heuristics can be given at all for the symmetric travelling salesman problem (1.1), unless $\mathcal{P} = \mathcal{N}\mathcal{P}$. But if the intercity distances c_{ij} satisfy the triangle inequality $c_{ij} + c_{jk} \geq c_{ik}$, $1 \leq i < j < k \leq n$, then the $0(n^3)$ algorithm of Christofides (1976) produces a tour which is at most 50 % longer than the optimal tour. It is unknown whether a smaller bound can be achieved in polynomial time.

There are hard problems for which heuristics with provably best possible polynomial time performance guarantee exist. For example, Hochbaum & Shmoys (1984) describe a polynomial time heuristic for the k-center problem (with triangle inequality) which gives a solution whose value is at most twice as large as the optimum value, and they prove that the existence of a polynomial time approximation algorithm with better performance guarantee would imply $\mathcal{P} = \mathcal{N}\mathcal{P}$.

Very few hard problems can be approximated up to any given accuracy. One such example is the knapsack problem. (Find a 0/1-vector maximizing $c^T x$ over all 0/1-vectors satisfying $a^T x \leq b$.) For this problem Ibarra & Kim (1975) have designed a so-called *fully polynomial approximation scheme* which is an algorithm that, given an instance of the knapsack problem and a rational $\varepsilon > 0$, produces a solution S such that the error of the value $c(S)$ of S relative to the optimum value is at most ε, and which has a running time which is polynomial in the input length of the instance and ε^{-1}.

Striking progress has been made with respect to the bin packing problem (pack n items into as few bins as possible). The First Fit heuristic (Pick any item and put it into the first bin into which it fits!) was shown by Garey et al. (1976) to give no more than $\frac{17}{10}$th of the optimum number of bins, D. Johnson (1973) proved that a variant of this, First Fit Decreasing, produces $\frac{11}{9}$ th of the optimum value, Yao (1980) improved this to $\frac{11}{9} - \varepsilon$ for some very small $\varepsilon > 0$. Finally Fernandez de la Vega & Lueker (1981) gave a linear time algorithm — based on a linear programming relaxation — which for any fixed $\varepsilon > 0$ gives no more than $(1 + \varepsilon)$ times the optimum number of bins. But the running time of this algorithm increases very quickly with ε getting small. Karmarkar & Karp (1982) considered the dual of the Fernandes de la Vega & Lueker linear program and applied the ellipsoid method to obtain a fully polynomial approximation scheme, see Coffman, Garey & Johnson (1983) for a survey.

For many hard combinatorial optimization problems no polynomial time heuristic algorithms with good (or any) performance guarantee are known. For example, for the acyclic subdigraph problem, see Section 3.2, a trivial algorithm gives a 50 % relative error (Take any linear ordering of the nodes of $D = (V, A)$, let B be the arcs of D which are consistent with this ordering and let $B' := A \setminus B$. Then clearly B and B' are acyclic and the value of B or B' is at least $1/2$ of the optimum value!), but nothing better is available. For the asymmetric travelling salesman problem (with triangle inequality) no constant bound is known. All known performance guarantees depend on the number of cities.

For problems, like the TSP, where no performance guarantee can be given at all, worst-case analysis is not an appropriate tool to judge a heuristic algorithm. More promising in such (but not only in these) cases is a *probabilistic analysis* of the performance of a heuristic. The idea here is the following. Given a combinatorial optimization problem, then with each $n \in \mathbb{N}$ a probability distribution over the instances of input size n is associated. Then one tries to prove that with probability tending to 1 (with growing input size) a certain heuristic produces a solution whose value is ε-close to the optimum solution.

Algorithms with such a good *average-case* behaviour have been designed for various hard combinatorial optimization problems, (e. g. the TSP and the acyclic subdigraph problem), but this area is still in its infancy. Most of the algorithms that have been analyzed are extremely simple (and it has empirically been observed that they perform rather poorly (within the practically relevant problem sizes) compared with other widely used heuristics). It seems, however, to be a very difficult problem to make the stochastic machinery work for more sophisticated algorithms.

Surveys of this approach are Karp (1976), Lueker (1979) and Weide (1980). Two interesting papers, for example, are Halton & Terada (1982) and Burkard & Finke (1984). An annotated bibliography of the probabilistic analysis of algorithms is Karp (1984).

(4.6)   Exact Optimization Algorithms for Hard Problems

In many real world situations it is necessary (or desirable or profitable) to know the true optimum solution of a problem instance and not only a "good" solution. Thus, algorithms have to be designed for hard problems which empirically show good running time performance. It would of course be even better to be able to prove that such algorithms run fast on the average (within some probability model), but to my knowledge no result of this type has been obtained so far for $\mathcal{NP}$-complete problems.

There are two principle methods available for solving hard problems exactly. One is the cutting plane technique described in Sections 3.3 and 3.5. The other is the branch & bound method.

The guiding idea of the branch & bound technique is to enumerate all feasible solutions in an "intelligent manner". The enumeration is organized in such a way that at every step the universe of feasible solutions is partitioned into disjoint subsets and that at every step lower and/or upper bounds for the value of the best solution within these subsets are computed. If (in case of a maximization problem) this upper bound for a certain subset is smaller than the present best known feasible solution or the best known lower bound, all solutions contained in this subset can be omitted from further considerations. Otherwise the subset is split into smaller pieces to obtain a finer partition of the set of all feasible solutions, and the procedure is continued.

It is apparent that the relative success of a branch & bound method heavily depends on the partitioning strategy and the quality of the bounds that are computed. The determination of bounds is the most important feature of such algorithms and a large part of the research effort in the last years has gone into finding methods for computing good bounds with reasonable computational effort.

It is impossible to compare these two approaches in general with respect to their quality. Historically the cutting plane methods came first. They were then superseded in the sixties by the branch & bound algorithms. In particular the technique of Lagrangean relaxation developed in the early seventies considerably improved the performance of these algorithms. Now there is a revival of cutting plane methods as described in Section 3.5. But still all such quality judgements are extremely problem specific. An approach, working well for some problem, might fail in another. This also shows that we do not know enough about the "character" of hard problems.

The current trend is to combine the two methods and enrich them with various heuristic features. These approaches and their combination are described in Grötschel (1982). Most of these techniques have in fact been developed in order to solve travelling salesman problems. This problem seems to have become a standard hard problem for which everybody tries to show the success of his new ideas. The branch & bound algorithms existing for the travelling salesman problem are surveyed in Balas & Toth (1983) and the cutting plane methods for this problem in Padberg & Grötschel (1984).

(4.7) **Adaptation to Technical Progress**

Without the rapid development of computers the explosive growth of combinatorial optimization in the last thirty years is unimaginable. It is absolutely hopeless to try to solve 15-city travelling salesman problems by hand. However, even the biggest computer cannot handle 30-city problems using brute force enumeration only. But the joint progress in mathematics, algorithm design techniques and computer hardware makes it possible to solve 150-city problems routinely, and even problems with more than 300 cities have been solved, cf. Crowder & Padberg (1980).

The last years brought two new technical developments: microcomputers and parallel computers. Both these two technologies have already influenced the research in combinatorial optimization.

With respect to microcomputers an attempt is made to design algorithms for combinatorial optimization problems which are particularly suited for these types of machines, especially to give small companies which have no access to large computing facilities the possibility to benefit from the results obtained in mathematical programming. The public interest in these developments is, for instance, shown by the growing number of technical sessions on this subject at meetings of the Operations Research Society of America. I have, however, the feeling that due to further technical progress this problem area will disappear within the next ten years, say. I believe that in a few years we shall have cheap desk calculators with 10 Megabyte or more central memory, and so there will be no need any more for special purpose algorithms for computers with small central or peripheral memory.

More significant is parallelization. Most of the existing algorithms cannot be parallelized, and so really new methods have to be developed to exploit the power of these new computers. Very interesting progress has been made in this area in the recent years. It is impossible to survey all the theoretical and practical aspects of parallelism here. We recommend the very up-to-date annotated bibliography Kindervater & Lenstra (1984) on this subject, and the $\mathcal{NP}$-completeness columns seven and eight by D. S. Johnson (1983) in the Journal of Algorithms.

4.3   *Some Concrete Open Problems*

Before, I have outlined general developments in combinatorial optimization which I expect or hope for. Now I would like to mention a few concrete open problems which I am interested in. Most of the problems are probably nontrivial. I do not consider this problem list representative for the field. The problems reflect my own research interests.

The first group of problems could be called "*combinatorialization of ellipsoidal results*". What I mean by this is the following. A host of combinatorial optimization problems was shown to be solvable in polynomial time using the ellipsoid method, cf. Section 3.4. The ellipsoid method is — for various reasons

— not a really good algorithm from the practical point of view. But for quite a number of combinatorial problems these ellipsoidal algorithms are the only polynomial ones known to date. Thus, I would like to see good combinatorial algorithms for these problems. The following problems are of particular interest.

(4.8)   **Problem.** *Find a polynomial time combinatorial algorithm to solve the weighted stable set, clique, clique covering and colouring problem on perfect graphs.*

There has been some progress recently by the Grenoble group (Burlet, Fonlupt, Uhry et al.) who gave good combinatorial algorithms for these problems for large classes of perfect graphs, but the general case seems to be hard. A side remark on perfect graphs! Maybe a proof of

(4.9)   **The Perfect Graph Conjecture.** *A graph is perfect if and only if it contains neither an odd chordless cycle of length at least five nor the complement of such an odd chordless cycle as an induced subgraph.*

could shed new light on the structure of perfect graphs and provide the tools for a good algorithm, see the books Golumbic (1980) and Berge & Chvátal (1984) for perfect graphs and conjecture (4.9).
The second problem of this ellipsoidal group is:

(4.10) **Problem.** *Find a polynomial time combinatorial algorithm to minimize submodular functions.*

Problem (4.10) has already been mentioned in Section (3.2), see Theorem (3.12). Such an algorithm could be derived from a positive solution of

(4.11) **Problem.** (Polymatroid Separation Problem). *Given a submodular function $f: 2^E \to \mathbb{Q}$ satisfying $f(\emptyset) = 0$ and $S \subseteq T \subseteq E \Rightarrow f(S) \leq f(T)$. Set*

$$P_f := \{x \in \mathbb{R}^E \mid \sum_{e \in F} x_e \leq f(F) \quad \text{for all } F \subseteq E$$
$$x_e \geq 0 \qquad \text{for all } e \in E\}.$$

Find a combinatorial polynomial time separation algorithm for P_f. (In fact it suffices to find an algorithm which checks whether the point $(1,1,\ldots,1) \in \mathbb{R}^E$ is contained in P_f.)

Cunningham (1984) has a good algorithm (based on network flow techniques) which solves (4.10) in case f satisfies in addition $f(S) \leq |S|$ for all $S \subseteq E$.
      In order to get efficient cutting plane algorithms for hard combinatorial optimization problems it is necessary to have polynomial time separation algorithms for large classes of facet defining inequalities of the associated polyhedra, see Sections 3.1, 3.2, 3.4 and 3.5. With respect to the travelling salesman polytope Q_T^n the following problem is unsolved, cf. Theorem (3.16).

(4.12) **Problem.** *Find polynomial time algorithms which check whether a given point $y \in \mathbb{Q}^E$ satisfies the comb inequalities (3.16) (4) resp. the clique tree inequalities (3.16) (5) and if not provide a violated inequality of this type.*
Padberg & Rao (1982) can handle a special case of the comb inequalities (the so-called 2-matching inequalities), but not more is known. Similarly for the acyclic subdigraph problem we have (cf. Theorem (3.17)).

(4.13) **Problem.** *Find polynomial time algorithms that check whether a given point $y \in \mathbb{Q}^A$ satisfies all k-fence inequalities (3.17) (3) resp. all Möbius ladder inequalities (3.17) (4) and if not yield a violated inequality of this type.*

It follows from results of Grötschel, Lovász & Schrijver (1981) that the facets of polyhedra associated with easy problems are algorithmically well-characterized in the following sense.

(4.14) **Theorem.** *If $P_{\mathscr{I}} \subseteq \mathbb{R}^n$ is a polytope associated with a combinatorial optimization problem (as described in 3.1) which is solvable in polynomial time and if $c^T x$ is an objective function such that $\gamma = max\{c^T x \mid x \in P_{\mathscr{I}}\}$, then one can find in polynomial time n facet defining inequalities $a_i^T x \leqq \alpha_i$ and nonnegative rationals λ_i, $i = 1, \ldots, n$ such that*

$$c = \lambda_1 a_1 + \ldots + \lambda_n a_n \text{ and } \gamma = \lambda_1 \alpha_1 + \ldots + \lambda_n \alpha_n. \quad \square$$

Theorem (4.14) shows that in some way one can "get his hand on the facets" of $P_{\mathscr{I}}$, which suggests, that it should be possible to find an explicit linear characterization of $P_{\mathscr{I}}$. However, there are some problems whose associated polytopes have resisted all characterization attacks so far.

A graph is called *claw-free* if it does not contain the graph $K_{1,3}$ as an induced subgraph.

(4.15) **Problem.** *Find a complete linear characterization for the convex hull of the incidence vectors of stable sets in claw-free graphs.*
Minty (1980) and Sbihi (1978) have shown that maximum stable sets in claw-free graphs can be found in polynomial time. Giles & Trotter (1981) discovered some "wild" facets of stable set polytopes of claw free graphs, and Chvátal (unpublished) showed that the Chvátal rank of this class of polyhedra is unbounded. So this is probably a tough problem.

For graphs with positive edge weights it is easy to find not only shortest paths but also shortest paths and cycles of even or odd lengths using matching techniques or modifications of Dijkstra's method. Similary, in digraphs shortest odd dicycles are easy to compute. The polytope $P_S(D)$ of shortest (s, t)-paths in a digraph D has the nice description given in Theorem (3.13). Nothing similar is known if a parity condition is added.

Let $G = (V, E)$ be a graph, $D = (V, A)$ be a digraph and let $s, t \in V$ be different nodes. Define

$$Q_1(G) := \text{conv}\,\{\chi^P \in \mathbb{R}^E \mid P \subseteq E \text{ contains an } (s,t)\text{-path of even length}\}$$

$$Q_2(G) := \text{conv}\,\{\chi^P \in \mathbb{R}^E \mid P \subseteq E \text{ contains an } (s,t)\text{-path of odd length}\}$$

$$Q_3(G) := \text{conv}\,\{\chi^C \in \mathbb{R}^E \mid C \subseteq E \text{ contains an even cycle}\}$$

$$Q_4(G) := \text{conv}\,\{\chi^C \in \mathbb{R}^E \mid C \subseteq E \text{ contains an odd cycle}\}$$

and let $P_1(D), \ldots, P_4(D)$ be defined analogously (replacing (s, t)-path by directed (s, t)-path and cycle by directed cycle).

(4.16) **Problem.** *Find complete linear characterizations of the polytopes $Q_1(G), \ldots, Q_4(G)$ and $P_1(G), \ldots, P_4(G)$.* $\square$

Finally I would like to repeat a very interesting and difficult problem mentioned in Section 3.2, whose solution probably also depends on a good characterization of a certain polyhedron.

(4.17) **Problem.** *Design a polynomial time algorithm for the weighted matroid matching problem for matric matroids.*

4.4   Further Information and Conclusion

The number of books on combinatorial optimization (and related areas) is not too large. A short list of relatively new books — each with a different emphasis — is: Christofides (1975), Garey & Johnson (1979), Garfinkel & Nemhauser (1972), Golumbic (1982), Grötschel, Lovász & Schrijver (1985), Lawler (1976), Lawler, Lenstra & Rinnooy Kan (1984), Lovász (1979), Lovász & Plummer (1984), Papadimitriou & Steiglitz (1982), Recski (1985), Schrijver (1984a), Tarjan (1983). Soon the book O'hEigeartaigh, Lenstra & Rinnooy Kan (1984) will appear that contains annotated bibliographies on various branches of combinatorial optimization. In these books more detailed information can be found about the concepts and problems whose developments have been surveyed in this paper.

In the thirty years of its existence combinatorial optimization has developed into a fertile and rapidly expanding field. It has many relations to other mathematical disciplines (I hope I could point this out), manifold nontrivial applications to areas like Physics, Management, Economics, Engineering etc. Fortunately, combinatorial optimization has not split into a pure, a computational, and an applied branch (yet), and a large part of the attraction of this field (at least to me) originates from the fact that still many new problems come into the field from (quite varying) issues of the real world which cannot be solved routinely.

References

Adleman L. M. (1983), *"On Breaking Generalized Knapsack Public Key Crypto-Systems"*, Proceedings of the 15th Annual ACM Symposium on Theory of Computing, ACM, 1983, 402—412.

Bachem A., M. Grötschel & B. Korte (eds.) (1983), *"Mathematical Programming — The State of the Art, Bonn 1982"*, Springer Verlag, Heidelberg, 1983.

Baker T., J. Gill & R. Solovay (1975), *"Relativizations of the $P = \omega NP$ Question"*, SIAM Journal on Computing **4** (1975) 431—442.

Balas E. & M. W. Padberg (1975), *"Set partitioning"* in: B. Roy (ed.), *"Combinatorial Programming: Methods and Applications"*, Reidel, Dordrecht, 1975, 205—258.

Balas E. & P. Toth (1983), *"Branch and Bound Methods for the Traveling Salesman Problem"*, Management Science, Research Report No. MSRR **488**, Carnegie-Mellon University, Pittsburgh, March 1983.

Barahona F. & E. Maccioni (1982), *"On the Exact Ground States of Three-Dimensional Ising Spin Glasses"*, Journal of Physics A: Math. Gen. **15** (1982) L611—L615.

Berge C. & V. Chvátal (1984), *"Topics in Perfect Graphs"*, North-Holland, Amsterdam, 1984, to appear.

Billera L. (1983), *"Polyhedral Theory and Commutative Algebra"* in: A. Bachem, M. Grötschel & B. Korte (eds.), *"Mathematical Programming — The State of the Art, Bonn 1982"*, Springer Verlag, Heidelberg, 1983, 57—77.

Birkhoff G. (1946), *"Tres observaciones sobre el algebra lineal"*, Rev. Univ. Nac. Tucuman, Ser. A, **5** (1946) 147—148.

Bloniarz P. A. (1983), *"A Shortest-Path Algorithm with Expected Time $0 (n^2 \log n \log^* n)$"*, SIAM Journal on Computing **12** (1983) 588—600.

Borgwardt K.-H. (1982), *"The average number of pivot steps required by the simplex-method is polynomial"*, Zeitschrift für Operations Research **26** (1982) 157—177.

Burkard R. E. & U. Finke (1984), *"Probabilistic Asymptotic Properties of Some Combinatorial Optimization Problems"*, Discrete Appl. Math. (1984) to appear.

Christofides N. (1975), *"Graph Theory, an Algorithmic Approach"*, Academic Press, New York, 1975.

Christofides N. (1976), *"Worst-Case Analysis of a New Heuristic for the Travelling Salesman Problem"*, Tech. Report, Grad. School Industrial Administration, Carnegie-Mellon University. Abstract in *"Algorithms and Complexity — New Directions and Recent Results"*, J. F. Traub (ed.), (1976) 441. Academic Press, New York.

Chvátal V. (1973), *"Edmonds Polytopes and a Hierarchy of Combinatorial Problems"*, Discrete Mathematics **4** (1973) 305—337.

Cobham A. (1965), *"The Intrinsic Computational Difficulty of Functions"*, Proc. 1964 International Congress for Logic, Methodology and Philosophy of Science, Y. Bar-Hillel (ed.), North-Holland, 1965, 24—30.

Coffman E. G., M. R. Garey & D. S. Johnson (1983), *"Approximation Algorithms for Bin-Packing — An Updated Survey"*, Preprint, Bell Laboratories, Murray Hill, 1983.

Cook S. A. (1971), *"The Complexity of Theorem-Proving Procedures"*, Proc. ACM Symp. Theory of Computing **3** (1971) 151—158.

Coppersmith D. & S. Winograd (1982), *"On the Asymptotic Complexity of Matrix Multiplication"*, SIAM Journal on Computing **11** (1982) 472—492.

Crowder H., E. L. Johnson & M. W. Padberg (1983), *"Solving Large-Scale Zero-One Linear Programming Problems"*, Operations Research **31** (1983) 803—834.

Crowder H. P. & M. W. Padberg (1980), *"Solving Large-Scale Symmetric Traveling Salesman Problems to Optimality"*, Management Science **26** (1980) 495—509.

Cunningham W. (1984), *"Testing Membership in Matroid Polyhedra"*, Journal of Combinatorial Theory (B) (1984) to appear.

Dantzig, G. B., D. R. Fulkerson & S. M. Johnson (1954), *"Solution of a Large-Scale Traveling Salesman Problem"*, Operations Research **2** (1954) 393—410.

Dijkstra E. W. (1959), *"A Note on two Problems in Connection with Graphs"*, Numerische Mathematik **1** (1959) 269—271.

Edmonds J. (1965a), *"Paths, Trees and Flowers"*, Canad. J. Math. **17** (1965) 449—467.

Edmonds J. (1965b), *"Maximum Matchings and a Polyhedron with 0 – 1 Vertices"*. J. Res. Nat. Bur. Standards Sect. B **69** (1965) 125—130.

Edmonds J. (1967), *"Optimum Branchings"*, J. Res. Nat. Bur. Standards Sect. B **71** (1967) 233—240.

Edmonds J. (1970), *"Submodular Functions, Matroids and Certain Polyhedra"*, in: R. Guy (ed.), *"Combinatorial Structures and Their Applications"*, Proc. Calgary International Conference. Gordon and Breach, New York, (1970) 69—87.

Edmonds J. (1971), *"Matroids and the Greedy Algorithm"*, Mathematical Programming **1** (1971) 127—136.

Edmonds J. & R. Giles (1977), *"A Min-Max Relation for Submodular Functions on Graphs"*, Annals of Discrete Mathematics **1** (1977) 185—204.

Edmonds J. & R. Giles (1984), *"Total Integrality of Linear Inequality Systems"*, Proceedings of the Silver Jubilee Conference on Combinatorics, held at the University of Waterloo, Waterloo, Canada, 1982 to appear.

Edmonds J. & R. M. Karp (1972), *"Theoretical Improvements in Algorithmic Efficiency for Network Flow Problems"*, J. ACM **19** (1972) 248—264.

Elias, P., A. Feinstein & C. E. Shannon (1956), *"A Note on the Maximum Flow through a Network"*, IRE Trans. Information Theory IT **2** (1956) 117—119.

Fernandez de la Vega W. & G. S. Lueker (1981), *"Bin Packing can be Solved within 1 + ε in Linear Time"*, Combinatorica **1** (1981) 349—355.

Fisher M. L. (1980), *"Worst-Case Analysis of Heuristic Algorithms"*, Management Science **26** (1980) 1—17.

Fisher M. L. (1981), *"Lagrangean Relaxation Methods for Combinatorial Optimization"*, Management Science **27** (1981) 1—18.

Ford L. R. & D. R. Fulkerson (1956), *"Maximum Flow through a Network"*, Canad. J. Math. **8** (1956) 399—404.

Frank A. (1979), *"Kernel System of Directed Graphs"*, Acta Sci. Math., Szeged, **41** (1979) 63—76.

Frank A. (1984), *"Generalized Polymatroids"*, Proceedings of the 6th Hungarian Combinatorial Colloquium, in: A. Hajnal. L. Lovász & V. T. Sôs (eds.) *"Finite and Infinite Sets"*, North-Holland, Amsterdam, 1984, to appear.

Frederickson G. N. (1983), *"Shortest Path Problems in Planar Graphs"*, 24th Annual Symposium on Foundations of Computer Science, IEEE, 1983, 242—247.

Fujishige S. (1983), *"A Characterization of a Base Polyhedron Associated with a Submodular System"*, Working Paper No **83283**-OR, Institut für Ökonometrie und Operations Research, Universität Bonn, 1983.

Fulkerson D. R. (1971), *"Blocking and Anti-Blocking Pairs of Polyhedra"*, Mathematical Programming **1** (1971) 168—194.

Gabow H. N. (1983), *"Scaling Algorithms for Network Problems"*, 24th Annual Symposium on Foundations of Computer Science, IEEE, 1983, 248—257.

Garey M. R., R. L. Graham, D. S. Johnson & A. C. Yao (1976), *"Resource Constrained Scheduling as Generalized Bin Packing"*, Journal of Combinatorial Theory A **21** (1976) 257—298.

Garey M. R. & D. S. Johnson (1979), *"Computers and Intractability: A Guide to the Theory of NP-Completeness"*, Freeman, San Francisco, 1979.

Garfinkel R. S. & G. L. Nemhauser (1972), *"Integer Programming"*, Wiley, London, 1972.

Geoffrion A. M. (1974), *"Lagrangean Relaxation for Integer Programming"*, Mathematical Programming Studies **2** (1974) 82—114.

Giles R. & L. E. Trotter (1981), "*On Stable Set Polyhedra of $K_{1,3}$-Free Graphs*", Journal of Combinatorial Theory B **31** (1981) 313—326.

Golumbic M. C. (1980), "*Algorithmic Graph Theory and Perfect Graphs*", Academic Press, New York, 1980.

Gomory R. E. (1958), "*Outline of an Algorithm for Integer Solutions to Linear Programs*", Bull. Amer. Math. Soc. **64** (1958) 275—278.

Gomory R. E. (1960), "*Solving Linear Programming Problems in Integers*", in: R. Bellmann & M. Hall (eds.), "*Combinatorial Analysis*", American Mathematical Society, Providence, 1960, 211—216.

Grötschel M. (1982), "*Approaches to Hard Combinatorial Optimization Problems*" in: B. Korte (ed.), "*Modern applied mathematics: Optimization and Operations Research*", North-Holland, Amsterdam, 1982, 437—515.

Grötschel M. & O. Holland (1984), "*A cutting Plane Algorithm for the Matching Problem*", 1984, to appear.

Grötschel M., M. Jünger & G. Reinelt (1982), "*On The Acyclic Subgraph Polytope*", Working Paper No. **82215**-OR, Institut für Okonometrie und Operations Research, Universität Bonn, 1982.

Grötschel M., M. Jünger & G. Reinelt (1983), "*Optimal Triangulation of Large Real World Input-Output Matrices*", Preprint No. **9**, Mathematisches Institut, Universität Augsburg, 1983.

Grötschel M., L. Lovász & A. Schrijver (1981), "*The Ellipsoid Method and its Consequences in Combinatorial Optimization*", Combinatorica **1** (1981) 169—197.

Grötschel M., L. Lovász & A. Schrijver (1985), "*The Ellipsoid Method and Combinatorial Optimization*", Springer, Berlin, 1985, to appear.

Grötschel M. & M. W. Padberg (1984), "*Polyhedral Aspects of the Traveling Salesman Problem I: Theory*" in: E. L. Lawler, J. K. Lenstra & A. H. G. Rinnooy Kan, (eds.), "*The Traveling Salesman Problem*", Wiley, 1984, to appear.

Grünbaum B. (1967), "*Convex Polytopes*", Wiley, London, 1967.

Halton J. H. & R. Terada (1982), "*A Fast Algorithm for the Euclidean Traveling Salesman Problem, Optimal with Probability One*", SIAM Journal on Computing **11** (1982) 28—46.

Hartmanis J. & J. E. Hopcroft (1976), "*Independence Results in Computer Science*", SIGACT News **8,4** (1976) 13—24.

Hassin R. (1978), "*On Network Flows*", Ph. D. Thesis, Yale University, Boston, 1978.

Hausmann D. (1978), "*Integer Programming and Related Areas: A Classified Bibliography 1976—1978*", Lecture Notes in Economics and Mathematical Systems **160**, Springer, Berlin, 1978.

Hochbaum D. & D. Shmoys (1984), "*A Best Possible Heuristic for the k-Center Problem*", Mathematics of Operations Research, 1984, to appear.

Hoffman A. J. & J. B. Kruskal (1956), "*Integral Boundary Points of Convex Polyhedra*", in: H. W. Kuhn & A. W. Trucker (eds.), "*Linear Inequalities and Related Systems*", Ann. of Math. Studies **38**, Princeton Univ. Press, Princeton, N. J., 1956, 233—246.

Hoffman A. J. & D. E. Schwartz (1978), "*On Lattice Polyhedra*", in: A. Hajnal & V. T. Sôs (eds.), "*Combinatorics*", North-Holland, Amsterdam, 1978, 593—598.

Hopcroft J. E. & R. E. Tarjan (1972), "*Efficient Planarity Testing*", SIAM Journal on Computing **2** (1973) 225—231.

Ibarra O. H. & C. E. Kim (1975), "*Fast Approximation Algorithms for the Knapsack and Subset Sum Problems*", J. Assoc. Comput. Mach. **22** (1975) 463—468.

Iri M. (1983), "*Applications of Matroid Theory*", in: A. Bachem, M. Grötschel & B. Korte (eds.), "*Mathematical Programming — The State of the Art, Bonn 1982*", Springer, Heidelberg, 1983, 158—201.

Johnson D. S. (1973), "*Near-Optimal Bin Packing Algorithms*", Ph. D. Thesis, Department of Mathematics, MIT, Cambridge, Massachussetts, 1973.

Johnson D. S. (1983), "The $\mathcal{NP}$-Completeness Column: An Ongoing Guide", Journal of Algorithms **4** (1983) 189—203 and 286—300.

Johnson D. S., A. Demers, J. D. Ullman, M. R. Garey & R. L. Graham (1974), *"Worst-Case Performance Bounds for Simple One-Dimensional Packing Algorithms"*, SIAM Journal on Computing **3** (1974) 299—325.

Kannan R. & A. Bachem (1979), *"Polynomial Algorithms for Computing the Smith and Hermite Normal Forms of an Integer Matrix"*, SIAM Journal on Computing **8** (1979) 499—507.

Karmarkar N. & R. M. Karp (1982), *"An Efficient Approximation Scheme for the One-Dimensional Bin Packing Problem"*, Proc. 23rd Annual Symposium on Foundations of Computer Science, 1982, 212—230.

Karp R. M. (1972), *"Reducibility Among Combinatorial Problems"*, in: R. E. Miller & J. W. Thatcher (eds.), *"Complexity of Computer Computations"*, Plenum Press, New York, 1972, 85—103.

Karp R. M. (1976), *"The Probabilistic Analysis of Some Combinatorial Search Algorithms"*, Memorandum No. ERL-M **581**, University of California, Berkeley 1976.

Karp R. M. (1984), *"Probabilistic Analysis of Deterministic and Probabilistic Algorithms: An Annotated Bibliography"*, in: O'hEigeartaigh et al. (eds.) (1984) to appear.

Karp R. M. & Ch. H. Papadimitriou (1982), *"On Linear Characterization of Combinatorial Optimization Problems"*, SIAM Journal on Computing **11** (1982) 620—632.

Kastning C. (1976), *"Integer Programming and Related Areas: A Classified Bibliography"*, Lecture Notes in Economics and Mathematical Systems **128**, Springer, Berlin, 1976.

Khachiyan L. G. (1979), *"A Polynomial Algorithm in Linear Programming"*, Soviet Math. Dokl. **20** (1979) 191—194.

Kindervater G. A. P. & J. K. Lenstra (1984), *"Parallel Algorithms in Combinatorial Optimization: An Annotated Bibliography"*, in: O'hEigeartaigh et al. (eds.) (1984) to appear.

Klee V. (1980), *"Combinatorial Optimization: What is the State of the Art"*, Mathematics of Operations Research **5** (1980) 1—26.

Klee V. & P. Kleinschmidt (1984), *"The d-Step Conjecture and its Relatives"*, SIAM, Philadelphia, 1984, to appear.

Korte B. (1979), *"Approximative Algorithms for Discrete Optimization Problems"*, Annals of Discrete Mathematics **4** (1979) 85—120.

Korte B. & L. Lovász (1983), *"Structural Properties of Greedoids"*, Combinatorica **3** (1983) 359—374.

Lagarias J. C. & A. M. Odlyzko (1983), *"Solving Low-Density Subset Sum Problems"*, 24th Annual Symposium on Foundations of Computer Science, IEEE, 1983, 1—10.

Lawler E. L. (1976), *"Combinatorial Optimization: Networks and Matroids"*, Holt, Rinehart & Winston, New York, 1976.

Lawler, E. L., J. K. Lenstra & A. H. G. Rinnooy Kan (1984), *"The Traveling Salesman Problem"*, Wiley, New York, 1984, to appear.

Lawler E. L. & C. U. Martel (1982), *"Flow Network Formulations of Polymatroid Optimization Problems"*, Annals of Discrete Mathematics **16** (1982) 189—200.

Lenstra A. K., H. W. Lenstra jun. & L. Lovász (1982) *"Factoring Polynomials with Rational Coefficients"*, Math. Ann. **261** (1982) 515—534.

Lenstra jun. H. W. (1983), *"Integer Programming with a Fixed Number of Variables"*, Mathematics of Operations Research **8** (1983) 538—548.

Lovász L. (1979), *"Combinatorial Problems and Exercises"*, North-Holland, Amsterdam, 1979.

Lovász L. (1981), *"The Matroid Matching Problem"*, in: L. Lovász & V. T. Sôs (eds.) *"Algebraic Methods in Graph Theory"*, North-Holland, Amsterdam, 1981, 495—517.

Lovász L. (1982), *"Bounding the Independence Number of a Graph"*, Annals of Discrete Mathematics **16** (1982) 213—223.

Lovász L. (1983), *"Submodular Functions and Convexity"*, in: A. Bachem, M. Grötschel & B. Korte (eds.), *"Mathematical Programming — The State of the Art, Bonn 1982"*, Springer, Heidelberg, 1983, 235—257.

Lovász L. & M. Plummer (1984), *"The Matching Structure of Graphs"*, Akademia Kiado, Budapest, 1984, to appear.

Lueker G. S. (1979), "*Maximization Problems on Graphs with Edge Weights Chosen from a Normal Distribution*", Proc. Xth Annual ACM Symp. on Theory of Computing 1978.

Mader W. (1978a), "*Über die Maximalzahl kantendisjunkter A-Wege*", Arch. Math., Basel, **30** (1978) 325—336.

Mader W. (1978b), "*Über die Maximalzahl kreuzungsfreier H-Wege*", Arch. Math., Basel, **31** (1978) 387—402.

Minty G. J. (1980), "*On Maximal Independent Sets of Vertices in a Claw-Free Graph*", J. Combinatorial Theory B **28** (1980) 284—304.

von Neumann J. (1953), "*A Certain Zero-Sum Two-Person Game Equivalent to the Optimum Assignment Problem*", in: W. Tucker & H. W. Kuhn (eds.), "*Contributions to the Theory of Games II*", Annals of Math. Studies **38**, Princeton Univ. Press, Princeton, N. J., 1953, 5—12.

O'hEigeartaigh M., J. K. Lenstra & A. H. G. Rinnooy Kan (1984), "*Combinatorial Optimization: Annotated Bibliographies*", Wiley, New York, 1984, to appear.

Padberg, M. W. & M. Grötschel (1984), "*Polyhedral Aspects of the Traveling Salesman Problem II: Computation*", in: E. L. Lawler, J. K. Lenstra & A. H. G. Rinnooy Kan (eds.), "*The Traveling Salesman Problem*", Wiley, 1984, to appear.

Padberg M. W. & M. R. Rao (1982), "*Odd Minimum Cut-Sets and b-Matchings*", Mathematics of Operations Research **7** (1982) 67—80.

Padberg M. W. & M. R. Rao (1984), "*The Russian Method for Linear Inequalities III: Bounded Integer Programming*", Mathematical Programming Studies, 1984, to appear.

Papadimitriou Ch. H. (1984), "*Polytopes and Complexity*", in:: Proceedings of the Silver Jubilee Conference on Combinatorics, held at the University of Waterloo, Waterloo, Canada, 1982, 1984, to appear.

Papadimitriou Ch. H. & K. Steiglitz (1982), "*Combinatorial Optimization: Algorithms and Complexity*", Prentice-Hall, Englewood Cliffs, 1982.

Pulleybank W. R. (1983), "*Polyhedral Combinatorics*", in: A. Bachem, M. Grötschel & B. Korte (eds.), "*Mathematical Programming — The State of the Art, Bonn 1982*", Springer, Heidelberg, (1983), 312—345.

von Randow R. (1982), "*Integer Programming and Related Areas: A Classified Bibliography 1978—1981*", Lecture Notes in Economics and Mathematical Systems **197**, Springer, Berlin 1982.

Recski A. (1985), "*Matroid Theory and its Applications*", Springer, Berlin 1985, to appear.

te Riele H. (1983), "*Mertens' Conjecture Disproved*", Research Announcement, CWI Newsletter, Amsterdam **1** (1983) 23—24.

Rivest R. & S. Vuillemin (1978), "*On Recognizing Graph Properties from Adjacency Matrices*", Theor. Comp. Sci. **3** (1978) 371—384.

Roberts F. S. (1978), "*Graph Theory and its Applications to Problems of Society*", SIAM, Philadelphia, 1978.

Sbihi N. (1978), "*Etudes des stables dans les graphes sans étoile*", M. Sc. Thesis, Univ. Sci. et Méd. Grenoble, 1978.

Schrijver A. (1980), "*On Cutting Planes*", Annals of Discrete Mathematics **9** (1980) 291—296.

Schrijver A. (1983), "*Min-Max Results in Combinatorial Optimization*", in:A. Bachem, M. Grötschel & B. Korte (eds.), "*Mathematical Programming — The State of the Art, Bonn 1982*", Springer, Heidelberg, 1983, 439—500.

Schrijver A. (1984a), "*Polyhedral Combinatorics*", Wiley, New York, 1984, to appear.

Schrijver A. (1984b), "*Total Dual Integrality from Cross-Free Families — a General Framework*", Mathematical Programming (1984) to appear.

Schrijver A. (1984c), "*Total Dual Integrality from Graphs, Crossing Families, and Sub- and Supermodular Functions*", Proceedings of the Silver Jubilee Conference on Combinatorics, Univ. of Waterloo, Academic Press, 1984, to appear.

Seymour P. D. (1977), "*The Matroids with the Max-Flow Min-Cut Property*", Journal of Combinatorial Theory B **23** (1977) 189—222.

Seymour P. D. (1980), *"Decomposition of Regular Matroids"*, Journal of Combinatorial Theory B, **28** (1980) 305—359.

Seymour P. D. (1981), *"Matroids and Multicommodity Flows"*, Europ. J. Comb. **2** (1981) 257—290.

Shor N. Z. (1970), *"Convergence Rate of the Gradient Descent Method with Dilatation of the Space"*, Cybernetics **6** (1970) 102—108.

Shor N. Z. (1977), *"Cut-Off Method with Space Extension in Convex Programming Problems"*, Cybernetics **13** (1977) 94—96.

Strassen V. (1969), *"Gaussian Elimination is Not Optimal"*, Numer. Math. **13** (1969) 354—356.

Tarjan R. E. (1978), *"Complexity of Combinatorial Algorithms"*, SIAM Rev. **20** (1978) 457—491.

Tarjan R. E. (1983), *"Data Structures and Network Algorithms"*, Society for Industrial and Applied Mathematics, Philadelphia, 1983.

Tarjan R. E. (1983a), *"Algorithms for Maximum Network Flow"*, Paper presented at the NET-FLOW **83** Workshop, Pisa, Italy, 1983.

Tutte W. T. (1965), *"Lectures on Matroids"*, J. Res. Nat. Bur. Standards **69** B (1965) 1—47.

Weide B. (1977), *"A Survey of Analysis Techniques for Discrete Algorithms"*, Computing Surveys **9** (1977) 291—313.

Weide B. (1980), *"Random Graphs and Graph Optimization Problems"*, SIAM J. Comput. **9** (1980) 552—557.

Welsh D. J. A. (1976), *"Matroid Theory"*, Academic Press, New York, 1976.

Yamnitsky B. & L. A. Levin (1982), *"An Old Linear Programming Algorithm Runs in Polynomial Time"*, Paper presented at the Silver Jubilee Conference, Waterloo, Ontario, Canada, 1982.

Yao A. C. (1980), *"New Algorithms for Bin Packing"*, J. ACM **27** (1980) 207—227.

Perspectives in Mathematics
Anniversary of Oberwolfach 1984
© Birkhäuser Verlag, Basel

Spread and Age Structure in Epidemic Models

K. P. HADELER

Lehrstuhl für Biomathematik, Universität Tübingen,
Auf der Morgenstelle 28, D-7400 Tübingen (FRG)

0   Introduction

The first attempts to describe mathematically the spread of infectious diseases go back to 1927 when Kermack and McKendrick designed an elementary model and exhibited the famous threshold theorem, based on the observation that essentially two parameters determine the behaviour of an infection in a population, namely the contact rate between infectives and susceptibles and the length of the period of infectiousness. If the initial number of susceptibles is lower than the threshold then the number of infectives will decrease, if it is larger than the threshold then an outbreak will occur.

It appears that the Kermack-McKendrick model is the nucleus of all later and more elaborate models. Many additional features have been incorporated, in particular finer subdivision of the host population, periods of incubation and of temporal immunity, age structure of the host population, and seasonal variation of parameters. Most of the earlier models, well-suited to describe bacterial or virus infections, are prevalence models where only the prevalence of the disease in an individual is counted. For many important diseases, in particular infections by helminths, it is appropriate to count individual parasites. Such models are closely related to birth and death processes.

The mathematically most interesting feature of epidimic models are the close interrelation of ordinary and partial differential equations, of integral equations and stochastic processes, the occurence of various bifurcation phenomena and of oscillations and travelling epidemic waves.

The organisation of the paper is as follows: First I discuss the Kermack-McKendrick model, then the classical Lotka model for a population with age-structure and its discrete analogon, the Leslie model. The results on spatial spread of epidemies are discussed in some detail. Next I describe autonomous and forced oscillations, finally recent developments in epidemic models for discrete parasites. In each of the sections I refer to recent literature, before all to the monography of Bailey [6] and to the Lecture Notes of Hoppensteadt [42], Frauenthal [24], and Waltman [78]. This review has of course the shortcoming that a field with extremely dispersed literature extending into the biological and medical sciences can hardly be presented on a few pages, also my own preferences introduce some bias in the presentation.

1   The Kermack-McKendrick model

In the classical epidemic model, originating from the work of Kermack and McKendrick [46], only the prevalence of the disease is described. The population is partitioned into the classes of susceptibles S, infectives I, and recovered R. The transition $S \to I \to R \to S$ is modeled by ordinary differential equations derived from a mass action law with constant population size being assumed,

$$\dot{S} = -\beta SI + \gamma R - \mu_1 S + (\mu_1 S + \mu_2 I + \mu_3 R)$$

(1.1) $$\dot{I} = \beta SI - \alpha I - \mu_2 I$$

$$\dot{R} = \alpha I - \gamma R - \mu_3 R.$$

The parameter $\beta > 0$ is the contact rate between infectives and susceptibles, $\alpha > 0$ is the rate at which infectives become recovered, and $\gamma \geq 0$ is the rate for recovered to become susceptible again. The parameters $\mu_1, \mu_2, \mu_3 \geq 0$ are the death rates for the three classes S, I, R. In fact the equations are independent of μ_1.

The quantity $P \equiv S + I + R$ is an invariant of motion. Thus the system is essentially two-dimensional,

(1.2) $$\dot{S} = -\beta SI + (\gamma + \mu_3)(P - S) - (\gamma - \mu_2 + \mu_3) I,$$
$$\dot{I} = \beta SI - (\alpha + \mu_2) I.$$

The state space of the system (1.2) is the triangle $T = \{S \geq 0, I \geq 0, S + I \leq P\}$ which is positively invariant with respect to the flow. First consider the endemic case $\gamma + \mu_3 > 0$. There are two stationary states

$$(S_0, I_0) = (P, 0)$$

(1.3) $$(S_1, I_1) = \left(\frac{\alpha + \mu_2}{\beta}, \frac{\gamma + \mu_3}{\alpha + \gamma + \mu_3} \left(P - \frac{\alpha + \mu_3}{\beta} \right) \right)$$

In the endemic case the threshold theorem can be expressed as follows: If the total population size P is less than the threshold value $S_1 = (\alpha + \mu_2)/\beta$ then the point (S_1, I_1) is not in the half-plane $I \geq 0$, the point (S_0, I_0) is the only stationary point and it is stable. On the other hand, if P exceeds S_1 then (S_1, I_1) is located in the interior of T, and it is the only stable stationary point.

Applying Dulac's criterion with weight function $1/I$ and the Poincaré-Bendixson theorem one sees that for $P < S_1$ all trajectories converge to $(P, 0)$, for

$P > S_1$ all trajectories except for (S_0, I_0) converge to (S_1, I_1). Thus an endemic level of the disease can be maintained only if the total population size is above the threshold level.

Conversely, consider the number of infectives $I(\beta)$ at the stable equilibrium as a function of the contact rate β. For small β one has $I = 0$. At $\beta_0 = (\alpha + \mu_2)/P$ there occurs a bifurcation with an exchange of stability, then $I(\beta)$ increases from 0 to $I_{max} = (\gamma + \mu_3) P/(\alpha + \gamma + \mu_3)$. We shall see that a similar diagram arises in the model for discrete parasites.

For a certain range of parameters the system (1.2) admits a Lyapunov function. Assume that in equations (1.2) one has $\gamma - \mu_2 + \mu_3 > 0$, i.e. an increase in the number of infectives leads to an immediate decrease in the number of susceptibles. Then

$$(1.4) \quad V(S, I) = S - \frac{\alpha + \gamma + \mu_3}{\beta} \log \left(\frac{\gamma - \mu_2 + \mu_3}{\beta} + S \right) + I - I_1 \log I$$

is a Lyapunov function, since

$$(1.5) \quad \frac{dV}{dt} = - \frac{\beta(\gamma + \mu_3)}{\alpha + \gamma + \mu_3} \cdot \frac{\gamma - \mu_2 + \mu_3 + \beta P}{\gamma - \mu_2 + \mu_3 + \beta S} (S - S_1)^2.$$

The limit case $\gamma = \mu_3 = 0$ has attracted greatest attention. Then the line $I = 0$ consists of stationary points. For the two-dimensional system

$$(1.6) \quad \begin{aligned} \dot{S} &= - \beta S I + \mu_2 I \\ \dot{I} &= \beta S I - (\alpha + \mu_2) I \end{aligned}$$

the total population size P or the point $(P, 0)$ do not play an exceptional rôle. The system has an invariant of motion

$$(1.7) \quad V(S, I) = S - \frac{\alpha}{\beta} \log \left(S - \frac{\mu_2}{\beta} \right) + I$$

Thus every trajectory approaches a stationary point with $I = 0$. The threshold theorem assumes the following form: Consider trajectories which start from points $(S(0), I(0))$ with $I(0) > 0$ close to zero. If $S(0) < (\alpha + \mu_2)/\beta$ then $I(t)$ decreases monotonely. If $S(0) > (\alpha + \mu_2)/\beta$ then $I(t)$ first increases to a maximum, then decreases to zero. For any trajectory in $I > 0$ then limits $S(-\infty)$ and $S(+\infty)$ of the S-coordinate are coupled through the equation

$$(1.8) \quad V(S(+\infty), 0) = V(S(-\infty), 0), \quad S(+\infty) < S(-\infty).$$

The classical model has been extended in various directions, in many cases without direct relation to epidemiological field data. A state of exposed individuals can be introduced as the transition between susceptibles and infectives. In the simplest case such models lead to ordinary differential equations in $I\!R^3$. Incubation periods can also be modeled by introducing delays into the equations, which, as can be expected, cause oscillations for certain choices of the parameters. Some diseases may be transmitted directly from mother to child (vertical transmission, see e.g. Busenberg, Cooke, Pozio [13] for further references). Longini [51] discusses populations which undergo genetic selection under the influence of an endemic disease.

2   Populations with age structure

To derive the classical type of renewal equation for age-structured populations we first follow the argument of Sharpe and Lotka [66]. Let $u(t, a)$ be the non-normalized age distribution of the population at time t, i.e. $u(t, a)\,\Delta a$ is the size of the cohort with age in $[a, a + \Delta a]$ at time t. Let $p(a)$ be the probability that an individual survives up to time a, and let $b(a)$ be the reproduction rate at age a. Let $N(t)$ be the birth rate at time t, i.e. $N(t)\,\Delta t$ is the number of births in $[t, t + \Delta t]$. Let $u(0, a) = u_0(a)$ give the state of the population at time $t = 0$.

Individuals with age $a > t$ have existed at time $t = 0$. They were born at time $t - a$ and have survived up to time zero, i.e. at time zero their age was $a - t$. For individuals of age $a - t$ at time zero the probability to survive up to time t is $p(a)/p(a - t)$.

Individuals with age $a < t$ have been born at $t - a > 0$. Since their age is a, the probability of survival has been $p(a)$. Hence the function u has a representation

$$(2.1) \quad u(t, a) = \begin{cases} u_0(a - t)\,\dfrac{p(a)}{p(a - t)} & t < a \\[2em] N(t - a)\,p(a) & t > a \end{cases}$$

On the other hand, the birth rate N satisfies

$$(2.2) \quad \begin{aligned} N(t) &= \int_0^\infty b(a)\,u(t, a)\,da \\ &= \int_0^t b(a)\,p(a)\,N(t - a)\,da + \int_t^\infty b(a)\,\frac{p(a)}{p(a - t)}\,u_0(a - t)\,da \end{aligned}$$

i.e. the birth rate satisfies a renewal equation or Volterra equation of convolution type

(2.3) $N(t) = \int\limits_0^t \phi(a)\, N(t-a)\, da + g(t)$

It is astonishing that in this context the notions of partial differential equations and of characteristics enter not earlier than 1959 in a paper by von Foerster [23] on cell proliferation. Let $\mu(a)$ be the mortality of individuals of age a. Since individuals age with chronological time the function u satisfies an equation

(2.4) $u_t + u_a + \mu(a)\, u = 0$

(subscripts denote partial derivatives). The characteristic differential equations $dt/ds = 1$, $da/ds = 1$, $du/ds = -\mu(a)\, u$ lead immediately to the solution

(2.5) $u(t, a) = \begin{cases} u_0(a-t)\, \exp\left(-M(a) + M(a-t)\right) \\ N(t-a)\, \exp\left(-M(a)\right) \end{cases}$

where

(2.6) $M(a) = \int\limits_0^a \mu(s)\, ds.$

Then the renewal equation assumes the form

(2.7)
$$N(t) = \int\limits_0^t b(a)\, e^{-M(a)}\, N(t-a)\, da$$
$$+ \int\limits_t^\infty b(a)\, e^{-M(a)+M(a-t)}\, u_0(a-t)\, da.$$

Thus the two approaches lead to the same result with

(2.8) $p(a) = \exp\left(-\int\limits_0^a \mu(s)\, ds \right)$

Lotka's idea was to look for persistent age distributions, i.e. for solutions of the homogeneous equation

(2.9) $N(t) = \int\limits_0^\infty b(a)\, e^{-M(a)}\, N(t-a)\, da$

in the form $N(t) = N_0 \exp(\lambda t)$. Then λ satisfies

(2.10) $\int\limits_0^\infty b(a)\, e^{-M(a)}\, e^{-\lambda a}\, da = 1$

and indeed, under mild assumptions, this equation has a single positive simple root λ, which is "Lotka's r".

A discrete version of the Lotka model has been presented only in 1945 by Leslie [49] although it could have been easily obtained by discretizing equations (2.4). The population is subdivided into discrete age classes $u_1, \ldots, u_n$. Let $v_i > 0$ be the probability to enter the class u_{i+1} from class u_i, let $b_i \geq 0$ the reproduction rate of individuals in class u_i. Suppose not all of the b_j vanish. Let u_i^t be the size of class u_i at time t. Then for the u_i^t one has the linear recursion

$$u_{i+1}^{t+1} = v_i u_i^t, \quad i = 1, \ldots, n-1$$

(2.11)

$$u_1^{t+1} = \sum_{j=1}^{n} b_j u_j^t.$$

In vector notation the recursion can be written

(2.12) $u^{t+1} = A u^t$,

where $u = (u_1, \ldots, u_n)$ and

$$(2.13) \quad A = \begin{bmatrix} b_1 & \cdots & & b_n \\ v_1 & & & \\ & \ddots & & \\ & & v_{n-1} & 0 \end{bmatrix}$$

Thus the evolution of the population is described by the iteration of the matrix A, which is diagonally similar to a Frobenius companion matrix. The characteristic polynomial is easily obtained as

(2.14) $p(\lambda) = v_1 \ldots v_{n-1} b_n + v_1 \ldots v_{n-2} b_{n-1} \lambda + \cdots + v_1 b_2 \lambda^{n-2} - b_1 \lambda^{n-1} - \lambda^n.$

There is a single positive root r, which is simple. At least when all b_j are positive, all other roots are less in modulus and the iterates behave asymptotically like

$$u^t = \text{const} \cdot r^t \cdot \bar{u}$$

where $\bar{u} = (1, v_1/r, v_2 v_1/r^2, \ldots, v_{n-1} \ldots v_1/r^{n-1})$ is a right eigenvector of A. The vector $\bar{u}$ describes the stationary ("persistent") age distribution, and r^t is the asymptotic growth rate.

The fundamental problem of the linear renewal equation (2.7) or of the discrete version (2.13) is to show that for some exponent λ which can be read off from the characteristic equation the function $e^{-\lambda t} u(t)$ assumes a limit (or $r^{-t} u^t$ assumes a limit in the discrete case). The first "renewal theorems" go back to

Feller [20]. Many results on renewal equations have been motivated by the theory of stochastic processes (see e. g. Jagers [43]). Thieme [74] has proved renewal theorems for discrete equations of Leslie type

$$u^t = \sum_{k=0}^{t} A_k u^{t-k} + u_0^t$$

where u assumes values in a partially ordered Banach space. An important reference is the recent monography by Webb [81], who develops the theory of the Lotka—von Foerster equation systematically within the frame-work of operator semigroups.

3   The Gurtin-McCamy model

A well-known paper by Gurtin and McCamy [27] has elicited a large number of successors. Since the Lotka model is linear, and populations either tend to extinction or grow exponentially, it is a natural idea, similar to the transition from exponential growth to the Verhulst law (1845!) to assume that the parameter functions of the Lotka model depend on the population. The simplest law of course is a dependence on the total population size

$$(3.1) \quad P(t) = \int_0^\infty u(t, a) \, \mathrm{d}a$$

Then the equations read

$$(3.2) \quad u_t + u_a + \mu(a, P) u = 0$$

$$(3.3) \quad u(t, 0) = \int_0^\infty b(a, P) u(t, a) \, \mathrm{d}a$$

If the functions $P(t)$ and $N(t) = u(t, 0)$ are known one can solve the linear partial differential equation in terms of P and N, then insert this solution in equations (3.1) and (3.3) to obtain a pair of coupled renewal equations.

$$(3.4) \quad \begin{aligned} N(t) &= \int_0^t b(t-a, P(t)) \, e^{-\int_0^{t-a} \mu(S, P(a+s)) \, \mathrm{d}s} \, N(a) \, \mathrm{d}a \\ &+ \int_0^\infty b(t+a, P(t)) \, e^{-\int_0^t \mu(a+s, P(s)) \, \mathrm{d}s} \, u_0(a) \, \mathrm{d}a \end{aligned}$$

$$P(t) = \int_0^t e^{-\int_0^{t-a} \mu(s,\, P(a+s))\, ds}\, N(a)\, da$$

(3.5)

$$+ \int_0^\infty e^{-\int_0^t \mu(a+s,\, P(s))\, ds}\, u_0(a)\, da$$

Under appropriate conditions on b, μ, and u_0 one can obtain local and global existence via Banach's contraction theorem, say.

The asymptotic behavior depends, of course, on the qualitative behavior of the functions b and μ. If, as a function of P, the birth rate decreases fast and the death rate increases fast, then solutions should stabilize to stationary states. For stationary solutions one has P and N constant, and these satisfy the equations (unless $N = P = 0$)

(3.6) $$\phi(P) \equiv \int_0^\infty b(a, P)\, e^{-M(a,\, P)}\, da - 1 = 0$$

and

(3.7) $$N \int_0^\infty e^{-M(a,\, P)}\, da = P$$

Here

(3.8) $$M(a, P) = \int_0^a \mu(s, P)\, ds.$$

Thus stationary solutions correspond to zeros of the function ϕ.

While the existence prblems are rather simple, the questions of stability and asymptotic behavior lead to difficult analytic problems. Various special cases have been discussed in the literature. In the original work of Gurtin and McCamy [27] only the birth rate was assumed to depend on P. In a sequel Sovunmi [71] has extended to results to a population where both sexes are counted separately. Rorres [64], Lamberti and Vernole [48], Di Blasio, Iannelli, Sinestrari [9] considered a birth rate depending on P. Marcati [55] studied the case $b = b(a)$, $\mu = \mu_0(a) + \gamma P$ in great detail. Prüß [60] studies a system of equations for several species where the death rate depends on the population sizes (and of more general functionals) of all species. Various extensions including the case of spatial diffusion one finds in recent papers of Webb, in particular in the monograph [81]. Special cases of the Lotka model and the Gurtin-McCamy model are equivalent to systems of ordinary differential equations [38].

4   An epidemic model with accumulation

This may be the appropriate place to mention a model introduced by Hoppensteadt and Waltman [40]. Although the concept of dose accumulation in a state of exposition to the disease may require further biological clarification (and possibly modification of the model) the equations are mathematically interesting. The basic idea is a transition $S \to E \to I \to R$. A susceptible individual becomes exposed by contact with infectious individuals, accumulates a dose of infecting agent, becomes infectious if the dose arrives at a critical level m, remains infectious for an interval ω and becomes immune. At each moment the state of the system is described by an element $(S, e, i, R) \in \mathbb{R} \times C[0, m] \times C[0, \omega] \times \mathbb{R}$. Here $S(t)$ counts susceptibles, $e(t, \delta)$ the exposed having accumulated a dose δ, $i(t, \tau)$ the infectious which became infected at $t - \tau$, $R(t)$ counts the immunes. The following equations are self-explanatory

$$\dot{S}(t) = -rI(t)\,S(t)$$

$$e(t, 0) = \frac{r}{\varrho}\,S(t)$$

(4.1)
$$e_t(t, \delta) + \varrho I(t)\,e_\delta(t, \delta) = 0$$

$$i(t, 0) = \varrho I(t)\,e(t, m)$$

$$i_t(t, \tau) + i_\tau(t, \tau) = 0, \quad I(t) = \int_0^\omega i(t, \tau)\,d\tau$$

$$\dot{R}(t) = i(t, \omega)$$

A detailed study of the characteristics shows the following: Let S_0, e_0, i_0, R_0 give the initial state of the population. If the equation

$$x(t) = \int_0^t \chi_{[t-\omega, t]}(\tau)\,e_0\left(m - \varrho \int_0^\tau x(s)\,ds\right) d\tau + f(t)$$

(4.3)
where $f(t) = \int_0^\infty \chi_{[0, \omega - t]}(\tau)\,i_0(\tau)\,d\tau$

has a solution $x(t)$ such that $\varrho \int_0^t x(t)\,dt < m$ for all $t < \infty$, then newly exposed become never infected, and the state of the system is entirely described in terms of $x(t)$, in particular, $I(t) = x(t)$, and $I(t) \to 0$ as $t \to \infty$, although $I(t) > 0$ may hold for all $t > 0$. On the other hand, if there is a (first) $t_1 < \infty$ such that the solution $x(t)$ satisfies $\varrho \int_0^t 1\,x(t)\,dt = m$, then for $t > t_1 + \omega$ the function $S(t)$ satisfies

(4.4) $\dot{S}(t) = -r\,e^{\frac{mr}{\varrho}}\,S(t)\,[S(t-\omega) - S(t)].$

Thus the system is determined by the history of the variable S. The equation (4.4) has been discussed in [82].

5   Spatial spread of epidemics

For several epidemic diseases, notable plague, flue, or rabies, propagation of epidemic waves is empirically known. Following the model of R. A. Fisher [22] for the advance of advantageous genes, studied in great detail by Kolmogorov, Petrovskij, Piskunov [47] and many followers ([21], [54], [65], [10], [11], [77], [28], [29]), one has explained such phenomena as resulting from the interaction of a Kermack-McKendrick kinetics and a diffusion-like process simulating migration of the population, of infectious individuals only, or of germs. A first discussion has been given by Kendall [45], who introduced a convolution operator into the Kermack-McKendrick model

$$S_t = -\beta S \bar{I}$$
$$(5.1) \quad I_t = \beta S \bar{I} - \alpha I$$
$$R_t = \alpha I$$

where

$$(5.2) \quad \bar{I}(t, x) = (k * I)(t, x) = \int_{\mathbb{R}} k(x + y) I(t, y) \, dy$$

and then replaced the convolution by a diffusion operator

$$(5.3) \quad \begin{aligned} S_t &= -\beta S(I + I_{xx}) \\ I_t &= \beta S(I + I_{xx}) - \alpha I \end{aligned}$$

Noble [59] has suggested to model the propagation of plague by a Kermack-McKendrick model where both susceptibles and infectious migrate with the same rate

$$(5.4) \quad \begin{aligned} \frac{\partial S}{\partial t} &= -\beta SI + S_{xx} \\ \frac{\partial I}{\partial t} &= \beta SI - \mu I + I_{xx}. \end{aligned}$$

He tried to fit this model to available data on the progress of the black death in 1347—50. More generally one could introduce a diffusion term into the general model (1.1) for each of the variables. (See also Capasso [14], Capasso and Maddalena [15], Di Blasio, Iannelli, Sinestrari [9], Webb [79], [80].)

The concept of travelling fronts or travelling pulses in reaction-diffusion equations has been rather successful. Assume $\dot{u} = f(u)$ is a system of ordinary differential equations in $\mathbb{R}^m$ describing some chemical or ecological interaction. Let u_1 and u_2 be stationary points. Then consider the reaction-diffusion equation $u_t = f(u) + Du_{xx}$ on the real axis, where D is a positive diagonal matrix of diffusion coefficients. If one imposes the boundary conditions

$$\lim_{x \to -\infty} u(t, x) = u_1, \quad \lim_{x \to +\infty} u(t, x) = u_2$$

then in many cases intuition tells that solutions of the initial-value problem of the parabolic equation approach for large t travelling fronts, i.e. solutions of the special form

$$u(t, x) = \phi(x - ct), \quad \text{with} \quad \lim_{x \to -\infty} \phi(x) = u_1, \quad \lim_{x \to +\infty} \phi(x) = u_2,$$

where the function of one variable ϕ describes the shape of the front, and the number c is the speed of propagation. A travelling front corresponds to a trajectory of the ordinary differential equation $-c\dot{\phi} = f(\phi) + D\ddot{\phi}$ or, equivalently

$$\dot{\phi} = \psi, \quad -c\psi = f(\phi) + D\dot{\psi}$$

in $\mathbb{R}^{2m}$ connecting the two stationary points $(u_1, 0)$, $(u_2, 0)$.

It is well known, e.g., that the one-dimensional problem

$$u_t = u(1 - u)(u - \alpha) + u_{xx}, \quad u_1 = 1, \quad u_2 = 0, \quad 0 < u < 1$$

admits a unique travelling front (up to translations, of course) if $0 < \alpha < 1$, and a continuum of wave speeds c, if $\alpha < 0$.

For the present epidemic model (1.1) one expects a front connecting (S_0, I_0) and (S_1, I_1) in the endemic case, and a family of fronts in the epidemic case, each of which connects two of the stationary states in (1.8).

For the model (5.3) Kendall has completely solved the existence problem. The ansatz

$$S(t, x) = u(x - ct), \quad I(t, x) = v(x - ct)$$

leads to the equations

$$(5.5) \quad \begin{aligned} -c\dot{u} &= -\beta u(v + \ddot{v}) \\ -c\dot{v} &= \beta u(v + \ddot{v}) - \alpha v \end{aligned}$$

for the functions of one variable u and v, which satisfy the boundary conditions

(5.6) $u(-\infty) = S_-, \quad u(+\infty) = S_+, \quad v(\pm\infty) = 0.$

This system can be integrated once and rescaled,

(5.7) $\dot{u} = \left(1 + \dfrac{\alpha^2}{c^2}\right) v - \left(\gamma - u + \dfrac{\alpha}{\beta} \log u\right)$

$\dot{v} = \gamma - u + \dfrac{\alpha}{\beta} \log u - v.$

From the boundary condition follows

(5.8) $\gamma = S_+ - \dfrac{\alpha}{\beta} \log S_+ = S_- - \dfrac{\alpha}{\beta} \log S_-.$

The system (5.7) can be treated by phase-plane methods. A travelling front corresponds to a trajectory connecting the stationary points $(S_-, 0)$, $(S_+, 0)$. One finds ([45], [28]) that there is a half-line of possible speeds $[c_0, \infty)$. The minimal speed is given by

(5.9) $c_0 = 2\sqrt{\beta S_- (\beta S_- - \alpha)}.$

Thus the minimal speed of the front is a monotone function of the initial number of susceptibles. It approaches zero, if this number is close to the threshold. The problem of the asymptotic behavior of the equations is more difficult and apparently it has not been dealt with.

The case of the convolution operator (5.1), (5.2) has been treated by several authors. Mollison [56] has considered special cases, Atkinson and Reuter [5] have transformed the equation for travelling fronts

(5.10) $\begin{aligned} -c\dot{u} &= -\beta u(k * v) \\ -c\dot{v} &= \beta u(k * v) - \alpha v \end{aligned}$

into an integral equation for the function $U = \log u$.

For $\alpha = 0$ it has the form

(5.11) $U(x) = -\displaystyle\int_{-\infty}^{x} (1 - e^{U(y)}) Q(y - x) \, dy$

where

(5.12) $Q(x) = \displaystyle\int_{-\infty}^{x} k(y) \, dy.$

Aronson [4] has treated the asymptotic behavior of the solutions of the problem (5.1), (5.2) with initial conditions

(5.13) $S(0, x) = S_0(x), \quad I(0, x) = I_0(x), \quad R(0, x) = R_0(x).$

He transforms the problem into an integral equation for the function R,

(5.14) $R_t + \dfrac{\alpha}{\beta} R = \dfrac{\alpha}{\beta} S_0 \left[1 - \exp \left\{ -\dfrac{\beta}{\alpha} k * (R - R_0) \right\} \right] + \dfrac{\alpha}{\beta} (I_0 + S_0)$

and considers the situation where a front of infectious travels along a susceptible population, i. e. where $S_0 > 0$ const., $R_0 \equiv 0$, and $I_0 \not\equiv 0$, but $I_0(x) = 0$ for $x > 0$. Of course these assumptions imply that the total population is unevenly distributed. The characteristic function

(5.15) $F(\lambda) = \dfrac{1}{\lambda} \left\{ \displaystyle\int_{-\infty}^{\infty} k(y)\, e^{\lambda y}\, dy - \dfrac{\alpha}{\beta S_0} \right\}$

appears in a natural way. In [5] it is shown with some further assumptions that solutions of the equation $F(\lambda) = c$ correspond to solutions U of the integral equation with $U(x) < 0$, and the desired asymptotic behavior. It can also be shown that there are no fronts for $c < \min F(\lambda)$.

From the results in [4] it follows that the minimal speed $c_0 = \min F(\lambda)$ in fact determines the behavior of the time-dependent solutions. If $S_0 < \alpha/\beta$, i.e. if susceptibles are below the threshold everywhere, then for every $x \in \mathbb{R}$ and $c > 0$ one has $\lim_{t \to \infty} R(t, x + ct) = 0$, i.e. an observer travelling with any positive speed will outrun the epidemy. On the other hand, if $S_0 > \alpha/\beta$ then $\lim_{t \to \infty} R(t, x + ct) = 0$ for $c > c_0$ and $\lim_{t \to \infty} R(t, x + ct) = R_\infty > 0$ for $c < c_0$ and all $x \in \mathbb{R}$.

If susceptibles are above the threshold everywhere then the epidemic front travels with asymptotic speed c_0. Here R_∞ is the solution of the equation $1 - R = \exp(-\beta R/\alpha)$. The hypothesis of [4] is satisfied for

$k(x) = (\sqrt{2\pi\sigma})^{-1} \exp(-x^2/2\sigma^2).$

In this case the mapping $\alpha/\beta \to c_0(\alpha/\beta)$ is implicitly given by

$\alpha/\beta = (1 - \lambda^2) \exp(\lambda^2 \sigma^2/2), \quad c_0 = \lambda/(1 - \lambda^2).$

Thus c_0 is very large for $S_0 \gg \alpha/\beta$ and decreases monotonely if S_0 approaches α/β.

Various approaches have been chosen to design models for the spatial spread of epidemics with general transmission laws. The model of Diekmann [17] is among the earliest. Let $S(x, t)$ be the density or susceptibles at time t and position x, and let $i(t, \tau, x)$ be the density of infectious which were infected at time $t - \tau$. Then

(5.16) $I(t, x) = \int\limits_0^\infty i(t, \tau, x)\, d\tau$

is the density of infectious at time t at position x. The kernel $A(\tau, x, \xi)$ describes the infectivity at position x of an infectious individual at position ξ which has been infected at time $t - \tau$. Then the function

(5.17) $B(t, x) = \int\limits_0^\infty \int\limits_\Omega A(\tau, x, \xi)\, i(t, \tau, \xi)\, d\xi\, d\tau$

measures the total infectivity at position x at time t. The evolution of the population is described by the equations

$$\frac{\partial S(t, x)}{\partial t} = - B(t, x)\, S(t, x)$$

(5.18) $i(t, 0, x) = \dfrac{\partial S(t, x)}{\partial t}$

$$i(t, \tau, x) = i(t - \tau, 0, x).$$

At time $t = 0$ initial conditions are required. Using the substitution (suppose $S_0(x) > 0$)

$$u(t, x) = - \log\big(S(t, x)/S_0(x)\big)$$
$$g(y) = 1 - e^{-y}$$
$$f(t, x) = \int\limits_0^t \int\limits_0^\infty \int\limits_\Omega A(s + \tau, x, \xi)\, i_0(\tau, \xi)\, d\xi\, d\tau\, ds$$

the problem can be transformed into a single integral equation

(5.19) $u(t, x) = \int\limits_0^t \int\limits_\Omega A(\tau, x, \xi)\, S_0(x)\, g\big(u(t - \tau, \xi)\big)\, d\xi\, d\tau + f(t, x).$

Diekmann shows global existence and uniqueness for the solution of equation (5.19) in an appropriate function space, provided the function g is locally Lipschitz, furthermore positivity results. Under appropriate conditions he can show that the solution $u(t, x)$ converges towards a limit function $\bar{u}$ which satisfies a stationary equation

$$(5.20) \quad \bar{u}(x) = \int_\Omega \int_0^\infty A(\tau, x, \xi) \, S_0(\xi) \, g(\bar{u}(\xi)) \, d\tau \, d\xi + \bar{f}(x).$$

The work of Barbour [7], Brown and Carr [12], is closely related to the approach of Atkinson and Reuter. Schumacher [69], [70] has studied travelling fronts in Fisher's equation with the Laplacian replaced by an integral operator, see also Thieme [76]. Radcliffe et al. [62], [63] have investigated the problem of travelling fronts for n types of population between which the disease is transmitted. The proofs follow the general line, but require much linear algebra and Perron-Frobenius theory. A somewhat different approach has been taken by Thieme [72], who studies the asymptotic behavior of the solutions of a general Hammerstein integral equation

$$w = A_0 w + A_1 F([w + w_0]_+)$$

where A_0, A_1 are linear integral operators and F is a Nemytskij operator.

6   Oscillations in epidemic models

In several infectious diseases regular periodic oscillations can be observed. Such oscillations have been explained either as periodic solutions of autonomous systems, typically originating from a Hopf bifurcation, or as subharmonics of forced oscillations caused by an annual cycle of climate, social behavior etc.

Several authors [16], [50], [19] have considered contact models with delay in which autonomous oscillations occur. If a susceptible individual is infected at time t then it becomes infective at time $t + \tau_1$, stays infective up to $t + \tau_2$, remains immune till $t + \tau_3$, and becomes susceptible again. For $0 \leq \tau \leq \tau_3$ let $i(t, \tau)$ be the density at time t of those individuals which have become infected at time $t - \tau$. Then

$$(6.1) \quad E(t) = \int_0^{\tau_1} i(t, \tau) \, d\tau, \quad I(t) = \int_{\tau_1}^{\tau_2} i(t, \tau) \, d\tau, \quad R(t) = \int_{\tau_2}^{\tau_3} i(t, \tau) \, d\tau$$

are the exposed, infected, and recovered, respectively. The susceptibles $S(t)$ satisfy

$$(6.2) \quad \dot{S}(t) = -i(t, 0) + i(t, \tau_3).$$

Let $A(\tau)$ be a measure for the infectivity of an individual at time t which has been infected at time $t - \tau$. Then

(6.3) $\displaystyle\int_{\tau_1}^{\tau_2} A(\tau)\, i(t, \tau)\, d\tau$

is the total infectivity at time t. By the contact law

(6.4) $i(t, 0) = S(t) \displaystyle\int_{\tau_1}^{\tau_2} A(\tau)\, i(t, \tau)\, d\tau$

and thus

(6.5) $i(t, 0) = \left(P - \displaystyle\int_{0}^{\tau_3} i(t, \tau)\, d\tau \right) \displaystyle\int_{\tau_1}^{\tau_2} A(\tau)\, i(t, \tau)\, d\tau$

where P is the total population size.

If the population has existed at least for $t \geq -\tau_3$, then for $t \geq 0$ one has $i(t, \tau) = i(t - \tau, 0)$. Then the equations can be transformed into an equation for $x(t) = i(t, 0)$

(6.6) $x(t) = \gamma \left(1 - \displaystyle\int_{0}^{1} x(t - \tau)\, d\tau \right) \displaystyle\int_{0}^{1} b(\tau)\, x(t - \tau)\, d\tau$

where $b \geq 0$, $\displaystyle\int_{0}^{1} b(\tau)\, d\tau = 1$, supp $b \subset [0, \tau_2/\tau_3]$. The parameter $\gamma > 0$ is essentially the population size.

The characteristic equations at the two constant solutions $x_1 \equiv 0$, $x_2 \equiv 1 - 1/\gamma$ read

(6.7) $\gamma \bar{b}(\lambda) = 1$

   $\bar{b}(\lambda) + (1 - \gamma)\dfrac{1}{\lambda}(1 - e^{-\lambda}) = 1$

respectively, where

(6.8) $\bar{b}(\lambda) = \displaystyle\int_{0}^{1} b(\lambda)\, e^{-\lambda \tau}\, d\tau.$

These equations have been discussed by Diekmann and Montijn [18] in detail. For small $\gamma > 0$ the behavior corresponds to that of the Kermack-McKendrick model. For $0 < \gamma < 1$ the only nonnegative solution x_1 is stable, at $\gamma = 1$ there is a bifurcation with an exchange of stability. x_1 is unstable for $\gamma > 1$, and x_2

is stable for $\gamma > 1$, $\gamma - 1$ small. However, x_2 may loose its stability in a Hopf-type bifurcation. One can show: The number of positive Fourier coefficients

$$b_n = 2 \int_0^1 b(\tau) \sin (2 \pi n \tau) \, d\tau$$

is exactly the number of complex-conjugate roots which pass the imaginary axis (always simple pairs, from left to right) as γ runs from 1 to ∞. If τ_2/τ_3 is small then the first few of the b_n are positive. In biological terms: If either the period of immunity or the population size is increased then the steady state can be destabilized leading to oscillator behavior.

    If in equation (6.1) one assumes that the transition times are distributed then one arrives at an equation

$$(6.9) \quad x(t) = \gamma \left(1 - \int_{-\infty}^t a(t - \tau) \, x(\tau) \, d\tau \right) \int_{-\infty}^t b(t - \tau) \, x(\tau) \, d\tau$$

for which Gripenberg [25] has shown a Hopf bifurcation theorem under the hypothesis that there is a first pair of conjugate roots which passes the imaginary axis with positive speed and is not in resonance with other roots.

    An extensive bibliography on periodic oscillations in epidemic models has been collected in [39].

    The spread of infectious deseases is influenced by the seasonal variation of parameters, such as contact rates changing with school terms. One has tried to explain biennial and other cycles as subharmonics of the annual cycle. Mathematical models have been designed by Cooke and Yorke [16], London and Yorke [50] in the form of delay equations, Dietz [19], [26] has proposed a simpler model, a $S \to I \to R$ model with periodic forcing

$$(6.10) \quad \dot{S} = \mu - \mu S - \beta (1 + \delta \cos 2 \pi t) \, I S$$
$$\dot{I} = \beta (1 + \delta \cos 2 \pi t) \, I S - (\gamma + \mu) \, I$$
$$\dot{R} = \gamma I - \mu R$$

The bifurcation of subharmonics, i. e. of periodic solutions with periods greater than 1, has been discussed by Smith [67], [68] for a certain restricted range of parameters.

7   Discrete Parasites

    Prevalence models as the Kermack-McKendrick model are well-suited to describe bacterial and virus diseases, where the host acquires the disease at one instant, typically from an infected individual, and then carries a large number of germs. On the other hand, in helminthic infections, which are of utmost medical and social importance particularly in tropical countries, the host acquires a small number of parasites at different times. The number of parasites in an individual host influences the severeness of the disease. In many cases there are intermediate hosts, where the parasite population multiplies, and larvae directly immigrate into the definitive host, or the disease is transmitted by vectors such as biting insects.

Nåsell and Hirsch [57] have designed a stochastic model for helminthic infections in which there is a constant population of N_1 definitive hosts and a constant population of N_2 intermediate hosts. The process ξ_k, $k = 1, \ldots, N_1$, counts the number of mature parasites in the k-th definitive host, and η is the number of infected intermediate hosts, $0 \leq \eta \leq N_2$. The problem can be stated in terms of a system of differential equations for the functions

$$
\begin{aligned}
P_{mn}(s, t) &= P\{\xi_k(t) = n \,|\, \xi_k(s) = m\} \\
\pi_{ij}(s, t) &= P\{\eta(t) = j \,|\, \eta(s) = i\}
\end{aligned}
\tag{7.1}
$$

$$
m, n \in \mathbb{N} \cup \{0\}; \quad k = 1, \ldots, N_1; \quad i, j = 0, \ldots, N_2; \quad 0 < s < t.
$$

The basic assumptions are as follows: a definitive host acquires a new parasite with a rate proportional to the expected number of infected intermediate hosts, and looses its parasites according to a death process,

$$
\begin{aligned}
P_{m,m+1}(t, t + \Delta t) &= \gamma_1 E[\eta(t)]\,\Delta t + o(\Delta t) \\
P_{m,m-1}(t, t + \Delta t) &= m\mu_1\,\Delta t + o(\Delta t) \\
P_{m,m+1}(t, t + \Delta t) &+ P_{mm}(t, t + \Delta t) + P_{m,m-1}(t, t + \Delta t) = 1 + o(\Delta t).
\end{aligned}
\tag{7.2}
$$

On the other hand, intermediate host become infected in proportion to the expected number of infected definitive hosts and are replaced by non-infected intermediate hosts according to a death process,

$$
\begin{aligned}
\pi_{i,i+1}(t, t + \Delta t) &= \gamma_2 \sum_{k=1}^{N_1} E[(\xi_k)(t)]\,(N_2 - i)\,\Delta t + o(\Delta t) \\
\pi_{i,i-1}(t, t + \Delta t) &= i\mu_2\,\Delta t + o(\Delta t) \\
\pi_{i,i+1}(t, t + \Delta t) &+ \pi_{ii}(t, t - \Delta t) + \pi_{i,i-1}(t, t + \Delta t) = 1 + o(\Delta t).
\end{aligned}
\tag{7.3}
$$

The authors give existence proofs and investigate in detail the asymptotic behaviour of the solutions, also they derive strategies for the control of parasite populations. Further work, mainly on the control of schistomiasis, is contained in [53], [58].

The following model makes use of the concept of a birth and death process with killing.

The classical birth and death process counts particles or individuals of a population. Thus the state space is $\mathbb{N} \cup \{0\}$. In an infinitesimal time interval only a birth of a particle or the death of a particle may occur. On the other hand, in the birth and death processes with killing which have been studied by Karlin and Tavaré [44] and earlier, though more implicit, by Puri [61], there is an additional state H, in which the population is absent or "killed". Such processes are suited to describe populations in which, by the population itself or by action from outside, the habitat is destroyed. Here we shall consider only processes with constant rates. Let $\varrho \geqq 0$ and $\sigma > 0$ be the birth rate and death rate of individuals, and let $\alpha > 0$ be the killing parameter: If r individuals are present, then the probability of transition to H during Δt is $\alpha r \, \Delta t + o(\Delta t)$. The probability generating functions $u(t, z)$ for this process satisfy the partial differential equation

$$(7.4) \quad u_t + g(z) \, u_z - \varphi(t) \, (z - 1) \, u = 0,$$

where $\varphi(t)$ is the time-dependent immigration rate and

$$(7.5) \quad g(z) = (\alpha + \sigma + \varrho) \, z - \sigma - \varrho z^2.$$

Using the method of characteristics, one can represent the solutions of these equations in terms of the solution operator G of the Riccati equation $\dot{z} = g(z)$,

$$(7.6) \quad u(t, z) = u_0 \, (G(-t, z)) \, \exp \left[\int_0^t G(s - t, z) - 1 \right] \varphi(s) \, ds.$$

Here

$$(7.7) \quad G(t, z) = \frac{z_1 (z - z_2) + z_2 (z_1 - z) e^{-\varkappa t}}{(z - z_2) + (z_1 - z) \, e^{-\varkappa t}}$$

where z_1, z_2 are the solutions of the equation $g(z) = 0$, $z_1 \geqq 1 \geqq z_2$, and $\varkappa = \varrho(z_1 - z_2)$. $(G(t, z) = 1 - (1 - z) \, e^{\varkappa t} - (\alpha/\varkappa) \, (1 - e^{\varkappa t})$ with $\varkappa = \alpha + \sigma$ in case $\varrho = 0$.)
Most properties of the process can be studied in terms of the function G.
Anderson and May [1], [2], [3] derived models in the form of ordinary differential equations for the densities of hosts and parasites from ad hoc assumptions on the parasitic distribution. Hadeler and Dietz [30] assume that the host population satisfies an equation of the form (2.4), and that in each individual host the parasite population is governed by a birth and death process with killing (of the host). Furthermore it is assumed that the acquisition rate of parasites (i.e. the immigration rate of parasites) is a nonlinear function of the average parasite load.
Let $\varrho \geqq 0$ and $\sigma > 0$ be the birth and death rate of parasites within a host, and $\alpha > 0$ is the differential mortality of hosts due to the presence of one parasite,

thus the mortality of a host of age a carrying r parasites is $\mu(a, P) + \alpha r$ where P is the total population size. If W is the average parasite load then

$$(7.8) \quad \varphi = \beta f(W, P)$$

is the parasite acquisition rate. Here $f: [0, \infty)^2 \rightarrow [0, \infty)$ is a nonlinear function with $f(0, P) \equiv 0$, $\partial f(0, P)/\partial W = 1$, and $\beta > 0$ is a normalization constant which may serve as a bifucation parameter. Let $n(t, a, r)\, da$ be the number of hosts at time t with age in $[a, a + da)$ carrying r parasites. These functions satisfy a system of infinitely many differential equations [30], which can be described in terms of the generating function

$$(7.9) \quad u(t, a, z) = \sum_{r=0}^{\infty} n(t, a, r)\, z^r.$$

This function satisfies the differential equation

$$(7.10) \quad u_t + u_a + g(z)\, u_z - [\varphi(z - 1) - \mu(a)]\, u = 0$$

where the function g is given by (7.5).
   In terms of the generating function the average parasite load is

$$(7.11) \quad W(t) = \frac{\int_0^{\infty} u_z(t, a, 1)\, da}{\int_0^{\infty} u(t, a, 1)\, da}.$$

The initial condition

$$(7.12) \quad u(0, a, z) = u_0(a, z)$$

describes the age structure of the host population and the distribution of parasites within the hosts at time 0. The side condition at $a = 0$ specifies the neonatals, either by a prescribed rate

$$(7.13) \quad u(t, 0, z) = N(t)$$

(neonatals are assumed not to be infected) or by a Lotka birth law similar to (3.3)

$$(7.14) \quad u(t, 0, z) = \int_0^{\infty} b(a)\, u(t, a, \omega)\, da.$$

Here $\omega \in [0,1]$ is a parameter which measures the geometric decrease in fertility due to the presence of parasites. For $\omega = 1$ there is no influence, for $\omega = 0$ only noninfected individuals reproduce. (Observe that the factor $n(t, a, r)$ in (7.9) is multiplied by ω^r.)

In several papers [30—37] the following approach has been choosen: Assume the function φ in (7.10) were known. Then the linear differential equation (7.10) can be solved explicitly by the method of characteristics. This solution can be introduced into the equations (7.11), (7.8). Then, in case of the boundary condition (7.13), one obtains an Volterra integral equation for the function $\varphi(t)$. In case of the boundary condition (7.14) one arrives at a system of two coupled integral equations for the functions $\varphi(t)$ and $N(t) = u(t, 0)$.

Finally, as in the Gurtin-McCamy model, the functions b and μ can depend on the age a and on the total population size P. Then one obtains a system of three coupled Volterra integral equations for the functions φ, N, P. In the following I discuss only the stationary case. Then φ, N, P are constants, which satisfy the equations

(7.15) $\varphi = \beta f(P, \tilde{W}(P, \varphi))$

(7.16) $\int_0^\infty b(a, P)\, e^{-M(a, P)}\, e^{-Q_\omega(a)\varphi}\, da = 1$

(7.17) $P = N \int_0^\infty e^{-Q(a)\varphi - M(a, P)}\, da$

(7.18) $\tilde{W}(P, \varphi) = \dfrac{\int_0^\infty e^{-Q(a)\varphi - M(a, P)}\, q(a)\, da\, \varphi}{\int_0^\infty e^{-Q(a)\varphi - M(a, P)}\, da}$

The coefficient functions Q_ω, q, $Q = Q_1$, and M are given as

$$Q_\omega(a) = -(z_1 - 1)\, a + \frac{1}{\varrho} \log \frac{(\omega - z_2) + (z_1 - \omega)\, e^{\varkappa a}}{z_1 - z_2} \qquad (\omega > 0)$$

(7.19) $q(a) = \dfrac{1}{\varrho} \dfrac{e^{\varkappa a} - 1}{z_2 + (z_1 - 1)\, e^a} \qquad (\omega > 0)$

$$M(a, P) = \int_0^a \mu(s, P)\, ds$$

There are several main cases:
1) f, b, μ do not depend on P. The boundary condition has the form (7.13) with $N(t) = \text{const}$. Then from equation (7.17) one obtains the trivial solution $\varphi = 0$ and a nontrivial branch given by the same equation,

$\beta = \varphi/f(\tilde{W}(\varphi))$. Then the total population size is given by (7.17). This branch need not be monotone, i.e. for a given value of β there may be several nontrivial stationary states, and there may be hysteresis phenomena. In [35] a rather detailed stability analysis has been given which shows, as could be expected, a strong connection between the slope of the bifurcating branch and the stability properties of the nontrivial solutions.

2) f, b, μ do not depend on P. The boundary condition has the form (7.14). In general stationary solutions do not exist. For persistent solutions

$$u(t, a, z) = u_\infty(a, z) e^{\lambda t}$$

one has to replace in equations (7.15), (7.16), (7.18) the function $M(a)$ by $M(a) + \lambda a$, then solve for φ and λ. Again one finds that there is a trivial branch $\varphi = 0$ and a nontrivial branch of persistent solutions with positive φ.

3) f, b, μ depend on P, and the boundary condition has the form (7.14). In general equations (7.15), (7.16) will admit several stationary solution branches with $\varphi = 0$ and such branches with $\varphi > 0$.

In all three cases certain monotonicity assumptions lead to uniqueness. Natural assumptions are that $b(a, P)$ is decreasing in P, $\mu(a, P)$ and $f(W, P)$ are increasing in P.

Finally, and most important let the function f be concave in the sense of Krasnoselskij with respect to the variable W

$$\frac{\partial f}{\partial W} > 0, \quad \frac{\partial}{\partial W}\left(\frac{f(W, P)}{W}\right) < 0, \quad \frac{\partial}{\partial P}\left(\frac{f(W, P)}{\partial P}\right) > 0$$

then, in addition to the trivial solution, for each β there is at the most one nontrivial stationary or persistent solution. This branch represents equilibria where the population is kept below the natural capacity of the biotope due to the action of parasites.

References

[1]   Anderson, R. M. and R. M. May, *Directly transmitted infectious diseases*, Control by vaccination. Science **215**, 1053—1060 (1982).

[2]   Anderson, R. M. and R. M. May, *Population biology of infectious diseases I*. Nature **280**, 361—367 (1979).

[3]   May, R. M. and R. M. Anderson, *Population biology of infectious diseases II*. Nature **280**, 455—461 (1979).

[4]   Aronson, D. G., *The asymptotic speed of propagation of a simple epidemic*. In: Nonlinear Diffusion, p. 1—23, W. E. Fitzgibbon, H. F. Walker Eds., Research Notes in Mathematics **19**, Pitman, London (1977).

[5]   Atkinson, C. and G. H. Reuter, *Deterministic epidemic waves*. Math. Proc. Cambr. Philos. Soc. **80**, 315—330 (1976).

[6] Bailey, N. T. J., *The mathematical theory of infectious diseases and its applications*, 2nd ed. London, Griffin (1975).

[7] Barbour, A. D., *The uniqueness of Atkinson and Reuter's epidemic waves*. Math. Proc. Cambr. Phil. Soc. **82**, 127—130 (1977).

[8] Di Blasio, G., *Non-linear age-dependent population diffusion*. J. Math. Biol. **8**, 265—284 (1979).

[9] Di Blasio, G., M. Iannelli, E. Sinestrari, *Approach to equilibrium in age structured population with an increasing recruitment process*, J. Math. Biol. **13**, 371—382 (1982).

[10] Bramson, M., *Maximal displacement of branching Brownian motion* Comm. Pure App. Math. **91**, 531—581 (1978).

[11] Bramson, M., *Minimal displacement of Branching Brownian motion*. Z. Wahrsch. Theor. verw. Gebiete **45**, 89—108 (1978).

[12] Brown, K. J. and J. Carr, *Deterministic epidemic waves of critical velocity*. Math. Proc. Cambr. Phil. Soc. **81**, 431—433 (1977).

[13] Busenberg, S., K. L. Cooke, M. A. Pozio, *Analysis of a model of a vertically transmitted disease*. J. Math. Biol. **17**, 305—329 (1983).

[14] Capasso, V., *Global solution for a diffusive nonlinear deterministic model SIAM*, J. Appl. Math. **35**, 274—284 (1978).

[15] Capasso, V. and L. Maddalena, *Convergence to equilibrium states for a reaction-diffusion system modelling the spatial spread of a class of bacterial and viral diseases*. J. Math. Biol. **13**, 173—184 (1981).

[16] Cooke, K. L. and J. A. Yorke, *Some equations modelling growth processes and gonorrhea epidemics*. Math. Biosc. **16**, 75—101 (1973).

[17] Diekmann, O., *Thresholds and travelling waves for the geographical spread of infection*. J. Math. Biol. **6**, 109—130 (1978).

[18] Diekmann, O. and R. Montijn, *Prelude to Hopf bifurcation in an epidemic model: Analysis of a characteristic equation associated with a nonlinear Volterra integral equation*. J. Math. Biol. **14**, 117—127 (1982).

[19] Dietz, K., *The incidence of infectious diseases under the influence of seasonal fluctuations*. Lecture Notes in Biomath. **11**, p. 1—15, Berlin-Heidelberg-New York, Springer (1976).

[20] Feller, W., *On the integral equation of renewal theory*. Ann. Math. Stat. **12**, 243—267 (1941).

[21] Fife, P. C. and J. B. Mc Leod, *The approach of solutions of nonlinear diffusion equations to travelling wave solutions*. Arch. Rat. Mech. Anal. **65**, 333—361 (1977).

[22] Fisher, R. A., *The advance of advantageous genes*. Ann. of Eugenics **7**, 355—369 (1937).

[23] Von Foerster, H., *Some remarks on changing populations*. In: The kinetics of cellular proliferation. p. 382—407. New York, Grune and Stratton (1959).

[24] Frauenthal, J. C., *Mathematical modeling in epidemiology*. Springer-Verlag (1980).

[25] Gripenberg, G., *Periodic solutions of an epidemic model*. J. Math. Biol. **10**, 271—280 (1980).

[26] Grossman, Z., I. Gumovski, K. Dietz, *The incidence of infectious diseases under the influence of seasonal fluctuations—analytic approach*. In: Nonlinear systems and applications to life sciences p. 525—546, Academic Press, New York (1977).

[27] Gurtin, M. E. and R. C. MacCamy, *Nonlinear age-dependent population dynamics*. Arch. Rat. Mech. Anal. **54**, 281—300 (1974).

[28] Hadeler, K. P. and F. Rothe, *Travelling fronts in nonlinear diffusion equations*. J. Math. Biol. **2**, 251—263 (1975).

[29] Hadeler, K. P., *Free boundary value problems in biological models*, In: Free boundary problems, theory and applications, Vol. II, 664—681, A. Fasano, M. Primicerio Eds., Pitman Boston-London-Melbourne (1983).

[30] Hadeler, K. P. and K. Dietz, *Nonlinear hyperbolic partial differential equations for the dynamics of parasite populations*. Comp. and Math. with Appl. **9**, 415—430 Pergamon Press (1983).

[31]   Hadeler, K. P., *An integral equation for helminthic infections: Stability of the non-infected population*. In: Trends in Theor. Pract, Nonl. Diff. Eq. p. 231 — 240 V. Lakshmikantham Ed., Lecture Notes in Pure Appl. Math. **90**, M. Dekker (1984).

[32]   Hadeler, K. P., *Integral equations with discrete parasites: Hosts with a Lotka birth law*. In: Conf. Proc. Autumn Course on Math. Ecology, Trieste 1982, S. Levin, T. Hallam Eds., Lect. Notes in Biomathematics **54**, Springer Verlag (1984).

[33]   Hadeler, K. P. and K. Dietz, *An integral equation for helminthic infectious: Global existence of solutions*. In: Recent Trends in Mathematics, Conf. Proc. Reinhardsbrunn, Teubner-Verlag, Leipzig (1982).

[34]   Hadeler, K. P., *Hysteresis in a model for parasitic infection*. In: Conf. Num. Math. for Bifurcation Problems, Dortmund 1983, H. Mittelmann, T. Küpper Eds., Birkhäuser Verlag, to appear.

[35]   Hadeler, K. P., *Vector models for Infectious diseases, Stability of the nontrivial stationary states*. Proc. Conf. on Ord. and Partial Diff. Equ., Dundee 1984.

[36]   Hadeler, K. P. and K. Dietz, *Population dynamics of killing parasites which reproduce in the host*. J. Math. Biol. in revision.

[37]   Hadeler, K. P. and K. Dietz, *Models for parasitic diseases: Transmission rate depending on total population size*. In preparation.

[38]   Hadeler, K. P., *Reduction of Lotka age structure models to ordinary differential equations*. In preparation.

[39]   Hethcote, H. W., H. W. Stech, P. van den Driessche, *Periodicity and stability in epidemic models:* a survey Diff. Eq. and Appl. in Ecology, Epidemics, and Population Problems. 1980.

[40]   Hoppensteadt, F. and P. Waltman, *A problem in the theory of epidemics*. Math. Biosc. **9**, 71 — 91 (1970).

[41]   Hoppensteadt, F. and P. Waltman, *A problem in the theory of epidemics II*. Math. Biosc. **12**, 133 — 145 (1971).

[42]   Hoppensteadt, F., *Mathematical Theories of Populations: Demographics, Genetics and Epidemics*, SIAM Reg. Conf. Series **20**, Philadelphia (1975).

[43]   Jagers, P., *Branching Processes with Biological Applications*. John Wiley 1975.

[44]   Karlin, S. and S. Tavaré, *Linear birth and death processes with killing*. J. Appl. Prob. **19**, 477 — 487 (1982).

[45]   Kendall, D. G., *Mathematical models of the spread of infection*. Mathematics and Computer Science in Biology and Medicine. p. 213 — 225. London H.M.S.O. 1965.

[46]   Kermack, W. O. and A. G. McKendrick, *Contributions to the mathematical theory of epidemics*, part I., Proc. Roy. Soc., Ser. A 115, 700 — 721 (1927).

[47]   Kolmogorov, A., I. Petrovskij, N. Piskunow, *Etude de la quantité de la matière et son application à un problème biologique*, Bull. Univ. Moscou Ser. Internation., Sec. A, 1, 6, 1 — 25 (1937).

[48]   Lamberti, L. and P. Vernole, *An age structured epidemic model-asymptotic behavior*, to appear.

[49]   Leslie, P. H., *On the use of matrices in certain population mathematics*. Biometrika **33**, 183 — 212 (1945).

[50]   London, W. P. and J. A. Yorke, *Recurrent outbreak of measles chickenpox, and mumps I:* seasonal variation in contact rates Amer. J. Epidem. **98**, 453 — 468 (1973).

[51]   Longini, I. M., *Models of epidemics and endemicity in genetically variable host populations*. J. Math. Biol. **17**, 289 — 304 (1983).

[52]   Lotka, A. J., *The stability of the normal age distribution*, Proc. Nat. Acad. Science **8**, 339 — 345 (1922).

[53]   MacDonald, G., *The dynamics of helminthic infections, with special reference to schistosomes*. Trans. Roy. Soc. Trop. Med. Hyg. **59**, 489 — 506 (1965).

[54]   McKean, H. P., *Application of Brownian motion to the equation of Kolmogorov-Petrovskii-Piskunov*. Comm. Pure Appl. Math. **28**, 323 — 331 (1975).

[55]   Marcati, P., *On the global stability of the logistic age dependent population growth*. J. Math. Biol. (1982).

[56]   Mollison, D., *Possible velocities for a simple epidemic*. Adv. Appl. Prob. **4**, 233—257 (1972).

[57]   Nåsell, I. and W. M. Hirsch, *A mathematical model of some helminthic infections*. Comm. Pure Appl. Math. **25**, 459—477 (1972).

[58]   Nåsell, I., *Mating models for schistosomes*, J. Math. Biol. **6**, 21—35 (1978).

[59]   Noble, J. V., *Geographical and temporal development of plagues*. Nature **250**, 726—729 (1974).

[60]   Prüß, J., *Equilibrium solutions of age-specific population dynamics of several species*, J. Math. Biol. **11**, 65—84 (1981).

[61]   Puri, P. S., *A method for studying the integral functionals of stochastic processes with applications III*. Proc. Sixth Berkeley Symp. Math. Stat. Prob. Vol. III, 481—500, UCLA Press 1972.

[62]   Radcliffe, J., L. Rass, W. D. Stirling, *Wave solutions for the deterministic host-vector epidemic*. Math. Proc. Cambr. Phil. Soc. **91**, 131—152 (1982).

[63]   Radcliffe, J. and L. Rass, *Wave solutions for the deterministic non-reducible n-type epidemic*. J. Math. Biol. **17**, 45—66 (1983).

[64]   Rorres, C., *Local stability of a population with density dependent fertility*, Theor. Pop. Biol. **10**, 26—46 (1976).

[65]   Rothe, F., *Convergence to travelling fronts in semilinear parabolic equations*, Proc. Roy. Soc. Edinburgh 80 A, 213—234 (1978).

[66]   Sharpe, F. R. and A. J. Lotka, *A problem in age distribution*, Phil. Mag. **21**, 435—438 (1911).

[67]   Smith, H. L., *Subharmonic bifurcation in an S-I-R epidemic model*, J. Math. Biol. 163—177 (1983).

[68]   Smith, H. L., *Multiple stable subharmonics for a periodic epidemic model*, J. Math. Biol. **17**, 179—190 (1983).

[69]   Schumacher, K., *Travelling fronts for integro-differential equations I*, J. Reine Angew. Math. **316**, 54—70 (1980).

[70]   Schumacher, K., *Travelling fronts for integro-differential equations II*, In: Conf. Proc. Biological Growth and Spread, Heidelberg 1979, p. 296—309, W. Jäger, H. Rost, P. Tautu, Eds., Lecture Notes in Biomathematics **38**, Springer-Verlag (1980).

[71]   Sovunmi, C. O. A., *Female dominant age-dependent deterministic population dynamics*. J. Math. Biology **3**, 9—17 (1976).

[72]   Thieme, H. R., *The asymptotic behaviour of solutions of nonlinear integral equations*, Math. Zeitschr. 157, 141—154 (1977).

[73]   Thieme, H. R., *A model for the spatial spread of an epidemic*, J. Math. Biol. **4**, 337—351 (1977).

[74]   Thieme, H. R., *Renewal theorems for linear discrete Volterra systems*, J. Reine Angew. Math. (to appear).

[75]   Thieme, H. R., *Renewal theorems for some mathematical models in epidemiology*, J. Integral Eqs. (to appear).

[76]   Thieme, H. R., *Asymptotic estimates of the solutions of nonlinear integral equations and asymptotic speeds for the spread of populations*, J. Reine Angew. Math. **306**, 94—121 (1979).

[77]   Uchiyama, K., *The behavior of solutions of some nonlinear differential equations for large time*, J. Math. Kyoto Univ. **18**, 453—508 (1978).

[78]   Waltman, P., *Deterministic threshold models in the theory of epidemics* (Lecture Notes in Biomathematics Vol. 1), Berlin-Heidelberg-New York, Springer (1974).

[79]   Webb, G. F., *An age-dependent epidemic model with spatial diffusion*, Arch. Rat. Mech. Anal. **75**, 91—102 (1980).

[80]   Webb, G. F., *A recovery-relapse epidemic model with spatial diffusion*, J. Math. Biol. **14**, 177—194 (1982).

[81]   Webb, G. F., *Theory of nonlinear age-dependent population dynamics*, Mscr. 330 pp. Nashville (1983).

[82]   Wilson, L. O., *An epidemic model involving a threshold*, Math. Biosc. **15**, 109—121 (1972).

[83]   Yorke, J. A. and W. P. London, *Recurrent outbreaks of measles, chickenpox, and mumps II: systematic differences in contact rates stochastic effects*, Amer. J. Epidem. **98**, 469—492 (1973).

Perspectives in Mathematics
Anniversary of Oberwolfach 1984
© Birkhäuser Verlag, Basel

Calculus of Variations Today, Reflected in the Oberwolfach Meetings

S. HILDEBRANDT

Mathematisches Institut der Universität,
Wegelerstraße 10, D-5300 Bonn 1 (FRG)

1   It is rather difficult, if not impossible, to explain the benefit and the importance of the Mathematical Research Institute in Oberwolfach to a non-mathematician. Whenever I went to another Oberwolfach meeting, my father suspected that I was leaving for some kind of vacation. A house in the Black Forest, long hikes through the woods, and extended discussions—this did not sound like respectable, serious work. Yet, in retrospect, I can say that the meetings at the Lorenzenhof have been of invaluable help to me when I was trying to become a mathematician, and certainly many of my colleagues, younger and older ones alike, have had similar experiences. I particularly well remember the two meetings on partial differential equations in 1961 and 1963 when, in addition to many other exciting lectures, Moser talked about Harnack's inequality for elliptic equations, Jörgens presented his new results on nonlinear wave equations, and Courant lectured on old and new experiments for minimal surfaces. There it became completely clear to me that I had to go to the Mecca for Analysis, the Courant Institute at New York, to complete my studies and to enlarge my view. Fortunately my teacher Ernst Hölder helped me to become acquainted with Courant and to obtain an invitation to his institute. This experience, I think, is quite typical of one role that Oberwolfach plays for mathematicians, in particular for German mathematicians who come from small departments. At a well organized meeting they have the chance to get a survey of their field of interest, to learn about new ideas, new results and important problems. Discussions during a whole week have initiated lasting collaborations, and many a young mathematician has made contacts which were to strongly influence his further scientific development.

The annual or biannual gatherings serve as information bazaars for the specialists in the principal mathematical fields. Of course, there arises the danger of inbreeding and sterility. This ever present danger can best be avoided by sufficient international cooperation, and, furthermore, a limited period of duty of the organizers would certainly be helpful.

Another function of the Oberwolfach Institute is to give German mathematicians the opportunity to leave the predominant areas of research and to pursue new promising tracks, either by organizing seminars with colleagues from other departments or by arranging meetings which are directed by experts from other countries. In some cases, new tracks might in fact be rather old ones, unjustly neglected in the past. For instance, in 1968, we tried to revitalize the calculus of variations and to draw again attention to this beautiful and important field which thrived so well in the United States, in Italy, and also in the Soviet Union. (Young mathematicians should feel encouraged to take the iniative and to propose new meetings if they see a need for them. I have found that the Director of the Institute and the Advisory Board support all interesting projects.)

A third and equally important objective of the Oberwolfach Institute is to serve the international mathematical community by providing an unexcelled location for scientific exchange. Where in the world can mathematicians meet and discuss, having an excellent library at hand both day and night?

One cannot separate the different functions and goals of the Institute since, obviously, scientific exchange means give and take. The renewal of mathematicial research in our country after the brown disaster would have been impossible without Oberwolfach and without the help of our foreign colleagues. While we celebrate the anniversary of Oberwolfach, I have no doubt that the Institute can keep its vitality and will, in its unique way, serve mathematics and mathematicians for many years to come.

2   I should like to discuss the development of the calculus of variations during the last fifteen years and to compare it with the program of the Oberwolfach meetings of the same name. In June 1968, the first variational meeting took place, organized by Erhard Heinz, Willi Jäger and the author. Besides Jens Frehse, Ernst Hölder and Johannes Nitsche, the other participants were assistants and students from Göttingen and Mainz. When we set up plans, we found the following situation: The theory of linear elliptic equations and systems with regular coefficients was more or less completed. Everything that was needed in the calculus of variations of multiple integrals had been presented by Morrey in his monumental treatise [12] which had just appeared. Over a period of thirty years, Morrey had developed the "*direct methods*" of the calculus of variations. The basic idea of this approach, due to Hilbert, Lebesgue, and Tonelli, is fairly simple. If one wants to minimize a multiple integral, say, of the kind

$$I(u) = \int_{\Omega} F\big(x, u(x), Du(x)\big) \, dx,$$

where the competing functions $u(x)$ are subjected to certain subsidiary conditions as boundary conditions, one makes a good guess what properties the minimum $v(x)$ may have. Then one sets up a class $\mathscr{C}$ of admissible functions that, as one hopes, will contain the solution $v(x)$, and minimizes the functional I within $\mathscr{C}$. After choosing a sequence of admissible functions u_n with $I(u_n)$ tending to the infimum e of I within the class $\mathscr{C}$, one tries to select a subsequence $\{v_n\}$ of $\{u_n\}$

which, in some sense, will converge to some function v of $\mathscr{C}$ whence

$$\lim_{n\to\infty} I(v_n) = e \leqq I(v). \tag{1}$$

This requires some compactness of the sequence $\{u_n\}$ which is the easier to prove the larger the class $\mathscr{C}$ and the coarser its topology is choosen. Yet simple examples show that one can no longer hope for continuity of the functional I with respect to this convergence but has to be content with lower semicontinuity, i.e., with the relation

$$\lim_{n\to\infty} \inf I(v_n) \geqq I(v).$$

But, on account of (1), this would suffice to prove $I(v) = e$.

Morrey, Sobolev, and L. Schwartz had developed a theory of function spaces suitable for this approach, and Morrey had obtained sufficiently general lower semicontinuity results. Thus the framework of an existence theory of so-called *weak solutions* was available. The main question to be solved was to prove regularity (in the sense of Hilbert's problem 19) of these weak solutions in the case of elliptic integrands F. This had been done by Morrey for double integrals, whereas De Giorgi and Nash had solved the basic problem for multiple integrals depending on scalar-valued functions $u(x)$. The general regularity problem, however, was still undecided. In 1967, Morrey proved a "partial regularity theorem" by showing that, under suitable restrictions on F, weak solutions were regular up to a locally compact set Σ of measure zero, and briefly afterwards Giusti and Miranda obtained a similar theorem. These results seemed incomplete and were therefore considered as unsatisfactory. Thus it was quite a surprise when, in 1967, De Giorgi, Giusti and M. Miranda found examples of elliptic variational problems with irregular weak solutions (the corresponding papers appeared in print in 1968).

We then decided to direct our attention to two-dimensional variational problems, where a technique valid for systems was available, for most questions in differential geometry lead unavoidably to general mappings, that is, to non-linear systems. Our choice therefore was quite natural. As topics for our seminar we selected Heinz's 1954-paper on surfaces of constant mean curvature, the theory of unstable minimal surfaces (following Courant and Morse-Tompkins), Morrey's regularity theory for double integrals, Morrey's variational approach to the global version of Lichtenstein's theorem that two-dimensional surfaces of the topological type of a planar domain bounded by k curves can be mapped conformally into the plane, Morrey's solution of Plateau's problem for minimal surfaces in a Riemannian manifold, Morrey's treatment of the general two-dimensional parametric problem, and Hermann's theorem from 1966 stating isolatedness of minimal surfaces spanned into a given boundary curve.

We moreover decided to leave aside the Morse theory developed by

Palais and Smale since there seemed to be no applications to interesting geometric problems except for those leading to one-dimensional variational problems. We had in fact to wait until 1982 when Struwe, Brezis and Coron saw suitable modifications of this theory which permitted to study unstable minimal surfaces, harmonic mappings, and surfaces of constant mean curvature. For similar reasons we excluded the very "general" theory of nonlinear equations based on the study of monotone operators and similar ideas which, a few years before, had been initiated by Minty. The study of obstacle problems and variational inequalities, related to monotone operators, appeared more promising but was left to the next meeting.

3   The seminar was started with lectures on surfaces of constant mean curvature ("H-surfaces") and on unstable minimal surfaces. The paper of Morse-Tompkins was found unsatisfactory in so far as we thought it hopeless to ever be able to check the conditions on which these authors had founded their theory. In addition, no generalization to H-surfaces was in sight. On the other hand, Courant's method was considered as very simple, elegant, and suitable to be generalized. This approach employed the "mountain pass lemma" which was applied to minimal surfaces within polygonal boundaries, and the isoperimetric inequality permitted to pass to general boundaries by an appropriate limit consideration. Heinz was able to show how Courant's ideas could be carried over to H-surfaces within contours Γ provided that $R \cdot |H| < 1/2$, where R is the radius of the smallest ball containing Γ. This was the first encouraging success of the meeting which, in later years, stimulated various related investigations. For instance Ströhmer carried over Courant's approach to Plateau's problem for minimal surfaces in Riemannian manifolds and to related questions. Jarausch used the same method to treat Plateau's problem numerically, and recently Wohlrab was able to solve in a similar way free boundary value problems for minimal surfaces. Finally, these first lectures, together with a rather cryptic paper by Marx and Shiffman, initiated the penetrating investigations of Heinz on minimal surfaces with polygonal boundaries which are still not entirely completed.

The lectures on the Lichtenstein theorem ended in a general turmoil since Morrey's proof was found to contain gaps which could not be filled by the auditorium (cf. [12], chapter 9). More than ten years passed by until Jost obtained a flawless modification of Morrey's approach. These ideas have led to remarkable new results on harmonic and on conformal mappings of surfaces.

The lecture on Hermann's paper was never given because of an incorrect minus sign, and the result was past saving. Unfortunately, the question as to whether minimal surfaces within a fixed contour are isolated is still unsolved, although it is of great importance if one wants to solve one of the most difficult open problems for minimal surfaces. This is the question whether every closed regular Jordan curve will bound only finitely many minimal surfaces. At least, the Oberwolfach meetings stimulated Tomi to prove a beautiful partial result:

"There are only finitely many disk-type minimal surfaces within a closed and real analytic Jordan curve which furnish an absolute minimum of area." (Beeson obtained some generalization to relative minima for special boundary contours.)

But let us return to the first variational meeting. The rest of the program was carried through except that the lecture on the parametric integrals was replaced by a talk on the boundary regularity result for minimal surfaces which had been found only a few weeks before the seminar started. This result and the lectures on Morrey's regularity theory for double integrals initiated a development which still leads to interesting results. We only mention the ramified theory of free boundary value problems which began with Jäger's result, found by the end of 1969.

4   Thus the first variational meeting was a good start. Further gatherings were conceived, and the participants set up a list of problems, a result of long discussions after the lectures and on hikes, and these were written on the black board. I should like to give the nearly complete list, using only a slight rewording:

(1)   Can one prove boundary regularity of H-surfaces spanning a regular boundary contour?

(2)   Can one prove boundary regularity for minimal surfaces with "free boundaries"? What about the length of the free trace?

(3)   Do there exist H-surfaces within a given boundary if $R|H| \geq 1/2$?

(4)   What about branch points for minimal surfaces? Are there curves which only bound (disk-type) minimal surfaces with branch points? Can one find conditions on the boundary Γ such that Γ only bounds minimal surfaces free from branch points? What is the situation for H-surfaces?

(5)   Is there a sharpening of Sasaki's inequality for knotted curves Γ? (Sasaki's formula estimates the number of branch points by the total curvature of Γ.)

(6)   Can one find an analogue of Sasaki's inequality for H-surfaces?

(7)   Solve the so-called "thread Problem" for minimal surfaces (in this problem, the contour consists in part of a fixed arc Γ and in part of a movable arc of given length the shape of which is to be determined). Can one prove analyticity of those parts of the free arc which do not attach to Γ?

(8)   Are minimal surfaces (H-surfaces) within a given contour Γ isolated, at least, if Γ is polygonal?

(9)   Estimate the number of minimal surfaces (H-surfaces) within a given curve from above and from below.

(10)   What can one say about branch points at the boundary?

(11)   Study Courant's function which appears in Courant's theory of unstable minimal surfaces.

(12)   Can one prove that there are always at least two solutions of the Plateau problem for the H-surface equation under suitable conditions on H ("Rellich's

conjecture")? What about the corresponding Dirichlet problem? Can one show that it never has more than two solutions?

(13)   Can one derive isoperimetric inequalities for H-surfaces and for minimal surfaces in Riemannian manifolds?

(14)   Let x be a solution of the Dirichlet problem for the H-surface equation, and suppose that $|x| \leq R$ and $|H| < 1/R$. Is the solution unique among all surfaces y with $|y| \leq R$?

(15)   Treat capillary problems (existence and regularity).

I still think that these problems were not so badly chosen. Here is a brief glimpse on the status of these first list of questions:

(1)   was, in 1969, solved by the author for the minima of the H-functional, and in full generality by Heinz who, in 1970, found another method which can hardly be simplified. It was carried over to minimal surfaces in Riemannian manifolds.

(2)   The first breakthrough was attained by Jäger who, under suitable side conditions on the supporting surface S, was able to prove regularity for area minimizing surfaces. This result was later improved and generalized, so that by now most of the interesting cases are covered, including minima of the functional

$$E(x) = \int \int \{|x_u|^2 + |x_v|^2 + Q(x) \cdot (x_u \wedge x_v)\} \, du \, dv$$

which, for instance, appears in the theory of capillarity. In 1981, almost simultaneously, the corresponding regularity theorem for stationary minimal surfaces was proved in two papers, the first one due to Grüter, Nitsche and the author, and the second due to Dziuk. The length of the free trace has been estimated by Nitsche and the author, and Küster has obtained the optimal estimate.

(3)   The answer is "yes" if $R \cdot |H| \leq 1$ as, in 1969, was proved by the author, and even variable H are permitted. Moreover, in the same year, Heinz proved that this result is optimal, at least, if H is constant. Yet, for complicated boundary curves, there might exist solutions even if the condition $R \cdot |H| \leq 1$ is violated. Thus one should look for other sufficient conditions which may guarantee existence. Such conditions were found by Wente, Gulliver-Spruck, and by Steffen between 1969 and 1975. Related results have also been proved for H-surfaces in Riemannian manifolds.

(4)   Ossermann (1970) announced that, for every Jordan curve Γ, the area minimizing minimal surfaces spanning Γ were free of interior branch points. The basic reasoning of Ossermann was very beautiful; yet his proof treated only one of two possible cases. The proof was completed independently by H. W. Alt and by Gulliver who extended their arguments even to H-surfaces and to minimal surfaces in Riemannian manifolds. Steffen and Wente also tackled volume constraints. However, nothing is known about curves which bound only surfaces without interior branch points, apart from the old results of Radó and from some simple cases.

(5)   Nothing is known for this question.

(6)   Such an estimate has been found by Heinz and the author in 1970; this result was improved by Kaul in 1972. However these results appear to be not optimal, and very likely interesting formulas remain to become discovered.

(7)   A satisfactory existence proof was obtained by H. W. Alt in 1973 whereas Nitsche (1971) established a regularity theorem for the free boundary which, however, did not yield the analyticity. The final result was found by Dierkes (1981), employing an idea suggested by Hans Lewy.

(8)   The answer is not known.

(9)   Ditto.

(10)   By applying a technique due to Carleman, Heinz (1969) was able to derive an asymptotic expansion for minimal surfaces near a boundary branch point. This kind of expansion has been carried over to free boundary value problems and to other related situations. Furthermore, Gulliver and Lesley showed in 1973 that, for real analytic boundaries, the area minimizing solutions of Plateau's problem have no boundary branch points.

(11)   was investigated by Heinz in a series of papers beginning in 1979.

(12)   A partial proof of "Rellich's conjecture" was given by Steffen in 1970. The complete proof was, almost simultaneously, given by Struwe, Steffen, and Brezis-Coron in 1982. Struwe had the first result, valid for a class $\mathscr{C}$ of curves defined by some implicit condition, and he showed that planar curves are contained in $\mathscr{C}$, whereas Steffen found that every reasonable curve belongs to $\mathscr{C}$. Brezis and Coron had the best, in fact the optimal result, by proving that, for every value of H in $0 < |H| < 1/R$, there are at least two H-surfaces bounded by Γ. All these authors employ results described in (3), but it should be mentioned that ideas and results due to Wente play a particularly important role. Without the work of Wente it would have hardly been possible to achieve these beautiful theorems. Similar results were derived for the Dirichlet problem, but the answer to the last question is not known to me.

(13)   Here the answer and the references are the same as in (6) and, again, I think that the problem is far from being exhausted.

(14)   A remarkable uniqueness theorem by Jäger (1975) shows that the answer is "yes" for constant H. For variable H, however, Ströhmer has shown that the condition $|xH| < 1$ will, in general, not be sufficient for uniqueness.

(15)   The progress made within 15 years cannot be described in a few lines. There occurred an explosion of new results, after the matter had more or less rested for nearly a hundred years. I particularly mention the work of Finn and Concus and, as one of their most beautiful results, their detailed study of the singular solution of the pending drop equation which, in the last century, had been discovered by Lord Kelvin and which, thereafter forgotten, had been rediscovered by numerical experiments around 1970. A detailed survey of all results will appear in a forthcoming monograph by Robert Finn.

5   I do not try to describe all further seven meetings (from 1970 till 1982) with the same completeness, indeed I have to excuse myself for having claimed your attention with so many details about a single seminar. What I hoped to show was how, by a week of work, one can set the stage for a successful development during the next one or even two decades, provided that the right people meet and have some luck. Oberwolfach, of course, is part of this luck.

The second calculus of variations-meeting was again organized as a seminar, with the four main themes:
(I)   Variational problems with inequalities as subsidiary conditions;
(II)   Surfaces of prescribed mean curvature with free boundaries;
(III)   Existence and regularity for surfaces of prescribed mean curvature;
(IV)   Regularity theorems.

This time we had invited four new foreign guests: Mario Miranda and Enrico Giusti from Pisa who reported on minimal surfaces with obstacles, and Kjell-Ove Widman from Uppsala who lectured on the singularity of Green's function for degenerated elliptic equations. Moreover, Courant participated in what probably was his last scientific meeting.

Section (I) was completed with lectures we had prepared on the work of Lewy and Stampacchia from 1969 and on an obstacle problem for H-surfaces treated by Tomi. Sections (II) and (III) mostly discussed results the participants had obtained since the first seminar, and in (IV) Frehse lectured on his regularity theorems for elliptic variational inequalities of higher order.

In the following years, the variational meeting lost more and more its character as a "seminar". This was regrettable in that the close cooperation between Göttingen and Mainz (later: Bonn) started to disappear. On the other hand, this development was for four reasons unavoidable. First of all, we wanted to learn what our foreign colleagues were presently working on. It therefore seemed useful to select "hot topics" and to invite experts to lecture on these areas. Secondly, colleagues from other countries increasingly became interested to see what we were doing, and so we invited visitors who wanted to become acquainted with our work. Thirdly, since our Oberwolfach work had inspired lectures and seminars at our home universities, the various groups became so strong and active that soon new teams at other universities formed. Thus the organization of a seminar was no longer a simple matter. Finally, collaboration began with some of our foreign colleagues so that interests were becoming more divergent. For these reasons, the variational meeting became more and more like one of the mammoth conventions, except for a few but important differences. We usually selected some guiding themes which would vary with different years. Correspondingly, guests were invited according to whether they fitted into the program, of course always with some exceptions, so that irrational sparkles, born out of the occasion, were admitted. Then we had the principle of "no lectures in the afternoons" (which should be free for hikes, private work, and discussions) and "spontaneous lectures after dinner", organized by the participants and not by the organizing committee. This left space for side tracks which would not be interesting for all, and allowed the organizers to stick closely to the program: no hodge-podge of lectures. It also meant that the main lectures could be carefully selected since everybody could speak in the evening—if he found somebody to listen.

6   Before I end my historical account I should like to discuss some of the main topics of the "variational mettings" between 1972 and 1982, because they reflect, at least to some extent, the development of the calculus of variations during this decade.

In 1972, Allard, Almgren, and Jean Taylor lectured on Geometrical Measure Theory, and it became clear that this was a very powerful technique. Steffen therefore organized a seminar on GMT, and the participants tried to understand Federer's book. For some time, only the habilitation thesis of Steffen reflected this meeting in our country, and the reception of GMT was limited to the results of DeGiorgi and his Italian school. I only mention the contributions of Gerhardt to the capillary problem. Lately, however, ideas from GMT increasingly enter the work of German analysts, for instance, in the important thesis by Grüter, but by no means as much as it should. For instance, the beautiful result by Leon Simon and Robert Hardt (on which Simon lectured at Oberwolfach in 1978) that one can span into every smooth curve Γ an area minimizing, smooth, compact, oriented and embedded minimal surface of genus g (which depends on Γ) shows impressively the power of these techniques.

Moreover, I mention the regularity theory for nonlinear elliptic systems and, in particular, for systems in diagonal form with "quadratic growth" of the right hand side, to which was devoted a large part of the time. Already Hans Lewy and Heinz had pointed out the utility of this kind of equations for differential geometry. Examples by Heinz, Frehse, Widman and the author, Meier, Struwe showed that, even in two dimensions, regularity can only be expected if the parameters of the solution satisfy certain inequalities, and it was investigated if these necessary conditions also are sufficient for regularity. This was carried out in the work of Widman and the author, and of Wiegner. In particular Widman's "hole filling technique" and the estimates of Green's function for elliptic operators became powerful tools. These ideas were applied by Kaul, Widman and the author to harmonic mappings of Riemannian manifolds. This problem had been somewhat neglected since the important paper of Eells and Sampson had appeared in 1964, because there seemed no way to admit target manifolds with positive curvature. The three authors mentioned before were able to treat the general case and to arrive at optimal conditions for existence.

A completely different approach was chosen by Giaquinta, Giusti, Modica and, to some extent, by Ivert. Giaquinta and Modica saw how the "Gehring lemma" had to be modified in order to become applicable to elliptic systems. By combining this method of the "reverse Hölder inequalities" with an iteration procedure developed by Morrey and Campanato, they obtained a powerful technique to get regularity or partial regularity for minima and "quasi-minima" of multiple integrals. Their approach probably is the most promising one for multiple integrals although, for harmonic mappings, the recently developed variational method of Schoen and Uhlenbeck presently yields stronger results. All these ideas have to be discussed in further meetings. In particular, the regularity theory of higher order variational problems should be further developed for interesting problems of differential geometry. The results of Frehse, Widman, Giaquinta/Modica should give a firm basis to proceed in this direction. The work of Frehse on regularity of solutions to "thin obstacle problems" and to obstacle problems for higher order integrals contains many interesting ideas.

Harmonic mappings were and probably will still be a privileged area for interesting geometrical applications of the calculus of variations. From the past, I mention the uniqueness theorem by Jäger and Kaul and the minimum property (for $n \geq 7$) of the mapping $u(x) = \left(\dfrac{x}{|x|}, 0 \right)$ discovered by these authors, which led to the non-existence theorem of Baldes. Moreover, Eells and Lemaire reported on their deformation theorem for harmonic maps and on harmonic mappings of two-dimensional manifolds. Jost (in part joint work with Schoen) lectured on very general new existence theorems for harmonic diffeomorphisms between surfaces. Finally, I mention Liouville theorems for harmonic mappings and their application to minimal submanifolds via the Gauss map which have been discussed during the last years. In the same spirit, Meier has proved a theorem concerning the removability of singularities.

Plateau's Problem for minimal surfaces continued to catch our interest. As most outstanding results, I recall Nitsche's uniqueness theorem which states that a closed curve Γ of total curvature less than 4π bounds only one minimal surface, the example of Frank Morgan (exhibiting a boundary contour that consists of closed curves which bound a continuum of minimal surfaces), Tomi's finiteness theorem and the regularity result of Hardt-Simon which were already mentioned, the Böhme-Tromba index theorem together with its various applications such as the generic finiteness theorem (simplified proofs of this theorem were recently obtained by Söllner and by Tomi/Schüffler; the techniques of these authors also work for H-surfaces, yet they only treat interior branch points), and Heinz's theory for minimal surfaces within polygonal boundaries.

Nonparametric minimal surfaces and H-surfaces as well as capillary surfaces have extensively been discussed by M. Miranda, Giusti, Gerhardt, Heinz, Spruck, Wente, and in particular by Finn, and very likely the interest in this field will remain. One of the challenging problems, raised by Bombieri, is to describe the growth at infinity of non-planar and non-parametric "entire" minimal hypersurfaces.

Karen Uhlenbeck lectured on her "removability of singularities" theorem for Yang-Mills fields but, unfortunately, although the Yang-Mills theory is presently one of the most interesting fields in the calculus of variation, that lecture has remained an isolated event in Oberwolfach.

Finally, I should like to touch on another field that over the years has found much interest at Oberwolfach. I refer to the "free boundary problems", by which we mean a whole collection of very different phenomena not having very much in common. Here one finds truly exciting questions, and a good idea of the problems and methods can be obtained from the recent monograph [5] by Avner Friedman and from [A].

As a first example, I cite the study of flows through porous media. Although such flows are not ruled by a variational principle, they can be treated by techniques which, in part, grew out of the study of variational inequalities. In this field, the work of H. W. Alt has initiated an interesting development.

Free boundary problems for fluids and gases pose many difficult and inspiring problems. These quite often allow a variational formulation, so that one can apply the "direct methods". Such methods usually yield much more solutions than the "implicit function theorems" which only catch solutions in the neighborhood of an equilibrium configuration.

Typical examples of such free boundary problems are jets and cavities of fluids, vortex rings of gases, equilibrium configurations of rotating fluids, and confined plasmas in magneto-hydrodynamics. Alt, Friedman, Caffarelli, Auchmuty, and Bemelmans have lectured in Oberwolfach on their work.

A fundamental contribution was given by Alt and Caffarelli to variational problems of the kind

$$\int_{\Omega \cap \{u > 0\}} F(x, u(x), Du(x)) \, dx \to \min$$

with the subsidiary conditions

$$u = 0 \quad \text{on} \quad \Omega \cap \partial \{u > 0\}.$$

Their methods ought to be better understood and simplified because they may be helpful analytical tools for various other free boundary problems. Ideas from GMT have been rather helpful for this paper.

Minimal surfaces lead to many interesting free boundary problems as, for instance, to systems of minimal surfaces which are spanned into a given framework and possess free boundary arcs at which three surfaces of the system always meet at angles of 120 degrees. Jean Taylor (1972) has proved existence of such systems and $C^{1,\alpha}$-regularity of the free arcs. Nitsche was able to show C^∞-regularity, and Kinderlehrer, Nirenberg, and Spruck have verified that the free arcs are real analytic (1978). All this work has been discussed in Oberwolfach lectures.

Moreover, Taylor has completed Lamarle's classification of the singular cones of such systems and has shown which three of the ten possible cones can only arise in area minimizing systems.

Another kind of free boundary appears for minimal surfaces above obstacles. In this case, one wants to determine the analytical and topological nature of the set of coincidence of the minimal surface with the obstacle. Here I refer to the work of H. Lewy, Nitsche, Giusti, M. Miranda, and in particular to the papers of Kinderlehrer.

Still another type of free boundary problems arises for minimal surfaces with boundaries on supporting surfaces which themselves have a nonvoid boundary. This boundary of the supporting surfaces can be viewed as a thin obstacle for the boundary values of the minimal surfaces. Nitsche and the author have in this case investigated the shape and the regularity of the "free trace".

A virtually untouched topic are stationary minimal surfaces with free boundaries on closed surfaces, say, on the surface of a convex body, although H. A. Schwarz in 1872 already exhibited a stationary minimal surface within a tetrahedron. Such surfaces within convex bodies can never minimize area so that a critical point theory is needed to prove existence. Recently Struwe has made some initial progress by demonstrating the existence of one stationary solution but, of course, one expects that three of them will exist.

Let me at this point close my account on the past although I realize that it is incomplete and biased. I will be content if I have been successful in demonstrating that most of the important progress in the calculus of variations during the last fifteen years has been reflected in the Oberwolfach meetings and, moreover, that these meetings have had a considerable influence on the development of nonlinear analysis. In my opinion, regularity theory for systems of Euler equations in particular and for nonlinear elliptic systems in general has been at the center of progress. On the other hand, the existence theory for multidimensional variational problems is still in the stone age, since only the direct methods are available as the tool which functions in virtually all cases. In the last years, however, there have been signs that, at last, a global theory of nonlinear analysis might be developing. Fortunately these global methods, which have proved their value for one-dimensional problems such as geodesics or Hamiltonian flows, are presently also used to attack concrete problems in geometry and physics. It is, for instance, a great success that we now have the beginnings of a structure theory for the set of solutions to Plateau's problem thanks to Böhme, Heinz, Tomi, Tromba, and to their students. In this context, I should also mention the interesting work by Brian White. It is to be hoped that, in the future, topological methods as those developed by Tromba and Paul Rabinowitz may play a similar role for multidimensional systems as they presently play for problems with one independent variable and for certain simple scalar problems with several variables.

Which international developments have been left out in the variational meetings? Control theory has completely been neglected because it is a field in its own right. The same holds for geodesics and for Hamiltonian theory (yet Rabinowitz has lectured in Oberwolfach on his work), and one might also say that the numerical treatment of variational problems is indeed a separate area. I however think that, at least occasionally, numerical analysts should be invited to report on the state of their art. The work of Concus and Finn shows how successful the collaboration between a numerical and a theoretical analyst can be.

Surely the progress in linear and nonlinear eigenvalue problems has been underrepresented, be it in differential geometry or in quantum physics (see, for instance the work of Brascamp, Brezis, E. H. Lieb, and of Gidas, Ni, and Nirenberg).

7   This leads me to the question as to which areas promise to be fruitful for further research, and in which way the calculus of variations-meetings could or should develop.

To begin with the second question, I should like to suggest some ideas:
(1)   Each meeting should focus on one or two main themes (which very likely will change with the years).
(2)   General progress in the calculus of variations, as far as it does not fit into the main program, should be left to afternoon and evening discussions, and to spontaneously organized lectures.
(3)   Therefore, the organizers should confine the principal lectures strictly to the mornings, *without any exception.*
(4)   Every meeting should end with a problem list, and all participants should, during the first session, be invited to contribute to that list.
(5)   It would also be desirable if, at each meeting, one or two experts were asked to give a survey on the themes under discussion, and each survey should end with a list of the significant problems.
(6)   The meetings should be open to as many (invited) young people as the capacities of the Institute permit, since Oberwolfach offers them a unique chance to become acquainted with the experts in the field. This does not mean that everyone should lecture, for this would be impossible since there are only $4 \cdot 5 = 20$ morning lectures available whereas the number of participants might range between forty and fifty.
(7)   In order to initiate the study of new areas, one could plan seminars (with about 20—30 participants) which should be organized by young scientists who want to explore new possibilities.

The question of what should be the directions of future research is more difficult to answer. None of the present fields of investigation shows any sign of exhaustion. Yet, since the time of the Bernoullis and of Euler, a special feature of the calculus of variations has been its wide range of applications. Thus I propose to enlarge the scope of our endeavours to include the following areas:
(I)   Variational problems in differential geometry. There exist, for instance, interesting relations between the spectrum of the Laplacian on a Riemannian manifold and isoperimetric inequalities. The latter should be systematically explored.

The critical points of curvature integrals will provide a rich field for investigations. Well known examples are Euler's elastic lines, which are the stationary curves of the Daniel Bernoulli integral

$$\int R^{-2} \, ds, \qquad (R = \text{curvature radius}),$$

the Willmore problem for surfaces in euclidean 3-space which is related to the integral

$$\int H^2 \, dA, \qquad (H = \text{mean curvature}),$$

and the Einstein equations which are the Euler equations of

$$\int R \, dA, \qquad (R = \text{scalar curvature}).$$

The survey of Yau and his problem list in [18] will yield further suggestions. I believe that the calculus of variations would miss many possibilities and might even become sterile, if it did not follow progress in differential geometry.

(II)   Equilibrium configurations in continuum mechanics. For instance, elasticity is a vast and largely unexplored field. It is well known that the standard regularity techniques fail in the theory of elasticity. First of all, the elastic energy does not lead to strictly elliptic equations so that even lower semicontinuity becomes a more difficult matter. This, however, has to some extent been clarified by Morrey and by John Ball. There are probably various reasons why one cannot expect that the standard regularity theory works. First of all, the elastic energy remains invariant with respect to a very large group of transformations. A similar situation arises if one attacks length and area by the mapping point of view. These are very unpleasant integrals which admit many singular solutions in which one is not interested. There are various procedures describing how to eliminate these solutions in order to obtain only "reasonable" ones. For instance, one can replace length or area by the Dirichlet integral, and Morrey devised a similar procedure for the general two-dimensional parametric variational integral (cf. [12], chapter 9). Another possibility is, as in GMT, to forget about the mappings and to treat the elastic bodies as sets. Then the real solution is selected from the nonsensical ones by density estimates. This procedure seems less desirable in elasticity since the mappings will also be interesting.

Secondly, as the examples of Nečas show there really will be singular solutions. Thus one problem would be to find out whether these singularities are mathematical phenomena with no significance for physics (since the physical model is no longer valid for this solution), or whether such a singularity explains a physical phenomenon such as tearing. One should then study equilibrium states for varying parameters. It would be interesting to follow the various branches of solutions and to study their bifurcation behaviour and the change of stability from one branch to another. Here one encounters a problem which one also finds in other areas of continuum mechanics. At bifurcation points (surfaces), the equilibrium states may alter their topological structure. Rotating fluids might, via pearshaped forms, change from one into two lumps, or a ring can dissolve from oblate spheroidical shapes which, in turn, may decompose into several simply connected drops. Plateau's experiments and star models in astrophysics are rich sources of models. To my knowledge, this change of the topological type of equilibrium configurations at bifurcations has never systematically been attacked. I think that it is one of the most interesting and most important questions in the calculus of variations, and all efforts should be made to solve problems of this type. A careful stability analysis should help to understand what happens, and the occurance of singular solutions might just be the consequence of the fact that the topological type of the stable solution changes. In fact, singular solutions might be the key to solve this problem. Henry Wente's penetrating stability analysis of the pendent drop [17] gives ideas of the direction in which the problem should be pursued.

An important model problem are closed surfaces of prescribed mean curvature which have been studied by E. Hölder (1926), Bakelman (1970), Treibergs and Wei (1982), and recently also by Bemelmans. A counterexample due to Wente shows that unexpected difficulties may occur.

Rotating fluids also show that one should not study equilibrium configurations independently from the dynamical problem because configurations can become unstable if one does not admit interior motions of the liquid. This phenomenon has been discovered by Dirichlet and Riemann [15] but not much has been done except for the work of Chandrasekhar which is mainly a second order stability analysis. Only very recently, Bemelmans has started to investigate more general equilibrium problems taking the Navier-Stokes equations into account. I think that people interested in continuum mechanics, in the calculus of variations, and in nonlinear hyperbolic systems should join forces in order to attack these beautiful and difficult mathematical problems. Eventually, the relativistic many body-problems should be approached. The report [1] shows which mathematical tools presently are vailable, and it becomes very clear that, despite all the successes of the past, we are only at the very beginning of a theory of continuum mechanics and of nonlinear analysis.

8.   I close my reflections on the calculus of variations with some *references to the literature*. I have only quoted some recent books and survey papers. There one can find detailed references to the literature and a description of many of the results mentioned in my account.

References

[1]   Ball, J. M., editor, *Systems of nonlinear partial differential equations*, NATO ASI series, Series C, No. 111, Proceedings of the NATO Advanced Study Institute Oxford 1982, Reidel Publ., Dordrecht–Boston–Lancaster 1983.

[2]   Böhme, R., *New results on the classical problem of Plateau on the existence of many solutions*, Séminaire Bourbaki 1981/82, no. 579.

[3]   Eells, J., and Lemaire, L., *A report on harmonic maps*, Bull. London Math. Soc. **10**, 1—68 (1978).

[4]   Finn, R., *Forthcoming treatise on capillary problems*, Springer.

[5]   Friedman, A., *Variational principles and free-boundary problems*, Wiley, New York 1982.

[6]   Giaquinta, M., *Multiple integrals in the calculus of variations and nonlinear elliptic systems*, Annals of Math. Studies, Princeton University Press 1983.

[7]   Gilbarg, D., and Trudinger, N., *Elliptic partial differential equations of second order*, Springer, Heidelberg–New York 1977.

[8]   Hildebrandt, S., *Nonlinear elliptic systems and harmonic mappings*, Proceedings of the 1980 Beijing Symposium on Diff. Geom. and Diff. Equ., Vol. **1**, 481—615, Science Press, Beijing 1982.

[9]   Hildebrandt, S., *Minimal surfaces with free boundaries and related problem*, (a survey, to appear in Astérisque).

[10]   Jost, J., *Harmonic maps between surfaces*, Vorlesungsreihe des SFB 72 Bonn, no. **15**, 1983; to appear in Springer Lecture Notes.

[11]   Ladyzenskaya, O. A., and Ural'ceva, N. N., *Linear and quasilinear elliptic equations*, Academic Press, New York and London 1968 (English translation of the first Russian edition 1964).

[12]   Morrey, C. B., *Multiple integrals in the calculus of variations*, Springer, Berlin–Heidelberg–New York 1966.

[13]   Nitsche, J. C. C., *Vorlesungen über Minimalflächen*, Springer, Berlin–Heidelberg–New York 1975.

[14]   Osserman, R., *Minimal surfaces, Gauss maps, total curvature, eigenvalue estimates and stability*, The Chern Symposium 1979, Springer 1980.

[15]   Riemann, B., *Ein Beitrag zu den Untersuchungen über die Bewegung eines flüssigen gleichartigen Ellipsoids*, Ges. Math. Werke, 168—197, Teubner 1876.

[16]   Tromba, A., *On the number of simply connected minimal surfaces spanning a curve*, Memoirs A. M. S. **12**, No. 194 (1977).

[17]   Wente, H. C., *The stability of the axially symmetric pendent drop*, Pacific J. Math. **88**, 421—470 (1980).

[18]   Yau, S. T., editor, *Seminar on Differential Geometry*, Annals of Math. Studies, No. **102**, Princeton University Press 1982.

In addition, I refer to:

[A]   Vol. 88, No. 2 of the Pacific J. of Math. (1980) which entirely is devoted to the study of free boundary problems;

[B]   Notes of the 1982 Seminar "Gauge theories and four-manifolds", organized by Mike Freedman and Karen Uhlenbeck at Berkeley, Fall 1982; Notes taken by Dan Freed;

[C]   Almgren, F., *Q valued functions minimizing Dirichlet's integral and the regularity of area minimizing rectifiable currents up to codimension two*, Preprint 1982, Princeton University, 3 vol.; these notes also contain a valuable bibliography;

[D]   Almgren, F., *Existence and regularity almost everywhere of solutions to elliptic variational problems with constraints*, Memoirs of the Amer. Math. Soc. **4**, No. 165 (1976).

Perspectives in Mathematics
Anniversary of Oberwolfach 1984
© Birkhäuser Verlag, Basel

Aspects of Modern Control Theory

H. W. KNOBLOCH and M. THOMA

Mathematisches Institut der Universität, Am Hubland,
D-8700 Würzburg (FRG) and
Institut für Regelungstechnik, Appelstraße 11,
D-3000 Hannover (FRG)

1   Introduction

Control theory is a relatively young branch of applied science. Its origin dates back to the end of the second world war when first attempts where undertaken to attack problems of guidance and control of aircraft via mathematical modeling. At that stage problems were treated within the context of existing mathematical theories: Stability was studied as a problem of differential equations and complex variables, filtering as a problem of probability theory and optimization as a problem of calculus.

A new situation arose at the end of the fifties when the state space approach became more popular among engineers at the expense of "classical" frequency domain techniques. However it was not so much the shift to a different setting which marked the beginning of modern control theory than the appearance of concepts and notions which were not borrowed from other mathematical areas. This was in part due to the rise of the idea of a dynamical system. But above all it was the notion of controllability as introduced first by Kalman in the beginning of the sixties which has contributed in an essential way to the shaping of modern control theory. Not only because it was appealing from the viewpoint of mathematical aesthetics, but mainly because it was and still is a valuable guide in finding mathematical answers to questions which are put forward by control engineers and which are centered around the following three basic themes.

  (i) Mathematical modeling of real systems.
 (ii) Analysis of the model (with respect to basic system properties as e. g. stability, controllability, observability).
(iii) Design of controllers (i. e. the construction of mathematical models for devices which influence a given system in a desired way and improve system properties with respect to inaccurate modeling and uncertain or varying data).

Historically optimal control emerged as the first and so far most known approach to the design problem. In its beginning it followed very much the lines of the classical calculus of variations. There are some important and lasting results from this initial period of optimal control theory, some of them actually being by-products of the analysis of optimal trajectories. We mention the solution of the linear-quadratic regulator problem (LQ- and LQG-problem) which is by now a standard tool for the construction of a stabilizing feedback control.

Presently optimal control comprises a much larger spectrum of optimization techniques than say 20 years ago. This is clearly due to the demand for increasing productivity and product quality of controlled processes and for minimizing side effects as environmental pollution. But it also reflects the impact which the rapid development of microelectronics has on control theory. The effort required to implement a mathematical procedure and the availability of standard program packages play an important role in choosing a specific approach. This tendency will certainly prevail in the future when more and more complex situations with physical, chemical, biological or economic background will be studied from the viewpoint of control theory.

In the second half of this survey (Section 4 and 5) we will try to give an impression of the present state of development in optimal control theory. We have chosen for this purpose two representative research areas where the wishes and needs of applications determine the frame of mathematical analysis.

The first is called large scale systems and is concerned basically with standard optimal control problems for lumped parameter systems (i. e. systems described by ordinary differential equations), even with linear-quadratic problems. The complexity of the problem is due to the high dimension of the underlying model and makes the standard approach inefficient. In fact, the terminology " large scale system" indicates that top priority in the list of design objectives is given to the problem of coping with the dimension. We will explain some ideas and proposals which have been set forward in order to meet this requirement. The importance of this type of problems and the need for developing specific mathematical tools has not been duly recognized so far.

A further example of temporary activities in optimal control theory is discussed in Section 5 and concerns distributed parameter systems i. e. systems described by partial differential equations or integral equations. We briefly examine three topics which are of general interest, namely

control, simulation, identification,

and we wish to demonstrate in this section where the essential effort presently and certainly in the foreseeable future will be devoted to: Namely to find out the appropriate methods within the full range of applied mathematics, notably in functional analysis and approximation theory, which can be of use in control theory, and adopt these methods to the particular problems in question.

What has been said about distributed parameter systems is of course true for lumped parameter systems. The topics listed above will also remain in the foreground of interest and a large portion of research will be devoted to the extraction of appropriate methods from the existing reservoir of mathematical techniques. However in the last years there has been a certain movement toward an alternative approach to the design problem. The underlying philosophy is to exploit more the qualitative nature of the problem which optimization techniques quite often fail to do, because of their generality and the tendency just to be independent from the specific type of problem. This setback of optimal control is in particular felt if uncertainties about the data of the problem play a major role in the formulation of design objectives. These uncertainties may occur with respect to the environment in which the system operates, with respect to parameters of the plant, or even with respect to the model as a whole.

The first part of this survey (Section 2 and 3) will therefore emphasize qualitative methods for lumped parameter systems. We wish to demonstrate how these methods can be applied in order to solve one of the basic design problems, namely the attenuation of the influence which external disturbances may exert on a given system. This is done separately and by different techniques for the deterministic case (Section 2) and the stochastic case (Section 3). There is good reason to make this distinction. In the analysis of deterministic systems the property of controllability plays the main role whereas the dominant viewpoint for the stochastic situation is stability. It has been well understood in recent years that controllability is not a very meaningful property for a stochastic system, whereas stability obviously is.

Sections 2 and 3 provide a first impression of the type of problems which probably will present the main incentive to develop new ideas within the field of control theory itself rather than to try to import them from the outside. The essential feature of these problems is that they are concerned with the inherent deficiencies of mathematical modeling. Progress in the handling of these problems will determine the extent to which mathematical control theory will become a useful tool also for workers in biological or social sciences who hitherto have to rely upon the somewhat vague and merely qualitative arguments from conventional "cybernetics".

This survey article has been written jointly by four participants of the conferences on control theory which have been held regularly at Oberwolfach since 1968. The group consists of

H. W. Knobloch, A. Munack, M. Thoma and J. L. Willems

and may be viewed as a representative cross-section of the audience of these conferences. It is not the intention of the authors to give a complete picture of the state of the art. Rather they wish to point out certain lines of development which demonstrate the fruitful interplay between engineering intuition and mathematical reasoning.

2   Finite dimensional deterministic control systems (H. W. Knobloch)

2.1   *The output regulator problem*

The situation underlying our considerations is depicted in Fig. 1. The plant represents the system which reacts in two ways with its environment

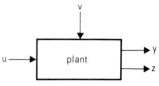

Fig. 1

namely through a pair of input variables (u, v) and a pair of output variables (y, z). u is the portion of the input which we have at our disposal. v represents the disturbance which influences the system and in most cases cannot be controlled directly nor is accessible to precise measurement. The output y is the observed variable (information available about the instantaneous situation of the plant) whereas z is the controlled variable (quantity which one really wants to control). For simplicity we will frequently assume that y and z are identical and that $z = 0$ in the absence of input (i.e. if $u = 0$, $v = 0$ then z will approach 0 in the course of the time). The main objective of the controller is then to bring the controlled variable close to the desired value zero or to keep it there in the presence of a disturbance signal $v(\cdot)$. This is, generally speaking, the output regulator problem. In more precise terms it can be formulated as follows.

Design a strategy for specializing the input variable u which is based on past observations (i.e. the value of $u = u(t)$ at some time t depends upon the values of $y(\tau)$ for $t_0 \leqq \tau \leqq t$ only) and which reduces the influence of v on the output z.

What "reduction of influence" means has to be specified. We will consider here two possible design objectives:

(i) Minimizing a performance criterion which depends upon the control effort $u(\cdot)$ and the achieved output $z(\cdot)$.
(ii) Bringing z asymptotically down to zero (disturbance attenuation).

In the first case one chooses a quantitative, in the second case a qualitative criterion. The second one aims at exploiting a more structural property of the plant (and the disturbance): The control mechanism of the plant is such that in principle one can eliminate the influence of the disturbance completely provided one has sufficient time. Such a property is not required if one tries to solve the optimal control problem (i). On the contrary one can expect less difficulties in proving the existence of a minimizing solution if the possibilities of reducing the output are a-priori limited. Such limitations are explicitly expressed as standard regularity conditions which appear in the theory of the optimal regulator and the optimal state estimator.

Next we wish to specify the type of mathematical model on which our further considerations will be based. We assume that a link between input and output can be established by writing down relations of the form

$$\dot{x} = f(t, x, u, v), \quad y = z = h(t, x). \tag{2.1}$$

x is a finite dimensional quantity, the state variable, which evolves in time according to a prescribed and explicitly given dynamical law.

This assumption of course implies among other things that we have a correct mathematical model of the plant, it may therefore appear somewhat unrealistic. Nevertheless a considerable portion of the most common problems in control engineering can be formulated within the frame of the output regulator problem as we have defined it so far. Note also, that parameters appearing in the dynamical equation of the plant formally can be "updated" to state variables. This — in principle — allows one to formulate the regulator problem also in case of parameter uncertainties in the mathematical model of the plant and one still remains within the scope of the methods which we will discuss in this survey (though not within the context of linear system theory). So the output regulator problem represents an illustrative example of problem development in control theory and certainly will remain so in the future.

If the disturbance is short-living then one need not worry about what v actually is. What counts is the displacement of z which has been caused by v, and this displacement will determine the correcting action of the controller. So handling momentary disturbances is actually nothing else than stabilizing the output of the plant. This is of course a classical problem which can be attacked by a variety of methods and not all of them require a full model description of the form (2.1). It is the case of persistent disturbances where the advantages of working with a complete state space model become apparent, provided one has in addition a dynamical model for the disturbance. Thereby one is in a position to exhibit in a quantitative way the interplay between the dynamics of the plant and the dynamics of the disturbance. This in turn prepares the stage for the construction of appropriate strategies. How this general line of approach can successfully be carried through is best demonstrated by the solution of the linear stochastic regulator problem. This notion refers to the situation where one chooses a linear model for the plant and assumes furthermore that v can be interpreted as output of a linear system driven by white noise, the so called shaping filter for the disturbance signal. The optimal strategy is then expressed in terms of the separation principle which will be explained in Section 3. According to this principle the solution of the output regulator problem can be inferred from the solutions of two standard optimization problems, namely the linear optimal regulator and the linear optimal state estimator. The solution to both problems is essentially known since the end of the sixties and nothing really new has been added to it since then except that one has succeeded to make the presentation more elegant and elementary. The popularity of the linear-quadratic theory stems from this fact rather than from the scope of the method as far as applications are concerned.

In a concrete situation it is quite often difficult to find a dynamical model which meets the requirements of the theory. Not only because one has in-sufficient information about the disturbance. Also in reality external distur-bances will frequently be composed of rapidly fluctuating signals as well as time functions which vary rather regularly though the pattern which they follow is not known beforehand. The source for the second type of disturbance may be non-stochastic physical phenomena (as elastic vibrations) which the mathematical model fails to explain and which therefore are regarded as external input signals. It is clear that the standard concept of the shaping filter is too narrow in order to generate a reasonable model for a more complex type of disturbance signals. There have been some attempts to modify this concept accordingly and then again proceed along the lines of the linear-quadratic regulator problem (see e. g. Johnson 1971), but the results are not as satisfactory as in the case of the standard stochastic regulator.

On the other hand there is a simple heuristic argument which suggests a treatment of the deterministic case under the qualitative viewpoint. Assume one knows that $v(\cdot)$ is a time function which is completely determined by a set of "hidden" parameters (amplitudes, phases, Fouriercoefficients) which do not change in time. If there is no time limit for the controller's action then he may try to learn from the systems's reaction to control inputs as much about these parameters as is necessary in order to outmanoeuvre the disturbance eventually. We will indicate in the next section how this type of argument can be converted into a constructive scheme for finding suitable strategies.

A somewhat simplified picture of the present state of development of the output regulator problem for linear systems may be described as follows. There are two alternative lines of approach, favouring different objectives and making different assumptions about the disturbance. Each assumption as such is more or less unrealistic from the practical point of view, a mixture of both would represent much better actual situations. A combination of the two methods therefore seems to serve best the needs of applications. How this can be achieved is unclear in the moment and is certainly a question of considerable significance even within the context of linear systems theory.

The mathematical background of the questions discussed in this and the following section can be found in advanced textbooks on linear control theory e. g. in Knobloch/Kwakernaak (1984).

2.2   Geometric theory, disturbance decoupling and the internal model principle

We consider the general configuration of Fig. 1, but will assume throug-hout this section that the model equations (2.1) are linear and time invariant and hence can be written in this form

$$\dot{x} = Ax + Bu + Gv, \quad y = z = Cx. \tag{2.2}$$

The first and most radical attempt to solve the output regulator problem for linear systems is commonly known as disturbance decoupling. It is concerned with the question whether one can disrupt the dynamic connection between v and z by means of state feedback $u = -Fx$. In precise terms this problem can be phrased as follows: Find a matrix F such that the transfer function from v to z of the linear system

$$\dot{x} = (A - BF)x + Gv, \quad z = Cx$$

is identically zero. That it can be reformulated as a standard problem of linear algebra is easy to demonstrate if one uses some basic facts about controlled invariance.

This notion is a refinement of the concept of controllability and was first discussed by Basile and Marro (1969). Later Wonham has build around it his "geometric theory" of linear systems (Wonham 1979).

If disturbance decoupling is possible then the matrix F in the state feedback law in general is not unique. How this freedom can be used in order to shift eigenvalues of $A - BF$ into the left half plane (pole placement) is an additional and very useful information which can be derived from geometric theory. Hence the role of geometric theory is more accurately described as follows: It is concerned with the question whether and how two important design objectives of linear control theory, namely disturbance decoupling and pole placement, can be achieved simultaneously by means of state feedback.

State feedback however is not an admissible strategy in the sense of our definition, since it requires full information about the state of the system. How to find a substitute $\hat{x}$ for the state x in case the latter one is not accessible is now a standard procedure in control theory: One uses as $\hat{x}$ the state of a linear system with input (u, y) (usually a so called dynamical observer). One speaks then of output feedback and this is a type of strategy which falls into the category described above. It is therefore of interest to learn the extent to which pole placement and disturbance decoupling can be achieved simultaneously by output feedback. A conclusive answer to this question can be found in the paper by Willems and Commault (1981) which also contains additional references and a short account of the history of the problem.

Disturbance decoupling is the best possible solution of the output regulator problem since it requires no a-priori knowledge about the disturbance. It is however questionable whether the results will play any significant role in practial applications since they are tied up with rather restrictive conditions for the matrix G (cf. (2.2)).

From the control engineering point of view one would therefore prefer a less radical solution in return for a wider range of applicability. A natural question is then how much could be gained if one would aim at disturbance attenuation (cf. Sec. 2.1) rather than disturbance decoupling. This modification of the design objective alone does not help much unless one is not willing to give up the idea of finding strategies which work simultaneously for all possible disturbances.

So we will now discuss the problem of disturbance attenuation under the assumption that v is element of a finite dimensional space of time dependent functions with known dynamics. To be more precise: v is regarded as output of a deterministic linear differential equation with a known coefficient matrix. We seek a strategy which is effective against any possible v out of this class of functions without the necessity of identifying it beforehand.

The approach to this problem is in its beginning similar to the stochastic case. The system description is written in such a way that x includes both the state of the system and the "state" of the disturbance. Thereby v formally disappears from the equation and the uncertainty about the disturbance is shifted to the uncertainty about the initial state. The problem of regulator design then becomes a special case of the problem of output stabilization for systems of the form

$$\dot{x} = Ax + Bu, \quad z = y = Cx. \tag{2.3}$$

Accordingly it can now be phrased as follows. Find strategies for the specialization of u which are based on past observations of z and which bring z asymptotically down to zero irrespective of the initial state of the system (2.3).

The existence of such strategies is a structural property of the given system and the disturbance model. If it is satisfied then one can always achieve output stabilization by means of a feedback-observer structure (Knobloch/Kwakernaak 1984, Ch. 7). Since the observer has to be chosen for a system which includes the disturbance model the "internal" dynamics of v will somehow enter into the regulator model. This is sometimes called the internal model principle (Wonham loc. cit, Ch. 8).

To summarize one can say that the problem of disturbance attenuation using a deterministic disturbance model can be completely settled within the frame of elementary control theory, at least from the theoretical viewpoint. In principle the scope of the method is larger than that of alternative methods (stochastic regulator, disturbance decoupling), since one has considerable freedom in choosing the disturbance model. So one has also an answer to the question how to meet uncertainties about the disturbance: Make the model rich enough, so that every possible disturbance mode is represented.

From the practical viewpoint however the internal model principle is a rather unsatisfactory answer to the design problem. Research effort is certainly required to improve the method in three respects.

(i) Introduction of design constraints in order to enforce uniqueness of the solution.

(ii) To remove the greatest obstacle to the effective use of rich (and therefore realistic) disturbance models, namely the necessity of employing a full order observer for a system which includes the state of the disturbance model.

(iii) A realistic assessment how much uncertainties about the disturbance really can be met by choosing "rich" models. This includes a discussion of the robustness and the possibilities of on-line-adjustments of the regulator.

2.3   *Affine systems: Controllability and Controlled Invariance*

We consider again a time-invariant control system of the form (2.1) and will assume for the remaining part of this section that f is linear with respect to the inputs $u := (u_1, \ldots, u_m)$ and $v := (v_1, \ldots, v_r)$. Hence it can be written in the form

$$f(x, u, v) = f_0(x) + \sum_{v=1}^{m} u_v g_v(x) + \sum_{\mu=1}^{r} v_\mu e_\mu(x). \tag{2.4}$$

Finite dimensional systems with this property are sometimes called affine. If f_0, g_v, e_μ are linear functions of x one speaks of a bilinear system.

The study of affine systems is an intermediate step on the way from linear to general nonlinear systems theory. In the last ten years successful attempts have been undertaken to generalize to affine systems some parts of qualitative linear theory which center around the notion of controllability. This brought out certain aspects of nonlinearity which had not been recognized during the area of optimal control theory. Their emergence went hand in hand with the rise of differential geometric methods in nonlinear systems theory. Conditions for controllability of bilinear systems (Brockett 1972) and the formulation of stability criteria for the Itô equation (Section 3) provide first hints at the role which Lie algebras may play in the study of affine systems. It is by now quite clear that for a system of the form

$$\dot{x} = f_0(x) + \sum_{v=1}^{m} u_v g_v(x) \tag{2.5}$$

a certain Lie-algebra — sometimes called accessibility distribution and denoted by $\mathcal{L}_0$ — has to be regarded as the analogue of the controllability space associated with a linear system

$$\dot{x} = Ax + \sum_{v=1}^{m} u_v b_v.$$

The elements of $\mathcal{L}_0$ are vectors h, k depending upon the state variable x. The definition of the product is the usual one: $[k, h] = h_x k - k_x h$. In precise terms $\mathcal{L}_0$ is then the algebra generated by the elements ad $^\varrho(f_0)g_v$, $\varrho = 0, 1, \ldots, v = 1, \ldots, m$ where the ad-operator has the usual meaning (i.e. $\text{ad}^0(h)k = k$

$\mathrm{ad}^{\varrho}(h)k = [h, \mathrm{ad}^{\varrho-1}(h)k])$. There does not exist so far a conclusive result (as in the linear case) which provides a common roof for the various statements which relate system theoretic properties of (2.5) to algebraic properties of $\mathscr{L}_0$. This is in part due to the fact that the notion of controllability for time invariant linear systems splits into several notions (controllability, reachability, accessibility, see e.g. Hermann and Krener (1977)), for nonlinear systems. The main reason however is of course the more complex situation which rarely allows to close the gap between necessary and sufficient conditions. A reasonable impression of the various steps which have been undertaken in order to understand more clearly the role of $\mathscr{L}_0$ can be obtained from these references: Hermann (1961), Sussmann and Jurdjevic (1972), Hermes (1974) and (1976), Sussmann (1983), Knobloch and Wagner (1984).

Knowing the appropriate generalization of controllability spaces it is somewhat natural to ask oneself whether and how those notions of linear control theory which rest directly upon controllability can be extended to affine systems. First answers to this question have been given only quite recently (cf. e.g. Hirschhorn (1981), Isidori et al. (1981), Nijmeijer (1982)). They concern the topics controlled invariance and disturbance decoupling and allow – at least in principle – applications to nonlinear design problems.

At this stage of the development it is too early to predict the impact which modern differential geometric thinking will eventually have on actual design practice in control engineering. It certainly will depend upon the extent to which the rather academic constructions of nonlinear feedback strategies which are the outcome of existence theorems can be replaced by efficient algorithms. This is probably the greatest challenge to future research in nonlinear control theory.

2.4   References

a) For a more detailed account of the literature on nonlinear geometric theory the reader is referred to:
Brockett, W., S. Millman, H. J. Sussmann, *Differential Geometric Control Theory*, eds., Progress in Mathematics, Vol. **27**, Birkhäuser, Boston, Basel, Stuttgart 1983.
b) The following references depict some general background of modern geometric thinking in systems theory and its relation to physics:
Brockett, R. W., *Control theory and analytical mechanics*, in: C. Martin and R. Hermann, eds., Geometric Control Theory, Lie Groups History, Frontiers and Applications, Vol. **II**, Mathem. Sci. Press, Brookline, MA, 1977.
Willems, J. C., *System theoretic models for the analysis of physical systems*, Richerche di Automatica, vol. **10**, 1977, pp. 71—106.
c) Literature quoted in Section 2:
Basile, G. and G. Marro, *Controlled and conditioned invariant subspaces in linear system theory*, J. Optimization Th. Appl., vol. **3**, 1969, pp. 306—315.
Brockett, R. W., *System Theory on Group Manifolds and Coset Space*, SIAM J. on Control, Vol. **10**, 1972, pp. 265—284.

Hermann, R., *On the Accessibility Problem in Control Theory*, International Symposium on Nonlinear Differential Equations and Nonlinear Mechanics, La Salle, Lefschetz eds., Academic Press, London 1963.

Hermann, R. and A. J. Krener, *Nonlinear controllability and observability*.

Hermes, H., *High Order Algebraic conditions for Controllability*, Lecture Notes in Economics and Mathematical Sciences, Vol. **131**, pp. 165—171. Springer-Verlag Berlin–Heidelberg–New York 1976.

Hermes, H., *On local and global controllability*, SIAM J. on Control, Vol. **12**, 1974, pp. 252—261.

Hirschhorn, R. M., *(A, B)-invariant distributions and disturbance decoupling of nonlinear systems*, SIAM J. on Control, Vol. **19**, 1981, pp. 1—19.

Isidori, A., A. J. Krener, C. Gori-Giorgi and S. Monaco, *Nonlinear decoupling via feedback: A differential geometric approach*, IEEE Trans. Autom. Contr., Vol. AC-**26**, 1981, pp. 331—345.

Johnson, C. D., *Accomodation of External Disturbances in Linear Regulator and Servomechanism Problems*, IEEE Trans. Autom. Contr., Vol. AC-**16**, 1971, pp. 635—644.

Knobloch, H. W. and H. Kwakernaak, *Lineare Kontrolltheorie*, To appear 1984, Springer-Verlag.

Knobloch, H. W. and K. Wagner, *On local controllability of nonlinear systems*. To appear in "Dynamical Systems and Microphysics: Control Theory and Mechanics", 1984, Academic Press.

Nijmeijer, H., *Controllability distributions for nonlinear control systems*, Systems & Control Letters, Vol. **2**, 1982, pp. 122—129.

Sussmann, H. J. and V. Jurdjevic, *Controllability of nonlinear systems*, J. Differential Equations, Vol. **12**, 1972, pp. 95—110.

Sussmann, H. J., *Lie Brackets and Local Controllability*, SIAM Journal on Control, Vol. **21**, 1983, pp. 686—713.

Wonham, W. M., *Linear Multivariable Control: A Geometric Approach*, 2nd. edn., Applic. of Math. Vol. **10**, Springer-Verlag, Berlin–New York, 1979.

Willems, J. C. and C. Commault, *Disturbance Decoupling by measurement feedback with stability or pole placement*, SIAM J. on Control, Vol. **19**, pp. 490—504.

3   A survey of some results in stochastic control theory (J. L. Willems)

Stochastic control theory is concerned with modeling, analysis and design of control systems in a random environment [17]. The stochastics are introduced to model random input components and external disturbances, measurement errors, and uncertainties or unknown variations of the parameters. In the present report some results and research directions in this area are reviewed. Only continuous-time stochastic systems are discussed; the developments for discrete-time stochastic control systems are very similar.

3.1   *Modeling of stochastic systems*

Let a deterministic control system be described by a set of first-order differential equations

$$\dot{x}(t) = f[x(t), u(t), v(t), t] \tag{3.1}$$

where $t \in \mathbb{R}$ denotes the time variable, $x(t) \in \mathbb{R}^n$ the state of the system, $u(t) \in \mathbb{R}^m$ the control input, $v(t) \in \mathbb{R}^k$ the system disturbances. The disturbances may represent either time-dependent inputs which are beyond the control of the operator, or variations of some parameters. In most applications such disturbances (or some of them) are unknown and immeasurable; thus it is quite natural to represent them as stochastic processes. The unpredictability of the disturbances leads to the use of white noise processes, that is processes which are completely unpredictable. Sometimes it may be more accurate to model the disturbances as processes with non-zero correlation between the values at different times; then they may be represented as outputs of dynamic systems excited by white noise. It can often be assumed that the random processes are Gaussian; the motivation is that Gaussian processes (and Gaussian random variables) are good models for events which depend on a large number of independent causes. Moreover for any set of random variables there exists a set of Gaussian random variables with the same first and second order moments. Hence if only these statistics are known or measured, one can as well use Gaussian random variables or processes. These considerations lead to the representation of system (3.1) by differential equations with white noise terms in the right hand side, hence to Itô equations [1]. An Itô differential equation is a stochastic differential equation of the type

$$dx(t) = f[x(t), u(t), t]\, dt + g[x(t), u(t)t]\, dW \tag{3.2}$$

with $f: \mathbb{R}^n \times \mathbb{R}^m \times \mathbb{R} \rightarrow \mathbb{R}^n$, $g: \mathbb{R}^n \times \mathbb{R}^m \times \mathbb{R} \rightarrow \mathbb{R}^{n \times k}$, where W is an $\mathbb{R}^k$-valued zero-mean Wiener process, that is a Gaussian random process with independent increments, also called Brownian motion. The "derivative" dW/dt can then be considered to be white noise. Under suitable technical assumptions the solution $x(t)$ of (3.2) is a Markov process [1].

In a control system description we add output measurements to the model; taking white noise measurement errors into consideration leads to

$$dy(t) = h(x(t), t)\, dt + r[x(t), t]\, dV \tag{3.3}$$

where $y \in \mathbb{R}^p$, $h: \mathbb{R}^n \times \mathbb{R} \rightarrow \mathbb{R}^p$, $r: \mathbb{R} \times \mathbb{R} \rightarrow \mathbb{R}^{p \times q}$ and V is an $\mathbb{R}^q$-valued zero-mean Wiener process.

An important question in the area of modeling stochastic dynamic systems is whether or not the Itô differential equation is a good approximation for the representation of systems with wide-band noise (the so-called physical noise); ideal white noise has infinite bandwidth. Consider the dynamic system equation

$$\dot{x}(t) = f[x(t), t] + g(x(t), t)\beta(t) \tag{3.4}$$

where $\beta(t)$ is a Gaussian $\mathbb{R}^k$-valued zero-mean random process with very large bandwidth; the control input is not explicitly included in (3.4): an open loop control is assumed to be expressed as a time function; a feedback control is assumed to be expressed in terms of the state x. Much research effort has been devoted to the study of the limiting case of equation (3.4) if the random process $\beta(t)$ has a constant spectral density and if the bandwidth tends to infinity. It was proved [23] that the solutions of (3.4) do *not* converge to the solutions of the Itô equation

$$d x(t) = f[x(t), t] \, dt + g[x(t), t] \, dW \tag{3.5}$$

where $W(t)$ is the Wiener process which corresponds to the limiting case of the integral of the random process $\beta(t)$. However it was shown that the solutions of (3.4) converge to the solutions of a different Itô equation [23]:

$$d x(t) = f_1[x(t), t] \, dt + g[x(t), t] \, dW \tag{3.6}$$

where

$$f_1(x, t) := f(x, t) + c(x, t)$$

with the correction term

$$c(x, t) := \frac{1}{2} \sum_{r=1}^{n} \sum_{s=1}^{k} a_r(x, t) \, g_{rs}(x, t).$$

In this expression $a_r(x, t)$ denotes the r-th column of the Jacobian matrix $\partial g(x, t)/\partial x$ and $g_{rs}(x, t) \in \mathbb{R}$ is the element on the r-th row and the s-th column of the matrix $g(x, t)$. This correction term is often referred to as the Stratonovitch correction. It vanishes for systems where the random disturbances in the system equation are independent of the state, that is the so-called additive noise case. In other cases the correction term should be taken into consideration, that is in the multiplicative noise case. The additive noise case is very important in control theory, since a linear stochastic differential equation

$$d x(t) = [A x(t) + B u(t)] \, dt + G \, dW(t) \tag{3.7}$$

with output equation

$$d y(t) = C x(t) + d V(t) \tag{3.8}$$

is often used to model linear dynamic systems with stochastic disturbances and random measurement errors. The analysis of Itô equations cannot be done by means of ordinary calculus, but requires special techniques, known as Itô calculus.

Considerable research effort has also been devoted to the study of control systems with non-Gaussian white noise disturbances. Then a stochastic differential equation of the type (3.2) is obtained where the Wiener process $W(t)$ is to be replaced by a non-Gaussian stochastic process with independent increments. It has been shown that *martingale* theory [13] is very well suited to attack such control theory problems. Because of space limitation the discussion in the present paper is restricted to the Gaussian case; it is this case which has most extensively been investigated and for which a fairly complete theory for control system design has been developed.

The derivation of stochastic system models for chemical engineering and computer applications is discussed in the contributions by Arnold and Boel in [11].

3.2   *Stability of stochastic systems*

As has been mentioned in the introduction stability is a much more interesting qualitative property than controllability for stochastic systems. It is one of the fundamental requirements in the design phase, and, even more, stability is often the basic reason for using feedback control. Much research has been concerned with the development of techniques for the analysis of stability of stochastic dynamic systems [7]. Here also various definitions of asymptotic stability for stochastic systems can be derived from the definition of asymptotic stability for deterministic systems, depending on the mode of convergence of the random trajectory to the equilibrium state. Let the desired equilibrium state be at the origin of the state space. The following convergence properties and corresponding asymptotic stability definitions have been considered in many publications:
— *almost sure convergence*: the trajectories (originating in some neighborhood of the origin) converge to the origin with probability one;
— *mean square convergence*: the trajectories (originating in some neighborhood of the origin) converge to the origin in the mean square; if $x(t)$ denotes the state, $\|x\|$ a norm of x, and E the average value, then this means

$$\lim_{t \to \infty} E[\|x(t)\|^2] = 0.$$

— *p-th mean convergence*: the trajectories (originating in some neighborhood of the origin) converge to the origin in the p-th mean (with $p > 0$):

$$\lim_{t \to \infty} E[\|x(t)\|^p] = 0.$$

Most research effort has been devoted to the analysis of almost sure asymptotic stability and mean square asymptotic stability. The former property is rather weak, probably too weak for control systems. Mean square asymptotic stability is much stronger, it is required for the solvability of optimal control problems with quadratic cost functions. Moreover, an interesting feature is that for linear stochastic systems with multiplicative white noise the mean square stability properties can be shown to be equivalent to the corresponding stability properties of a deterministic linear dynamic system. Indeed, consider the Itô differential equation ($x \in \mathbb{R}^n$):

$$dx(t) = Ax(t) \, dt + \sum_{i=1}^{k} \sigma_i G_i x(t) \, dW_i \qquad (3.9)$$

where, for simplicity, only constant matrices A and G_i are considered, and where the Wiener processes W_i are assumed to be independent and normalized:

$$E(dW_i \, dW_j) = \delta_{ij} \, dt$$

with δ_{ij} denoting the Kronecker delta. The second moment matrix $M(t) := E[x(t) \, x(t)']$ satisfies the deterministic matrix differential equation [1]

$$\dot{M}(t) = AM(t) + M(t)A' + \sum_{i=1}^{k} \sigma_i^2 G_i M(t) G_i' \qquad (3.10)$$

This is equivalent to a set of $n(n+1)/2$ linear differential equations. The mean square stability properties of (3.9) can hence be derived from the properties of the linear operator L on $\mathbb{R}^{n(n+1)/2}$, the space of symmetric matrices of order n, defined by

$$P \to L(P) = AP + PA' + \sum_{i=1}^{k} \sigma_i^2 G_i P G_i' \qquad (3.11)$$

An important characteristic [7] for the analysis of the mean square stability properties of (3.9) and also for the p-th mean stability properties, with $p \neq 2$, is the Lie algebra generated by the set of matrices $\{A, G_1, G_2, \ldots, G_k\}$; that is the smallest matrix Lie algebra containing these matrices. A subspace of square matrices of order n is called a matrix Lie algebra if it is closed with respect to the commutator product $[A, B] := AB - BA$. A Lie algebra is solvable if its derived series defined by

$$L^{(0)} := L, \; L^{(1)} := [L, L], \; L^{(2)} = [L^{(1)}, L^{(1)}], \; \ldots$$

satisfies

$$L^{(k)} = \{0\}$$

for some integer k. A rather complete stability analysis of the Itô equation (3.9) with a solvable Lie algebra is given in Willems' contribution in [7]. For these systems criteria are obtained for almost sure asymptotic stability, mean square asymptotic stability, and p-th mean asymptotic stability. Also explicit criteria for the same stability properties for the colored noise system

$$\dot{x}(t) = Ax(t) + \sum_{i=1}^{k} f_i(t) \, G_i x(t) \tag{3.12}$$

where $f_i(t)$ are stochastic processes and the same Lie algebra is solvable, are given in that paper.

Except for the case of a solvable Lie algebra, the analysis of almost sure asymptotic stability of systems described by the linear Itô equations (3.9) is very difficult; some interesting results have been obtained by Arnold and co-workers [2]. The problem associated with the analysis of the mean square stability properties for system (3.12) is that either the convergence should be derived from the partial differential equation governing the probability density of the state, or from an infinite set of ordinary differential equations governing the moments. Indeed the moments of different orders are not uncoupled. Approximation algorithms have been developed for the analysis of the moments by truncating this set of equations by neglecting some coupling terms [7].

3.3   Synthesis of stochastic control systems: Open loop versus closed loop

In an idealized situation where the mathematical model of a deterministic control system is an exact description of its physical behavior, open loop control (using knowledge of only the initial state) and closed loop control (using knowledge of all past states) are completely equivalent. It is however well known by control engineers that closed loop control is less sensitive to and more robust against uncertainties and inaccuracies in the system model; this is the main reason for using feedback control. From this observation it is clear that open loop and closed loop control strategies are not at all equivalent for stochastic control systems [8]. Since the behavior of the system variables is random, it cannot exactly be predicted. Closed loop control uses on-line measurements of the system variables and hence more information is available to the controller than in the case of open loop control where only some a priori information is known.

Obviously for practical applications only *causal* controls are realizable: the control at time t is constrained to depend only on the observations that have been obtained prior to or at time t. However one may assume that the controller has knowledge of the future observation *program* and of the statistics of the future observation errors and system disturbances. Hence the following essentially different control policies can be defined [3]:

Open loop control. The controller has no knowledge of any observation. It only knows the system model, the statistics of the disturbances, and the initial state or its probability model.

Feedback control. The observations up to time t are available for the controller at time t, and also the mathematical model, the statistics of the disturbances, the previous inputs, and the statistics of the available measurements, but not the future measurement policy or the statistics of the future measurement errors.

Closed loop control. The observations up to time t are available for the controller at time t, and also the mathematical model, the statistics of the disturbances, the previous inputs, and the statistics of the available measurements, and also the future measurement policy and the statistics of the future measurement errors.

Note that the distinction between feedback and closed loop control is not standard. With respect to this distinction, an interesting feature of control in stochastic systems is that the control action, which obviously affects the behavior of the state, may also affect the future uncertainty on the state; this is called the *dual effect* of the control. In fact, the future disturbances, parameter uncertainties, and measurement errors may depend on the present control action. Closed loop control, but not feedback control, as defined above, takes this dual effect completely into account. One may distinguish two different aspects, called *caution* and *probing*:

(i) Due to the uncertainties inherently associated with a stochastic control problem, the controller must be cautious not to increase unfavourably the effect of future uncertainties on the control system performance; this caution can be reduced if the future observation program is known to the controller.
(ii) When the dual effect exists, the control can be designed partly to improve the estimation of the future states or of the parameters by decreasing the future uncertainty. In such cases closed loop control has the capability of *active learning*. The lack of dual effect has been called *neutrality*. An interesting discussion of the dual effect and its implications was given by Bar-Shalom and Tse [3].

3.4   *Optimal control of linear stochastic systems*

One of the most interesting and appealing results in deterministic control theory is the optimal linear regulator. Consider the linear finite-dimensional control system described by

$$\dot{x}(t) = A(t)\, x(t) + B(t)\, u(t) \tag{3.13}$$

For a given initial state x_0 at the initial tissue t_0, the control in (t_0, t_1) has to be designed to minimize the cost function

$$\eta := \int_{t_0}^{t_1} [x(t)' Q(t) x(t) + u(t)' R(t) u(t)] \, dt + x(t_1)' S x(t_1) \qquad (3.14)$$

with $Q(t) \geq 0$, $R(t) > 0$ in (t_0, t_1) and $S \geq 0$. This problem is commonly called the LQ-problem. Its stochastic counterpart, the LQG-problem, is obtained by assuming

(i)   that some additive Gaussian white noise disturbances act on the system,
(ii) that not the state of the system is measured, but only linear output measurements are available, where the measurement errors are modelled as Gaussian white noise processes,
(iii) that the average value of the cost function is to be minimized.

This optimal control problem is undoubtedly the most studied problem in stochastic control theory. Its complete solution is known and can be obtained as follows [8]: First we compute the optimal state feedback control for the deterministic version of the stochastic problem (by replacing the stochastic processes by their average values); then the state is replaced by its optimal estimate from the available measurements. This property is called the *certainty equivalence* property. This algorithm is also used in other stochastic optimal control problems, where it yields an ad hoc suboptimal control strategy. Weaker than certainty equivalence is the *separation property* where the optimal control depends on the measurement data via the estimate of the state, but where the dependence is not necessarily the same as in the deterministic version of the stochastic control problem. It is intuitively clear that both properties are closely related to neutrality and absence of the dual effect; a nice discussion of the relationship has been given by Bar-Shalom and Tse [3] and by Witsenhausen [22].

The LQG-problem is one of the exceptional cases where a stochastic optimal control problem has been solved completely. Some other cases are reported by Beneš [4]. The dynamic programming algorithm, using Bellman's principle of optimality, is very well suited to deal with stochastic control problems [6, 10]. As far as optimal filtering or estimation of the state variables or parameters of a stochastic system from the observations is concerned, the main problem is that the optimal estimation algorithm does not correspond to a finite dimensional linear system; this is discussed by Beneš [5] and also in some papers in [11] and is the subject of a lot of current research [19].

Next we consider the optimal control of a linear stochastic system with *multiplicative* Gaussian white noise and exact state measurements. The system is described by the Itô equation

$$dx(t) = [A(t)x(t) + B(t)u(t)]dt + \sum_{i=1}^{k} \sigma_i F_i(t)x(t)dW_i(t)$$

$$+ \sum_{j=1}^{k} \varrho_j G_j(t)u(t)dV_j(t) \tag{3.15}$$

where the Wiener processes $W_1(t), \ldots, W_k(t), V_1(t), \ldots, V_l(t)$ are zero mean and independent with

$$E[dW_i(t)^2] = dt, \quad E[dV_j(t)^2] = dt.$$

The cost function is the same as for the LQG-problem. The optimal control strategy in (t_0, t_1) is given by

$$u(t) = K^*(t)\,x(t)$$

where

$$K^*(t) = -\tilde{R}(t)^{-1} B(t)'\, P(t)$$

and

$$\tilde{R}(t) = R(t) + \sum_{j=1}^{l} \varrho_j^2 G_j(t)'\, P(t)\, G_j(t)$$

The matrix $P(t)$ is the solution of the matrix differential equation

$$\dot{P}(t) = -A(t)'\, P(t) - P(t)\, A(t) - \sum_{i=1}^{k} \sigma_i^2 F_i(t)'\, P(t)\, F_i(t)$$

$$+ P(t)\, B(t)\, \tilde{R}(t)^{-1} B(t)'\, P(t) - Q(t) \tag{3.16}$$

with

$$P(t_1) = S$$

The solution of this control problem does not possess the certainty equivalence property; here the dual effect exists. Considering the case $n=1$, where (3.15) is a first-order system, we see that the effect of the state dependent noise in (3.15) is to increase the feedback, and effect of the control dependent noise in (3.15) is to decrease the feedback. Both effects tend to decrease the future uncertainty in the system equation.

3.5   *Stabilization of linear stochastic systems with multiplicative noise*

We have already pointed out that stability is one of the fundamental requirements for the design of control systems. Many classical design techniques in
deterministic control theory are aimed at stabilizing rather than optimally
controlling a plant by feedback. In the present section we discuss the stabilization of a simple stochastic control with multiplicative noise by state feedback. Consider the stochastic system

$$\mathrm{d}x(t) = [Ax(t) + Bu(t)]\,\mathrm{d}t + \sum_{i=1}^{k} \sigma_i F_i x(t)\,\mathrm{d}Wi \tag{3.17}$$

where for simplicity the control dependent noise terms in (3.15) have been
deleted; $x \in \mathbb{R}^n$ and $u \in \mathbb{R}^m$. The matrices $A, B, F_1, \ldots, F_k$ are assumed to be
constant and to have appropriate dimensions. The Wiener processes W_i are
assumed to be independent and normalized as in (3.15). The noise intensities are
explicitly expressed by the positive constants σ_i. The question considered is
whether there there exists a state feedback control

$$u(t) = Kx(t)$$

with K constant, such that the closed loop system

$$\mathrm{d}x(t) = (A + BK)\,x(t)\,\mathrm{d}t + \sum_{i=1}^{k} \sigma_i F_i x(t)\,\mathrm{d}W_i \tag{3.18}$$

is mean square asymptotically stable. The criteria for this kind of stability have
been discussed in 3.2. Moreover it is desired that the stability property be *robust*
with respect to uncertainties in the model, in particular with respect to changes in
the noise intensities σ_i; it is quite logical to assume that these intensities are not
known exactly. Note also that mean square stabilizability of (3.17) is a necessary
condition for the solvability of the infinite horizon case of the optimal control
problem considered in 3.4, that is the case $t_1 \rightarrow \infty$. The following results on this
problem have been obtained [20].
    Criterion. The stochastic system (3.17) is stabilizable in the mean square
by state feedback for the noise intensities σ_i if there exist square positive definite
matrices Q and R of order n and m respectively such that the equation

$$SA + A'S - SBR^{-1}B'S + \sum_{i=1}^{k} \sigma_i^2 F_i' S F_i = -Q \tag{3.19}$$

has a positive definite solution S.
    Note that the selection of the positive definite matrices Q and R does not
pose any problems; indeed if the above criterion is satisfied for some positive

definite matrices Q and R, it is satisfied for all such matrices. Note also that the stabilizing feedback matrices can directly be derived from the positive definite solution S of (3.19), indeed a suitable feedback matrix is

$$K = -R^{-1}B'S.$$

This feedback is robust in the sense that a feedback matrix which stabilizes (3.17) for some noise intensities σ_i also stabilizes the system for all smaller noise intensities. The stabilizability of the deterministic system

$$\dot{x}(t) = Ax(t) + Bu(t) \tag{3.20}$$

is a necessary condition for mean square stabilizability of (3.18). In general when (3.20) is stabilizable, the stochastic system (3.18) is only mean square stabilizable up to some maximum noise intensities. For larger noise intensities the system cannot be stabilized; this is entirely due to the stochastic elements. This phenomenon is called the *uncertainty treshold principle* [9].

One may ask if there are cases where stabilizability of (3.18) is perfectly robust, that is where system (3.18) can be stabilized for all noise intensities by means of the same state feedback; of course then there is no uncertainty treshold. A criterion for perfect robust stabilizability and the computation of a suitable feedback matrix has been derived by means of the concepts of geometrical theory of linear systems [21].

3.6   References

Only a few references on stochastic control theory are indicated below. For further references the reader should in particular consult the bibliographies of the conference proceedings [11] and [13], where he may find extensive lists of relevant papers and books.

[1]   Arnold, L., *Stochastische Differentialgleichungen*, Oldenbourg, München, 1973.

[2]   Anold, L. and W. Kliemann, *Qualitative theory of stochastic systems*, in: *Probabilistic Analysis and Related Topics* (Editor: A. T. Bharucha-Reid), Academic Press, New York, 1983.

[3]   Bar-Shalom, Y. and E. Tse, *Dual effect, certainty equivalence, and separation in stochastic control*, IEEE Transactions on Automatic Control, Vol. AC-**19**, 1974, pp. 494—500.

[4]   Beneš, V. E., L. A. Shepp and H. S. Witsenhausen, *Some solvable stochastic control problems*, Stochastics, Vol. **4**, 1980, pp. 39—83.

[5]   Beneš, V. E., *Exact finite-dimensional filters for certain diffusions with nonlinear drift*, Stochastics, Vol. **5**, 1981, pp. 65—92.

[6]   Bertsekas, D. P., *Dynamic Programming and Stochastic Control*, Academic Press, New York, 1976.

[7]   Clarkson, D. L., (Editor), *Stochastic Problems in Dynamics*, Pitman, London, 1970.

[8]   Davis, M. H. A., *Linear Estimation and Stochastic Control*, Chapman and Hall, London 1977.

[9]   Dersin, P. L., M. Athans and D. A. Kendrick, *Some properties of the dual adaptive stochastic control algorithm*, IEEE Transactions on Automatic Control, Vol. AC-**26**, 1981, pp. 1001—1008.
[10]   Fleming, W. H. and R. W. Rishel, *Deterministic and Stochastic Optimal Control*, Springer Verlag, Berlin, 1972.
[11]   Hazewinkel, M. and J. C. Willems (Editors), *Stochastic Systems: The Mathematics of Filtering and Identification and Applications*, D. Reidel, Dordrecht, 1981.
[12]   Kailath, T., *A view of three decades of linear filtering theory*, IEEE Transactions on Information Theory, Vol. IT-**20**, 1974, pp. 145—181.
[13]   Kohlmann, M. and W. Vogel (Editors), *Stochastic Control Theory and Stochastic Differential Equations*, Springer Verlag Lecture Notes in Control and Information Sciences **16**, Berlin, 1979.
[14]   Kushner, H., *Stochastic Stability and Control*, Academic Press, New York, 1967.
[15]   Kwakernaak, H. and R. Sivan, *Linear Optimal Control Systems*, Wiley Interscience, New York, 1972.
[16]   Striebel, C., *Optimal Control of Discrete Time Stochastic Systems*, Springer Verlag Lecture Notes in Economics and Mathematical Systems **110**, Berlin, 1975.
[17]   Schweppe, F., *Uncertain Dynamic Systems*, Prentice Hall, Englewood Cliffs, N. J., 1973.
[18]   Van de Water, H. and J. C. Willems, *The certainty equivalence property in stochastic control theory*, IEEE Transactions on Automatic Control, Vol. AC-**26**, 1981, pp. 1080—1087.
[19]   Sussmann, H. J., *Approximate finite-dimensional filters for some nonlinear problems*, Systems and Control Letters, to appear.
[20]   Willems, J. L. and J. C. Willems, *Feedback stabilizability for stochastic systems with state and control dependent noise*, Automatica, Vol. **12**, 1976, pp. 277—283.
[21]   Willems, J. L. and J. C. Willems, *Robust stabilization of uncertain systems*, SIAM Journal of Control and Optimization, Vol. **21**, 1983, pp. 352—374.
[22]   Witsenhausen, H. S., *Separation of estimation and control for discrete time systems*, Proceedings IEEE, Vol. **59**, 1971, pp. 1557—1566.
[23]   Wong, E., *Stochastic Processes in Information and Dynamical Systems*, McGraw-Hill, New York, 1971.
[24]   Zabczyk, J., *Controllability of linear stochastic systems*, Systems and Control Letters, Vol. **1**, 1981, pp. 25—31.

4   Large scale systems (M. Thoma)

4.1   *Complexity and decentralization*

The complexity of present-day technological, environmental and other processes is one of the foremost challenges to system theory. Complexity is, however, a subjective notion; but we consider besides others the *dimensionality* as an important source (Siljak 1983). For large size systems it is either impossible or uneconomical to analyse the system as a whole. Very often real large size systems are by their nature composed by partly autonomous subsystems with more or less interconnections. A standard approach therefore is to use decomposition techniques and gain both a conceptual insight and a numerical simplicity in solving the design problem. This procedure leads in many cases to a decentralized control concept; they are more reliable, safer, and give a better (physical) understanding of the system behaviour.

An interesting possibility is the decentralized control by decomposing the overall system into hierarchical composed subsystems. A control system of hierarchical structure is a system in which there are several controllers of which some override or supervise the actions of others. Two fundamental concepts have up to now developed which are called *Multilayer Systems*, where the control of an object is split into algorithms, or layers, each of which acts at different time intervals, and *Multilevel Systems*, where control of an interconnected, complex system is divided into local control units (lower level) and their action is coordinated by a coordinator (higher level). However, both have in common that the decision making has been achieved in a hierarchical dependence. This means, that there exist several decision units (controller) in the structure, but only some of them have direct access to the controlled system. The others are at a higher level and define the tasks and coordinate the lower level units, but they do not override their decisions (Findeisen et al. 1980, Wilson 1979, Siljak 1978).

4.2   Multilayer systems

In this case the control is split up in superimposed layers, which in general have different time horizons. In most cases the time horizons increase by going from lower to higher layers. This so called vertical decomposition by different goals or time horizons can be found in a number of real systems. However, up to now there is no explicit mathematical design method existing. Questions like what number of layers and how should their time horizon be chosen, questions of stability and suboptimality etc. are not yet solved. The existing procedures are more or less heuristical and they depend on the considered problem (Lefkowitz 1965, Findeisen et al. 1980).

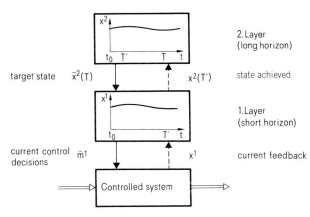

Fig. 2: Two-layer control with different time horizon

In order to demonstrate the above mentioned idea, we choose a production process. The monthly production goal depends on the longer-term marketing strategies, which are not so much dependent on disturbances (long horizon). While the daily, short-term marketing strategies are taken into account by the lower layer with a short time horizon. In other words the dynamic optimization is subdivided as shown in figure 2.

— Each layer has a different time horizon; the highest layer has the longest horizon.

— The model used at each layer, or the degree to which details of the problem are considered, is also different—the top level is the least detailed.

The hierarchical structure can be mathematical formulated in the following way (Findeisen et al. 1980). Assume the optimal production problem was to determine $m^1(t)$ such that the cost functional

$$\int_{t_0}^{T} g^1 [x^1(t), m^1(t), z^1(t)] \, dt \tag{4.1}$$

is maximized subject to constraints of the form

$$\dot{x}^1(t) = f^1 [x^1(t), m^1(t), z^1(t)] \tag{4.2}$$

where the state $x^1(t_0)$ is given and $x^1(T)$ is free or specified at $t = T$ and

$$\begin{aligned}
& x^1(t) \text{ designates the state vector} \\
& m^1(t) \text{ designates the control (decision) vector} \\
& z^1(t) \text{ designates the disturbance vector.}
\end{aligned} \tag{4.3}$$

For simplicity we will assume in the sequel that the problem is divided among two layers only.

Top layer (long horizon). Here the problem is formulated as follows:

$$\underset{m^2(t)}{\text{Max}} \int_{t_0}^{T} g^2 [x^2(t), m^2(t), z^2(t)] \, dt \tag{4.4}$$

subject ot the constraints

$$\dot{x}^2(t) = f^2 [x^2(t), m^2(t), z^2(t)], \tag{4.5}$$

$x^2(t_0)$ given and $x^2(T)$ free or specified.

The variables $x^2(t)$, $m^2(t)$, $z^2(t)$ are simplified or aggregated variables of the same meaning as defined above. g^2 is a performance function of the same type as in the original formulation (cf. (4.1)) but dependent on the aggregated variables. The solution of the long horizon problem as described by equations (4.4), (4.5) determines among other things the state variable $\hat{x}^2(T')$ at time T' < T which will serve as target condition for the lower layer (cf. (4.7)).

Lower layer (short horizon).
The description of the problem is as detailed as for the original problem, so the symbols x^1, etc. have the same meaning as in (4.1)—(4.3). Note however, that the time horizon T' on the lower layer is smaller than T and that there is an additional boundary condition which represents the short horizon target set by the top layer. So far the lower layer the optimization problem assumes this form

$$\underset{m^1(t)}{\text{Max}} \int_{t_0}^{T'} g^1[x^1(t), m^1(t), z^1(t)]\, dt \tag{4.6}$$

subject to these constraints

$$\dot{x}^1(t) = f^1[x^1(t), m^1(t), z^1(t)], \tag{4.7}$$
$$x^1(t_0) \text{ given}, \quad q^2[x(T')] = x^2(T')$$

q^2 is a mapping from the x^1-space into the x^2-space (note that these spaces in general have different dimensions).
One can improve the multi layer concept by providing an additional feedback from the lower to the upper layer (dashed line in figure 2). The actual state of the lower layer which is achieved at time T' is updated and is used for the policy makers in the upper layer in order to fix their initial data at time T':

$$x^2(T') = q^1[x^1(T')], \tag{4.8}$$

and consequently at time $2\,T'$ (after two days)

$$x^2(2\,T') = q^1[x^1(2\,T')].$$

and so on. This leads to a repetitive optimization problem (OLFO-Algorithm).

4.3   Multilevel System

In this case the overall system is decomposed into more or less independent subsystems (parallel or spatial decomposition), whereby each subsystem is controlled or optimized separately. Of course, the structures can be very different. However, an important class is defined by systems with hierarchical structures. The control action applied to the plant by the local controller (first level) is dependent upon information both fed back from the plant and

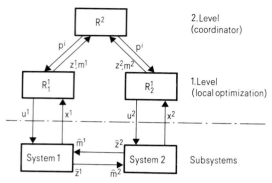

Fig. 3: Two-level control with goal coordination

passed down from the higher-level controller (coordinator). The lower-level controller is dependent on the coordination information and must act in accordance with this information (priority of the coordinator).

Actually this kind of hierarchical control includes three main problems:

(i) *Decomposition*: The overall system and overall performance criteria has to be decomposed in subsystems (local systems) and sub-criteria (modification).

(ii) *Solution of the parametrical subproblems*: These subproblems arise by the modification process.

(iii) *Coordination of the subproblems*: If coordinability is assumed this process leads in general to an iterative coordination algorithm.

An important class are the linear-quadratic optimal control problems. The n-dimensional overall system (plant)

$$\dot{x}(t) = Ax(t) + Bu(t), \quad y(t) = Cx(t) \tag{S}$$

where $x(t) \in \mathbb{R}^n$ state vector, $u(t) \in \mathbb{R}^m$ control vector, $y(t) \in \mathbb{R}^r$ output vector, and A, B, C, D are constant matrices of appropriate dimensions, is assumed to be completely controllable and observable. Besides the performance index

$$J = \frac{1}{2} \int_{t_0}^{t_1} [x^T(t) \, Qx(t) + u^T(t) \, Ru(t)] \, dt \tag{4.9}$$

is given; the constant matrices Q and R are assumed to be positive semidefinite respectively positive definite.

Let the overall system (S) now be decomposed into N interconnected subsystems of dimensions n_i, so that

$$\sum_{i=1}^{N} n_i = n. \tag{4.10}$$

It is assumed that the subsystems are just coupled dynamically through the state variables or components of the state vector $x(t)$, but that they enter uncoupled into the performance index. We partition the state vector into subvectors x_i and the Matrix A into submatrices A_{ij}, $i, j = 1, \ldots, N$. According to our assumption we can then write $B = \operatorname{diag}(B_i)$, $C = \operatorname{diag}(C_i)$, $Q = \operatorname{diag}(Q_i)$ and $R = \operatorname{diag}(R_i)$. We thereby arrive at a system description of this form

$$\dot{x}_i(t) = A_{ii}x_i(t) + B_iu_i(t) + \sum_{\substack{j \neq i}}^{N} A_{ij}x_j(t),$$

$$x_i(t_0) = x_{i0}, \quad (i = 1, 2, \ldots, N) \tag{4.11}$$

and

$$J = \sum_{i=1}^{N} \frac{1}{2} \int_{t_0}^{t_1} [x_i(t)^T Q_i x_i(t) + u_i(t)^T R_i u_i(t)] \, dt. \tag{4.12}$$

Again it is assumed that the constant matrices Q_i and R_i are positive semidefinite respectively positive definite, and all subsystems are controllable and observable. In case of linear time invariant overall systems (S) directed graphs (digraphs) are used for the decomposition process. In context of graph theory it is natural to replace that properties by their structural counter parts, structural controllability and observability. If both hold we speak of a (completely) structured system (Thoma 1980, Söte 1980, Lin 1974).

By the way, the digraph representation is for linear systems very important not only for decomposition purposes but also for system analysis, synthesis modelling etc. because it is closely related to the real system behaviour.

The following applied decomposition and coordination is based on the *coupling balance principle* (goal coordination) (Lasdon 1965, Mesarovic et al. 1970). In order to explain this principle we first rewrite the equation (4.11) as follows

$$\dot{x}_i(t) = A_{ii}x_i(t) + B_iu_i(t) + m_i(t) \tag{4.13}$$

where

$$m_i(t) = \sum_{\substack{j \neq i}}^{N} A_{ij}x_j(t) \qquad (i = 1, \ldots, N) \tag{4.14}$$

Next we introduce formally the $m_i(t)$ into the performance index using an arbitrary set of parameters $v_i(t)$:

$$J = \sum_{i=1}^{N} \left\{ \int_{t_0}^{t_1} \frac{1}{2} [x_i(t)^T Q_i x_i(t) + u_i(t)^T R_i u_i(t)] \right.$$

$$+ v_i(t)^T \left[\sum_{j \neq i}^{N} A_{ij} x_j(t) - m_i(t) \right] \right\} dt . \tag{4.15}$$

The coupling balance principle now proposes the following decoupling procedure. We dispense with the relation (4.14) and regard instead m_i as additional control variable (pseudo-variable) whereas the v_i play the role of parameters which are chosen by the coordinator (coordination variables). The problem of maximizing J subject to the constraints (4.13) then splits into N "local" problems, each depending upon x_i, u_i, m_i, v_i only. A conformity of the sum of the local optimal performance criteria with the original global performance criteria is only given if the parameter $v_i(t)$ is chosen by the coordinator in a way that the coupling conditions (4.14) are met by the optimal solution for the pseudo-variable $m_i(t)$ and the state variable $x_i(t)$.

Each local problem is of the linear quadratic type, the performance index however is linear in the pseudo-variable m_i. This leads to a singular control problem which should be avoided, in particular because the coordination procedure is in most cases based on an iterative process. Therefore the overall performance index is extended by including the coupling variables $m_i(t)$ in

$$J_i = \frac{1}{2} \int_{t_0}^{t_1} [x_i(t)^T Q_i x_i(t) + u_i(t)^T R_i u_i(t) + m_i(t)^T S_i m_i(t)] dt$$
$$+ \int_{t_0}^{t_1} \left[x_i(t)^T \sum_{j \neq i}^{N} A_{ji}^T v_j(t) - m_i(t)^T v_i(t) \right] dt , \tag{4.16}$$

whereby the performance matrices S_i $(i = 1, 2, \ldots, N)$ are also assumed as positive semidefinite.

Each local problem is now a standard LQ-problem. We introduce the Hamiltonian

$$H_i = \frac{1}{2} [x_i(t)^T Q_i x_i(t) + u_i(t)^T R_i u_i(t) + m_i(t)^T S_i m_i(t)]$$
$$+ x_i(t)^T \sum_{j \neq i}^{N} A_{ji}^T v_i(t) - m_i(t)^T v_i(t) \tag{4.17}$$
$$+ p_i(t)^T [A_{ii} x_i(t) + B_i u_i(t) + m_i(t)] .$$

and determine the solution from the stationarity conditions

$$\dot{x}_i(t) = \frac{\partial H_i}{\partial p_i} = A_{ii} x_i(t) + B_i u_i(t) + m_i(t) \tag{4.18}$$

$$\dot{p}_i(t) = \frac{-\partial H_i}{\partial x_i} = - A_{ii}^T p_i(t) - Q_i x_i(t) - \sum_{j \neq i}^{N} A_{ji}^T v_j(t) \tag{4.19}$$

$$0 = \frac{\partial H_i}{\partial u_i} = R_i u_i(t) + B_i^T p_i(t) \tag{4.20}$$

$$0 = \frac{\partial H_i}{\partial m_i} = S_i m_i(t) + p_i(t) - v_i(t) \tag{4.21}$$

together with the boundary conditions

$$x_i(t_0) = x_{i0} \quad \text{und} \quad p_i(t_1) = 0. \tag{4.22}$$

The solution of the stationarity equations (4.19)—(4.22) leads to a control law

$$u_i(t) = - R_i^{-1} B_i^T K_i(t) x_i(t) + R_i^{-1} B_i^T g_i(t), \tag{4.23}$$

where $K_i(t)$ is given by the well-known Matrix-Riccati equation

$$\dot{K}_i(t) = - K_i(t) A_{ii}^T K_i(t) + K_i(t) [B_i R_i^{-1} B_i^T + S_i^{-1}] K_i(t) - Q_i \tag{4.24}$$

$$K_i(t_1) = 0.$$

At the same time the vectors $g_i(t)$ can be calculated from

$$\dot{g}_i(t) = [K_i(t)(B_i R_i^{-1} B_i^T + S_i^{-1} - A_{ii}^T)] g_i(t) + K_i(t) S_i^{-1} v_i(t) +$$

$$+ \sum_{\substack{j=1 \\ j \neq i}}^{N} A_{ji}^T v_j(t), \quad g_i(t_1) = 0. \tag{4.25}$$

Observe that eqs. (4.23) and (4.24) contain only variables which are related to the i-th subsystem; in other words the solution of the Riccati-equation (4.24) is independent of the coordination variable. The control of each subsystem for each fixed $g_i(t)$ is based on the well-known linear-quadratic optimization strategy. This is also advantageous for the interative numerical solution because the solution has to be calculated just once and can be afterwards stored in the computer memory. However, the vector $g_i(t)$ is dependent on the coordination variable $v_i(t)$ and has to be calculated at each iteration. The target of the coordinator is to determine the optimal coordination variable $v_i^0(t)$; under this condition the term of the optimal solutions of the subproblems with the modified performance criteria (4.16) fulfills also the solution of the global problem (necessary condition). Different coordinator strategies like gradient methods, contraction mapping and others can be used. In fig. 4 the local diagram of a two-level control system consisting of two interconnected subsystems is shown. It leads concerning the suboptimal control to a closed loop structure (Wend 1980).

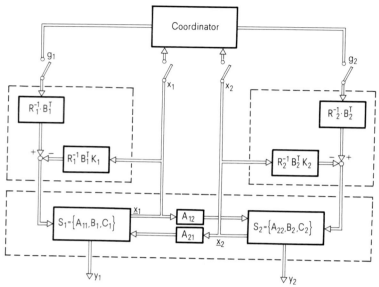

Fig. 4: Local diagram of a two-level control system consisting
of two interconnected subsystems

4.4   *Conclusions*

Both presented hierarchical concepts are, of course, of a special structure but they indicate the general idea, namely to split up a complex dynamical system in a structure with several decision units. This gives a better understanding or intuition of the system behaviour and how to control it. Besides the fact that real systems are sometimes hierarchically structured, a decentralized control concept is applicable which has several advantages. Due to the development of microelectronics the economic implementation of several (parallel operating) process computers is possible.

A number of contributions are dealing with very different decomposition- and coordination procedures and the extension of multi-level (-layer) system structures to different classes of (nonlinear) systems, see e. g. Findeisen et al. (1970), Singh and Titli (1978) and Siljak (1978).

In order to reduce for linear systems the necessary expenditure one could think of using order-reduction (model reduction) principles like e. g. Litz (1981). Another idea is to use the inclusion principle which has recently been proposed by Ikeda and Siljak (1980). It is based on the mathematical framework for expansions and contractions of dynamic systems. Since the subsystems share a common part of the overall system (overlapping subsystems), the original system is mathematical expanded into a larger-dimensional space where the overlapping subsystems appear as disjoint. In the expanded space the identity of the subsystems is preserved and their interconnections identified, so that the standard methods can be used for system analysis and design.

By looking at fig. 4 one can see that the two-level control concept has a so called "open-closed loop structure". The system is open loop concerning the coordination, but closed loop concerning the local optimization at the lower level. The switchs indicate on one hand that the optimal vector function $g_i(t)$ is calculated off-line and are then passed down to the lower layer. This is necessary because the interconnections in real systems cannot be disconnected. In this case the coordination procedure is a non feasible method which has to be realized with the help of a model for the subsystems (model-reference coordination). The modelling mainly with the help of process computers is by no means specific for multi-level control but it is a necessary condition for the implementation of modern control strategies (algorithms).

The optimal control of the subsystems leads in case of fig. 4 to a closed loop structure which is, of course, advantageous concerning disturbances etc. of the subsystem. By the way, the constant feedback $R_i^{-1} B_i^T K_i$ ($i = 1, 2$) is due to the assumption $t_1 = \infty$.

Compared with the orthodox design techniques for a centralized control strategy a number of additional questions arise like applicability of decomposition, coordinability and stability not only of the overall system and subsystems, but also in the model domain (process computer) and numerical stability for the coordination procedure; another stability concept, called connective stability has been considered by Siljak (1978) which is concerned with failures of connection paths in the real system. This all has to do with the *dynamic reliability* when hardware and software failures can occur either in parts of the plant or the decision units (controllers). Therefore reliability has become a major concern. This indicates that research into complex dynamic systems is wide open. However, without additional rigorous theoretical development, no real progress can be achieved.

4.5   References

Findeisen, W. et al., *Control and Coordination in Hierarchical Systems*, John Wiley & Sons, Chichester, New York, Brisbane, Toronto, 1980.

Ikeda, M., D. D. Siljak, *Overlapping Decompositions, expansions and contractions of dynamic systems*, Large Scale Systems **1**, (1980), pp. 29—38.

Lasdon, L. S., J. D. Schoeffler, *A multi-level technique for optimization*, Proc. JACC 1965, Troy, N.Y., pp. 85—92.

Lefkowitz, I., *Multilevel approach applied to control system design*, Proc. JACC, 1965, Troy, N.Y., pp. 100—109.

Lin, C. T., *Structural controllability*, IEEE Transactions **19**, (1974), pp. 201—208.

Litz, L., *Order reduction of linear state-space models via optimal approximation of the nondominant modes*, J. Large Scale Systems **2**, (1981), pp. 171—184.

Mesarovic, M. D. et al., *Theory of hierarchical multilevel systems*, Academic Press, New York, 1970.

Siljak, D. D., *Large-Scale Dynamic Systems—Stability and Structure*, North-Holland Publ. Co., New York, (1978).

Siljak, D. D., *Complex dynamic systems: Dimensionality, structure, and uncertainty*, Large Scale Systems **4**, (1983), pp. 279—294.

Singh, M. G., A. Titli, (eds.), *Handbook of large scale systems engineering*, North-Holland Publ., Co., Amsterdam, 1978.

Söte, W., *Strukturelle Methoden zur Dekomposition von Großsystemen*, Regelungstechnik **28**, (1980), pp. 37—44.

Thoma, M., *Verfahren zur Beschreibung und Optimierung hierarchischer Automatisierungssysteme*. In: Ernst, D., Thoma, M.: Meß- und Automatisierungstechnik, INTERKAMA-Kongreß 1980, Düsseldorf. Fachberichte Messen — Steuern — Regeln, Band 5, Springer-Verlag Berlin, Heidelberg, New York, 1980, pp. 380—409.

Wend, H. D., *On hierarchical control of complex technological systems*, J. Large Scale Systems — Theory and Applications **1**, (1980), pp. 63—75.

Wilson, I. S., *Foundations of Hierarchical Control*, Intern. J. Control, Vol. **29**, (1979), pp. 899—933.

5   Control of distributed-parameter systems (A. Munack)

5.1   *Modeling*

Distributed parameter systems (DPSs) are a class of systems which cannot be described exactly by ordinary differential equations of finite order, since the systems are continuously distributed in space as well as evolving in time. Therefore, infinite dimensional state spaces are used to describe these processes. Whereas classical problems of mechanics (beams, strings, membranes) were treated quite early, several technical applications have arisen recently which have increased the motivation to treat this class of systems from an engineering point of view. Moreover, the modern large computer centers offer the opportunity to carry out extensive simulation studies and investigations of optimal design and optimal control of these plants. In addition, the powerful modern microcomputers provide the tool to implement the sophisticated control algorithms based on the results of control theory for DPSs. To mention only a few applications which have been treated recently, we refer to modelling and control of river systems, flow in long pipes, and various types of reactors for nuclear, chemical, or biotechnological purposes. Large furnaces for the steel producing and steel processing industry have been treated as well as problems of semiconductor dotation.

In the following we will treat several aspects of control theory for DPSs which are relevant for applications.—A fundamental step in system and control engineering is the construction of a mathematical model of the plant. In case of a DPS, this usually leads either to partial differential equations (PDEs) or to integral equations. A PDE model is normally derived by writing down balance equations for (in the limit) infinitely small elements of the spatial domain. These balance equations are formulated using the physical principles of conservation of mass, momentum, and energy. In most practical cases, however, one has several unknown parameters in these equations; therefore parameter identification techniques must be considered in order to determine these coefficients.

A further treatment requires the solution of the system equations. If possible, an analytical expression is sought; if not, simulations will be tried. Therefore we will first discuss the possibilities of simulation using digital computers. This leads to the problem of finding adequate numerical approximation procedures, since for computation the DPS has to be approximated by a finite-order system of equations.

Next stages may be an optimal design study and the determination of a controller. Depending on the type of system involved, here nearly all the control problems and methods come into consideration which are correspondingly known from control of lumped-parameter systems. So assurance of stability is a fundamental task, and for systems with temporally varying parameters robust or adaptive control algorithms are desirable. Systems which are described by a set of coupled PDEs may be treated by methods developed in the framework of large scale systems theory. In case of incomplete information about the system's state, or noise perturbing the system or the measurements, observers or filters are applied.

In the following, we will treat theoretical results of *simulation, optimal control* and *parameter identification* in some more detail. These topics form the basis for other modern control techniques for DPSs, which we will outline briefly at the end of this chapter.

In the presentation we will restrict ourselves to a treatment of *parabolic PDEs*, since this type of equation is apparently most often found in applications. Systems incorporating the effects of diffusion, dispersion, and heat conduction can be modelled by this class of PDEs.

5.2   Simulation techniques

Various simulation procedures for parabolic PDEs are known from the literature; a classification of these can be made by dividing them into classes, where space- and time-domain are treated as continuous, discretized, or transformed. This leads to a maximum of nine classes, of which we will treat the most relevant for applications. This means that the transformed time methods will not be considered, since an on-line application of these or an interactive simulation is not feasible. Each of the methods will be evaluated by two considerations: Off-line simulations, i. e. calculations without temporal requirements and with the facilities of large computing centers, usually lead to the demand of a very *flexible* and *easy to use* program package. However, in on-line applications, e. g. needed for adaptive control or decision planning, or in case studies where the problem has to be solved frequently with different sets of parameters, *fast* routines are required. Under these two considerations, valuations of the available simulation techniques will be given in the following.

The *DSCT*-method (discrete space, continuous time) is the classical analog-computer method for simulation of PDEs, cf. e.g. Bekey/Karplus (1968). The spatial domain of the system is divided into a certain number of intervals, and at the boundaries between the intervals the spatial differential operator is substituted by a difference operator, which only uses values of the solution at the location under consideration and the adjacent node points. After inclusion of the boundary conditions one gets a coupled system of ordinary differential equations in time-domain, the solution of which approximates the solution of the PDE at the node points. — On analog computers the solution of the entire system of ordinary differential equations can be performed in parallel. For digital computers a lot of approximative procedures are known. A valuation with respect to the above formulated criteria yields the following results:

1. clear and flexible in the programming structure; digital simulation packages are available,
2. on analog computers extremely fast, but with fine discretization relatively complicated to patch, and on digital computers relatively slow.

The *CSDT*-method (continuous space, discrete time) is applicable on pure analog computers only if an analog memory is installed. Here the temporal differential operator is approximated by a difference operator. In this way, a sequence of spatial boundary value problems is formed, the solutions of which are approximations for the solution of the PDE at the corresponding time-instants. Due to the continuous treatment of the spatial variable the procedure is very well suited for problems with spatially varying parameters. The algorithm is readily implemented on a hybrid computer; for pure digital realizations approximation routines for ordinary differential equations are used as in the DSCT-method. However, for problems of parabolic type direct integration of the two-point boundary value problems usually leads to severe difficulties, since the equation is not integrable in the forward nor in the backward direction without unstable error-propagation. A method proposed by Vichnevetsky (1968) circumvents this difficulty by decomposing the problem into one forward integrable and one backward integrable subsystem. There the final solution in each time step is formed by a superposition of three partial solutions, two of which can be precomputed for standard (linear, constant parameters) cases. Coupled equations and equations with nonlinear or spatially varying parameters usually require iterative methods for solution.

Valuation:
1. programming relatively complicated; simulation package not known,
2. on a hybrid computer very fast, particularly in the case of coupled PDEs; on digital computers relatively slow.

As *DSDT*-method (discrete space, discrete time) the commonly used difference-approximations are known. There both the temporal as well as the spatial differential operator are approximated by a suitable difference operator. At every time-step one gets a system of algebraic equations, the solution of which approximates the solution of the PDE at this time-instant at the node points. Detailed information is provided by the book of Richtmyer/Morton (1967).

Valuation:
1. programming very easy and clear; programming packages available,
2. relatively slow, particularly in the case of systems of equations.

In applications relatively restricted are the Monte-Carlo-methods which generally also use a grid-like discretized time-space-domain. Due to the complete different basic idea of this method it is not classified as a DSDT-method.

TSCT/TSDT-methods (transformed space, continuous/discrete time) are the last methods considered here. The spatial profile of the solution of the PDE is approximated by a time-dependent (linear) combination of given coordinate or basis functions, which are usually members of a set of orthogonal functions in space domain. Insertion of these basis functions into the differential equation results in a spatial error function, the so-called equation residual. Correspondingly a boundary and an initial residual can be defined. In most cases the so-called "interior method" is used, where the boundary residual is identical to zero by appropriate construction of the basis. The inner product of the equation residual with a suitable spatial weighting function is finally used to determine the temporal derivatives of the coefficient functions for the corresponding basis functions. In the case of discretized time domain, a system of algebraic equations for the coefficients of the next time step is formed in the same way. —A lot of well-known and widely used algorithms can be assigned to this class of "functional approximation methods". So the Galerkin-methods use as weighting functions just the basis functions, and taking δ-functions one gets the collocation methods. Other basis functions lead to the spline- or finite-element-approximation and to the modal simulation.

Having in mind this variety of possibilities it is impossible to give a valuation in general; so that following statements may certainly not be true in every special case.

Valuation:
1. programming simple, if simulation packages are used, otherwise except for collocation methods relatively time-consuming,
2. on analog- or hybrid computers extremely fast if implementable (normalization problems!); on digital computers relatively fast; in the case of nonlinear or coupled problems sometimes great reduction in velocity.

From the above stated valuations one can draw the conclusion, that for a fixed problem the TSCT- or TSDT-method would be preferable for our purposes. Compared with the discretizing methods, the lower flexibility then will be no

disadvantage, whereas the low computation time offers great advantages. —
During tests with a lot of modifications in the used model however, it may be
preferable to take the simple and very flexible DSDT-methods which have
usually automatic mesh size control and some other very pleasant characteristics
if a simulation package is available.

5.3   *Optimal control*

Contributions to this field have been made by various scientists. Among
these are several who treat the problem using integral equations, cf. Butkovskiy
(1969) or the semigroup approach developed by Curtain/Pritchard (1978).
Widely used results using differential equations have been obtained by Lions
(1971) and his coworkers.

These methods apply to cases, where the system description can be made
in the form (the temporal derivative being a partial derivative)

$$\frac{dy(t)}{dt} + A(t)\,y(t) = f(t) + B(t)\,v(t), \quad \text{in} \quad]0,\,T[, \tag{5.1}$$

$$y(0) = y_0,$$

where A is the system operator, which means that Ay contains all spatial
derivatives of the system state y. A is an elliptic differential operator, which
assures that the whole PDE is of parabolic type. The action of the control v on
the system is performed through the input operator B, and all disturbances are
combined in the disturbance function f. Several further assumptions have to be
made concerning the solution spaces and the operators involved. We will not go
into detail here and refer to the literature.

Now the optimal control problem will be defined. In most applications,
the system states should follow almost precisely a prescribed trajectory, and —
on the other hand — this aim should be achieved using only a small amount of
energy. This leads to the commonly used quadratic functional

$$J(v) = \int_0^T \|Cy(v) - z_s\|_F^2 \, dt + \int_0^T (Gv, v)_E \, dt, \tag{5.2}$$

where C is an output operator and G gives the weighting of control energy; z_s
denotes the desired trajectory. The aim is to find an optimal control u which
minimizes the cost functional, such that

$$J(u) \leqq J(v) \quad \forall\, v \in U_{ad}. \tag{5.3}$$

The solution of this problem is given in Lions (1971) by means of a variational inequality which characterizes the minimizing element u. After definition of the adjoint state by

$$-\frac{dp(v)}{dt} + A'p(v) = C'\Lambda_F(Cy(v) - z_s) \quad \text{in} \quad]0, T[,$$ (5.4)

$$p(t = T, v) = 0,$$

this variational inequality reads

$$\int_0^T (\Lambda_E^{-1} B'p(u) + Gu, v - u)_E \, dt \geq 0 \quad \forall v \in U_{ad}; \quad u \in U_{ad}.$$ (5.5)

If there are no constraints for the control function u and, if furthermore, some smoothness conditions are fulfilled for the right-hand side of equation (5.1), then the optimal control function can be computed by a feedback law

$$u = -G^{-1}\Lambda_E^{-1} B'p(u) = -G^{-1}\Lambda_E^{-1} B'(Py + r),$$ (5.6)

where $P(t)$ is the solution of the operator Riccati differential equation

$$-\frac{dP}{dt} + PA + A'P + PBG^{-1}\Lambda_E^{-1} B'P = C'\Lambda_F C \quad \text{in} \quad]0, T[,$$ (5.7)

$$P(T) = 0,$$

and $r(t)$ is given by

$$-\frac{dr}{dt} + A'r + PBG^{-1}\Lambda_E^{-1} B'r = Pf - C'\Lambda_F z_s \quad \text{in} \quad]0, T[,$$ (5.8)

$$r(T) = 0.$$

With this, we have stated some basic results of optimization theory from the work of Lions.

The solution of this classical linear-quadratic optimal control problem has been widely used and/or adapted to more specialized cases. Various numerical methods have been proposed to solve the optimality system (which consists of the state equation (5.1) and the adjoint equation (5.4)) or the operator Riccati differential equation. Also extensions to the case where there are constraints on the state or constraints on the control have been made. These problems may be solved by duality or penalization methods.—Some nonlinear problems have been attacked. It could be shown that an optimality system also exists in some of these cases, which leads to computational algorithms for minimization of the cost functional. In contrast to the linear case, however, the optimal control may be non-unique. — Further extensions of the classical situation are problems of optimal design. Here the control variable is a domain and therefore does not belong to a Hilbert or Banach space. Another situation occurs e. g. in melting processes, where the functional may be non-differentiable with respect to the control. For some special biochemical processes, problems arise where the state equation does not admit a unique solution. However, also in these cases an optimality system may be formulated which allows the minimization of the cost functional.

5.4   Parameter identification

The determination of unknown system parameters as the last stage of system modelling is an essential step in the process of designing adequate control strategies for a given system. A very rigorous approach to this problem was derived by Chavent using optimal control theory. To apply these results, the parameter identification problem has to be formulated in terms of optimal control.

Let the system be described by

$$\frac{d y_S}{d t} + A_S y_S = f_S + B_S u \quad \text{in} \quad]0, T[, \tag{5.9}$$

$$y_S(0) = y_{S_o}.$$

All assumptions on operators and spaces are the same as in the foregoing section. In order to measure the system's state, k sensors may be connected to it, which yield measurements

$$z_j(t) = \int_0^1 S_j(x, t) \, y_S(x, t) \, dx, \quad j = 1, \ldots, k \tag{5.10}$$

where the (in this case one-dimensional) spatial domain has been normalized to $[0, 1]$, and $S_j(x, t)$ are the sensor characteristics. Identification is performed by means of a system model given by

$$\frac{d y_M}{d t} + \hat{A} y_M = f + \hat{B} u \quad \text{in} \quad]0, T[, \tag{5.11}$$

$$y_M(0) = y_{M_o}.$$

In most cases it is assumed that $y_{M_o} = y_{S_o}$ holds; otherwise one can either identify the initial state, too, or one can use a time-variant weighting. Anyway, if the system is stable and the observation time T is long enough, then the effect of the initial state will diminish with increasing time.

After application of the sensor characteristics onto the model's state, it is possible to define the errors between system and model as

$$e_j(t) = z_j(t) - \int_0^1 S_j(x, t) \cdot y_M(x, t)\, dx, \quad j = 1, \ldots, k.$$ (5.12)

A common weighting of all errors is given by the quadratic functional

$$J_I = \int_0^T \sum_{j=1}^k W_j(t) \cdot e_j^2(t)\, dt,$$ (5.13)

where

$$W_j(t) \geqq 0 \quad \forall\, t \in [0, T].$$ (5.14)

The parameter identification problem is now formulated as an optimization problem. We want to determine estimated parameters $\hat{A}_{opt}, \hat{B}_{opt}, \hat{f}_{opt}$, such that

$$J_I(\hat{A}_{opt}, \hat{B}_{opt}, \hat{f}_{opt}) \leqq J_I(\hat{A}, \hat{B}, \hat{f}) \quad \forall\, \hat{A}, \hat{B}, \hat{f}.$$ (5.15)

A necessary condition for optimality is that the first variation of the functional becomes zero, that is

$$\delta J_I(\hat{A}_{opt}, \hat{B}_{opt}, \hat{f}_{opt}) = 0.$$ (5.16)

Using formal arguments this first variation is calculated as follows

$$\delta J_I(\hat{A}, \hat{B}, \hat{f}) = \int_0^T (p, \delta \hat{A} y - \delta \hat{B} u - \delta f)\, dt,$$ (5.17)

with the adjoint state p given by

$$-\frac{dp}{dt} + \hat{A}'p = 2 \cdot \sum_{j=1}^k W_j S_j e_j \quad \text{in} \quad]0, T[.$$ (5.18)

$$p(T) = 0.$$

In practical applications it is advisable to calculate the variation of the functional (5.17) directly. The adjoint state used for solution of the parameter identification problem has the same left-hand side of the underlying PDE as the adjoint state in the optimal control problem. Therefore both equations are of the same type and can be solved by the same simulation procedure.

Since the (infinite dimensional) gradient of the functional with respect to the parameters (or coefficient functions) is given by (5.17), highly efficient gradient methods can be used to compute the optimal parameters. Using numerical approximations of the second derivatives, very satisfactory results have been obtained with Newton techniques.

The approach presented above uses results of optimal control theory for DPS and therefore stays in the original state space as long as possible. Approximations need not be introduced until solving the optimality system. Other parameter identification methods introduce approximations in a far earlier stage, i.e. a finite order representation is sought both for system and model. Then minimization of the functional is carried out in the framework of optimization theory for lumped-parameter systems. A rigorous mathematical treatment of such approximations has been published recently by Banks and Kunisch. The original problem is projected onto a sequence of subspaces which are generated by spline functions. These subspaces may be contained in the domain of the system operator; however, also schemes where this condition is not satisfied have been proven to converge and to work highly satisfactory.

In the previous paragraphs we have discussed very briefly three fundamental topics of scientific research work on control of distributed-parameter systems. Deep theoretical work, like stability results, the design of finite-dimensional compensators, the theory of stochastic systems with distributed parameters, the attempts to establish a structural theory, the use of differential geometrical methods, and other fields of research could not be included in this overview.

5.5   References

The following references only contain some fundamental work on control of DPS. For an overview on actual research in this discipline the reader is referred to

Babary, J.-P., Le Letty, L. (eds.), *Proceedings of the 3rd IFAC Symposium on Control of Distributed Parameter Systems*. Pergamon, Oxford and New York, 1983,

and the literature cited therein. The articles of Banks and Kunisch may also be found in these proceedings.

Bekey, G. A., W. J. Karplus, *Hybrid Computation*, Wiley & Sons, New York, 1968.
Butkovskiy, A. G., *Distributed Control Systems*, Elsevier, New York, 1969.
Chavent, G., *Identification of Functional Parameters in Partial Differential Equations*. In: *Identification of Parameters in Distributed Systems*, ASME, New York, 1974.
Curtain, R. F., A. J. Pritchard, *Infinite Dimensional Linear Systems Theory*, Lecture Notes in Control and Information Sciences **8**, Springer, Berlin, 1978.
Lions, J. L., *Optimal Control of Systems Governed by Partial Differential Equations*, Springer, Berlin, 1971.
Richtmyer, R. D., K. W. Morton, *Difference Methods for Initial-Value Problems*. Wiley & Sons, New York, 1967.
Vichnevetsky, R., *A New Stable Computing Method for the Serial Hybrid Computer Integration of Partial Differential Equations*, Proceedings of the SJCC, 1968, pp. 143—150.

Perspectives in Mathematics
Anniversary of Oberwolfach 1984
© Birkhäuser Verlag, Basel

Geometry in Total Absolute Curvature Theory[1])

N. H. KUIPER

Institut des Hautes Etudes Scientifiques, 35, route de Chartres,
F-91440 Bures-sur-Yvette (France)

Summary

This is a (incomplete) report on geometrical results of the last twenty five years in the theory of total absolute curvature, in particular concerning its minimal value in certain classes of embeddings of manifolds.

1   Definitions and General Problems

The analytic definition of total curvature. The unit normal vectors of a connected closed smooth n-manifold $M, n \geq 0$ embedded by $f : M \to \mathbb{R}^N$ into euclidean N-space, form a smooth $N-1$-manifold Q of dimension $N-1$. Parallel transport of these unitvectors to the origin $0 \in \mathbb{R}^N$ yields a map

$$v : Q \to S^{N-1} \subset \mathbb{R}^N$$

into the unit $N-1$-sphere. If its volume-form is denoted ω_{N-1}, then the total absolute curvature is defined by

$$\tau_a(f) = c \cdot \int |v^*(\omega_{N-1})|, \ c^{-1} = \text{vol } S^{N-1}. \tag{1.1}$$

where $v^*(\omega_{N-1})$ is the pullback of ω_{N-1} by v. The constant c is such that $\tau(\text{point}) = 1$.

An equivalent *geometric definition*, denoted $\tau(f)$, which however works for a broader class of subsets of $\mathbb{R}^N$ is as follows:

For any oriented line L through $0 \in \mathbb{R}^N$, let π_L denote orthogonal projection into L. The function $\pi_L : \mathbb{R}^N \to L$, where L is identified with $\mathbb{R}$ by an oriented euclidean coordinate (scale) on L, can be considered as *one* coordinate z of a system of euclidean coordinates in $\mathbb{R}^N$. We will often call a line L *vertical* and the corresponding function z *height*. For almost all lines L the function

---

[1]) Gratefully presented for the Oberwolfach anniversary volume 1984.

$\pi_L \circ f : M \to L$ is non-degenerate. Let the number of critical points be $\mu_L = \mu(\pi_L \circ f)$. Then the total (absolute) curvature is the *mean value* (expectation value $\mathscr{E}$):

$$\tau(f) = \mathscr{E}_L \mu_L. \tag{1.2}$$

For smooth manifolds

$$\tau(f) = \tau_a(f).$$

In particular we have in customary notations:

$$\begin{aligned}
\tau(f) &= \int |\varrho \, ds|/\pi && \text{for } \textit{curves} \text{ in } \mathbb{R}^N \\
&= \int |K \, d\sigma|/2\pi && \text{for } \textit{surfaces} \text{ in } \mathbb{R}^3 \\
&= c \int |K^*(p) \, d\sigma| && \text{for smooth manifolds,}
\end{aligned}$$

where $K^*(p)$ is a curvature density related to Lipshitz-Killing curvature and $d\sigma$ the volume element of $M \subset \mathbb{R}^N$. We see that this definition applies to other subsets $f : X \to \mathbb{R}^N$ of $\mathbb{R}^N$ under suitable definitions of critical point and its multiplicity as follows.

Consider, $\varphi = z \circ f : M \to \mathbb{R}^N$, a nondegenerate smooth function as before, but with any two critical points assumed at different levels. If we let

$$M_t = \{q \in M : \varphi(q) \leq t\}, \tag{1.3}$$

then by the geometry of the situation a point $p \in M$ is *a critical point if and only if* in *Čech-homology over* $\mathbb{Z}_2$:

$$H_*((M_t, M_{t-\varepsilon}; \mathbb{Z}_2) = \mathbb{Z}_2$$

for $t = \varphi(p)$ and small $\varepsilon > 0$. Equivalently the sum of the Bettinumbers $\beta(M_t)$ changes by *one* (plus or minus) at the critical value $t = \varphi(p)$ and does not change at non-critical values, for increasing t.

Taking *this property as definition* of critical point we see immediately that for example for an embedded compact *convex set* $f : X \to \mathbb{R}^N$ one has $\mu_L(f) = 1$ for all L, and so $\tau(f) = \mathscr{E}_L \mu_L(F) = 1$ is well defined. The class of compact connected sets for which the definition can apply includes also all finite polyeders. See [23] for precise definitions.

Returning to smooth n-manifolds M and smooth embeddings, we denote the *minimal number of critical points* a smooth nondegenerate function $\varphi : M \to \mathbb{R}$ can have by $\gamma = \gamma(M)$, and the sum of the $\mathbb{Z}_2$-*Bettinumbers* by $\beta = \beta(M) = \sum_i \beta_i(M)$.

By definition (1.2), and by the Morse inequalities, we conclude

$$\tau(f) \geqq \gamma \geqq \beta.\tag{1.4}$$

The infimum of $\tau(g)$ for all embeddings g of M into euclidean spaces is known to be[1])

$$\gamma(M) = \gamma \geqq \beta.\tag{1.5}$$

For closed curves and surfaces $\gamma = \beta$.
    The infimum of $\tau(g)$ for all embeddings g that are isotopic to f ($g \in [f]$, the isotopy class of f) is denoted

$$\tau[f] = \inf_{g \in [f]} \tau(g).$$

Our general problem is whether these infimums can be obtained under various conditions and how the embeddings then look.

Variants of this problem are expressed with the following definitions. The embedding f has *minimal total absolute curvature* if

$$\tau(f) = \gamma(M) = \gamma\tag{1.6}$$

Tight submanifolds and sets
    The embedding f is called *tight* if $\gamma = \beta$ *and* the mean value attains the infimum:

$$\mathscr{E}_L(\mu_L) = \tau(f) = \gamma(M) = \gamma = \beta.\tag{1.7}$$

If f is tight, then it does have minimal total absolute curvature, but not vice versa (See §9). Recall that the sum $\beta(M_t)$ of the $\mathbb{Z}_2$-Bettinumbers of the subspace

$$M_t = \{q \in M : z \circ f(q) \leqq t\},\tag{1.3}$$

for nondegenerate $z \circ f = \pi_L \circ f$, changes with increasing t only at the β critical points and for each by *one*. In order to attain the number $\beta = \beta(M)$ in β steps, each change must be an *increase*! This implies that the inclusion $M_t \subset M$ de-

---

[1]) *Question.* Is the infimum the same for embeddings of M into the euclidean space $\mathbb{R}^N$, for exactly the smallest value N, for which embeddings exist? Not so for exotic spheres!

termines injectivity for the homomorphism in homology:

$$H_*(M_t; \mathbb{Z}_2) \to H_*(M; \mathbb{Z}_2).$$

Observe also that $M_t = M \cap h$, is the intersection of $M \subset \mathbb{R}^N$ with the *half space* $h = \{r \in \mathbb{R}^N : z(r) \leqq t\}$. These observations lead to the *alternative*
    Definition (theorem). The embedding $f: M \subset \mathbb{R}^N$ is *tight* if and only if in Čech-homology

$$H_*(M \cap h; \mathbb{Z}_2) \to H_*(M; \mathbb{Z}_2) \text{ is injective} \tag{1.8}$$

for *every* half space $h \subset \mathbb{R}^N$. Then $\tau(f) = \beta(M)$.

    Remark. *For   closed   surfaces   tightness   is   already   sure   if* $H_0(M \cap h; \mathbb{Z}_2) \to H_0(M; \mathbb{Z}_2) = \mathbb{Z}_2$ *is injective, that is if and only if $M \cap h$ is connected, for all half spaces h.*
    The euclidean space $\mathbb{R}^N$ can be considered as the complement $\mathbb{R}P^N \setminus \mathbb{R}P^{N-1}$ of a hyperplane $\mathbb{R}P^{N-1}$ in the real projective space $\mathbb{R}P^N$. If a projective transformation of $\mathbb{R}P^N$ onto itself sends M back into $\mathbb{R}^N$, we call this a projective transformation of the embedding. *Tightness* is, by definition (1.8), a *projective property* in the sense that it is invariant under projective transformations.
    Tightness by its definition makes sense for compact subsets $f: X \subset \mathbb{R}^N$ which are not necessarily smooth manifolds. For example, every compact convex set X is clearly *tight with* $\tau(f) = 1$. Also, the boundary of any convex set $X \subset \mathbb{R}^k \subset \mathbb{R}^N$ with interior points in $\mathbb{R}^k$ is a tight topological embedding of a $k - 1$-sphere ∂X, with $\tau(f) = 2$.

Taut submanifolds and sets
    A tight submanifold (compact set) of the round (unit) N-sphere S^N in $\mathbb{R}^{N+1}$, is called *taut*. Intersections of half spaces of $\mathbb{R}^{N+1}$ with S^N are round N-balls in S^N. The projective transformations of $\mathbb{R}P^{N+1}$ which leave invariant S^N, $S^N \subset \mathbb{R}^{N+1} = \mathbb{R}P^{N+1} \setminus \mathbb{R}P^N$ induce conformal or Moebius transformations for S^N, and all are so obtained. With the definition (1.7), we get the

    Alternative definition. The embedding of a connected compact set, $f: X \subset S^N$, is *taut* if and only if

$$H_*(X \cap b) \to H_*(X)$$

is injective for every round ball $b \subset S^N$.

    Corollary. *Tautness is a conformal property.* It is invariant under Moebius transformations.

Remark. Euclidean space $\mathbb{R}^N$ is conformally equivalent (stereographic projection) to $S^N \setminus \{\text{point}\}$. Compactify $\mathbb{R}^N$ with one point ∞:

$$S^N = \mathbb{R}^N \cup \infty.$$

A compact set in $\mathbb{R}^N$ is called taut in case it is taut in $\mathbb{R}^N \cup \infty = S^N$. *Taut sets in $\mathbb{R}^N$ are* clearly *tight in $\mathbb{R}^N$,* because half spaces compactified by the point ∞, are round balls in S^N.

Isotopy tight submanifolds
The embedding $f: M \to \mathbb{R}^N$ is called *isotopy tight* in case

$$\tau(f) = \tau[f].$$

This concerns knotted submanifolds. For closed surfaces in $\mathbb{R}^3$ isotopy tightness is again a *projective property*, as we will see in §9.

2   The Theorem of Chern-Lashof

Theorem [8]. *If $f: M \subset \mathbb{R}^N$, a smooth embedding of a closed n-manifold, $n \geq 1$, in euclidean N-space, has total curvature $\tau(f) = 2$, then M is an n-sphere, boundary of a convex body in an $n+1$-plane of $\mathbb{R}^N$.*

The original proof used the calculus of differential forms. Our proof [20, 24] leads us outside differential analysis, with respect to the methods (we use convex sets for example) and with respect to *the conclusion, which holds equally well for topological embeddings of n-manifolds.*

Proof. As $\tau(f) = 2$ is a mean value, $\mathscr{E}_L \mu_L$, and as $\pi_L \circ f$ has at least two critical points for almost all L, we conclude $\mu_L = \mu(\pi_L \circ f) = 2$, for almost all lines L; $\tau(f) = \beta = 2$, and f is *tight*. As M admits a function with exactly two critical points, it is homeomorphic to S^n.

By tightness any half space h induces injectivity in

$$H_*(h \cap M) \to H_*(M).$$

As $H_i(M, \mathbb{Z}_2) = \mathbb{Z}_2$ for $i = 0, n$, and equals 0 otherwise, then $H_*(h \cap M) = H_0^\circ(h \cap M) = \mathbb{Z}_2$, and $\beta(h \cap M) = 1$, if $\emptyset \neq h \cap M \neq M$. Now take for h an M-*supporting half space*. This means *by definition*:

$$\emptyset \neq M' = h \cap M = \partial h \cap M, \quad \mathring{h} \cap M = \emptyset$$

where ∂h is the hyperplane bounding h and $\mathring{h}$ is the interior of h. The set M' is called a *top-set* of M. By studying limits of half-space intersections with M, one proves the

Fundamental lemma [24]. *Top-sets of compact tight sets in euclidean space are themselves tight.*

So M' is *tight*, and

$$\tau(M') = \beta(M') = \beta(h \cap M) = 1, \quad \text{if } M' \neq M.$$

Assume $M \subset \mathbb{R}^N$ spans $\mathbb{R}^N$. Then *every* top-set M' of M is tight with $\tau(M') = 1$, and so is every top-set of M' etc. Assume by induction that all top-sets M'' of a set M' with $\tau(M') = 1$ are *convex*. These top-sets M'' then *fill* the boundary (∂) of the *convex hull* $(\mathcal{H})$ of M' in the (say) k-plane $\xi(M')$ which spans M'. So M' contains the $k-1$-sphere

$$\partial \mathcal{H} M' \subset M' \subset \xi(M') \subset \mathbb{R}^N.$$

As $\tau(M') = \beta(M') = 1$, the $k-1$-sphere $(k \geq 1)$ $\partial \mathcal{H} M'$ must be bounding in $M' \subset \mathcal{H}(M')$. Then every point of $\mathcal{H}(M')$ must belong to (M') and M' is convex.

Applying this method to M, we find a convex hypersurface contained in M:

$$\partial \mathcal{H} M \subset M \subset \xi(M) = \mathbb{R}^N.$$

Then $\partial \mathcal{H} M$ has dimension $N-1 \leq n$ and so $N = n+1$. The n-sphere $\partial \mathcal{H} M \subset M$ must coincide with M, which is the required conclusion. Observe that we proved also the

Lemma. *A compact subset $f: X \subset \mathbb{R}^N$ is convex if and only if $\tau(f) = 1$ or equivalently $\mu(\pi_L \circ f) = 1$ for all lines L.*

Remark. A bounded set in Hilbert space need not be convex if $\mu(\pi_L \circ f) = 1$ for all lines L. Take a convex set with interior points and delete some disjoint open convex sets from the interior.

3   Smooth Tight Submanifolds Have Small Substantial Codimension

The highest possible codimension is only attained for particular unique submanifolds. Here the embedding $f: M \to \mathbb{R}^N$ is called *substantial* if $f(M)$ spans $\mathbb{R}^N$.

Theorem. *If $f: M^n \to \mathbb{R}^N$ is a smooth tight, substantial embedding of a closed n-manifold, then $N \leq n(n+3)/2$.*

Equality is only obtained if $M = \mathbb{R}P^n$, $n \geq 0$, the real projective n-space and $f(M)$ is the Veronese manifold $V \subset E \subset \mathbb{R}(n+1)^2$ (See [24]) (unique up to projective transformations) of E defined by

$$V = \{A = \{a_{ij} \in \mathbb{R}, \, i, j = 0, \ldots, n\} :$$
$${}^t A = A = A^2, \text{ trace } A = 1 = \text{rank } A\} \tag{3.1}$$
$$E = \{A : {}^t A = A, \text{ trace } A = 1\} \text{ of dimension } n(n+3)/2.$$

If the sequence of Betti numbers $\{\beta_0, \ldots, \beta_n\}$ has lacuna, we can do better in the first part [20]. See §2 where $\beta_1 = \cdots = \beta_{n-1} = 0$ and then $N \leq n+1$!

On the proofs: The first part of the theorem (Kuiper [18]) rests on a simple local observation: For any normal vector v at a maximal height point $p \in M \subset \mathbb{R}^N$, consider the inner product $\langle v, q \rangle$ for $q \in M$. Call the 2-jet at p of this function $J^2(\langle v, q \rangle)$. It is a quadratic form at p in n local coordinates. The relation

$$v \to J^2(\langle v, q \rangle)$$

is linear and by local arguments for tightness must be injective:

$$N \leq n + n(n+1)/2 = n(n+3)/2.$$

The second part is a considerable geometrical piece of work using local differential analysis. The author obtained this for surfaces, $n = 2$ [19], and Little and Pohl [29] for n-manifolds in general, $n \geq 2$.

Important remark. Full tightness is not needed for the proof. If $h \cap M$ is connected for all h then the same conclusions hold already!

Conclusion. Tightness with high codimension can have dramatic consequences of uniqueness and projective rigidity.

Problem. *Are these conclusions false or true for embeddings of differentiability class C^1?* This would need a truly new approach.

4   Tight Piecewise Linear Surfaces Can Have High Codimension

T. Banchoff studied curvature for piecewise linear submanifolds of $\mathbb{R}^N$ and found many tight surfaces of high codimension. Here are examples [1, 2]:

a) projective
plane

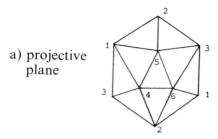

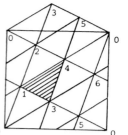

b) torus

Fig. 1

Take a triangulation of $\mathbb{R}P^2$ with six vertices. Place the vertices in general position in $\mathbb{R}^5$ and fill in the plane triangles. This is Banchoff's tight PL-projective plane subtantially in $\mathbb{R}^5$.

In the flat *torus* $\{(x, y),\ x \bmod 1,\ y \bmod 1\}$ take the seven points

$$(x, y) = (k/7,\ 3k/7)\ \ k = 0,\ \ldots, 6.$$

They are the vertices $e_0, \ldots, e_6$ of a triangulation by flat triangles. Place $e_0, \ldots, e_6$ in general position in $\mathbb{R}^6$ and fill in the plane triangles. This is Banchoff's tight PL-torus in dimension 6.

Let M_g be a surface of genus $g = (p-2)(p-3)/12$. Then there is a triangulation of M_g with $p+1$ vertices (Ringel [35]). Map these vertices in those of a p-simplex σ_p in $\mathbb{R}^p$ and fill in the plane triangles. You get a tight imbedding $f: M_g \to \mathbb{R}^p$ (Banchoff) and $f(M_g)$ contains the 1-skeleton $Sk_1(\sigma_p)$ and is contained in the 2-skeleton $Sk_2(\sigma_p)$:

$$Sk_1(\sigma_p) \subset M_g \subset Sk_2(\sigma_p) \subset \sigma_p$$

$(g, p) = (1,6)$ yields the torus again
$(g, p) = (6,10)$ is the next example, namely of M_6 in $\sigma_{10} \subset \mathbb{R}^{10}$.

Kühnel [16] obtained the highest codimension for tight PL-embeddings of all closed surfaces, and related results for surfaces with boundary.

5   Tightness Can Force Topologically Embedded Surfaces to Be Algebraic or Piecewise Linear and very Special

We mention two theorems

Theorem (Kuiper-Pohl [27]). *Let $f: M \subset \mathbb{R}^5$ be a substantial topological embedding of the real projective plane in $\mathbb{R}^5$, such that $h \cap M$ is connected for every halfspace $h \subset \mathbb{R}^5$. (tightness!). Then M is either the algebraic Veronese surface (§3) or Banchoff's PL-model with six vertices. (§4).*

Theorem (Pohl [34]). *Let $f: M \subset \mathbb{R}^6$ be a substantial topological embedding of the 2-torus in $\mathbb{R}^6$, such that $h \cap M$ is connected for every half space $h \subset \mathbb{R}^6$ (tightness!). Then M is Banchoff's PL-model with seven vertices.* (§4).

The proofs are hard and long. They rest on the analysis of top-sets of M. These top-sets are essential cycles if not convex, and then certain intersections are nonempty by homology.

Question. Are the tight examples for $M_g \subset \mathbb{R}^p$ with $(p-2)(p-3) = 12g$ in §4 unique for topological tight embeddings?

6   Kühnel's 9-Vertex Complex Projective Plane

The author proved in [24] among others

Theorem. *Let $f: \mathbb{C}P^2 \to \mathbb{R}^8$ be a smooth tight substantial embedding of the complex projective plane. Then $f(\mathbb{C}P^2)$ is the standard Veronese algebraic model for $\mathbb{C}P^2$ in $\mathbb{R}^8$, unique but for real projective transformations in $\mathbb{R}^8$.* (Take $a_{ij} \in \mathbb{C}$, $n = 2$, and $^tA = \bar{A}$ instead of $^tA = A$ in (3.1).)

The author conjectured that topological embeddings would give the same conclusion. But what a great surprise it was when Kühnel [17] found a triangulation of $\mathbb{C}P^2$ with 9 vertices that embeds tightly and affine on each simplex, into the 4-skeleton of the simplex $\sigma_8 \subset \mathbb{R}^8$, and contains the 2-skeleton:

$$Sk_2(\sigma_8) \subset \mathbb{C}P_k^2 \subset Sk_4(\sigma_8);$$

Banchoff and Kühnel wrote an excellent article on this phenomenon in the Intelligencer [17]. The subject developed much during differential geometry meetings in Oberwolfach.

Brehm and Kühnel [3] also studied and obtained approximation of polyhedral surfaces M_{PL} in euclidean three space by smooth surfaces M_t such that $\tau(M_t)$ converges to $\tau(M_{PL})$ for $t \to \infty$. It would be interesting to have such approximations for manifolds in $\mathbb{R}^N$. *Can Kühnel's tight $\mathbb{C}P_k^2$ in $\mathbb{R}^8$ be approximated by smooth manifolds M_t with $\tau(M_t) > 3$ converging to the limit infimum $\tau(\mathbb{C}P_k^2) = 3$?* There also remains the interesting

Question. Is any topological tight substantial embedding $f: \mathbb{C}P^2 \to \mathbb{R}^8$ either onto the Veronese model or onto Kühnel's model?

7   Existence of Tight and Taut Submanifolds

Given a closed manifold M one cannot in general expect it to admit a tight embedding or immersion, or even less a taut embedding. All orientable surfaces have tight embeddings in $\mathbb{R}^3$. The non-orientable ones have tight immersions in $\mathbb{R}^3$, except for a) the projective plane $\mathbb{R}P^2$ ($\chi = 1$) and b) the Klein bottle $K(\chi = 0)$, not even if one permits the immersion to be topological [22], and c) the surface with Euler characteristic $\chi = -1$, for which the question has been completely open for long.

There is a piecewise linear (Banchoff [2]) but so far no smooth tight embedding of K in $\mathbb{R}^4$. There is a piecewise linear and there should be a smooth tight embedding of the surface with $\chi = -1$ in $\mathbb{R}^4$ but no proof is written.

In higher dimensions as soon as the fundamental group $\pi_1(M)$ (resp $\pi_i(M)$) has more generators then $H_1(M, \mathbb{Z}_2)$ (resp. $H_i(M, \mathbb{Z}_2)$), for example for a homology 3-sphere different from S^3, then $\gamma > \beta$ and we can at most hope for embeddings of minimal total curvature but not for tightness. It seems likely that also "most" simply connected manifolds of dimension $n > 2$ have no tight embeddings.

Even so, there is a wealth of examples of tight manifolds, namely by homogeneous spaces like $SO(n) \subset \mathbb{R}^{n^2}$, all projective spaces (real, complex etc.), all compact homogeneous Kähler manifolds, and "R-spaces" (Kobayashi-Takeuchi [15]). All these models are placed in a round hypersphere and are therefore taut as well.

8   Taut Submanifolds and Taut Sets

Tautness is a much more restrictive property than tightness. For *closed surfaces* M ($n = 2$) in $\mathbb{R}^N$ it means that any round ball in $\mathbb{R}^N \cup \infty = S^N$ meets M in a connected set. Equivalently $M \setminus S^{N-1}$ has at most two components for any round $N-1$-sphere S^{N-1}.

This is Banchoff's spherical two pieces property. He proved that a smooth taut closed surface in $\mathbb{R}^N$ (which is then tight as well!) is a round 2-sphere, a (round) Veronese surface ($\subset S^4 \subset \mathbb{R}^5$), or a Dupin cyclide surface ($\subset S^3 \subset \mathbb{R}^4$).

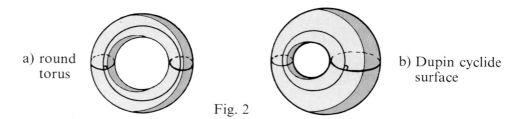

a) round torus   b) Dupin cyclide surface

Fig. 2

The last is an embedded torus with two orthogonal transversal foliations by round circles. It is a Moebius group transform of a standard torus example in

$\mathbb{R}^3$. *Thus among closed surfaces only the projective plane* $\mathbb{R}P^2$, *the sphere and the torus admit taut embeddings.* [2a]

The author [21] studied *connected compact taut* ANR-*sets* in $\mathbb{R}^3$ and found the

Theorem: The *only case is again the Dupin cyclide surface* if one does not count the trivial examples: point, round circle and round two sphere.

If a round N-ball in S^N meets a given set X only on its boundary,

$$\emptyset \neq X' = X \cap b = X \cap \partial b,$$

then X' is called a spherical top-set. *Spherical top-sets of taut sets are taut.* This information could help for deciding the

Question: Are there taut sub-manifolds which are not real algebraic?

Taut higher dimensional sub-manifolds, compact or closed in $\mathbb{R}^N$, have been studied by Carter-West [4, 6], Cecyl-Ryan [7], (also taut embeddings in hyperbolic space); Thorbergsson [41] and Pinkal [33]. Cecyl and Ryan [7] proved that every isoparametric hypersurface in the sense of E. Cartan (see Ferus, Karcher, Münzner [11]) is taut. They, as well as Carter and West, studied the relation (equality) with totally focal hypersurfaces in $\mathbb{R}^N$. Thorbergsson [41] shows that a taut $n-1$-connected $2n$-dimensional manifold is diffeomorphic to a projective plane (real, complex etc.), to $S^n \times S^n$ (generalized Dupin cyclide) or to S^{2n} (and then round). See also Hebda [14] who obtains a tight embedding for a connected sum of $k \geq 2$ copies of $S^n \times S^n$ which cannot be taut.

9   Isotopy Tight Surfaces

If a knot, that is a *closed curve* $\gamma : S^1 \to \mathbb{R}^3$, can be isotoped in a situation where some height function $z = \pi_L \circ \gamma$ has $B(\gamma)$ relative maxima, and such that this is the minimal number possible, then $B(\gamma)$ is called the *bridge number* of this "knot". For the unknot it is $B(\gamma) = 1$. By the work of Fenchel, Fary, Fox and Milnor, the total curvature of a knot $\gamma : S^1 \to \mathbb{R}^3$ is

$$\tau(\gamma) \geq \tau[\gamma] = 2B(\gamma), \tag{9.1}$$

and this infimum is *not attained,*

$$\tau(\gamma) = \tau[\gamma] = 2B(\gamma),$$

but for the unknot

$$\tau(\gamma) = \tau[\gamma] = 2$$

and *by the plane convex curves. No knot in* $\mathbb{R}^3$ *can be isotopy tight.*

Langevin and Rosenberg [28] started the study of $\tau(f)$ for *knotted surfaces.* With later work of Morton [32] and Meeks [30] it was found that an embedding $f: M_g \to \mathbb{R}^3$ of an orientable surface of genus g with

$$\tau(f) < \beta + 4 = 2g + 2 + 4$$

is unknotted.

Meeks and the author found [25], that if the two components of $\mathbb{R}^3 \setminus M_g$ have fundamental groups with at least $g + \sigma_1^{int}$ and $g + \sigma_1^{ext}$ generators, then

$$\tau(f) \geq 2g + 2 + 4(\sigma_1^{int} + \sigma_1^{ext}). \tag{9.2}$$

In Fig. 3c, $g = 2$, $\sigma_1^{int} = \sigma_1^{ext} = 1$, $\tau > 14$.

There are analogous formulas [26, 40] for hypersurfaces $M^n \subset \mathbb{R}^{n+1}$, in which also higher homotopy groups can play a role.

We also studied the possibility of isotopy tight surfaces:

$$\tau(f) = \tau[f].$$

By a fundamental lemma, then $\tau[f] = \beta + 4k$, $k \geq 0$ an integer, there must exist $1 + k$ convex bodies in $\mathbb{R}^3$ with boundaries $\partial B_0, \ldots, \partial B_k$, and the set of points with positive Gauss curvature on M coincides exactly with the union of the analogous sets on $\partial B_0, \ldots, \partial B_k$:

$$M_{K > 0} = \bigcup_{i=0}^{k} (\partial B_i)_{K > 0}.$$

For $k = 0$ this concerns tight surfaces and was known, also for immersed surfaces, possibly non orientable.

By this description it follows that for embedded surfaces in $\mathbb{R}^3$ *isotopy tightness is a projective property.* The following is a surprising

Theorem [25]. *For surfaces* M_g *of genus* $g \geq 3$ *there do exist isotopy tight surfaces in* $\mathbb{R}^3$ *with*

$$\tau(f) = \beta + 4 = 2g + 2 + 4,$$

(see Fig. 3b) *but none for* $g = 1$ *and* $g = 2$.

For the torus Meeks and the author recently found a complete analogue of the theorem of Fary-Fenchel-Fox-Milnor for curves as follows.

Theorem [26]. *Let* $f: T \to \mathbb{R}^3 \subset \mathbb{R}^3 \subset \infty = S^3$ *be an embedding of a torus. At least one component of* $S^3 \setminus T$ *is known to be (and it is proved again) isotopic in* S^3 *to a tubular neighborhood of a closed curve* $\gamma : S^1 \to \mathbb{R}^3 \subset S^3$. *Let* $B(\gamma)$ *be the bridgenumber of* γ. *Then the infimum* $\tau[f]$ *is* $4B(\gamma)$ *and it is only attained for the unknotted torus and by the classical tight embeddings of the torus:* $\tau(f) = \tau[f] = 4$: *No knotted torus can be isotopy tight.*

A relatively isotopy tight torus embedding (suitably defined) with $\tau(f) = \beta + 4 = 8$ is shown in its "unique" form in Fig. 3a.

The closure of the bounded component of $\mathbb{R}^3 \setminus M_{g=3}$ of the embedded surface in Fig. 3b, is a 3-*manifold with boundary in* $\mathbb{R}^3$ *with minimal total absolute curvature, but not tight.* We denote the space by F and have

$$\tau(F) = \gamma(F) = \gamma = 6 > \beta = 4. \tag{9.4}$$

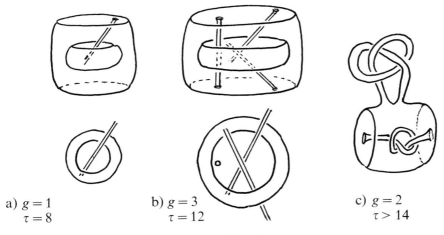

a) $g = 1$ b) $g = 3$ c) $g = 2$
$\tau = 8$ $\tau = 12$ $\tau > 14$

Fig. 3

Here is an example of a *closed* manifold with *minimal total curvature but not tight* [25]. Take for small $\varepsilon > 0$ the boundary W^n of an ε-neighborhood of $F(M_3) \subset \mathbb{R}^3 \subset \mathbb{R}^{n+1}$ in euclidean space $\mathbb{R}^{n+1}$, $n \geq 4$. The embedding $g: W^n \subset \mathbb{R}^{n+1}$ has minimal total curvature

$$\tau(g) = \gamma(W^n) = 12 > 8 = \beta(W^n)$$

but is not tight: $\tau(g) \neq \beta(W^n)$.

10   Total Curvature of (smooth) n-Knots

A n-knot is a smooth embedding $f: M \to \mathbb{R}^{n+2}$ of a closed n-manifold M homeomorphic to S^n. Let $M^c = (\mathbb{R}^{n+2} \cup \infty) \setminus M$ be the complement of M in $\mathbb{R}^{n+2} \cup \infty = S^{n+2}$. If z is a heightfunction for which $z \circ f$ is non degenerate and has less than 4 critical points for n odd, or less than 6 critical points for n even, then it can be shown that M^c has the homotopy groups of the circle S^1 and M^c is homotopy equivalent ($\cong$) to S^1. Furthermore if $M^c \cong S^1$ and $n \neq 2$, then M^c is isotopic to a standard embedding of S^n in $\mathbb{R}^{n+2}$. (By hard results of J. Levine for $n \geq 4$; and C. T. C. Wall and J. Shaneson for $n = 3$.) For $n = 2$ this is not known. However M. Scharlemann [37] recently proved for $n = 2$ that if $z \circ f$ has four critical points, then f is in fact unknotted, that is isotopic to any standard embedding. This is an outstanding result in lower dimensional topology. With the definition

$$\tau(f) = \mathscr{E}_L \mu_L$$

we then deduce immediately the

Theorem. Let $f: M^n \to \mathbb{R}^{n+2}$ be a n-knot. If n is odd and $\tau(f) < 4$, or if n is even and $\tau(f) < 6$, then f is unknotted.

We conjecture that also for equality $\tau(f) = 4$, for n odd and $\tau(f) = 6$ for n even, the embedding is unknotted. This would generalize the case $n = 1$ for which we saw this already. The spun trefoil n-knot [36] is an embedding $f: S^n \to \mathbb{R}^{n+2}$ with $\tau(f) > 6$, and (infimum) $\tau[f] = 6$.

Ferus [9] studied the exotic n-spheres M, $n = 2q - 1 = 4m + 1$ given in complex variables $z_0, \ldots, z_q$ by Brieskorn's equations:

$$M : z_0^d + z_1^2 + \cdots + z_q^2 = 0, \quad \sum_{i=0}^{q} z_i \bar{z}_i = 2, \quad d \text{ odd} \geq 3, \quad q = 2m + 1 \text{ is odd}.$$

M lies in the $4m + 3$-sphere $S^{n+2} : \sum_{i=0}^{q} z_i \bar{z}_i = 2$ in $\mathbb{C}^{q+1} = \mathbb{R}^{4m+4}$.

Here $\pi_i(M^c) = \pi_i(S^1)$ for $i < q = 2m + 1$ and $\pi_q(M^c) = \mathbb{Z}^{d-1} \neq 0$. M is Kervaire's exotic 9-sphere for $m = 2$. It is a non trivial n-knot in S^{n+2} for all odd $q \geq 1$. Ferus obtained a model for this knot of $f: M \subset \mathbb{R}^{n+2}$ in $\mathbb{R}^{n+2}$ with a height function with 4 critical points. Then one can find an embedding f with $\tau(f) < 4 + \varepsilon$, for any $\varepsilon > 0$. Thus the infimum is

$$\tau[f] = 4.$$

Whether such knotted n-knots (with $\tau[f] = 4$) exist for dimensions $n = 4m + 3$, $m \geq 0$ is an interesting open question.

"*Polyhedral n-knots*" Wintgen [40] observed that the "polyhedral $n+1$-knot" which is the double cone $\Sigma f: S^{n+1} \to \mathbb{R}^{n+3}$ on any piecewise linear locally unknotted n-knot $f: S^n \to \mathbb{R}^{n+2}$, can have $\tau(\Sigma f) < 2 + \varepsilon$ for any $\varepsilon > 0$. It is locally knotted at the two cone vertices.

Exercise: Find a piecewise linear knotted two sphere in $\mathbb{R}^4$ with $\tau < 6$, but locally knotted at exactly one vertex.

References

[1]   Banchoff, T., *Tightly embedded 2-dimensional polyhedral manifolds*, Amer. J. Math. **87** (1965), pp. 462—472.

[2]   Banchoff, T., *Tight polyhedral Klein bottles, projective planes and Moebius bands*, Math. Ann. **207** (1974), pp. 233—243.

[2a] Banchoff, T., *The spherical two-piece property and tight surfaces in spheres*, J. Diff. Geom. **4** (1970), pp. 193—205.

[3]   Brehm, U., and Kühnel, W., *Smooth approximation of polyhedral surfaces regarding curvatures*, Geometria Dedicata **12** (1982), pp. 435—461.

[4]   Carter, S., and West, A., *A characterization of isoparametric hypersurfaces in spheres*, J. London Math. Soc. (2), (1982), pp. 183—192.

[5]   Carter, S., and West, A., *Isoparametric systems and transnormality*, Preprint, Leeds 1983.

[6]   Carter, S., and West, A., *Tight and taut immersions*, Proc. London Math. Soc. **25** (1972), pp. 701—720.

[7]   Cecil, T., and Ryan, P. J., *Focal sets, taut embeddings and the cyclides of Dupin*, Math. Ann. **236** (1978), pp. 177—190.

[8]   Chern, S. S., and Lashof, R. K., *On the total curvature of immersed manifolds*, I, II, Amer. J. Math. **79**, (1957), pp. 306—313. Mich. Math. J. (1958), pp. 5—12.

[9]   Ferus, D., *Über die absolute Totalkrümmung höher-dimensionaler Knoten*, Math. Ann. **171** (1971), pp. 81—86.

[10]* Ferus, D., *Totale Absolutkrümmung in Differentialgeometrie und Topologie*, Springer Lecture Notes **66** (1968).

[11] Ferus, D., Karcher, H., and Münzner, H. F., *Clifford algebren und neue isoparametrische Hyperflächen*, Math. Zeitschrift **177** (1981), pp. 479—502.

[12] Fox, R. H., *On the total curvature of some tame knots*, Ann. Math. **52** (1950), pp. 258—262.

[13] Hausmann, J. C., (editor), *Knot theory*, Proc. Plans-sur-Bex (1977), Springer Lecture Notes **685**.

[14] Hebda, J. J., *Some new tight imbeddings which cannot be made taut*, Geometria Dedicata.

[15] Kobayashi, S., and Takeuchi, M., *Minimal imbeddings of R-spaces*, J. Differential Geometry 2 (1968), pp. 215—230.

[16] Kühnel, W., *Tight and 0-tight polyhedral embeddings of surfaces*, Invent. Math. **58** (1980), pp. 161—177.

[17] Kühnel, W., and Banchoff, T., *The 9-vertex complex projective plane*, Intelligencer Volume 5, N⁰ 3 (1983), pp. 11—22.

[18] Kuiper, N. H., *Immersions with minimal total absolute curvature*, In Coll. de Géométrie Diff., Centre Belge de Recherches Math., Bruxelles (1958), pp. 75—88.

[19] Kuiper, N. H., *On convex maps*, Nieuw Archief voor Wisk **10** (1962), pp. 147—164.

[20] Kuiper, N. H., *Minimal total absolute curvature for immersions*, Invent. Math. **10** (1970), pp. 209—238.

[21]* Kuiper, N. H., *Taut sets in three space are very special*. To appear in Topology.

[22] Kuiper, N. H., *There is no tight topological immersion of the Klein bottle in* $\mathbb{R}^3$. Preprint IHES (1983).

[23] Kuiper, N. H., *Morse relations for curvature and tightness*, Proc. Liverpool Singularities Symposium II, Springer Lecture Notes in Mathematics **209** (1971), pp. 77—89, (C. T. C. Wall, Editor).

[24]* Kuiper, N. H., *Tight imbeddings and maps*, The Chern Symposium 1979, Springer-Verlag 1980, pp. 97—145.

[25] Kuiper, N. H., and Meeks III, W. H., *Total curvature for knotted surfaces*, to appear in Invent. Math.

[26] Kuiper, N. H., and Meeks III, W. H., *The total curvature of a knotted torus*, in preparation.

[27] Kuiper, N. H., and Pohl, W. F., *Tight topological embeddings of the real projective plane in* E^5, Invent. Math. **42** (1977), pp. 177—199.

[28] Langevin, R., and Rosenberg, H., *On curvature integrals and knots*, Topology **15** (1976), pp. 405—416.

[29] Little, J. A., and Pohl, W. F., *On tight immersions of maximal codimension*, Invent. Math. **13** (1971), pp. 179—204.

[30] Meeks III, W. H., *The topological uniqueness of minimal surfaces in three dimensional euclidean space*, Topology **20** (1981), pp. 389—411.

[31] Milnor, J. W., *On the total curvature of knots*, Ann. Math. **52** (1950), pp. 248—257.

[32] Morton, H. R., *A criterion for an embedded surface to be unknotted*, Springer Lecture, Notes in Mathematics **722**, pp. 92—98.

[33] Pinkal, U., *Curvature properties of taut submanifolds*. Preprint (1983), Freiburg, Germany.

[34] Pohl, W. F., In preparation.

[35]* Ringel, G., *Map colour theorem*, Grundlehren der Math. Wiss., Band 209, Springer Verlag (1974).

[36]* Rolfsen, D., *Knots and links*, Math. Lecture Series **7**, Publish or Perish (1976).

[37] Scharlemann, M., *Smooth spheres in* $\mathbb{R}^4$ *with four critical points are standard*, To appear.

[38]* Willmore, T. J., *Total curvature in Riemannian geometry*, Ellis Horwood Limited, Chichester, England (1982), pp. 1—168.

[39]* Willmore, T. J., *Tight immersions and total absolute curvature*, Bull. London Math. Soc. **3** (1971), pp. 129—151.

[40] Wintgen, P., *Totale absolutkrümmung von Hyperflächen*, Beiträge zur Algebra und Geometrie **10** (1980), pp. 87—96.

[41] Thorbergsson, G., *Highly connected taut submanifolds*, Math. Ann. **265** (1983), pp. 399—405.

* Survey aspects, books.

Perspectives in Mathematics
Anniversary of Oberwolfach 1984
© Birkhäuser Verlag, Basel

Some Tendencies in Contemporary Algebra

PETER ROQUETTE

Mathematisches Institut, Im Neuenheimer Feld 288,
D-6900 Heidelberg (FRG)

*Dem Andenken an **Wilhelm Süss** gewidmet, den Gründer und
spiritus rector des Mathematischen Instituts in Oberwolfach*

1   Introduction

Mathematics is, on the one hand, a *cumulative science*. Once a mathematical theorem has been proved to be true then it remains true forever; it is added to the stock of mathematical discoveries which has piled up through the centuries and it can be used to proceed still further in our pursuit of knowledge.

On the other hand, the mere proof of validity of a theorem is in general not satisfactory to mathematicians. We also want to know "why" the theorem is true, we strive to gain a better understanding of the situation than was possible for previous generations. Consequently, although a mathematical theorem never changes its content, we can observe a continuous change of the *form of presentation*, in the course of history of our science. Sometimes a result seems to be better understood if it is generalized, or if it is looked at from a different point of view, or if it is embedded into a general theory which opens analogies to other fields of mathematics. Also, in order to make further progress possible it is often convenient and sometimes necessary to develop a framework, conceptual and notational, in which the known results become trivial and almost self-evident at least from a formalistic point of view. So when we look at the history of mathematics we indeed observe a change, not in the nature of mathematical truth but in *the attitude of mathematicians* towards it. It may well be that sometimes a new theory is but the response to a current fashion, and sometimes it may be mere fun to derive a theorem by unconventional means. But mostly the changes in attitude reflect a serious effort towards a better understanding of the mathematical universe.

It is fascinating to observe such trends in the past and see how they have led to the picture of today's mathematics. But likewise it is not without interest to search for trends in *contemporary* mathematics since they will shape the future of our science.

In this article we shall restrict our discussion to *algebra*. But even in algebra the reader cannot expect a full account of all the many recognizable threads of development. Some of these are quite obvious to the attentive observer: for instance the tendency towards *geometrization of algebra*, which means adopting the language of geometry and its way of arguing. This has had striking successes, the most recent one being the proof of Mordell's conjecture by Faltings, and it has found its due recognition by the contemporary mathematical public. Another obvious trend in present-day algebra is its *algorithmization*, i.e. the desire to supplement every existence proof and actually all arguments, by an effective algorithm if this is possible at all. Again, this is not exactly new and algorithms have been pursued throughout the history of algebra. But during this century it often seemed, or it was at least proclaimed that the *structural viewpoint* is to be dominating, and that in this framework computational or algorithmical considerations are at most of secondary interest. Today, since the structural viewpoint has become firmly inserted in contemporary mathematical thinking it is realized again that algorithms are essential, not only for computers but also theoretically, as part of our understanding of mathematics.

In this article we propose to discuss some other tendency of contemporary algebra, not yet quite as auspicious as the ones mentioned above and not even known to many, but nevertheless somewhat remarkable. This is the *intrusion of model theoretic notions and arguments into algebra*. Here, "model theory" is used in the sense of mathematical logic. The field of mathematical logic was formerly regarded as pertaining mainly to the foundations of mathematics only, giving the mathematician a (hopefully) solid base for his work but otherwise not influencing actual mathematical research activity. Some mathematicians had even voiced their opinion that mathematical logic does not properly belong to mathematics but should be considered as part of philosophy. This has completely changed, in the meantime. Model theoretic notions and results have intruded heavily into mathematics and in particular into algebra. Model theory provides the algebraist with a new way of reasoning which was not available before. This is possible if he (the algebraist) remains conscious of *the formal language with which algebraic structures are described*. Perhaps a good illustration will be the discussion of the model theoretic notion of *elementary equivalence* which is much more adapted to the investigation of algebraic structures than the notion of *isomorphism*, contrary to what is claimed in many textbooks of algebra. The notion of isomorphism is of set-theoretical nature and introduces set-theoretical difficulties into algebra which are not inherent in the algebraic problem itself. Whereas the notion of elementary equivalence permits to deduce the same consequences as from isomorphism, but it also allows to change sets without disturbing algebraic properties.

The idea to make use of model theory in this way is originally due to *Abraham Robinson*.

When I first planned to write this article I meant to give a review about all applications of model theory in algebra, or at least about the most important ones. But soon it turned out that this task would have required more space than available here, and more time for preparation. Also, in view of the large amount of literature is has become doubtful whether a comprehensive report about model theory in algebra would be of much value today. As said above, model theory provides us with a new method, a new way of mathematical reasoning. In order to explain this method it is perhaps sufficient to discuss particular examples where it has been applied.

Accordingly this article will be restricted to certain parts of algebra. The examples selected for our discussion come from field theory; they are all connected with the work of the algebra group in Heidelberg.*) Although we realize that this selection is somewhat arbitrary we hope that it will be sufficient for our purpose: to acquaint the reader with some of the *basic principles* of model theoretic methods in algebra.

2   Elementary Equivalence

Let us discuss the model theoretic notion of *elementary equivalence* in relation to the classical algebraic notion of *isomorphism*. Modern algebra deals with *algebraic structures* as its basic objects. Examples of algebraic structures are groups, rings, fields, ordered fields etc. Along with the notion of structure there comes the notion of *isomorphism* between structures of the same kind: group isomorphism, ring isomorphism etc. If we have established that two algebraic structures K and L, say fields, are isomorphic then we know that *every algebraic property of K is shared by L, and vice versa*. This is precisely the reason why isomorphism theorems are of fundamental importance in algebra.

But is isomorphism between K and L really necessary to draw the above conclusion? *Are there non-isomorphic algebraic structures which do have the same algebraic properties?*

In order to discuss this question it is necessary to specify what is meant by "algebraic property" of a given algebraic structure. As a rule, an algebraist does not bother to make this precise because in concrete situations it always seems to be evident which properties are admissible as "algebraic properties" in the above sense. For instance, if we consider rings then the property of being commutative is certainly admissible: if a ring K is commutative then every isomorphic copy of K is commutative too. What is the general description of "algebraic properties" which are admissible in this context?

We consider algebraic structures of a given type, say fields or groups. These structures are defined by *axioms*. The axioms contain references to certain functions and relations which are defined on structures of the type considered. For instance, the group axioms refer to a 2-variable function $x \cdot y$ which denotes the group operation. The field axioms refer to two 2-variable functions $x + y$ and

---

*) We have *excluded* the work with *nonstandard methods* in algebra and number theory. For this we refer to our report [Rq3].

$x \cdot y$, denoting addition and multiplication in the field, etc. Now an "algebraic statement" with respect to the given type of structure must be expressible *solely* in terms of these functions and relations which appear in the axioms.

We envisage the axioms written in a *formal language* $\mathscr{L}$ whose vocabulary is adapted to the structure type considered; thus the vocabulary contains the appropriate function symbols and relation symbols (including constant symbols). It is always understood that $\mathscr{L}$ is an *elementary* (or first order) language. This means that the variables in $\mathscr{L}$ denote individuals only; there are no set variables or function variables. Accordingly, quantification ($\forall$ or $\exists$) is permitted with respect to individuals only. Historical experience shows that for most of the relevant algebraic structures, the axioms can indeed be stated in an elementary language. (This has led some authors to *define* algebraic structures as being those whose axioms can be stated in an elementary language.)

We now can give the definition of an admissible "algebraic property" of a structure: *such property should be expressible by a sentence φ in the language $\mathscr{L}$.* In other words: a structure K has this property if and only if φ holds in K. For instance in ring theory, the property of a ring K to be commutative means that the following sentence holds in K:

$$(\forall x)\,(\forall y)\,(xy = yx).$$

The precise definition of the notion of *sentence* in a formal language $\mathscr{L}$ can be found in any introductory treatise on mathematical logic [Po], [Bar], [CK].

Now let us review the situation: We have a formal (elementary) *language* $\mathscr{L}$, with a vocabulary consisting of certain function symbols and/or relation symbols (including perhaps some constant symbols, e. g. 0 and 1 in case of field theory). We consider *structures of type* $\mathscr{L}$, which means that the function symbols and relation symbols in $\mathscr{L}$ should have an interpretation by means of functions resp. relations in the structure. Any set of sentences of $\mathscr{L}$ defines a *theory*; the defining sentences are called *axioms* of that theory. An $\mathscr{L}$-structure K is called a *model* of the theory if all the axioms hold in K. If φ is any sentence in $\mathscr{L}$, and if φ holds in all models then φ is said to be a *theorem* of the theory. In general, if φ is an arbitrary sentence then φ may hold in some models but perhaps not in all models of the theory. If φ holds in K then it is customary to write $K \models \varphi$. We then say that φ defines an "algebraic property" of K or that K has the property φ.

Definition: Two models K and L are said to be *elementary equivalent* if every algebraic property of K is shared by L, and vice versa. In other words: $K \models \varphi$ iff $L \models \varphi$. If this is so then we write $K \equiv L$.

If K and L are isomorphic then we write $K \approx L$. Clearly, isomorphic models are elementary equivalent. Now our above question can be formulated, more precisely, as follows:

Do there exist non-isomorphic models which are elementary equivalent?

Model theory provides the answer to this question: Yes, *to every infinite model K there exists a non-isomorphic, elementary equivalent model L.* Indeed, for every sufficiently large cardinal number α there exists a model $L \equiv K$ such that $|L| = \alpha$. "Sufficiently large" means that α should be infinite and greater or equal to the cardinality of the vocabulary of the language $\mathscr{L}$. This is the theorem of Löwenheim-Skolem-Tarski. If we are dealing with group theory, or ring theory, or field theory etc. then there are only finitely many function symbols and relation symbols in the vocabulary and hence we can find a model $L \equiv K$ in *every* infinite cardinality α; in particular L can be countable.

In case of a *finite* model K it is easily seen that every elementary equivalent model $L \equiv K$ is isomorphic: $L \approx K$.

We conclude that, for the investigation of properties of algebraic structures, *the notion of isomorphism* is adequate in the case of finite structures, but it is *inadequate for infinite structures.* The adequate notion is that of elementary equivalence. If we want to know whether an algebraic structure K has a certain property φ then it suffices to find a structure $L \equiv K$ for which the validity of φ can be checked; L need not be isomorphic to K.

Hence a *new way of mathematical reasoning is introduced into classical algebra, with the notion of isomorphism being replaced by elementary equivalence.* The success of this new method depends on how one is able to handle, algebraically, the notion of elementary equivalence. Let us look at some examples, classical ones and some that are more recent.

3   Algebraically Closed Fields

We refer to the celebrated paper of *E. Steinitz* [St] which appeared in volume 137 of *Crelle's Journal* (1910). This paper contains the first systematic study of fields from the "algebraic" point of view, i.e. solely as models of the field axioms. Today we are so used to this kind of viewpoint that we can hardly imagine the impact and the source of inspiration which Steinitz' paper generated for the mathematicians of his time. It became a "classic" and, because of its fundamental importance it was reprinted in book form some twenty years after its first appearance (and later again in one of the post-war Chelsea editions). Steinitz' program was to give a constructive description of all fields, up to isomorphisms. For our purpose his results on *algebraically closed* fields are relevant. Steinitz showed that an algebraically closed field K is completely determined, up to isomorphisms, by two invariants: its *characteristic p* $= \text{char}(K)$ which is either 0 or a prime number, and its *transcendence degree t* $= \text{tr}(K)$ which is a cardinal number. Both p and t can be arbitrarily prescribed: there exists an algebraically closed field with given characteristic and given transcendence degree.

If $\text{tr}(K)$ is finite then the algebraically closed field is denumerable, $|K| = \aleph_0$. If $\text{tr}(K)$ is infinite then $|K| = \text{tr}(K)$. Consequently, if $|K| > \aleph_0$ then $|K| = \text{tr}(K)$; in this case Steinitz' theorem says that for fixed characteristic

$p \geqq 0$, K is *uniquely determined* (up to isomorphisms) *by its cardinality*. This fact is expressed by saying that *the theory of algebraically closed fields of characteristic p is categorical in every uncountable cardinality*.

Now there is a general theorem of model theory, called *Łos-Vaught test*, to the following effect: For any theory T, if T is categorical in but one infinite cardinality greater or equal to the cardinality of the vocabulary of the language, then all infinite models of T are elementary equivalent [Bar]. We conclude that *all algebraically closed fields of characteristic p are elementary equivalent*. Thus if we want to know whether an algebraically closed field K has a certain algebraic property φ then the transcendence degree $\operatorname{tr}(K)$ is of no importance; if φ holds in any other algebraically closed field of the same characteristic then φ also holds in K. In characteristic zero this fact is commonly known as the *Lefschetz principle*. Historically, this name is not exactly correct since Lefschetz in his book on Algebraic Geometry [Le] stated a somewhat different principle, related but not quite identical to the above. To explain this we need the notion of *elementary extension*.

Quite generally let T be an arbitrary theory, formulated in an elementary language $\mathscr{L}$, and let K, L be models of T. We suppose that $K \subset L$, which means that L should be an extension of K. The notion of elementary equivalence $K \equiv L$ has been defined above already: every sentence φ of $\mathscr{L}$ which holds in K, should also hold in L. Now let us consider formulas $\varphi(x_1, \ldots, x_n)$ depending on free parameters $x_1, \ldots, x_n$. Let $c_1, \ldots, c_n \in K$. Then $\varphi(c_1, \ldots, c_n)$ repesents a property of K in whose formulation the elements $c_1, \ldots, c_n \in K$ are involved, besides of the other constants, functions and relations which are defined in K as an $\mathscr{L}$-structure. More precisely, $\varphi(c_1, \ldots, c_n)$ is a sentence in the extended language $\mathscr{L}_K$ whose vocabulary contains constants to denote the individual elements $c \in K$, besides of the constant symbols, function symbols and relation symbols in the vocabulary of $\mathscr{L}$. Now if every sentence $\varphi(c_1, \ldots, c_n) \in \mathscr{L}_K$ which holds in K does also hold in L (and vice versa), then L *is said to be an elementary extension of* K. Notation: $K \prec L$. Clearly, $K \prec L$ implies $K \equiv L$. (But not conversely.)

In the theory of algebraically closed fields, it is immediate from Steinitz' work (this time applying the Łos-Vaught test to $\mathscr{L}_K$):

Every extension of algebraically closed fields is an elementary extension.

This is the "Lefschetz principle" as stated by Lefschetz in his book. Indeed, he considers an algebraically closed field K of characteristic zero. Let V denote an irreducible variety, defined in n-dimensional affine space by a system of polynomial equations

(E) $f_1(x_1, \ldots, x_n) = 0, \ldots, f_n(x_1, \ldots, x_n) = 0.$

Let K_0 be an algebraically closed subfield of K, containing all the finitely many coefficients $c_1, \ldots, c_r$ of $f_1, \ldots, f_m$, and such that K_0 is of finite transcendence degree. Then K_0 may be isomorphically embedded into the complex number field; let us identify K_0 with its isomorphic image: $K_0 \subset \mathbb{C}$. Then the equations

(E) define also a complex variety V^* in n-dimensional affine space over $\mathbb{C}$. Now the important fact, Lefschetz continues, is that *in the passage from V to V* all the strictly algebraic properties of V are preserved.* Lefschetz does not explain what is meant by "strictly algebraic properties of V". If we interpret this as those properties which can be expressed by sentences of the form $\varphi(c_1, \ldots, c_r)$ in the elementary language $\mathscr{L}_{K_0}$ then indeed, they are preserved because $K_0 \prec K$ and $K_0 \prec \mathbb{C}$. (It is conceivable, though, that the Lefschetz principle applies also to other "strictly algebraic properties" which can be expressed in a higher order language only. See [Ek] and the literature cited there.)

In model theory the following terminology is used: A theory T is called *complete* if all models of T are elementary equivalent. T is called *model-complete* if every extension of models of T is an elementary extension. Thus the result of our above discussion can be briefly stated as follows:

The theory of algebraically closed fields of fixed characteristic $p \geq 0$ is complete, and it is also model-complete.

Before proceeding to the next example, let us briefly indicate how this fact may be used as a tool in various arguments of geometric nature.

Consider a system of polynomial equations of the form (E) where the $f_1, \ldots, f_m$ are polynomials over an algebraically closed field K. The *Hilbert Nullstellensatz* gives a necessary and sufficient condition for the existence of a solution $x_1, \ldots, x_n \in K$ of the equation (E). Namely, the ideal generated by $f_1, \ldots, f_m$ in the polynomial ring $K[X_1, \ldots, X_n]$ should not be trivial; it should be a *proper* ideal. Clearly, this condition is necessary. Conversely, if the condition is satisfied let P be a maximal ideal of $K[X_1, \ldots, X_n]$ containing $f_1, \ldots, f_m$ and consider the residue class ring $K[X_1, \ldots, X_n]/P = R$. This is an integral domain containing K, and the residue classes $x_i = X_i$ mod P satisfy (E). Now R can be embedded into its field of quotients which in turn can be embedded into an algebraically closed field L, by the work of Steinitz. By construction, the statement

$$(\exists x_1)(\exists x_2) \ldots (\exists x_n)(f_1(x_1, \ldots, x_n) = 0 \wedge \ldots \wedge f_m(x_1, \ldots, x_n) = 0)$$

holds in L. Note that this statement is expressible by a sentence in $\mathscr{L}_K$, the constant parameters being the coefficients $c_1, \ldots, c_r \in K$ of the polynomials $f_1, \ldots, f_m$. Since $K \prec L$ we conclude that the above statement holds in K too, which is to say that there exist $x_1, \ldots, x_n \in K$ satisfying (E).

Thus we see that the Hilbert Nullstellensatz is *almost trivial* in view of the model-completeness of the theory of algebraically closed fields.

A less direct example of an application of model-completeness for algebraically closed fields is the following. Let K be an algebraically closed field of characteristic 0 and $K(t)$ the rational function field of one variable over K. Given any finite subset $S \subset K \cup \{\infty\}$, consider the maximal algebraic field extension F_S of $K(t)$ which is unramified outside of S. Then F_S is a Galois extension of $K(t)$. What is the structure of its Galois group G_S? The answer is that G_S is generated, as a profinite group, by $s = |S|$ elements $\sigma_1, \ldots, \sigma_s$ with the *defining relation* $\prod_{1 \leq i \leq s} \sigma_i = 1$. *Hence* G_S *is a free profinite group in* $s - 1$ *generators.* In the case $K = \mathbb{C}$ a proof of this assertion is well known from the analytic theory of holomorphic functions, using the Riemann existence theorem. By means of model-completeness we would like to deduce that the above assertion holds for an arbitrary algebraically closed field of characteristic 0. In trying to apply model-completeness we try to express the above assertion by a sentence $\varphi \in \mathcal{L}_S$, the language of fields extended by s constant symbols to denote the elements of S. However, such sentence φ does not exist. But it is possible to find a sequence $\varphi_1, \varphi_2, \ldots$ of sentences in $\mathcal{L}_S$ such that the above assertion is equivalent to that all φ_n hold ($n = 1, 2, 3, \ldots$). Each φ_n describes the factor groups of G_S of order n. For details of the construction of φ_n we refer to the literature [DrR2]. By model-completeness we can now deduce that each φ_n holds in every algebraically closed field K of characteristic 0 with a prescribed finite set $S \subset K \cup \{\infty\}$. Hence indeed, G_s is free profinite on $s - 1$ generators.

No algebraic proof, without the use of analysis and topology, has been found for this structure theorem for G_S. However it follows from model theoretic fundamentals that such an algebraic proof exists, i.e. each φ_n can be deduced within $\mathcal{L}_S$ from the axioms of algebraically closed fields of characteristic 0.

If K is an algebraically closed field of characteristic $p > 0$ then the structure of G_S as a profinite group is still unknown.

Let us close this section with the following remark. While talking about the "theory of algebraically closed fields" we have envisaged that theory as given by certain axioms. The reader should keep in mind, however, that these axioms are not finite in number. Indeed, to the finitely many axioms of field theory there is to be added for each $n = 1, 2, 3, \ldots$ an axiom which says that every polynomial of degree n has a root. It is not possible to find finitely many axioms in $\mathcal{L}$ for the theory of algebraically closed fields.

Similarly, the axioms expressing the property of characteristic 0 are not finite in number: for each $n = 1, 2, 3, \ldots$ one has the axiom that 1 added n times to itself does not give 0. It is not possible to find finitely many axioms in $\mathcal{L}$ which imply characteristic zero.

4   Real Closed Fields

The algebraic theory of real fields was originated by E. Artin and O. Schreier [AS] with their three papers in volume 5 of the *Hamburger Abhandlungen* (1926). It is noteworthy that Artin and Schreier start by mentioning the paper of Steinitz. They point out its importance for the development of "abstract" algebra and then they put forward a program how to include real algebra into the abstract framework of Steinitz. A closer look at the Artin-Schreier program reveals that one problem is left out and not mentioned at all, namely the classification problem for real closed fields, up to isomorphisms. This would be the analog of Steinitz' classification for algebraically closed fields as discussed above. Later authors have tried to fill this gap and to arrive at a classification for real closed fields; partial results were obtained [EGH] but no general satisfactory solution of the problem is known today. The transcendence degree (resp. cardinality) of the real closed field is *not* sufficient for classification up to isomorphism; one has to add more subtle invariants, as e. g. the order type of the field with respect to its (unique) ordering. But the order type is not sufficient either [Ro3]. The problem seems to be difficult.

In any case, if the classification problem were important then it should have been at least mentioned in the general program of Artin-Schreier, even though the solution could not be given. Why are Artin and Schreier silent on this problem? The answer is easily perceived: because the classification up to isomorphism is *not* important; it is of little significance with respect to algebraic problems. Instead, the *classification up to elementary equivalence* is the proper problem to be considered.

This broader classification problem is indeed mentioned, and also solved, in the Artin-Schreier papers. It follows from their work that *all real closed fields are elementary equivalent*. In other words: *The theory of real closed fields is complete*.

To be sure, the completeness theorem is not formally considered in the Artin-Schreier papers; this was done later by Tarski [TM]. But all the essential ingredients of the completeness proof can be found in the Artin-Schreier papers. The authors recognized quite clearly the significance of their work for the elementary-equivalence classification problem. This can be inferred from the statement, to be found in their paper, that *"the theorems of real algebra hold in any real closed field"*. Properly interpreted, this means for every sentence φ in the language $\mathscr{L}$ of fields, if φ holds in the ordinary real number field $\mathbb{R}$ then φ holds in every real closed field. Indeed: this means completeness.

In addition to completeness, Artin-Schreier also tacitly proved *model-completeness* for the theory of real closed fields; again this was later formally verified by Tarski [TM]. Today this is known as the *Tarski principle* for real closed fields, and it is the real analog to the Lefschetz principle for algebraically closed fields.

The main motivation for the introduction and study of real closed fields was Artin's solution of the 17th Hilbert problem. Artin's theorem says that every positive definite rational function $f(X_1, \ldots, X_n) \in \mathbb{R}(X_1, \ldots, X_n)$ is a sum of squares of rational functions.*) In other words: if f is not a sum of squares in $\mathbb{R}(X_1, \ldots, X_n)$ then f is not positive definite, i. e. there exist $a_1, \ldots, a_n \in \mathbb{R}$ such

---

*) Strictly speaking, Hilbert formulated this problem over $\mathbb{Q}$ instead of $\mathbb{R}$. But today it has become customary to speak of Hilbert's 17th problem with reference to $\mathbb{R}$ (or to any real closed base field). From this the corresponding problem for $\mathbb{Q}$ can be deduced by simple density arguments.

that $f(a_1, \ldots, a_n)$ is defined and $f(a_1, \ldots, a_n) < 0$. As A. Robinson [Ro2] has observed, this theorem is an *almost trivial* consequence of model-completeness for real closed fields. The argument is as follows.

Since f is not a sum of squares in $\mathbb{R}(X_1, \ldots, X_n)$ it follows from Artin's theory [AS] that there exists an ordering of the field $\mathbb{R}(X_1, \ldots, X_n)$ such that $f < 0$. Let L denote the real closure of $\mathbb{R}(X_1, \ldots, X_n)$ with respect to this ordering. Then $\mathbb{R} \subset L$ and hence, by model-completeness, L is an *elementary* extension of $\mathbb{R}$. Consequently, if the sentence

$$(\exists x_1) \ldots (\exists x_n) : f(x_1, \ldots, x_n) < 0 \tag{S}$$

holds in L then it also holds in $\mathbb{R}$. But in L this sentence holds by construction; one may take $x_i = X_i (1 \leqq i \leqq n)$. Thus the above sentence holds in $\mathbb{R}$ which means that there exist $a_1, \ldots, a_n \in \mathbb{R}$ such that $f(a_1, \ldots, a_n) < 0$.   Q.E.D.

The beauty and simplicity of this (A. Robinson's) proof for Artin's theorem is apparent.

Note that in the sentence (S) above, an inequality sign is used which *a priori* does not belong to the vocabulary of the language of fields. The use of this inequality sign can be avoided since the ordering relation in a real closed field is uniquely determined: the positive elements are precisely the non-zero squares. Consequently, the above sentence (S) can be replaced by the following sentence in the language of fields:

$$(\exists z) (\exists x_1) \ldots (\exists x_n) \left(z \neq 0 \wedge f(x_1, \ldots, x_n) = -z^2 \right). \tag{S'}$$

However, for the validity of A. Robinson's argument it does not matter whether we use (S) or (S'). For the model-completeness theorem or real closed fields holds in either case: with respect to the language of fields, and with respect to the language of ordered fields.

As to the proof of model-completeness, we have said above already that all its essential ingredients are to be found in the Artin-Schreier papers. However, by using basic fundamentals of model theory the proof can be considerably simplified, and in particular Artin's specialization arguments can be avoided, i.e. replaced by embedding arguments. Since this is perhaps not yet widely known among algebraists, let us briefly sketch this proof:

Let $K \subset L$ be an extension of real closed fields. We have to show that L is an *elementary* extension of K. By general model theoretic principles it suffices to show that L can be K-isomorphically embedded into a sufficiently saturated non-standard model (or ultrapower) K^* of K. More precisely, K^* should be α-saturated for some cardinal number $\alpha > |L|$. By general induction (Zorn's lemma) the problem is reduced to the case where $L|K$ is of transcendence degree 1. Let $x \in L$ be transcendental over K; then L is the real closure of the ordered field $K(x)$. By the functorial property of the real closure, every order-preserving K-embedding $K(x) \hookrightarrow K^*$ extends (uniquely) to an embedding $L \hookrightarrow K^*$ (note that K^* is real closed too). Hence it suffices to construct an order-preserving K-embedding $K(x) \hookrightarrow K^*$. Now the ordering of $K(x)$ is uniquely determined by the *Dedekind cut* which x determines in K. The cut consists of the two disjoint sets $C_x, D_x \subset K$:

$$C_x = \{c \in K : c < x\}$$
$$D_x = \{d \in K : d > x\}.$$

The saturation property of K^* guarantees the existence of $\xi \in K^*$ which determines the same cut:

$$C_\xi = C_x, \quad D_\xi = D_x.$$

Then the substitution $x \mapsto \xi$ yields the required order-preserving K-embedding $K(x) \approx K(\xi) \subset K^*$.

We have given this proof because the same typical pattern of argument applies also to other situations, yielding model-completeness results for other classes of fields. See below for various classes of valued fields. In all those cases the problem is reduced to an embedding problem for a purely transcendental extension $K(x)$ into a saturated model K^* of K, the embedding preserving certain additional structure which is connected with the class of fields considered. By the way, in the case of algebraically closed fields as discussed in section 3, the proof of model completeness can also be made to follow the same pattern. In this case there is no additional structure given, and thus we are faced with the K-embedding of $K(x)$, where K is algebraically closed, into a saturated ultrapower K^*. But this is trivially achieved by the substitution $x \mapsto \xi$ where $\xi \in K^*$ is an arbitrary transcendental over K.

Hilbert's 17th problem yields but one example for the application of the model-completeness theorem for real closed fields. There are many other examples: for instance, the *real Nullstellensatz* [Kri], [Du], giving conditions for a variety to have *real* points. It is also possible to generalize the Krull-Neukirch theorem [KrN] about the structure of the absolute Galois group of $\mathbb{R}(t)$, to arbitrary real closed base fields K instead of $\mathbb{R}$ [DrR1]. Various other applications have been given: e.g. to real algebraic function theory or to real algebraic geometry [DK]. A very interesting development concerning sums of $2n$-th powers (instead of sums of squares as in Hilbert's 17th problem) has recently been started by E. Becker; see [BeJ], [Pr].

Perhaps it is not superfluous to state explicitly the *axioms for real closed fields*:

(1)   the field axioms;

(2)   every sum of two squares is a square;

(3)   if a is not a square then $-a$ is a square;

$(4)_n$ every polynomial of odd degree $2n + 1$ has a root.

Note that $(4)_n$ is an infinite set of axioms, one axiom each for $n = 1, 2, 3, \ldots$ It is not possible to find finitely many axioms for the theory of real closed fields.

The completeness result of Tarski says that these axioms describe the algebraic theory of $\mathbb{R}$. In other words: If φ is a sentence in the language of fields then φ *holds in* $\mathbb{R}$ *if and only if* φ *can be derived from the above axioms.*

5   *p*-adically Closed Fields

In the foregoing section we have discussed the algebraic theory of $\mathbb{R}$. It seems natural to ask, analogously, for the algebraic theory of the Hensel field $\mathbb{Q}_p$. Here and in the following, p denotes a prime number and $\mathbb{Q}_p$ is the completion of the rational number field $\mathbb{Q}$ with respect to the p-adic valuation.

Hensel's book [He] containing his discovery of the fields $\mathbb{Q}_p$ had appeared in 1908, two years before the Steinitz' paper. Steinitz informs us that it was mainly Hensel's discovery which induced him to write his article. It seems that in those times, the Hensel fields were somehow regarded to be strangers in the mathematical world; they had never been encountered before, at least not explicitly. Hence there was a desire for a general, axiomatic, abstract field theory into which the p-adic fields would fit naturally. According to Steinitz, his article was to be regarded as a first step in this direction containing the *foundations* only of a general field theory. He announced further investigations including, he said, applications to geometry, number theory and analytic theory of functions. It is not quite clear what he had in mind because none of those announced applications were ever published. His reference to *geometry* might perhaps indicate that he had envisaged an abstract theory of *real fields*, of the kind which Artin and Schreier presented 16 years later. When he mentions *number theory* then we may assume that he included an abstract theory of *p-adic fields;* this seems quite probable since he knew about Hensel's discovery and explicitly mentions it as a source of inspiration for his work.

In any case, Steinitz did never return to his announced applications, and his work was continued by Artin-Schreier [AS] in the case of real fields. It is natural to expect that soon after the appearance of the Artin-Schreier papers, the analogous theory for the p-adic fields would have been developed. Strangely enough, this was not the case. The algebraic theory of p-adic fields was given only recently by Kochen [Ko1,2]. This delay is surprising because during the twenties and thirties, the use of p-adic fields in number theory had led to striking results: for quadratic forms, for simple algebras, in class field theory etc. These successes created an atmosphere quite favorable for the so-called p-adic methods; the Hensel fields $\mathbb{Q}_p$ were now accepted, not only by the specialists, as belonging to the fundamentally important mathematical structures, similar in importance to the field $\mathbb{R}$. What, then, were the historical reasons that

Artin-Schreier's real algebra was not immediately matched by an analogous *p*-adic algebra?

Perhaps one of the reasons may be seen in the fact that the notion of *elementary equivalence* had *not yet been recognized* in its fundamental importance for algebra and number theory. If it would have been, then the search for the fields elementary equivalent to Q_p would probably have started, or at least have been mentioned as being desirable. But this was not the case and it seems that the Artin-Schreier theory was considered to be of singular nature, not being transferable to the *p*-adics.

In some cases, though, elementary equivalence for Q_p was tacitly mentioned and used. For instance, it was observed that the field $Q_p^{(0)}$, of algebraic numbers within Q_p, has for all practical purposes the same algebraic properties as Q_p itself. For instance, the absolute Galois group over $Q_p^{(0)}$ is the same as the Galois group over Q_p, and local class field theory also holds over $Q_p^{(0)}$, etc. Later, it was observed that *every* subfield $K \subset Q_p$ which is algebraically closed within Q_p, can be used equally well as a base field for *p*-adic algebraic geometry [La2]. These arguments were in fact of the nature of elementary equivalence arguments, without however being explicitly stated that way.

A second reason for the delay in the systematic development of *p*-adic algebra was perhaps the lack of suitable prominent problems. We remember that one of the driving forces for the Artin-Schreier theory was the solution of Hilbert's 17th problem. This was concerned with *positive definite* functions. The notion of positive definite function involves the ordering relation $<$ and hence belongs to *real* algebra. But Hilbert did not state a corresponding problem belonging to *p*-adic algebra.

What, then, would be the *p*-adic analog to Hilbert's 17th problem?

Instead of positive definite functions it seems natural to consider, in the *p*-adic case, the *p-integral definite functions*. Let $\mathbb{Z}_p$ denote the ring of *p*-integers in Q_p; it can also be described as the *p*-adic completion of $\mathbb{Z}$. A rational function

$$f(X_1, \ldots, X_n) \in Q_p(X_1, \ldots, X_n)$$

is called *p-integral definite* if

$$f(a_1, \ldots, a_n) \in \mathbb{Z}_p$$

for all $a_1, \ldots, a_n \in Q_p$, provided $f(a_1, \ldots, a_n)$ is defined which means that $a_1, \ldots, a_n$ is not a zero of the denominator of f. (By continuity, it suffices that $f(a_1, \ldots, a_n) \in \mathbb{Z}_p$ on some Zariski-open subset of Q_p^n.)

Problem: To describe all *p*-integral definite rational functions $f \in Q_p(X_1, \ldots, X_n)$. Let us call this the Problem $(17)_p$—it is the p-adic analog to Hilbert's 17th problem.

In the real case, positive definite functions are sums of squares. In the *p*-adic case, the square operator x^2 has to be replaced by some other operator $\gamma_p(x)$

which is adapted to p-adic theory. Kochen [Ko1] has found such an operator, which in p-adic algebra plays a similar role as does the square operator in real algebra:

$$\gamma_p(x) = \frac{1}{p}\left(\wp(x) - \frac{1}{\wp(x)}\right)^{-1}$$

with $\wp(x) = x^p - x$ the Artin-Schreier operator. This operator looks somewhat complicated; no wonder that the development of p-adic algebra was delayed. Merckel in his thesis [Me] gave a detailed study of those rational functions $\eta(x) \in \mathbb{Q}_p(x)$ which may serve, in p-adic algebra, in the same way as does $\gamma_p(x)$. It is clear that $\eta(x)$ cannot be a polynomial because polynomials have a p-adic pole (at infinity) and hence cannot be p-integral definite. Merckel has found certain admissible functions $\eta(x)$ of smaller denominator degree than $\gamma_p(x)$, but it seems that $\gamma_p(x)$ is the most "natural" operator. In the following we write $\gamma(x)$ instead of $\gamma_p(x)$, since p remains fixed in the discussion. $\gamma(x)$ is called the *p-adic Kochen operator*.

It is easily verified that $\gamma(x)$ is p-integral definite, i.e. $\gamma(a) \in \mathbb{Z}_p$ for every $a \in \mathbb{Q}_p$. Consequently, for arbitrary

$$g = g(X_1, \ldots, X_n) \in \mathbb{Q}_p(X_1, \ldots, X_n)$$

the function

$$\gamma(g) \in \mathbb{Q}_p(X_1, \ldots, X_n)$$

is p-integral definite too. Hence so is every expression of the form

$$\Phi\big(\gamma(g_1), \ldots, \gamma(g_r)\big)$$

where $\Phi(Y_1, \ldots, Y_r)$ is a polynomial over $\mathbb{Z}$ and

$$g_1, \ldots, g_r \in \mathbb{Q}_p(X_1, \ldots, X_n).$$

Note that

$$1 + p \cdot \Phi\big(\gamma(g_1), \ldots, \gamma(g_r)\big)$$

assumes values in $1 + p\mathbb{Z}_p$; these values are *units* in $\mathbb{Z}_p$. It follows that every rational function of the form

$$f = \frac{\Psi\big(\gamma(g_1), \ldots, \gamma(g_r)\big)}{1 + p\,\Phi\big(\gamma(g_1), \ldots, \gamma(g_r)\big)} \tag{F}$$

is p-integral definite, Φ and Ψ being arbitrary polynomials in $\mathbb{Z}[Y_1, \ldots, Y_r]$ (for some $r \in \mathbb{N}$) and $g_1, \ldots, g_r \in \mathbb{Q}_p(X_1, \ldots, X_n)$.

As said above, γ is the p-adic analog of the square operator. The expressions of the form (F) are the p-adic analogs of the "sum of squares" in the real theory.

Now Kochen [Ko1] and the author [Rq1] have proved that *every p-integral definite rational function can be put into the form* (F). This result can be viewed as the p-adic analog of Artin's theorem in the real case.

Its proof is a mere copy of A. Robinson's proof as given in the preceding section, *once* the proper notions and facts of p-adic algebra are established. These are as follows (for the proofs we refer e.g. to [PRq]).

Let K be a field. K is called *p-adic* if $\frac{1}{p}$ cannot be written in the form (F) with $g_1, \ldots, g_r \in K$. (This is the analog of the definition of *real* field: there, -1 cannot be written as a sum of squares). Every p-adic field K admits a maximal algebraic p-adic extension field L; the latter is called a *p-adic closure* of K. If $K = L$ then K is called *p-adically closed*. In a p-adically closed field K the elements of the form $\gamma(g)$ with $g \in K$ form a ring $\mathcal{O}_K$ which is in fact a valuation ring of the field K. The maximal ideal of $\mathcal{O}_K$ is generated by p, and the residue class field $\mathcal{O}_K/p$ is of order p. In general, for arbitrary field K, valuation rings with these two properties are called *p-valuation rings*. K admits a p-valuation ring if and only if K is a p-adic field. (Thus in p-adic theory, the p-valuations are the analogs of the ordering relations in the real theory.) Given any p-valuation ring $\mathcal{O}$ of K there exists a p-adic closure L of K whose canonical p-valuation ring $\mathcal{O}_L = \gamma(L)$ lies above $\mathcal{O}$, i.e. $\mathcal{O}_L \cap K = \mathcal{O}$. If an element $f \in K$ is contained in all p-valuation rings of K then f is called *totally p-integral*. This is the case if and only if f is of the form (F) as explained above. (The last statement is not quite straightforward to prove. It is easy to see that every totally p-integral $f \in K$ is a root of a monic polynomial whose coefficients are of the form (F). In order to show that f itself is actually of the form (F), one has to rely more heavily on commutative algebra; see [Rq1].)

All the above mentioned facts from p-adic algebra are quite analogous to the corresponding facts in real algebra. So is the following

Model Completeness Theorem: Every extension $K \subset L$ of p-adically closed fields is an elementary extension, i.e. the theory of p-adically closed fields is model-complete.

This is the Lefschetz-Tarski principle for p-adically closed fields. Based on this principle, the solution of "Hilbert's problem $(17)_p$" (i.e. the proof of the above mentioned Kochen-Roquette theorem) can be given straightforward, copying A. Robinson's proof of Artin's theorem. This will be left to the reader.

As to the proof of the p-adic model-completeness theorem, it is obtained by a similar pattern as described above in the real case. Firstly, general model theoretic considerations permit the reduction to the following embedding problem: Let $K(x)$ be a rational function field over a p-adically closed field K, and suppose that $K(x)$ is equipped with a p-valuation ring $\mathcal{O}$, i.e. $K(x)$ is *p*-valued. *Then $K(x)$ admits a K-isomorphic, valuation-preserving embedding into every sufficiently saturated non-standard model K^* of K.*

Secondly, in order to solve this embedding problem one has to give a description of the possible *p*-valuations of $K(x)$. It turns out that only two cases are possible:

(1) *Either $K(x)$ is ramified over K,* i. e. the value group of $K(x)$ is a proper extension of the value group of K. In this case, after suitable change of the generator x, the value group $v(K(x))$ is generated by $v(x)$ over $v(K)$. (We use v to denote the *p*-valuation of $K(x)$.) $v(x)$ is of infinite order modulo $v(K)$ and *the given p-valuation of $K(x)$ is uniquely determined by the cut which $v(x)$ determines in the (totally ordered) value group $v(K)$.* Using saturation property, we find $\xi \in K^*$ such that $v(\xi)$ determines the same cut in $v(K)$; hence the substitution $x \mapsto \xi$ yields the desired embedding $K(x) \approx K(\xi) \subset K^*$.

(2) *Or $K(x)$ is an immediate extension of K,* i. e. $K(x)$ and K have the same value groups and the same residue fields. In this case we consider the *neighborhood filter* which x determines on K, as follows. For each $c \in K$ consider the distance $v(x - c)$ from c to x; let U_c denote the set of those $a \in K$ which lie in the disc around x with radius $v(x - c)$. These sets U_c then form the neighborhood filter of x on K. It turns out that *the p-valuation of $K(x)$ is uniquely determined by this neighborhood filter.* Using saturation property we can find $\xi \in K^*$ which determines the same neighborhood filter on K; hence the substitution $x \mapsto \xi$ yields the desired embedding $K(x) \approx K(\xi) \subset K^*$.

Let us add some more words about the problem $(17)_p$. Its solution, as stated above, says that every *p*-integral definite function is representable in the form (F). A glance at (F) will convince the reader that this kind of representation looks much more involved than the corresponding "sum of squares"-representation in the real case. Hence one would like to have more information about (F). There is an effective bound for the number r of $g_i's$ and for the degrees of their numerators and denominators, the bound depending on the degrees of numerator and denominator of f (and of course on the number n of variables). The existence of such effective bound follows from general principles of model theory. No explicit form for this bound has yet been obtained; probably it would be of no great value for use in computations. In the real case Pfister [Pf] has shown that for n variables, any sum of squares is already the sum of 2^n squares. Is there a corresponding result for the *p*-adics, giving a bound for r in terms of the number of variables only? This is not known. While in the real case the theory of quadratic forms is available for the study of "sums of squares", no equivalent in the *p*-adics is known to investigate the structure of the Kochen operator. The only structure theorem known in this direction is the *principal ideal theorem* [Rq2] which says that the ring of elements of the form (F) is in fact a *Bezout ring:* every finitely generated ideal is principal. We have already mentioned Merckel's thesis [Me] searching for simpler operators $\eta(x)$ which can replace $\gamma(x)$. Perhaps this search should be extended to operators $\eta(x_1, \ldots, x_s)$ of several variables, and also to finite or infinite systems $\eta_1, \eta_2, \ldots$ which simultaneously can replace

$\gamma(x)$. We also mention the thesis by J. Unruh [Un] who investigated systematically whether the g_i in (F) can be restricted to a proper subset $E \subset \mathbb{Q}_p(X_1, \ldots, X_n)$, such that the validity of the main theorem is preserved. Such a set E must necessarily be infinite. More precisely, there must be infinitely many irreducible polynomials appearing in the numerator of elements of E, or likewise in the denominators of elements in E (if $n \geq 2$). Many similar results in [Un] show that such admissible E is fairly big; nevertheless it is possible to construct admissible E which are considerably smaller than the whole $K(X_1, \ldots, X_n)$. Here again, the situation is not yet fully clear.

Perhaps it is more suitable to investigate the problem from the *multiplicative* point of view; this would amount to the study of the *unit group* of the Kochen ring. (For quadratic forms, the multiplicative theory had led to Pfister's result mentioned above.) The above remarks have been inserted to point out that the *p*-adic Kochen operator is not yet fully understood and that more research in this direction would be desirable.

We have arranged the above discussion of *p*-adic algebra such as to match the foregoing discussion of real algebra. In particular, our discussion was focused around the proof of "Hilbert's problem $(17)_p$", the *p*-adic analog of the Hilbert problem 17 in the proper sense. We should mention, however, that historically the "problem $(17)_p$" did not play any role in the development of *p*-adic algebra. This problem $(17)_p$ was stated by Kochen [Ko1] only *after p*-adic algebra had been properly formulated. Independent of special problems to be solved, the formulation of *p*-adic algebra was given in full recognition of the fact that real algebra should have a counterpart in the *p*-adics. This recognition came about as a fall out from the work of Ax-Kochen on another prominent problem, namely Artin's conjecture about the C_2-property of $\mathbb{Q}_p$. (See also section 7.)

The model completeness theorem has been used for various other applications. For instance: the *p*-adic *Nullstellensatz*, the question about *p*-adic rational points in algebraic geometry, *p*-adic places and holomorphy rings in *p*-adic function fields etc. For details we refer e.g. to [PRq].

Let us explicitly state the *axioms for p-adically closed fields*:

(1)   the field axioms;

(2)   the elements of the form $\gamma_p(a)$ form a *p*-valuation ring $\mathcal{O}$ of the field;

$(3)_n$ Hensel's lemma with respect to $\mathcal{O}$, for polynomials of degree n.

Note that $(3)_n$ are infinitely many axioms; one for each $n = 1, 2, 3, \ldots$ Axioms $(3)_n$ express the fact that a *p*-adically closed field K is *Henselian* with respect to its canonical valuation. But the Henselian property does *not* yet suffice to characterize *p*-adically closed fields. There is another set of axioms which express the fact that the value group $v(K)$ is a $\mathbb{Z}$-group. Recall that a totally ordered abelian group Γ is called a $\mathbb{Z}$-group if it is elementary equivalent to $\mathbb{Z}$. This means that Γ contains a smallest positive element ε and for each n, the factor group Γ/n has exactly n cosets, represented by $0, \varepsilon, 2\varepsilon, \ldots, (n-1)\varepsilon$. In the case of a *p*-valued field K with valuation ring $\mathcal{O}$, the value group is isomorphic to the multiplicative group $K^{\times}$ modulo the units $\mathcal{O}^{\times}$. Its smallest positive element is the value of p. This gives rise to the following set of axioms:

(4)$_n$ Every field element a can be written in the form

$$a = p^i b^n u$$

for some $i = 0, 1, \ldots, n-1$, where b is a field element and u is a unit in the valuation ring $\mathcal{O}$.

A field is p-adically closed if and only if it satisfies the above axioms. The theory of p-adically closed fields is complete [PRq], and hence *the above axioms describe the algebraic theory of the Hensel field* $\mathbb{Q}_p$.

So again the situation is quite analogous to the situation in case of real algebra.

In number theory, not only the fields $\mathbb{Q}_p$ are of interest but also their finite extensions. These can be dealt with in the same way; the results are quite similar, the differences to the case of $\mathbb{Q}_p$ are of technical nature only. See [PRq].

6   Elimination of Quantifiers

Let T be a theory and $\varphi(x_1, \ldots, x_n)$ a formula in the language of T, where $x_1, \ldots, x_n$ are free parameters in $\mathscr{L}$. Given a model K of T and $a_1, \ldots, a_n \in K$ we want to know whether $\varphi(a_1, \ldots, a_n)$ holds in K. Usually a mathematician will ask for a necessary and sufficient criterion $\psi(x_1, \ldots, x_n)$, i.e. $\varphi(a_1, \ldots, a_n)$ holds in K if and only if $\psi(a_1, \ldots, a_n)$ does. $\psi(x_1, \ldots, x_n)$ should also be a formula in $\mathscr{L}$ and it should be independent of K, i.e. the criterion should hold in every model of T. If the criterion is to be of interest then $\psi(x_1, \ldots, x_n)$ should be somehow "simpler" than $\varphi(x_1, \ldots, x_n)$. If possible then $\psi(x_1, \ldots, x_n)$ *should be free of quantifiers*, so that $\psi(a_1, \ldots, a_n)$ can be checked directly by looking at $a_1, \ldots, a_n$ and the algebraic relations between them, without referring to the whole structure K.

The classical example of such problem is *elimination theory* in the theory of algebraically closed fields, formulated within the language of fields. In its simplest case, we are given two homogeneous forms $f(X, Y)$, $g(X, Y)$ in two variables. The resultant $R(f, g)$ is a certain polynomial in the coefficients of f and g. The vanishing of the resultant: $R(f, g) = 0$, is a quantifier-free criterion for the existence of a non-trivial common zero in an algebraically closed field. The parameters of the general set-up are here the coefficients of $f(X, Y)$ and $g(X, Y)$.

An arbitrary theory T is said to admit *quantifier elimination* in the language $\mathscr{L}$, if every formula $\varphi(x_1, \ldots, x_n)$ in $\mathscr{L}$ is T-equivalent, in the sense as explained above, to a quantifier-free formula $\psi(x_1, \ldots, x_n)$ in $\mathscr{L}$.

General model theory provides us with a very useful condition for the existence of quantifier elimination. Namely, a theory T admits elimination of quantifiers in the language $\mathscr{L}$ if and only if *every two models L_1, L_2 of T are substructure equivalent* in the following sense: If K is a common substructure, $K \subset L_1$ and $K \subset L_2$, then L_1 and L_2 should be elementary equivalent over K. This means that L_1 and L_2 satisfy the same sentences in $\mathscr{L}_K$, the language $\mathscr{L}$ augmented by constants to denote the elements of K. Note that in this condition, K need not be a model of T; the axioms of T need not be satisfied in K. The only requirement is that K is a substructure of L_1 and of L_2 with respect to all the relations and functions belonging to the vocabulary of the language $\mathscr{L}$.

For instance, consider the theory of algebraically closed fields within the language of fields. To check the above condition, consider two algebraically closed fields L_1 and L_2 which contain a common subfield*) K, not necessarily algebraically closed. But then L_1, L_2 each contain an algebraic closure $\tilde{K}_1$ resp. $\tilde{K}_2$ of K. By Steinitz' theory it follows that any two algebraic closures of K are K-isomorphic: $\tilde{K}_1 \underset{K}{\approx} \tilde{K}_2$. After identifying $\tilde{K}_1 = \tilde{K}_2 = \tilde{K}$, both L_1 and L_2 now appear as extensions of the algebraically closed field $\tilde{K}$. By model completeness: $\tilde{K} \prec L_1$, $\tilde{K} \prec L_2$, hence L_1 and L_2 are elementary equivalent over $\tilde{K}$ and *a fortiori* over K. *We conclude that the theory of algebraically closed fields admits elimination of quantifiers.*

This has been first proved by Tarski [TM], and today it constitutes one of the prominent class-room examples for quantifier elimination. It explains why one has to expect a condition of the type of the resultant: $R(f, g) = 0$, for the existence of common non-trivial zeros of f and g. But of course the explicit form of the resultant as given in algebra textbooks will *not* come out of such general considerations. In each special case, the explicit form of the quantifier-free ψ equivalent to φ is to be the object of detailed investigation.

The theory of real closed fields admits elimination of quantifiers in the language of ordered fields. The ordinary language of fields does not suffice in this case. This is clearly seen by testing the above criterion of substructure equivalence: L_1 and L_2 are now real closed fields, containing the subfield K. Hence they contain real closures $\tilde{K}_1, \tilde{K}_2$ respectively, of K. Now if we work in the language of ordered fields then K is an *ordered* subfield of L_1 and L_2; this means that L_1 and L_2 induce the same ordering in K. It follows that $\tilde{K}_1$ and $\tilde{K}_2$ are real closures *with respect to the same ordering* of K. Hence by Artin-Schreier [AS] they are K-isomorphic. The rest of proof can proceed in the same way as in the case of algebraically closed fields, this time using model-completeness for the theory of real closed fields. On the other hand, if we work in the language of fields then K is just a subfield of L_1 and of L_2. The orderings of K induced by L_1, L_2 may be

---

*) A substructure of a field is an integral domain (provided the vocabulary of field theory is taken to be the same as for ring theory, with no specific symbol for division). Now if a field L contains an integral domain D then L contains, canonically, the quotient field K of D.

distinct and, in that case, $\tilde{K}_1$ and $\tilde{K}_2$ are *not* K-isomorphic. The proof breaks down and, in fact, we have counterexamples showing that there are models of the theory which are *not* substructure-equivalent.

The most prominent example for elimination of quantifiers in real algebra is given by *Sturm's theorem*. For given n, consider monic polynomials $f(X)$ of degree n. Sturm's theorem gives a quantifier free criterion for $f(X)$ to admit a root ϑ in a given interval $a \leq \vartheta \leq b$. (The parameters in this problem are the coefficients of f and the end points of the interval.) The Sturm criterion is formulated in the language of *ordered* fields; it is *not* possible to formulate it without quantifiers in the theory of fields only, in the absence of the relation symbol $\leq$. (Of course, every inequality $x \leq y$ may be replaced in real closed fields by $(\exists z)(y - x = z^2)$. But the latter formula contains an existential quantifier.)

According to Tarski, the elimination of quantifiers for the algebraic theory of $\mathbb{R}$ implies elimination of quantifiers for Euclidean geometry. Was this what Steinitz had in mind when he announced applications of his algebra to geometry? Most probably we shall never know.

Now let us discuss elimination of quantifiers in p-adic algebra. Due to our experience in real algebra, we work p-adically in the *language of valued fields*, not just in the language of fields. But even in the language of valued fields, elimination of quantifiers does *not* hold for the theory of p-adically closed fields. For if we try to verify the condition of substructure equivalence in the same manner as in the above two cases, then we arrive at the following problem: Let K be a p-valued field and consider the p-adic closures $\tilde{K}$ whose canonical p-valuation induces the given valuation in K. Are all such p-adic closures K-isomorphic? (As it is the case in the corresponding problem for real closures.) The answer is: *No*, there are counterexamples. This implies that quantifier elimination fails.

In view of this situation there arises the problem of classification of the various p-adic closures $\tilde{K}$ of a p-valued field K. It is no restriction to assume K to be Henselian, for in any case, the Henselization K^h of K is contained in $\tilde{K}$. Now the classification problem for $\tilde{K}$ leads to certain invariants of cohomological nature which, as it turns out, can be explicitly identified. There are infinitely many, in fact uncountably many non-isomorphic p-adic closures $\tilde{K}$ of K (except in the trivial case when K itself is already p-adically closed). In any case, it can be proved that *every p-adic closure $\tilde{K}$ can be obtained from K by the adjunction of radicals* [PRq]. (An n-th radical ϑ over K is the form $\vartheta = \sqrt[n]{a}$ with $a \in K$; in other words, ϑ is the root of the binomial $X^n - a$.) As a consequence, *it can be shown* [PRq] *that two p-adic closures $\tilde{K}_1$, $\tilde{K}_2$ of a p-valued field K are isomorphic if and only if*

$$\tilde{K}_1^n \cap K = \tilde{K}_2^n \cap K$$

for each $n = 1, 2, \ldots$ Here, $\tilde{K}_1^n$ denotes the set of n-th powers of elements in $\tilde{K}_1$;

hence $\tilde{K}_1^n \cap K$ is the set of those elements $a \in K$ which admit n-th radicals $\sqrt[n]{a}$ in $\tilde{K}_1$. (Similarly for $\tilde{K}_2^n \cap K$.)

The above results imply that elimination of quantifiers can be achieved, also in the *p*-adic case, in a certain extended language. Let us extend the language of valued fields by adjoining predicate symbols $P_1, P_2, P_3, \ldots$ In *p*-adically closed fields $P_n(x)$ it to be interpreted as x being an *n*-th power. Now if L is *p*-adically closed and K is a *substructure of L with respect to this extended language*, then:

(1) K is a subfield of L,

(2) K carries a valuation ring $\mathcal{O}$ which is induced by the canonical valuation ring $\mathcal{O}_L$, i.e. $\mathcal{O} = K \cap \mathcal{O}_L$.

(3)   For each n, K carries a certain distinguished subset $P_n(K)$ which is induced by the set of n-th powers of L, i.e. $P_n(K) = K \cap L^n$.

Consequently, if K is a substructure (in the extended sense) of two *p*-adically closed fields L_1 and L_2 then $K \cap L_1^n = K \cap L_2^n$. If $L_1 = \tilde{K}_1$ and $L_2 = \tilde{K}_2$ are *p*-adic closures of K we conclude from the above theorem that $\tilde{K}_1$ and $\tilde{K}_2$ are isomorphic. Consequently, if we work in the extended language then we can argue in the same way as in the real closed case or in the algebraically closed case. Hence, any two *p*-adically closed fields are substructure equivalent with respect to the extended language. Hence, *the theory of p-adically closed fields admits quantifier elimination in the extended language.*

According to our above definition, the "extended language" is obtained from the language of valued fields by adjunction of the new predicate symbols $P_1, P_2, \ldots$ It can be shown that *the above result remains valid if we start from the language of fields, not valued fields.*

It would be interesting to know explicitly *the p-adic analog of Sturm's theorem*, giving a criterion for a monic polynomials $f(X)$ to have a *p*-integral root. This criterion should be quantifier-free in the extended language. In other words: the criterion should be expressible in the language $\mathscr{L}$ of fields, with quantifiers permitted in the form $(\exists y)(x = y^n)$ only, where x is a term. No such *p*-adic analog of Sturm's theorem is known, except in special cases when the polynomial satisfies in addition the conditions of Hensel's lemma (or of related lemmas).

Elimination of quantifiers in the extended language has been discovered by Macintyre [Ma]. As a consequence he proved that every infinite, definable subset of $\mathbb{Q}_p^n$ (the *n*-dimensional vector space over $\mathbb{Q}_p$) has a non-empty interior. For real closed fields, this was known before, in view of Tarski's elimination of quantifier theorem.

Another interesting application of elimination of quantifiers in the algebraic theory of $\mathbb{Q}_p$, has recently been given by J. Denef [Den]. The problem was to prove the rationality of the Poincaré series associated to the *p*-adic points of a variety, and of related series. In the course of proof a certain integral has to be evaluated over a certain subset D of $\mathbb{Z}_p^n$; this subset is definable and can be shown (by quantifier elimination) to be a Boolean combination of rather simple sets, for which the corresponding integral can indeed be evaluated, giving the desired information.

Quantifier elimination in the extended language can be carried over *mutatis mutandis* to finite extensions of $\mathbb{Q}_p$. See [PRq].

7   Diophantine Problems and Valued Fields

We have seen above that the theory of p-adically closed fields is complete, and that it admits a recursive set of axioms. Consequently, it follows that the theory is *decidable*. That is, there exists an effective procedure to determine for each sentence φ in the language of valued fields, whether φ does or does not hold in all p-adically closed fields. By completeness, this is equivalent to saying that φ holds (resp. does not hold) in $\mathbb{Q}_p$.

Now let V be an affine variety in n-space, given by polynomial equations of the form

$$f_1(x_1,\ldots,x_n)=0,\ldots,f_r(x_1,\ldots,x_n)=0. \tag{D}$$

We assume that the coefficients of these polynomials are integers in $\mathbb{Z}$. The *Diophantine problem* in $\mathbb{Z}$ asks whether (D) has a solution in integers $x_1,\ldots,x_n \in \mathbb{Z}$, i.e. whether V as a $\mathbb{Z}$-rational point. The corresponding problem for $x_1,\ldots,x_n \in \mathbb{Z}_p$ is *decidable*, by what we have said above. It has been shown by Weispfenning [We1] that the decision procedure can be made to be *uniform* in p, in a certain sense. This implies that *it is decidable whether the Diophantine problem (D) has a solution everywhere locally* or, in other words, whether there is a solution $x_1,\ldots,x_n$ whose coordinates are contained in the adele ring $A(\mathbb{Z}) = \prod_p \mathbb{Z}_p$ (direct product).

Now if we knew that the variety V satisfies Hasse's *local-global principle* (for integer points) then we could conclude that the original Diophantine problem (D) over $\mathbb{Z}$ is decidable. However, varieties with local-global principle are very rare. It has been shown by Julia Robinson and Matijasevic [DMR] that the *general* Diophantine problem (D) over $\mathbb{Z}$ is in fact *not* decidable. There does *not* exist an effective algorithm which permits for every system of equations of the form (D), to decide whether there is a solution $x_1,\ldots,x_n \in \mathbb{Z}$.

The situation does not improve if we replace $\mathbb{Q}$ by an algebraic number field K of finite degree and, accordingly, $\mathbb{Z}$ by the ring of algebraic integers of K. But we may go to the limit, arriving at the algebraic closure $\mathbb{Q}^a$ of $\mathbb{Q}$, the field of all algebraic numbers, and correspondingly at the integral closure $\mathbb{Z}^a$ of $\mathbb{Z}$, the ring of all algebraic integers. Does there exist a decision procedure for solutions of (D) with $x_1, \ldots, x_n \in \mathbb{Z}^a$? In other words: can Hilbert's 10th problem be positively solved for $\mathbb{Z}^a$ instead of $\mathbb{Z}$?

The conjecture that this may perhaps be the case had been voiced by Skolem [Sk]. He said that by general mathematical experience, many algebraic problems become decidable if one considers structures satisfying *closure properties* which are connected with the particular problem. This rather vague heuristic principle can be made precise, as we have seen, with respect to *algebraic closure, real closure, p-adic closure*. These closure properties render the respective theories decidable. Skolem's question is whether the algebraic theory of $\mathbb{Q}^a$, with distinguished subring $\mathbb{Z}^a$, is decidable? The proper language for this theory would be the theory of fields augmented by one predicate symbol to identify the elements in $\mathbb{Z}^a$.

The answer to the above general Skolem question is not yet known. However one may restrict this question and ask for the decidability at least of the Diophantine problems of the form (D), with solutions $x_1, \ldots, x_n \in \mathbb{Z}^a$.

This restricted question has recently been answered affirmatively by Rumely in a yet unpublished paper [Ru], based on former work by D. Cantor and the author [CRq]. The main point is *the local-global principle for Diophantine problems of the form* (D). This means the following: let $\mathbb{Q}_p^a$ denote the algebraic closure of $\mathbb{Q}_p$ and $\mathbb{Z}_p^a$ the integral closure of $\mathbb{Z}_p$ in $\mathbb{Q}_p^a$. Then $\mathbb{Z}_p^a$ is called the ring of algebraic p-adic integers. Now the local-global principle can be stated as follows: *If for each p, the Diophantine problem* (D) *admits a solution in* $\mathbb{Z}_p^a$ *then* (D) *admits solution* $x_1, \ldots, x_n \in \mathbb{Z}^a$.

In [CRq] this has been proved in the case when the variety V is unirational, i.e. parametrizable via rational functions. (More precisely, [CRq] covers the problem for *simple* points on unirational varieties.) The generalization to arbitrary varieties has been announced by Rumely, based on a detailed study of local heights on algebraic curves of higher genus.

In the course of proof of this local-global principle it turns out that for given V, only *finitely many* prime numbers are critical, in the sense that for the other primes the existence of a local solution in $\mathbb{Z}_p^a$ can easily be ascertained by general theorems. Consequently, the decision problem for global solutions is reduced to the local decision problem for a finite set of primes given in advance; by induction this is reduced to one single prime.

Now we recall that $\mathbb{Q}_p^a$ is an *algebraically closed field*, and $\mathbb{Z}_p^a$ is a *valuation ring* of $\mathbb{Q}_p^a$. We see that in the local case, we are studying *algebraically closed valued fields*. In this situation there is an old theorem of A. Robinson, saying that *the theory of algebraically closed valued fields is model-complete*, with respect to the language of valued fields. (It is understood that the valuation is non-trivial.) This, together with the fact that the axioms of algebraically closed fields are recursively enumerable, shows that *the algebraic theory of* $\mathbb{Q}_p^a$, *valued by* $\mathbb{Z}_p^a$, *is decidable*. In this respect we have for $\mathbb{Q}_p^a$, $\mathbb{Z}_p^a$ the same situation as described above for $\mathbb{Q}_p$, $\mathbb{Z}_p$.

In particular we see that, indeed, Diophantine problems (D) over $\mathbb{Z}^a$ are decidable.

Besides of its application to Diophantine problems over $\mathbb{Z}^a$, the theorem of A. Robinson is of importance also on its own. It admits a *classification of algebraically closed valued fields, up to elementary equivalence*, by two invariants: the characteristic p_1 of the field and the characteristic p_2 of the residue field.

For given invariants p_1, p_2 there exists a *prime model*, i. e. an algebraically closed valued field with invariants p_1, p_2 which is isomorphically contained in *every* algebraically closed valued field with the same invariants. If $p_1 = 0, p_2 = p$ > 0 then the prime model is $\mathbb{Q}^a$, the field of algebraic numbers, equipped with a valuation extending the p-adic valuation of $\mathbb{Q}$. (It does not matter which extension is chosen because all such extensions are conjugate over $\mathbb{Q}$.) If $p_1 = p_2 = 0$ then the prime model is $\mathbb{Q}(x)^a$, the field of algebraic functions, equipped with any valuation over $\mathbb{Q}$ (it does not matter which). Similarly, if $p_1 = p_2 = p > 0$ then the prime model is $\mathbb{F}_p(x)^a$ with any valuation ($\mathbb{F}_p$ denoting the field with p elements). By general model theory, the existence of prime models together with model-completeness implies completeness. Hence the theory of algebraically closed valued fields with given invariants (p_1, p_2) is complete.

By the way, the theory of algebraically closed valued fields admits *elimination of quantifiers*, because any two models are substructure-equivalent. Indeed, if K is a valued field contained in two algebraically closed valued fields L_1 and L_2 then the algebraic closure K^a, as a field, is isomorphically contained in both L_1 and L_2. The valuations of K^a induced by L_1 and by L_2 are both extensions of the given valuation of K. Hence they are *conjugate* over K. Hence after applying a suitable K-isomorphism we may assume that both L_1 and L_2 induce in K^a the same valuation. By model-completeness, L_1 and L_2 are now elementary equivalent over K^a, hence a priori over K. So the argument is completely analogous to the corresponding argument in the case of p-adically closed fields, or real closed fields, or algebraically closed fields. Model theory provides the proper framework for questions of this kind.

A. Robinson published his theorem about algebraically closed valued fields already in the year 1956 [Ro 1]. But it seems that it did not become widely known among algebraists, at least not as much as it would deserve in view of its importance. This is somewhat curious because algebraic geometry over *valued* fields has been quite well recognized as an important object of study, via Hensel's p-adic theory.

There have been several generalizations of Robinson's above theorem, concerning valued fields which are not algebraically closed. The first motivation came from the work of Ersov [Er 1], Ax and Kochen [AK] about Artin's conjecture on the C_2-property of $\mathbb{Q}_p$. Let us briefly discuss the model theoretic framework of the problem, without going into the detail about the content of the C_2-property.

Let φ be a sentence in the language of valued fields, and suppose we want to know whether φ holds in the fields $\mathbb{Q}_p$. If the problem is difficult then we may first try the power series fields $\mathbb{F}_p((x))$ in one variable x over the finite field $\mathbb{F}_p$ with p elements. $\mathbb{F}_p((x))$ is a complete valued field with the same residue field as $\mathbb{Q}_p$ (viz. $\mathbb{F}_p$) and the same value group (viz. $\mathbb{Z}$). Mathematical experience shows that often (though not always) the field $\mathbb{F}_p((x))$ is easier to investigate than $\mathbb{Q}_p$. Suppose then that we have found that φ holds in $\mathbb{F}_p((x))$, for all prime numbers p. Can we infer from this that φ holds in every $\mathbb{Q}_p$?

Consider the ultraproduct $K = \prod_p \mathbb{F}_p((x))/\mathscr{D}$ of the fields $\mathbb{F}_p((x))$ *modulo a non-principal ultrafilter $\mathscr{D}$ on the set of all prime numbers p.* This ultraproduct is a valued field, and the algebraic properties of K are precisely those which hold in $\mathscr{D}$-almost all factors $\mathbb{F}_p((x))$. (That is, there should exist a set D of the ultrafilter such that the property holds in $\mathbb{F}_p((x))$ for all $p \in D$.) In particular, φ holds in K. We now compare K with the corresponding ultraproduct of p-adic fields: $K' = \prod_p \mathbb{Q}_p/\mathscr{D}$. We note that K, K' both have the same residue field, namely $\prod_p \mathbb{F}_p/\mathscr{D}$, which is of characteristic zero. Also, they have the same value group, namely $\prod_p \mathbb{Z}/\mathscr{D}$; this is an ultrapower of $\mathbb{Z}$ and hence elementary equivalent to $\mathbb{Z}$. Both K and K' are Henselian. This is because the Henselian property can be described by (infinitely many) sentences in the language of valued fields; since each factor $\mathbb{F}_p((x))$ resp. $\mathbb{Q}_p$ is Henselian it follows that K and K' are indeed Henselian. From the above it follows that K, K' are elementary equivalent via the following

Theorem of Ersov [Er 1]. *Consider two Henselian valued fields K, K' whose residue fields are of characteristic 0. If the residue fields of K and K' are elementary equivalent in the language of ordered groups, then K and K' are elementary equivalent in the language of valued fields.*

It follows that φ holds in K', hence φ holds for $\mathbb{Q}_p$ if p belongs to a certain set D of the ultrafilter $\mathscr{D}$. Since $\mathscr{D}$ is an *arbitrary* non-principal ultrafilter on the set of primes we conclude: φ holds in almost all $\mathbb{Q}_p$ (i.e. all but for a finite number of primes p). Consequently, by means of Ersov's theorem we have obtained the following important *transfer principle: If φ holds in almost all power series fields $\mathbb{F}_p((x))$ then φ holds in almost all $\mathbb{Q}^p$, and vice versa.*

This transfer principle applies to all sentences φ in the language of valued fields, not just to those sentences which make up the C_2-property.

Note that Ersov's theorem contains Robinson's completeness theorem in case of residue characteristic 0. For if K, K' are algebraically closed valued fields then their residue field $\bar{K}, \bar{K'}$ are algebraically closed too; by the Lefschetz principle it follows that $\bar{K}, \bar{K'}$ are elementary equivalent. Moreover, their value groups $v(K)$ and $v(K')$ are *divisible*. It is known that all divisible, totally ordered abelian groups are elementary equivalent. Thus indeed, Ersov's theorem shows that K, K' are elementary equivalent. Hence Robinson's completeness theorem.

There arises the question whether Ersov's theorem can be generalized to the case of residue characteristic $p > 0$. This is true if the following additional condition is satisfied: K is of characteristic 0 and is absolutely unramified, i. e. the value $v(p)$ is the smallest positive element in the value group. In addition, the residue field is to be perfect. It is clear that this result includes e. g. the completeness theorem for p-adically closed fields.

If K too is of characteristic $p > 0$ then the Henselian property of K is not sufficient in this context. Itn seems natural to replace it by the property of K to be *immediately closed* This means that K (as a valued field) should not admit any proper algebraic immediate extension. Every immediately closed field is Henselian, but not conversely.

Ersov's theorem now holds for char $(K) = p > 0$, if K is *immediately closed* and, moreover, K satisfies the so-called Kaplansky hypothesis (A). This means, firstly, that the residue field $\bar{K}$ does not admit any algebraic extension of finite degree divisible by p; equivalently, every additive polynomial with coefficients in $\bar{K}$ should have a root in $\bar{K}$. Secondly, the value group $v(K)$ should not admit, in its divisible hull, any extension of finite degree divisible by p; equivalently, every equation $p\xi = \alpha$ with $\alpha \in v(K)$ should have a solution $\xi \in v(K)$. See [Ka], [Wh].

The above generalizations of what we called Ersov's theorem have also been proved by Ersov himself [Er1], and independently by Ax-Kochen [AK].

Kaplansky's hypothesis (A) does not seem unnatural in this context, but many fields appearing in mathematical nature do *not* satisfy it. Some effort has been spent to investigate other cases where hypothesis (A) does not apply. See e. g. Delon [Del]. Pank [Pa] has studied the Galois theoretic interpretation of hypothesis (A). Kuhlmann [Ku] has studied Hensel fields where hypothesis (A) is *not* satisfied. But the results obtained so far are not yet conclusive.

There have been generalizations of the theorems of Ersov and Ax-Kochen. An extensive literature is centered around it. A thorough and quite general study has been made by Basarab [Bas]. See also Transier [Tr].

8   Concluding Remarks

As pointed out already, our report is not meant to be comprehensive. There are many more instances where the intrusion of model theoretic concepts and methods into algebra could have been demonstrated. The reader may consult the papers of our list of references, or the literature cited in those papers. Here it was our purpose to get the reader interested in this new kind of algebraic reasoning. But perhaps we should avoid the word "new" in this context because as we have seen the roots of model theory can be traced right back to the beginning of "Modern Algebra" (in the sense of van der Waerden's book [vdW]). In fact, *model theory is naturally and inherently an offspring of the axiomatic viewpoint of contemporary mathematics.* So when today we observe model theory playing an important role in algebra, then this is not surprising and not exactly new. But now mathematicians are becoming more and more conscious about the *close connection between mathematical structures and the formal language $\mathscr{L}$ which is used to describe those structures.*

Of course there is no inherent reason why to restrict the language to be elementary (or first order), as we did in this article. For even when the axioms can be stated in an elementary language (as is the case for most algebraic theories) then one may be interested in properties which are of higher order. For instance in ring theory, the property of a ring to be *Noetherian* cannot be expressed in the elementary language of ring theory. In the theory of valued fields, *topological properties* are in general not of elementary nature. Hence if one wants to include such properties into the investigation then one has to work with higher order languages, adapted to the specific purpose.

A beautiful example in this direction is the work of Prestel and Ziegler [PrZ] about the so-called *local theory of topological fields.* In this theory, the language $\mathscr{L}$ contains (besides of the ordinary vocabulary of field theory) a set of additional variables U to denote neighborhoods of 0. Accordingly, the language contains a relation symbol $t \in U$ to denote that t belongs to U; here t may be any term in the language of fields. A sentence φ in this language is called *local* if, for every topological field K in which φ holds, φ remains true if the range of the set variables is restricted to a *basis* of neighborhoods around 0. Similarly if φ does not hold in K then it is required that φ remains false if the range of the set variables is restricted to a *basis* around 0. For instance the Hausdorff axiom is local in this sense: it requires that for $a \neq b$ there exist neighborhoods U, V such that $(a + U) \cap (b + V) = \emptyset$; clearly this can be expressed in the above language $\mathscr{L}$ and, if it holds in K then it also holds if U, V are restricted to a basis, and conversely. By inspection one verifies that all the axioms of topological fields, if stated in terms of neighborhoods around 0, are local.

The above definition of local sentences is semantic; they can also be defined syntactically: see [PrZ].

Two topological fields are called *locally equivalent if they satisfy the same local sentences.*

Now it is shown in [PrZ] that *the complex number field $\mathbb{C}$ is locally equivalent to each $\mathbb{Q}_p^a$, the algebraic closure of the p-adic Hensel field $\mathbb{Q}_p$.* In fact: all algebraically closed topological fields of characteristic 0 are locally equivalent, provided the topology is Hausdorff and non-discrete. This is a "*Lefschetz principle*" *for topological algebraically closed fields.* It seems remarkable that the field $\mathbb{C}$ with the ordinary archimedean absolute value, and the fields $\mathbb{Q}_p^a$ with their non-archimedean valuations referring to different primes p, are all equivalent with respect to their local topological properties. In a way this corresponds to the experience gained in developing p-adic analysis. Although the fields $\mathbb{C}$ and $\mathbb{Q}_p^a$ are essentially different as valued fields, experience has shown that they show a similar behavior with respect to *local algebro-topological properties.* Perhaps it will be possible to extend the Prestel-Ziegler result also to local *analytic* properties, if the $\mathbb{Q}_p^a$ are replaced by their completions.

As to the fields $\mathbb{R}$ and $\mathbb{Q}_p$, they are *not* locally equivalent because they are not even elementary equivalent in the language of fields. On the other hand, these fields do share certain local properties, for instance the following: For every n, the monic polynomials in $K[X]$ of degree n which have a *simple* zero in K form an *open set*, in the space of all monic polynomials of degree n. It is shown [PrZ] that this is equivalent to the validity of the *Implicit Function Theorem* (for polynomials) to hold in the field. Every such field is locally equivalent to a topological field L which admits a basis of neighborhoods around 0 consisting of *Henselian valuation ideals* (i.e. maximal ideals of Henselian valuation rings of L). The nice thing about this is, that e.g. the Implicit Function Theorem for $\mathbb{R}$ (resp. for $\mathbb{Q}_p$) can be reduced to the ordinary Hensel's Lemma in such field L. Since the original work by Hensel [He], Rychlik [Ry] and Ostrowski [Os] it has become common knowledge that there is a great similarity between Hensel's lemma and the classical analytic theorem for implicit functions (or continuity of roots etc.). Here too, model theory provides the proper framework to give this experience a precise meaning and, moreover, model theory gives us new ways of mathematical reasoning: e.g. to *deduce* analytic theorems for $\mathbb{R}$ or $\mathbb{C}$ from properties of non-archimedean valued fields.

It is to be hoped that further investigations can provide us with even more detailed information about the similarities in local analytic behavior of the fields $\mathbb{R}$ and $\mathbb{Q}_p$.

References

[AS]   Artin, E., O. Schreier, *Algebraische Konstruktion reeller Körper. Über die Zerlegung definiter Funktionen in Quadrate. Eine Kennzeichnung der reell abgeschlossenen Körper.* Hamburger Abhandlungen **5** (1926), 85—231.

[AK]   Ax, J., S. Kochen, *Diophantine problems over local fields*, I, II. Amer. J. Math. **87** (1965), 605—648. III. Annals of Math. **83** (1966), 437—456.

[Bar]   Barwise, J. (ed.), *Handbook of Mathematical Logic.* North Holland (1978).

[Bas]   Basarab, S., *A model theoretic transfer theorem for henselian valued fields.* Crelle's Journal 311/312 (1979), 1—30.

[BeJ]   Becker, E., B. Jacob, *Rational points on algebraic varieties over a generalized real closed field.* A model theoretic approach. To appear in Crelle's Journal (1984).

[CRq]   Cantor, D., P. Roquette, *On diophantine equations over the ring of all algebraic integers.* Journ. Number Theory **18** (1984), 1—26.

[CK]   Chang, C., J. Keisler, *Model Theory.* North Holland (1973).

[Ch]   Cherlin, G., *Model theoretic algebra.* Selected topics. Springer Lecture Notes **521** (1976).

[DMR]   Davis, M., Yu. Matijasevic, J. Robinson, *Hilbert's tenth problem. Diophantine equations: positive aspects of a negative solution.* Proc. Symp. Pure Math. **28** (1976), 223—378.

[Del]   Delon, F., *Quelques propriétés des corps valués en théorie des modèles.*Thèse, Paris (1981).

[Den]   Denef, J., *The rationality of the Poincaré series associated to the p-adic points on a variety* (Second version). Preprint (1983).

[DK]   Delfs, H., M. Knebusch, *On the homology of algebraic varieties over real closed fields.* Crelle's Journal **335** (1982), 122—163.
[Dr1]   van den Dries, L., *Model theory of fields.* Thesis, Utrecht (1978).
[Dr2]   van den Dries, L., *Reducing to prime characteristic, by means of Artin approximation and constructible properties, and applied to Hochster algebras.* Comm. Math. Inst. Utrecht **16** (1983).
[DrR1]   van den Dries, L., P. Ribenboim, *Lefschetz principle in Galois theory.* Queen's Math. preprint **5** (1976).
[DrR2]   van den Dries, L., P. Ribenboim, *Application de la théorie des modèles aux groupes de Galois de corps de fonctions.* C. R. Acad. Sciences Paris **288** (1979), série A, 789—792.
[Du]   Dubois, D., *A Nullstellensatz for ordered fields.* Ark. Math. **8** (1969), 111—114.
[Ek]   Eklof, P., *Lefschetz's principle and local functors.* Proc. AMS **37** (1973), 333—339.
[EGH]   Erdös, P., L. Gillman, M. Hendrikson, *An isomorphism theorem for real closed fields.* Annals of Math. **61** (1955), 542—554.
[Er1]   Ersov, Yu., *On the elementary theory of maximal valued fields* (russian). Algebra i Logika I: **4** (1965), 31—69. II: **5** (1966), 8—40. III: **6** (1967), 31—73.
[Er2]   Ersov, Yu., *Decision problems in constructible models* (russian). Moscow (1980) (translation to appear in Springer Verlag).
[He]   Hensel, K., *Theorie der Algebraischen Zahlen I.* Teubner Leipzig (1908).
[Ka]   Kaplansky, I., *Maximal fields with valuations.* Duke Math. J. **9** (1942), 303—321.
[Ko1]   Kochen, S., *Integer valued rational functions over the p-adic numbers.* In: Number Theory, Proc. Symp. Pure Math. XII. Houston (1967), 57—73
[Ko2]   Kochen, S., *The model theory of local fields.* In: Logic Conference Kiel (1974), Springer Lecture Notes **499**.
[Kri]   Krivine, J. L., *Anneaux préordonnés.* Journ. d'Analyse Math. **12** (1964), 307—326.
[KrN]   Krull, W., J. Neukirch, *Die Struktur der absoluten Galoisgruppe über dem Körper $R(t)$.* Math. Annalen **193** (1971), 197—209.
[Ku]   Kuhlmann, F. V. (to appear).
[La1]   Lang, S., *On quasi algebraic closure.* Annals of Math. **55** (1952), 373—390.
[La2]   Lang, S., *The theory of real places.* Annals of Math. **57** (1953), 378—391.
[La3]   Lang, S., *Some applications of the local uniformization theorem.* Amer. Journ. Math. **76** (1954), 362—374.
[Le]   Lefschetz, S., *Algebraic Geometry.* Princeton (1953).
[Ma]   Macintyre, A., *On definable sets of p-adic numbers.* J. Symb. Logic **41** (1976), 605—610.
[MaMD]   Macintyre, A., K. McKenna, L. van den Dries, *Elimination of quantifiers in algebraic structures.* Advances in Math. **47** (1983), 74—87.
[Me]   Merckel, M., *Darstellung des Ringes der total-p-adisch ganzen Elemente eines formal p-adischen Körpers als ganze Hülle von Quotientenringen von Wertbereichen geeigneter Funktionen.* Dissertation Heidelberg (1978).
[Os]   Ostrowski, A., *Untersuchungen zur arithmetischen Theorie der Körper.* Math. Z. **39** (1935), 269—404.
[Pa]   Pank, M., *Beiträge zur reinen und angewandten Bewertungstheorie.* Dissertation Heidelberg (1976).
[Pf]   Pfister, A., *Zur Darstellung definiter Funktionen als Summe von Quadraten.* Inventiones math. **4** (1967), 229—237.
[Po]   Potthoff, K., *Einführung in die Modelltheorie und ihre Anwendungen.* Wiss. Buchges. Darmstadt (1981).
[Pr]   Prestel, A., *Model theory of fields. An application to positive definite polynomials over* ℝ. To appear: Mémoirs de la Societé Mathématique de France.
[PRq]   Prestel, A., P. Roquette, *Formally p-adic fields.* Springer Lecture Notes 1050 (1984).
[PrZ]   Prestel, A., M. Ziegler, *Model theoretic methods in the theory of topological fields.* Crelle's Journal 299/300 (1978), 318—341.

[Ro1]   Robinson, A., *Complete theories*. North Holland (1956).
[Ro2]   Robinson, A., *On ordered fields and definite functions*. Math. Ann. **130** (1955), 257—271.
[Ro3]   Robinson, A., *Solution of a problem by Erdös-Gillman-Henrikson*. Proc. Amer. Math. Soc. **7** (1956), 908—909.
[Ro4]   Robinson, A., *Introduction to model theory and the metamathematics of algebra*. North Holland (1963).
[RoG]   Robinson, A., P. C. Gilmore, *Metamathematical considerations and the relative irreducibility of polynomials*. Canad. Journ. Math. **7** (1955), 483—489.
[RoRq]   Robinson, A., P. Roquette, *On the finiteness theorem of Siegel and Mahler concerning diophantine equations*. Journ. Number Theory **7** (1975), 121—176.
[Rq1]   Roquette, P., *Bemerkungen zur Theorie der formal p-adischen Körper*. Beitr. Algebra Geometrie **1** (1971), 177—193.
[Rq2]   Roquette, P., *Principal ideal theorems for holomorphy rings in fields*. Crelle's Journal 262/263 (1973), 361—374.
[Rq3]   Roquette, P., *p-adische und saturierte Körper*. Neue Variationen zu einem alten Thema von Hasse. Mitteilungen Math. Gesellschaft Hamburg **11**, Heft 1 (1982), 25—45.
[Ru]   Rumely, R., *Capacity theory on algebraic curves and canonical heights*. Manuscript (Sept. 8, 1982).
[Ry]   Rychlik, K., *Zur Bewertungstheorie der algebraischen Körper*. Crelle's Journal 153 (1924), 94—107.
[Sk]   Skolem, Th., *Lösung gewisser Gleichungen in ganzen algebraischen Zahlen, insbesondere in Einheiten*. Skrifter Norske Videnskap-Akad. Oslo I. Mat. Nat. Kl. No. 10 (1934).
[St]   Steinitz, E., *Algebraische Theorie der Körper*. Crelle's Journal 137 (1910), 167—309.
[TM]   Tarski, A., J. McKinsey, *A decision method for elementary algebra and geometry*. Univ. California Press Berkeley (1951).
[Tr]   Transier, R., *Verallgemeinerte formal p-adische Körper*. Archiv d. Math. **32** (1979), 572—584.
[Un]   Unruh, J., Dissertation Heidelberg (1984).
[vdW]   van der Waerden, B., *Moderne Algebra* (Erster Teil). Springer Verlag (1. Aufl. 1936).
[We1]   Weispfenning, V., *Die Entscheidbarkeit des Adelringes eines algebraischen Zahlkörpers*. Manuskript (1976). Vgl. auch Habilitationsschrift Heidelberg (1978).
[We2]   Weispfenning, V., *Nullstellensätze — A model theoretic framework*. Zeitschrift f. math. Logik und Grundlagen d. Math. **23** (1977), 539—545.
[We3]   Weispfenning, V., *Aspects of quantifier elimination in algebra*. Preprint (revised Nov. 1983).
[Wh]   Whaples, G., *Galois cohomology of additive polynomial and n-th power mappings of fields*. Duke Math. Journ. **24** (1957), 143—150

Perspectives in Mathematics
Anniversary of Oberwolfach 1984
© Birkhäuser Verlag, Basel

Fifteen Problems in Mathematical Physics

BARRY SIMON*

Departments of Mathematics and Physics,
California Institute of Technology, Pasadena, CA 91125 (USA)

Abstract   Presentation and discussion of a number of important open problems in mathematical physics.

0   Introduction

When the editors of this volume asked me to contribute, I had mixed feelings. Since I had recently written several long review articles, I was very reluctant to write another. One the other hand, I had fond remembrances of the scattering theory meetings I attended at Oberwolfach in 1971, 1974 and 1977, meetings which clearly had an important positive influence on the field. In thinking of the rather special character of Oberwolfach and its vitality, I realized an article which looks towards the future belonged among those rigthfully celebrating the past. The editors responded very warmly to my suggestion of an article on open problems in mathematical physics: hence this article. By looking towards the future, I also was able to survey broad areas of mathematical physics; unfortunately, Oberwolfach has intersected mathematical physics mainly in scattering theory and in classical mechanics, but I hope the future sees conferences in areas like quantum field theory, statistical mechanics and mathematical aspects of condensed matter physics!

It is with some misgivings that I set out in writing this article. Broad problem survey articles bring to mind Hilbert's famous article [1]. I am no Hilbert, and I certainly don't want anyone to think I feel any comparison is possible except using Lev Landau's logarithmic scale. Nevertheless, I have borrowed some of Hilbert's devices. While many of the problems stated are quite explicit and precise, some are so vague as to be close to ludicrous. Also, even more than Hilbert, I use the device of grouping several problems into "one", but when I do that, I have labeled them A, B, . . . Indeed, my 15 problems are really 32, explicitly:

*   Research partially supported by USNSF under Grant MCS-81-20833.

Problem 1 A: Almost always global existence for Newton's equations
         1 B: Existence of non-collisional singularities in the Newtonian N-body problem
Problem 2 A: Ergodicity of gases with soft cores
         2 B: Approach to equilibrium
         2 C: Asymptotic abelianness for the quantum Heisenberg dynamics
Problem 3   : Turbulence and all that
Problem 4 A: Fourier's heat law
         4 B: Kubo formula
Problem 5 A: Exponential decay of $v = 2$ classical Heisenberg correlations
         5 B: Pure phases at low temperatures in the $v \geqq 3$ classical Heisenberg model
         5 C: GKS for classical Heisenberg models
         5 D: Phase transitions in the quantum Heisenberg model
Problem 6   : Existence of ferromagnetism
Problem 7   : Existence of continuum phase transitions
Problem 8 A: Formulation of the renormalization group
         8 B: Proof of universality
Problem 9 A: Asymptotic completeness for short range N-body quantum systems
         9 B: Asymptotic completeness for Coulomb potentials
Problem 10 A: Monotonicity of ionization energy
        10 B: The Scott correction
        10 C: Asymptotic ionization
        10 D: Asymptotics of maximal ionized charge
        10 E: Rate of collapse of Bose matter
Problem 11   : Existence of crystals
Problem 12 A: Existence of extended states in the Anderson model
        12 B: Diffusive bound on "transport" in random potentials
        12 C: Smoothness of $k(E)$ through the mobility edge in the Anderson model
        12 D: Analysis of the almost Mathieu equation
        12 E: Point spectrum in a continuous almost periodic model
Problem 13   : Critical exponent for self-avoiding walks
Problem 14 A: Construct QCD
        14 B: Renormalizable QFT
        14 C: Inconsistency of QED
        14 D: Inconsistency of φ_4^4
Problem 15   : Cosmic censorship

   In deciding what is mathematical physics, I have generally tried to follow two basic rules: (1) Problems like "quantize gravity", where it is clear that the basic underlying physics is not understood, have not been included even if their solution is likely to involve a lot of mathematics. (2) Problems in "pure mathematics", even quite close to mathematical physics (like operator algebras) have generally not been included (which forces me to give some explanation in connection with Problem 13).

In an undertaking like this, I have benefited greatly from advice and information I received from a number of colleagues whom I consulted. I would like to thank Jürg Fröhlich, Bob Geroch, Jim Glimm, Anatoly Katok, Joel Lebowitz, Elliott Lieb, John Mather, Roger Penrose, Derek Robinson, Don Saari, Alan Sokal, Arthur Wightman and most especially, Tom Spencer, for their aid.

1   Existence for Newtonian Gravitating Particles

Newton's equations for N-particles of masses $m_1, \ldots, m_N$ interacting gravitationally in units where $G = 1$ are

$$m_i \ddot{\vec{r}}_i = \sum_{j \neq i} m_i m_j (\vec{r}_j - \vec{r}_i) |\vec{r}_i - \vec{r}_j|^{-3} \tag{1.1}$$

It is obvious that already for $N = 2$, (1.1) can fail to have solutions global in time for suitable initial conditions, e.g. $\dot{r}_1 = \dot{r}_2 = 0$. For $N = 2$, it is easy to see that the set of initial conditions leading to a collision is a subset of those conditions of total angular momentum zero, so the set of initial conditions for which global existence fails has measure zero if $N = 2$.

Problem 1 A (*Almost always global existence for Newton's equations*). Prove that the set of initial conditions for which (1.1) fails to have global solutions has measure zero in $\mathbb{R}^{6N}$.

We show our general feeling for what we believe is the answer, but we should emphasize that some excellent mathematicians believe that there may be an open set of initial conditions leading to non-global solutions.

To be more precise, the problem of singularities of (1.1) is connected with some pair colliding, i.e. we say a global solution fails to exist if at some finite time, T, $\lim_{\substack{t \uparrow T \\ i \neq j}} [\min |\vec{r}_i(t) - \vec{r}_j(t)|] = 0$. It is easy to see that the set of initial conditions, NE, leading to this is an F_σ so if NE has measure zero, its complement is automatically a dense G_δ.

We call a singular time, T, a *collision* if, for each i, $\lim_{t \uparrow T} \vec{r}_i(t)$ exists (and is a finite point). A *binary collision* is one where only pairs of $\vec{r}_i(T)$ are equal. The set of all initial conditions leading to a collision we will call C, and its complement in NE we call NC. The subset of C leading to binary collisions is denoted BC.

Painlevé [2] appears to have been the first person to have seriously discussed these questions and, in particular, he proved that NC is empty if $N = 3$. Much more recently, Saari [3] proved that NC has measure zero if $N = 4$. The analogous problem is open for $N \geq 5$ and, as we shall see, is the key question.

Birkhoff [4], applying a result of Sundman [5], showed that BC has measure zero, and in 1972—73, Saari proved [6]:

Theorem 1.1 C has measure zero and is Baire first category for any N.

This result does not immediately imply the same for various invariant subsets of lower dimension (i.e. $I \cap C$ has zero measure in the appropriate measure on an invariant subset of lower dimension). For I the set of configurations lying in a fixed plane, this is proven by Saari in the same references, and for the manifold of fixed angular momentum, it is a result of Urenko [7]. Saari's proof depends on an interesting and detailed analysis of precisely what happens at a collision.

Theorem 1.1 reduces an affirmative solution to problem 1 A to showing that NC has measure zero, and in particular, Saari's later result that NC has measure zero if $N = 4$ settles Problem 1 A in that case.

One general fact is known about NC, namely:

Theorem 1.2 (Sperling [8], based on ideas of von Zeipel [9]). For a solution whose initial conditions is in NC,

$$\lim_{t \uparrow T} \sum |\vec{r}_i(t)|^2 = \infty \qquad (1.2)$$

As we have remarked, it is known (Painlevé) that NC is empty when $N = 3$. There is no proof that it is not always empty, but there are strong indications it is not. First, for particles on a line there is an obvious way continuing through a binary collision (have the particles bounce off each other in their mutual center of mass frame). Mather and McGehee [10] found an initial configuration of 4 particles on the line which, if continued through binary collisions by this rule, have a time T which is an accumulation point of binary collisions, and (1.2) holds.

Recently, J. Gerver [11] produced a simple mechanism for non-collisional singularity in $N = 5$. He imagines a situation of 3 very massive particles at the edges of an isosceles triangle. A light "moon" is rotating about the particle, S, at the distinguished vertex and it is the "falling" of this moon into S that serves as the "engine" pumping energy into the system. A fifth particle travels more or less around this triangle. It moves essentially in a hyperbola as it swings around each vertex with the edges of the triangle being the asymptotes of the hyperbola. As it passes each vertex is gives an "outwards" kick to each particle. As it passes by S it picks up enough energy from the "engine" (i.e. the moon of S ends up in a smaller orbit after the passage of particle 5 trough the area of S) to enable it to continue its circuit around the enlarged triangle. Scaling arguments show that as the triangle gets bigger the circuit time of particle 5 decreases geometrically, and in finite time the triangle becomes infinitely larger. Gerver presents a number of detailed calculations to support this picture. Since he doesn't present a complete proof, we have

Problem 1 B (*Existence of non-collisional singularities in the Newtonian N-body problem*) Show NC is non-empty for some N and suitable m_i.

We caution the reader that Mather has made significant progress on making a rigorous proof of Gerver's scenario.

Next, a reason we tend to believe NE has measure zero. In quantum mechanics, it is a theorem of Kato [12] that global solutions of the Schrodinger equation with Coulomb potentials exist. Since quantum mechanics tends to only smooth out sets of measure zero, one expects that NE has measure zero. No doubt this reasoning will infuriate classical mechanics. In any event, if NE turns out to have an open subset, the *classical* limit of the corresponding quantum theory will be very interesting.

The quantum analog of Problem 1 A is, as we have noted, solved. Indeed, there is an enormous and more or less complete literature on the solubility of the Schrodinger equation summarized in Reed-Simon, Vol. II [13]. With the recent paper of Leinfelder-Simader [14] who solved one interesting open question in this area, only one basic selfadjointness question remains:

Jörgen's Conjecture Let $W(x) \geqq V(x)$ on R^ν and let M be a finite union of closed submanifolds in R^ν. Suppose that $-\varDelta + V$ is essentially selfadjoint on $C_0^\infty (R^\nu \smallsetminus M)$ and bounded below. Then $-\varDelta + W$ is essentially selfadjoint on $C_0^\infty (R^\nu \smallsetminus M)$.

We note there are counterexamples if the assumption the $-\varDelta + V$ is bounded below is dropped (see Pg. 155—156 [13]).

We have been careful not to include this among our list of problems. It has intrinsic interest but its importance is primarily technical. It is significant in part because it is over 10 years old and several technically strong people have worked on it without success.

2   Open Questions in Ergodic Theory

The founding fathers of statistical mechanics, especially Boltzmann and Gibbs, realized that the deepest aspect of thermodynamics from a microscopic point of view was the "zeroth law", that bulk systems rapidly approach equilibrium states parametrized by a few macroscopic parameters. By 1930, the standard wisdom was that the key notion is a proof that the classical dynamics on the constant energy manifolds of phase space is ergodic (see e. g. Avez-Arnold [15] for a discussion of the basic notions of ergodic theory). It is ironic that Sinai's celebrated result that the hard sphere gas is ergodic was announced [16] at approximately the same time that the KAM theory developed, for one important consequence of KAM is that many classical systems will not be ergodic: There will be an invariant subset of phase space consisting of a union of invariant tori of positive total measure.

It has been 20 years since Sinai's announcement, and a complete, detailed proof has not yet appeared except for the (nontrivial) case of two particles [17].

A partial sketch for $N = 3, 4, 5$ appears in [18]. Recently, Sinai and Chernoff [19] have proven that the Kolomogorov-Sinai entropy of a hard sphere gas is positive, and even that the entropy per particle is positive in the thermodynamic limit. (While these results are mathematically independent of ergodicity, the ideas in their proof are presumably an important aspect of a possible proof of ergodicity.)

Despite the blow that KAM gives to the 1930's wisdom, it is an interesting question to extend Sinai's proof beyound the hard sphere gas: His system in its simplest form involves N particles in a cubic box bouncing elastically off the walls and each other.

*Problem 2*A (*Ergodicity of gases with soft cores*) Find a class of repulsive smooth potentials for which the N-particle dynamics in a box (with, say, smooth wall potentials) is ergodic.

The expected lack of ergodicity for systems with interacting potentials which are not strictly repulsive requires a convincing revised standard wisdom to explain the approach to equilibrium. One idea advocated by Wightman [20] among others is that there is one ergodic component of such systems which, in the limit as the volume goes to infinity (with constant particle density), occupies a larger and larger fraction of phase space.

Problem 2 B (*Approach to equilibrium*) Verify the above scenario to justify approach to equilibrium of large systems with forces which are attractive at suitable distances, or else find an alternate scenario which doesn't rely on strict ergodicity in finite volume.

We want to emphasize that the studies of ergodicity of the dynamics of infinite partial systems (see e. g. [21]), while interesting mathematically, does *not*, in our opinion, address this issue. The ergodicity of the infinite particle non-interacting gas shows that this kind of ergodicity comes from the fact that one puts equilibrium into the system at infinity by the choice of underlying measure and that equilibrium "diffuses" into finite regions.

We also note that neither the standard wisdom or the above candidate for a revised standard wisdom addresses the basic question of why the approach to equilibrium in the real world is on a time scale so short compared to typical recurrence times in the system.

Finally, we should say a few words about approach to equilibrium in quantum systems which is very difficult for many reasons, e. g. in finite volume the systems tend to have discrete spectrum and thus almost periodic behavior in time. There has been some interesting study of approach to equilibrium in infinite quantum lattice systems, but even here much more is unknown than known. One basic question involves the notion of asymptotic abelianness under time translation. The notion was originally introduced under space translation where it is an obvious feature very useful in the abstract study of such systems (see e. g. Ruelle [22] and Bratelli-Robinson [23]). The relevance of these ideas in quantum systems is discussed in Chapter 6 of Bratelli-Robinson [23]. Unfortunately, the only examples where it is known there is asymptotic abelianness are quasi-free states and the closely related one-dimensional $X - Y$ model (on the even algebra). For simplicity we state the next problem for a definite model, but any non quasi-free, intrinsically non-abelian multidimensional model would be interesting.

Problem 2 C (*Asymptotic Abelianness for the Quantum Heisenberg Dynamics*) Prove (or disprove) that the multidimensional quantum Heisenberg model has asymptotically abelian dynamics.

3   Long Time Behavior of Dynamical Systems

Problem 3 (*Turbulence and all that*) Develop a comprehensive theory of the long time behavior of dynamical systems including a theory of the onset of, and of fully developed turbulence.
  This problem is so general as to be verging on the absurd. We include it in part to indicate our strong feeling that this is an area which is not only fashionable but important as well. We state it in this form because it seems the field is not yet at a level of maturity where one can focus on certain crucial questions; rather, the first problem is to formulate the really significant questions. For some recent reviews of some of the more spectacular developments in the area, see Feigenbaum [24] or the book of Collet-Eckmann [25].
  As for the question of turbulence, there has been considerable progress in understanding the onset of turbulence (see e. g. Ruelle [26] or Eckmann [27]), but our understanding of fully developed turbulence is far from fully developed.
  The connection between turbulence and the Navier-Stokes equation is not clear, but there may well be one. In this regard, we should note that the existence theory for this important equation is not completely satisfactory; see Foias-Tenam [28] for a review.

4   Transport Theory

  At some level, the fundamental difficulty of transport theory is that it is a steady state rather than equilibrium problem, so that the powerful formalism of equilibrium statistical mechanics is unavailable, and one does not have any way of precisely identifying the steady state and thereby computing things in it.
  A second difficulty concerns the fact that most transport is a diffusion phenomena and there is no satisfactory derivation of diffusion from an underlying microscopic dynamics except in some limit in which a physically fixed scale (like particle sizes) is varied rather than a physically varied scale (like system sizes).

To explain this diffusion remark in an example, consider a very crude model of a linear system with particles moving between a wall at 0, and another at L. We characterize the fact that we imagine the wall at 0 having temperature T_0 and the one at L having temperature T_1 by saying that upon collision with the wall, all 0 the particles always come off with velocity $v_0 \sim \sqrt{T_0}$ and upon collision with the wall at T_1 with velocity $v_1 \sim \sqrt{T_1}$. As L varies we imagine increasing the number of particles moving back and forth to keep their density fixed. If we change L and assume the particles are non-interacting, a simple calculation shows that the rate of energy transport between the two walls is unchanged although Fourier's heat law says that it should go as $(\varDelta T) L^{-1}$. If one, by fiat, imagines that particle interaction causes a diffusion of heat so that transit times go as L^2 not L, then the rate of heat transfer has the proper L^{-1} behavior.

The connection with diffusion links these transport questions with the material discussed in Section 12.

In the problem below, we would allow a model which brought temperature in even with as bad a caricature as the above crude model.

Problem 4 A (Fourier's Heat Law) Find a mechanical model in which a system of size L has a temperature difference $\varDelta T$ between its ends and in which the rate of heat transfer in the infinite L limit goes as L^{-1}.

There are also serious foundational questions in quantum transport. A basic formula in condensed matter physics is the Kubo formula for conduction; see e. g. [29] for discussion. Not only are the usual derivations suspect, but van Kampen [30], among others, has seriously questioned its validity on physical grounds.

Problem 4 B (Kubo Formula) Either justify Kubo's formula in a quantum model, or else find an alternate theory of conductivity.

5   Heisenberg Models

Lattice models of statistical mechanics have been fruitful testing grounds for ideas in the theory of phase transitions. The last 15 years have seen remarkable progress in the rigorous study of these models, especially the Ising model. For each site α in $\mathbb{Z}^v$ we imagine a spin $\vec{\sigma}_\alpha$ taking values in S^{D-1}, the unit sphere in D-dimensions. Given $\varLambda \subset \mathbb{Z}^v$, a finite subset, we define

$$H_\varLambda = - \sum_{<\alpha\gamma> \in \varLambda} \vec{\sigma}_\alpha \cdot \vec{\sigma}_\gamma \tag{5.1}$$

the sum being over all nearest neighbor pairs in $\varLambda$. Given a parameter $\beta =$ inverse temperatures, we form a probability measure on $(S^{D-1})^\varLambda$ by

$$\langle f \rangle_{\Lambda,\beta} = \int f(\sigma_\alpha) \, e^{-\beta H_\Lambda(\sigma_\alpha)} \prod_{\alpha \in \Lambda} \mathrm{d}\mu_0(\sigma_\alpha)/Z_\Lambda \qquad (5.2\,\mathrm{a})$$

$$Z_\Lambda = \int e^{-\beta H_\Lambda(\sigma_\alpha)} \prod_{\alpha \in \Lambda} \mathrm{d}\mu_0(\sigma_\alpha) \qquad (5.2\,\mathrm{b})$$

where $\mathrm{d}\mu_0(\sigma_\alpha)$ is the usual invariant measure on S^{D-1} (if $D = 1$, so $S^{D-1} = \{\pm 1\}$, $\mathrm{d}\mu_0(\sigma_\alpha) = \frac{1}{2}[\delta(\sigma_\alpha + 1) + \delta(\sigma_\alpha - 1)]$. By $\langle f \rangle_\beta$ we mean suitable limits as Λ approaches Z^ν. $D = 1$ is called the Ising model, $D = 2$ the plane rotor and $D = 3$ the classical Heisenberg model. These models are quite different because their symmetry groups are quite distinct: In $D = 1$ a discrete group, in $D = 2$ an abelian continuous group and in $D = 3$ a non-abelian continuous group.

Problem 5A (Exponential decay of $\nu = 2$, $D = 3$ correlations). Consider the two dimensional classical Heisenberg model ($\nu = 2$, $D = 3$). Prove that for any β, $\langle \sigma_\alpha \cdot \sigma_\gamma \rangle_\beta$ decays exponentially as $|\alpha - \gamma| \to \infty$.

Here is some background on this problem: If $\lim_{|\alpha - \gamma| \to \infty} \langle \sigma_\alpha \cdot \sigma_\gamma \rangle \neq 0$, one says the model has long range order (LRO), an indication of multiple phases (see Ruelle [22], Griffiths [31], Israel [32], or Simon [33]). For $D = 1$ (Ising), there is LRO when β is sufficiently large so long as $\nu \geq 2$ (Peierls [34]); for $D \geq 2$, it is known there is LRO for β large if $\nu \geq 3$ (Fröhlich et al. [35]), but if $\nu = 2$, there is no LRO for any β (Mermin-Wagner [36]). Dyson [37] gave an intuitive argument that when $\nu = 2$, $D \geq 2$ and β is large, $\langle \sigma_\alpha \cdot \sigma_\gamma \rangle_\beta$ should only have power decay; in the '70's it was realized in the non-rigorous theoretical physics literature (see e.g. [38]) that due to a renormalization group intuition, one should expect that there is this power decay when $D = 2$ but not if $D \geq 3$. Recently, Fröhlich-Spencer [39] have proven that if $D = 2$, $\nu = 2$ there is only power decay if β is large. The important open question above concerns whether the situation is different if $D \geq 3$, $\nu = 2$. Because of the connection with "infrared freedom", an important notion in Q.C.D., this problem has importance in quantum field theory.

The next problem concerns the structure of the set of "pure phases" ($\equiv$ extreme points of the set of translation invariant DLR states); we will not give the precise definition on this notion: See Ruelle [22], Israel [32] or Simon [33]. The symmetry group acts on the set of equilibrium states.

Problem 5B (Pure phase at low temperatures). Prove that at large β and $\nu \geq 3$, the set of equilibrium states for the $D = 3$ model forms a single orbit under $SO(3)$ which is the sphere S^2.

This result says that at fixed low temperature, the phases are characterized by a single unit vector describing, say, the direction of the magnetization. The analogous result for $D = 1$ was proven by Gallavotti-Miracle Sole [40] and for $D = 2$ by Fröhlich-Pfister [41]. It is likely that a solution of this problem will either involve developing new correlation inequalities for these types of models, or else one will understand these phenomena without correlation inequalities.

Problem 5C (GKS for classical Heisenberg models). Consider the model with $D = 3$, arbitrary v. Let f, g be finite products of the form $(\sigma_\alpha \cdot \sigma_\gamma)$. Is it true that

$$\langle fg \rangle_{\Lambda,\beta} \geq \langle f \rangle_{\Lambda,\beta} \langle g \rangle_{\Lambda,\beta} \tag{5.3}$$

for all Λ, β.

Actually, one wants this for more general ferromagnets than nearest neighbor coupling. (5.3) for $D = 1$ is the famous inequality of Griffiths [42], Kelly and Sherman [43]. It was extended to $D = 2$ by Ginibre [44] who obtained it from a generalized set of inequalities (Ginibre's inequalities). Shortly before his death, Sherman announced a proof of Ginibre's inequality, and therefore GKS for general D, but his notes seemed to contain an error. In fact, Sylvester [45] has recently proven that Ginibre's inequality is false for $D \geq 3$. This leaves the GKS situation open; it is generally believed they are true. Many other inequalities and many applications would immediately follow.

The final of our Heisenberg-model problems involves the quantum model. The phase space $(S^{D-1})^\Lambda$ is replaced by a Hilbert space $\mathbb{C}^{2^{|\Lambda|}}$ thought of as $\mathbb{C}^2 \otimes \cdots \otimes \mathbb{C}^2$ ($|\Lambda|$ times). $\sigma_{\alpha i}$ is the operator which is a tensor product of τ_i in the α factor and 1 in the others. Here τ_i are the standard Pauli matrices

$$\tau_1 = \begin{pmatrix} 0 & 1 \\ 1 & 0 \end{pmatrix} \qquad \tau_2 = \begin{pmatrix} 0 & -i \\ i & 0 \end{pmatrix} \qquad \tau_3 = \begin{pmatrix} 1 & 0 \\ 0 & -1 \end{pmatrix}$$

H_Λ is still given by (5.1) but (5.2) is replaced by

$$E(f) = \text{Tr}(f e^{-\beta H_\Lambda})/Z_\Lambda \tag{5.4a}$$

$$Z_\Lambda = \text{Tr}(e^{-\beta H_\Lambda}). \tag{5.4b}$$

Problem 5D (Phase transition in the quantum Heisenberg model). Prove that for $v \geq 3$ and β large, the quantum Heisenberg model has LRO in the sense that

$$\lim_{|\alpha - \gamma| \to \infty} \langle \sigma_\alpha \cdot \sigma_\gamma \rangle_\beta \neq 0.$$

A positive solution of this problem was announced by Dyson, Lieb and Simon [46], but they made an error. For the antiferromagnet, i.e. $\beta < 0$ and $|\beta|$ very large, Dyson, Lieb and Simon [47] prove that

$$\lim_{|\alpha - \gamma| \to \infty} |\langle \sigma_\alpha \cdot \sigma_\gamma \rangle| \neq 0.$$

Quite likely, this problem is connected with problem (vi) below.

Here are some other interesting open questions in lattice models: we suppose some familiarity with terminology (see [22, 31, 32, 33]).

(i) Let J be a non-negative function on Z^v. The model given by (5.2) with

$$H_\Lambda = - \sum_{\alpha\gamma \in \Lambda} J(\alpha - \gamma)\, \sigma_\alpha \cdot \sigma_\gamma$$

and $D = 1$ is called the general ferromagnetic Ising model. If $J(\alpha) = 0$ for all but finitely many α, one calls the model finite range. One defines

$$p(\beta) = \lim_{|\Lambda| \to \infty} |\Lambda|^{-1} \ln Z_\Lambda(\beta).$$

$p(\beta)$ is convex and so automatically differentiable for all but finitely many β. One expects that $p(\beta)$ is actually C^1 for these models. Prove it. This is important because of results of Lebowitz [48].

(ii) The one dimensional Ising model with $J(\alpha) = |\alpha|^{-2}$ for $\alpha \neq 0$ is especially interesting. An argument of Thouless [49] (made partially rigorous by Simon-Sokal [50]) suggests that the magnetization of this model is discontinuous in β. It is known (Fröhlich-Spencer [51]) that the magnetization is nonzero for β large. Prove the magnetization is discontinuous.

(iii) Consider the basic nearest neighbor model with $D = 1$, $v \geq 3$. Define

$$\beta_c^{(1)}(v) = \inf \{\beta | \lim_{|\alpha - \gamma| \to \infty} \langle \sigma_\alpha \cdot \sigma_\gamma \rangle \neq 0\} \quad \text{and}$$

$$\beta_c^{(2)}(v) = \sup \{\beta | \langle \sigma_\alpha \cdot \sigma_\gamma \rangle \leq C_1\, e^{-C_2 |\alpha - \gamma|} \text{ for some } C_1, C_2\}.$$

Clearly $\beta^{(1)} \geq \beta^{(2)}$. Prove they are equal.

(iv) There are interesting questions concerning the existence of equilibrium states ($\equiv$ DLR states) which are not translation invariant. For $D = 1$, $v = 2$, Aizenman [52] proved these don't occur. Dobrushin [53] proved for $D = 1$, $v \geq 3$, there are such states. Define $\beta_r(v)$, the roughening temperature, to be the inf over all β for which there exist nontranslation invariant states; van Beijeren [54] proved that $\beta_r(v) \geq \beta_c^{(1)}(v - 1)$. A basic question is that a "roughening transition occurs", i.e. $\beta_r(3) < \beta_c^{(1)}(3)$. There is reason to believe (see Fröhlich et al. [55]) that $\beta_r(v) = \beta_c^{(1)}(v)$ if $v \geq 4$. Prove or disprove this.

(v) Do plane rotors have nontranslation invariant states? If they do, the states will be quite different from those in the Ising case. See [55, 41] for further discussion.

(vi) Find additional methods for proving phase transitions occur when there is continuous symmetry ("spontaneously broken continuous symmetry"). At this point, all we have are reflection positivity methods (Fröhlich, Spencer, Simon [35]; Fröhlich et al. [56]) which are quite rigid in terms of when they apply and the "scales of contours" method of Fröhlich-Spencer [39] which seems to be restricted to $D = 2$.

6   Ferromagnetism

Mathematicians who have been exposed to the Ising model often delude themselves that in understanding that, they have understood the reason for magnetism. While it is true that the lesson of that model, namely that local interactions can cooperatively produce long range order, is an important aspect of ferromagnetism, it is not the only one nor the most mysterious one.

The point is that the Ising model postulates an interaction which tends to make neighboring spins point parallel. These spins which are associated with the magnetic moments of neighboring atoms produce bulk magnets by aligning in parallel. It is true that magnetic dipoles have direct interactions with each other, but the magnitude of such interactions is so small that they would set temperature scales much lower than those associated with real magnets (and they don't have the proper $\sigma_1 \cdot \sigma_2$ form to boot!).

The mysterious aspect of magnetism is what produces the strong effective spin aligning interaction. There is a standard explanation due to Heisenberg based on the Pauli principle: Since electron-electron interactions are repulsive, their spatial wave function wants to be as antisymmetric as possible (tending to keep them apart), so by the Pauli principle, their spin wave function is as symmetric as possible, which produces a tendency for parallel spins.

While this picture is quite possibly the correct one, it is far from proven: Indeed, in one space dimension, it is false! Lieb and Mattis [57] have shown that the total electron spin of the ground state of an even number of electrons in one dimension is zero!

Problem 6 (Explanation of Ferromagnetism). Verify the Heisenberg picture of the origin of ferromagnetism (or an alternative) in a realistic quantum system or in a suitable model.

7   Continuum Phase Transitions

Phase transitions are one of the more striking phenomena in nature. While there has been considerable rigorous understanding in the case of lattice systems, there has been virtually none on continuum models—a phase transition has been proven in only one rather artificial model [58].

To state the problem precisely, we quickly review some basic statistical mechanics. Because it uses more familiar quantities, we work in the canonical ensemble; technically, the grand canonical ensemble is often easier to deal with (see Ruelle [22]). We fix a pair potential, v obeying

(1)   (stability) For some C and all $x_1, \ldots, x_N \in R^3$:

$$\sum_{1 \leq i < j \leq N} v(x_i - x_j) \geq -CN$$

(2)   (temperedness) $|v(x)| \leq C(1 + |x|)^{-3-\varepsilon}$

Given a finite volume, Λ, in R^3, a number N and an inverse temperature β, we define the partition function, Z_Λ and free energy F_Λ by

$$F_\Lambda(\beta, N) = -\ln Z_\Lambda$$

$$Z_\Lambda = \int_{x_i \in \Lambda} \prod_{i=1}^{N} d^3 p_i \, d^3 x_i \, e^{-\beta H(p_i, \, x_i)}$$

$$H(p_i, x_i) = \sum_{i=1}^{N} \frac{p_i^2}{2} + \sum_{i<j} v(x_i - x_j).$$

One can show that if ϱ, a value of density, is fixed and if Λ approaches R^3 in a suitable way and $N/\Lambda \to \varrho$, then $|\Lambda|^{-1} F_\Lambda(\beta, N)$ has a limit, called $f(\beta, \varrho)$. This function is concave in β, so the one-sided derivatives exist at all points. A (first order in β) phase transition corresponds to f failing to be C^1:

Problem 7 (Existence of Continuum Phase Transitions). Show that for suitable choices of v, and for ϱ sufficiently large, f is non-C^1 at some β.
    A reasonable v to think about is a function like the Lenard-Jones potential $v(r) = ar^{-12} - br^{-6}$ which gets very large and positive for r small but has a small negative well in which particles can stick.
    Alternatively, instead of looking for a phase transition in β, one can pass to grand canonical ensemble and look for a transition in fugacity where the density jumps.

8   Rigorous Renormalization Group

One of the most celebrated developments in theoretical physics during the past 15 years is surely the "renormalization group theory of critical phenomena" of Fisher, Kadanoff and Wilson (see e. g. [59]). The basic idea of shifting scales as one approaches a critical point via a nonlinear map of Hamiltonians and obtaining information from the fixed points of that map is being applied in a variety of situations, e. g. the work of Feigenbaum [24] and parts of the philosophy are often present in work which doesn't embrace the full machine, e. g. the spirit of the renormalization group hovers over the recent work of Fröhlich-Spencer [39].
    In some of these analog studies, the nonlinear maps are on well defined spaces and there has been considerable progress on a rigorous mathematical analysis, e. g. the work of Collet, Eckmann and Landford [60] on the Feigenbaum theory. The original Wilson theory is on functions of inifinitely many variables and it is far from clear how to formulate the maps in a mathematically precise way (let alone then analyze their fixed point structure); indeed, there are various no-go theorems [61] to certain obvious ways one might try to make a

precise formulation. To make the following problem precise, we specialize to lattice systems:

Problem 8A (Formulation of the Renormalization Group). Develop a mathematically precise version of the renormalization transformations for v-dimensional Ising-type systems.

It may turn out that this problem can be finessed and one can get out renormalization group type results without a complete formalism. In this regard, see the work of Gawedski-Kupiainen [62].

It is often claimed that the renormalization group "explains" universality. It seems to me that it does not; rather, it assumes universality! For the kind of local analysis done in the renormalization group framework doesn't explain why the fixed points found seem to have "basins of attraction" which are all (or at least most) of the space of interactions. Thus:

Problem 8B (Proof of Universality). Show that the critical exponents in the three dimensional Ising models with nearest neighbor coupling but different bond strengths in the three directions are independent of the ratios of these bond strengths.

9   Asymptotic Completeness for Atomic Scattering

We begin with a brief description of multiparticle systems. See [63], Section XI.5, for more details. Consider n quantum mechanical particles of masses $m_1, \ldots, m_n$ moving in v-dimensions. After removing the center of mass motion, the wave functions live naturally on $\{(x_1, \ldots, x_n) \in R^{vn} | \sum m_i x_i = 0\}$ $= X$ (isomorphic to $R^{v(n-1)}$). Place the metric $d(x, y) = [\sum m_i (x_i - y_i)^2]^{\frac{1}{2}}$ on X and let H_0 be $(-\frac{1}{2})$ times the Laplace Beltrami operator in this metric. Here is an equivalent definition: Given any $(x_1, \ldots, x_n) \in R^{vn}$, let $R(x) = (\sum m_i)^{-1} \sum m_i x_i$ and let $\pi(x) = x - (R, R, \ldots, R)$. Given f a function on X, let $\pi^*(f)$ be the function on R^{vn}, given by $\pi^*(f)(x) = f(\pi(x))$. Then H_0 can be defined by

$$[-\sum (2m_i)^{-1} \Delta_{x_i}] \pi^*(f) = \pi^*(H_0 f)$$

H_0 is the kinetic energy of these particles.

Pick functions V_{ij} on R^v. For the time being, let us suppose that

$$|V_{ij}(x)| \leqq C(1 + |x|)^{-1-\varepsilon} \tag{9.1}$$

for some $\varepsilon > 0$. We write V_{ij} for the function on X given by $V_{ij}(x_i - x_j)$. On X we define

$$H = H_0 + \sum_{i,j} V_{ij}.$$

Let a be a partition of $\{1, \ldots, n\}$, i.e. a family of $\#(a)$ disjoint subsets whose union is all of $\{1, \ldots, n\}$. If i and j are in the same subset of a, we write $(ij) \subset a$. If these are in destinct subsets, we write $(ij) \notin a$. Elements of a are called clusters: Define

$$H(a) = H_0 + \sum_{(ij) \subset a} V_{ij}$$

$H(a)$ describes a situation of particles interacting within clusters but not between cluster. Given a, we can pick coordinates for X in two classes, x^a and x_a. The x^a describe difference of centers of mass of different clusters and the x_a coordinate differences within a cluster. Corresponding to such a decomposition, $L^2(X) = \mathscr{H}_a \otimes \mathscr{H}^a$ where $\mathscr{H}_a$ is functions of the x_a. $H(a)$ then decomposes to

$$H(a) = H_a \otimes I + I \otimes T^a$$

T^a is independent of V and describes the kinetic energy of relative motion of the clusters. H_a describes internal motion of the clusters. Let P_a denote the projection in $\mathscr{H}_a$ onto the point spectrum of H_a. Ran P_a is the sum of products of bound states for each cluster. Let $P(a) = P_a \otimes I$. Thus $P(a)$ describes the projection in $L^2(X)$ onto functions which are sums of products of bound states in the clusters and free motion of their centers of mass.

Theorem 9.1 Suppose $v \geq 3$ and that (9.1) holds. Then

$$s - \lim_{t \to \mp \infty} e^{itH} e^{-itH(a)} P(a) = \Omega_a^\pm \tag{9.2}$$

exist. Moreover, if $a \neq b$, then Ran $\Omega_a^+ \perp$ Ran Ω_b^+. $\varphi \in$ Ran Ω_a^+ if and only if there exists η so that

$$\|e^{-itH} \varphi - e^{-itH(a)} P(a) \eta\| \to 0 \text{ as } t \to -\infty \tag{9.3}$$

Remarks 1. See [63; XI. 5] for a proof. The result is claimed there for $v = 1,2$ also but the proof is in error; in $v = 1,2$ one requires some information on decay of the eigenfunctions of the H_a.

2. (9.3) says that as $t \to -\infty$, the interacting state $e^{-itH} \varphi$ looks asymptotically like bound clusters moving relatively freely.

3. We have included the case a_1 where a has one cluster for which $\Omega_{a_1}^\pm = P(a_1) =$ projection onto the bound clusters of H.

Problem 9A—1st Form (Asymptotic Completeness for Short Range N-body Systems).

Under the hypotheses $v \geq 3$, (9.1) proves that

$$\bigoplus_a \operatorname{Ran} \Omega_a^+ = L^2(X)$$

As already remarked, $\operatorname{Ran} \Omega_{a_1}^+ = \mathscr{H}_{p.p.}$, the point spectral subspace for H, and it can be shown that if $a \neq a_1$, $\operatorname{Ran} \Omega_a^+ \subset \mathscr{H}_{a.c.}$ the absolutely continuous space for H. Thus problem 9A is often stated as

Problem 9A—2nd Form

(i) Prove $\mathscr{H}_{s.c.}$, the singular continuous space is empty.

(ii) Prove $\displaystyle\bigoplus_{a \neq a_1} \operatorname{Ran} \Omega_a^+ = \mathscr{H}_{a.c.}$.

The limits in Thm. 9.1 fail to exist in the case where $V_{ij}(x) \sim |x|^{-1}$ at ∞. There is a modification of the wave operators (9.2) due to Dollard, for which the limits $\Omega_a^{D, \pm}$ exist. These are described, e. g. in [63], Section XI.9.

Problem 9B (Asymptotic Completeness for Coulomb Potentials). Under the hypotheses,

$v = 3$, $V_{ij}(x) = e_{ij} |x|^{-1}$, prove

$$\bigoplus_a \operatorname{Ran} \Omega_a^{D, +} = L^2(X).$$

Of course, one wants to allow sums of Coulomb and short range potentials.

For $n = 2$, these problems were solved over 20 years ago at least if $|V(x)| = 0(|x|^{-v-\varepsilon})$ with the sharpest results due to Agmon-Kuroda and Enss (see [64] and [65]). For $n = 3$ and $|V(x)| = 0(|x|^{-2-\varepsilon})$ and an extra assumption (no resonances in two body subsystems), Faddeev [66] solved the problem; see Ginibre-Moulin [67], Thomas [68], Howland [69], Kato [70], Yajima [71], Sigal [72] and Hagedorn-Perry [73] for additional information. The Coulomb 3-body problem was solved by Mercuriev [74] and by Enss [75]. Enss also treated the general $(0(|x|^{-1-\varepsilon}))$ 3-body problem. Mourre [76] has announced general 3-body results also. For a suitable class of analytic potentials and a suitable sense of genericity, Hagedorn [77] (for $n = 3,4$) and Sigal [78] for general n have solved the problem, but genericity plays a central role.

The "half" of asymptotic completeness that requires $\mathscr{H}_{s.c.}$ is empty was solved by Balslev-Combes [79] for Coulomb potentials and for a wide class of short range V's by Perry et al. [80] using ideas of Mourre [81]. The basic open question concerns completeness without requiring genericity or analyticity.

This important open question has been studied to some extent since 1960, and very actively for the last ten years. At the risk of jinxing the solution, it seems to me like a good bet that it will be solved in the next five years and probably sooner: Both Enss' and Mourre's three-body methods appear promising for N-bodies and Mourre's long range two-body work [82] may be useful in the N-body problem.

10   Quantum Potential Theory

Basic to atomic and molecular physics is the binding energy of a quantum mechanical system of electrons interacting with one or more nuclei. To be explicit, fix N and consider two classes of operators on $L^2(\mathbb{R}^{3N})$, with $x \in \mathbb{R}^{3N}$ written as $x = (x_1, \ldots, x_N)$. For any fixed Z define

$$H_N(Z) = \sum_{i=1}^{n} \left(-\Delta_i - \frac{Z}{|x_i|} \right) + \sum_{1 \leq i < j \leq N} 1/|x_i - x_j| \tag{10.1}$$

and for Z_0, k and $R_1, \ldots, R_k \in R^3$ we define

$$H_N^{(k)}(R_1, \ldots, R_k; Z_0) = \sum_{i=1}^{N} -\Delta_i + \sum_{1 \leq i < j \leq N} \frac{1}{|x_i - x_j|}$$

$$+ \sum_{1 \leq \alpha, \beta \leq k} \frac{Z_0^2}{|R_\alpha - R_\beta|} - \sum_{\substack{1 \leq \alpha \leq k \\ 1 \leq i \leq N}} \frac{Z_0}{|x_i - R_\alpha|} \tag{10.2}$$

Note that the 3rd term in (10.2) is a constant depending only on the parameters R_i and not an operator on $L^2(\mathbb{R}^{3N})$. (10.1) is the Hamiltonian of an atom in the approximation of infinite nuclear mass and (10.2) that of a molecule in Born-Oppenheimer approximation. We define

$$E_B(N; Z) = \inf \operatorname{spec}(H_N(Z))$$
$$E_B^{(k)}(N; R_1, \ldots, R_k; Z_0) = \inf \operatorname{spec}(H_N^{(k)}(R_1, \ldots, R_k; Z_0)).$$

The B stands for "Boson" since the operators are taken on $L^2(\mathbb{R}^{3N})$ and ignore the Pauli principle. For fermion electrons one should restrict $H_N(Z)$ (and $H_N^{(k)} \ldots$) to $\mathscr{H}_{\text{phys}}$ the subset of $L^2(\mathbb{R}^{3N})$ of all functions $f(x_1, \ldots, x_N)$ antisymmetric under interchange of the coordinates (actually, because of the fact that electrons have two spin states, we should take f to be a sum of functions transforming under permutations as representations with at most two columns in their Young tableaux). The inf of the spectrum of restricted operators we will call E without any subscript. These are the physically relevant objects so we do not give them a subscript F.

The total binding energies are basic physical objects. While several significant properties are known (see especially Thms. 10.1, 2, 3 below), it is shocking how little we know about $E(N; Z)$ and $E^{(k)}(N; R_1, \ldots, R_k; z_0)$. This is shown by the first open problem. Define

$$(\Delta E) (N, Z) = E(N - 1, Z) - E(N, Z)$$

the energy it takes to remove electron N. It is a consequence of the HVZ theorem ([83], Section XIII.5) that $(\Delta E) (N, Z) \geqq 0$.

Problem 10A (Monotonicity of the Ionization Energy). Prove that

$$(\Delta E) (N - 1, Z) \geqq (\Delta E) (N, Z)$$

for all N, Z.

This is just the fact, almost obvious, that it takes more energy to remove inner electrons than outer ones. Since in removing electron $(N - 1)$ there is one fewer electron to repel, and since the Pauli principle only makes things better this should be true. It seems to be remarkably difficult to prove. Indeed, it is false if one requires it for nuclei with all possible finite masses (rather than our infinite mass assumption) and one allows for electron spin [84]. The inequality to be proved says that $E(N, Z)$ for Z fixed is convex in N.

A weaker result that would be of interest would be to prove: "If $\Delta E(N, Z) = 0$, then $\Delta E(N + 1, Z) = 0$". This result would be relevant in connection with the Ruskai-Sigal theorem (Thm. 10.2 below).

To state the next open problem, we need to recall

Theorem 10.1 (Lieb-Simon [85]) $\lim_{Z \to \infty} E(Z, Z)/Z^{7/3}$ exists.

In fact, the limit is given by a "Thomas-Fermi" energy, e_{TF}. See Lieb [86] for further information and insight.

Problem 10B (The Scott Correction). Prove that $\lim_{Z \to \infty} \left(E(Z, Z) - e_{TF} Z^{7/3}\right)/Z^2$ exists and is the constant found by Scott [87].

If one drops the electron-electron repulsion, one can find the new $E(Z, Z)$ exactly and compute the $0(Z^2)$ term exactly and see it corresponds to the fact that Thomas-Fermi fails to get the inner electrons correctly. Since the electron repulsion shouldn't matter for the inner electrons, Scott [87] conjectured that the Z^2 term is the same as for the non-interacting case. Recent physicists' arguments which seem difficult to make rigorous for the Scott correction can be found in Bander [88] and Schwinger [89]. We remark that the obvious asymptotic series one might conjecture on the basis of the last theorem and problem, namely

$$E(Z, Z) \sim a_1 Z^{7/3} + a_2 Z^2 + a_3 Z^{5/3} + a_4 Z^{4/3} + \cdots$$

is almost surely *not* correct: There may be a $Z^{5/3}$ term but after that there are almost surely oscillations at the $Z^{4/3}$ level.

Physically, the quantity $E(Z, Z)$ as a total binding energy is not so interesting. The ionization energy $(\Delta E)(Z, Z)$ is much more interesting.

Problem 10C (Asymptotics of Ionization Energy). Find the leading asymptotics of $(\Delta E)(Z, Z)$ for large Z.

Lieb-Simon [85] suggest that $(\Delta E)(Z, Z)$ goes to a constant, but even on an intuitive level, that is not clear. Indeed, it isn't clear to me whether the leading power $(\alpha = \lim\limits_{Z \to \infty} \ln \Delta E(Z, Z)/\ln Z)$ is 0, positive or negative!

For the next problem, we need to recall

Theorem 10.2 (Ruskai [90], Sigal [91]). For every Z, there is an N_0 so that $(\Delta E)(N, Z) = 0$ if $N \geq N_0$.

This says that one cannot bind arbitrarily many electrons to a nucleus. Let $N(Z)$ be the smallest N_0 for which the above is true. Zhislin [92] (see also Simon [93]), showed that $(\Delta E)(N, Z) > 0$ if $N \leq Z$, so $N(Z) \geq Z$ and thus

$$\varliminf_{Z \to \infty} N(Z)/Z \geq 1.$$

Moreover, Sigal [91] has proven that

$$\varlimsup_{Z \to \infty} N(Z)/Z \leq 2 \tag{10.3}$$

It is quite reasonable to think this 2 can be replaced by 1.

Problem 10D (Asymptotics of Maximal Ionized Charge). Prove that

$$\lim_{Z \to \infty} N(Z)/Z = 1.^{*)}$$

Sigal's argument (for 10.3) in [91] uses the Pauli principle. In fact, if Problem 10D has a positive solution, it must use the Pauli principle, since Benguria-Lieb [94] have proven that if $N_B(Z)$ is defined using E_B in place of E, then $\lim\limits_{Z \to \infty} N_B(Z)/Z > 1$.

For the last formal problem, we need to recall what is the most significant result known about Coulomb energies, "the stability of matter".

Theorem 10.3 (Dyson-Lenard [95], Lieb-Thirring [96]). For a universal constant:

$$E^{(k)}(N, R_1, \ldots, R_k; Z_0) \geq - C(1 + Z_0^{7/3})[N + k].$$

---

*) See Note added in proof.

For Z_0 bounded this was first proven by Dyson-Lenard [95]; Lieb-Thirring [96] not only simplified the proof considerable, but found a value of C within the best possible value by an order and half in magnitude (Dyson-Lenard [95] had a constant off by many orders). This result is important since it implies that bulk matter doen't contract as more particles are added (see e. g. Lieb [86]); it is the starting point of a proof of the existence of good thermodynamics for Coulomb systems (Lebowitz-Lieb [97]).

This result depends critically on the Fermi nature of the electrons. Indeed, define

$$E_B(k, N; z_0) \equiv \inf_{R_i} E_B^{(k)}(N, R_1, \ldots, R_k; z_0).$$

Then Lieb [98] has proven that

$$-DN^{5/3} \leqq E_B(N, N; 1) \leqq -CN^{5/3}.$$

Let $\tilde{E}_B(k, N, Z_0)$ be the analogous object where now the "protons" are given a finite mass and so both "electrons" and "protons" (viewed as bosons) are treated quantum mechanically. Then, Dyson [99] has proven that for a suitable $C > 0$:

$$\tilde{E}_B(N, N; 1) \leqq -CN^{7/5}$$

The best lower bound known is

$$\tilde{E}_B(N, N; 1) \geqq -DN^{5/3}.$$

Problem 10E (Rate of Collapse of Bose Matter). Find suitable C_1, C_2, and α so that

$$-C_1 N^\alpha \leqq \tilde{E}_B(N, N; 1) \leqq -C_2 N^\alpha.$$

One suspects that $\alpha = 7/5$. Since electrons in nature are not bosons, one might think that this problem is of purely mathematical interest. In fact, since Dyson's trial function is of BCS type, a real understanding of this problem could improve our understanding of superconductivity.

The reader can consult Lieb's Lausanne lecture [100] for a list of other interesting open Coulomb problems. In connection with the Lieb-Thirring proof of Thm. 10.3, we should mention the open question of finding the best constant in the Cwickel-Lieb-Rosenbljum bound: See Simon [101], pgs. 96—97 and Glaser-Martin [102] for further discussion of this and related problems.

11   Existence of Crystals

It is an observed fact of nature that most materials occur in a crystalline state at low temperatures. Yet there is no proof or even a very convincing argument to show that even at zero temperatures ensembles of quantum mechanical atoms want to form crystals. Clearly to avoid boundary effects, one must take an infinite system; any finite system will not be a strictly crystalline form. Moreover, as a first problem, one should imagine infinite nuclear masses.

Thus, we should fix an integer, z_0 (a nuclear charge), and take $N = k z_0$ and consider the function $E^{(k)}(N; \vec{R}_1, \ldots, \vec{R}_k; z_0)$. We denote by $\{\vec{R}_i^{(k)}\}_{i=1}^{(k)}$ a minimizing configuration for this function (it is not automatic, indeed, not proven that such a minimum exists, i.e. that the minimum isn't taken for some $|\vec{R}_i - \vec{R}_j| = \infty$; presumably, for suitable z_0, such a minimum does exist).

Of course, the minimizing configurations is not unique; it is invariant under a common Euclidean motion of the nuclei or under permutation of indices and there could be additional non-uniqueness. For this reason, we are careful to deal with "a choice" below.

Here is one possible statement which would show at zero temperature atoms with atomic number z_0 form a crystal. There is a choice of minimizing configurations $R^{(k)}$, so that (i) $R_j^{(k)}$ converges to some $R_j^{(\infty)}$ as $k \to \infty$ for each fixed j. (ii) For any R_0, there is a J so that $|R_j^{(k)}| \geq R_0$ if $j > J$. (iii) The $R_j^{(\infty)}$ lie in a lattice, i.e. a subset of R^3 left invariant by a subgroup of translations isomorphic to Z^3. Condition (ii) is included to prevent one nucleus from "getting lost" in the limit due to mislabeling.

Problem 11 (Existence of Crystals). Prove the above statement or another suitable version of the existence of crystals for some z_0.

We note that the classical analog of this result is unknown. There is, however, an interesting series of papers on this classical question by Radin [103] and a paper of Duneau-Katz [104].

Of course, if one solves Problem 11, the next thing is to worry about finite but low temperature, then melting, then...

12   Random and Almost Periodic Potentials

In this section we want to discuss $-\Delta + V$ on $L^2(R^\nu)$ and its discrete analog

$$((hu)(n) = \sum_{|\delta|=1} u(n+\delta) + V(n) u(n))$$

on $l^2(Z^\nu)$ where V is either a stochastic process with strong mixing properties ("random potentials") or an almost periodic function. This is an area of considerable current interest to me, and so I may be accused of lacking perspective in including the five problems listed here. It seems to me that the first problem below is very significant from any viewpoint; perhaps (but I think not) the second, third and fourth are too specialized; the fifth is included so the reader can help me win a bet.

To be precise about random potentials, one can consider a particular model known as the Anderson model. Choose the $V(n)$ to be independent, identically distributed random variable with distribution uniformly in $(-\lambda, \lambda)$ (λ is a number known as the coupling constant). The following results are proven: In any dimension, v, the spectrum $\sigma(h)$ is almost surely $[-2v - \lambda, 2v + \lambda]$ (e. g. Kunz-Souillard [105]) and if $v = 1$, almost surely h has only dense pure point spectrum ("localized states") (see [105] and Delyon et al. [106]; also Goldshade et al. [107] for the case of $-d^2/dx^2 + V(x)$ with suitable random V). In the physics literature the belief is that the same result holds if $v = 2$ (although until roughly 5 years ago this was not the belief) but when $v \geq 3$ it is believed that one has only dense point spectrum when $\lambda \geq \lambda_0 > 0$, but for $\lambda < \lambda_0$ there is a region $[-a(\lambda), a(\lambda)]$ of absolutely continuous spectrum ("extended states") with dense point spectrum in $\pm [a(\lambda), 2v + \lambda]$. Fröhlich and Spencer [108] have recently obtained some results in the region where there is supposed to be localized states and they will probably succeed in proving dense point spectrum soon when either λ is large or $|e|$ is near $2v + \lambda$. This leaves the region of extended states.

Problem 12A (Existence of Extended States in the Anderson Model). Prove that in $v \geq 3$, for λ small, there is a region with absolutely continuous spectrum, and determine whether this is false when $v = 2$.

We mention here the interesting results of Kunz-Souillard on extended states in the Anderson model on a Bethe lattice (which in some sense has $v = \infty$) of which so far only an announcement exists [109].

At first sight, one might think that since the $\lambda = 0$ operator has a. c. spectrum, extended states shouldn't be so hard since one just has to find a simple perturbation argument. That this is not the case is shown by the expectation that when $\lambda \neq 0$ the a. c. spectrum should be associated with diffusive motion; explicitly, when $\lambda = 0$, $(\delta_0, (e^{itH_0} \vec{N} e^{-itH_0})^2 \delta_0) \sim 0(t^2)$ where δ_0 is the element of $l^2(Z^v)$ which is 1 at $\vec{0}$ and 0 elsewhere and $(\vec{N}u)(\vec{n}) = \vec{n}u(\vec{n})$ while we expect that:

Problem 12B (Diffusive Bound on "Transport" in Random Potentials). For the Anderson model (and more general random potentials) prove that

$$\text{Exp}(\delta_0, (e^{itH} \vec{N} e^{-itH})^2 \delta_0) \leq c(1 + |t|)$$

This result is clearly connected with our discussion in Section 4. We note that it is easy to prove the analogous bound if ct is ct^2; indeed, that is true for any bounded V (see e. g. Radin-Simon [110]) and that when there are extended states, it is believed that expectations grow as $D_{\pm} t$ for t large.

There is one last aspect of the Anderson model we want to mention. A basic object is the integrated density of states (e. g. [111]), $k(E)$. In the Anderson model, there are two places we might worry about lack of smoothness in k; at the edges of the spectrum where k is certainly non-analytic and at the mobility edge (indicated as $\pm a(\lambda)$ above). At the edges, a result known as Lifschitz tails (see

e. g. [112]) suggests k is C^∞ so the following problem is really about the mobility edge:

Problem 12C (Smoothness of k (E) Through the Mobility Edge in the Anderson Model).
Is $k(E)$ a C^∞ function of E in the Anderson model at all couplings?
　　Of course, there are a myriad of other questions about the mobility edge with more physics (e. g. behavior of the diffusion constant); we list the above as the simplest one.
　　Our last pair of problems involves the case of almost periodic potentials. The simplest example in many ways is the almost Mathieu equation on $l^2(Z)$

$$(hu)(n) = u(n+1) + u(n-1) + \lambda \cos(2\pi\alpha n + \theta) u(n) \tag{12.1}$$

where λ, θ and α are parameters. It is an idea of Sarnak [113] that the spectral properties should depend on Diophantine properties of α: if

$$\left| \alpha - \frac{p}{q} \right| \geq C q^{-k}$$

for some C and k, we call α a Roth number and if there is an infinite sequence q_k with

$$\left| \alpha - \frac{p_k}{q_k} \right| \leq \exp(-k q_k),$$

we call α a Liouville number. (The Roth numbers have full Lebesgue measure while the Liouville numbers are a dense G_δ!) Here is the belief about the spectrum of (12.1) (see e. g. [114]):
　　(a) If α is a Liouville number and $\lambda \neq 0$, then for a. e. θ, the spectrum is purely singular continuous.
　　(b) If α is a Roth number and $|\lambda| < 2$, the spectrum is purely absolutely continuous for a. e. θ.
　　(c) If α is a Roth number and $|\lambda| > 2$, the spectrum is purely dense pure point.
　　(d) If α is a Roth number and $|\lambda| = 2$, $\sigma(h)$ has Lebesgue measure zero and the spectrum is purely singular continuous.
　　All that has been proven about this model is: (i) (a) is true if $|\lambda| > 2$ [111] (ii) In case (a), there is at least no point spectrum [115, 111] (iii) When α is Roth there is at least some a.c. spectrum when $|\lambda|$ is very small and some point spectrum when $|\lambda|$ is very large [116] (iv) If $\lambda > 2$, there is at least no a.c. spectrum [111].

Problem 12D (Analysis of the Almost Mathieu Equation). Verify the picture (a)—(d) above.

Our final problem is the only one involving the continuum case $-\varDelta + V$. As noted above, it is known that for λ large and α suitable, (12.1) has some point spectrum. Here is a continuous analog of that:

Problem 12E (Point Spectrum in a Continuum Almost Periodic Model). Show that for α, λ, μ suitable

$$-\frac{d^2}{dx^2} + \lambda \cos(2\pi x) + \mu \cos(2\pi \alpha x + \theta) \tag{12.2}$$

has some point spectrum for a.e. θ.

I pick this problem among all possible continuum problems because two excellent mathematicians have bet me that (12.2) has no point spectrum. I don't give their names to spare them public embarrassment (not caused by their choosing to disagree with me, but by the fact that, in this case, they are wrong!).

13   Self-Avoiding Random Walks

We want to first describe a mathematical problem which is easy to describe, and then we will briefly explain why it is included in a list of problems in mathematical physics. Consider the lattice Z^d of integral points in d-dimensions. (We abandon our usual v here because in this subject v is usually used for the object in (13.1)). A self-avoiding walk (SAW) of length n is a sequence of $n+1$ *distinct* points $R(0), \ldots, R(n) \in Z^d$ so that $R(0) = 0$ and $|R(i+1) - R(i)| = 1$. This differs from ordinary random walks in the requirement that the R's be distinct (hence self-avoiding). Let $k(n)$ denote the number of SAW of length n, labeled

$$\{R_i^{(n)}(j) | j = 0, \ldots, n; \, i = 1, \ldots, k(n)\}$$

and we define the mean displacement by:

$$D(n) = \langle R^{(n)}(n)^2 \rangle^{\frac{1}{2}} \equiv [k(n)^{-1} \sum_i R_i^{(n)}(n)^2]^{\frac{1}{2}}.$$

One expects that (more or less) $D(n) \sim C n^v$ as $n \sim \infty$. Essentially nothing is known about v which we might define by

$$v = \lim_{n \to \infty} n^{-1} \ln D(n) \tag{13.1}$$

(it is *not* known that the limit exists). Intuitively, the self-avoiding property should force the path to grow faster than in ordinary random walks where $v = \frac{1}{2}$

so one certainly expects that

$$v \geqq \tfrac{1}{2} \tag{13.2}$$

but even this is unknown. Indeed, a few moment reflection on "trapping in cul de sacs" will indicate the problems. A proof of (13.2) would be *very* interesting and probably represent real progress.

This subject is reviewed in [117, 118].

Computer calculations suggest that if $d = 2$, $v = 3/4$, if $d = 3$, $v \cong .59$ (some prefer $v = 3/5$ and don't believe .59), and if $d \geqq 4$, $v = 1/2$. That v seems to be $1/2$ if $d \geqq 4$ is believed connected with the fact that Brownian motion is non-selfinter-secting if $d \geqq 5$ and has only "logarithmic" selfintersections if $d = 4$ (see e.g. [101] and reference therein).

Problem 13 (Critical Exponents for self-Avoiding Walks). Prove that $v = 1/2$ for $d \geqq 4$ and $v > 1/2$ for $d \leqq 3$.

So much for the simple statement of this problem. Why is this problem here? There are many reasons:

(1) v is in many ways the most elementary example of a critical exponent and problem 13 is an expression of the fact that in high dimension these exponents are supposed to agree with mean field theory, which in this case is the Gaussian value $v = 1/2$. Critical exponents are important in the theory of phase transitions so this section is related to Section 8; indeed, there is an analog of universality; v is supposed to be dependent only on dimension of the lattice and not on its exact form (e.g. in 3-dimensions, the SAW on the cubic lattice and the face centered lattice are believed to be the same).

(2)   The SAW model is related to elementary models of polymers; indeed, much of the work on SAW has been done by polymer people and both review articles mentioned above appear in Advances in Chemical Physics.

(3)   There is supposed to be connection between SAW and the Ising model. Actually, it seems to me that this is at a deep level only through Fisher's bounds [119] and, in particular, the real relevance to the Ising model is only the analogy.

(4)   Symanzik [120] had a vision of φ^4 field theories which have been a fertile source of intuition and which relates φ^4 field theories to SAW. Indeed, Brydges et al. [121] have made use of a random walk expansion of φ^4 theories which relates them to random walks in which self-intersections are not forbidden but are suppressed relative to SAW's. If one writes out the formalism for n-compo-nent φ^4 and then formally sets n to 0, SAW's result! (This is a remark of de Gennes [122]). Progress on understanding SAW could help us understand quantum field theory.

See Westwater [123] for additional information on this and related problems.

14   Quantum Field Theory

A list of problems like this one written 10 or even 30 years ago would surely have included the mathematically consistent construction of quantum electrodynamics (QED). We will see what happened to that problem below, but it is clear that quantum field theory remains a basic element of fundamental physics and a continual source of inspiration to mathematicians.

The most spectacular development in theoretical physics of the past 10 years has been the formulation of a generally accepted model of strong interaction physics, a model of quarks interacting through a non-abelian Yang-Mills gauge fields (whose quanta are called gluons). This model is normally called quantum chromodynamics (QCD). As with most quantum field theories, the theory is written down by physicists by giving a formal Lagrangian and there are numerous infinities only eliminated formally; that is, one is quite far from a mathematically precise set of objects.

Problem 14A (Construct QCD). Give a precise mathematical construction of quantum chromodynamics.

For a discussion of the model formally (actual class of models depending on the number of quarks and of various groups), see e. g. [124].

The past 15 years have seen the development of the first mathematically consistent quantum field theories in two and three space-time dimensions. This area, known as constructive quantum field theory, is nicely summarized in the book of Glimm and Jaffe [125]; see Seiler [126] for a discussion of mathematical aspects of Yang-Mills field theories. All the models constructed lie in a class known as "super renormalizable" since their infinities are rather mild. There is another class of formal field theories known as "renormalizable", of which QCD is the most interesting but also one of the more complex technically. It is possible to imagine someone constructing a renormalizable theory but being unable to handle QCD because of some of the difficulties intrinsic to Yang-Mills fields or to fermions. Thus, the following is interesting:

Problem 14B (Renormalizable QFT). Construct any non-trivial renormalizable but not superrenormalizable quantum field theory.

With regard to QED, for many years this was believed to be the fundamental theory of electrons and photons. The impressive agreement between experiment and QED was used as an argument that the formal theory had an underlying mathematically consistent formulation. This is no longer believed to be the case, at least among an overwhelming majority of the theoretical high energy physics community. Rather, it is believed that QED by itself is *not* consistent; rather, there is a consistent (non-abelian gauge) unified theory of weak and electromagnetic interactions, but the differences of the perturbation series of this consistent theory and QED are very small at low energies, explaining the agreement with experiment. This should be taken as a warning to those who argue that a theory that seems to agree with nature must be mathematically consistent and it is pointless to prove such an "obvious" fact. In any event, this leads to:

Problems 14C (Inconsistency of QED). Prove that QED is not a consistent theory.

Alan Sokal has dubbed the discipline of proving certain field theories are not consistent "destructive field theory". There are some results in this new area. Fröhlich [127] (using ideas from Brydges et al. [121]) and Aizenman [128] have shown that there is no non-trivial limit of lattice cutoff φ^4 theories in space time dimension $d > 4$, if one only renormalizes with mass and coupling constant renormalization. This is *not* a verification of the phenomena responsible for the putative inconsistency of QED, where it is believed perturbation theory is misleading because of lack of infrated stability. For φ_d^4, $d > 4$, formal perturbation theory suggests that renormalization of higher degree than four will be required. Thus an analogous result for $d = 4$ where the heuristics for QED are also valid, would be especially interesting. Fröhlich and Aizenman have results in $d = 4$ but they suffer from various loopholes, e. g. at this point a φ_4^4 with finite field strength renormalization has not been ruled out. There is also a loophole suggested by Gallavotti-Rivasseau [129] which, while an intriguing possibility, is probably not going to save φ_4^4. This leads us to propose:

Problem 14D (Inconsistency of φ_4^4). Prove that a non-trivial φ_4^4 theory does not exist.

15   Cosmic Censorship

The reader who has tired at the length of this article may well wish that the title of this section had been applied sooner.

Classical general relativity is a discipline whose death has been prematurely claimed by too many theoretical physicists. It remains healthy and vigorous, in part because of input from astrophysics (such as the identification of probable black holes and the identification of an effect of gravitational radiation) and, in part, due to a frequent injection of fertile mathematical ideas (such as those of Hawking and Penrose and, more recently, of Schoen-Yau and Witten). By any reasonable definition of the term, it is clear that much of classical general relativity is "mathematical physics". It is unfortunate that a horizon seems to separate general relativists and other mathematical physicists. I am not alone among mathematical physicists in knowing less about the subject than I should. I have included a problem from general relativity here to express my belief in the unity of mathematical physics, but I must confess a feeling of "I hope I got it right". In any event, the reader should consult various articles of Penrose [130] for more information, and his article [131] for additional problems in general relativity.

Problem 15 (Cosmic Censorship). Formulate and then prove or disprove a suitable form of cosmic censorship.

Very roughly speaking, cosmic censorship says that for Einstein's equations coupled to matter obeying "realistic" evolution equations (such as Maxwell's equations or suitable Yang-Mills equations), "naked singularities" do not "generically" occur. It would be interesting to prove the result even for vacuum solutions of the Einstein's equations (i. e. those with no matter).

Cosmic censorship deals with the deep and thorny issue of singularities in general relativity. The first "singularity" in general relativity was the Schwarzschild singularity: If the Schwarzschild solution (the field of a static, spherically symmetric source) is continued, in the absence of matter, to a distance (which, for usual bodies, is far within the matter producing the field) called the Schwarzschild radius, there appears to be a singularity. We say "appears" because it was realized later (by Eddington, Lemaitre and Synge) that the singularity was not one of the geometry but rather of the coordinate system used: e. g., in another coordinate system found by Kruskal, one can continue past the Schwarzschild radius until a true singularity appears. We say "true singularity" because a suitable curvature scalar, a coordinate independent object, diverges there.

While the Schwarzschild "singularity" is not a singularity of the geometry, it has important geometric and physical significance: It is a horizon in that no light rays from inside it can pass out to infinity. In this way, it prevents us from "seeing" the true singularity which would presumably be the ultimate psychedelic experience.

It is not easy to get explicit solutions of the Einstein equations because of the many components and variables, and for that reason, most known solutions have very high symmetry. For a while, there was a belief that the true singularity in the Schwarzschild solution (which can arise in finite "time" from non-singular Cauchy data if matter collapses to a point) might be an artifact of the symmetry, and that most solutions very near to Cauchy data leading to a singularity might well be free of singularities. A basic discovery of Penrose and Hawking [132] was that this was not the case but that solutions near one with a Schwarzschild-solution-type (true) singularity have some type of singularity. This stability result for black holes is very significant, given the apparent occurence of black holes in the cosmos.

The Hawking-Penrose theorem says we must learn to live with singularities or else rely on some quantum effect to save us. Upon some reflection, a singularity like that in the Schwarzschild solution is not so difficult to live with because it doesn't live next door, i. e. we don't see it. A "naked singularity" is, roughly speaking, one with the property that light rays form points arbitrarily near it can escape to infinity. These are much more disturbing from a physical point of view. One cannot conjecture that naked singularities never occur, since one does in a solution called the Taub-NUT solution, which Misner [133] has dubbed a "counterexample to almost anything". However, this solution has a high symmetry and one can conjecture that naked singularities tend to become clothed by horizons under most small perturbations. This is the content of Cosmic Censorship. Given the history of the Hawking-Penrose theorem, one might well suspect that the idea that naked singularities are associated with symmetries is wrong; however, recent results of Isenberg-Moncreif [134] tend to support the notion that naked singularities imply symmetry.

There are other examples of naked singularities among the Weyl axisymmetric class. To eliminate such examples, it may be necessary to make some kind of hypothesis of initial conditions which are non-singular and "realistic".

Note added in proof

Problem 10D has been solved by E. Lieb, I. Sigal, B. Simon and W. Thirring (in prep.), who prove that $\lim\limits_{Z \to \infty} N(Z)/Z = 1$ but no effective control on the rate of convergence is obtained. Can one prove that $N(Z) - Z$ is bounded?

References

[1] Hilbert, D., Bull. Am. Math. Soc. **8** (1902), 437
[2] Painlévé, P., *Lecons sur la théorie analytique des equations différentielles*, Herman, Paris, 1897.
[3] Saari, D., J. Diff. Eqns. **26** (1977), 80.
[4] Birkhoff, G. D., *Dynamical systems*, A.M.S. Colloq. Publ. **9** (1927).
[5] Sundman, K., Acta Soc. Sci. Fenn. **34** (1907); **35** (1909).
[6] Saari, D., Trans. A.M.S. **162** (1971), 267; 181 (1973), 351; Proc. A.M.S. **47** (1975), 442.
[7] Urenko, J., SIAM J. App. Math. **36** (1979), 123.
[8] Sperling, H. J. Reine Ang. Math. **245** (1970), 15.
[9] von Zeipel, H., Ark. Mat. Astr. Fys. **4** No. 32 (1908).
[10] Mather, J. and R. Mc Gehee, Springer Lecture Notes in Physics **38** (1975), 573.
[11] Gerver, J., J. Diff. Eqn., to appear.
[12] Kato, T., Trans. A.M.S. **70** (1951), 195.
[13] Reed, M. and B. Simon, *Methods of modern mathematical physics, II. Fourier analysis. Self-adjointness*, Academic Press, 1975.
[14] Leinfelder, H. and C. Simader, Math. Zeit. **176** (1981).
[15] Arnold, V. and A. Avez, *Ergodic problems of classical mechanics*, Benjamin, 1968.
[16] Sinai, J., Sov. Math. Dokl. **4** (1963), 1818.
[17] Sinai, J., Usp. Akad. Nauk. SSSR **25** (1970), No. 2, 141—192; L. Bunimovich and J. Sinai, Math. Sb. **90** (1970), No. 3, 415—431.
[18] Sinai, J., Appendix to English translation of S. Krylov, *Works on the Foundations of Statistical Physics*, Princeton University Press, 1980.
[19] Chernoff, N. and J. Sinai, Proc. Petr. Sem. **8** (1982), 218—238.
[20] Wightman, A. S., Proc. Symp. Pure Math. **28** (1976), 147.

[21] Volkovysski, K. and J. Sinai, Funk. Analiz. **5** (1971), (3) 19; 0. de Pazzis, Commun. Math. Phys. **22** (1971), 121; S. Goldstein and J. Lebowitz, Commun. Math. Phys. **22** (1974), 1; M. Aizenman et al., Springer Lecture Notes in Physics **38** (1975), M. Aizenman et al., Commun. Math. Phys. **39** (1975), 289.

[22] Ruelle, D., *Statistical mechanics*, Benjamin, 1969.

[23] Bratteli, O. and D. Robinson, *Operator algebras and quantum statistical mechanics*, 2 Vols., Springer, 1979; 1981.

[24] Feigenbaum, M., Los Alamos Science, Summer 1980; **4**.

[25] Collet, P. and J. P. Eckman, *Iterated maps on the interval as dynamical systems*, Birkhäuser, 1980.

[26] Ruelle, D., Math. Intell. **2** (1980), 126; Bull. Am. Math. Soc. **5** (1981), 29.

[27] Eckmann, J. P., Rev. Mod. Phys. **53** (1981), 643.

[28] Foias, C. and R. Tenam, J. Math. Pures Appl. **58** (1979), 339.

[29] Kubo, R., *Synergetics*, (ed. H. Haken), Teubner, 1973; R. Kubo, K. Matsuo and K. Katahara, J. Stat. Phys. **9** (1973), 51.

[30] van Kampen, N., Physica Norvegica **5** (1971), 279—284; in *Fluctuation phenomena in solids* (ed. R. Burgess), Academic Press, 1965; *A discussion on linear response theory*, Symposium in Commemoration of Ohm's Law, Cologne (1976).

[31] Griffiths, R., in *Phase transitions and critical phenomena*, Vol. 1 (ed. C. Domb and M. Green), Academic Press, 1972, 7—110.

[32] Israel, R., *Convexity in the theory of lattice gases*, Princeton Univ. Press, 1979.

[33] Simon, B., *The statistical mechanics of lattice gases*, Princeton Univ. Press, in prep.

[34] Peierls, R., Proc. Camb. Phil. Soc. **32** (1936), 477; R. Griffiths, Phys. Rev. **136A** (1964), 437; R. Dobrushin, Theo. Prob. Appl. **10** (1965), 193.

[35] Fröhlich, J., B. Simon and T. Spencer, Commun. Math. Phys. **50** (1976), 79.

[36] Mermin, N. and H. Wagner, Phys. Rev. Lett. **17** (1966), 1133, 1307.

[37] Dyson, F., Phys. Rev. **102** (1956), 1217; 1230.

[38] Kogut, J., Rev. Mod. Phys. **51** (1979), 659.

[39] Fröhlich, J. and T. Spencer, Commun. Math. Phys. **81** (1981), 527.

[40] Gallavotti, G. and S. Miracle-Sole, Phys. Rev. **5B** (1972), 2555; see also Ref. 48.

[41] Fröhlich, J. and C. Pfister, Commun. Math. Phys. **89** (1983), 303.

[42] Griffiths, R., J. Math. Phys. **8** (1967), 478; 484.

[43] Kelly, D. and S. Sherman, J. Math. Phys. **9** (1968), 466.

[44] Ginibre, J., Commun. Phys. **16** (1970), 310.

[45] Sylvester, G., Commun. Math. Phys. **73** (1980), 105.

[46] Dyson, F., E. Lieb and B. Simon, Phys. Rev. Lett. **37** (1976), 120.

[47] Dyson, F., E. Lieb and B. Simon, J. Stat. Phys. **18** (1978), 335.

[48] Lebowitz, J., J. Stat. Phys. **16** (1977), 463.

[49] Thouless, D., Phys. Rev. **187** (1969), 732.

[50] Simon, B. and A. Sokal, J. Stat. Phys. **25** (1981), 679.

[51] Fröhlich, J. and T. Spencer, Commun. Math. Phys. **84** (1982), 87.

[52] Aizenman, M., Commun. Math. Phys. **73** (1980), 83.

[53] Dobrushin, R., Theo. Prob. Appl. **17** (1972), 582.

[54] van Beijeren, H., Commun. Math. Phys. **40** (1975), 1.

[55] Fröhlich, J., C. Pfister and T. Spencer, IHES preprint, 1982.

[56] Fröhlich, J., R. Israel, E. Lieb and B. Simon, Commun. Math. Phys. **62** (1978), 1; J. Stat. Phys. **22** (1980), 297.

[57] Lieb, E. and D. Mattis, Phys. Rev. **125** (1962), 164; J. Math. Phys. **3** (1962), 749.

[58] Ruelle, D., Phys. Rev. Lett. **27** (1971), 1040.

[59] Wilson, K. and J. Kogut, Phys. Rep. **12C** (1974), 75; D. Walker and R. Zia, Rep. Prog. Phys. **41** (1978), 1.

[60] Collet, P., J. P. Eckmann and O. Lanford, Commun. Math. Phys. **76** (1981), 211.

[61] Griffiths, R. and P. Pearce, J. Stat. Phys. **20** (1979), 499.

[62]   Gawedski, K. and A. Kupiainen, Commun. Math. Phys. **82** (1981), 407; **83** (1982), 469; **88** (1983), 77.
[63]   Reed, M. and B. Simon, *Methods of modern mathematical physics, III. Scattering theory*, Academic Press, 1979.
[64]   Agmon, S., Ann. Scuola Norm. Sup. Pisa **II,2** (1975), 151; S. Kuroda, J. Math. Soc. Japan **25** (1973), 75.
[65]   Enss, V., Commun. Math. Phys. **61** (1978), 285; Ann. Phys. **119** (1979), 117.
[66]   Fadeev, L., *Mathematical aspects of the three body problem in quantum scattering theory*, Steklov Institute, 1963.
[67]   Ginibre, J. and M. Moulin, Ann. Inst. H. Poincaré, **A21** (1974), 97.
[68]   Thomas, L., Ann. Phys. **90** (1975), 127.
[69]   Howland, J., J. Func. Anal. **22** (1976), 250.
[70]   Kato, T., J. Fac. Sci. Univ. Tokyo **24** (1977), 503.
[71]   Yajima, K., J. Fac. Sci. Univ. Tokyo **25** (1978), 109.
[72]   Sigal, I., Mem. A.M.S. **209** (1978); M. Loss and I. Sigal, preprint, 1982.
[73]   Hagedorn, G. and P. Perry, Comm. Pure Appl. Math. **36** (1983), 213.
[74]   Merkuriev, S., Ann. Phys. **130** (1980), 395; Acta Phys. Aust. Supp. **23** (1981), 65.
[75]   Enss, V., J. Func. Anal., to appear; Commun. Math. Phys. **89** (1983), 245 and in prep.
[76]   Mourre, E., CNRS preprint, 1982.
[77]   Hagedorn, G., Trans. Am. Math. Soc. **258** (1980), 1.
[78]   Sigal, I., Bull. Am. Math. Soc. **84** (1978), 152.
[79]   Balslev, E. and J. Combes, Commun. Math. Phys. **22** (1971), 280.
[80]   Perry, P., I. Sigal and B. Simon, Ann. Math. **114** (1981), 519; see also, R. Froese and I. Herbst, Duke Math. J. **49** (1982), 1075.
[81]   Mourre, E., Commun. Math. Phys. **78** (1981), 391.
[82]   Agmon, S., Sem. Goulauic. Schwartz 1978—79 and in prep?
[83]   Reed, M. and B. Simon, *Methods of modern mathematical physics, IV. Analysis of operators*, Academic Press, 1978.
[84]   Morgan, J. and B. Simon, unpublished.
[85]   Leib, E. and B. Simon, Adv. Math. **23** (1977), 22.
[86]   Lieb, E., Rev. Mod. Phys. **48** (1976), 553; Rev. Mod. Phys. **53** (1981), 603.
[87]   Scott, J., Philos. Mag. **43** (1952), 899.
[88]   Bander, M., Ann. Phys. **144** (1982), 1.
[89]   Schwinger, J., Phys. Rev. **22A** (1980), 1827; **24A** (1981), 2353.
[90]   Ruskai, M. B., Commun. Math. Phys. **82** (1982), 457; **85** (1982), 325.
[91]   Sigal, I., Commun. Math. Phys. **85** (1982), 309.
[92]   Zhislin, G., Tr. Mosk. Mat. Obs. **9** (1960), 81.
[93]   Simon, B., Helv. Phys. Acta **43** (1970), 607.
[94]   Benguria, R. and E. Lieb, Phys. Rev. Lett. **50** (1983), 50.
[95]   Dyson, F. and A. Lenard, J. Math. Phys. **8** (1967), 423; **9** (1968), 698.
[96]   Lieb, E. and W. Thirring, Phys. Rev. Lett. **35** (1975), 687.
[97]   Lebowitz, J. and E. Lieb, Adv. in Math. **9** (1972), 316.
[98]   Lieb, E., Phys. Lett. **70A** (1969), 71.
[99]   Dyson, F., J. Math. Phys. **8** (1967), 1538.
[100]   Lieb, E., Springer Lecture Notes in Phys. **116** (1980), 91.
[101]   Simon, B., *Functional Integration*, Academic Press, 1979.
[102]   Glaser, V. and A. Martin, CERN preprint, 1983.
[103]   Radin, C. and C. Gardner, J. Stat. Phys. **20** (1979), 719; C. Radin and G. Hamrick, J. Stat. Phys. **21** (1979), 601; C. Radin and R. Heltman, J. Stat. Phys. **22** (1980), 281; C. Radin, J. Stat. Phys. **26** (1981), 365; Physica **113A** (1982), 338.
[104]   Duneau, M. and A. Katz, Ann. Inst. H. Poincaré **37A** (1982), 249.
[105]   Kunz, H. and B. Souillard, Commun. Math. Phys. **78** (1980), 201.
[106]   Delyon, F., H. Kunz and B. Souillard, J. Phys. **A16** (1983), 25.

[107]   Goldshade, I., S. Molchanov and L. Pastur, Funkt. Anal. Pril. **11** (1977), 1.
[108]   Fröhlich, J. and T. Spencer, Commun. Math. Phys. **88** (1983), 151.
[109]   Kunz, H. and B. Souillard, J. Phys. Lett. **44** (1983), L 411 and in prep.
[110]   Radin, C. and B. Simon, J. Diff. Eqn. **29** (1978), 289.
[111]   Avron, J. and B. Simon, Duke Math. J. **50** (1983), 369.
[112]   Romerio, M. and W. Wreszinski, J. Stat. Phys. **21** (1979), 169 and refs. therein.
[113]   Sarnak, P., Commun. Math. Phys. **84** (1982), 377.
[114]   Simon, B., Adv. Appl. Math. **3** (1982), 463.
[115]   Gordon, A., Usp. Math. Nauk. **31** (1976), 257.
[116]   Bellissard, J., C. Lima and D. Testard, Commun. Math. Phys. **88** (1983), 207.
[117]   Domb, C., Adv. Chem. Phys. **15** (1969), 229.
[118]   Whittington, S., Adv. Chem. Phys. **51** (1982), 1.
[119]   Fisher, M., Phys. Rev. **162** (1967), 480.
[120]   Symanzik, K., in *Local quantum theory*, ed. R. Jost, Academic Press, 1969.
[121]   Brydges, D., J. Fröhlich and T. Spencer, Commun. Math. Phys. **83** (1982), 123.
[122]   de Gennes, P., Phys. Lett. **38A** (1972), 339.
[123]   Westwater, M., Proc. 4th Bielefeld (Zif) Conf. in Math. Phys., to appear.
[124]   Itzykson, C. and J. Zuber, *Quantum field theory*, McGraw-Hill, 1980; P. Raymond, *Field theory—A modern primer*, Benjamin, 1981.
[125]   Glimm, J. and A. Jaffe, *Quantum physics*, Springer, 1981.
[126]   Seiler, E., Springer Lecture Notes in Physics **159** (1982).
[127]   Fröhlich, J., Nuc. Phys. **B200** (1982), 281.
[128]   Aizenman, M., Phys. Rev. Lett. **47** (1981), 1; Commun. Math. Phys. **86** (1982), 1.
[129]   Gallavotti, G. and V. Rivasseau, preprints, 1983.
[130]   Penrose, R., in *Theoretical principles in astrophysics and relativity* (eds. N. Lebovitz et al.) U. Chi. Press (1978), and in *General relativity and Einstein centenary survey* (eds. S. Hawking and W. Israel), Cambridge Univ. Press (1979).
[131]   Penrose, R., in *Seminar on differential geometry*, Princeton Univ. Press, 1982.
[132]   Penrose, R., Phys. Rev. Lett. **14** (1965), 57; S. Hawking and O. Penrose, Proc. Roy. Soc. London **A314** (1970), 529.
[133]   Misner, C., in *Relativity theory and astrophysics*, *I* (ed. J. Ehler), A.M.S., 1967.
[134]   Isenberg, J. and V. Moncrief, preprint, 1982.

Perspectives in Mathematics
Anniversary of Oberwolfach 1984
© Birkhäuser Verlag, Basel

Linear Algebraic Groups

T. A. SPRINGER

Rijksuniversiteit Utrecht, Mathematisch Instituut,
Budapestlaan 6, Utrecht 2506 (The Netherlands)

1   Introduction

The theory of linear algebraic groups is a creation of the last thirty years. It has proved its viability, for example by its applications, made in various parts of mathematics. Several Oberwolfach conferences have been devoted to the theory of algebraic groups and its ramifications.

The reader will find below a survey of the theory of linear algebraic groups. The central part is section 5, which contains a review of the basic results of the theory of linear algebraic groups over algebraically closed ground fields, completed in section 6 by a review of rationality results, involving an arbitrary ground field. These sections are preceded by a survey of some relevant anterior results from 19th and 20th century mathematics.

The later sections 7 and 8 briefly discuss some applications of the theory. I have not aimed at completeness, my purpose was only to indicate the scope of these applications.

I am grateful to A. Borel for a number of critical remarks.

2   The definition of linear algebraic groups

We shall first give a definition of the notion of linear algebraic group. It will be useful to have it in mind in the sequel.

Let K be an algebraically closed field. We denote by $SL_n(K)$ the group of $n \times n$-matrices $x = (x_{ij})$ with entries in K and determinant 1.

A *linear algebraic group over K* is a subgroup G of some $SL_n(K)$ such that the following holds:

(2.1)   *There is a set S of polynomials in $K[X_{ij}]_{1 \leq i,j \leq n}$ such that $x = (x_{ij})$ lies in G if and only if $P(x_{ij}) = 0$ for all $P \in S$.*

We shall abbreviate "linear algebraic group over K" to "K-group". If $G \subset SL_n(K)$ is a K-group we denote by $K[G]$ the K-algebra of functions f on G of the form $f(x) = F(x_{ji})$, where $F \in K[X_{ij}]_{1 \leq i,j \leq n}$. We call $K[G]$ the *algebra of regular functions* on G.

Let $G \subset SL_m(K)$ and $H \subset SL_n(K)$ be two K-groups. A *homomorphism of*

K-groups is a homomorphism of abstract groups $\varphi : G \to H$ with the following property:

(2.2)   *There are polynomials F_{ij} in the indeterminates $X_{hl}(1 \leqq h,j \leqq m, 1 \leqq i,j \leqq n)$ such that for $x = (x_{hl}) \in G$ the (i,j)-matrix entry of $\varphi(x)$ is given by $F_{ij}(x_{hl})$.*

An equivalent requirement is: $\varphi^*(f) = f \circ \varphi$ defines a homomorphism of $K[H]$ to $K[G]$.

It is clear how to define notions like isomorphism of *K*-groups and algebraic subgroup (or *K*-subgroup). It is also clear that a *K*-group is an affine algebraic variety, so the notions of algebraic geometry can be applied to it.

The preceding definitions of *K*-groups and their homomorphisms are not intrinsic. Better definitions can be given (see (4.5)). For example, one could define a *K*-group to be a group object in the category of affine algebraic varieties over *K*.

(2.3)   *Examples* of *K*-groups. (a) The group $GL_n(K)$ of all nonsingular $n \times n$-matrices can be viewed as a *K*-group, contained in $SL_{n+1}(K)$ via the imbedding $x \mapsto \begin{pmatrix} x & 0 \\ 0 & (\det x)^{-1} \end{pmatrix}$. We also write GL_n for this *K*-group, (b) SL_n, (c) the orthogonal group $O_n = \{x \in GL_n | x \cdot {}^t x = 1\}$, and the special orthogonal group $SO_n = \{x \in SL_n | x \cdot {}^t x = 1\}$, (d) $T_n \subset GL_n$, the group of diagonal matrices, (e) $B_n \subset GL_n$, the group of upper triangular matrices, (f) the additive group $G_a = \left\{ \begin{pmatrix} 1 & \xi \\ 0 & 1 \end{pmatrix} \in SL_2(K) \,\middle|\, \xi \in K \right\}$.

3   Linear algebraic groups in 19th century mathematics

Nowadays we are inclined to view the theory of linear algebraic groups as an algebraization of the theory of Lie groups. We could therefore expect to find the beginnings of the theory of linear algebraic groups in Lie's work. But this is not so. Certainly Lie's work contains results which, in hindsight, can be viewed as pertaining to the theory of linear algebraic groups, but he did not mention them explicitly (as far as I can see).

(3.1)   However there is another subject initiated at the end of the 19th century, where linear algebraic groups do appear naturally, namely the Galois theory of linear diffential equations, initiated by E. Picard in 1883 (nowadays called Picard-Vessiot theory). A typical example of the problems studied by Picard is the following.

Consider an nth order linear differential equation with polynomial coefficients in the complex plane:

$$\frac{d^n f}{dz^n} + a_{n-1}(z) \frac{d^{n-1} f}{dz^{n-1}} + \ldots + a_0(z) f = 0. \tag{*}$$

The problem is to develop a Galois theory for such equations, on the model of the Galois theory of algebraic equations. In present-day language, the Galois group of (*) is introduced in the following manner. One knows that (*) has n holomorphic solutions which are linearly dependent over $\mathbb{C}$, say $f_1, \ldots, f_n$. Let L be the field obtained by adjoining the f_i and all their derivatives to $\mathbb{C}(z)$ and denote by G the group of $\mathbb{C}(z)$-linear automorphisms of L which commute with derivation. This is the Galois group of (*).

If $x \in G$ then

$$x \cdot f_i = \sum_{j=1}^{n} x_{ji} f_j$$

and one shows that the matrices $(x_{ij}) \in GL_n(\mathbb{C})$, for $x \in G$, form a linear algebraic group, isomorphic to G as an abstract group. Hence $\mathbb{C}$-groups appear here as Galois groups of linear differential equations. It seems that Picard, in his early work on this subject, announced in 1882 and published in detail in 1887 [62], was fully aware of the fact that these Galois groups are linear groups described by algebraic relations between matrix elements.

He speaks explicitly of "groupes algébriques". Picard's work was completed by Vessiot (see [89]). A brief modern exposition of Picard-Vessiot theory is given in [39]. More recent and more complete is Kolchin's book [46].

A little later one encounters the notion of algebraic group in the work of Maurer, where invariant theory seems to be in the background (see [55]). The problem he is interested in in [56], is the following one (in a modern paraphrase). Let G be a closed subgroup of $GL_n(\mathbb{C})$, in the ordinary topology. Give conditions for G to be an algebraic group.

(3.2)   A few other developments ought to be mentioned, which were initiated in the 19th century, and which influenced later developments in the theory of linear algebraic groups.

First and foremost there is the theory of Lie groups, started in Lie's monumental work [50]. Of particular importance is what is nowadays called the classification of simple complex Lie algebras, initiated by Lie and completed by Killing [41] and E. Cartan (in his thesis, see [15, vol. 1, p. 137—287]). Lie's work also contains detailed studies of particular matrix groups over $\mathbb{C}$.

Coming from a different direction, namely from the theory of finite groups (and in the last resort from the Galois theory of algebraic equations), C. Jordan was also led to study such special groups, but not over $\mathbb{C}$. In his "Traité des Substitutions" (1870) one finds an elaborate study of the general linear group over a finite prime field, and of subgroups like the orthogonal groups over such a field. Here one also finds a version of what is now called "Jordan's normal form", over a finite field. Similar results over $\mathbb{C}$ are contained in the work of Weierstrass on elementary divisors [91] (but this similarity is clear only in hindsight).

4   Linear algebraic groups in the first half of the 20th century

We shall now follow briefly later developments which have a bearing on the theory of linear algebraic groups.

(4.1)   Picard-Vessiot theory was taken up by Ritt around 1930. His work aimed at an algebraization of the theory of differential equations. The group-theoretical aspects of this work were taken up by Kolchin in 1948 (see his papers [44], [45]). Kolchin's work is the starting point of the subsequent development of the theory of linear algebraic groups. We shall review some of his results.

Let K be an arbitrary algebraically closed field and G a K-group. We say that G is *connected* if G has no algebraic subgroup of finite index (this is tantamount to saying that G is connected in the Zariski topology). Connectedness of G is also equivalent to irreducibility of G, in the sense of algebraic geometry. Then Kolchin proves [44, p. 10]:

(4.2)   *G has a unique connected normal algebraic subgroup $G°$ of finite index.* We call $G°$ *the identity component of G.*

Example. Let char $(K) \neq 2$. The group O_n (example (c) of (2.3)) is not connected. Its identity component is $SO_n = O_n \cap SL_n$.

Another basic theorem is the *Lie-Kolchin theorem* [loc. cit., p. 19]:

(4.3)   *Let $G \subset GL_n(K)$ be a K-group which is connected and which is solvable as an abstract group. There is $x \in GL_n(K)$ such that xGx^{-1} is contained in the group B_n of invertible upper triangular matrices.*

The importance of this theorem for the Galois theory of linear differential equations is that it implies that the solutions of a linear homogeneous differential equation are "elementary" functions if and only if its Galois group is solvable [loc. cit. p. 38]. Lie already proved an analogue of (4.3) for a complex solvable Lie algebra $\mathfrak{g}$ of $n \times n$-matrices. This result about solvable Lie algebras is not true, however, over fields of positive characteristic.

It is therefore noteworthy that the "global" result about algebraic groups does carry over to positive characteristics. It should also be noted that the geometric assumption of connectedness in (4.3) is essential, as easy counterexamples of finite solvable groups show.

(4.4)   In the beginnings of the theory of Lie groups (starting with Lie himself) the "infinitesimal" approach was paramount. A Lie group G, which in present-day language is a C^∞- or analytic manifold equipped with a compatible group structure, was studied via its "infinitesimal group", i. e. the Lie algebra $\mathfrak{g}$ of the vector fields on G which are invariant by left translations with elements of G ($\mathfrak{g}$ can be identified with the tangent space of G at the neutral element). The Lie algebra product is given by the Poisson bracket of vector fields.

In the later developments, the "global" geometric aspects become more and more important. For example, in H. Weyl's work on the representation theory of compact Lie groups in the twenties [96], and in particular in his proof of the character formula, the global approach is essential. He uses integration on the group manifold, and he makes a start with the study of the topology of compact Lie groups.

In E. Cartan's work on symmetric spaces and on the topology of Lie groups, which is a little later (see for example [15, vol. 2, p. 1165—1227]) we meet the same interest in the global geometric aspects of the theory of Lie groups.

When the basic notions of topology and differential topology were well understood, an exposition of the foundations of the theory of Lie groups could be given from a global point of view. This was done by Chevalley in 1946 in his book [16]. Here algebraic groups also make an appearance [loc. cit., p. 198]. Chevalley associates to a compact Lie group G a $\mathbb{C}$-group, the complexification of G.

(4.5)   A. Weil's work on Jacobians of algebraic curves (published in 1948 in [92]) required an algebro-geometric study of abelian varieties, i.e. projective algebraic varieties, over any algebraically closed field K, which have a compatible group structure (which then is automatically commutative). Classically, abelian varieties over $\mathbb{C}$ were studied by transcendental methods which go back to Riemann.

In [loc. cit.] Weil also starts the study of general algebraic groups. An algebraic group G over the algebraically closed field K is an algebraic variety over K provided with a group structure such that the product map $G \times G \to G$ and the inversion map $G \to G$ are morphisms of algebraic varieties. This notion comprises both the notion of abelian variety (G a projective variety) and of a linear algebraic group (G an affine variety). In two later papers ([93] and [94]) Weil continues the study of general algebraic groups. The following basic facts about quotients are contained in these papers.

(4.6)   *Let G be an algebraic group over the algebraically closed field K, let H be an algebraic subgroup of G. There exists a unique structure of algebraic variety on the set G/H of cosets gH such that*
(a) the translations by elements of G are isomorphisms of G/H;
(b) the canonical map $\pi: G \to G/H$ is a morphism of algebraic varieties which is separable;
(c) for any morphism of algebraic varieties $f: G \to V$ such that $f(gh) = f(g)$ for all $g \in G$, $h \in H$ there is a unique morphism of varieties $\varphi: G/H \to V$ such that $f = \varphi \circ \pi$;
(d) if H is a normal algebraic subgroup of G then G/H has a structure of algebraic group, with the properties familiar from group theory.

This is the theorem about existence of quotients. A first version of it was proved by S. Nakano [61], and similar results were also proved by Rosenlicht [63]. Notice that the theorem is an analogue of an easy result on topological groups. The technical condition of separability of (b) is needed if char(K) > 0. It is equivalent to the more geometric condition that in some point g of G the linear tangent map $(d\pi)_g$ is a surjective map of the tangent space of G at g onto the tangent space of G/H at πg. The results proved in [63] and [94] are more precise, in that there algebraic groups over not necessarily algebraically closed fields are studied. Our formulation is somewhat different from the one in these papers, but is the accepted one nowadays. (4.6) is really a result pertaining to algebraic geometry, but it is quite essential for the theory of linear algebraic groups. As a complement to (d), one has that if G is a linear algebraic group and H a normal algebraic subgroup of G, then G/H is also a linear algebraic group [2, p. 183].

(4.7)   C. Jordan's study of special finite linear groups, mentioned in (3.2), was further developed by E. Dickson around 1910, and is described in his book [24]. This contains a thorough study of what is now called (after H. Weyl) the "classical" groups (general and special linear groups, orthogonal and symplectic groups) over finite fields. Dickson worked with matrices. Later on, Dieudonné took up the classical groups again, and studied them over more general fields, in the context of linear algebra (see [27] and [28]).

Dickson was clearly aware of the fact that the classical groups over finite fields are finite analogues of the simple complex Lie groups of types A, B, C, D. He also constructed analogues over finite fields of two of the five exceptional types of the classification of Killing-Cartan, namely the types G_2 and E_6 ([25], [26]). This work, and also Dieudonné's work is oriented towards the group-theoretical study of the groups in question, in particular it contains proofs of simplicity for certain groups.

A decisive progress was made by Chevally in 1955, in [18]. He constructs for *any* type of the Killing-Cartan classification of complex simple Lie groups, an analoguous linear group over any field k, and he proves simplicity theorems. Of particular interest are the finite groups which one obtains by taking k to be finite. His construction is based on the general theory of semi-simple Lie algebras over $\mathbb{C}$. In particular, it exploits the fact that such a Lie algebra has a basis whose structure constants are *integers* (a fact which is already contained in E. Cartan's thesis of 1894 [15, p. 137—287]). [18] also contains simplicity proofs.

5   Basic results on linear algebraic groups

The developments described in the previous section were brought together by Borel and Chevalley, who in the years 1956—1958 founded the theory of linear algebraic groups over algebraically closed fields, in the form we know-nowadays. In [1] Borel established a number of basic general results on linear algebraic groups. In his seminar [19] Chevalley brought out the analogy with the theory of Lie groups, culminating in a classification of simple linear algebraic groups in any characteristic, which is "the same" as the classification of complex simple Lie groups.

Both Borel and Chevalley make an essential use of results from algebraic geometry. One can—with some exaggeration—claim that the theory of linear algebraic groups is an application of algebraic geometry.

We shall now review some basic facts. Most of what follows is already contained in [19]. More recent expositions of the theory of linear algebraic groups are [2], [38], [74]. We denote by K an algebraically closed field, of arbitrary characteristic, and by $G \subset GL_n(K)$ a linear algebraic group over K, as defined in (2.1).

(5.1)   *Jordan decomposition.*

It is an easy consequence of the results about Jordan normal forms of matrices that any $g \in GL_n(K)$ can be written in a unique manner as a product $g = g_s g_u$ of two commuting matrices, such that g_s is semi-simple (i.e. diagonalizable) and g_u is unipotent (i.e. all its eigenvalues are 1). Then:

(5.1.1) (a) *If $g \in G$ then $g_s, g_u \in G$;*
(b) *If $f: G \to H$ is a homomorphism of K-groups (see (2.2)) then $f(g_s) = f(g)_s, f(g_u) = f(g)_u$.*
(5.1.1) implies that we have in G an intrinsic notion of semisimple and unipotent elements, and a *Jordan decomposition* $g = g_s g_u = g_u g_s$. This is an elementary, but important result. In this generality it was first stated in [1], but it is essentially already contained in [45, §3].

G is called *unipotent* if all its elements are unipotent. The following result was already proved by Kolchin [45, §1].

(5.1.2) *If G is any subgroup of $GL_n(K)$ (not necessarily algebraic) consisting of unipotent elements then there is $x \in GL_n(K)$ such that xGx^{-1} is contained in the group B_n of invertible upper triangular matrices.*

This implies that a unipotent K-group is nilpotent, as an abstract group.

The *radical $R(G)$* is the maximal connected, normal, solvable subgroup of G and the *unipotent radical $R_u(G)$* is the maximal connected, normal, unipotent subgroup. Clearly $R_u(G) \subset R(G)$. The existence proof of these radicals is easy. G is *semi-simple* if $R(G) = \{e\}$ and *reductive* if $R_u(G) = \{e\}$.

(5.2)   *Tori.*

A K-group G isomorphic to a group of diagonal matrices T_m (see example (2.3)(d)) is called a *torus*. The name comes from the fact that the tori play a role similar to the one of the usual tori in the theory of compact Lie groups.

(5.2.1) *If G is a torus there is $x \in GL_n(K)$ such that xGx^{-1} lies in the group $T_n(K)$ of invertible diagonal matrices.*

Another way of stating this is that any representation in the sense of algebraic groups $T_m \to GL_N(K)$ is equivalent to a direct sum of 1-dimensional representations.

Assume that T is a torus. Its *character group* $X^*(T)$ is the abelian abstract group of all homomorphisms of K-groups $\chi: T \to GL_1(K)$; its group of *1-parameter subgroups* $X_*(T)$ is the abelian abstract group of all homomorphisms of K-groups $u: GL_1(K) \to T$.

(5.2.2) *Let T be a torus.*
(a) *$X^*(T)$ and $X_*(T)$ are free abelian groups of finite rank, and there is a duality pairing $p: X_*(T) \times X^*(T) \to \mathbf{Z}$;*
(b) *$T \mapsto X_*(T)$ defines an equivalence between the category of tori over K and the category of free abelian groups of finite rank.*

The pairing p of (a) is defined as follows: if $\chi \in X^*(T)$, $u \in X_*(T)$ then $\chi(u(t)) = t^{p(\chi, u)}$, for all $t \in K^*$.

These results about tori are elementary, but quite useful. They are contained essentially in [2, §7].

(5.3)   *Connected solvable groups.*
Here the tools of algebraic geometry come into use. The proof of the following results on solvable K-groups involves an induction on dimension, which exploits the existence of quotients (4.6). They are due to Borel [1, §12—13].

(5.3.1) *Let G be a connected solvable K-group.*
(a) *The set G_u of unipotent elements of G is a connected, normal, unipotent algebraic subgroup;*
(b) *All maximal subtori of G are conjugate. If T is a maximal torus, then G is the semi-direct product $T \times G_u$, in the sense of algebraic groups;*
(c) *Let $s \in G$ be semi-simple. Then s is contained in a maximal torus and its centralizer is a connected algebraic subgroup.*

(5.3.2) (*Borel's fixed point theorem.*) *Let G be a connected solvable K-group. Assume that it operates on a non-empty projective algebraic variety V over K. Then G has a fixed point in V.*

We sketch a proof. First one shows that G has a Zariski-closed orbit in V (this is elementary), hence we may assume that V itself is a G-orbit. Then one shows by an induction that we may reduce to the case that G is abelian. Let $v \in V$ and denote by G_v the isotropy subgroup of v in G. Then $g \mapsto g \cdot v$ induces an injective morphism of the affine algebraic variety G/G_v onto the projective variety V. It follows that both varieties must be reduced to a point, whence the statement. Although intuitively clear, the algebro-geometric details require some care. The proof sketched here is Borel's original one [1, §15].

One can view Borel's theorem as a generalization of the Lie-Kolchin theorem (4.3). In fact, if G is as in (4.3), we take V to be the variety of all complete flags in the n-dimensional vector space K^n, a flag F being an ordered set $F = (F_1, F_2, \ldots, F_{n-1})$ of subspaces of K^n, with dim $F_i = i$, $F_i \subset F_{i+1}$ $(1 \leq i \leq n-2)$. Classically, the set of flags is a projective algebraic variety V, on which $GL_n(K)$ acts. Applying (5.3.2) to G and V, we obtain (4.3). The "flag manifold" V can be identified with the quotient variety GL_n/B_n (where B_n is the group of invertible upper triangular matrices).

(5.4)   *Borel subgroups.*
Now let G be any connected K-group. A *Borel subgroup* B of G is a connected, solvable, algebraic subgroup of G, which is maximal for these properties (their existence is trivial). A *parabolic subgroup* P of G is an algebraic subgroup of G which contains a Borel subgroup.

(5.4.1) (a) *P is parabolic if and only if G/P is a projective variety;*
(b) *All Borel subgroups of G are conjugate.*
      The fixed point theorem (5.3.2) implies the if-part of (a), it also shows that (a) implies (b). It remains to prove that for any Borel subgroup B the quotient variety G/B is projective. This follows by using the Lie-Kolchin theorem (4.3) ([1, §16], for another proof of (5.4.1) see [74, p. 159]).
      It should be remarked that parabolic groups do not yet appear in [2] and [19], they were first systematically introduced in [8].

(5.4.2) *Let P be a parabolic subgroup of G. Then P equals its normalizer in G and P is connected.*
      The crucial result is that a Borel subgroup equals its normalizer. This was first proved by Chevalley [19, exp. 9]. A simpler proof was given by Borel (published in [38, no. 23]). This is an important result, needed for results like (5.4.6).
      As a consequence of (5.3.1) (b) and (5.4.1) (b) we have the following analogue of a well-known theorem about compact Lie groups. It is due to Borel [1, §16].

(5.4.3) *All maximal tori of G are conjugate.*
      The dimension of a maximal torus of G is the *rank* of G. Another basic result is

(5.4.4) *Any element of G lies in a Borel subgroup.*
Let B be a Borel subgroup. By (5.4.1) (b), an equivalent statement is $G = \bigcup_{g \in G} gBg^{-1}$. To prove this, an algebro-geometric construction is used [1, §17]. We follow here [19]. Let H be any algebraic subgroup of G and denote by $N = \{g \in G | gHg^{-1} = H\}$ its normalizer in G, this is an algebraic subgroup. Consider the subset of $G/N \times G$

$$X = \{(gN, x) | g^{-1}xg \in H\}.$$

One shows that X is an algebraic subvariety of $G/N \times G$. In fact, X is isomorphic to the associated fiber bundle $G \times^N H$, for the quotient map $G \to G/N$ i.e. the quotient of $G \times H$ by the N-action $n(g, h) = (gn^{-1}, nhn^{-1})$. If $\pi \colon X \to G$ is the projection, then clearly $\pi X = \bigcup_{g \in G} g H g^{-1}$. So a geometric study of the morphism π will lead to information about the union of the conjugates of H.

In the particular case $H = B$ the completeness of projective varieties implies that πX is an algebraic subvariety of G and (5.4.4) is tantamount to $\pi X = G$. The proof of this fact uses once again the preceding construction, with H the centralizer of a maximal torus.

A consequence of (5.4.4) and the fixed point theorem is the following connectedness theorem.

(5.4.5) *Let S be a subtorus of G. Then its centralizer in G is connected.*

We finally come to another basic result, which characterizes the radicals of G (see 5.1). It is due to Chevalley [19, exp. 12]. Its proof (but not its formulation) requires some work on roots, which we have not yet mentioned.

(5.4.6) *Let T be a maximal torus of G. The radical $R(G)$ is the identity component of the intersection of all Borel subgroups B containing T. The unipotent radical $R_u(G)$ is the identity component of the intersection of all B_u, where B runs through the Borel subgroups containing T.*

This has the following consequence:

(5.4.7) *Assume G to be reductive (i.e. $R_u(G) = \{e\}$). The centralizer of any subtorus of G is reductive. The centralizer of a maximal torus T coincides with T.*

(5.5)   *The Lie algebra of G.*

Before proceeding with the description of the structure of reductive K-groups, we shall make a digression about the Lie algebra of G, which is a useful ingredient of the theory.

It was already remarked before (in 4.4) that in the original development of the theory of Lie groups an infinitesimal approach was used, studying a Lie group G via its Lie algebra $\mathfrak{g}$ and deducing properties of G via an integration process from those of $\mathfrak{g}$.

If now G is an algebraic group over K, one can give the following definition of its Lie algebra $\mathfrak{g}$, which is modeled on the definition of the Lie algebra of a Lie group: $\mathfrak{g}$ is the space of vector fields on G, in the sense of algebraic geometry, which are invariant under left translations by elements of G, the Lie algebra product being given by the Poisson bracket. Then $\mathfrak{g}$ can be identified as a vector space with the tangent space $T_e G$ of G at the neutral element.

The details, in the case of linear algebraic groups, are given in [2, I.3] or [74, 3.3] (in these references the language is less geometric). In particular, one obtains a functor $G \mapsto \mathfrak{g}$ of the category of K-groups to the category of finite dimensional Lie algebras over K.

In [17], Chevalley started, in the case that $\operatorname{char}(K) = 0$, an approach to the theory of K-groups via their Lie algebras, but he did not get very far. It seems that for some time the Lie algebras of K-groups, in the case $\operatorname{char}(K) > 0$ were thought to be rather pathological objects. One will not come across Lie algebras in [1] and [19]. But the respectability of these Lie algebras was restored subsequently, also because of the use made of them by Grothendieck in [23, exp. XIV], and in the present expositions ([2], [38], [74]) they are quite useful.

One of the uses made of the Lie algebra $\mathfrak{g}$ of the K-group G is in the introduction of the root system of G with respect to a maximal torus. First let G be an arbitrary K-group with Lie algebra $\mathfrak{g}$. The right translation by an element $x \in G$ defines an automorphism $\operatorname{Ad}(x)$ of the vector space $\mathfrak{g}$. The map $x \mapsto \operatorname{Ad}(x)$ is a representation in the sense of algebraic groups, the *adjoint representation* of G. Denote by T a maximal torus of G. The restriction of the adjoint representation to T is a direct sum of 1-dimensional representations of T (5.2.1). A nontrivial character $\alpha \in X^*(T)$ is a *root* of G with respect to T if (a) it occurs in this direct sum, (b) if S^α is the identity component of the kernel of α (which is a subtorus of T of codimension 1) then the centralizer of S is non-solvable (this centralizer is connected by (5.4.5)). We denote by $\Phi = \Phi(T, G)$ the set of roots. Notice that if $\alpha \in \Phi$ there is $X \in \mathfrak{g}$ such that $\operatorname{Ad}(t) X = \alpha(t) X$ for $t \in T$. The roots were first defined in [19], the definition here given is equivalent.

(5.6)   *Some classification results.*

We next review some results about classification of "small" K-groups. The first one is the classification of 1-dimensional groups.

(5.6.1) *Let G be a connected K-group of dimension 1. Then G is isomorphic to either GL_1 of G_a.*

These groups are defined in (2.3).

In spite of its simplicity, this is not a trivial result. It is explained by the following facts from the theory of algebraic curves: (a) an irreducible smooth projective curve over K which has an infinite group of automorphisms fixing one point is isomorphic to the projective line $\mathbb{P}^1(K)$, (b) (Lüroth's theorem) the automorphism group of the projective line $\mathbb{P}^1(K)$ is the group $PGL_2(K)$ of all fractional linear transformations

$$\xi \mapsto (a\xi + b)(c\xi + d)^{-1} \left(\xi \in \mathbb{P}^1(K), \begin{pmatrix} a & b \\ c & d \end{pmatrix} \in GL_2(K) \right).$$

For a proof using these facts see [2, p. 257]. One can give more elementary proofs (see [38, no. 20], [74, 2.6]). The first proof in the literature seems to be the one by Grothendieck [19, exp. 7], which uses properties of unramified coverings of the algebraic variety K.

Another classification result describes the "minimal" semi-simple K-groups.

(5.6.2) *Let G be a connected semi-simple K-group of rank 1. Then G is isomorphic either to $SL_2(K)$ or to $PGL_2(K)$.*
   $PGL_2(K)$, defined above, can also be viewed as the quotient of $GL_2(K)$ by its centre.
   In [19] there is a somewhat weaker result (in exp. 12 by M. Lazard), see also [2, p. 319]. In these references it is first established that, under these assumptions on G, we have (B being a Borel subgroup) that G/B is isomorphic to $\mathbb{P}^1(K)$. Then Lüroth's theorem is used to obtain a surjective homomorphism $G \to PGL_2(K)$. A more group-theoretical proof of the stronger statement (5.6.2) is given in [74, 8.2].

(5.6.3) The following use is made of (5.6.2). Let G be any connected K-group and T a maximal torus. Let $\alpha \in \Phi = \Phi(T, G)$ be a root (see (5.5)), denote by S_α the identity component of the kernel of α and by H the centralizer of S_α. Then H is a connected algebraic subgroup which is not solvable. One easily checks that the quotient $H/R(H)$ of H by its radical is connected semi-simple of rank 1, so (5.6.2) can be applied to it.
   Let $X = X^*(T)$ be the character group of T and $X^v = X_*(T)$ its group of 1-parameter subgroups, then X and X^v are in duality via a pairing $< , >$ (see (5.2.2)). Using (5.6.2) we deduce that for any $\alpha \in \Phi$ there exists $\alpha^v \in X^v$ such that $\langle \alpha, \alpha^v \rangle = 2$. We call α^v the *coroot* associated to α.

Examples.   (a) $G = SL_2(K)$, $T = \left\{ \begin{pmatrix} t & 0 \\ 0 & t^{-1} \end{pmatrix} \middle| t \in K^* \right\}$.   We   have $X \cong X^v \cong \mathbb{Z}$. The roots (written additively) are α and $-\alpha$, with $\alpha \begin{pmatrix} t & 0 \\ 0 & t^{-1} \end{pmatrix} = t^2$.
We have $\alpha^v(t) = \begin{pmatrix} t & 0 \\ 0 & t^{-1} \end{pmatrix}$.
(b) $G = GL_n(K)$, $T = T_n$, the subgroup of diagonal matrices. Now $X \cong X^v \cong \mathbb{Z}^n$, the pairing being the canonical one. The roots $\alpha \in \Phi$ are given by

$$\alpha(\mathrm{diag}(t_1, \ldots, t_n)) = t_i t_j^{-1} \ (i \neq j, \ 1 \leq i, j \leq n)$$

and $\alpha^v(t)$ is the diagonal matrix with t in the ith diagonal entry, t^{-1} in the jth, and 1 elsewhere.
(c) $G = SL_n$, T the subgroup of its diagonal matrices. Now $X = \mathbb{Z}^n/\mathbb{Z}u$, where

$$u = (1, \ldots, 1) \quad \text{and} \quad X^v = \left\{ (x_1, \ldots, x_n) \in \mathbb{Z}^n \middle| \sum_{i=1}^{n} x_i = 0 \right\}.$$

The $\alpha \in \Phi$ are as before, likewise the α^v.

(5.6.4) A *root datum* is a quadruple $\Psi = (X, \Phi, X^v, \Phi^v)$ where X and X^v are free abelian groups of finite rank, in duality by a pairing $< , >$, with Φ and Φ^v finite subsets of X and X^v. Assume given a bijection $\alpha \mapsto \alpha^v$ of Φ onto Φ^v. Define automorphisms s_α of X and X^v by

$$s_\alpha x = x - \langle x, \alpha^v \rangle \alpha, \ s_\alpha x^v = x^v - \langle \alpha, x^v \rangle \alpha^v.$$

We impose the following axioms:
(RD 1) *If* $\alpha \in \Phi$ *then* $\langle \alpha, \alpha^v \rangle = 2$;
(RD 2) *If* $\alpha \in \Phi$ *then* $s_\alpha \Phi \subset \Phi, s_\alpha \Phi^v \subset \Phi^v$.
Then the $s_\alpha (\alpha \in \Phi)$ generate a finite group of automorphisms of X, the *Weyl group* $W(\Psi)$ of the root datum Ψ.

It now follows from the results of (5.6.3) that we can associate with the connected K-group G and a maximal torus T of G a *root datum* $\Psi = \Psi(T, G)$. Its Weyl group $W(T, G)$ can naturally be viewed as a group of automorphisms of T, induced by inner automorphisms of G. In fact one has:

(5.6.5) *Let N be the normalizer of T in G and Z the centralizer. Then* $W(T, G) \cong N/Z$.

We have followed here [74], where more details can be found. We mention some additional facts and results. Let $\Psi = (X, \Phi, X^v, \Phi^v)$ be a root datum. Denote by $Q \subset X$ the subgroup spanned by Φ and put $V = Q \otimes_\mathbb{Z} \mathbb{Q}$. Then Φ (identified with $\Phi \otimes 1$) is a root system in V, in the sense of [11, p. 142].

Now let $\Psi = \Psi(T, G)$. Then Φ is a reduced root system (i.e. if $\alpha \in \Phi$ then $\mathbb{Z}\alpha \cap \Phi = \{\alpha, -\alpha\}$, see [74, p. 190]). Denote by B a Borel subgroup containing T. Then B defines an ordering of the roots of Φ, and determines, in particular, a *basis* Δ of Φ (i.e. a subset of Φ such that any element in Φ is an integral linear combination of the elements of Δ, with coefficients which are either all $\geqq 0$ or all $\leqq 0$). Put $\Delta^v = \{\alpha^v \in \Phi^v | \alpha \in \Delta\}$. Then we call the quadruple $(X, \Delta, X^v, \Delta^v)$ the *based root datum* $\tilde{\Psi} = \tilde{\Psi}(T, B, G)$ defined by G, B, T. Since Δ defines Φ uniquely, we can deduce Ψ from $\tilde{\Psi}$. The basis Δ defines a *Dynkin diagram*, according to [11, p. 195]. We denote it also by Δ.

We further note the following result. The notations are as above.

(5.6.6) *Let $\Psi = \Psi(T, G)$. Then G is semi-simple if and only if Q has finite index in X.* Let G be semi-simple. Then G is *adjoint* if $X = Q$ and is *simply connected* if (with obvious notation) $X^v = Q^v$. One shows that if G is adjoint its center is trivial and if G is simply connected its center is "as large as possible".

Examples. (a) $G = SL_n$. Then G is semi-simple and simply connected.
(b) $G = PGL_n$ (the quotient of $GL_n(K)$ by its center). Then G is semi-simple and adjoint.

(5.7)   *The existence and uniqueness theorems.*
We come now to the results on the classification of reductive K-groups.

Let G be a connected K-group and T a maximal torus of G. Denote by Ψ $=(X, \Phi, X^v, \Phi^v)$ the corresponding root datum. Let (G_1, T_1) be another such pair, with root datum Ψ_1. If $\varphi : G \to G_1$ is an isomorphism of algebraic groups with $\varphi T = T_1$, then there is an induced isomorphism of root data $f(\varphi) : \Psi \cong \Psi_1$ (it is clear how to define this last notion). The converse of this fact is the *uniqueness theorem* for root data:

(5.7.1) *Let* $f : \Psi \cong \Psi_1$ *be an isomorphism of root data. There is an isomorphism* $\varphi :$ $G \cong G_1$ *with* $\varphi T = T_1$ *such that* $f = f(\varphi)$. *If* φ' *is another such isomorphism there is* $t \in T$ *such that* $\varphi'(x) = \varphi(t x t^{-1})$ *for* $x \in G$.

In less precise terms: G is determined by the root datum up to isomorphism. A result of this kind for semi-simple groups was first proved by Chevalley [19, exp. 24]. The notion of root datum is due to Demazure [23, exp. XXI]. In [loc. cit., exp. XXIV] a similar theorem is proved for group schemes. We have followed here the exposition in [74, Ch. 11].

The proof of (5.7.1) is lengthy. We mention a few aspects.

With the notations introduced above let $\alpha \in \Phi$. Using (5.6.2) one associates with α an *additive* 1-*parameter subgroup* x_α of G, i.e. an isomorphism of G_a onto a 1-dimensional algebraic subgroup X_α of G, such that

$$t x_\alpha(\xi) t^{-1} = x_\alpha(\alpha(t) \xi) \ (\xi \in K, t \in T).$$

If α is given then X_α is unique. In fact, if x'_α is another such 1-parameter subgroup there is $c \in K$ with $x'_\alpha(\xi) = x_\alpha(c \xi) \ (\xi \in K)$. We fix x_α for each $\alpha \in \Phi$. It follows from (5.6.2) that we may assume that

$$n_\alpha = x_\alpha(1) x_{-\alpha}(-1) x_\alpha(1)$$

lies in the normalizer N of T. We then have

$$x_\alpha(\xi) x_{-\alpha}(-\xi^{-1}) \ x_\alpha(\xi) = \alpha^v(\xi) n_\alpha, \tag{*}$$

if $\xi \in K^*$, which shows that the coroot α^v is determined by x_α and $x_{-\alpha}$. Now let $\alpha, \beta \in \Phi, \alpha + \beta \neq 0$. We then have a commutator formula

$$x_\alpha(\xi) x_\beta(\eta) x_\alpha(-\xi) x_\beta(-\eta) = \prod_{\substack{i\alpha + j\beta \in \Phi \\ i,j > 0}} x_{i\alpha + j\beta}(c_{\alpha,\beta;i,j} \xi^i \eta^i) \tag{**}$$

The product is taken in some preassigned order, and the $c_{\alpha,\beta;i,j}$ are elements of K depending on the order (the proof of (**) is easy, see [74, 10.1.4]).

One of the ingredients of the proof of (5.7.1) is an analysis of the commutator formulas (**). Another ingredient of the proof given in [74] is a study of abstract groups with generators $\tilde{x}_\alpha(\xi) \ (\alpha \in \Phi, \xi \in K)$, and relations similar to relations satisfied by the $x_\alpha(\xi)$. For example:

$\tilde{x}_\alpha(\xi + \eta) = \tilde{x}_\alpha(\xi) x_\alpha(\eta)$, if $\xi, \eta \in K$; the analogue of (*) and certain relations of the form (**) (the $c_{\alpha, \beta; i, j}$ being as in (**)). The analysis of such abstract groups was initiated by Steinberg [78], and has led for example to the study of the "Steinberg groups" from algebraic K-theory (see [57, § 5]). In this analysis one also encounters "Tits systems", about which more will be said below.

Another approach to the uniqueness theorem, based on the theory of hyperalgebras was given by Takeuchi [84].

The *existence theorem* which follows completes the theory.

(5.7.2) *Let* Ψ *be a root datum. There exists a connected reductive K-group G, with a maximal torus T such that* $\Psi(G, T)$ *is isomorphic to* Ψ.

Chevalley's Tôhoku paper [18] which we mentioned in (4.7), gives the ingredients for the proof of this theorem. It contains, essentially, a proof of (5.7.2) for an important special case (G semi-simple and adjoint). The theorem can be viewed as a consequence of the existence theorem for simple Lie algebras over $\mathbb{C}$ [37, Ch. V]. For a proof requiring a minimum of Lie algebra theory see [74, Ch. 12].

The preceding results show that connected reductive K-groups are classified by root data, and that this classification is independent of K. It is also to be noted that it follows from the theory of Lie groups that both compact connected Lie groups and connected complex reductive Lie groups are also classified by root data.

(5.8)   *The Bruhat decomposition, Tits systems.*

Let G be a connected reductive K-group. Denote by B a Borel subgroup and by T a maximal torus contained in B. Let N be the normalizer of T in G and $W = N/T$ the Weyl group (see (5.6.5) and (5.4.7)). If $w \in W$ the double coset BwB is well defined.

(5.8.1) (*Bruhat decomposition*) (a) $G = \coprod_{w \in W} BwB$ (*disjoint union*);

(b) *The decomposition of* (a) *induces a decomposition* $G/B = \coprod_{w \in W} X(w)$. *Each* $X(w)$ *is a locally closed algebraic subvariety of the algebraic variety G/B, isomorphic to an affine space* $K^{l(w)}$.

These results are contained in [19, exp. 13]. The Bruhat decomposition was used for general Lie groups by Bruhat (see [12, p. 187]), it has been introduced for $GL_n(K)$ by Gelfand-Naimark [29, p. 88—90] and Steinberg [77, p. 275]. In Chevalley's Tôhoku paper [18] a decomposition like (a) was quite important. We shall come back to the geometric result (b) later in (8.2).

In 1961, Tits gave an axiomatic treatment of the Bruhat decompositions (theory of BN-pairs, (see [85])), in the context of abstract group theory. We follow Bourbaki [11] in adopting the name "Tits system" instead of "BN-pair".

A *Tits system* is a quadruple (G, B, N, S), where G is a group, B and N two subgroups and S a subset of $W = N/(B \cap N)$, satisfying the following axioms:

(T1) $B \cup N$ generates G and $B \cap N$ is a normal subgroup of N;

(T2) S generates W, it consists of elements of order 2;

(T3) If $s \in S$, $w \in W$ then $sBw \subset BwB \cup BswB$;

(T4) If $s \in S$ then $sBs \not\subset B$.

One proves that these axioms imply a Bruhat decomposition $G = \coprod_{w \in W} BwB$ [11, p. 25]. Also, one can establish a quite general simplicity theorem, which can be viewed as a generalization of the simplicity theorems alluded to in 4.7 [loc. cit., p. 30].

(5.8.2) *Example.* $G = GL_n(K)$, $B = B_n$ (see (2.3)(e)), $N =$ the subgroup of G of the matrices which have only one nonzero element in each row and column. If $g = (g_{ij}) \in N$, let s be the permutation of the symmetric group S_n such that $g_{s(i),i} \neq 0$ $(1 \leq i \leq n)$. Then $g \mapsto s$ induces an isomorphism $W = N/(B \cap N) \to S_n$. We take for $S \subset W$ the set of elements corresponding to the transpositions $(i, i+1)$ of S_n $(1 \leq i \leq n-1)$. Then (G, B, N, S) is a Tits system. The proof of the axioms is elementary (see [11, p. 24]).

In the situation of (5.8.1) we have indeed a Tits system. Now $B \cap N = T$, so W is as before. The elements of S correspond (via 5.6.5) to certain reflections s_α, namely those defined by the so-called simple roots $\alpha \in \Phi$ defined by B. They are such that $X_\alpha \subset B$ (see 5.7) and that $s_\alpha B s_\alpha \cap B$ is a subgroup of codimension 1.

The verification of the axioms is made in [38, no. 29], it is also contained in [19, exp. 13] (although Tits systems are not mentioned there).

It also should be remarked that the theory of Tits systems leads to an explicit description of all subgroups of G containing the group B [11, p. 27]. In the case of K-groups this gives an explicit description of the parabolic subgroups containing a Borel group, in terms of the root system of G.

(5.9)   *Representation theory.*

Let G be a connected semi-simple K-group. In [19, exp. 15] Chevalley established the basic facts about the representation theory of G (in the sense of algebraic groups). They are quite similar to the results about representations of complex semi-simple Lie algebras. Fix a Borel subgroup B of G and a maximal torus $T \subset B$.

Let $\varrho : G \to GL_m(K)$ be a homomorphism of algebraic groups, defining a representation of G in $V = K^m$.

(5.9.1) *If ϱ is irreducible there is a unique one-dimensional subspace Kv of V which is fixed by B.*

There is a character λ of T such that $\varrho(t)v = \lambda(t)v$ $(t \in T)$, this character is the *highest weight* of ϱ, with respect to T and B.

(5.9.2) *Two irreducible representations with the same highest weight are equivalent.*

To prove existence of irreducible representationed with a given highest weight the following construction is used. Let μ be a character of T, extend it to a homomorphism of algebraic groups $\mu: B \to GL_1$ which is trivial on unipotent elements. Now consider the subspace V_μ of the algebra of regular functions $K[G]$ (see no. 2) formed by the functions f such that $f(gb) = \mu(b)f(g)$ $(g \in G, b \in B)$. Then V_μ is finite dimensional and we have a representation ϱ_μ of the algebraic group G in V defined by left translations:

$$(\varrho_\mu(x)f)(g) = f(x^{-1}g) \quad \text{if} \quad g, x \in G.$$

If $V_\mu \neq 0$ there is a unique irreducible subspace of V_μ, whose highest weight is related to μ in an easy manner, and one shows that one obtains thus all irreducible representations of G. If char $(K) = 0$ then V_μ itself is irreducible (or 0), but this need not be so in characteristic $p > 0$.

For a more complete discussion of these matters we can refer to [38, Ch. X]. The construction of V_μ can be given in a more geometric manner. The character μ defines a line bundle L_μ over the flag manifold $X = G/B$: L_μ is the quotient of $G \times K$ by the B-action given by $b(g, \xi) = (gb^{-1}, \mu(b) \xi)$ (this quotient exists by general theorems, (see [2, II.6]) and the projection on the first factor defines a line bundle structure $L_\mu \to X$, which is locally trivial for the Zariski topology, as a consequence of Bruhat's lemma. Now V_μ is the space of sections of the line bundle. We thus see that the construction of V_μ is the analogue in algebraic geometry of the Borel-Weil construction of the irreducible representations of a compact Lie group. This point of view was already taken by Chevally [19].

The representation theory in characteristic $p > 0$ is by no means well understood. For example, an analogue of Weyl's character formula, for a representation with given highest weight, has not yet been established.

6   Linear algebraic groups over arbitrary fields

Let k be a subfield of an algebraically closed field K. We denote by $\bar{k}$ the algebraic closure of k in K and by k_s its separable closure. An example is $K = \mathbb{C}$, $k = \mathbb{Q}$.

(6.1)   *k-groups.*

Let $G \subset GL_n(K)$ be a K-group, with algebra of regular functions $A = K[G]$ (see no. 2). The group structure is defined by morphisms $G \times G \to G$ (product), $G \to G$ (inversion), $\{point\} \to G$ (unit element), which are given by algebra homomorphisms $\Delta: A \to A \otimes_K A$, $\iota: A \to A$, $\varepsilon: A \to K$. We say G is provided with a *k-structure* if we are given a k-subalgebra A_0 of A such that A

$= A_0 \otimes_k K$ and that the maps $\varDelta$, $\imath$, ε, all come, by extension of the ground field, from corresponding homomorphisms $A_0 \to A_0 \otimes_k A_0$, $A_0 \to A_0$, $A_0 \to k$. A K-group with a k-structure is a k-group $((k, K)$-group would be more correct). In that case we write $k[G] = A_0$. It is clear how to define homomorphisms of k-groups, k-subgroups etc. It is also clear that if G is a k-group and l is an extension of k contained in K, then G is canonically an l-group.

The group $GL_n(K)$ has an obvious k-structure, given by the algebra $k[X_{ij}, \det(X_{ij})^{-1}]$. The same is true for the examples of (2.3). One can show that if $G \subset GL_n(K)$ has a k-structure, it is k-isomorphic to a k-subgroup of some $GL_m(K)$ (see [2, p. 101]).

Let G be a k-group. The group $G(k)$ of its k-*rational points* is the group of homomorphisms $k[G] \to k$. If G is a k-subgroup of $GL_n(K)$, then $G(k)$ is the intersection $G \cap GL_n(k)$.

One has, more generally, the notions of *algebraic variety over k*, or k-*variety*, and its set $V(k)$ of k-rational points. A first fact which we mention is that the result 4.6 about quotients can be refined: If G is a k-group and H a k-subgroup then the quotient space G/H of 4.6 is a k-variety, such that π: $G \to G/H$ is a k-morphism of varieties (for a proof see [2, II.6]).

We shall now review some refinements for k-groups of results discussed in no. 5.

(6.2) k-*tori*.

A k-*torus* T is a torus with a k-structure. T is called k-*split* if it is k-isomorphic to a group of diagonal matrices T_m.

Example of a non-split torus: Let $K = \mathbb{C}$, $k = \mathbb{R}$ and $T \subset SL_2(\mathbb{C})$ the $\mathbb{R}$-group of all matrices $\begin{pmatrix} x & y \\ -y & x \end{pmatrix}$ with $x^2 + y^2 = 1$.

We have the following results.

(6.2.1) *Let T be a k-torus. There is a finite separable extension $l \subset k_s$ of T such that T is l-split.*

Denote by $\varGamma = \varGamma_k$ the Galois group of k_s/k, a profinite group provided with the Krull topology. Because of (6.2.1), the group $\varGamma$ acts on the character group $X^*(T)$ and its dual $X_*(T)$. The action is continuous with respect to the discrete topology on $X^*(T)$ (resp. $X_*(T)$), i.e. factors through an open normal subgroup of $\varGamma$ (with finite quotient).

(6.2.2) $T \mapsto X_*(T)$ *defines an equivalence of categories between the category of k-tori and the category of free abelian groups of finite rank with a continuous $\varGamma$-action.*

The k-torus T is *anisotropic* if it does not contain a nontrivial split k-subtorus.

(6.2.3) *Let T be a k-torus. There is a unique maximal k-split k-subtorus T_s of T and* *a unique maximal anisotropic k-subtorus T_a. We have $T = T_s T_a$ and $T_s \cap T_a$ is finite.*

For these results see [2, p. 219].

(6.3)   *Connected solvable k-groups.*

Let G be a connected solvable k-group. The results of (5.3.1) do not all carry over to k-groups. The group G_u of 5.3.1 (a) need not be a subgroup of G. But 5.3.1 (b) carries over, in the following form.

(6.3.1) *G contains a maximal torus which is a k-subgroup. Two such tori are conjugate by an element of $G(k)$.*

This is due to Rosenlicht [64, p. 110]. For a proof see [8, no. 11].

G is called *k-solvable* if it has a composition series $G = G_0 \supset G_1 \supset \ldots \supset G_s = \{e\}$ consisting of connected k-groups such that G_i/G_{i+1} is k-isomorphic to G_a or GL_1 ($0 \leq i \leq s - 1$). It follows from the Lie-Kolchin theorem (4.3) and (5.6.1) that G is K-solvable. Also, if k is perfect and G is unipotent then G is k-solvable ([2, p. 362]). This result is not true if k is non-perfect, see counterexamples in [64, p. 114]).

Borel's fixed point theorem has the following version over k.

(6.3.2) *Let G be k-solvable. Assume that it operates on a non-empty projective k-variety V, the action being defined over k. If $V(k) \neq \phi$ then G has a fixed point in $V(k)$.*

For a proof see [2, p. 358].

Now let G be an arbitrary connected k-group. It is a very special property of G to have a Borel subgroup which is defined over k (a reductive G with this property is called quasi-split, see (6.4)).

For maximal tori, we have the following result.

(6.3.3) *Let G be a connected k-group. Then G contains a maximal torus which is defined over k.*

Two such tori need not be conjugate by an element of $G(k)$. (Example: $k = \mathbb{R}$, $G = SL_2$. Let $T_1 \subset G$ be the torus described in (6.2) and T_2 the diagonal subtorus of G. Then T_1 and T_2 are maximal tori in SL_2, defined over $\mathbb{R}$, but not conjugate by an element of $SL_2(\mathbb{R})$).

(6.3.3) for arbitrary fields is contained in results of Grothendieck [23, exp. XIV, no. 6]. For another proof see [2, p. 382].

(6.4)   *Connected reductive k-groups.*

Assume G to be a connected reductive k-group. We then have the following density theorem:

(6.4.1) *Let k be infinite. Then $G(k)$ is dense in G for the Zariski topology.*

This follows from the fact that G is a unirational variety over k (proved first by Grothendieck [23, exp. XIV, no. 6], see also [2, p. 385]). If G is viewed as a k-subgroup of GL_n, then (6.4.1) shows that the algebraic group $G \subset GL_n(K)$ is completely determined by its group of k-rational points $G(k) \subset GL_n(k)$. This motivates the use made by some authors of locutions like "G is an algebraic subgroup of $GL_n(k)$". However it is advisable to avoid these, and adhere to the terminology adopted here. For one thing, (6.4.1) is not true for arbitrary G (the first example in [64, p. 114] provides a counterexample). The result is true if k is perfect and infinite (see [2, p. 385]).

It is unreasonable to expect, over an arbitrary ground field k, classification results like (5.7.1) and (5.7.2) for connected reductive k-groups. To indicate the complications inherent in a classification we give a few examples.

(6.4.2) *Examples.* (a) Assume K infinite. Let D be a central simple algebra over k, of dimension $d = n^2$. Then $D \otimes_k K$ is isomorphic to the matrix algebra $M_n(K)$, and $GL_n(K)$ can be viewed as the group of invertible elements of $D \otimes_k K$. Now define a k-structure on $GL_n(K)$ as follows. Let Nr be the reduced norm of D. If $(e_1, \ldots, e_d)$ is a k-basis of D, then there is a polynomial $F \in k[X_1, \ldots, X_d]$ such that $Nr(x_1 e_1 + \ldots + x_d e_d) = F(x_1, \ldots, x_d)$, if $x_i \in k$. The k-structure is now defined by the algebra $k[X_1, \ldots, X_d, F^{-1}]$, which becomes isomorphic to $K[GL_n]$ after extension of the base field. If G is the corresponding k-group, then $G(k)$ is the group D^* of invertible elements of D. In particular, any division algebra D leads to a k-form of GL_n.

(b) Let $\mathrm{char}(k) \neq 2$ and let $f = \sum\limits_{1 \le i, j \le n} a_{ij} X_i X_j$ be a non-degenerate quadratic form over k. Assume $a_{ij} = a_{ji}$ and let S be the symmetric matrix $(a_{ij})_{1 \le i, j \le n}$. The group G of linear transformation over K of determinant 1 fixing f, which is isomorphic to the group of $X \in SL_n(K)$ such that

$$({}^t X) S X = S,$$

is a k-form of SO_n.

However (5.7.1) and (5.7.2) carry over to a special case. The connected reductive k-group G is said to be k-*split* if G has a maximal torus T defined over k which is k-split (see (6.2)). In that case all ingredients connected with the root datum of (G, T) (such as the 1-parameter subgroups x_α of (5.7)) are defined over k. One can then refine the proofs of (5.7.1) and (5.7.2) to obtain:

(6.4.3) *The k-isomorphism classes of k-split reductive connected groups are parametrized by the isomorphism classes of root data, as in* (5.7.1) *and* (5.7.2).

The proof in [74, Ch. 11, 12] for k algebraically closed carries over; see also [2, p. 388—390].

One can generalize this a bit. G is said to be *quasi-split over* k if G has a

Borel subgroup B which is defined over k (it follows from (6.4.3) that "split" implies "quasi-split"). In that case let T be a maximal torus of B defined over k (see (6.3.1)). The based root datum $\tilde{\Psi} = \tilde{\Psi}(T, B, G)$ (see 5.6) has some extra structure. In fact, it follows from (6.2.2) that there is a homomorphism α of the Galois group $\Gamma = \mathrm{Gal}(k_s/k)$ into the automorphism group of $\tilde{\Psi}$ (with finite image). Then it can be shown that the k-isomorphism classes of quasi-split reductive connected groups over k are described by (isomorphism classes) of pairs (Ψ, α). See also (6.6.1).

A thorough study of parabolic k-subgroups of G has been made by Borel and Tits [8]. We shall now review some of their results.

(6.5)   *Parabolic k-subgroups.*

We assume that G is a connected reductive k-group.

(6.5.1) *Let P be a parabolic k-subgroup of G.*
(a) *The unipotent radical $R_u P$ is a k-subgroup, which is unipotent and k-solvable;*
(b) *There exist reductive k-subgroups L of P such that P is the semi-direct product of k-groups $P = L \times R_u P$. Two such groups are conjugate by an element of $R_u P(k)$. L is called a* Levi *subgroup of P. See [8, §3—4].*

(6.5.2) *Let S be a split k-subtorus of G, let L be its centralizer in G, this is a reductive k-subgroup. There exists a parabolic k-subgroup P of G with Levi subgroup L. If S is a maximal split k-subtorus then P is a minimal parabolic k-subgroup.*

In the case that $k = K$ a maximal k-split K-subtorus S is just a maximal torus. Then $L = S$ and P is a Borel subgroup. The conjugacy of Borel subgroups and maximal tori has the following counterpart over k.

(6.5.3) (a) *Two minimal parabolic k-subgroups are conjugate by an element of $G(k)$;*
(b) *Two maximal split k-subtori are conjugate by an element of $G(k)$.*
See [8, p. 83].

Now let S be a maximal split k-subtorus and let $T \supset S$ be a maximal torus of G which is defined over k. Such a torus exists, by an application of 6.3.3 to the centralizer of S. A k-*root* of G with respect to S is a nontrivial character of S which is the restriction to S of a root of G with respect to T. We write $_k\Phi = {_k\Phi}(G, S)$ for the set of k-roots. Let $V \subset X^*(S) \otimes_{\mathbb{Z}} \mathbb{Q}$ *be the subspace spanned by* $_k\Phi \otimes 1$.

(6.5.4) *If $_k\Phi \neq \emptyset$ it is a root system in V.*

This is proved in [8, §5]. Compare with the results of (5.6). $_k\Phi$ is the *relative root system* with respect to k. By (6.5.3) (b) it is uniquely determined, up to isomorphism. The relative root system $_k\Phi$ need not be reduced, i.e. there may be $\alpha \in {_k\Phi}$ such that $2\alpha \in {_k\Phi}$.

The Weyl group $_kW$ of the $_k\Phi$ is the *relative Weyl group* of G. If L is the centralizer of S and N the normalizer, then these groups are k-subgroups of G and $_kW$ is isomorphic to $N(k)/L(k)$ ([loc. cit.]). If P is as in (6.5.3) (a) then it defines an ordering of $_k\Phi$ and a set of involutorial generators Σ of $_kW$.

(6.5.5) $(G(k),\ P(k),\ N(k),\ \Sigma)$ *is a Tits system.*
See [loc. cit.].
    The parabolic k-subgroups containing S can now be described in terms of the relative root system and any parabolic k-subgroup is conjugate to one containing S by (6.5.3).

    G is *anisotropic* over k if it does not contain proper parabolic k-subgroups of equivalently (by (6.5.2)) if the split k-subtori of G all lie in the center of G.

(6.5.6) *Let S be a maximal k-split subtorus. Then its centralizer is a (connected reductive) k-group which is k-anisotropic.*
    This follows from the description of parabolic subgroups.

(6.5.7) *Examples* (a) If $k = \mathbb{R}$ the relative root system $_k\Phi$ is the root system of the symmetric space defined by $G(\mathbb{R})_0$ in the sense of E. Cartan (see [loc. cit., p. 148]). In that case G is anisotropic if and only if $G(\mathbb{R})$ is compact modulo its center.
(b) Let G be as in (6.4.2) (a), with D a division algebra. Then G is anisotropic. Likewise in the case of (6.4.2) (b), if f is anisotropic.

    If G is anisotropic over k there are no proper parabolic k-subgroups. However if P is a parabolic subgroup of G over K, the algebraic variety G/P has in many cases a structure of k-variety. This is so, for example, if P is a Borel subgroup. In that case the k-variety G/P has a k-rational point if and only if G is quasi-split over k.
    As another case, let us take the example of (6.4.2) (b). The quadratic polynomial f defines a hypersurface in $\mathbb{P}^{n-1}(K)$, which is a projective k-variety, isomorphic over K to a variety G/P as above. It has k-rational points if and only if f is isotropic.
    We also mention the *Severi-Brauer varieties.* These one obtains in the case of example 6.4.2 (a), as the projective k-varieties defined by the elements of rank 1 in the algebra $D \otimes_k K$.

(6.6)   *Classification of reductive k-groups.*
    Let G be a connected reductive k-group. Fix a maximal split k-subtorus S of G and a minimal parabolic k-subgroup containing S. We denote $L \subset P$ the centralizer of S in G, this is an anisotropic k-group. We also fix a maximal torus T of G which is defined over k and contains S (hence lies in L), together with a Borel subgroup B of G and P containing T, which in general will not be a k-subgroup. Then the based root datum defined by T, B, G, say $\tilde{\Psi} = (X,\ \Delta,\ X^v,\ \Delta^v)$ (see 5.6) has some extra structure:
(a) there is a homomorphism α of the Galois group $\Gamma = \mathrm{Gal}(k_s/k)$ into the

automorphism group of $\tilde{\Psi}$, with finite image. In particular, Γ defines a permutation group of the Dynkin diagram Δ;

(b) We have a subset Δ_0 of Δ consisting of the characters $\alpha \in \Delta$ which are trivial on the subtorus S of T. The elements of Δ_0 are fixed by the action of Γ (A special case is that of a quasi-split G, mentioned in 6.4). Then $I = (\tilde{\Psi}, \alpha, \Delta_0)$ is called the *index* of the k-group G. This notion was introduced by Tits (see [86]), our notion is slighly more general. He also proved the following uniqueness theorem:

(6.6.1) *G is determined up to k-isomorphism by the index I and the anisotropic group L.* [loc. cit. p. 43].

This may be viewed as a counterpart for k-groups to (5.7.1).

Tits' paper also contains tables of the possible Δ_0, for a given Dynkin diagram Δ, The most difficult part of a classification of reductive k-groups is the classification of the semi-simple *anisotropic* k-groups. This depends on the field k, and may be quite complicated, as the examples at the end of (6.5.7) show.

A complete classification of all anisotropic k-groups seems out of reach. On a more modest level, one would like to have some general results about anisotropic groups. For example, it would be interesting to know whether the following is true: Let G be a semi-simple anisotropic k-group whose root system is irreducible and not of type A. Then $G(k)$ contains a semi-simple element which is non-central and non-regular (i. e. such that the identity component of its centralizer is not a maximal torus). This result would imply the, still improved, "Hasse principle" for groups over number fields (see (7.4) for more details). I conjectured the result several years ago, but have not been able to make any headway.

We shall not go further into the details of this general classification. We must mention, however, some general notions about Galois cohomology of k-groups, which give a convenient framework for dealing with classification problems.

(6.6.2) *Galois cohomology.* Let G and G' be k-groups. We say that G' is a *k-form* of G, if G' is K-isomorphic to G. Now, k_s denoting the separable closure of k in K, G is completely determined by the group $G(k_s)$ of its k_s-rational points (as follows from 6.4.1), and we have, moreover, an action of $\Gamma = \mathrm{Gal}(k_s/k)$ on $G(k_s)$ which completely determines the k-group G. One can then describe all k-forms G' up to k-isomorphism in the following manner: we have $G'(k_s) = G(k_s)$, and there is a continuous function $c : s \mapsto c_s$ of Γ in the group of k_s-automorphisms of G (Γ having the Krull topology and the second group the discrete topology) satisfying

$$c_{st} = c_s \cdot s(c_t) \ (s, t \in \Gamma), \tag{*}$$

such that the action of Γ on $G'(k_s)$, denoted by $(s, g) \mapsto s * g$, is obtained by "twisting" the original action with c: i.e. $(s * g) = c_s(s \cdot g)$. Then G' is k-isomorphic to G if and only if there is an automorphism c with $c_s = c^{-1}(s(c))$. We say that G' is an *inner form* of G is c_s an inner automorphism.

If C is a group on which Γ acts (with the discrete topology), the continuous functions $s \mapsto c_s$ of Γ in which satisfy (*) are called 1-cocycles of Γ with values in C. On this set there is an equivalence relation: $(c_s) \sim (c'_s)$ if there is $c \in C$ with $c'_s = c^{-1} c_s(s(c))$, and the equivalence classes form the 1-*cohomology set* $H^1(k, C)$. It is a pointed set, with special element 1, coming from the constant function $c_s = e$.

If G is a k-group, we write $H^1(k, G) = H^1(k, G(k_s))$. What we mentioned before implies that the k-forms of the k-group G are classified by $H^1(k, \text{Aut } G)$ (provided Aut G can be viewed as a k-group, which is so if G is semi-simple, for example). The classification problem can thus be viewed as the determination of this Galois cohomology set. For more details see [66, Ch. III]. We shall use this language in a next section.

(6.7) Group schemes.

Recall that a K-group G can be completely described by its ring of regular functions A, together with homomorphism Δ, ι, ε as in (6.1). They satisfy certain relations, reflecting the group axioms (see [23, exp. I]). This description suggests a generalization, where K is replaced by an arbitrary ring.

Let R be a ring and A an R-algebra. We suppose given algebra homomorphisms $\Delta: A \to A \otimes_R A$, $\iota: A \to A$, $\varepsilon: A \to R$ as in (6.1), satisfying relations as before. These imply that for any S-algebra R, the set $G(S)$ of S-homomorphisms $A \otimes_R S \to S$ has a canonical group structure, defining a functor G of the category of R-algebras to the category of groups. G is called a *group scheme* over R. If S is an R-algebra, we denote by $G \times_R S$ the group scheme over S defined by $A \otimes_R S$.

If G is a k-group, and k is an R-algebra, we say that G has an R-*structure* if there is a group scheme G_0 over R such that $G = G_0 \times_R k$ (in the special case where $k = K$ and R is a subfield of K we recover the notion of k-group of (6.1)).

The notion of group scheme introduced above is really that of an affine group scheme. A thorough treatment of group schemes over rings, or more generally over schemes was given by Grothendieck and his collaborators in [23]. One of the features of [loc. cit.] is the systematic study and use of group schemes over a non-reduced ring R (i.e. a ring with nilpotent elements). As we mentioned already, Grothendieck proved here some basic results about k-groups (such as (6.3.3) and (6.4.1)). One also finds in [loc. cit.] a thorough treatment by M. Demazure of reductive groups over $\mathbb{Z}$. The basic result here is the "Kroneckerian" result that "all reductive groups come from $\mathbb{Z}$". The precise result is as follows [23, exp. XXV].

(6.7.1) *Let* $\Psi = (X, \Phi, X^v, \Phi^v)$ *be a root datum. There exists a group scheme* G

over $\mathbb{Z}$, *together with a subgroup scheme T, such that for any field k, the k-group* $G \times_{\mathbb{Z}} k$ *has* $T \times_{\mathbb{Z}} k$ *as a maximal torus which is k-split and* $\Psi(G \times_{\mathbb{Z}} k, T \times_{\mathbb{Z}} k) = \Psi$.

For G semi-simple this was first proved by Chevalley ([20], see also [5, part A]). This proof can certainly be extended to prove (6.7.1). The proof involves representation theory of Lie algebras, and is simpler then the one of [23], which uses more algebraic geometry. In [23] one also finds precise information about the "automorphisms group scheme" of the group G of 6.7.1.

One cannot say that the theory of group schemes developed in [23] has been widely used. But it is a recognized fact that the group schemes are useful and even indispensable in the theory of algebraic groups over fields. This is illustrated by the basic result (6.7.1). Also, in the theory of algebraic groups over global fields one comes naturally across group schemes over their rings of integers (see (7.4)). The same is true for groups over local fields (see for example [14]).

Of a different nature are the infinitesimal finite group schemes. For simplicity, assume that we work over the algebraically closed field K. A group scheme G over K is finite if its K-algebra $K[G]$ (denoted by A in the beginning of (6.7)) is finite-dimensional and infinitesimal if $K[G]$ is a local ring, with maximal ideal Ker ε. If G is finite we call dim $_k K[G]$ its order.

Example. The group μ_n of nth roots of unity. We have $K[\mu_n] = K[T]/(T^n - 1)$ and $\Delta(T) = T \otimes T$, $\iota(T) = T^{n-1}$, $\varepsilon(T) = 1$. This is a group scheme of order n. If char$(K) = p > 0$ and n is a power of p, it is infinitesimal.

One encounters infinitesimal group schemes naturally in the theory of algebraic groups over K. For example, if char $K = 2$, one can define a kernel of the canonical map

$$SL_2(K) \rightarrow PGL_2(K),$$

which is an infinitesimal central subgroup of order 2. This, and similar results show that it is useful to have results about quotients of the type (4.6) for somewhat more general situations. Such results are contained in [23].

One can get around the use of group schemes here, by using quotients of an algebraic group by a subalgebra of its Lie algebra (see [2, §17]).

In another situation, which has been studied in recent years, one cannot get around the use of group schemes. Let $G \subset GL_n(K)$ be an algebraic group over the algebraically closed field of characteristic $p > 0$. Let $F_n: G \rightarrow GL_n(K)$ be the homomorphism of algebraic groups with $F(x_{ij}) = (x_{ij}^{p^n})$. Then F_n is a surjective homomorphism, with infinitesimal kernel K_n. There are homomorphisms $K_1 \rightarrow K_2 \rightarrow K_3 \dots$. The "Frobenius kernels" K_n have turned out to be quite useful in representation theory (see [21], an application is described in [loc. cit., p. 267]).

The preceding remarks will have shown the importance of the notion of group scheme for the theory of linear algebraic groups over fields.

We have reviewed above some of the basic concepts and results from the theory of linear algebraic groups over fields. Most of these were already obtained some twenty years ago. Subsequent developments have demonstrated the usefulness of these concepts and results in various parts of mathematics. We shall discuss below briefly some of these developments.

7   Special ground fields

We first discuss uses of the theory of k-groups for special fields k.

(7.1)   *Finite fields.*

Let k be a finite field, with q elements. We take K to be an algebraic closure $\bar{k}$ of k, then $k_s = \bar{k}$. The Galois group $\Gamma = \mathrm{Gal}(\bar{k}/k)$ has a topological generator F, the *Frobenius* (automorphism) with $F\xi = \xi^q$ for all $\xi \in K$.

Let G be a k-group, viewed as a k-subgroup of $GL_n(\bar{k})$. If $x = (x_{ij}) \in G$, then $Fx = (x_{ij}^q)$. The basic theorem about algebraic groups over finite fields is *Lang's theorem*:

(7.1.1) *Let G be a connected k-group. The morphism $x \mapsto x(Fx)^{-1}$ of G is surjective.*

This is a quite early result [48]. Actually, it is true for any connected algebraic k-group (not necessarily linear), and in fact Lang applied it to abelian varieties. In the language of Galois cohomology, (7.1.1) states that $H^1(k, G) = 1$ (see (6.6.2)). The geometric fact explaining the theorem is that the differential of the morphism in question is everywhere bijective. See [74, p. 102—103] for a precise proof. The implications of Lang's theorem are discussed in [5, Part E]. We mention in particular [loc. cit. p. 175]:

(7.1.2) *Let G be a connected reductive k-group. Then G is quasi-split, i.e. contains a Borel subgroup which is defined over k.*

This enables us to classify completely these groups G. In particular, one can classify them in the case that G is semi-simple. A thorough discussion can be found in [81].

If G is as in (7.1.2), the group $G(k)$ is a finite group, called a *finite group of Lie type*. These finite groups should be viewed as the finite analogues of the reductive Lie groups. The finite Chevalley groups mentioned in (4.7) are important special cases. Lang's theorem is an indispensable tool for the study of the properties of the finite groups of Lie type. As an example of its use, we prove the following result.

(7.1.3) *Let G be a k-group. Assume a and b are elements of $G(k)$ which are conjugate in G. If the centralizer of a is connected then a and b are conjugate in $G(k)$.*

Let Z be the centralizer of a, it is a connected k-subgroup. Let $b = gag^{-1}$. Since $Fa = a$, $Fb = b$, we have that $g^{-1}Fg \in Z$. By Lang's theorem, applied to Z, there is a $z \in Z$ with $g^{-1}Fg = z(Fz)^{-1}$. Hence $F(gz) = gz$, so $h = gz \in G(k)$ and $b = hah^{-1}$.

The last result indicates that the study of the geometric property of connectednes in G is of importance for the determination of the conjugacy classes of the finite group $G(k)$. This can be reduced to the determination of the conjugacy classes of semi-simple and unipotent elements, respectively. The semi-simple classes are discussed in [5, Part E]. There is an extensive literature on unipotent classes in reductive groups and finite groups of Lie type. References can be found in [69] and [73]. We shall come back to them below (8.3).

As another application of Lang's theorem we mention the following classification result on maximal tori.

(7.1.4) *Let G be a connected reductive k-group which is k-split. These is a bijection of the set of $G(k)$-conjugacy classes of maximal tori of G which are defined over k onto the set of conjugacy classes of the Weyl group W of the root system of G.* See [5, p. 186].

In the representation theory of finite groups of Lie type the general notions of the theory of linear algebraic groups are widely used. Steinberg [79] has shown how one can determine the irreducible representations of $G(k)$ over the field $\bar{k}$ (the p-modular representations, where $p = \mathrm{char}(k)$) from the irreducible representations of the algebraic group G, in the sense of 5.9.

Great progress has been made in recent years in the ordinary representation theory of finite groups of Lie type. We mention, in particular, the work of G. Lusztig. We shall get not go into this here and we only refer to [54] and [76]. A very interesting aspect of this work is the exploitation of algebraic topology in characteristic p (l-adic cohomology and intersection cohomology). The theory of algebraic groups is used here together with algebraic topology, to obtain deep results about finite groups of Lie type.

(7.2) $\mathbb{R}$.

Take $k = \mathbb{R}$, $\bar{k} = \mathbb{C}$. If G is an $\mathbb{R}$-group then its group $G(\mathbb{R})$ of rational points is canonically a Lie group. If G is connected in the Zariski topology then $G(\mathbb{R})$ need not be connected in the ordinary topology (example: GL_n). At any rate, if G is semi-simple, the topological identity component $G(\mathbb{R})^0$ is a connected semi-simple Lie group with finite center. Any connected semi-simple Lie group is a covering group of one of this kind. The connection between semi-simple $\mathbb{R}$-groups and semi-simple Lie groups is sufficiently close, so as to be usable for deducing results about Lie groups from the theory of algebraic groups. In my opinion, one should view the theory of semi-simple Lie groups as much as possible as a consequence of the theory of algebraic groups, via this correspondence. In [72, no. 5] one finds, for example, a treatment of the

Iwasawa decomposition along these lines, for a fairly extensive class of Lie groups. One also finds there a brief elementary discussion of the symmetric spaces.

We now mention a few other results which fit into the context of algebraic groups over $\mathbb{R}$. The first one is the existence and uniqueness theorem for anisotropic groups.

(7.2.1) *Let Ψ be a root datum. There exists a connected reductive $\mathbb{R}$-group G, together with a maximal torus T of G defined over $\mathbb{R}$ such that $\Psi(T, G) = \Psi$ and that T and G are anisotropic over $\mathbb{R}$. The $\mathbb{R}$-group G is unique up to $\mathbb{R}$-isomorphism. $G(\mathbb{R})$ is a compact connected Lie group and any compact connected Lie group can be so obtained.*

This collects well-known results. However, as far as I know a proof from the point of view of algebraic groups is nowhere in the literature.

(7.2.2) *Let G be as in (7.2.1). Any two maximal tori of G defined over $\mathbb{R}$ are conjugate by an element of $G(\mathbb{R})$.*

This is the conjugacy theorem for maximal tori of a compact Lie group. For an "algebraic" proof see [32].

(7.2.3) *Let G be a $\mathbb{R}$-group whose identity component is as in (7.2.1). There is a bijection of $H^1(\mathbb{R}, G)$ onto the set of conjugacy classes in $G(\mathbb{R})$ of elements x with $x^2 = 1$.*

This was proved by Borel and Serre (see [66, III, 4.5], but the proof is not completely algebraic). Let G be as in (7.2.1) and apply (7.2.3) to the automorphism group Aut G of G. Since $H^1(\mathbb{R}, \text{Aut } G)$ classifies the $\mathbb{R}$-forms of G (see (6.6.2)) we see that these $\mathbb{R}$-forms are described by involutorial automorphisms of $G(\mathbb{R})$ of order ≤ 2, a result due to E. Cartan. Using the explicit description of the twisting procedure given in (6.6.2), one deduces from these facts the existence of Cartan involutions. This is the following result.

(7.2.4) *Let G be a connected reductive $\mathbb{R}$-group, let σ denote the complex conjugation automorphism of $G = G(\mathbb{C})$. There exists an automorphism θ of the $\mathbb{R}$-group G which commutes with σ such that $\theta^2 = 1$ and that the fixed point set of $\theta \circ \sigma$ in $G(\mathbb{C})$ is a compact Lie group.*

Such a θ is called a *Cartan involution* of G (these are usually introduced via the Lie algebra of $G(\mathbb{R})$). Problems about the $\mathbb{R}$-group G can sometimes be reduced to problems about the algebraic group G (over $\mathbb{C}$) and the automorphism θ. This is the case, for example, in some problems about infinite dimensional representations of $G(\mathbb{R})$ (study of Harish-Chandra modules, see [90]). As another instance we mention the following result (see [loc. cit. p. 107—108]). The notations are as in (7.2.4).

(7.2.5) *There is a bijection of the set of $G(\mathbb{R})$-conjugacy classes of maximal tori of*

G defined over ℝ *onto the set of classes of θ-stable maximal tori of G, under
conjugation by the group K of fixed points of θ.*

It would be desirable to have an exposition of the basic facts of the theory
of semi-simple Lie groups, as consequence of the theory of linear algebraic
groups. So far, the relevant facts are only scattered in the literature.

(7.3)   *Local fields.*

Next let k be a local field, i.e. a field with a nontrivial discrete valuation,
which is complete for the valuation and has finite residue field. Let G be a
semi-simple and simply connected k-group.

A profound study of these groups has been made by Bruhat and Tits
([13], [14] a review is given in [86]), for more general fields. In their work, group
schemes over the ring of integers of k play an essential role. One of the results of
their work is that the group $G(k)$ of k-rational points carries a structure of Tits
system, with an infinite Weyl group (namely an affine Weyl group, in the sense of
[11, p. 173]). Closely connected is the construction of a space on which $G(k)$
operates, the *building* of G, which is an analogue of the symmetric space from the
theory of Lie groups.

To give an idea of the Tits system in the groups $G(k)$ and of the buildings
we give two examples. The field k is as above. We denote by $\mathfrak{o}$ its ring of integers
and by π a generator of the maximal ideal of $\mathfrak{o}$. We denote by $|\cdot|$ a non-
archimedean absolute value on k corresponding to the valuation of k.

(a) Let $G = SL_2(k)$. We denote by B the subgroup of G of the matrices

$$\begin{pmatrix} a & b \\ \pi c & d \end{pmatrix},$$

with $a, b, c, d \in \mathfrak{o}$, $ad - \pi bc = 1$. Let $N \subset G$ be the subgroup of matrices which
have only one nonzero element in each row and column (compare (5.8.2)). Then
$B \cap N$ is the group of diagonal matrices contained in B, and $W = N/B \cap N$ is
isomorphic to the infinite dihedral group, generated by the images s and s' of

$$\begin{pmatrix} 0 & 1 \\ -1 & 0 \end{pmatrix} \quad \text{and} \quad \begin{pmatrix} 0 & \pi \\ -\pi^{-1} & 0 \end{pmatrix} \text{ in } W. \text{ Let } S = \{s, s'\}. \text{ Then } (G, B, N, S) \text{ is a Tits}$$

system (see [11, p. 54, ex. 21]).

(b) Let $G = GL_n(k)$, $V = k^n$. Then G operates on V. Let φ be a non-archimedean
norm on the k-vector space V. It is known that there exists a basis $(e_1, \ldots, e_n)$ of
V and real numbers $\lambda_i > 0$ such that

$$\varphi\left(\sum_{i=1}^{n} x_i e_i\right) = \underset{1 \le i \le n}{\text{Max}} \ \lambda_i |e_i|.$$

Let X be the set of all such norms. We define a distance on X by

$$d(\varphi, \varphi') = \log \sup_{\substack{v \in V \\ v \neq 0}} \left(\frac{\varphi(v)}{\varphi'(v)}, \frac{\varphi'(v)}{\varphi(v)} \right).$$

Provided with the corresponding metric, X is a complete metric space, on which G acts. It is the building of $GL_n(k)$, (see [13, p. 238—239]).

The work of Bruhat and Tits also leads to a classification of simple G. One basic result here is

(7.3.1) *Let G be semi-simple and simply connected. Then $H^1(k, G) = 1$.*
This was first proved by M. Kneser if $\operatorname{char}(k) = 0$ [43], via a case by case analysis. We refer to [86, no. 4] for the classification.

The results of Bruhat and Tits have been instrumental for the infinite dimensional representation theory of the locally compact group $G(k)$.

(7.4)   *Global fields.*
We now assume that k is a global field, i.e. a finite extension of either $\mathbb{Q}$ or a field of rational functions $F(t)$, with F finite. If v is a nontrivial valuation of k we denote by k_v the corresponding completion and if v is non-archimedean by $\mathfrak{o}_v$ the ring of integers.

Let A be the adele ring of k. Recall that A is the direct limit of the locally compact rings $\prod_{v \in S} k_v \times \prod_{v \notin S} \mathfrak{o}_v$, as S runs over the finite sets of valuations containing the archimedean ones (see [95, p. 60]). A is a locally compact k-algebra, containing k as a discrete subfield, and A/k is compact.

Now let G be a k-group, assume G is a k-subgroup of GL_n. Since A is a k-algebra, the group $G(A)$ of A-points of G is defined. Viewing GL_n as a closed subset of $(n^2 + 1)$-dimensional affine space (compare (2.3) (a)) we obtain an imbedding $g \mapsto (g, (\det g)^{-1})$ of $G(A)$ onto a closed subset of A^{n^2+1}. Provided with the induced topology, $G(A)$ is a locally compact group, the *adele group* of G (this is independent of the choice of the imbedding $G \hookrightarrow GL_n$). $G(k)$ is a discrete subgroup of $G(A)$.

(7.4.1) *Example.* Let $k = \mathbb{Q}$, $G = SL_n$. One then checks, in an elementary way, that

$$G(A) = SL_n(\mathbb{R}) \cdot \prod_p SL_n(\mathbb{Z}_p) \cdot G(\mathbb{Q}),$$

where $\mathbb{Z}_p$ is the ring of p-adic integers. In the product p runs over the primes. Let N be a positive integer and put

$$\Gamma(N) = \{\gamma \in SL_n(\mathbb{Z}) | \gamma \equiv 1 \pmod{N}\},$$

then

$$G(A)/G(\mathbb{Q}) = \varprojlim SL_n(\mathbb{R})/\Gamma(N).$$

In particular, there is a surjective continous map

$$G(A)/G(\mathbb{Q}) \to SL_n(\mathbb{R})/SL_n(\mathbb{Z}).$$

The right-hand side is a space which was already encountered by Minkowski, in his reduction theory of quadratic forms. He computed its volume, for a natural measure [58].

A reduction theory for k-groups was developed by Borel and Harish-Chandra in the case of number fields ([7], [3]) and by Harder for function fields [34]. Actually, using the elementary procedure of "restriction of the base field", one needs only to consider $k = \mathbb{Q}$ and $k = F(t)$, which is a considerable technical simplification. As a result of reduction theory one has the following general theorems about the quotient $G(A)/G(k)$.

(7.4.2) *Assume G to be connected and semi-simple.*
(a) $G(A)/G(k)$ *has finite invariant volume;*
(b) $G(A)/G(k)$ *is compact if and only if G is anisotropic over k.*
     The invariant measure involved in (a) is deduced from the Haar measure on $G(A)$.
     For $k = \mathbb{Q}$, (7.4.2) (a) is equivalent to the following, more concrete result. We assume G to be a connected semi-simple $\mathbb{Q}$-subgroup of SL_n. Let L be a lattice in $\mathbb{R}^n$ and put $\Gamma = \{\gamma \in G(\mathbb{R}) | \gamma L = L\}$.

(7.4.3) *Γ is a discrete subgroup of $G(\mathbb{R})$ and $G(\mathbb{R})/\Gamma$ has finite invariant volume.*
     For the proof of such arithmetic results the theory of linear algebraic groups is absolutely essential. That it has been possible to prove them must be viewed as one of the successes of the theory.
     In the situation of (7.4.2) (a) one can introduce the Hilbert space of square integrable functions $L^2(G(A)/G(k))$, for the invariant measure. One has a natural (infinite dimensional) representation of the locally compact group $G(A)$ in this Hilbert space. The study of this representation is the object of the theory of *automorphic forms*, generalizing the classical theory (which one recovers for $G = SL_2$). This is a subject which has led to vast developments (reported on in [6]). Again, the theory of k-groups is an indispensable tool. For example, in the study of the "continuous" part of the above representation via the theory of Eisenstein series by Langlands [49], the theory of parabolic k-subgroups is an essential ingredient. See also [4].
     To end our brief remarks about global fields we must say something about the classification of semi-simple k-groups. The striking thing is that over

number fields, and even over $\mathbb{Q}$, the classification problem has not yet been solved. The same is true for the (practically equivalent) problem of classifying the simple Lie algebras over $\mathbb{Q}$.

Let G be a connected semi-simple k-group. There is a canonical map

$$\gamma : H^1(k, G) \rightarrow \prod_v H^1(k_v, G_v),$$

where v runs over all valuations of k and $G_v = G \times_k k_v$ (see 6.6). One has the following *conjecture*:

(7.4.4) (*"Hasse principle"*). *If G is simply connected then γ is bijective.*

(7.4.4) would give a description of $H^1(k, G)$. In fact, by (7.3.1), we have $H^1(k_v, G_v) = 1$ if v is non-archimedean, and also of course if v is complex archimedean. So, for $k = \mathbb{Q}$, we would have $H^1(k, G) \cong H^1(\mathbb{R}, G_{\mathbb{R}})$. A proof of (7.4.4) is the major step in the classification of k-groups. But even assuming (7.4.4), the full classification of simple k-groups would require more work, which is not in the literature, as far as I know. The Hasse principle (7.4.4) is almost proved. In fact Harder proved it for number fields for all simple types of the Killing-Cartan classification *except* type E_8, via a case by case analysis [33]. Subsequently, he gave a proof for function fields based on general principles, by making use of the theory of Eisenstein series [35].

8   Some geometric applications

The results and the language of the theory of algebraic groups have been used in the study of geometric questions. We shall review a few of them in this section.

(8.1)   *Invariant theory.*
   As was mentioned in (3.1), invariant theory was one of the parts of 19th century mathematics where algebraic groups were encountered. The theory of linear algebraic groups has been applied in recent years to invariant theory, and has been instrumental in the new developments of the latter.
   It is best to describe the subject matter of invariant theory in geometric terms, as the study of the actions linear algebraic groups on algebraic varieties. A typical example is the following.
   Let $G \subset G L_n(K)$ be a linear algebraic group (K algebraically closed). Then G acts linearly in $V = K^n$. A geometric study of this action leads one to questions about G-orbits in V, for example the question whether one can assemble the orbits in a quotient variety V/G. A way of doing this is to consider the algebra $K[V]$ of polynomial functions on V. The group G acts on it, let R be the subalgebra of G-invariant functions in $K[V]$ (the algebra of invariants).

If $G = SL_2(\mathbb{C})$, acting on the homogeneous polynomials of a given degree d in two indeterminates, the elements of R are the "invariants of binary forms", the study of which initiated invariant theory in the 19th century.

If R is a K-algebra of finite type then it defines an affine algebraic variety, which we view as the desired quotient V/G. The question whether R is always of finite type is *Hilbert's* 14th *problem*, which was answered negatively by a counter-example of Nagata [60]. But for reductive groups the answer is positive. The crucial result for the proof of this finiteness theorem is the following.

(8.1.1) *Let G be reductive. Assume that v is a nonzero vector such that $G \cdot v = v$. There is a nonconstant homogeneous $f \in R$ such that $f(v) \neq 0$.*

If $\operatorname{char}(K) = 0$ one can show that this property implies that any rational representation of G is fully reducible. A proof (due to M. Schiffer) that the property of full reducibility implies the finite generation of R can already be found in H. Weyl's book "The classical groups" [97, p. 100—103]. If $\operatorname{char}(K) > 0$ the only connected reductive groups G having this property of full reducibility are the tori. That, in any characteristic, reductive groups have the property of 8.1.1 was conjectured by Mumford [59, Preface], and proved by Haboush [31], using facts from the representation theory of semi-simple groups.

As an example from invariant theory where the use of notions from the theory of linear algebraic groups clarify a "classical" result we mention the *theorem of Hilbert-Mumford:*

(8.1.2) *Let G be connected, reductive. If $v \in V$ is such that $f(v) = 0$ for all nonconstant homogeneous $f \in R$, there is a homomorphism of algebraic groups $\lambda: GL_1 \to G$ such that the morphism $\varphi: GL_1 \to V$ with $\varphi(\xi) = \lambda(\xi) \cdot v$ extends to a morphism $\varphi: K \to V$ with $\varphi(0) = 0$.*

If v is as in this theorem, one shows that the Zariski closure of the orbit $G \cdot v$ contains 0. The theorem then asserts that there is then already a 1-dimensional subtorus S of G such that the closure of $S \cdot v$ contains 0. For $G = GL_n$ the theorem is due to Hilbert [36, V]. For a proof if $\operatorname{char}(K) = 0$ see Mumford [59, Ch. 2]. A general proof was given by Kempf. Hilbert already applied the theorem in concrete situations, see also [loc. cit., Ch. 4].

Finally, we mention another important recent result. If $v \in V$ is as before, we denote by G_v its isotropy group in G, i.e. $G_v = \{g \in G | g \cdot v = v\}$.

(8.1.3) *Assume G reductive and $\operatorname{char}(K) = 0$. Let $v \in V$ be such that its orbit $G \cdot v$ is closed.*
(a) *G_v is reductive;*
(b) *There exists an affine subvariety U of V, passing through v and stable under G_v such that the induced morphism $G \times_{G_v} U \to V$ is an étale morphism onto an open neighbourhood of v in V.*
(b) is *Luna's slice theorem*, proved in [51], for (a) see [loc. cit., p. 84]. $G \times_{G_v} U$ is the quotient of $G \times U$ by the G_v-action $x(g, u) = (g x^{-1}, xu)$. Recall that an étale

morphism is the analogue in algebraic geometry of an unramified covering of finite degree.

(8.2)   *Schubert varieties.*
         Schubert varieties were introduced by H. Schubert in 1886, for use in his calculus of enumerative geometry. We can define them as follows.
         Let K be an algebraically closed field, and put $V = K^n$. Fix a flag in V, i.e. a sequence of nested subspaces

$$V_0 \subset V_1 \subset \ldots \subset V_d. \tag{*}$$

Consider the d-dimensional spaces W of V such that

$$\dim(V_i \cap W) \geq i.$$

They form a subvariety of the Grassman variety $G_{d,n}$ of all d-dimensional subspaces of V; these subvarieties are the Schubert varieties (see [42] for further details about the use of Schubert varieties in enumerative problems).
         The language of the theory of algebraic groups provides a simple manner to describe Schubert varieties. Let G be a connected reductive K-group, and let P and Q be two parabolic subgroups. Then $X = G/P$ is a projective variety (5.4.1). It follows from the Bruhat decomposition (5.8.1) that Q has finitely many orbits in X. We call *Schubert variety* S the closure of such an orbit. One recovers the original ones for $K = \mathbb{C}$, $G = GL_n(\mathbb{C})$, P the stabilizer of a fixed d-dimensional subspace, and Q the stabilizer of the flag (*). From our general definition it is clear that S is a projective variety.
         In particular, we might take $P = Q = B$, a Borel subgroup. In that case $X = G/B$, the "flag manifold" of G, and it follows from the Bruhat decomposition (5.8.1) that the Schubert varieties are the closures $S(w)$ of the Bruhat cells $X(w)$. So they are parametrized by Weyl group elements. It also follows that $S(w)$ is a union of Bruhat cells $X(w')$. Define an ordering on the Weyl group W by:

$$w \leq w' \Leftrightarrow X(w) \subset S(w').$$

This order can be described algebraically (see [9, p. 267]).
         The Schubert varieties $S(w)$ are quite interesting algebraic varieties (in general with singularities) which have attracted a great deal of attention in recent years. We mention:
(a) The study and apllication of the "Bott-Samelson" resolution of $S(w)$ by Demazure [22];
(b) The determination of their "intersection cohomology" by Kazhdan-Lusztig [40]. There are very striking applications to problems about infinite dimensional representations of Lie algebras (reported on in [75]);

(c) Quite recently, the singular locus of the $S(w)$ has been determined by Lakshmibai and Seshadri, in the case of classical groups [47]. This is an application of their "standard monomial theory", in which bases are constructed for spaces of sections like $H^0(G/P, L)$, L being an ample line bundle.

(8.3)   *The unipotent variety.*

Let $G \subset GL_n(K)$ be a connected reductive K-group. Denote by X the set of its unipotent element (5.1). Clearly, X is an algebraic variety, the *unipotent variety* of G. Its basic properties are as follows.

(8.3.1) (a) *X is an irreducible normal variety, whose dimension is* dim *G*-rank *G* ; (b) *G acts on X by conjugation, and has finitely many orbits in X.*

(a) is proved in [80]. The result proved there also show that X is singular (if not reduced to a point). (b) states that the number of unipotent conjugacy classes in G is finite. This is easy if $G = GL_n(K)$ (using Jordan normal forms), and under some restrictions on char (K) (b) can be deduced from this, by a geometric argument due to Richardson (see [5, p. 182]). The general statement (b) was proved by Lusztig [52]. His proof requires *l*-adic cohomology.

There is a great deal of detailed information about the unipotent conjugacy classes, and the structure of the centralizer of a unipotent element. We refer to [69] and [73] for further details and references.

There is an interesting resolution of singularities of X, which we now describe. Fix a Borel subgroup B of G and put

$$\tilde{X} = \{(x, gB) \in X \times G/B \,|\, g^{-1}xg \in B\}.$$

One shows that $\tilde{X}$ is an algebraic variety (it is a variety of the type encountered in the proof of (5.4.4)), on which G acts by $h(x, gB) = (hxh^{-1}, hgB)$.

(8.3.2) *Assume G semi-simple and simply connected. Then the projection* π: $\tilde{X} \to X$ *is a G-equivariant resolution of X.*

The G-equivariance is obvious. The fact that π is a resolution means that $\tilde{X}$ is smooth ($=$ non-singular) and that π is birational and proper. For a proof see [83]. The smoothness of $\tilde{X}$ follows by considering the projection $\tilde{X} \to G/B$. In fact, $\tilde{X}$ can be viewed as the cotangent bundle of G/B.

If $x \in X$, the fiber $\pi^{-1}x$ is isomorphic to the subvariety $\mathscr{B}_x$ of $\mathscr{B} = G/B$ of the points fixed under the action by x. The resolution π and the fibers $\mathscr{B}_x$ have been much studied recently. We mention a few results.

(8.3.3) $\mathscr{B}_x$ *is connected. All its irreducible components have dimension* $e(x) = \frac{1}{2}$ (dim $Z_G(x) -$ rank G).

Here $Z_G(x)$ is the centralizer of x in G. For a proof see [69] (the proof of the statement about dimensions, for fields of arbitrary characteristic, is quite involved).

Assume now, for simplicity, that $K = \mathbb{C}$. Then the varieties $\mathcal{B}_x$ are compact complex spaces. Their cohomology groups $H(\mathcal{B}_x, \mathbb{Q})$ are quite interesting. For example, the Weyl group W of G (with respect to a maximal torus $T \subset B$) operates on them and the study of the representation of W in the top cohomology group $H^{2e(x)}(\mathcal{B}_x, \mathbb{Q})$ leads to a parametrization of the irreducible representations of W in terms of unipotent conjugacy classes of G. These matters were first taken up in [70] and [71]. The recently developed "intersection cohomology" of singular spaces (see [30]) has shed more light on the construction of the operation of W on $H(\mathcal{B}_x, Q)$. In [10], Borho-MacPherson make a study of the resolution π in the context of intersection homology and give an explanation of the operation of W, in this context (in the quoted references one works with the variety N of nilpotent elements of the Lie algebra of G, rather than with the variety N of unipotent elements of G. But if $K = \mathbb{C}$ it is easy to see that N and X are isomorphic).

For review of the matters briefly mentioned here we refer to [75].

We next mention a geometric interpretation of a combinatorial result, involving the varieties $\mathcal{B}_x$. We assume $G = GL_n(\mathbb{C})$. In that case the $\mathcal{B}_x$ have been studied by Spaltenstein [69, II.5]. He showed, in particular, that then the irreducible components of $\mathcal{B}_x$ can be parametrized by standard Young tableaux, whose shape is determined by the Jordan normal form of the unipotent matrix x.

Let X and B be as before. Consider the subvariety Y of $X \times G/B \times G/B$ of the triples (x, B', B'') such that $B', B'' \in \mathcal{B}_x$. One can count the irreducible components of Y in two ways, first by looking at B' and B'' and using the Bruhat decomposition (5.8.1). One then sees that all components of Y have the same dimension and that they can be parametrized by the elements of the Weyl group, which in the case $G = GL_n(\mathbb{C})$ is the symmetric group S_n. Secondly, by looking at irreducible components of the $\mathcal{B}_x$ one sees that the components of Y can also be parametrized by representatives x of the unipotent conjugacy classes of G, together with ordered pairs of components of $\mathcal{B}_x$. Translating this into combinatorial terms via the result of Spaltenstein mentioned above one obtains a bijection of S_n onto the set of ordered pairs of standard Young tableaux of the same shape. This is the correspondence of Robinson-Schensted from combinatorics [65]. One has thus a geometric description of this correspondence, can also be generalized to other groups (see [69, II.2], [73] for more details). As far as I know, the combinatorics of the generalizations have not yet been worked out.

It follows from (8.3.3) that $\dim Z_G(x) - \operatorname{rank} G$ is $\geqq 0$ and even, for $x \in X$. If $\dim Z_G(x) = \operatorname{rank} G$, the element x is called *regular* unipotent. Those elements are studied in [80]. In particular, they form one conjugacy class, and the regular elements are precisely the smooth points of X.

If $\dim Z_G(x) = \operatorname{rank} G + 2$ then x is called *subregular*. There is a very interesting connection between subregular elements and *rational double points*. Let Γ be a finite subgroup of $SL_2(\mathbb{C})$. It acts on $\mathbb{C}^2$, and there is a quotient variety $\mathbb{C}^2/\Gamma$, which has a singular point at the image of 0 (also denoted by 0). An isolated singular point of an algebraic surface whose local ring is isomorphic to that of one of the quotient varieties $\mathbb{C}^2/\Gamma$ in 0 is called a *rational double point*.

The possible Γ are well-known, they are (up to isomorphism) the cyclic groups $\mathbb{Z}_n$, the dihedral groups $\mathbb{D}_n$, the tetrahedral group $\mathbb{T}$, the octahedral group $\mathbb{O}$, the icosahedral group $\mathbb{I}$. So for each of these types one has a type of rational double point.

Assume (for simplicity) that $K = \mathbb{C}$ and let G be a simple $\mathbb{C}$-group. We use the previous notations and fix a subregular element $x \in X$. Let S be a suitable transversal section in G to the orbit $G(x)$ of x (i.e. the conjugacy class of x). The intersection $X \cap S$ has dimension 2.

(8.3.4) *Let Γ be a finite subgroup of $SL_2(\mathbb{C})$. There is a simple $\mathbb{C}$-group G such that the singularity $(C^2/\Gamma, 0)$ is isomorphic to $(X \cap S, x)$. The type of the root system of G, for the various Γ, is as follows:*

$$\mathbb{Z}_n : A_{n-1},$$
$$\mathbb{D}_n : D_{n+2},$$
$$\mathbb{T} \ : E_6,$$
$$\mathbb{O} \ : E_7,$$
$$\mathbb{I} \ : E_8.$$

Here isomorphism of singularities means isomorphism of the corresponding analytic local rings.

This connection between singularities and simple $\mathbb{C}$-groups was established by Brieskorn (and was first conjectured by Grothendieck). A detailed treatment and references can be found in [68, no. 6]. The theorem can be refined so as to describe the "semi-universal deformation" of the singularity $(C^2/\Gamma, 0)$ in terms of G [loc. cit., p. 136].

9   Concluding remarks

The brief review of results in the previous sections will have shown that the theory of linear algebraic groups can be used in various parts of mathematics and provides sometimes a way of looking at classical questions which sheds now light on them.

It can be expected that applications of this nature will be further developed in the future; algebraic groups might turn up in some unexpected places. But I don't want to enter here into speculations on future applications.

I shall conclude by mentioning some questions and problems which should be dealt with in the (near?) future. First there is the problem of the representation theory of algebraic groups in characteristic $p > 0$, of which our understanding is still far from perfect (see (5.9) and also (7.1)). One of the results lacking here is an analogue of Weyl's character formula. A conjectural result of Lusztig [53, Problem IV] seems to indicate a connection with the singularities of

Schubert varieties of certain infinite dimensional flag manifolds (associated to affine Weyl groups). This indication comes from the analogy with the theory of Verma modules, where the Schubert cells $S(w)$ of (8.2) enter the picture. However in the last case one makes uses of the machinary of the theory of $\mathcal{D}$-modules ("algebraic theory of differential equations"), which is a characteristic 0 theory.

A related problem is that of the study of the cohomology of line bundles on the flag manifold G/B in characteristic $p > 0$.

The second problem I want to mention is the proof of the, still conjectural, Hasse principle for number fields (7.4). It is quite vexing that this conjecture still has not been proved. Perhaps a further exploitation of the analytic machinery of Eisenstein series should be made. Another conjecture, which might be related (also settled over function fields by Harder [35]) is Weil's conjecture about the canonical volume of $G(A)/G(k)$, in the situation of (7.4.2) (determination of the "Tamagawa number").

As remarked in (6.6), further progress in the theory of k-groups might also lead to a proof of the Hasse principle.

Finally I would like to mention infinite dimensional groups. In recent years there has been much interest in certain infinite dimensional Lie algebras, the *Kac-Moody algebras*, and corresponding groups have been constructed as abstract groups (see for example [88]). Some of these groups can be endowed with a structure of ind-algebraic group. i.e. group object of the category of inductive limits of algebraic groups (such groups are introduced by Shafarevich in [67]).

As an example of the sort of infinite dimensional groups which appear here I mention the *loop groups*. Let G be a connected reductive $\mathbb{C}$-group, and sonsider the group $G(\mathbb{C}[T, T^{-1}])$ of its points with coordinates in the ring of Laurent polynominals $\mathbb{C}[T, T^{-1}]$. This is an ind-algebraic group over $\mathbb{C}$. It is the algebraic version of the topological loop group of the compact Lie group associated with G.

A further study of ind-algebraic groups and their homogeneous spaces, or of certain types of these, might be of interest.

References

[1]   Borel, A., *Groupes linéaires algébriques*, Ann. of Math. 64 (1956), 20—82.
[2]   Borel, A., *Linear algebraic groups*, Benjamin, New York, 1969.
[3]   Borel, A., *Introduction aux groupes arithmétiques*, Hermann, Paris, 1969.
[4]   Borel, A., *Automorphic L-functions*, Proc. Symp. Pure Math. vol. XXXIII part 2, 27—61, Amer. Math. Soc., 1979.
[5]   Borel, A. et al., *Seminar on algebraic groups and related finite groups*, Lect. Notes in Math. no. 131, Springer Verlag, 1970.
[6]   Borel, A. and W. Casselman (ed.), *Automorphic forms, representations and L-functions*, Proc. Symp. Pure Math. vol. XXXIII (2 parts), Amer. Math. Soc. 1979.

[7]   Borel A. and Harish-Chandra, *Arithmetic subgroups of algebraic groups*, Ann. of Math. 75 (1962), 485—535.

[8]   Borel, A. and J. Tits, *Groupes réductifs*, Publ. Math. I.H.E.S. 27 (1965), 55—150.

[9]   Borel, A., and J. Tits, *Compléments à l'article "Groupes réductifs"*, ibid. 41 (1972), 253—276.

[10]   Borho, W. and R. MacPherson, *Représentations de groupes de Weyl et homologie d'intersection pour les variétés nilpotentes*, C. R. Acad. Sci. Paris t. 292, sér. I (1981), 707—710.

[11]   Bourbaki, N., *Groupes et algèbres de Lie*, Ch. 4, 5, 6, Hermann, Paris, 1968.

[12]   Bruhat, F., *Sur les représentations induites des groupes de Lie*, Bull. Soc. Math. France 84 (1956), 97—205.

[13]   Bruhat, F. and J. Tits, *Groupes réductifs sur un corps local I*, Publ. Math. I.H.E.S. no. 41 (1972), 5—251.

[14]   Bruhat, F. and J. Tits, *Groupes réductifs sur un corps local II*, to be published.

[15]   Cartan, E., *Œuvres complètes, Partie I*, Gauthier-Villars, Paris, 1952.

[16]   Chevalley, C., *Theory of Lie groups I*, Princeton University Press, 1946.

[17]   Chevalley, C., *Théorie des groupes de Lie II*, Hermann, Paris, 1951.

[18]   Chevalley, C., *Sur certains groupes simples*, Tôhoku Math. J. 7 (1955), 14—66.

[19]   Chevalley, C., *Classification des groupes de Lie algébriques*, Séminaire Ec. Norm. Sup., Paris, 1956—1958.

[20]   Chevalley, C., *Certains schémas de groupes semi-simples*, Sém. Bourbaki no. 219, Paris 1960—1961.

[21]   Cline, E., B. Parshall, L. Scott, *On the tensor product theorem for algebraic groups*, J. Alg. 63 (1980), 264—267.

[22]   Demazure, M., *Désingularisation des variétés de Schubert*, Ann. Ec. Norm. Sup. (4) 7 (1974), 53—88.

[23]   Demazure, M. and A. Grothendieck, *Schémas en groupes*, Lect. Notes in Math. nos. 151, 152, 153, Springer Verlag, 1970.

[24]   Dickson, L. E., *Linear groups*, Teubner, Leipzig, 1901.

[25]   Dickson, L. E., *Theory of linear groups in an arbitrary field*, in: Coll. Math. Papers II, p. 54—74, Chelsea, New York, 1975.

[26]   Dickson, L. E., *A class of groups in an arbitrary realm connected with the configuration of the 27 lines on a cubic surface*, in: Coll. Math. Papers V, p. 385—413, Chelsea, New York, 1975.

[27]   Dieudonné, J., *Sur les groupes classiques*, Act. Scient. et Ind., no. 1040, Hermann, Paris, 1948.

[28]   Dieudonné, J., *La géometrie des groupes classiques*, Erg. d. Math. (Neue Folge), Bd. 5, Springer Verlag, 1963.

[29]   Gelfand, I. M. and M. A. Neumark, *Unitäre Darstellungen der klassischen Gruppen*, Akademie-Verlag, Berlin, 1957.

[30]   Goresky, M. and R. MacPherson, *Intersection homology II*, Inv. Math. 71 (1983), 77—129.

[31]   Haboush, W. J., *Reductive groups are geometrically reductive*, Ann. of Math. 102 (1975), 67—83.

[32]   Harder, G., *Über einen Satz von E. Cartan*, Abh. Math. Sem. Univ. Hamburg 28 (1965), 208—214.

[33]   Harder, G., *Über die Galoishomologie halbeinfacher Matrizengruppen I*, Math. Z. 90 (1965), 404—428, II, ibid. 92 (1966), 396—415.

[34]   Harder, G., *Minkowskische Reduktionstheorie über Funktionenkörpern*, Inv. Math. 7 (1969), 33—54.

[35]   Harder, G., *Chevalley groups over function fields and automorphic forms*, Ann. of Math. 100 (1974), 249—306.

[36]   Hilbert, D., *Über die vollen Invariantensysteme*, Ges. Abh., II², 287—347, Springer Verlag, 1970.

[37]   Humphreys, J. E., *Introduction to Lie algebras and representation theory*, Graduate Texts in Math., no. 9, Springer Verlag, 1972.

[38]   Humphreys, J. E., *Linear algebraic groups*, Graduate Texts in Math. no. 21, Springer Verlag, 1975.

[39]   Kaplansky, I., *An introduction to differential algebra*, Hermann, Paris, 1957.

[40]   Kazhdan D. and G. Lusztig, *Schubert varieties and Poincaré duality*, Proc. Symp. Pure Math. vol. XXXVI, p. 185—203, Amer. Math. Soc., 1980.

[41]   Killing, W., *Die Zusammensetzung der stetigen endlichen Transformationsgruppen II*, Math. Ann. 33 (1889), 1—48.

[42]   Kleiman, S. L., *Problem* 15. *Rigorous foundation of Schubert's enumerative calculus*, in: Proc. Symp. Pure Math., vol. XXVIII, 2, p. 445—482, Amer. Math. Soc., 1976.

[43]   Kneser, M., *Galois-Kohomologie halbeinfacher algebraischer Gruppen über p-adischen Körpern I*, Math. Z. 88 (1965), 40—47; II, ibid. 89 (1965), 250—272.

[44]   Kolchin, E. R., *Algebraic matric groups and the Picard-Vessiot theory of homogeneous linear ordinary differential equations*, Ann. of Math. 49 (1948), 1—42.

[45]   Kolchin, E. R., *On certain concepts in the theory of algebraic matric groups*, Ann. of Math. 49 (1948), 774—789.

[46]   Kolchin, E. R., *Differential algebra and algebraic groups*, Academic Press, New York, 1973.

[47]   Lakshmibai, V. and C. S. Seshadri, *Singular locus of a Schubert variety*, to appear.

[48]   Lang, S., *Algebraic groups over finite fields*, Amer. J. Math. 78 (1956), 555–563.

[49]   Langlands, R. P., *On the functional equations satisfied by Eisenstein series*, Lect. Notes in Math. no. 544, Springer Verlag, 1976.

[50]   Lie, S., *Theorie der Transformationsgruppen* (3 vol.), Teubner, Leipzig, 1930.

[51]   Luna, D., *Slices étales*, Bull. Soc. Math. France (supplément), Mém. no. 33 (1973), 81—105.

[52]   Lusztig, G., *On the finiteness of the number of unipotent classes*, Inv. Math. 34 (1976), 201—213.

[53]   Lusztig, G., *Some problems in the representation theory of finite Chevalley groups*, in: Proc. Symp. Pure Math. vol. XXXVII, p. 313—317, Amer. Math. Soc., 1980.

[54]   Lusztig, G., *Characters of reductive groups over a finite field*, Princeton Univ. Press, 1984.

[55]   Maurer, L., *Über Invariantentheorie*, J. f. d. reine u. angew. Math. 107 (1891), 89—116.

[56]   Maurer, L., *Zur Theorie der continuierlichen homogenen und linearen Gruppen*, Sitz. Ber. Math. Phys. Classe d. k. bayer. Akademie der Wiss., XXIV (1894), 297—341.

[57]   Milnor, J., *Introduction to algebraic K-theory*, Ann. of Math. Studies, no. 72 (1971).

[58]   H. Minkowski, *Diskontinuitätsbereich für arithmetische Äquivalenz*, in: Ges. Abh. Bd. 2, p. 53—100, Teubner, Leipzig—Berlin, 1911.

[59]   Mumford, D., *Geometric invariant theory*, Erg. d. Math. Bd. 34, Springer Verlag, 1965.

[60]   Nagata, M., *On the* 14*th problem of Hilbert*, Am. J. Math. 81 (1959), 766—772.

[61]   Nakano, S., *Note on group varieties*, Mem. Coll. Sc., Univ. of Kyoto A 27 (1952), Math. 55—66.

[62]   Picard, E., *Equations différentielles linéaires et les groupes algébriques de transformations*, in: Œuvres, t. II, p. 117—131, Ed. C.N.R.S., 1979.

[63]   Rosenlicht, M., *Some basic theorems on algebraic groups*, Amer. J. of Math. 78 (1956), 401—443.

[64]   Rosenlicht, M., *Questions of rationality for solvable algebraic groups over nonperfect fields*, Annali di Math. (IV), vol. LXI (1963), 97—120.

[65]   Schützenberger, M.-P., *La correspondence de Robinson*, in: Combinatoire et représentation du groupe symétrique, Lect. Notes in Math. no. 579, Springer Verlag, 1977.

[66]   Serre, J.-P., *Cohomologie galoisienne*, Lect. Notes in Math. no. 5, Springer Verlag, 1964.

[67]   Shafarevich, I. R., *On certain infinite-dimensional groups II*, Math. USSR—Izvestija 18 (1981), 185—194.

[68]   Slodowy, P., *Simple singularities and simple algebraic groups*, Lect. Notes in Math. no. 815, Springer Verlag, 1980.

[69] Spaltenstein, N., *Classes unipotentes et sous-groupes de Borel*, Lect. Notes in Math. no. 946, Springer Verlag, 1982.

[70] Springer, T. A., *Trigonometric sums, Green functions of finite groups and representations of Weyl groups*, Inv. Math. 36 (1976), 173—207.

[71] Springer, T. A., *A construction of representations of Weyl groups*, ibid. 44 (1978), 279—293.

[72] Springer, T. A., *Reductive groups*, in: Proc. Symp. Pure Math. vol. XXXIII 1, p. 3—27, Amer. Math. Soc., 1979.

[73] Springer, T. A., *Geometric questions arising in the study of unipotent elements*, in: Proc. Symp. Pure Math. vol. XXXVII, p. 255—264, Amer. Math. Soc., 1980.

[74] Springer, T. A., *Linear algebraic groups*, Birkhäuser, Boston—Basel—Stuttgart, 1981.

[75] Springer, T. A., *Quelques applications de la cohomologie d'intersection* (Sém. Bourbaki no. 582), Astérisque no. 92—93, p. 249—273, Soc. Math. France, 1982.

[76] Srinivasan, B., *Representations of finite Chevalley groups*, Lect. Notes in Math. no. 764, Springer Verlag, 1979.

[77] Steinberg, R., *A geometric approach to the representations of the full linear group over a Galois field*, Trans. Amer. Math. Soc. 71 (1951), 274—282.

[78] Steinberg, R., *Générateurs, relations et revêtements de groupes algébriques*, in: Colloq. théorie des groupes algébriques (Bruxelles, 1962), Gauthier-Villars, Paris, 1962.

[79] Steinberg, R., *Representations of algebraic groups*, Nagoya Math. J. 22 (1963), 33—56.

[80] Steinberg, R., *Regular elements of semisimple algebraic groups*, Publ. Math. I.H.E.S. 25 (1965), 49—80.

[81] Steinberg, R., *Endomorphisms of linear algebraic groups*, Mem. Amer. Math. Soc. no. 80 (1968).

[82] Steinberg, R., *Conjugacy classes in algebraic groups*, Lect. Notes in Math. no. 366, Springer Verlag, 1974.

[83] Steinberg, R., *On the desingularization of the unipotent variety*, Inv. Math. 36 (1976), 209—224.

[84] Takeuchi, M., *A hyperalgebraic proof of the isomorphism and isogeny theorems for reductive groups*, to appear.

[85] Tits, J., *Théorème de Bruhat et sous-groupes paraboliques*, C. R. Acad. Sci. Paris 254 (1962), 2910—2912.

[86] Tits, J., *Classification of algebraic semisimple groups*, in: Proc. Symp. Pure Math. vol. IX, p. 33—62, Amer. Math. Soc., 1966.

[87] Tits, J., *Reductive groups over local fields*, in: Proc. Symp. Pure Math. vol. XXXIII,1 p. 29—69, Amer. Math. Soc., 1979.

[88] Tits, J., *Résumé de cours* (1980—81), Collège de France, Paris.

[89] Vessiot, E., *Sur les intégrations des équations différentielles linéaires*, Ann. Sci. Ec. Norm. Sup. (3) 9 (1892), 192—280.

[90] Vogan, D. A., *Representations of real reductive Lie groups*, Birkhäuser, Boston—Basel—Stuttgart, 1981.

[91] Weierstrass, K., *Zur Theorie der bilinearen und quadratischen Formen*, in: Math. Werke, Bd. II, p. 19—44, Mayer & Müller, Berlin, 1895.

[92] Weil, A., *Variétiés abéliennes et courbes algébriques*, Act. Sci. et Ind. no. 1064, Hermann, Paris, 1948.

[93] Weil, A., *On algebraic groups of transformations*, Am. J. Math. 77 (1955), 355—391.

[94] Weil, A., *On algebraic groups and homogeneous spaces*, ibid. (1955), 493—512.

[95] Weil, A., *Basic number theory*, Grundlehren d. math. Wiss. Bd. 144, Springer Verlag, 1967.

[96] Weyl, H., *Theorie der Darstellung kontinuierlicher halbeinfacher Gruppen durch lineare Transformationen*, Ges. Abh. Bd. II, p. 543—647, Springer Verlag, 1968.

[97] Weyl, H., *The classical groups*, Princeton Univ. Press, 1946.

Perspectives in Mathematics
Anniversary of Oberwolfach 1984
© Birkhäuser Verlag, Basel

Gauge Theories as a Tool for Low Dimensional Topologists*

RONALD J. STERN

Department of Mathematics, State University,
Salt Lake City, Ut. (USA)

The 1980s have experienced a tremendous growth in our understanding of four dimensional smooth manifolds. This has been principally through the work of Simon Donaldson [D] with his celebrated theorem.

Theorem: Let M be a smooth closed oriented simply-connected 4-manifold with positive definite intersection form Φ. Then Φ is "standard", i.e. over the integers.

$$\Phi \cong (1) \oplus \cdots \oplus (1)$$

Although this is a theorem about 4-dimensional topology, its proof is differential geometric and analytical in nature. The main theme of Donaldson's work is to study the space of solutions of the self-dual Yang-Mills equations on an $SU(2)$ bundle over the Riemannian manifold M and relate it to the topology of M.

At first (or even second) glance, the use of techniques from Mathematical Physics to solve an important problem in topology must be ad hoc and there must be a more "topological" proof! The purpose of this note is to argue that gauge theories and Yang-Mills connections *naturally* arise in the study of smooth 4-manifolds. (Why didn't low dimensional topologists discover them sooner?)

1   The Intersection Pairing

The traditional goal of geometric topology is to discover algebraic invariants which classify (at least partially) all manifolds in a given dimension. Historically, one of the most important of these invariants has been the intersection form.

---

* Partially supported by National Science Foundation Grant MCS 8002843A01

Perhaps it's best to start with two-dimensional manifolds where the intersection form and intersection numbers are more familiar. We can represent one-dimensional homology classes on a smooth surface S by smooth oriented curves. Suppose $\alpha, \beta \in H_1(S, \mathbb{Z})$ are represented by curves A and B. By slightly perturbing the curves we can assume that they intersect transversally in isolated points. This means that at each point of

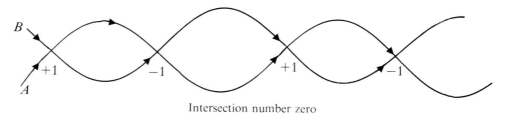

Intersection number zero

intersection a tangent vector to A, together with a tangent vector for B (in that order), form a basis for the tangent space of S. To each point of intersection we assign $+1$ if the orientation of this basis agrees with the orientation of S; otherwise we assign -1. The (oriented) intersection number $A \cdot B$ is defined to be the algebraic sum of these numbers over all points of intersection, and the intersection form is the induced bilinear pairing defined by $I_s(\alpha, \beta) = A \cdot B$. It's easy to see that I_s is skew-symmetric $[I_s(\alpha, \beta) = -I_s(\beta, \alpha)]$ and unimodular. In fact, for any such form we can choose a basis so that the matrix of the form is

$$\begin{pmatrix} 0 & I \\ -I & 0 \end{pmatrix}.$$

Intersection numbers and the intersection form for a smooth 4-manifold M are defined similarily. This time we suppose 2-dimensional homology classes $\alpha, \beta \in H_2(M, \mathbb{Z})$ are represented by smooth oriented surfaces A and B and that the surfaces intersect transversally in isolated points. Again we assign $+1$ to a point of intersection if an (oriented) basis for the tangent space of A together with an (oriented) basis for the tangent space of B agrees with the orientation for M; otherwise we assign -1. The intersection number $A \cdot B$ is the algebraic sum of these over all points of intersection, and the intersection form is the bilinear pairing $I_M(\alpha, \beta) = A \cdot B$. This time, however, the pairing is symmetric $[I_M(\alpha, \beta) = I_M(\beta, \alpha)]$. It is still unimodular – the matrix for the form has determinant ± 1.

For smooth manifolds there is another way to define the intersection form. By Poincaré duality we can define the pairing in cohomology rather than homology. If we use DeRham cohomology $H^*_{DR}(M)$, then $\alpha, \beta \in H^2_{DR}(M)$ can be represented by 2-forms a and b. We simply let

$$I_M(\alpha, \beta) = \int_M a \wedge b.$$

Defining the intersection pairing on cohomology allows us to extend the definition to all 4-manifolds, smooth or not. If $\alpha, \beta \in H_2(M; \mathbb{Z})$ and $[M] \in H_4(M; \mathbb{Z})$ is the fundamental class of M (given by the orientation on M), then $I_M(\alpha, \beta) = \alpha \cup \beta$, where "$\cup$" is the cup product in cohomology.

Here are some examples.

1. The 4-sphere S^4. Since $H_2(S^4, \mathbb{Z}) = 0$, the intersection form is trivial $I_{s^4} = 0$.

2. The complex projective plane $\mathbb{C}P^2$. Here $H_2(\mathbb{C}P^2; \mathbb{Z}) = \mathbb{Z}$, and so the matrix for $I_{\mathbb{C}P^2}$ is (1).

3. The product of spheres $S^2 \times S^2$. In this case $H_2(S^2 \times S^2; \mathbb{Z}) = \mathbb{Z} \oplus \mathbb{Z}$, and we can represent generators by the embedded surfaces $A = S^2 \times \{pt\}$ and $B = \{pt\} \times S^2$. Since A and B intersect in a single point, and each of them can be "pushed off" themselves, the matrix for $I_{s^2 \times s^2}$ is $\begin{pmatrix} 0 & 1 \\ 1 & 0 \end{pmatrix}$.

4. The Kummer surface $K = \{[Z_0, Z_1, Z_2, Z_3] \in \mathbb{C}P^3 \mid Z_0{}^4 + Z_1{}^4 + Z_2{}^4 + Z_3{}^4 = 0\}$. This time things are much more complicated. The rank of $H_2(K; \mathbb{Z})$ is 22, and one can show that the matrix for I_k is given by $E_8 + E_8 + 3 \begin{pmatrix} 0 & 1 \\ 1 & 0 \end{pmatrix}$, where

$$E_8 = \begin{pmatrix} 2 & -1 & 0 & 0 & 0 & 0 & 0 & 0 \\ -1 & 2 & -1 & 0 & 0 & 0 & 0 & 0 \\ 0 & -1 & 2 & -1 & 0 & 0 & 0 & 0 \\ 0 & 0 & -1 & 2 & -1 & 0 & 0 & 0 \\ 0 & 0 & 0 & -1 & 2 & -1 & 0 & -1 \\ 0 & 0 & 0 & 0 & -1 & 2 & -1 & 0 \\ 0 & 0 & 0 & 0 & 0 & -1 & 2 & 0 \\ 0 & 0 & 0 & 0 & -1 & 0 & 0 & 2 \end{pmatrix}$$

(In fact, E_8 is the Cartan matrix for the exceptional Lie algebra e_8.)

The intersection form is indeed a basic invariant for closed 4-manifolds. In 1949 Whitehead [W] showed that the homotopy type of a closed, simply-connected 4-manifold is completely determined by the isomorphism class of the intersection form.

The classification (up to isomorphism) of integral unimodular symmectic bilinear forms starts with three things: the rank (the dimension of the space on which the form is defined), the signature (the number of positive eigenvalues minus the number of negative eigenvalues considered as a real, rather than an integral form), and the type (the form is even if all the diagonal entries in its matrix are even, otherwise it's odd). A form is positive (negative) definite if all eigenvalues are positive (negative); otherwise it's indefinite.

For indefinite forms the rank, signature, and type form a complete set of invariants [MH]. The classification of definite forms, however, is much more difficult. There is only one nontrivial restriction on an even form—it's signature must be divisible by 8. In fact it is known that E_8 (mentioned above) is the unique positive definite form of rank 8; there are two even positive definite forms of rank 16 ($E_8 \oplus E_8$ and Γ_{16}); 24 such forms of rank 24; $\geqslant 10^7$ such forms of rank 32; $\geqslant 10^{51}$ such forms of rank 40.

The first indication that the intersection form of a smooth 4-manifold had more than algebraic restrictions was Rohlin's Theorem [R] which asserts that any even form coming from a smooth simply-connected 4-manifold has signature divisible by 16. In particular the form E_8 cannot occur as the intersection pairing of a simply-connected smooth 4-manifold.

2   A Study of $H^2(M;\mathbb{Z})$

In order to motivate the use of gauge theories to better understand the intersection form on a smooth 4-manifold M, we begin by studying the second cohomology group $H^2(M;\mathbb{Z})$ from the vantage point of an algebraic topologist, differential topologist, differential geometer, and then an analyist. This material is classical and appears in various (although rarely one) standard graduate courses. For simplicity we assume that $H_1(M;\mathbb{R})=0$.

When an algebraic topologist is confronted with the group $H^2(M;\mathbb{Z})$, obstruction theory comes to mind, whence $H^2(M;\mathbb{Z})=[M,CP^\infty]$. But CP^∞ is the classifying space for $SO(2)=U(1)$ bundles, so that the have a $1-1$ correspondence

$$H^2(M;\mathbb{Z}) \leftrightarrow \{\text{Isomorphism classes of } SO(2) \text{ bundles over } M\}$$

where $x \in H^2(M;\mathbb{Z})$ corresponds to that $SO(2)$ bundle L over M with Euler class $= e(L) = x$.

Now assuming that M is smooth, a differential topologist would consider $H^2(M;\mathbb{R})$ rather than $H^2(M;\mathbb{Z})$ and then use the DeRham theorem to identify $H^*(M;\mathbb{R})$ with $H^*_{DR}(M)$ i.e., the homology of the DeRham complex

$$(*)\ 0 \to \Omega^0 \xrightarrow{d} \Omega^1 \xrightarrow{d} \Omega^2 \to \cdots$$

where Ω^p are the p-forms on M and d is the exterior derivative. Since $d^2=0$, $(*)$ is indeed a complex and we can form its homology groups $H^*_{DR}(M)$. Thus any element $x \in H^2(M;\mathbb{R})=H^2{}_{DR}(M)=\ker d/\operatorname{im} d$ is represented by an element $x \in \Omega^2$ with $dx=0$.

Let's now assume that M is endowed with a Riemannian metric. This induces a metric $\langle,\rangle$ on p-forms, so that the differential operator d in $(*)$ has a formal adjoint δ; i.e., $\langle \delta a, b \rangle = \langle a, db \rangle$. An analyst would form the Laplacian $\Delta = d\delta + \delta d$ and then use Hodge theory to identify $H^2_{DR}(M)$ with the space of harmonic 2-forms, i.e., those 2-forms x with $\Delta x = 0$.

At this point, we can summarize our discussion with the following diagram

$$H^2(M;\mathbb{Z}) \xleftarrow{\text{bundle theory}} \{\text{Isomorphism classes of } SO(2) \text{ bundles over } M\}$$
$$\downarrow$$
$$H^2(M;\mathbb{R}) \xleftarrow{\text{DeRham theorem}} H^2_{DR}(M) \xleftarrow{\text{Hodge theory}} \{\text{harmonic 2-forms}\}.$$

We would like to make this into a "commutative" diagram by associating to an $SO(2)$ bundle over M a DeRham 2-form and to make this association unique. This is the realm of differential geometry through the study of connections and their curvatures!

Let E be a vector bundle over M. A *connection* ∇ on E is merely a rule which allows one to take derivatives of sections of E in the direction of tangent vectors of M. So, given a section $\sigma \in \Gamma(E)$ and a tangent vector field X on M, then $\nabla_X \sigma$ is the derivative of σ in the direction of X and it must satisfy a Leibniz rule. In other words, a *connection* on E is a linear differential operator $\nabla : \Gamma(E) \to \Gamma(T^*(M) \otimes E)$ such that

$$\nabla(f\sigma) = df \otimes \sigma + f\nabla\sigma$$

where $f : M \to R$. If E is endowed with a fiberwise metric $\langle,\rangle$ we then require ∇ to be Riemannian, that is

$$d\langle \sigma_1, \sigma_2 \rangle = \langle \nabla\sigma_1, \sigma_2 \rangle + \langle \sigma_1, \nabla\sigma_2 \rangle$$

It will be convenient for us to set the notation

$$\Omega^k(F) = \Gamma(\Lambda^k T^* M \otimes F)$$

for any bundle F over M. So, for example, $\nabla : \Omega^0(E) \to \Omega^1(E)$. A connection ∇ has a natural extension

$$d^\nabla : \Omega^k(E) \to \Omega^{k+1}(E)$$

defined by $d^\nabla(\alpha \otimes \sigma) = d\alpha \otimes \sigma + (-1)^k \alpha \wedge \nabla\sigma$, where $\alpha \in \Omega^k$ and $\sigma \in \Omega^0(E)$.

The *curvature* R^∇ of a connection ∇ on E is a 2-form with values in $\mathfrak{g}_E$

where $\mathfrak{g}_E \subset \operatorname{Hom}(E, E)$ is the subbundle of endomorphisms which are skew-symmetric on each fiber. In other words $R^\nabla \in \Omega^2(\mathfrak{g}_E)$. It is defined by

$$R^\nabla_{X,Y} = \nabla_X \nabla_Y - \nabla_Y \nabla_X - \nabla_{[X,Y]}.$$

Also, we have $R^\nabla = d^\nabla \circ \nabla$.

In the case at hand, E is an $SO(2)$ bundle, so that $\mathfrak{g}_E$ is the trivial real line bundle over M. Thus if ∇ is a connection in an $SO(2)$ bundle L over M, then its curvature $R^\nabla \in \Omega^2(\mathfrak{g}_E) = \Omega^2$ is a real 2-form! It is a consequence of the Bianchi identities that R^∇ is a closed 2-form, i.e., $dR^\nabla = 0$. Furthermore, it is a fundamental result of Chern that the real Euler class $e(L)_R \in H^2_{DR}(M)$ is represented by $\frac{1}{2\pi} R^\nabla$. We could now complete our diagram by associating to each $SO(2)$ bundle L over M the curvature 2-form of a connection on L. Any two such connections determine the same DeRham cohomology class.

But let's be greedy. There are many connections on L and many representatives for the real Euler class $e(L)$ of L. Given any closed 2-form α representing $e(L)$ there is a connection ∇_α whose curvature is $2\pi\alpha$ (for fix any connection ∇ and note $\frac{1}{2\pi} R^\nabla = \alpha + dw$ for some $w \in \Omega^1$. Set $\nabla_\alpha = \nabla - 2\pi w$, and then $\frac{1}{2\pi} R^\nabla \alpha = \frac{1}{2\pi} R^\nabla - dw = \alpha$). In particular there is a connection ∇_θ whose curvature is the unique harmonic 2-form θ representing $e(L)$. But there are many such connections ∇_θ! To see this, let ∇ be a connection in a vector bundle E over M and let $G_E \subset \operatorname{Hom}(E, E)$ be the bundle of orthogonal endomorphisms of E. If $g \in \Gamma(G_E)$, then $\nabla^g = g \circ \nabla \circ g^{-1}$ is a new connection whose curvature $R^{\nabla^g} = g \circ R^\nabla \circ g^{-1}$. ∇^g is said to be *gauge equivalent* to ∇ and $\mathcal{G}_E = \Gamma(G_E)$ is called the *gauge group*. In our case of an $SO(2)$ bundle L over M, $R^{\nabla^g} = g \circ R^\nabla \circ g^{-1} = R^\nabla$. In fact, $R^\nabla = R^{\nabla'}$ if and only if there exists a $g \in \mathcal{G}_E$ with $\nabla' = \nabla^g$! (If $R^\nabla = R^{\nabla'}$, then $\nabla' = \nabla + w$ where $dw = 0$. Since $H_1(M;\mathbb{R}) = 0$, $w = ds$ for some $s \in \Omega^0$. So $\nabla' = \nabla + ds = e^{-s}(e^s ds + e^s \nabla) = e^{-s} \nabla e^s$; hence ∇ and ∇' are gauge equivalent.)

We now associate to each $SO(2)$ bundle L over M the unique gauge equivalence class of connections whose curvatures are harmonic. This completes our commutative diagram.

$$H^2(M;\mathbb{Z}) \xleftarrow{\text{bundle theory}} \{\text{Isomorphism classes of } SO(2) \text{ bundles over } M\} \xleftarrow{\substack{\text{differential}\\ \text{geometry}}} \{\text{gauge equivalence classes of connections with harmonic curvature}\}$$

(2.1)

$$H^2(M;\mathbb{R}) \xleftarrow{\text{DeRham theory}} H^2_{DR}(M) \xleftarrow{\text{Hodge theory}} \{\text{harmonic 2-forms}\}$$

curvature/2π

Now that we have a cosmopolitan description of the free abelian group $H^2(M;\mathbb{Z})$ which utilizes material from (what should be) a standard graduate curriculum, so what?

3   Why Yang-Mills?

As is pointed out in the first section, in order to gain some understanding of 4-manifolds, the intersection form should come into play. This did not happen in the second section. As an attempt to introduce the intersection form into our scheme, a topologist might consider stable isomorphism classes of $SO(2)$ bundles over M rather than just isomorphism classes. That is, put an equivalence relation $\sim$ on $SO(2)$ bundles by declaring that $L \sim L'$ if and only if $L \oplus \varepsilon$ and $L' \oplus \varepsilon$ are isomorphic as $SO(3)$ bundles, where ε is a trivial $\mathbb{R}^1$ bundle over M.

To see if we have accomplished anything, what equivalence relation have we induced on $H^2(M;\mathbb{Z})$? By the classification of $SO(3)$ bundles over a 4-complex, due to Dold and Whitney [DW], $L \sim L'$ if and only if $e(L)_{(2)} = e(L')_{(2)}$ and $(e(L))^2 = (e(L'))^2$, where, for $a \in H^2(M;\mathbb{Z})$, $a_{(2)} \in H(M;\mathbb{Z}_2)$ is the mod 2 reduction of a, and a^2 is $(a \vee a)$ evaluated on the fundamental class of M. So by introducing the equivalence relation $\sim$ on $H^2(M;\mathbb{Z})$ given by $a \sim b$ if and only if $a_{(2)} = b_{(2)}$ and $a^2 = b^2$, we have the 1-1 correspondence

$$H(M;\mathbb{Z})/\sim \; \leftrightarrow \; \{\text{isomorphism classes of } SO(2) \text{ bundles over } M\} \Big/ \sim .$$

Let's now attempt to complete the picture.

The novel (at least for a topologist) viewpoint in the previous section was the study of connections on L. So we now study connections on $E = L \oplus \varepsilon$; in particular, we should study those connections whose curvature forms are harmonic. But what does this mean, since $R^\nabla \in \Omega^2(\mathfrak{g}_E)$ and $\mathfrak{g}_E$ is no longer trivial (in fact, for $SO(3)$ bundles E, $\mathfrak{g}_E \simeq E$). As mentioned above, we have the sequence

$$\Omega^0(E) \xrightarrow{\nabla} \Omega^1(E) \xrightarrow{d^\nabla} \Omega^2(E) \xrightarrow{d^\nabla} \cdots .$$

However, $R^\nabla \in \Omega^2(\mathfrak{g}_E)$, so we should look for a sequence involving forms with values in $\mathfrak{g}_E$. Given a connection ∇ in E, it induces a connection ∇ in $\mathfrak{g}_E$ given by $\nabla(\theta) = [\nabla, \theta]$ where $\theta \in \Omega^0(\mathfrak{g}_E)$, i.e., $\nabla(\theta)(\sigma) = \nabla(\theta(\sigma)) - \theta(\nabla\sigma)$ for any section σ of E. We then have the sequence

$$\Omega^0(\mathfrak{g}_E) \xrightarrow{\nabla = d^\nabla} \Omega^1(\mathfrak{g}_E) \xrightarrow{d^\nabla} \Omega^2(\mathfrak{g}_E) \xrightarrow{d^\nabla} \cdots$$

and, as in the real case, the Bianchi identities translate to the fact that $d^\nabla R^\nabla = 0$. Again, each d^∇ has a formal adjoint δ^∇, and we can form the Laplacian $\Delta^\nabla = d^\nabla \delta^\nabla$

$+ \delta^\nabla d^\nabla$. We then wish to study those connections ∇ in $E = L \oplus \varepsilon$ whose curvatures are harmonic, i.e., $\Delta^\nabla R^\nabla = 0$. If M is compact, this translates into two equations, $d^\nabla R^\nabla = 0$ and $\delta^\nabla R^\nabla = 0$, which by the Bianchi identities reduces to $\delta^\nabla R^\nabla = 0$. This is nothing more than the Yang-Mills equation!! A *Yang-Mills connection* is a connection whose curvature is harmonic.

As we saw in the previous section, there is the action of the gauge group on the space of connections which takes a Yang-Mills connection to a Yang-Mills connection. We are now led to the study of gauge equivalence classes of Yang-Mills connections on $E = L \oplus \varepsilon$, i.e., the moduli space $\mathcal{M}$ of Yang-Mills connections on E.

Note that $\mathcal{M} \neq \emptyset$, since $E = L \oplus \varepsilon$ has Yang-Mills connections arising from the unique gauge equivalence class of Yang-Mills connections on L direct summed with the trivial connection on E. Such connections are called *reducible* connections. The number of gauge equivalence classes of reducible Yang-Mills connections is, then, just the number m of distinct (up to orientation) splitting of E as $L' \oplus \varepsilon$ for some $SO(2)$ bundle L'. This number, as we saw above, is half the number of solutions to the equations

(i) $a^2 = (e(L))^2$

(ii) $a_{(2)} = e(L)_{(2)}$,

for $a \in H^2(M; \mathbb{Z})$. (ii) says that $a = e(L) + 2b$ for some $b \in H^2(M; \mathbb{Z})$, so m is half the number of solutions to the equation

(iii) $(e(L) + 2b)^2 = (e(L))^2$

which is equivalent to the equation

(iii)' $b \cdot (e(L) + b) = 0$.

Perhaps by studying M, the irreducible Yang-Mills connections will provide a cobordism between the reducible solutions which are completely determined by the intersection form on M.

4   Why Self-Dual Connections?

In order to complete the lower row of (2.1), we would like to relate harmonic forms with cohomology. Unfortunately, the sequence

(4.1) $\Omega^0(\mathfrak{g}_E) \xrightarrow{d^\nabla} \Omega^1(\mathfrak{g}_E) \to \Omega^2(\mathfrak{g}_E) \to \cdots$

is not a complex, since $d^\nabla \circ d^\nabla = R^\nabla$, which may not vanish. Don't despair!

Differential geometers have long been aware that dimension four has a property that distinguishes itself from other dimensions. The rotation group $SO(n)$ is a simple Lie group for all $n \neq 4$ and $SO(4)$ double covers $SO(3) \times SO(3)$, so that the Lie algebra $so(4)$ of $SO(4)$ is isomorphic to $so(3) \times so(3)$. Thus, since the six dimensional space $\Lambda^2(\mathbb{R}^4)$ of 2-forms on the inner product space $\mathbb{R}^4$ is isomorphic to $so(4)$, $\Lambda^2(\mathbb{R}^4)$ decomposes as the sum of 3-dimensional spaces $\Lambda^2_+ + \Lambda^2_-$. An alternate description of this decomposition is given in terms of the Hodge star operator $*: \Lambda^2(\mathbb{R}^4) \to \Lambda^2(\mathbb{R}^4)$. If $(e_1, \ldots, e_4)$ is an oriented basis for $\mathbb{R}^4$, then $*(e_i \vee e_j) = e_k \vee e_l$, where (i, j, k, l) is an even permutation of $(1, 2, 3, 4)$. As $(*)^2 = 1$, $\Lambda^2(\mathbb{R}^4)$ decomposes as the ± 1 eigenspaces $\Lambda^2_\pm$ of $*$. Thus, if M admits a Riemannian metric, $\Lambda^2(T^*M) = \Lambda^2_+(M) \oplus \Lambda^2_-(M)$, and this decomposition is an invariant of the conformal class of the metric on M. An element of $\Lambda^2_+(M)(\Lambda^2_-(M))$ is called a *self dual (anti-self dual) 2-form*.

Since $\Omega^2(\mathfrak{g}_E) = \Gamma(\Lambda^2(T^*(M)) \otimes \mathfrak{g}_E)$, $*$ extends to an operator $*: \Omega^2(\mathfrak{g}_E) \to \Omega^2(\mathfrak{g}_E)$ given by $* \otimes \mathrm{id}$. Thus $\Omega^2(\mathfrak{g}_E) \cong \Omega^2_+(\mathfrak{g}_E) \oplus \Omega^2_-(\mathfrak{g}_E)$. But $R^\nabla \in \Omega^2(\mathfrak{g}_E)$, so $R^\nabla = R^\nabla_+ + R^\nabla_-$. This is a *very* special property of 4-dimensional geometry—the curvature decomposes into its self dual and anti-self dual components.

The adjoint $\delta^\nabla: \Omega^p(\mathfrak{g}_E) \to \Omega^{p-1}(\mathfrak{g}_E)$ can be given by $\delta^\nabla = (-1)^{p+1} * d^\nabla *$. If $R^\nabla_- \equiv 0$, then $\delta^\nabla R^\nabla = -*d^\nabla R^\nabla = 0$. Thus *self dual (anti-self dual) connections*, i.e., connections ∇ for which $R^\nabla_-(R^\nabla_+)$ vanishes, are Yang-Mills connections.

We now obtain a complex from (4.1) as follows. Suppose ∇ is self dual connection. Then the sequence

$$(4.2) \quad \Omega^0(\mathfrak{g}_E) \xrightarrow{d^\nabla} \Omega^1(\mathfrak{g}_E) \xrightarrow{d^\nabla_-} \Omega^2_-(\mathfrak{g}_E) \to 0$$

with d^∇_- the orthogonal projection of d^∇ onto $\Omega^2_-(\mathfrak{g}^E)$, is a complex since $d^\nabla_- \cdot d^\nabla(\sigma) = [R^\nabla_-, \sigma] = 0$ for $\sigma \in \Omega^0(\mathfrak{g}_E)$. So, by considering self dual connections (which are Yang-Mills connections), we can extract from (4.1) a complex, hence consider its cohomology groups H^0_∇, H^1_∇ and H^2_∇.

The complex (4.1) and its cohomology groups contain much information: First, if ∇ and ∇' are Riemannian connections in E, $\nabla - \nabla' \in \Omega^1(\mathfrak{g}_E)$, so that, as an affine space, the space $\mathscr{C}_E$ of Riemannian connections on E is isomorphic to $\Omega^1(\mathfrak{g}_E)$. Furthermore, if $\nabla' = \nabla + A$ for some $A \in \Omega^1(\mathfrak{g}_E)$, $R^{\nabla'} = R^\nabla + d^\nabla A + [A, A]$, where $[A, A]_{X, Y} \equiv [A_X, A_Y]$. Second, the tangent space to the orbit of the gauge group $\mathscr{G} = \Gamma(G_E)$ at ∇, considered as a subspace of $\Omega^1(\mathfrak{g}_E) \cong T_\nabla \mathscr{C}_E$, is the image $d^\nabla(\Omega^0(\mathfrak{g}_E))$. To see this, we can view $\Omega^0(\mathfrak{g}_E) = \Gamma(\mathfrak{g}_E)$ as the infinitesimal gauge transformations. So given an element $\sigma \in \Omega^0(\mathfrak{g}_E)$, consider the corresponding curve $g_t = \exp(t\sigma)$ in G_E and note that $(d/dt)\nabla^{g_t}|_{t=0} = [\nabla, \sigma] = \nabla(\sigma) = d^\nabla(\sigma)$. Thus $\ker \delta^\nabla$ can be thought of as the tangent space of $\mathscr{C}/\mathscr{G}$ at $[\nabla]$.

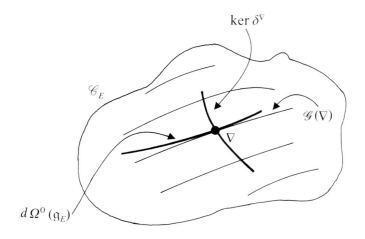

Third, if ∇ is self dual and $A \in \Omega^1(\mathfrak{g}_E)$, $\nabla + A$ is self dual if and only if $0 = R_{-}^{\nabla+A} = R_{-}^{\nabla} + d_{-}^{\nabla} A + [A, A]_{-} = d_{-}^{\nabla} A + [A, A]_{-}$. The linear part of this equation is $d_{-}^{\nabla} A = 0$. So, if we only consider linear information, a neighborhood of $[\nabla]$ in the moduli space $\mathscr{S}$ of gauge equivalence classes of self dual connections on M should be $\{A \in \Omega^1(\mathfrak{g}_E) | \delta^{\nabla} A = 0 \text{ and } d^{\nabla} A = 0\}$, that is, by Hodge theory, a neighborhood of 0 in H_{∇}^1.

What about the reducible Yang-Mills connections in $E = L \oplus \varepsilon$. It certainly is not the case that every harmonic 2-form is self dual. However, since the intersection pairing is positive (negative) definite on the self dual (anti-self dual) 2-forms, if the intersection form on M is positive definite, *every* harmonic 2-form is self dual. Thus under the assumption that the intersection form on M is positive definite (and $H^1(M; \mathbb{R}) = 0$, a fact we used in $\varepsilon 2$), we have that if m is half the number of solutions to

$$(e(L) + b) \cdot b = 0 \text{ for } b \in H^2(M; \mathbb{Z})$$

then contains m reducible connections.

It is from this point of view that in [FS1] we show, for instance, that $E_8 \oplus \Phi, \Phi$ any positive definite symmetric unimodular form, cannot occur as the intersection form on any closed smooth 4-manifold M with $H_1(M; \mathbb{Z})$ containing no 2-torsion. The outline of the proof is simple. Suppose such an M exists. We can assume $H_1(M; \mathbb{R}) = 0$, for surger out the free part of $H_1(M; \mathbb{Z})$ and note that the intersection form is unaffected. There exists an element $x \in H^2(M; \mathbb{Z})$ with $x^2 = 2$. Let L be the $SO(2)$ bundle over M with $e(L) = x$ and consider the $SO(3)$ bundle $E \cong L \oplus \varepsilon$. The work of K. Uhlenbeck ([U1], [U2]) can be used to show that $\mathscr{S}$ is *compact*. Then, using the work of Atiyah-Hitchin-Singer [AHS], we show that $\mathscr{S}$ is a manifold of dimension $2p_1(\mathfrak{g}_E) - 3 = 2p_1(E) - 3 = 2x^2 - 3 = 1$. $\mathscr{S}$ is then a disjoint union of circles and intervals whose endpoints correspond to the reducible self dual connections. But the solutions to $b \cdot (x + b) = 0$ is just the order of the torsion subgroup of $H^2(M; \mathbb{Z})$ (for $b \cdot (x + b) = 0$ if and only if $(x + 2b)^2 = x^2$ and $(x + 2b)^2 = (x + b + b)^2 = (x + b)^2 + (b)^2$. But then $b = \pm x$ or b is torsion since x is minimal, i.e., x cannot be written as $c + d$ with $c^2 < x^2$ or $d^2 < x^2$). Thus, by the universal coefficient theorem $m = |\text{tor } H^2(M; \mathbb{Z})| = |\text{tor } H_1(M; \mathbb{Z})|$. If $H_1(M; \mathbb{Z})$ has no 2-torsion, m is then odd. But intervals have an even number of end points, a contradiction!

References

[AHS]. M. F. Atiyah, N. J. Hitchin, and I. M. Singer, *Self-duality in 4-dimensional Riemannian Geometry*, Proc. Roy. Soc. London Ser. A. 362 (1978), 425—461.

[DW]. A. Dold and H. Whitney, *Classification of oriented sphere bundles over a 4-complex*, Ann. of Math. 69 (1959), 667—677.

[D]. S. Donaldson, *An application of gauge theory to four dimensional topology*, J. Diff. Geom. 18 (1983), 279—315.

[FS1]. R. Fintushel and R. J. Stern, *SO (3)-connections and the topology of 4-manifolds*, preprint.

[FS2]. R. Fintushel and R. J. Stern, *New invariants for homology 3-spheres*, preprint.

[MH]. J. Milnor and D. Husemoller, *Symmetric Bilinear Forms*, Springer-Verlag, New York 1973.

[R]. V. A. Rochlin, (1952) *New results in the theory of 4-dimensional manifolds* (Russian). Dokl. Akad. Nauk. SSSR, 84, 221—224.

[U1]. K. K. Uhlenbeck, *Removable singularities in Yang-Mills fields*, Comm. Math. Phys. 83 (1982), 11—30.

[U2]. K. K. Uhlenbeck, *Connections with L^p bounds on curvature*, Comm. Math. Phys. 83 (1982), 31—42.

[W]. J. H. C. Whitehead, *On simply connected 4-dimensional polyhedra*, Comment. Math. Helv. 22 (1949), 48—92.

Perspectives in Mathematics
Anniversary of Oberwolfach 1984
© Birkhäuser Verlag, Basel

Algebraische Berechnungskomplexität

VOLKER STRASSEN

Institut für Angewandte Mathematik der Universität Zürich,
Rämistrasse 74, CH-8001 Zürich (Switzerland)

1. Einleitung

Unter der Komplexität eines algorithmischen Problems versteht man im einfachsten Fall die minimale zu seiner Lösung hinreichende Anzahl von Rechenschritten. Ihr Studium ist der Gegenstand der Komplexitätstheorie.

Jeder Algorithmus liefert eine obere Schranke für die Komplexität der von ihm gelösten Aufgabe. Daher besitzt das Gebiet keine scharfe Grenze zur Numerik und zu denjenigen Teilen der Informatik, die sich mit dem Entwurf und der Analyse von Algorithmen befassen. Es gewinnt seine Identität erst durch die Betonung von unteren Komplexitätsschranken (bis hin zu Optimalitätsbeweisen für manche Verfahren).

Die Komplexitätstheorie vereint zwei ganz verschiedene Traditionen. Die eine entstammt der mathematischen Logik und der Theorie der berechenbaren Funktionen. Das grundlegende Berechnungsmodell ist hier die Turingmaschine, und man zeigt, daß z.B. gewisse Entscheidungsprobleme inhärent schwierig sind, auf ganz ähnliche Art, wie man in der mathematischen Logik die Unentscheidbarkeit von Theorien beweist: durch Diagonalisierung und durch Reduktion eines Problems auf ein anderes. Diese Turing-Komplexität wird z.B. in den Büchern Aho-Hopcroft-Ullman [2], Specker-Strassen [139], Mehlhorn [95], Paul [111], Machtey-Young [92] und Garey-Johnson [49] behandelt.

Die zweite Tradition hat sich aus Fragestellungen der numerischen Mathematik entwickelt. Das Berechnungsmodell für die hier betrachteten Probleme ist so etwas wie ein gewöhnlicher (endlicher) Computer, der allerdings mit der Fähigkeit ausgestattet ist, eine arithmetische Grundoperation in einem Schritt mit absoluter Genauigkeit auszuführen, und von dem verlangt wird, das Ergebnis ebenfalls ohne Exaktheitseinbuße abzuliefern. (Eine solche Idealisierung scheint für das Rechnen in endlichen Körpern ohne weiteres vernünftig. Sie liegt aber auch dem üblichen Konzept des numerischen Rechnens zugrunde, bei dem die mit der endlichen Genauigkeit zusammenhängenden Probleme durch zusätzliche Stabilitätsforderungen berücksichtigt werden. Die Korrektheit von unteren Schranken für die Komplexität wird durch diese Forderungen natürlich nicht berührt. Einen anderen, eher zur Turing-Tradition gehörenden Standpunkt nimmt Schönhage [133] ein.) Weil bei der Konstruktion von Algorithmen ebenso wie beim Beweis von unteren Schranken vor allem algebraische Methoden zur Anwendung kommen, spricht man von „algebraischer" Komplexitätstheorie. Sie ist Gegenstand des Buches Borodin-Munro [24]. Auch Knuth [81] und Aho-Hopcroft-Ullman [2] enthalten einschlägige Kapitel.

Der vorliegende Bericht ist hauptsächlich der zweiten Tradition gewidmet. Wir geben in den Abschnitten 16–18 zwar einen Überblick über denjenigen Teil der Turing-Komplexitätstheorie, der seinen Schatten auf das algebraische Gebiet zu werfen beginnt (Abschnitt 19); unsere Darstellung ist in jenen Abschnitten aber viel großzügiger als an anderen Stellen.

Die Begrenzung von Zeit, Platz und Kompetenz (welche unseren Gegenstand ja auch inhaltlich prägt) hat dazu geführt, daß wir nicht einmal die algebraische Komplexität mit einiger Vollständigkeit behandeln konnten. So fehlt eine Diskussion Boolescher Funktionen und ihrer Realisierung durch logische Netze (siehe Savage [122]) ebenso wie eine Behandlung des parallelen Modells. Auf der anderen Seite haben wir eine Reihe von ungelösten Problemen eingefügt. Sie richten sich in erster Linie an den Fachmann, sollen aber auch dem Außenstehenden einen Eindruck von der Lebendigkeit unseres Gebietes vermitteln.

Beim Leser wird nicht mehr als eine gute algebraische Allgemeinbildung vorausgesetzt, stellenweise eine gewisse Vertrautheit mit der Sprache der klassischen algebraischen Geometrie.

Die nun folgenden Abschnitte 2–19 erlauben eine natürliche Gliederung in weitgehend voneinander unabhängig lesbare Teile: Probleme mit allgemeinen Koeffizienten 2–5, Probleme hohen Grades 6–9, Probleme niedrigen Grades 10–15, vollständige Probleme 16–19. Bei Verweisungen wird die Nummer des Ziel-Abschnitts beigefügt: Satz 15 bezeichnet also den Satz des Abschnitts 15 (einen Satz, den es bedauerlicherweise nicht gibt).

Michael Clausen danke ich herzlich für seine Hilfe bei der Fertigstellung des Manuskripts.

2.   Additionsketten

1937 erschien eine kleine Note von A. Scholz [126] im Jahresbericht der DMV. Darin nennt Scholz eine Folge $1 = a_0 < \cdots < a_r = n$ ganzer Zahlen eine Additionskette für n, gewenn*) jedes a_i mit $i \geqq 1$ Summe zweier vorhergehender Elemente der Folge ist: $a_i = a_j + a_k$ für geeignete $j \leqq k < i$. Er bezeichnet mit $l(n)$ die kleinste Länge r einer Additionskette für n. $l(n)$ wird motiviert als der optimale Rechenaufwand zur Berechnung der Potenz x^n, wenn nur Multiplikationen zugelassen sind.

Diese Arbeit scheint der erste Beleg für den Begriff der Berechnungsfolge im Zusammenhang mit einer numerischen Fragestellung zu sein. A. Scholz war Zahlentheoretiker. Sein Göschenbändchen „Einführung in die Zahlentheorie"

---

*) ‚gewenn = genau dann, wenn (E. Specker).

kündet vom ungewöhnlichen Interesse des Autors an algorithmischen Problemen. Wir wollen gleich bemerken, daß der Berechnungsbegriff bereits 1931 von Turing und anderen auf eine viel umfassendere und grundsätzlichere Weise als in der Arbeit von Scholz analysiert worden war, allerdings auch weiter entfernt von den Bedürfnissen des numerischen Rechnens.

Kehren wir für den Augenblick zur Funktion l zurück. Die entscheidende Einsicht war bald gewonnen: $l(n) \sim \log n$*) (Scholz [126], A. Brauer [25]). Die Feinstruktur von l eröffnet aber dem mathematischen Tüftler eine Fülle von reizvollen und schwierigen Problemen. (Siehe Knuth [81], Schönhage [129].)

3.   Berechnungsfolgen

17 Jahre nach Scholz – das Zeitalter der programmierbaren Computer hatte bereits begonnen – veröffentlichte A. Ostrowski seine Arbeit [104] „On two problems in abstract algebra connected with Horner's rule", die zum Ausgangspunkt zahlreicher Untersuchungen über die Komplexität algebraischer Probleme wurde.

Bekanntlich läßt sich ein Polynom $a_0 x^n + a_1 x^{n-1} + \cdots + a_n$ nach Horner mit n Additionen und n Multiplikationen auswerten. Ostrowski stellt nun die Frage nach der Optimalität der Hornerschen Regel und äußert unter anderem die Vermutung, daß es kein allgemeines Auswertungsverfahren gibt, welches weniger als n Multiplikationen/Divisionen benötigt. Additionen und Subtraktionen werden dabei nicht gezählt. Von guter mathematischer Intuition geleitet erlaubt Ostrowski darüber hinaus die kostenlose Multiplikation mit Skalaren. (Skalare denke man sich als im Computerprogramm gespeicherte Konstante.)

Eine genaue Formulierung der obigen Vermutung erfordert natürlich die Angabe eines Berechnungsmodells. Wir folgen Ostrowskis Gedankengang, indem wir die Zwischenresultate eines Berechnungsverfahrens als Funktionen der Inputvariablen ansehen (in unserem Fall der Berechnung eines „allgemeinen" Polynoms also von $a_0, \ldots, a_n, x$) und nur die Ergebnisse zählender Rechenschritte registrieren.

Definition: Seien k ein unendlicher**) Körper, $x_1, \ldots x_m$ Unbestimmte über k. Eine Folge $(g_1, \ldots, g_r)$ aus $k(x_1, \ldots, x_m)$ heißt Berechnungsfolge, gewenn es zu jedem $i \leq r$ eine Darstellung

$$g_i = u_i v_i \quad \text{oder} \quad g_i = u_i / v_i$$

mit $u_i, v_i \in k + kx_1 + \cdots + kx_m + kg_1 + \cdots + kg_{i-1}$ gibt.

---

*) log bedeute in der vorliegenden Arbeit stets $\log_2$.
**) Diese Annahme wird hier und im weiteren hauptsächlich aus Gründen der Bequemlichkeit gemacht. Siehe auch [146], [148].

Für Horners Regel ist also $r = n$, $g_i = a_0 x^i + \cdots + a_{i-1} x \in k(a_0, \ldots, a_n, x)$, $u_i = g_{i-1} + a_{i-1}$ und $v_i = x$.

Definition: Seien $f_1, \ldots, f_p \in k(x_1, \ldots, x_m)$. Die (nichtskalare) Komplexität $L(f_1, \ldots, f_p)$ ist die kleinste natürliche Zahl r, für die es eine Berechnungsfolge $(g_1, \ldots, g_r)$ gibt mit

$$f_1, \ldots, f_p \in k + k x_1 + \cdots + k x_m + k g_1 + \cdots + k g_r.$$

Horners Regel liefert eine obere Schranke für die Komplexität:

$$L(a_0 x^n + \cdots + a_n) \leqq n$$

Ostrowskis Vermutung lautet

$$L(a_0 x^n + \cdots + a_n) = n.$$

4.   Substitution

Ein Beweis dieser Vermutung gelang zwölf Jahre später V. Pan [105]. Die dabei verwendete Substitutionsmethode wurde von Winograd [157], Strassen [143] und Hartmann-Schuster [63] weiter ausgestaltet. Wir erläutern sie am Beispiel der Quadratsumme $f = x_1^2 + \cdots + x_m^2 \in \mathbb{R}[x_1, \ldots, x_m]$, bei deren Berechnung wir der Einfachheit halber keine Divisionen zulassen wollen. Ist $(g_1, \ldots, g_r)$ eine optimale Berechnungsfolge, so sind u_1 und v_1 nichtkonstante lineare Polynome. Kommt z. B. x_m in u_1 vor, so gibt es eine affin lineare Substitution $x_m \mapsto \sum_{j=1}^{m-1} \lambda_j x_j + \lambda$, die u_1 und damit g_1 annulliert. Die Substitution ist ein Endomorphismus von $\mathbb{R}[x_1, \ldots, x_m]$, welcher $(g_1, \ldots, g_r)$ in eine Berechnungsfolge abbildet, deren erstes Element überflüssig ist. f wird damit in ein Polynom kleinerer Komplexität verwandelt. Nun sieht man leicht, daß die Quadratsumme nicht mit weniger als m solcher Substitutionen linearisiert werden kann, woraus zusammen mit der trivialen Abschätzung nach oben

$$L_{\mathbb{R}}(x_1^2 + \cdots + x_m^2) = m \tag{4.1}$$

folgt. Das Beispiel läßt sich verallgemeinern: Ist $\operatorname{char}(k) \neq 2$ und $f \in k(x_1, \ldots, x_m)$ eine beliebige quadratische Form, so gilt

$$L(f) = n - q,$$

wo n der Rang von f und q die Dimension der maximalen Nullräume von f ist. So haben wir z. B.

$$L_{\mathbb{C}}(x_1^2 + \cdots + x_m^2) = \lceil m/2 \rceil.$$

Der Vergleich mit (4.1) zeigt, daß sich bei Erweiterung des Grundkörpers k die Komplexität verkleinern kann. (Dieser Effekt tritt allerdings bei algebraisch abgeschlossenem Grundkörper nicht auf. Hingegen genügt die relative algebraische Abgeschlossenheit des Körpers in seiner Erweiterung nicht!) Die Anwendung der Substitutionsmethode auf das Polynom $a_0 x^n + \cdots + a_n \in k(a_0, \ldots, a_n, x)$ trifft noch auf die Schwierigkeit, daß eine Substitution von x offenbar alles verscherzen würde. Pan umgeht dieses Problem durch eine besondere Behandlung des Zwischenkörpers $k(x)$.

Als weiteres Beispiel betrachten wir die Auswertung einer „allgemeinen" rationalen Funktion vom Grad n

$$f_1 = \frac{a_0 x^n + \cdots + a_n}{x^n + b_1 x^{n-1} + \cdots + b_n} \in k(a_0, \ldots, a_n, b_1, \ldots, b_n, x). \tag{4.2}$$

Man erhält hier wieder das zu erwartende Resultat

$$L(f_1) = 2n.$$

Eine rationale Funktion läßt sich aber auch in einen Partialbruch oder in einen Kettenbruch entwickeln. Beschränken wir uns der Einfachheit halber auf die über algebraisch abgeschlossenen Grundkörpern im Normalfall zu erwartenden Formen solcher Entwicklungen, nämlich

$$f_2 = a_0 + \frac{a_1}{x - b_1} + \cdots + \frac{a_n}{x - b_n}$$

und

$$f_3 = a_0 + \cfrac{a_1}{(x + b_1) + \cfrac{a_2}{(x + b_2) + \cdots + \cfrac{a_n}{x + b_n}}},$$

so liefert die Substitutionsmethode

$$L(f_2) = L(f_3) = n.$$

Bei diesen Entwicklungen genügt zur Auswertung also genau die halbe Anzahl von Multiplikationen/Divisionen wie bei der natürlichen Darstellung (4.2).

Folgende Fragen bleiben zunächst unbeantwortet:

Gibt es noch auswertungseffizientere Darstellungen von rationalen

Funktionen als die Partial- und Kettenbruchentwicklung, oder von Polynomen als die durch ihren Koeffizientenvektor?

Wie verhält es sich in den betrachteten Fällen mit der Anzahl der Additionen/Subtraktionen?

Problem: Man bestimme die Komplexität eines Newtonschritts für die Nullstellenapproximation eines allgemeinen Polynoms:

$$L_{\mathbb{C}}\left(x - \frac{a_0 x^n + \cdots + a_n}{n a_0 x^{n-1} + \cdots + a_{n-1}}\right).$$

5.   Transzendenzgrad

Überraschenderweise wurden die obigen Fragen schon früher geklärt als die Vermutung von Ostrowski (Motzkin [102]; siehe auch Belaga [11], Reingold-Stocks [117]).

Sei K ein unendlicher Körper und $f \in K(x)$. Wir bezeichnen mit $L_+(f)$ die minimale Anzahl von Additionen/Subtraktionen, die zur Auswertung von f hinreicht. Multiplikationen, Divisionen und der Zugriff auf beliebige Konstante aus K gelten dabei als kostenlos. $L_*(f)$ ist analog definiert: Hier werden Multiplikationen/Divisionen einschließlich Skalarmultiplikationen gezählt. $L_+(f)$ heißt additive, $L_*(f)$ multiplikative Komplexität von f.

Für $u, v \in K[x]$ sei $T(u, v)$ der Transzendenzgrad der Menge der Koeffizienten von u und v über dem Primkörper von K.

Satz: Sind $u, v \in K[x]$ teilerfremd und hat v den führenden Koeffizienten 1, so gilt

$$L_+(u/v) \geqq T(u, v) - 1, \quad L_*(u/v) \geqq \frac{1}{2}\left(T(u, v) - 1\right).$$

Hiermit lassen sich unsere Fragen leicht beantworten. Um z. B. die additive Komplexität von $a_0 x^n + \cdots + a_n \in k(a_0, \ldots, a_n, x)$ nach unten abzuschätzen, vergrößern wir den Grundkörper k zu $K = (a_0, \ldots, a_n)$ (wodurch sich die Komplexität höchstens verkleinert) und wenden Satz 5 auf $a_0 x^n + \cdots + a_n \in K(x)$ an. Zusammen mit der trivialen oberen Schranke ergibt sich

$$L_+(a_0 x^n + \cdots + a_n) = n$$

und damit die vollständige Optimalität von Horners Regel.

Für die drei oben betrachteten Darstellungen rationaler Funktionen vom Grad n erhalten wir

$$L_+(f_i) = 2n.$$

Zum Beweis setzen wir $K = k(a_0, \ldots, a_n, b_1, \ldots, b_n)$ und bemerken, daß die Koeffizientenmenge des als Polynomquotient geschriebenen f_i mindestens den Transzendenzgrad $2n+1$ über k hat. Ebenso ergibt sich

$$L_*(f_1) \geq L_*(f_2) = L_*(f_3) = n,$$

was freilich schon aus den Resultaten des letzten Abschnitts zusammen mit $L_* \geq L$ folgt. Die jetzigen Beweise sind allerdings robuster als die dort verwendeten, da sie (wegen der Vergrößerung des Grundkörpers) eine „Koeffizientenvorbereitung" zulassen. Nun kann man jede Darstellung von Polynomen oder rationalen Funktionen als eine solche Koeffizientenvorbereitung interpretieren. Aus diesem Grunde besitzen Partial- und Kettenbruchentwicklung tatsächlich optimale Effizienz. Dagegen ist die natürliche Darstellung nicht einmal für Polynome auswertungsoptimal. (Siehe Belaga [11], Pan [105].)

Der Satz dieses Abschnitts läßt sich ohne weiteres auf mehrere Unbestimmte $x_1, \ldots, x_m$ ausdehnen. Eine wesentlichere Verallgemeinerung stammt von Baur-Rabin [9]: Ist k irgendein Unterkörper von K (z. B. der Primkörper), so sind K und $k(x)$ linear disjunkt über k. In der genannten Arbeit wird $K(x)$ durch eine beliebige Körpererweiterung $k \subset \Omega$ ersetzt, die mit zwei über k linear disjunkten Zwischenkörpern ausgestattet ist.

6.   Geometrischer Grad

Wir wollen als nächstes eine Methode beschreiben, welche es erlaubt, die nichtskalare Komplexität komplizierterer Berechnungsaufgaben brauchbar nach unten abzuschätzen.

Wir nehmen an, k sei algebraisch abgeschlossen und gehen aus von rationalen Funktionen $f_1, \ldots, f_p \in k(x_1, \ldots, x_m)$. Mit W bezeichnen wir den Graph der zugehörigen rationalen Abbildung von k^m nach k^p. W ist eine im Sinne der Zariski-Topologie lokal abgeschlossene irreduzible Teilmenge von k^{m+p} und besitzt als solche einen Grad deg W. (deg W ist die maximale endliche Schnittpunktzahl von W mit einem affin-linearen Raum der Dimension p.)

Satz: $L(f_1, \ldots, f_p) \geq \log \deg W$.

Sind z. B. $f_1, \ldots, f_m$ Polynome in $x_1, \ldots, x_m$ (also $p = m$) und hat das Gleichungssystem

$$f_1(\xi) = \cdots = f_m(\xi) = 0$$

genau N Lösungen $\xi \in k^m$, so gilt $L(f_1, \ldots, f_m) \geq \log N$.

Zum Beweis des Satzes verfolgt man eine hypothetische Berechnung von $f_1, \ldots, f_p$ sozusagen graphisch, deutet die arithmetischen Operationen als

Schnittbildungen mit Hyperflächen vom Grad 1 oder 2 nebst nachfolgender Projektion und wendet den Satz von Bézout an (Strassen [145], siehe auch Schönhage [130]).

Wie es scheint, ist das natürliche Anwendungsgebiet von Satz 6 die algebraische Manipulation von Polynomen und rationalen Funktionen. Da diese auf verschiedene Weise gegeben sein können, spielt auch der Übergang von einer Darstellung zu einer andern eine Rolle. Nur die einfachsten solcher Aufgaben haben eine (nicht-skalare) Komplexität, die sich durch eine lineare Funktion des Grades der betrachteten Polynome abschätzen läßt: Die Multiplikation (d. h. die Berechnung der Koeffizienten des Produktpolynoms), die Division mit Rest (Sieveking [137], Kung [82], Strassen [145]), die Entwicklung nach Potenzen von $x - a$ (Shaw-Traub [136]).

Betrachten wir nun die Berechnung der Koeffizienten eines Polynoms in einer Variablen aus seinen Nullstellen, d. h. bis auf das Vorzeichen die Berechnung der elementarsymmetrischen Funktionen $\sigma_1, \ldots, \sigma_m \in k(x_1, \ldots, x_m)$. Sind $\lambda_1, \ldots, \lambda_m \in k$, so ist das Gleichungssystem

$$\sigma_1(\xi) - \lambda_1 = \cdots = \sigma_m(\xi) - \lambda_m = 0$$

äquivalent zu der einen Gleichung

$$(t - \xi_1) \cdots (t - \xi_m) = t^m - \lambda_1 t^{m-1} + \cdots + (-1)^m \lambda_m$$

in $k[t]$, hat also bei geeigneter Wahl der λ_i genau $m!$ Lösungen $\xi \in k^m$. Es folgt $L(\sigma_1, \ldots, \sigma_m) = L(\sigma_1 - \lambda_1, \ldots, \sigma_m - \lambda_m) \geqq \log m!$ und damit die untere Schranke in

$$m(\log m - 2) \leqq L(\sigma_1, \ldots, \sigma_m) \leqq m \log m. \tag{6.1}$$

(Die obere Schranke ergibt sich aus einer induktiven Halbierung von $(t - x_1) \cdots (t - x_m)$.)

Ein anderes häufig auftretendes Problem ist die Auswertung eines Polynoms $f = a_0 t^n + \cdots + a_n$ vom Grad n an vielen (etwa $n+1$) Stellen, d. h. die Berechnung von $f(x_0), \ldots, f(x_n) \in k(a_0, \ldots, a_n, x_0, \ldots, x_n)$. Man möchte vermuten, daß die separate Auswertung des Polynoms an den einzelnen x_i mittels der Hornerschen Regel optimal ist. Ein solches Verfahren hat den nichtskalaren Aufwand $n(n+1)$. Überraschenderweise gibt es viel schnellere Algorithmen (Borodin-Munro [23], Fiduccia [42], Moenck-Borodin [100], Strassen [145]): Man denke sich die Inputvariablen $a_0, \ldots, a_n, x_0, \ldots, x_n$ als konkrete Zahlen aus k. Wir nehmen an, n sei ungerade, etwa $n = 2q+1$ und versuchen, f zunächst nur an den Stellen $x_0, \ldots, x_q$ auszuwerten. Division mit Rest

$$f(t) = g(t)(t - x_0) \cdots (t - x_q) + h(t)$$

liefert ein Polynom h vom Grad $q = \dfrac{n-1}{2}$, welches für $x_0, \ldots, x_q$ die gleichen Werte annimmt wie f. Wir können also zunächst die Koeffizienten von $(t-x_0)\cdots(t-x_q)$ berechnen (elementarsymmetrische Funktionen!), dann die Division mit Rest ausführen und schließlich das Polynom h an den Stellen $x_0, \ldots, x_q$ auswerten. Gehen wir bei den übrigen Stellen $x_{q+1}, \ldots, x_n$ analog vor, so haben wir die ursprüngliche Aufgabe zurückgeführt auf 2 Probleme des gleichen Typs, aber halber Größe. Man kann das Verfahren rekursiv anwenden und erhält für den nichtskalaren Gesamtaufwand γ_n die Rekursionsformel $\gamma_{2q+1} \leq \mathrm{const}\, q \log q + 2\gamma_q$, also $\gamma_n = 0\big(n(\log n)^2\big)$. Ein etwas sorgfältigeres Vorgehen ergibt die obere Schranke in

$$(n+1) \log n \leq L\big(f(x_0), \ldots, f(x_n)\big) = 0(n \log n). \tag{6.2}$$

Wie zeigt man die untere Schranke? Das Gleichungssystem

$$f(x_0) = \cdots = f(x_n) = a_0 - 1 = a_1 = \cdots = a_{n-2} = a_{n-1} - \alpha = a_n - \beta = 0$$

mit geeigneten $\alpha, \beta \in k$ hat n^{n+1} Lösungen in k^{2n+2}, also $L\big(f(x_0), \ldots, f(x_n)\big)$ $= L\big(f(x_0), \ldots, f(x_n), a_0 - 1, a_1, \ldots, a_{n-2}, a_{n-1} - \alpha, a_n - \beta\big) \geq \log(n^{n+1})$.

Kehrt man das gerade behandelte Problem um, so erhält man die Interpolationsaufgabe: Gegeben sind Stützstellen $x_0, \ldots, x_n$ und Werte $y_0, \ldots, y_n$. Zu bestimmen sind die Koeffizienten des Polynoms p vom Grad n mit $p(x_i) = y_i$ für alle i. Da dieses Problem invers zum vorherigen ist, hat es im wesentlichen den gleichen Graphen wie jenes. Seine Komplexität ist also mindestens $(n+1)\log n$. Andererseits kann man die Koeffizienten des Interpolationspolynoms mit nichtskalarem Aufwand $0(n \log n)$ berechnen (Horowitz [69]). Ein ähnliches Ergebnis erhält man, wenn man statt der Koeffizienten die baryzentrischen Koordinaten $w_0, \ldots, w_n$ von p im Sinne von Rutishauser [121] zu berechnen sucht. Auf die Frage einer möglichst effizienten Darstellung des Interpolationspolynoms gehen wir am Ende des nächsten Abschnitts ein.

Die Voraussetzung der algebraischen Abgeschlossenheit, welche wir bei der Formulierung von Satz 6 gemacht haben, können wir bei seinen Anwendungen wieder fallen lassen, da sich die Komplexität durch eine Körpererweiterung höchstens verkleinert.

Läßt sich die Gradmethode auf die additive Komplexität anwenden? Das Polynom $x^n - 1 \in \mathbb{C}[x]$ scheint diese Hoffnung zu zerstören. Für $f \in \mathbb{R}[x]$ haben jedoch Borodin-Cook [22], Grigoryev [53] und Risler [119]

$$L_+(f) \geq c \sqrt{\log N} \tag{6.3}$$

bewiesen, wo N die Anzahl der reellen Nullstellen von f und $c > 0$ eine universelle Konstante ist. (6.3) beruht auf einem Satz von Hovanskii [70] über die Nullstellen eines reellen Gleichungssystems mit wenigen Monomen.

Durch Anwendung einer Möbiustransformation auf $x^n - 1 \in \mathbb{C}[x]$ hat Van de Wiele [156] ein reelles Polynom f vom Grade n konstruiert, welches sich über $\mathbb{C}$ mit 3 Additionen/Subtraktionen berechnen läßt, jedoch n verschiedene reelle Nullstellen besitzt. Nach (6.3) ist $L_+(f) \geq c \sqrt{\log n}$ über $\mathbb{R}$. Eine Körpererweiterung vom Grad 2 kann also katastrophale Folgen für die additive Komplexität haben.

Problem 1: Man entscheide, ob mit den obigen Bezeichnungen und einer universellen Konstante $c_1 > 0$

$$L_+(f) \geq c_1 \log N \tag{6.4}$$

gilt. Diese Ungleichung ist eine Konsequenz aus der Richtigkeit der folgenden Vermutung von Kusnirenko (siehe Hovanskii [70]):

Die Anzahl der nicht entarteten Nullstellen mit positiven Koordinaten eines Systems $p_1, \ldots, p_n \in \mathbb{R}[x_1, \ldots, x_n]$ ist höchstens $\prod_{i=1}^{n} (m_i - 1)$, wo m_i die Anzahl der Monome von p_i bezeichnet.

Die Tschebyschew-Polynome T_{3^k} zeigen, daß (6.4) das beste Ergebnis seiner Art wäre, welches man erwarten kann (Borodin-Cook [22], Van de Wiele [156]). Das Interesse an der immer noch recht kleinen unteren Schranke (6.4) rührt daher, daß ihre Gültigkeit auf die Möglichkeit hindeuten würde, in Analogie zur Gradmethode $n \log n$-Schranken für die additive Komplexität von Systemen reeller Polynome in n Variablen zu beweisen.

Problem 2: Man finde heraus, wieviele ganzzahlige Nullstellen ein Polynom $f \in \mathbb{Q}[x]$ mit $L_+(f) \leq r$ (oder $L(f) \leq r$) höchstens besitzen kann.

7.   Ableitungen

Die Methode des letzten Abschnitts ergibt für die Komplexität eines einzelnen Polynoms $f \in k(x_1, \ldots, x_m)$ nur die triviale untere Schranke $L(f) \geq \log \deg f$. Läßt sich die Gradmethode hier auch auf eine wirksamere Art ausnutzen? Diese Fragestellung wurde zuerst in Schnorr [124] untersucht. Die folgende Ungleichung (Baur-Strassen [10]) liefert zusammen mit dem Satz des letzten Abschnitts in vielen Fällen untere Schranken in der richtigen Größenordnung.

Satz: Für $f \in k(x_1, \ldots, x_m)$ gilt

$$L\left(f, \frac{\partial f}{\partial x_1}, \ldots, \frac{\partial f}{\partial x_m}\right) \leq 3 L(f).$$

Der Beweis dieses Satzes ist konstruktiv. Man beachte, daß die Form der Ungleichung unabhängig von der Anzahl der Variablen ist und daß an den Grundkörper keine Voraussetzungen gemacht werden. Ähnliche Ungleichungen gelten auch bei Zählung aller Multiplikationen/Divisionen oder aller arithmetischen Operationen.

Als Anwendung betrachten wir zunächst die Potenzsumme $f = x_1^n + \cdots + x_m^n$, wobei wir annehmen wollen, daß n kein Vielfaches von $\mathrm{char}(k)$ sei. Offenbar ist $L(f) \leq m\,l(n)$ mit der Scholz-Funktion l (Abschnitt 2). Andererseits haben wir $\dfrac{\partial f}{\partial x_i} = n x_i^{n-1}$, also $L(f) \geq \dfrac{1}{3} L(x_1^{n-1}, \ldots, x_m^{n-1}) = \dfrac{1}{3} L(x_1^{n-1} + \alpha x_1 + \beta,$

$\ldots, x_m^{n-1} + \alpha x_m + \beta) \geq \dfrac{1}{3} m \log(n-1)$ (wie im Beweis von (6.2)). Damit ist

$$L(x_1^n + \cdots + x_m^n) \asymp m \log n$$

gezeigt. ($\asymp$ bedeute Gleichheit der Größenordnungen.)

Etwas mühsamer ist der Beweis von

$$L(\sigma_q) \asymp m \log \min\{q, m-q\},$$

wo σ_q die q-te elementarsymmetrische Funktion in m Variablen und $q < m-1$ ist. Die Berechnung eines „mittleren" Koeffizienten aus den Nullstellen eines Polynoms braucht also ungefähr so lange wie die Berechnung sämtlicher Koeffizienten.

Auch Resultante und Diskriminante als Funktionen der Wurzeln der beteiligten Polynome fallen in den Anwendungsbereich der beschriebenen Methoden:

$$L\left(\prod_{i,\,j=1}^{m} (x_i - y_j)\right) \asymp L\left(\prod_{i \neq j} (x_i - x_j)\right) \asymp m \log m. \qquad (7.1)$$

Insbesondere hat die Vandermonde-Determinante $\det[x_i^j]_{0 \leq i,\,j \leq m-1}$ eine Komplexität wenigstens der Größenordnung $m \log m$, da ihr Quadrat die Diskriminante ist. Der beste uns bekannte Algorithmus liefert freilich nur $L(\det[x_i^j]) = 0 \, (m(\log m)^2)$.

Schließlich noch ein schönes Anwendungsbeispiel von Stoss [141]: Sei $\mathrm{char}\,k = 0$. Sind $x_0, \ldots, x_n, y_0, \ldots, y_n, x$ Unbestimmte über k und ist $p(x)$ das Interpolationspolynom zu den (x_i, y_i), so gilt

$$L(p(x)) \geq \frac{1}{3} n \log n - n.$$

Die Komplexität der Auswertung des durch die x_i, y_i gegebenen Interpolationspolynoms an einer einzigen weiteren Stelle hat also schon die Größenordnung $n \log n$. Daraus folgt: Jede in linearer Zeit auswertbare Darstellung von p benötigt zu ihrer Herstellung einen Aufwand der Größenordnung $n \log n$. Koeffizientendarstellung und baryzentrische Darstellung haben also in diesem Sinne optimale Effizienz.

Problem 1: Man bestimme die Größenordnung der Komplexität von Resultante und Diskriminante (als Funktionen der Polynomkoeffizienten).

Problem 2: Man finde eine brauchbare Verallgemeinerung von Satz 7 auf mehrere Funktionen oder höhere (etwa zweite) Ableitungen. Diese beiden Fragen sind äquivalent: Die zweiten Ableitungen von f sind die ersten Ableitungen von $\dfrac{\partial f}{\partial x_1}, \ldots, \dfrac{\partial f}{\partial x_m}$, und die ersten Ableitungen von $f_1, \ldots, f_p$ kommen unter den zweiten Ableitungen von $a_1 f_1 + \cdots + a_p f_p$ vor. (Die a_i sind zusätzliche Unbestimmte.) Man beachte: Für $f = x_1 \cdots x_m$ ist $L(f) = m - 1$ (Substitution), aber

$$L\left(\left\{\frac{\partial^2 f}{\partial x_i \partial x_j} : 1 \leqq i, j \leqq m\right\}\right) \geqq \binom{m}{2},$$

wie ein einfaches Dimensionsargument zeigt.

8.   Verzweigungen

Übersetzt man eine Berechnungsfolge (im Sinne von Abschnitt 3) in ein Computerprogramm und wendet dieses auf einen konkreten Input ξ an, so ist es möglich, daß die Rechnung wegen einer verlangten Division durch 0 blockiert wird, obschon die zu berechnenden Funktionen $f_1, \ldots, f_p$ an der Stelle ξ definiert sind. Will man sich nicht damit zufrieden geben, daß dies nur auf einer „dünnen" Menge von Inputs vorkommen kann, so muß man das Programm mit Hilfe von Verzweigungen so vervollständigen, daß es für jeden Input im Definitionsbereich der f_i das richtige Ergebnis liefert (und sonst etwa $\emptyset$ ausdruckt). Um die Betrachtungen vom Grundkörper unabhängig zu machen, kommen als Verzweigungen nur Nulltests in Frage, d. h. Anweisungen der Form

$$if \ a = 0 \ then \ goto \ i \ else \ goto \ j. \tag{8.1}$$

(Dabei bezeichnet a ein jeweils schon berechnetes Zwischenergebnis.)
Es ist vernünftig, die Möglichkeit solcher Verzweigungen auch ins abstrakte Berechnungsmodell zu übernehmen. Aus einer Berechnungsfolge wird so ein „Berechnungsbaum". Das ist ein (endlicher) binärer Baum, an dessen einfachen Knoten arithmetische Anweisungen und an dessen Verzweigungsknoten Anweisungen der Form (8.1) stehen. (Für eine genaue Definition siehe z. B.

[148].) Ein Input ξ gibt Anlaß zu einem Weg von der Wurzel bis zu einem Blatt des Baumes, wo angegeben ist, welche Zwischenresultate den Output bilden. Wir verlangen jetzt natürlich, daß eine Division durch 0 auch für spezielle Inputwerte ausgeschlossen ist. Als Kosten des Berechnungsbaumes für den Input ξ bezeichnen wir die Anzahl der nichtskalaren Operationen und der Verzweigungen, welche längs dem durch ξ bestimmten Weg auszuführen sind. Das Maximum der Kostenfunktion heißt Aufwand des Baumes. Damit können wir ein neues Komplexitätsmaß $C(f_1, \ldots, f_p)$ als den minimalen Aufwand von Berechnungsbäumen zur Berechnung von $f_1, \ldots, f_p$ einführen. Wir nennen C die Verzweigungskomplexität.

Man sieht leicht, daß

$$L(f_1, \ldots, f_p) \leqq C(f_1, \ldots, f_p) \tag{8.2}$$

gilt. (Von einem Berechnungsbaum für $f_1, \ldots, f_p$ darf man annehmen, daß keine Verzweigung überflüssig ist. Wählt man aus einem solchen Baum den „dicken" Weg aus, in dem jede Verzweigung (8.1) im Sinne von $a \neq 0$ durchlaufen wird, so erhält man wegen der „Irrelevanz algebraischer Ungleichungen" eine Berechnungsfolge für $f_1, \ldots, f_p$.) Aus (8.2) ergibt sich, daß alle bisher für die nichtskalare Komplexität bewiesenen unteren Schranken auch für die Verzweigungskomplexität gültig bleiben.

Es scheint, daß C in konkreten Fällen wesentlich größer sein kann als L (siehe z. B. Keller [78]). Das liegt daran, daß vom Berechnungsbaum die lückenlose Auswertung der Funktionen f_i verlangt wird, von der Berechnungsfolge nicht. (Natürlich fällt L mit C zusammen, wenn keine Divisionen zugelassen sind.)

Die größere Leistungsfähigkeit des in diesem Abschnitt eingeführten Berechnungsmodells zeigt sich darin, daß es nicht nur auf die Auswertung endlicher Folgen von rationalen Funktionen zugeschnitten, sondern auf eine größere Klasse von Berechnungsaufgaben anwendbar ist. Hierzu ein Beispiel:

Wir haben in Abschnitt 4 die (nichtskalare) Komplexität der Auswertung eines Kettenbruches für den Fall angegeben, daß dieser die über algebraisch abgeschlossenen Grundkörpern zu erwartende Normalgestalt besitzt. (Ein entsprechendes Resultat läßt sich mit der dort verwendeten Substitutionsmethode auch für die Auswertung eines ganz beliebigen Kettenbruchs beweisen.) Wir wollen jetzt die Entwicklung eines Polynomquotienten in einen Kettenbruch untersuchen. (Über die algorithmische Bedeutung der Kettenbruchentwicklung und der eng damit zusammenhängenden euklidschen Entwicklung geben Collins [32] und Loos [89] Auskunft.) Sind $A_1, A_2 \in k[x]$ und ist $A_2 \neq 0$, so erhält man die Kettenbruchentwicklung von A_1/A_2 bekanntlich als Folge $(Q_1, \ldots, Q_t)$ der Quotienten im euklidschen Algorithmus

$$A_1 = Q_1 A_2 + A_3$$
$$A_2 = Q_2 A_3 + A_4$$
$$\dotsb\dotsb\dotsb\dotsb\dotsb$$
$$A_t = Q_t A_{t+1}.$$

Berechnet man die Koeffizientenfolge der Polynome $Q_1, \ldots, Q_t$ aus den Koeffizienten von A_1, A_2 durch Ausführen des euklidschen Algorithmus, so ergibt sich im Normalfall ein nichtskalarer Aufwand der Größenordnung n^2. (Dabei sei deg $A_1 = n$, deg $A_2 \leqq n$.) Knuth [80] und Schönhage [127] haben ein raffiniertes rekursives Verfahren zur Berechnung der Kettenbruchentwicklung gefunden, bei dem nur geeignete Anfangsstücke der A_i verwendet werden und welches mit $0(n \log n)$ nichtskalaren Operationen zum Ziel kommt.

Man beachte, daß die Anzahl der zu berechnenden Koeffizienten eine nichtkonstante Funktion der Inputpolynome A_1, A_2 ist. Das bedeutet, daß das Modell der Berechnungsfolge im vorliegenden Fall versagt und durch dasjenige des Berechnungsbaumes zu ersetzen ist. Formuliert man den Knuth-Schönhage-Algorithmus als Berechnungsbaum, so ergibt sich eine $0(n \log n)$-Abschätzung für die (wie oben definierte) Verzweigungskomplexität der Kettenbruchentwicklung. Daß dies die optimale Größenordnung ist, kann man wieder mit der Gradmethode des Abschnitts 6 zeigen. Tatsächlich erhält man eine viel genauere Auskunft [148]:

Der Knuth-Schönhage-Algorithmus besitzt eine Kostenfunktion der Größenordnung $n(H+1)$, wo H die Entropie des auf Koeffizientensumme 1 normierten „Output-Formats" (deg $Q_1, \ldots,$ deg Q_t) ist.

Die Kostenfunktion eines jeden Berechnungsbaumes für die Kettenbruchentwicklung ist $\geqq n(H+1)$ außerhalb einer „gleichmäßig dünnen" Menge von Inputs. Der Knuth-Schönhage-Algorithmus ist also „uniformly most powerful", um einen Ausdruck der Statistik zu gebrauchen.

Im folgenden wollen wir $k = \mathbb{R}$ annehmen und neben den Nulltests $a \overset{?}{=} 0$ auch Größenvergleiche $a \overset{?}{\geqq} 0$ zulassen. Wir bezeichnen die entsprechende Verzweigungskomplexität wieder mit C. Die Ungleichung (8.2) bleibt richtig, die in früheren Abschnitten bewiesenen unteren Schranken für die nichtskalare Komplexität gelten also weiterhin, wenn man L durch C ersetzt.

Die Verzweigungskomplexität von Entscheidungsproblemen $W \subset \mathbb{R}^n$ (zu berechnen ist hier der Indikator von W) ist im vergangenen Jahrzehnt von einer Reihe von Autoren untersucht worden. Erst in jüngster Zeit ist es gelungen, nichtlineare untere Schranken zu beweisen.

Satz (Ben-Or [12]): $W \subset \mathbb{R}^n$ besitze N Zusammenhangskomponenten. Dann ist

$$C(W) \geqq \frac{1}{1 + \log 3} (\log N - n \log 3).$$

Als Anwendung erhält man Abschätzungen der Größenordnung $n \log n$ für die Komplexität einer Fülle von natürlichen Entscheidungsproblemen. Hier sind einige Beispiele:

A: Gegeben $(x_1, \ldots, x_n) \in \mathbb{R}^n$. Zu entscheiden, ob die x_i paarweise verschieden sind.

B: Gegeben $(x_1, \ldots, x_n, y_1, \ldots, y_n) \in \mathbb{R}^{2n}$. Zu entscheiden, ob $\{x_1, \ldots, x_n\} = \{y_1, \ldots, y_n\}$.

C: Gegeben $(x_1, \ldots, x_n, y_1, \ldots, y_n) \in \mathbb{R}^{2n}$. Zu entscheiben, ob $\{x_1, \ldots, x_n\} \cap \{y_1, \ldots, y_n\} = \emptyset$.

Im Fall des „Rucksackproblems"

D: Gegeben $(y_1, \ldots, y_n) \in \mathbb{R}^n$. Zu entscheiden, ob $\sum_{i \in I} y_i = 1$ für ein $I \subset \{1, \ldots, n\}$

liefert Satz 8 sogar $C(D) \geq \text{const} \cdot n^2$. Dieses Resultat ist für restriktivere Berechnungsmodelle bereits von Dobkin-Lipton [40] und Steele-Yao [140] bewiesen worden. Die letzteren Autoren haben auch schon die wesentliche Ingredienz des Beweises von Satz 8 verwendet, eine Abschätzung von Milnor [99] und Thom [150] für die Anzahl der Zusammenhangskomponenten einer reellen algebraischen Varietät.

Im Fall $k = \mathbb{C}$ läßt sich aus Satz 8 mit einer elementaren Schlußweise Satz 6 (um den Faktor $\frac{1}{1 + \log 3}$ abgeschwächt) herleiten. Die direkten Anwendungen von Satz 8 auf Resultante und Diskriminante (vgl. (7.1)), wie sie Ben-Or [12] vorschlägt, sind wohl nur bei Ausschluß von Divisionen stichhaltig.

Problem 1: Man entscheide, ob die gleichmäßige Optimalitätseigenschaft des Knuth-Schönhage-Algorithmus erhalten bleibt, wenn man Größenvergleiche als Verzweigungen zuläßt.

Problem 2: Man finde heraus, ob die angegebenen unteren Schranken für die Verzweigungskomplexität von $A - D$ bei Grundkörpern endlicher Charakteristik Gültigkeit behalten.

9. Komplexitätsmengen

Unsere bisherigen Methoden liefern nur logarithmische untere Schranken für die Komplexität von Polynomen in einer Variablen, deren Koeffizienten algebraisch über dem Primkörper sind (z. B. von Polynomen in $\mathbb{Q}[x]$). Um diesen Mangel zu beheben, greifen wir zunächst auf die Ideen von Motzkin [102] und Belaga [11] zurück, und zwar in ihrer von Paterson-Stockmeyer [110] für die nichtskalare Komplexität entwickelten Gestalt. Der Einfachheit halber seien keine Divisionen zugelassen (obwohl die Resultate auch ohne diese Voraussetzung gültig bleiben) und der Grundkörper sei der Körper der komplexen Zahlen.

Es handelt sich darum, die Abhängigkeit der Koeffizienten zu berechnender Polynome von den in einer Berechnung auftretenden Parametern zu untersuchen. Dazu eignet sich der Begriff der „generischen Berechnung": Seien x, c_{ij}, d_{ij} ($1 \leqq i \leqq r$, $0 \leqq j \leqq i$) und e_j ($0 \leqq j \leqq r+1$) Unbestimmte über $\mathbb{C}$. Wir definieren U_i, V_i, $G_i \in \mathbb{C}[c, d, e, x]$ für $1 \leqq i \leqq r$ induktiv durch

$$U_i = c_{i0} + c_{i1}x + \sum_{j=1}^{i-1} c_{i,j+1}G_j, \quad V_i = d_{i0} + d_{i1}x + \sum_{j=1}^{i-1} d_{i,j+1}G_j, \quad G_i = U_iV_i$$

(vgl. die Definitionen 3). Dann setzen wir

$$F = e_0 + e_1 x + \sum_{j=1}^{r} e_{j+1}G_j \in \mathbb{C}[c, d, e, x].$$

Für $f \in \mathbb{C}[x]$ ist nun klar:

$$L(f) \leqq r \Leftrightarrow \exists\ \gamma_{ij}, \delta_{ij}, \varepsilon_j \in \mathbb{C}\quad f(x) = F(\gamma, \delta, \varepsilon, x). \tag{9.1}$$

Bezeichnen wir also mit $\mathbb{C}[x]_n$ den linearen Raum aller Polynome in $\mathbb{C}[x]$ vom Grad $\leqq n$, so ist die „Komplexitätsmenge" $\{f \in \mathbb{C}[x]_n : L(f) \leqq r\}$ konstruktibel und ihr Zariski-Abschluß X_{nr} ist eine Untervarietät von $\mathbb{C}[x]_n$ einer Dimension $< (r+2)^2$ (denn es gibt weniger als $(r+2)^2$ Unbestimmte c, d, e). Nimmt man hier $r = \lfloor \sqrt{n} - 2 \rfloor$, so ergibt sich, daß fast alle Polynome vom Grad n eine Komplexität $\geqq \sqrt{n} - 1$ haben. (Andererseits läßt sich nach Paterson-Stockmeyer [110] jedes Polynom vom Grad n mit nichtskalarem Aufwand $0(\sqrt{n})$ auswerten.)

Eine „ideale" Methode, die Komplexität konkreter Polynome nach unten abzuschätzen, bestünde darin, das Verschwindungsideal der affinen Varietät X_{nr} zu bestimmen. Seine Elemente sind diejenigen komplexen Polynome in $n+1$ Variablen, die auf den Koeffizientenvektoren aller $f \in X_{nr}$ (oder äquivalenterweise auf den Koeffizienten von $F \in \mathbb{C}[c, d, e][x]$) verschwinden. Wir wollen sie Resultanten nennen. Leider scheint das Auffinden interessanter Resultanten schwierig zu sein. Nun zeigt sich aber, daß schon eine ganz bruchstückhafte Information wie die Existenz von ganzzahligen Resultanten kleinen Grades und kleiner Höhe ausreicht, um brauchbare untere Schranken für die Komplexität konkreter Polynome zu gewinnen. (Strassen [147], Schnorr [123].) Hier ist ein Beispiel:

$$L\left(\sum_{j=0}^{n} 2^{2^j} x^j\right) \geqq \text{const}\ \sqrt{n/\log n}.$$

(Man beachte, daß die Schwierigkeit bei der Auswertung von $\sum_0^n 2^{2^j} x^j$ keineswegs darin liegt, die riesigen Koeffizienten herzustellen, da ja beliebige komplexe Zahlen kostenlos zur Verfügung stehen.)

Sucht man untere Schranken für die Komplexität von Polynomen, deren Koeffizienten zwar algebraisch, aber nicht notwendig rational sind, so kann man die oben angezeigte Methode deutlich verbessern, indem man statt der Existenz von Resultanten kleinen Grades die Tatsache benutzt, daß X_{nr} selbst einen kleinen Grad hat (Heintz-Sieveking [66]). So erhält man folgendes schöne und allgemeine Ergebnis.

Satz: Seien $\alpha_1, \ldots, \alpha_n \in \mathbb{C}$ algebraisch über einem Unterkörper k_0 von $\mathbb{C}$ und sei N die Größe der Bahn von $(\alpha_1, \ldots, \alpha_n)$ unter der Galoisgruppe von $\mathbb{C}$ über k_0. Gibt es Polynome $h_1, \ldots, h_n \in k_0 [a_1, \ldots, a_n]$ vom Grad $\leq M$, deren Nullstellenmenge in $\mathbb{C}^n$ endlich ist und $(\alpha_1, \ldots, \alpha_n)$ enthält, so gilt

$$L(\alpha_n x^n + \cdots + \alpha_1 x) \geqq \left(\frac{\log N}{24 \log (nM)} \right)^{1/2}$$

Obwohl beim Beweis wieder der Satz von Bézout ins Spiel kommt, besteht kein erkennbarer Zusammenhang mit der Methode von Abschnitt 6.

Als Anwendungsbeispiele haben wir

$$L \left(\sum_{j=1}^{n} \alpha^{1/j} x^j \right) \geqq \text{const } \sqrt{n/\log n}$$

für positiv reelles $\alpha \neq 1$, und

$$L \left(\sum_{j=1}^{n} j^q x^j \right) \geqq \text{const } \frac{\sqrt{n}}{\log n}$$

für $q \in \mathbb{Q} \setminus \mathbb{Z}$ (von zur Gathen-Strassen [51]). Für $q \in \mathbb{N}$ ist $L \left(\sum_{j=1}^{n} j^q x^j \right) = 0 (\log n)$.

Eine interessante Anwendung geht auf Lipton-Stockmeyer [88] zurück (siehe auch Schnorr [123]): Jedes Polynom $f \in \mathbb{C}[x]$ vom Grad n mit lauter einfachen Nullstellen besitzt Teiler, deren Komplexität die Größenordnung $\sqrt{n/\log n}$ hat (obwohl es sehr einfach zu berechnende solche f gibt, z.B. $f = x^n - 1$).

Die Methoden dieses Abschnitts sind auch auf andere Komplexitätsmaße anwendbar (Zählung der Additionen/Subtraktionen oder Zählung sämtlicher arithmetischen Operationen). Wir verweisen auf Schnorr-Van de Wiele [125] und die dort angegebene Literatur.

Problem 1: Für die „typische" nichtskalare Komplexität

$$\lambda_n = \min \{r: X_{nr} = \mathbb{C}[x]_n\}$$

von Polynomen vom Grad $\leq n$ folgt aus Paterson-Stockmeyer [110]

$$\sqrt{n-1} \leq \lambda_n \leq \sqrt{2n} + 0\,(\log n).$$

Man verbessere diese Abschätzung. (Vgl. auch Abschnitt 15.)

Problem 2: Man entscheide, ob $L\left(\sum_{j=0}^{n} x^j/j!\right)$ bzw. $L\left(\sum_{j=1}^{n} x^j/j\right)$ die Größenordnung $\log n$ hat. (Vgl. auch Brent [26].)

Problem 3: Man finde ein konkretes Polynom $f \in \mathbb{C}[x]$ mit Koeffizienten aus $\{0, 1\}$, dessen Komplexität nicht die Größenordnung $\log n$ hat. Nach Lipton [87] gibt es solche Polynome. (Siehe auch Schnorr Van de Wiele [125].)

10.   Multiplikation von Matrizen

Der Standardalgorithmus zur Multiplikation zweier n-reihiger Matrizen benötigt $2n^3 - n^2$ arithmetische Operationen. Die Einfachheit und Natürlichkeit dieses Verfahrens ließen seine Optimalität lange als selbstverständlich erscheinen. So verwundert es nicht, daß gerade die Suche nach einem strengen Optimalitätsbeweis zur Entdeckung eines asymptotisch schnelleren Algorithmus geführt hat [142]. Definieren wir (nach Wahl eines Grundkörpers) den Exponenten ω der Matrixmultiplikation als das infimum derjenigen c, für die es ein Multiplikationsverfahren vom Aufwand $0\,(n^c)$ gibt, so liefert dieser Algorithmus die Abschätzung

$$\omega \leq \log 7 < 2{,}808. \tag{10.1}$$

Die Forschungen der letzten Jahre zielen einerseits auf ein besseres Verständnis der Komplexität der Matrixmultiplikation und verwandter Probleme (Abschnitte 11–14), andererseits auf die möglichen Anwendungen solcher asymptotisch schnellen Algorithmen.

Die Bedeutung der Matrixmultiplikation liegt vor allem in ihrer Schlüsselrolle für die numerische lineare Algebra: So haben z. B. die folgenden Probleme alle den Exponenten ω: Matrixinversion, LR-Zerlegung, Berechnung der Determinante oder sämtlicher Koeffizienten des charakteristischen Polynoms, im Fall $k = \mathbb{C}$ auch QR-Zerlegung und unitäre Ähnlichkeitstransformation auf obere Hessenbergform (Strassen [142], [144], Bunch-Hopcroft [29], Schönhage [128], Baur-Strassen [10], Keller [78]). Ein Beispiel für die in den Beweisen auftretende Problemreduktion: Jedes schnelle Verfahren zur Berechnung der Determinante einer n-reihigen Matrix liefert ein solches für das System aller Minoren (Satz 7), also nach Cramer für die Inverse.

Weitere Anwendungen, wie das Bestimmen der Zusammenhangskomponenten gerichteter Graphen oder das Erkennen kontextfreier Sprachen, findet man im Übersichtsartikel von Paterson [109].

Problem: Sei η der Exponent der Auflösung quadratischer n-reihiger linearer Gleichungssysteme. Klar ist $\eta \leqq \omega$. Man entscheide $\omega \leqq \eta$.

11.   Bilineare Abbildungen

Der Exponent der Matrixmultiplikation ändert sich nicht, wenn man statt aller arithmetischen Operationen nur die nichtskalaren zählt. Wir wollen im folgenden deshalb stets mit der technisch bequemeren nichtskalaren Komplexität arbeiten. Ist

$$f: A \times B \to W$$

eine bilineare Abbildung endlich dimensionaler Vektorräume, so bezeichne $L(f)$ die Komplexität der Aufgabe, die Koordinaten des Vektors $f(a, b)$ aus den Koordinaten von a und b zu berechnen. (Hierzu muß man in A, B, W zunächst Basen wählen, $L(f)$ erweist sich aber als basisunabhängig.)

Bilineare Abbildungen treten z. B. als Multiplikationen in endlich dimensionalen (nicht notwendig assoziativen) Algebren A auf. Wir schreiben dann $L(A)$ für $L(f)$ und sprechen von der Komplexität der Algebra A. Speziell ist $L(k^{n \times n})$ die Komplexität der Matrixmultiplikation.

Wir geben eine einfache und koordinatenfreie Beschreibung von $L(f)$.

Satz: $L(f) \leqq r$, gewenn es Linearformen $u_1, \ldots, u_r,\ v_1, \ldots, v_r \in (A \oplus B)^*$ und Vektoren $w_1, \ldots, w_r \in W$ gibt, so daß für alle $a \in A,\ b \in B$

$$f(a, b) = \sum_{\varrho = 1}^{r} u_\varrho (a \oplus b)\, v_\varrho (a \oplus b)\, w_\varrho.$$

Der Beweis läuft im wesentlichen auf die Argumentation hinaus, daß bei der Berechnung einer Menge von quadratischen Formen ohne Effizienzverlust auf Divisionen verzichtet werden kann [144].

Die Komplexität bilinearer Abbildungen hat gute algebraische Eigenschaften, die sich am einfachsten für Algebren formulieren lassen: $L(A) \geqq L(A')$, falls A' Unteralgebra oder Quotient von A ist, $L(A \oplus A') \leqq L(A) + L(A')$, wo $A \oplus A'$ das direkte Produkt bezeichnet.

Eine kleine Modifikation der obigen Charakterisierung von $L(f)$ führt uns auf den Begriff des Rangs $R(f)$: Wir verlangen, daß die u_ϱ nur von a, die v_ϱ nur von b abhängen, so daß $u_\varrho \in A^*$, $v_\varrho \in B^*$. Um die Symmetrieeigenschaften des Rangs sichtbar zu machen, gehen wir von der bilinearen Abbildung f zu ihrem Strukturtensor $t \in U \otimes V \otimes W$ über, wo $U = A^*$, $V = B^*$.

Definition [143], [144]: Seien U, V, W endlich dimensionale Vektorräume, $t \in U \otimes V \otimes W$. Der Rang $R(t)$ ist die kleinste natürliche Zahl r, zu der es Vektoren $u_1, \ldots, u_r \in U$, $v_1, \ldots, v_r \in V$ und $w_1, \ldots, w_r \in W$ gibt mit

$$t = \sum_{\varrho=1}^{r} u_\varrho \otimes v_\varrho \otimes w_\varrho.$$

Eine solche Liste von u_ϱ, v_ϱ, w_ϱ heißt eine bilineare Berechnung von t.

Wir identifizieren im folgenden bilineare Abbildungen mit ihren Strukturtensoren und verwenden sinngemäß die Bezeichnungen $R(f)$ und $R(A)$. Es ist

$$\tfrac{1}{2}R(f) \leqq L(f) \leqq R(f).$$

Der Rang besitzt die oben für die Komplexität formulierten einfachen Eigenschaften, insbesondere

$$R(A \oplus A') \leqq R(A) + R(A').$$

Hinzu kommt die Ungleichung

$$R(A \otimes A') \leqq R(A)\, R(A'), \tag{11.1}$$

deren Gültigkeit die hauptsächliche Motivation für den Rangbegriff ist. Mit ihrer Hilfe können wir die Organisation des auf (10.1) führenden Algorithmus folgendermaßen beschreiben: Man zeigt zunächst $R(k^{2 \times 2}) \leqq 7$ durch Angabe einer bilinearen Berechnung, dann wendet man (11.1) an und findet

$$R(k^{2^q}\Sigma^{2^q}) = R(k^2\Sigma^2 \otimes k^{2^{q-1}}\Sigma^{2^{q-1}}) \leqq 7\, R(k^{2^{q-1}}\Sigma^{2^{q-1}}),$$

also

$$L(k^{2^q}\Sigma^{2^q}) \leqq R(k^{2^q}\Sigma^{2^q}) \leqq 7^q,$$

also $L(k^n\Sigma^n) = 0\,(n^{\log 7})$ durch Einbettung und damit $\omega \leqq \log 7$.

Es ist klar, daß man aus jeder Rangabschätzung für eine Matrixmultiplikation bestimmter Größe in analoger Weise eine Abschätzung von ω gewinnen kann. Auf Grund eines Resultats von Hopcroft-Musinski [68] ist dies sogar möglich, wenn man von der Multiplikation rechteckiger Matrizen ausgeht. Bezeichnen wir, wie es seit Schönhage [132] üblich ist, den Tensor der Matrixmultiplikation $k^{m \times n} \times k^{n \times p} \to k^{m \times p}$ mit $\langle m, n, p \rangle$, so gilt

$$R(\langle m, n, p \rangle) \leqq r \Rightarrow (mnp)^{\omega/3} \leqq r. \tag{11.2}$$

Problem: Man entscheide die Additivitätsvermutung

$$R(A \oplus A') = R(A) + R(A'). \tag{11.3}$$

(Strassen [144], Winograd [160], Schönhage [132].) Diese Vermutung ist wohl ein Ausdruck unseres Wunschdenkens: Ihre Richtigkeit würde nach Wedderburn den Rang beliebiger halbeinfacher $\mathbb{C}$-Algebren auf den Rang von Matrixalgebren zurückführen. In diesem Zusammenhang siehe auch Riffelmacher [118].

12.   Algebren

Die sich durch die Diskussion des vorangehenden Abschnitts aufdrängende Frage nach dem genauen Rang der Multiplikation zweireihiger Matrizen wurde von Hopcroft-Kerr [67] und Winograd [158] beantwortet:

$$L(k^{2 \times 2}) = R(k^{2 \times 2}) = 7.$$

Etwas allgemeiner haben Brockett-Dobkin [28] und Lafon-Winograd [83]

$$L(k^{n \times n}) \geqq 2n^2 - 1$$

gezeigt, Auch für die Komplexität (und damit den Rang) einiger anderer Klassen von assoziativen Algebren liegen nichttriviale untere Schranken vor. So hat man z. B. für Divisionsalgebren A der Dimension n

$$L(A) \geqq 2n - 1 \qquad (12.1)$$

(Fiduccia-Zalcstein [43]). Nach de Groote [57] steht hier das Gleichheitszeichen, gewenn A eine einfache Körpererweiterung ist. Angewandt auf die reelle Quaternionenalgebra $\mathbb{H}$ liefert dies die eine Hälfte eines schon früher bekannten Resultats

$$L(\mathbb{H}) = R(\mathbb{H}) = 8 \qquad (12.2)$$

(Dobkin [39], de Groote [54], Howell-Lafon [71], Feig [41]). Fiduccia-Zalcstein [43] und Winograd [160] haben die Multiplikation zweier Polynome modulo einem vorgegebenen Polynom f untersucht und

$$L(k[x]/(f)) = 2n - t$$

gefunden, wobei n der Grad und t die Anzahl der verschiedenen Primfaktoren von f ist. Hier wird wieder die Abhängigkeit der Komplexität vom Grundkörper augenfällig. (Z. B. hat die Gruppenalgebra $k[x]/(x^n - 1)$ der zyklischen Gruppe C_n für $k = \mathbb{Q}$ die Komplexität $2n - d(n)$, für $k = \mathbb{C}$ aber nur die Komplexität n.) Die bisher genannten unteren Schranken mit Ausnahme von (12.2) sind unmittelbare Korollare aus der Ungleichung

$$L(A) \geqq 2n - t, \tag{12.3}$$

wo $n = \dim A$ und t die Anzahl der maximalen zweiseitigen Ideale von A bedeutet (Alder-Strassen [4]). Diese Ungleichung eignet sich auch zur Bestimmung der Komplexität gewisser nicht-kommutativer Gruppenalgebren, zum Beispiel der Diedergruppe.

Wie bei (12.1) kann man sich auch bei (12.3) fragen, wann Gleichheit gilt. Ersetzt man zur Vereinfachung L durch R, so wird man auf den Begriff der „Algebra minimalen Ranges" geführt. De Groote-Heintz [58] haben die kommutativen Algebren minimalen Ranges vollständig bestimmt: Es sind dies die endlichen Produkte von einfach erzeugten Algebren (also Algebren der Form $k[x]/(f)$) und sogenannten verallgemeinerten Nullalgebren (d.h. Algebren $k[\omega_1, \ldots, \omega_q]$, wo alle ω_i nilpotent sind und sich gegenseitig annullieren).

In einer weiteren Arbeit [59] untersuchen Heintz und de Groote die Komplexität und den Rang einfacher Lie-Algebren. Der Rang von Lie-Algebren verdient u.a. deshalb Interesse, weil die Komplexität der Multiplikation in einer zusammenhängenden linearen algebraischen Gruppe nach unten bis auf einen konstanten Faktor durch den Rang der zugehörigen Lie-Algebra abgeschätzt wird [144]. Für die Komplexität von Moduln über assoziativen Algebren verweisen wir auf Hartmann [62] und die dort angegebene Literatur.

Kennt man den Rang einer Algebra A, so ist es natürlich, nach der Struktur der Menge der optimalen bilinearen Berechnungen zu fragen. Auf dieser Menge operiert die Symmetriegruppe Γ der Algebra. (Γ ist eine Gruppe verallgemeinerter Automorphismen.) Von de Groote [55] stammt das schöne Resultat, daß die Operation im Fall $A = k^{2 \times 2}$ transitiv ist. Dagegen besitzen einfache Körpererweiterungen i.a. mehrdimensionale Bahnenräume optimaler Algorithmen (Winograd [160], siehe auch de Groote [57]).

Problem 1: Man entscheide, ob $R(k^{n \times n})$ die Größenordnung n^2 hat.

Problem 2: Man bestimme $R(K^{2 \times 2})$, wo K eine einfache algebraische Körpererweiterung von k ist und $K^{2 \times 2}$ als k-Algebra aufgefaßt wird.

Problem 3: Man bestimme die Algebren minimalen Ranges. (Zunächst wird man wie oben einschränkende Voraussetzungen an eine solche Algebra A machen, z.B. A einfach, oder $k = \mathbb{C}$ und A sauber.)

Problem 4: Man verallgemeinere den Satz von Feig [41], daß eine Divisionsalgebra minimaler Komplexität minimalen Rang besitzt.

Problem 5: Die Inversion von Einheiten in $k[x]/(f)$ mit $\deg f = n$ hat eine nichtskalare Komplexität $0(n \log n)$, wie man mit Hilfe der euklidschen Entwicklung [148] sieht. Man verbessere diese Abschätzung. Dabei beachte man, daß z.B. für $f = x^n$ die Komplexität nach Sieveking [137], Kung [82] nur $0(n)$ ist.

13.   Der Exponent der Matrixmultiplikation

1978 gelang es Pan [106], [107], durch eine raffinierte Konstruktion

$$R(\langle n, n, n \rangle) = R(k^{n \times n}) \leqq \frac{1}{3}(n^3 - n) + \frac{9}{2}n^2$$

zu zeigen. Nimmt man $n = 48$ und wendet (11.2) an, so ergibt sich

$$\omega < 2{,}781\,,$$

eine Verschärfung von (10.1).

Wenig später und unabhängig von Pan fanden Bini-Capovani-Lotti-Romani [19] und Bini [16] die noch etwas bessere Abschätzung

$$\omega < 2{,}780\,. \tag{13.1}$$

Wichtiger als das Resultat ist hier die Methode, die in einer neuartigen Verwendung approximativer (bilinearer) Algorithmen besteht: Wir nehmen an, der Grundkörper k sei algebraisch abgeschlossen und betrachten die polynomiale Abbildung

$$\varphi_r : (U \times V \times W)^r \to U \otimes V \otimes W \tag{13.2}$$

$$(u_\varrho, v_\varrho, w_\varrho)_{\varrho \leqq r} \mapsto \sum_{\varrho = 1}^{r} u_\varrho \otimes v_\varrho \otimes w_\varrho.$$

Das Bild von φ_r besteht nach Definition von R genau aus den Tensoren $t \in U \otimes V \otimes W$, deren Rang $\leqq r$ ist. (Vgl. auch Abschnitt 10.) Die Menge dieser Tensoren ist also irreduzibel und konstruktibel. Ihren Zariski-Abschluß bezeichnen wir mit X_r.

Definition: Seit $t \in U \otimes V \otimes W$. Der Grenzrang $\underline{R}(t)$ ist die kleinste natürliche Zahl r mit $t \in X_r$.

$\underline{R}(t)$ ist also klein, gewenn t sich beliebig genau durch Tensoren von kleinem Rang approximieren läßt. Bini-Capovani-Lotti-Romani [19] zeigen nun

$$\underline{R}(\langle 3, 2, 2 \rangle) \leqq 10,$$

und Bini [16] überträgt (11.2) auf den Grenzrang:

$$\underline{R}(\langle m, n, p \rangle) \leqq r \Rightarrow (mnp)^{\omega/3} \leqq r. \tag{13.3}$$

Insgesamt ergibt sich (13.1). (Tatsächlich arbeiten die genannten Autoren mit einer anderen Definition des Grenzrangs, welche aber nach Alder [3] äquivalent zu der hier angegebenen ist.)

In seiner Arbeit [132] macht Schönhage zunächst effizienten Gebrauch von der ebenfalls auf Bini-Capovani-Lotti-Romani [19] zurückgehenden Idee der partiellen Matrixmultiplikation (mit dem Ergebnis $\omega < 2{,}609$). Diese Untersuchungen führen ihn dann zu einer weitreichenden Verallgemeinerung von (13.3).

Satz (Schönhage [132]): Aus

folgt
$$\underline{R}(\langle m_1, n_1, p_1 \rangle \oplus \cdots \oplus \langle m_l, n_l, p_l \rangle) \leqq r$$
$$(m_1 n_1 p_1)^{\omega/3} + \cdots + (m_l n_l p_l)^{\omega/3} \leqq r.$$

Der Satz ist verhältnismäßig leicht zu beweisen, wenn man die Richtigkeit der Additivitätsvermutung (11.3) annimmt. Schönhage zeigt durch eine neuartige Rekursionstechnik, daß diese Annahme überflüssig ist.

Man kann den Sachverhalt auch umkehren und den Satz als eine asymptotische Bestätigung von (11.3) auffassen. In der gleichen Arbeit bringt Schönhage freilich die Additivitätsvermutung ins Wanken, indem er zeigt, daß ihre naheliegende Übertragung auf den Grenzrang falsch ist.

$$\underline{R}(\langle m, 1, p \rangle \oplus \langle 1, (m-1)(p-1), 1 \rangle) = mp + 1, \tag{13.4}$$

obwohl $\underline{R}(\langle m, 1, p \rangle) = mp$ und $\underline{R}(\langle 1, (m-1)(p-1), 1 \rangle) = (m-1)(p-1)$. Wendet man schließlich den obigen Satz auf (13.4) an, so erhält man

$$\omega < 2{,}548,$$

eine erhebliche Verbesserung der früheren Abschätzungen.

Schönhage entwickelte diese Ideen im Sommer 1979 und trug darüber im Oktober des gleichen Jahres auf der Komplexitätskonferenz in Oberwolfach vor. Unter den Zuhörern war Pan, der bald bemerkte, daß sich seine und Schönhages Konstruktionen kombinieren lassen, mit dem Ergebnis [108]

$$\omega < 2{,}522.$$

Das letzte Wort haben bisher Coppersmith-Winograd [37], denen es gelang, durch eine dynamische Verallgemeinerung von (13.4) (und unter Verwendung von Satz 13)

$$\omega < 2{,}496 \tag{13.5}$$

zu beweisen. Ein interessantes Nebenprodukt dieser Arbeit ist die Erkenntnis, daß man durch einfache Anwendung von (11.2), (13.3) oder Satz 13 den Exponenten ω nicht treffen kann, daß also in der Konklusion dieser Aussagen stets $<$ steht.

Es soll nicht verschwiegen werden, daß die zu den obigen Exponentenabschätzungen führenden Algorithmen kein praktisches Interesse beanspruchen können, da sich der asymptotische Gewinn erst bei riesigen Matrixformaten bemerkbar macht.

Für die Multiplikation schmaler rechteckiger Matrizen verweisen wir auf Brockett-Dobkin [27], Coppersmith [36] und Lotti-Romani [90].

Problem 1: Man verbessere die Abschätzung (13.5) deutlich.

Problem 2: Man bestimme $\underline{R}(k^{2 \times 2})$ für algebraisch abgeschlossenes k. (Bini [18] hat $\underline{L}(k^{2 \times 2}) = 6$ gezeigt.)

Problem 3: Man entscheide, ob das folgende Analogon von (12.3) richtig ist:

$$\underline{R}(A) \geqq 2 \dim A - \dim(\operatorname{rad} A) - t \tag{13.6}$$

wo t wieder die Anzahl der maximalen Ideale von A bedeutet und k als algebraisch abgeschlossen angenommen wird. Die Ungleichung stimmt, falls man sich bei der Definition von $\underline{R}$ auf die Approximation von A durch assoziative Algebren beschränkt, statt beliebige bilineare Abbildungen zuzulassen:
$l(A)$ sei die gemeinsame Länge von Kompositionsreihen $0 = A_0 \subseteq \cdots \subseteq A_l = A$, wo A_i zweiseitiges Ideal von A_{i+1} ist. Durch Verfeinerung von $0 \subset \operatorname{rad} A \subset A$ sieht man $l(A) = \dim(\operatorname{rad} A) + t$. Wegen (13.6) gilt also

$$l(A) \geqq 2n - r \tag{13.7}$$

auf der Menge der assoziativen Algebrenstrukturen A vom Rang $\leqq r$ über einem festen Vektorraum der Dimension n. (13.7) ist aber äquivalent zur Existenz einer Normalreihe der Länge $2n - r$, also wegen der Vollständigkeit der Fahnenvarietäten eine abgeschlossene Bedingung, die ihrerseits $r \geqq 2n - \dim(\operatorname{rad} A) - t$ impliziert.

Freilich trifft die gleiche Schlußweise auf $\underline{L}$ an Stelle von $\underline{R}$ zu, wofür (13.6) falsch ist. (Siehe Problem 13.2.)

14.   Rang und Grenzrang

Wir tragen hier einige Ergebnisse über Rang und Grenzrang von Tensoren zusammen, die nicht speziell auf Strukturtensoren assoziativer Algebren zugeschnitten sind. Sei k algebraisch abgeschlossen und seien U, V, W k-Vektorräume der Dimensionen m, n, p. Als Format eines Tensors $t \in U \otimes V$

$\otimes W$ bezeichnen wir das Tripel (m, n, p). (Man beachte, daß das Format des Matrixtensors $\langle m, n, p \rangle$ nicht $(\langle m, n, p \rangle)$, sondern (mn, np, mp) ist.) Wegen der Symmetrie von R und $\underline{R}$ dürfen wir $m \geq n \geq p$ annehmen. Für $p = 1$ bedeuten Rang und Grenzrang einfach Matrixrang, weshalb wir im folgenden $p \geq 2$ voraussetzen wollen.

Für $p = 2$ läßt sich ein Tensor $t \in U \otimes V \otimes W$ nach Wahl von Basen durch zwei Matrizen $A, B \in k^{m \times n}$ beschreiben. Ist dann z. B. $m = n$ und A invertierbar, so gilt $\underline{R}(t) = n$ und $R(t) = n + l$, wo l die maximale Anzahl nichttrivialer Jordankästchen zu einem Eigenwert der Matrix $A^{-1}B$ bezeichnet. Dies folgt aus einem allgemeinen Resultat von Grigoryev [52] und Ja'Ja' [72], welches $R(t)$ durch die in der Kroneckerschen Normalform des Matrizenpaars (A, B) auftretenden Invarianten ausdrückt. Der Fall $p = 2$ kann damit als erledigt gelten. (In analoger Weise behandelt Bini [17] den Grenzrang.)

Für $p = 3$ wird $t \in U \otimes V \otimes W$ durch ein Tripel von Matrizen $A, B, C \in k^{m \times n}$ beschrieben. Nimmt man $m = n$ ungerade an und setzt $r = (3n + 1)/2$, so ist $X_r = U \otimes V \otimes W$, X_{r-1} eine irreduzible Hyperfläche in $U \otimes V \otimes W$. Insbesondere haben fast alle Tensoren den Rang und den Grenzrang r. In diesem Fall ist die Erzeugende des Verschwindungsideals von X_{r-1} bekannt [149]: Man ersetze die Koeffizienten von A, B, C durch Unbestimmte und bilde die rationale Funktion

$$F = (\det A)^2 \det (BA^{-1}C - CA^{-1}B).$$

Dann ist F ein irreduzibles Polynom, welches auf X_{r-1} verschwindet. Als Anwendung hiervon und einer entsprechenden Aussage für gerade n läßt sich z. B. der Grenzrang eines beliebigen $sl_2(\mathbb{C})$-Moduls berechnen.

Die Bestimmung der Dimensionen der X_r ist nicht nur für schmale Formate möglich. Aus der Definition von X_r folgt unter Berücksichtigung trivialer Fasern in (13.2)

$$\dim X_r \leq \min \{r(m + n + p - 2), mnp\}.$$

Ein Format (m, n, p) heiße gut, gewenn hier stets Gleichheit gilt. (Das oben für ungerade n betrachtete Format $(n, n, 3)$ ist also nicht gut.) Man kann eine relativ dichte Menge von guten Formaten angeben [149]. Der Nachweis, daß das besonders interessante kubische Format (n, n, n) für $n \neq 3$ gut ist, gelang Lickteig [86]. Bezeichnen wir den typischen Rang ($=$ maximalen Grenzrang) von Tensoren $t \in U \otimes V \otimes W$ mit $\underline{R}(m, n, p)$, so haben wir für gute Formate offenbar

$$\underline{R}(m, n, p) = \lceil mnp/(m + n + p - 2) \rceil.$$

Allgemein gilt immerhin

$$\underline{R}(m, n, p) \sim mnp/(m + n + p - 2)$$

(Atkinson-Lloyd [5], Strassen [149], Lickteig [86]).

Problem 1: Man bestimme alle guten Formate.

Problem 2: Sei $P = P(U \otimes V \otimes W)$ der projektive Raum der Strahlen von $U \otimes V \otimes W$, S die Segre-Varietät (d. h. das Bild der Segre-Einbettung von $P(U) \times P(V) \times P(W)$ in P). Eine r-Sekante von S ist ein projektiver Teilraum von P, der von r linear unabhängigen Punkten von S aufgespannt wird. Sei $Y \subset \mathrm{Grass}\,(r-1, P)$ der Abschluß der Menge der r-Sekanten von S. Dann gilt für $t \in (U \otimes V \otimes W) \setminus \{0\}$

$$\underline{R}(t) \leq r \Leftrightarrow \exists\, y \in Y \quad kt \in y.$$

Man bestimme die Schubertkoeffizienten von Y in konkreten Fällen. (Bei festem t ist $\{y \in \mathrm{Grass}\,(r-1, P): kt \in y\}$ eine Schubertvarietät.)

15.   Fouriertransformation

Ähnlich wie bei den bilinearen Abbildungen (siehe Satz 11) kann man sich bei der Auswertung eines Systems von Linearformen

$$\sum_{j=1}^{n} \alpha_{ij} x_j, \ldots, \sum_{j=1}^{n} \alpha_{mj} x_j \tag{15.1}$$

ohne wesentlichen Effizienzverlust auf die Verwendung von Linearkombinationen $(g, h) \mapsto \beta g + \gamma h$ (mit $\beta, \gamma \in k$) als Rechenoperationen beschränken. Die mit Hilfe eines solchen linearen Berechnungsmodells definierte Komplexität von (15.1) nennen wir die Komplexität der Matrix $a = [\alpha_{ij}] \in k^{m \times n}$ und bezeichnen sie mit $L_s(a)$. Sie ist die minimale Anzahl von Linearkombinationen, welche ausreicht, um die durch a gegebene lineare Abbildung $k^m \to k^n$ an einem Vektor x auszuführen. (Wir können auch von der Komplexität einer linearen Abbildung abstrakter Vektorräume sprechen, müssen aber im Auge behalten, daß dieser Begriff die Wahl von Basen voraussetzt.)

Von großer praktischer Bedeutung ist der durch den chinesischen Restsatz vermittelte Algebrenisomorphismus

$$D_n: \mathbb{C}[x]/(x^n - 1) \to \mathbb{C}^n,$$

die diskrete Fouriertransformation (DFT). Bezüglich der natürlichen Basen besitzen sowohl D_n wie $\frac{1}{n} D_n^{-1}$ die Matrix $[\zeta^{ij}] \in \mathbb{C}^{n \times n}$, wobei ζ eine primitive n-te Einheitswurzel ist. Der 1965 von Cooley-Tukey [35] (wieder-)entdeckte Algorithmus der „schnellen Fouriertransformation" (FFT) liefert zusammen mit

Bluestein [20] (oder Rader [116] und Winograd [159]) die Komplexitätsabschätzung

$$L_s(D_n) = 0(n \log n).$$

Von den vielfältigen Anwendungen der FFT sei hier nur die Multiplikation von Zahlen oder Polynomen erwähnt (Schönhage-Strassen [134], Schönhage [131]).

Was ist an unteren Komplexitätsschranken bekannt? Morgenstern [101] hat auf überraschend einfache Weise gezeigt, daß ein Algorithmus für die DFT, welcher bei festem $c \geqq 2$ nur Linearkombinationen $\beta g + \gamma h$ mit $|\beta| + |\gamma| \leqq c$ verwendet, wenigstens den Aufwand $(n \log n)/(2 \log c)$ besitzt.

Der Versuch, untere Schranken für $L_s(D_n)$ ohne Einschränkung zu beweisen, hat die Aufmerksamkeit auf eine Klasse von interessanten Graphen gelenkt: den n-Superkonzentratoren. Das sind Graphen mit je n ausgezeichneten blauen bzw. weißen Knoten, so daß man für jedes $q \leqq n$ jede Menge von q blauen mit jeder Menge von q weißen Knoten durch q disjunkte Wege verbinden kann. Ein Algorithmus zur Berechnung der DFT mit r Linearkombinationen gibt Anlaß zu einem n-Superkonzentrator mit $4r$ Kanten. Die minimale Kantenzahl von n-Superkonzentratoren ist also eine untere Schranke für $4L_s(D_n)$. Pinsker [112] und Valiant [151] haben das janusköpfige Ergebnis erzielt, daß es n-Superkonzentratoren mit $0(n)$ Kanten gibt. Die effektive Konstruktion solcher Graphen gelang Margulis [94] (siehe auch Gabber-Galil [48], Pippenger [113], Klawe [79]). Das Konzept des Superkonzentrators hat zahlreiche Anwendungen gefunden, über die man sich ebenfalls in der obigen Literatur informieren kann.

Die diskrete Fouriertransformation ist ein Spezialfall des Wedderburn-isomorphismus

$$\mathbb{C}[G] \to \mathbb{C}^{n_1 \times n_1} \times \cdots \times \mathbb{C}^{n_t \times n_t}$$

$$a \mapsto (\varrho_1(a), \ldots, \varrho_t(a))$$

von der Gruppenalgebra einer endlichen Gruppe G auf ein Produkt voller Matrixalgebren (Die ϱ_i repräsentieren die Äquivalenzklassen irreduzibler Matrixdarstellungen von G, sie sind folglich nur bis auf Konjugation bestimmt). Die algorithmischen Aspekte dieses Isomorphismus werden in Beth [15] für den Fall auflösbarer Gruppen untersucht.

Problem: Man bestimme die Größenordnung von $L_s(D_n)$.

16.   Berechenbarkeit

Der Begriff der berechenbaren zahlentheoretischen Funktion $f: \mathbb{N} \to \mathbb{N}$ wurde in den dreißiger Jahren von Herbrand, Gödel, Church, Kleene, Turing und Post auf verschiedene, aber äquivalente Weise formuliert. Für unsere Zwecke eignet sich die von Turing vorgeschlagene „maschinelle" Definition besonders gut. Eine Turingmaschine ist ein endliches Computerprogramm, das einen Lese-Schreib-Kopf auf einem potentiell unendlichen externen Bandspeicher schrittweise hin und her bewegt. Das Band ist in Felder gleicher Größe eingeteilt, deren jedes die Symbole 0, 1 oder * speichern kann. Auf das Band wird die binär codierte Inputzahl x geschrieben, und auf ihm steht am Schluß der Rechnung der Funktionswert $f(x)$.

Der Begriff der berechenbaren Funktion zieht den der entscheidbaren Menge nach sich: Eine Menge heißt entscheidbar, gewenn ihr Indikator berechenbar ist.

Die beiden Begriffe lassen sich leicht von Zahlen auf andere diskrete mathematische Objekte übertragen: Zahlenpaare, endliche Zahlenfolgen, ganzzahlige Polynome, formalisierte Aussagen der elementaren Zahlentheorie, endliche Graphen und Relationalstrukturen, ja sogar Turingmaschinenprogramme können auf natürliche und effiziente Weise durch Zahlen kodiert werden. Da es nicht auf die genaue Art der Codierung ankommt, können wir die genannten Objektbereiche (also z. B. $\mathbb{Z}^n$, $\mathbb{Z}[x_1, \ldots, x_n]$ oder die Menge der Turingmaschinenprogramme) jeweils mit $\mathbb{N}$ identifizieren.

Der Wert einer strengen Berechnungsdefinition liegt natürlich vor allem in der Möglichkeit, die Nichtberechenbarkeit von Funktionen bzw. die Nichtentscheidbarkeit von Mengen nachzuweisen. Sehen wir uns einige Beispiele von Entscheidungsproblemen an:

A: Zu entscheiden, ob ein elementarer zahlentheoretischer Satz aus den Peano-Axiomen folgt.

B: Zu entscheiden, ob ein ganzzahliges Polynom in mehreren Variablen eine ganzzahlige Nullstelle besitzt (Hilberts zehntes Problem).

C: Zu entscheiden, ob ein als Produkt von Erzeugenden geschriebenes Element einer durch endlich viele Erzeugende und Relationen gegebenen Gruppe das Einselement darstellt (Wortproblem der Gruppentheorie).

Diesen drei Problemen ist gemeinsam, daß sie zwar vielleicht nicht entscheidbar, aber jedenfalls nicht weit davon entfernt sind, es zu sein. Eine Menge von natürlichen Zahlen heißt verifizierbar*) (oder aufzählbar), gewenn sie die Projektion einer entscheidbaren Menge $E \subset \mathbb{N} \times \mathbb{N}$ auf die erste Komponente ist. Daß unsere Entscheidungsprobleme verifizierbar sind, sieht man am leichtesten im Fall B: Die Menge E besteht einfach aus allen Paaren (Polynom p, Nullstelle von p). Durch Angabe einer Nullstelle „verifiziert" man die Zugehörigkeit von p zur Menge B. (Bei A betrachtet man die Paare (Satz s, Beweis von s).)

Nicht jedes Turingmaschinenprogramm definiert eine berechenbare Funktion. Das liegt daran, daß die Berechnung einer Turingmaschine nicht auf jedem Inputwert schließlich zum Stillstand kommen muß, so daß die dargestellte

---

*) In der Wortwahl lehnen wir uns an die Einleitung von [139] an.

Funktion i. a. nur partiell definiert ist. Die Definitionsbereiche dieser „partiell berechenbaren" Funktionen sind wiederum gerade die verifizierbaren Mengen.

Wie kann man von einer solchen Menge zeigen, daß sie unentscheidbar ist? Die Antwort wurde schon von Gödel vorgezeichnet und beruht auf einer Anwendung des Cantorschen Diagonalverfahrens. Es ist plausibel, daß man ein Turingmaschinenprogramm finden kann, welches angesetzt auf den Input (x, e) im wesentlichen die gleiche Rechnung durchführt, wie das Turingmaschinenprogramm mit der Codenummer e angesetzt auf den Input x. (Ein solches universelles Programm muß nur in der Lage sein, die in den Codenummern steckende Information zu entschlüsseln und auszuführen.) Definieren wir also $U \subset \mathbb{N} \times \mathbb{N} \cong \mathbb{N}$ als die Menge der Paare (x, e), so daß das Maschinenprogramm e auf dem Input x schließlich zum Stillstand kommt, so ist U als Definitionsbereich einer partiell berechenbaren Funktion verifizierbar. U ist aber nicht entscheidbar, denn sonst wäre auch

$$U' = \{x \in \mathbb{N} : (x, x) \notin U\}$$

entscheidbar, also erst recht verifizierbar, im Widerspruch dazu, daß jede verifizierbare Menge als eine „Zeile" von U auftritt.

Wir können die verifizierbaren Mengen nach ihrer Schwierigkeit ordnen: $X \leqq Y$ (X ist transformierbar in Y), gewenn es eine berechenbare Funktion f mit

$$\forall x \in \mathbb{N} \quad (x \in X \Leftrightarrow f(x) \in Y)$$

gibt. Ist Y entscheidbar und gilt $X \leqq Y$, so ist auch X entscheidbar.

Eine verifizierbare Menge heißt vollständig, gewenn sie jede andere im Sinne dieser Präordnung dominiert. U ist vollständig (weil jede verifizierbare Menge eine „Zeile" von U ist). Vollständige Mengen sind unentscheidbar (weil sonst jede verifizierbare Menge entscheidbar wäre). Will man von irgendeiner verifizierbaren Menge zeigen, daß sie vollständig ist, so genügt der Nachweis von $U \leqq X$.

Die oben betrachteten drei Entscheidungsprobleme sind vollständig und damit unentscheidbar. Für A wurde dies von Church gezeigt. Analoge Unentscheidbarkeitsresultate hat man heute für die meisten interessanten axiomatischen Theorien. Bemerkenswerte Ausnahmen bilden nach Tarski die elementaren Theorien reell oder algebraisch abgeschlossener Körper. Die Vollständigkeit von B wurde nach Vorarbeiten von Julia Robinson und Davis-Putnam-Robinson von Matjasevitsch bewiesen und die von C verdankt man Novikov und Boone.

17.   Die Cooksche Hypothese

Der zeitliche Aufwand einer Turingmaschinenberechnung wird gewöhnlich als die Anzahl der Bewegungen des Lese-Schreib-Kopfes definiert. Wir nennen eine zahlentheoretische Funktion P-berechenbar, gewenn sie mit einem Aufwand berechnet werden kann, der durch ein Polynom in der Länge des Inputs beschränkt ist. Analog wird P-Entscheidbarkeit erklärt. (Die Klasse der P-entscheidbaren Mengen bezeichnet man mit **P**.) Beide Begriffe erweisen sich als robust gegenüber Variationen des Maschinenmodells, und ihre naheliegende Ausdehnung auf andere diskrete mathematische Bereiche hängt nicht von der Wahl der Codierungen ab, solange man dabei einigermaßen die Vernunft walten läßt.

Schon in den Anfängen der Komplexitätstheorie setzte sich die Erkenntnis durch, daß die P-Berechenbarkeit einer Funktion f die praktische Berechenbarkeit von f auf einem Computer mit recht guter Annäherung beschreibt.

In seiner bahnbrechenden Arbeit [34] nimmt Cook den Begriff der verifizierbaren Menge zum Vorbild für die Einführung der Mengenklasse **NP**. (Siehe auch Levin [85].) Wir nennen die Elemente von **NP** die P-verifizierbaren Mengen. Die Definition dieses Begriffs unterscheidet sich von der in Abschnitt 16 gegebenen Verifizierbarkeits-Definition durch die zusätzliche Forderung, daß die Zugehörigkeit eines (x, y) zur Menge E in polynomialer Zeit in der Länge von x entscheidbar sei.

In den letzten Jahren hat sich gezeigt, daß die meisten praktisch interessanten Entscheidungsprobleme P-verifizierbar sind. Wir geben hier nur eine kleine Auswahl:

A: Zu entscheiden, ob eine aussagenlogische Formel erfüllbar ist.

B: Zu entscheiden, ob sich ein (endlicher) Graph in einen andern isomorph einbetten läßt.

C: Zu entscheiden, ob zwei Graphen isomorph sind.

D: Zu entscheiden, ob ein Graph einen Hamiltonzyklus besitzt. (Dies ist ein Spezialfall des Problems vom „travelling salesman".)

E: Zu entscheiden, ob ein Graph dreifärbbar ist.

F: Zu entscheiden, ob ein System von linearen Ungleichungen über $\mathbb{Q}$ eine Lösung besitzt („Lineare Optimierung").

G: Zu entscheiden, ob eine natürliche Zahl prim ist.

H: Zu entscheiden, ob eine Zahl a einen Primfaktor $< b$ besitzt. (Der Input ist also das Zahlenpaar (a, b). Hier wird die Aufgabe, eine Zahl in Primfaktoren zu zerlegen, als Entscheidungsproblem formuliert: Ein schnelles Entscheidungsverfahren für H würde mit Hilfe einer Halbierungsmethode zum schnellen Auffinden des kleinsten Primfaktors von a führen.)

I: Zu entscheiden, ob die lineare diophantische Gleichung $a_1 x_1 + \cdots + a_n x_n = b$ eine Lösung in $\{0, 1\}^n$ besitzt. (Dies ist das für $\mathbb{Q}$ an Stelle von $\mathbb{R}$ formulierte Rucksackproblem aus Abschnitt 8.)

J: Zu entscheiden, ob die quadratische diophantische Gleichung $ax^2 + by - c = 0$ (mit $a, b\,c \in \mathbb{N}$) eine Lösung in $\mathbb{N}^2$ besitzt.

K: Zu entscheiden, ob die Kongruenz $x^2 \equiv a \bmod b$ eine Lösung x mit $0 \leqq x$ $< c$ besitzt. (Hier wird die Aufgabe, die Kongruenz wenn möglich zu lösen, als Entscheidungsproblem formuliert.)

Die Analogie zwischen verifizierbaren und P-verifizierbaren Mengen ist unvollkommen: So kann man die relative Schwierigkeit P-verifizierbarer Mengen zwar wiederum durch eine Präordnung beschreiben ($X \leqq_P Y$, gewenn X mit Hilfe einer P-berechenbaren Funktion in Y transformiert werden kann), und es gibt bezüglich dieser Ordnung wieder vollständige Mengen (diese heißen NP-vollständig), aber das Diagonalverfahren versagt und es ist nicht auszuschließen, daß jede P-verifizierbare Menge P-entscheidbar ist. An der Richtigkeit der berühmten Cookschen Hypothese $P \neq NP$ ist freilich kaum zu zweifeln. Aus ihr folgt, daß sich kein NP-vollständiges Problem in polynomialer Zeit entscheiden läßt.

Cook selbst hat die Vollständigkeit von A und B in der obigen Liste bewiesen. In seiner brillanten Arbeit [77] gibt Karp etwa 20 weitere NP-vollständige Probleme an, darunter D, E und I in unserer Liste. J und K sind ebenfalls vollständig (Manders-Adleman [93]). Dagegen liegt F in P (Schor [135], Judin-Nemirovskii [73], Kachian [74]). Der Status des Graphenisomorphieproblems C ist nicht geklärt. Immerhin gibt es polynomiale Algorithmen für spezielle Klassen von Graphen (Babai [6], Miller [98], Filotti-Mayer [44], Luks [91], Babai-Grigoryev-Mount [7]). G ist jedenfalls in einem praktischen Sinn schnell entscheidbar (Solovay-Strassen [138]*), Miller [97], Rabin [114], Adleman [1], Cohen-Lenstra [31]). Andererseits hat man bis heute keinen brauchbaren Algorithmus, um große (etwa 100-stellige) Zahlen in Primfaktoren zu zerlegen (siehe Knuth [81]). Diese Diskrepanz zwischen der (heutigen Beurteilung der) Faktorzerlegung und der Primzahlerkennung ist die Grundlage für eine revolutionäre Entwicklung in der Kryptographie (Diffie-Hellman [38], Rivest-Shamir-Adleman [120]; siehe auch Rabin [115], Ong-Schnorr-Shamir [103]). Wir bemerken, daß man über die Faktorzerlegung von Polynomen viel besser Bescheid weiß (Berlekamp [13], [14], Lenstra-Lenstra-Lovász [84],

---

*) Eine ursprüngliche Fassung dieser Arbeit gab Anlaß zu einer Wette, deren Text im Vortragsbuch 1974 des Mathematischen Instituts Oberwolfach niedergelegt ist und hier wiedergegeben sei:
Zwischen Ernst Specker und Volker Strassen, beide Zürich, wird eine Wette um folgende Aussage A abgeschlossen:
    Die Menge der Primzahlen ist in P
d. h. es gibt eine deterministische Turingmaschine im Sinne von Turing, welche für jede dezimal codierte Eingabe n, $n \in \mathbb{N}$, die Primheit von n in einer Schrittzahl entscheidet, welche durch ein Polynom in $\log n$ nach oben beschränkt ist.
Volker Strassen gewinnt die Wette, falls bis zum 1. November 1984 ein Beweis für A publiziert ist, der im Prinzip in ZF durchführbar ist. Andernfalls ist Ernst Specker Gewinner.
Der Verlierer lädt den Gewinner alsbald zu einer Ballonfahrt ein oder entschädigt ihn durch 50 Gramm Gold.

Oberwolfach, den 31. Oktober 1974

Chistov-Grigoryev [30], Kaltofen [75], [76], von zur Gathen [50] und die dort angegebene Literatur, Schönhage [133]).

Problem 1: Entscheide die Cooksche Hypothese.

Problem 2: Entscheide, ob C, G bzw. H in P liegen.

18.   Schwierige Entscheidungsprobleme

Wie wir im letzten Abschnitt bemerkt haben, versagt das Diagonalverfahren für die Klasse NP. (In gewissem Sinne läßt sich dies sogar beweisen, siehe Baker-Gill-Solovay [8].) Das bedeutet freilich nicht, daß Diagonalargumente in der Komplexitätstheorie überhaupt keine Rolle spielen. Schon lange vor Cook [34] haben Hartmanis-Stearns [61] mit derartigen Schlußweisen gezeigt, daß es unter den entscheidbaren Problemen solche von beliebig hohem Schwierigkeitsgrad gibt. (Ein definitives Resultat über die Feinstruktur dieser Schwierigkeitsgrade wurde kürzlich von Fürer [46] erzielt; siehe auch die dort angegebene Literatur. Andererseits hat Blum [21] auf erstaunliche pathologische Eigenschaften der Schwierigkeitsgrade aufmerksam gemacht.)

Es ist ein Verdienst von Meyer-Stockmeyer [96], natürliche Entscheidungsprobleme von nachweisbar hoher Komplexität entdeckt zu haben. Unter den vielen Resultaten, die in der Folge dieses methodischen Durchbruchs erzielt wurden, wollen wir nur einen besonders schönen Satz von Fischer und Rabin zitieren.

Satz (Fischer-Rabin [45]): Es gibt eine Konstante $c > 0$, so daß jedes Entscheidungsverfahren für die elementare Theorie der reellen Zahlen für jedes n_0 auf mindestens einer geschlossenen Formel einer Länge $n \geq n_0$ mehr als 2^{cn} Schritte benötigt.

Kurz: Die elementare Theorie der reellen Zahlen ist wenigstens exponentiell kompliziert. Andererseits ist diese Theorie nach Tarski entscheidbar. Das Tarskische Verfahren hat allerdings einen Aufwand von $2^{2^{\cdot^{\cdot^2}}}$ mit n übereinandergeschachtelten Zweiern. Vor einigen Jahren haben Collins [33] und etwas später Monck-Solovay (siehe Wüthrichs Beitrag in [139]) Entscheidungsverfahren angegeben, welche in vergleichsweise kurzer, nämlich nur doppelt exponentieller Zeit zum Ziele kommen. Entsprechende Resultate wurden von Wüthrich [161] und Heintz [64], [65] auf ganz anderem Wege für algebraisch abgeschlossene Körper beliebiger Charakteristik gewonnen.

19.   Die Permanente

Kehren wir zu den NP-vollständigen Problemen zurück. Zu entscheiden sei etwa, ob ein Graph mit den Knoten $1, \ldots, n$ und der Inzidenzmatrix

$[\alpha_{ij}]$ ($\alpha_{ij} = 1$ falls i und j durch eine Kante verbunden sind, sonst $\alpha_{ij} = 0$) einen Hamiltonzyklus besitzt. Es liegt nahe, hierzu den Enumerator

$$HC_{n \times n}([a_{ij}]) = \sum_{\sigma \in S_n \text{ Zykel der Länge } n} a_{1\sigma(1)} \cdots a_{n\sigma(n)} \in \mathbb{Q}[a_{ij}]$$

heranzuziehen: Wie man leicht sieht, ist $HC_{n \times n}[\alpha_{ij}]$ die Anzahl der Hamiltonschen Zyklen des Graphen. Ein schnelles Verfahren zur Auswertung des Polynoms $HC_{n \times n}$ auf einer Turingmaschine würde also ein schnelles Entscheidungsverfahren für die Existenz von Hamilton-Zyklen bewirken.

Auch andere NP-vollständige Probleme besitzen natürliche Enumeratoren, und es ist plausibel, daß sich diese in ihrer Berechnungskomplexität untereinander nicht sehr wesentlich unterscheiden (Valiant [152]). Überraschend ist hingegen die Entdeckung Valiant's, daß der Enumerator des bekannten Heiratsproblems in die gleiche Komplexitätsklasse gehört: n Knaben und n Mädchen sind teilweise miteinander befreundet. Kann jeder Knabe ein ihm befreundetes Mädchen heiraten? Setzen wir $\alpha_{ij} = 1$, wenn Knabe i mit Mädchen j befreundet ist, sonst $\alpha_{ij} = 0$, so können wir eine Verheiratung aller Knaben als eine Permutation $\sigma \in S_n$ mit der Eigenschaft $\alpha_{i\sigma(i)} = 1$ für $1 \leq i \leq n$ auffassen. Die Frage, ob eine Matrix $[\alpha_{ij}]$ mit Koeffizienten 0 oder 1 eine Verheiratung zuläßt, ist nach M. Hall [60] P-entscheidbar (siehe auch Specker's Beitrag V in [139] und die dort angegebene Literatur). Valiant [154] zeigt nun, daß der natürliche Enumerator des Heiratsproblems, die Permanente

$$\text{Perm}_{n \times n}([a_{ij}]) = \sum_{\sigma \in S_n} a_{1\sigma(1)} \cdots a_{n\sigma(n)},$$

die Komplexität von Enumeratoren NP-vollständiger Probleme besitzt. Unter der Annahme der Cookschen Hypothese sind also weder $HC_{n \times n}$ noch $\text{Perm}_{n \times n}$ in polynomialer Zeit berechenbar.

Leider ist es nicht ohne weiteres möglich, diese Aussagen ins algebraische Berechnungsmodell der Abschnitte 2–15 zu übernehmen. Valiant [153], [155] hat aber ein überzeugendes Analogon der Theorie von Cook und Karp ganz im Rahmen der algebraischen Komplexität entwickelt.

Die Gegenstände seiner Untersuchung sind nicht Entscheidungsprobleme, sondern Folgen $(F_m)_{m \geq 1}$ von Polynomen $F_m \in k[x_1, \ldots, x_m]$ mit der Eigenschaft, daß $\deg(F_m)$ P-beschränkt in m ist. (Eine Funktion $t : \mathbb{N} \to \mathbb{N}$ heiße P-beschränkt, gewenn sie von einem Polynom majorisiert wird.) Die Permanente führt auf eine solche Polynomfolge, indem man z.B. $F_m = 0$ setzt, falls m keine Quadratzahl ist, und $F_m = \text{Perm}_{n \times n}([a_{ij}])$ mit $a_{ij} = x_{(i-1)n+j}$, falls $m = n^2$.

Den P-entscheidbaren Mengen entsprechen die „P-berechenbaren" Polynomfolgen (F_m), bei denen $L(F_m)$ P-beschränkt in m ist. (Von einem endlichen Turingmaschinenprogramm, welches die Berechnung sämtlicher F_m steuert, ist hier nicht die Rede.)

Valiant nennt eine Folge (F_m) P-definierbar, gewenn es eine P-berechenbare Folge (G_m) und ein P-beschränktes t gibt, so daß für alle m

$$F_m(x_1, \ldots, x_m) = \sum_{\xi_{m+1}, \ldots, \xi_{t(m)} \in \{0, 1\}} G_{t(m)}(x_1, \ldots, x_m, \xi_{m+1}, \ldots, \xi_{t(m)})$$

gilt. Die P-definierbaren Folgen entsprechen den P-verifizierbaren Mengen. (Die Summation ersetzt die Projektion.) Die meisten natürlich auftretenden Polynomfolgen erweisen sich als P-definierbar.

Der Transformationsbegriff schließlich ist sehr einfach: $(F_m) \leqq (G_m)$, gewenn es ein P-beschränktes t gibt, so daß für alle m

$$F_m(x_1, \ldots, x_m) = G_{t(m)}(c_1, \ldots, c_{t(m)})$$

mit $c_i \in \{x_1, \ldots, x_m\} \cup k$ gilt. $(F_m) \leqq (G_m)$ besagt also, daß jedes F_m aus einem nicht gar zu fernen G_n durch eine Substitution mit Unbestimmten und Konstanten hervorgeht. (Daß Berechnen so etwas wie Substituieren bedeutet, haben wir ja schon in (9.1) gesehen.)

Satz (Valiant [153]): $(HC_{n \times n})$ ist vollständig bezüglich $\leqq$ in der Klasse der P-definierbaren Polynomfolgen. Ein Gleiches gilt für $(\text{Perm}_{n \times n})$, falls $\text{char}(k) \neq 2$.

Das Analogon der Cookschen Hypothese lautet jetzt: Die Folge der Permanenten ist nicht polynomial berechenbar, vorausgesetzt $\text{char}(k) \neq 2$. (Über Körpern der Charakteristik 2 ist die Permanente nichts anderes als die Determinante, also P-berechenbar.)

Problem: Eine Funktion $t: \mathbb{N} \to \mathbb{N}$ heiße QP-beschränkt, gewenn $t(n) = 2^{(\log n)^{0(1)}}$. Sei $\text{char } k \neq 2$. Man entscheide, ob es ein QP-beschränktes t gibt, so daß $\text{Perm}_{n \times n}[x_{ij}]$ aus $\text{Det}_{t(n) \times t(n)}[y_{rs}]$ durch eine Substitution mit Unbestimmten und Konstanten hervorgeht. (Gibt es kein solches t, so folgt aus Valiant [153], daß $L(\text{Perm}_{n \times n}[x_{ij}])$ nicht QP-beschränkt in n ist. Erst recht sind dann $(\text{Perm}_{n \times n})$, $(HC_{n \times n})$ und die anderen in Valiant [152] angegebenen vollständigen Polynomfolgen nicht P-berechenbar.)

20. Ausblick

Nimmt man Ostrowskis Arbeit [104] als Geburtsurkunde der algebraischen Komplexitätstheorie, so steht dieses Gebiet im dreißigsten Lebensjahr. Allerdings kann erst für die letzten 15 Jahre von kontinuierlicher Forschung die Rede sein. Naturgemäß ging die Entwicklung zunächst mehr in die Breite als in die Tiefe.

Herausragende Probleme wie die Matrixmultiplikation, die diskrete

Fouriertransformation und die Berechnung der Permanente sind immer noch ungelöst. Man hat aber heute eine viel klarere Vorstellung von ihrer Bedeutung und von den bei ihrer Untersuchung zu erwartenden Schwierigkeiten. Die Prognose, daß die Lösung jedes dieser Probleme vor dem Hintergrund der algebraischen Geometrie gelingen werde, ist gewiß kontrovers. Vor 15 Jahren wäre sie undenkbar gewesen.

Von den vorhandenen Techniken zum Beweis unterer Schranken für die Komplexität scheinen mir vor allem Gradmethoden und die direkte oder indirekte Verwendung von Resultanten Zukunftspotential zu besitzen. (Im Zusammenhang mit Abschnitt 9 wird man vielleicht eine stärkere Anlehnung an Ideen der diophantischen Approximation und der Transzendenzbeweise suchen. Andererseits führt die Konstruktion von Resultanten für das Tensorrangproblem in natürlicher Weise auf Fragen der Darstellungstheorie der symmetrischen Gruppen.)

Wie kaum ein anderes mathematisches Gebiet wird das hier dargestellte durch die Spannung zwischen algebraischem und algorithmischem Denken geprägt. Wie kaum ein anderes widerstrebt es der landläufigen Einteilung in „reine" und „angewandte" Mathematik: Bezieht es seine Methoden vor allem aus der Algebra, so verdankt es der Numerik einen Schatz von schönen und wichtigen Problemen neben vielen Einsichten in deren Struktur.

Literatur

[1]   Adleman, L. M., *On distinguishing prime numbers from composite numbers*, 21st Annual Symp. on Found. of Computer Science (1980), 387–406.

[2]   Aho, A. V., Hopcroft, J. E., Ullman, J. C., *The design and analysis of computer algorithms*, Addison-Wesley, Reading, 1974.

[3]   Alder, A., *Grenzrang und Grenzkomplexität aus algebraischer und topologischer Sicht*, Dissertation, Universität Zürich, 1983.

[4]   Alder, A., Strassen, V., *On the algorithmic complexity of associative algebras*, Theor. Computer Science **15** (1981), 201–211.

[5]   Atkinson, M. D., and Lloyd, S., *Bounds on the ranks of some 3-tensors*, Linear Algebra and its Applications **31** (1980), 19–31.

[6]   Babai, L., *Monte Carlo algorithms in graph isomorphism testing*, Preprint, 1979.

[7]   Babai, L., Grigoryev, D. Yu., and Mount, D. M., *Isomorphism of graphs with bounded eigenvalue multiplicity*, Proc. 14th Annual ACM Symposium on Theory of Computing, San Francisco (1982), 310–324.

[8]   Baker, T., Gill, J., and Solovay, R., *Relativizations of the P = ? NP question*, SIAM J. Comput. **4** (1975), 431–442.

[9]   Baur, W., and Rabin, M. O., *Linear disjointness and algebraic complexity*, in: Logic and Algorithmic: An international Symp. held in honour of Ernst Specker, Monogr. No. 30 de l'Enseign. Math. (1982), 35–46.

[10]   Baur, W., and Strassen, V., *The complexity of partial derivatives*, Theor. Comp. Science **22** (1982), 317–330.

[11]   Belaga, E. G., *Evaluation of polynomials of one variable with preliminary processing of the coefficients*, Problemy Kibernetiki **5** (1961), 7–15.

[12]   Ben-Or, B., *Lower bounds for algebraic computation trees* (Preliminary Report), Proc. of the 15th Annual ACM Symp. on Theory of Computing, Boston (1983), 80–86.

[13]   Berlekamp, E. R., *Factoring polynomials over finite fields*, Bell System Techn. J. **46** (1967), 1853–1859.

[14]   Berlekamp, E. R., *Factoring polynomials over large finite fields*, Math. Comp. **24/111** (1970), 713–735.

[15]   Beth, T., *On the complexity of group algebras*, Preprint Universität Erlangen, 1983.

[16]   Bini, D., *Relations between EC-algorithms and APA-algorithms, applications*, Nota interna B79/8 (March 1979) I. E. I. Pisa.

[17]   Bini, D., *Border rank of a pxqx2 tensor and the optimal approximation of a pair of bilinear forms*, Lecture Notes on Comp. Sci. **85** (1980), 98–108.

[18]   Bini, D., *A note on commutativity and approximation*, Nota interna B81/9 (1981) I. E. I. Pisa.

[19]   Bini, D., Capovani, M., Lotti, G., and Romani, F., $0(n^{2.7799})$ *complexity for matrix multiplication*, Inform. Proc. Letters **8** (1979), 234–235.

[20]   Bluestein, L. I., *A linear filtering approach to the computation of the discrete Fourier transform*, IEEE Trans. AU-**18** (1970), 451–455.

[21]   Blum, M., *A machine-independent theory of the complexity of recursive functions*, J. of the ACM (Assoc. for Comp. Mach.) **14** (1967), 322–336.

[22]   Borodin, A., and Cook, S., *On the number of additions to compute specific polynomials*, SIAM J. Comput. **5** (1976), 146–157.

[23]   Borodin, A., and Munro, I., *Evaluating polynomials at many points*, Inform. Proc. Letters **1** (1971), 66–68.

[24]   Borodin, A., and Munro, I., *Computational complexity of algebraic and numeric problems*, American Elsevier, New York, 1975.

[25]   Brauer, A., *On addition chains*, Bulletin AMS **45** (1939), 736–739.

[26]   Brent, R. P., *Multiple-precision zero-finding methods and the complexity of elementary function evaluation*, in: J. F. Traub (Ed.), Analytic computational complexity, Academic Press, 1976.

[27]   Brockett, R. W., and Dobkin, D., *On the number of multiplications required for matrix multiplication*, SIAM J. Comput. **5** (1976), 624–628.

[28]   Brockett, R. W., and Dobkin, D., *On the optimal evaluation of a set of bilinear forms*, Linear Algebra and Appl. **19** (1978), 207–235.

[29]   Bunch, J., and Hopcroft, J., *Triangular factorization and inversion by fast matrix multiplication*, Mathematics of Computation **28** (1974), 231–236.

[30]   Chistov, A. L., and Grigoryev, D. Yu., *Polynomial-time factoring of the multivariable polynomials over a global field*, LOMI preprint E-5-82 (1982) Leningrad.

[31]   Cohen, H., and Lenstra, H. W., *Primality testing and Jacobi sums*, Report 2–18, University of Amsterdam.

[32]   Collins, G. E., *Computer algebra of polynomials and rational functions*, American Math. Monthly **80/7** (1973), 725–755.

[33]   Collins, G. E., *Quantifier elimination for real closed fields by cylindrical algebraic decomposition*, Lecture Notes in Computer Science **33** (1975), 134–183.

[34]   Cook, St. A., *The complexity of theorem-proving procedures*, Proc. 3rd Ann. ACM Symp. on Theory of Comput., Shaker Heights (1971), 151–158.

[35]   Cooley, J. W., and Tukey, J. W., *An algorithm for the machine calculation of complex Fourier series*, Math. of Comput. **19/90** (1965), 297–301.

[36]   Coppersmith, D., *Rapid multiplication of rectangular matrices*, SIAM J. Comput. **11/3** (1982), 467–471.

[37]   Coppersmith, D., and Winograd, S., *On the asymptotic complexity of matrix multiplication*, SIAM J. Comput. **11** (1982), 472–492.

[38]   Diffie, W., and Hellman, M., *New directions in cryptography*, IEEE Trans. Inform. Theory IT-**22/6** (1976), 644–654.

[39]   Dobkin, D., *On the arithmetic complexity of a class of arithmetic computations*, Thesis, Harvard University, 1973.

[40]   Dobkin, D., and Lipton, R. J., *A lower bound of $\frac{1}{2}n^2$ on linear search programs for the Knapsack problem*, J. of Comput. and System Sci. **16** (1978), 413–417.

[41]   Feig, E., *On systems of bilinear forms whose minimal divisor-free algorithms are all bilinear*, J. of Algorithms **2** (1981), 261–281.

[42]   Fiduccia, C., *Polynomial evaluation via the division algorithm: the fast Fourier transform revisited*, Proc. 4th Symp. on the Theory of Comp. (May 1972), 88–93.

[43]   Fiduccia, C., and Zalcstein, I., *Algebras having linear multiplicative complexity*, J. of the ACM **24** (1977), 311–331.

[44]   Filotti, I., and Mayer, J., *A polynomial time algorithm for determining isomorphism of graphs of fixed genus*, 12th Annual ACM Symp. on Theory of Computing (1980), 236–243.

[45]   Fischer, M. J., and Rabin, M. O., *Super-exponential complexity of Presburger arithmetic*, in: R. M. Karp (Ed.), Complexity of Computation, American Math. Society (1974), 27–41.

[46]   Fürer, M., *The tight deterministic time hierarchy*, Proc. 14th Annual ACM Symp. on Theory of Computing, San Francisco (1982).

[47]   Fürer, M., Schnyder, W., and Specker, E., *Normal forms for trivalent graphs and graphs of bounded valence*, Proc. 15th Annual ACM Symp. on Theory of Computing, Boston (1983), 161–179.

[48]   Gabber, O., and Galil, Z., *Explicit constructions of linear size superconcentrators*, J. of Comput. and System Sci. **22** (1981), 407–420.

[49]   Garey, M. R., and Johnson, D. S., *Computers and Intractibility*, W. H. Freeman, San Francisco, 1979.

[50]   Von zur Gathen, J., *Factoring sparse multivariate polynomials*, Preprint, Dept. of Computer Science, University of Toronto, 1983.

[51]   Von zur Gathen, J., and Strassen, V., *Some polynomials that are hard to compute*, Theor. Computer Science **11/3** (1980), 331–336.

[52]   Grigoryev, D. Yu., *Some new bounds on tensor rank*, LOMI preprint E-2-1978, Leningrad (1978).

[53]   Grigoryev, D. Yu., *Notes of Scientific Seminars of LOMI*, **118** (1982), 25–82.

[54]   de Groote, H. F., *On the complexity of quaternion multiplication*, Inform. Proc. Letters **3** (1975), 177–179.

[55]   de Groote, H. F., *On varieties of optimal algorithms for the computation of bilinear mappings. II. Optimal algorithms for 2×2-matrix multiplication*, Theor. Computer Science **7** (1978), 127–148.

[56]   de Groote, H. F., *Lectures on the complexity of bilinear problems* (1982), to be published.

[57]   de Groote, H. F., *Characterization of division algebras of minimal rank and the structure of their algorithm varieties*, SIAM J. Comput. **12/1** (1983), 101–117.

[58]   de Groote, H. F., and Heintz, J., *Commutative algebras of minimal rank*, Linear Algebra and Its Applications **55** (1983), 37–68.

[59]   de Groote, H. F., and Heintz, J., *The isotropy group and a lower bound for the complexity of semisimple Lie algebras*, preprint Jan. 1983, Fachbereich Mathematik, Universität Frankfurt.

[60]   Hall, M., Jr., *An algorithm for distinct representatives*, American Math. Monthly **63** (1956), 716–717.

[61]   Hartmanis, J., and Stearns, R., *On the computational complexity of algorithms*, Trans. AMS **117** (1965), 285–306.

[62]   Hartmann, W., *Zwei Probleme aus der Berechnungskomplexität*, Dissertation, Universität Zürich, 1983, to appear in: SIAM J. Comput.

[63]   Hartmann, W., and Schuster, P., *Multiplicative complexity of some rational functions*, Theor. Computer Science **10** (1980), 53–61.

[64]   Heintz, J., *Definability bounds of first order theories of algebraically closed fields (abstract)*, Proc. Fundamentals of Computation Theory FCT'79 (1979), 160–166.

[65]   Heintz, J., *Definability and fast quantifier elimination in algebraically closed fields*, Dissertation, Universität Zürich, 1982.

[66]   Heintz, J., and Sieveking, M., *Lower bounds for polynomials with algebraic coefficients*, Theor. Computer Science **11** (1980), 321–330.

[67]   Hopcroft, J., and Kerr, L., *On minimizing the number of multiplications necessary for matrix multiplication*, SIAM J. Appl. Math. **20** (1971), 30–36.

[68]   Hopcroft, J. E., and Musinski, J., *Duality applied to the complexity of matrix multiplications and other bilinear forms*, SIAM J. Comput. **2** (1973), 159–173.

[69]   Horowitz, E., *A fast method for interpolation using preconditioning*, Inform. Processing Letters **1** (1972), 157–163.

[70]   Hovanskii, A. G., *On a class of systems of transcendental equations*, Soviet Math. Dokl. **22/3** (1980), 762–765.

[71]   Howell, T. D., and Lafon, J. C., *The complexity of the quaternion product*, Cornell University TR (1975), 75–245.

[72]   Ja'Ja', J., *Optimal evaluation of pairs of bilinear forms*, Proc. 10th Annual ACM Symp. on Theory of Computing (1978), 173–183.

[73]   Judin, D. B., and Nemirovskii, A. S., *Informational complexity and effective methods for the solution of convex extremal problems*, Ekonomika Matematicheskie Metody **12/2** (1976), 357–369 (Russisch).

[74]   Kachian, L. G., *A polynomial algorithm in linear programming*, Dokl. Akad. Nauk SSSR **244:5** (1979), 1093–1096, Soviet Math. Dokl. **20** (1979), 191–194.

[75]   Kaltofen, E., *A polynomial-time reduction from bivariate to univariate integral polynomial factorization*, Proc. 23rd Annual Symp. FOCS, Chicago (1982), 57–64.

[76]   Kaltofen, E., *Polynomial-time reduction from multivariate to bivariate and univariate integer polynomial factorization*, Manuscript, 1983, submitted to SIAM J. Comput.

[77]   Karp, R. M., *Reducibility among combinatorial problems*, IBM Symposium New York, 1972: *Complexity of Computer Computations*, Plenum Press, New York, 1972, 85–103.

[78]   Keller, W., *Asymptotisch schnelle Algorithmen der linearen Algebra*, Diplomarbeit, Universität Zürich, 1982.

[79]   Klawe, M., *Limitations on explicit constructions of expanding graphs*, Preprint, Computer Science Dept., IBM Research, San José/USA, 1983.

[80]   Knuth, D. E., *The analysis of algorithms*, Actes du congrès international des Mathématiciens 1970 Nice, tome **3**, 269–274.

[81]   Knuth, D. E., *The art of computer programming*, Vol. II: *Seminumerical algorithms*, Addison-Wesley, 2. Aufl., 1980.

[82]   Kung, H. T., *On computing reciprocals of power series*, Numer. Math. **22** (1974), 341–348.

[83]   Lafon, J. C., and Winograd, S., *A lower bound for the multiplicative complexity of the product of two matrices*, (Unpubliziertes) Manuskript, 1978.

[84]   Lenstra, A. K., Lenstra, H. W., and Lovász, L., *Factoring polynomials with rational coefficients*, Math. Ann. **261** (1982), 515–534.

[85]   Levin, L. A., *Universal Sorting Problems*, engl. translation in: Problems of Information Transmission **9** (1973), 265–266.

[86]   Lickteig, T., *Typical tensorial rank*, Preprint, Universität Tübingen, 1983.

[87]   Lipton, R. J., *Polynomials with 0 − 1 coefficients that are hard to evaluate*, 16th Annual Symp. on Found. of Computer Science, Univ. of California, Berkeley (1975), 6–10.

[88]   Lipton, R. J., and Stockmeyer, L. J., *Evaluation of polynomials with superpreconditioning*, Proc. 8th Annual ACM Symp. on Theory of Computing (1976), 174–180.

[89]   Loos, R., *Generalized polynomial remainder sequences*, Computing, Suppl. **4** (1982), 115–137.

[90]   Lotti, G., and Romani, F., *On the asymptotic complexity of rectangular matrix multiplication*, Theor. Computer Science **23** (1982), 1–15.

[91]   Luks, E., *Isomorphism of graphs of bounded valence can be tested in polynomial time*, J. of Comput. and System Sci. **25** (1982), 42–65.

[92]   Machtey, M., and Young, P., *An introduction to the general theory of algorithms*, Elsevier, 1978.

[93]   Manders, K. L., and Adleman, L., *NP-complete decision problems for binary quadratics*, J. of Comput. and System Sci. **16** (1978), 168–184.

[94]   Margulis, G., *Explicit constructions of concentrators*, engl. translation in: *Problems of Information Transmission*, Plenum Press, 1975.

[95]   Mehlhorn, K., *Effiziente Algorithmen*, Teubner, 1977.

[96]   Meyer, A. R., and Stockmeyer, L. J., *The equivalence problem for regular expressions with squaring requires exponential time*, Proc. 13th Annual Symp. on Switching and Automata Theory, IEEE, Long Beach (1972), 125–129.

[97]   Miller, G. L., *Riemann's hypothesis and tests for primality*, Proc. 7th Annual ACM Symp. on Theory of Computing, Albuquerque (1975), 234–239.

[98]   Miller, G. L., *Isomorphism testing for graphs bounded genus*, Proc. 12th Annual ACM Symp. on Theory of Computing (1980), 218–224.

[99]   Milnor, J., *On the betti numbers of real varieties*, Proc. AMS **15** (1964), 275–280.

[100] Moenck, R., and Borodin, A., *Fast modular transform via division*, Proc. 13th Annual Symp. on Switching and Automata Theory, IEEE (1972), 90–96.

[101] Morgenstern, J., *Note on a lower bound of the linear complexity of the fast Fourier transform*, J. ACM **20/2** (1973), 305–306.

[102] Motzkin, T. S., *Evaluation of polynomials and evaluation of rational functions*, Bull. AMS **61** (1955), 163.

[103] Ong, H., Schnorr, C. P., and Shamir, A., *An efficient signature scheme based on quadratic equations*, (extended abstract) Mathematischer Fachbereich, Universität Frankfurt, 1983.

[104] Ostrowski, A. M., *On two problems in abstract algebra connected with Horner's rule*, Studies in Math. and Mech. presented to Richard von Mises, Academic Press, New York (1954), 40–48.

[105] Pan, V. Ya., *Methods of computing values of polynomials*, Russian Math. Surveys **21/1** (1966), 105–136.

[106] Pan, V. Ya., *Strassen's algorithm is not optimal*, Proc. 19th IEEE Symp. on Foundations of Computer Science (1978), 166–176.

[107] Pan, V. Ya., *New fast algorithms for matrix operations*, SIAM J. Comput. **9** (1980), 321–342.

[108] Pan, V. Ya., *New combinations of methods for the acceleration of matrix multiplication*, Comput. Math. with Appl. **7** (1981), 73–125.

[109] Paterson, M. S., *Complexity of product and closure algorithms for matrices*, Proc. of the International Congress of Math., Vancouver, 1974.

[110] Paterson, M. S., and Stockmeyer, L., *Bounds of the evaluation time for rational polynomials*, IEEE Conference Record of the 12th Annual Symp. on Switching and Automata Theory (1971), 140–143.

[111] Paul, W., *Komplexitätstheorie*, Teubner, 1978.

[112] Pinsker, M., *On the complexity of a concentrator*, 7th International Teletraffic Conference, Stockholm (1973), 318/1–318/4.

[113] Pippenger, N., *Superconcentrators*, SIAM J. Comput. **6** (1978), 298–304.

[114] Rabin, M. O., *Probabilistic algorithms*, in: *Algorithms and Complexity*, J. F. Traub (Ed.), Academic Press, New York (1976), 21–40.

[115] Rabin, M. O., *Digitalized signatures and public-key functions as intractable as factorization*, MIT/LCS/TR-212, 1979.

[116]   Rader, C. M., *Discrete Fourier transforms when the number of data samples is prime*, Proc. IEEE **56** (1968), 1107–1108.

[117]   Reingold, E. M., and Stocks, I., *Simple proofs of lower bounds for polynomial evaluation*, in: *Complexity of Computer Computations*, R. Miller and J. Thatcher (Eds.), Plenum Press (1972), 21–30.

[118]   Riffelmacher, D., *Multiplicative complexity and algebraic structure*, J. of Comput. and System Sci. **26** (1983), 92–106.

[119]   Risler, J. J., *Additive complexity and zeros of real polynomials*, Jan. 1983 (to appear in SIAM J. Computing).

[120]   Rivest, R. L., Shamir, A., and Adleman, L., *A method for obtaining digital signatures and public-key cryptosystems*, Communications of the ACM **21** (1978), 120–126.

[121]   Rutishauser, H., *Vorlesungen über Numerische Mathematik*, Band 1, Birkhäuser, 1976.

[122]   Savage, J. E., *The complexity of computing*, Wiley, 1976.

[123]   Schnorr, C. P., *Improved lower bounds on the number of multiplications/divisions which are necessary to evaluate polynomials*, Theor. Computer Science **7** (1978), 251–261.

[124]   Schnorr, C. P., *An extension of Strassen's degree bound*, SIAM J. Comput. **10** (1981), 371–382.

[125]   Schnorr, C. P., and Van de Wiele, J. P., *On the additive complexity of polynomials*, Theor. Computer Science **10** (1980), 1–18.

[126]   Scholz, A., Jahresber. der DMV, class II, **47** (1937), 41/43.

[127]   Schönhage, A., *Schnelle Berechnung von Kettenbruchentwicklungen*, Acta Informatica **1/1** (1971), 139–144.

[128]   Schönhage, A., *Unitäre Transformationen großer Matrizen*, Numer. Mathematik **20** (1973), 409–417.

[129]   Schönhage, A., *A lower bound for the length of addition chains*, Theor. Computer Science **1** (1975), 1–12.

[130]   Schönhage, A., *An elementary proof for Strassen's degree bound*, Theor. Computer Science **3** (1976), 267–272.

[131]   Schönhage, A., *Schnelle Multiplikation von Polynomen über Körpern der Charakteristik 2*, Acta Informatica **7** (1977), 395–398.

[132]   Schönhage, A., *Partial and total matrix multiplication*, SIAM J. Comput. **10** (1981), 434–455.

[133]   Schönhage, A., *The fundamental theorem of algebra in terms of computational complexity*, Preliminary report, Univ. Tübingen, 1982.

[134]   Schönhage, A., and Strassen, V., *Schnelle Multiplikation großer Zahlen*, Computing **7** (1971), 281–292.

[135]   Schor, N. Z., *Cut-off method with space extension in convex programming problems*, Translation in: Cybernetics **13** (1977), 94–96.

[136]   Shaw, M., and Traub, J. F., *On the number of multiplications for the evaluation of a polynomial and some of its derivatives*, J. ACM **21** (1974), 161–167.

[137]   Sieveking, M., *An algorithm for division of power series*, Computing **10** (1972), 153–156.

[138]   Solovay, R., and Strassen, V., *A fast Monte-Carlo test for primality*, SIAM J. Comput. **6/1** (1977), 84–85.

[139]   Specker, E., and Strassen, V., *Komplexität von Entscheidungsproblemen*, Lecture Notes in Computer Science **43**, Springer, 1976.

[140]   Steele, J., and Yao, A., *Lower bounds for algebraic decision trees*, J. of Algorithms **3** (1982), 1–8.

[141]   Stoss, J., *Die Komplexität der Auswertung von Interpolationspolynomen*, Preprint 1983, Universität Konstanz.

[142]   Strassen, V., *Gaussian elimination is not optimal*, Numer. Mathematik **13** (1969), 354–356.

[143]   Strassen, V., *Evaluation of rational functions*, in: *Complexity of Computer Computations*, R. Miller and J. Thatcher (Eds.), Plenum Press, (1972), 1–10.

[144] Strassen, V., *Vermeidung von Divisionen*, Crelles Journal für die reine und angew. Mathematik **264** (1973), 184–202.

[145] Strassen, V., *Die Berechnungskomplexität von elementarsymmetrischen Funktionen und von Interpolationskoeffizienten*, Numer. Mathematik **20/3** (1973), 238–251.

[146] Strassen, V., *Some results in algebraic complexity theory*, Proc. of the International Congress of Math. 1974, Vancouver, 1974.

[147] Strassen, V., *Polynomials with rational coefficients which are hard to compute*, SIAM J. Comput. **3/2** (1974), 128–149.

[148] Strassen, V., *The computational complexity of continued fractions*, SIAM J. Comput. **12/1** (1983), 1–27.

[149] Strassen, V., *Rank and optimal computation of generic tensors*, Linear Algebra and Appl. **52** (1983), 645–685.

[150] Thom, R., *Sur l'homologie des variétés algébriques réelles*, Ed. S. S. Cairns, Princeton Univ. Press (1965), 255–265.

[151] Valiant, L. G., *On non-linear lower bounds in computational complexity*, Proc. 7th Annual ACM Symp. on Theory of Computing, Albuquerque (May 1975), 45–53.

[152] Valiant, L. G., *The complexity of enumeration and reliability problems*, SIAM J. Comput. **8/3** (1979), 410–421.

[153] Valiant, L. G., *Completeness classes in algebra*, Proc. 11th Annual ACM Symp. on Theory of Computing (1979), 259–261.

[154] Valiant, L. G., *The complexity of computing the permanent*, to appear in: Theor. Computer Science.

[155] Valiant, L. G., *Reducibility by algebraic projections*, in: *Logic and Algorithmic: An international Symp. held in honour of Ernst Specker*, Monogr. 30 de L'Ens. Math. (1982), 365–380.

[156] Van de Wiele, J.-P., *Complexité additive et zéros des polynômes à coefficients réels et complexes*, Rapport IRIA No. **292**, 1978.

[157] Winograd, S., *On the number of multiplications necessary to compute certain functions*, Comm. Pure & Appl. Math. **23** (1970), 165–179.

[158] Winograd, S., *On multiplication of 2×2 matrices*, Linear Algebra and Appl. **4** (1971), 381–388.

[159] Winograd, S., *On computing the discrete Fourier transform*, Proc. Nat. Acad. Sci. USA **73** (1976), 1005–1006.

[160] Winograd, S., *Some bilinear forms whose multiplicative complexity depends on the field of constants*, Math. Systems Theory **10** (1977), 169–180.

[161] Wüthrich, H.-R., *Ein schnelles Quantoreneliminationsverfahren für die Theorie der algebraischen abgeschlossenen Körper*, Dissertation, Universität Zürich, 1977.

Perspectives in Mathematics
Anniversary of Oberwolfach 1984
© Birkhäuser Verlag, Basel

Algebraic Independence of Transcendental Numbers. Gel'fond's Method and Its Developments

MICHEL WALDSCHMIDT

Institut Henri Poincaré 11, rue Pierre et Marie Curie,
F-75231 Paris Cedex 05 (France)

The first result on algebraic independence of transcendental numbers was proved one century ago by Lindemann and Weierstrass: if $\alpha_1, \ldots, \alpha_n$ are algebraic numbers which are linearly independent over $\mathbb{Q}$, then $e^{\alpha_1}, \ldots, e^{\alpha_n}$ are algebraically independent.

One knows four methods to derive the algebraic independence of transcendental numbers.

1   The construction, by Liouville, of transcendental numbers can be generalized to the construction of algebraically independent numbers. Several examples of algebraically free subsets of $\mathbb{C}$ with the power of continuum have been exhibited (J. von Neumann, O. Perron, H. Kneser, W. M. Schmidt, F. Kuiper and J. Popken, A. S. Fraenkel, M. G. de Bruin, A. Durand, P. Bundschuh and R. Wallisser, W. W. Adams, F. J. Wylegala, I. Shiokawa, Zhu Yao Chen, ...).

On the other hand, quantitative estimates (transcendence, linear independence or algebraic independence measures) enable one also to construct algebraically independent numbers (D. D. Mordoukhay-Boltovskoy, K. Mahler, A. O. Gel'fond, N. I. Fel'dman, S. Lang, A. A. Shmelev, A. I. Galochkin, W. D. Brownawell, K. Väänänen, M. Laurent, A. Bijlsma, F. J. Wylegala, ...).

2   In 1929, C. L. Siegel introduced a new method which enabled him to generalize the Lindemann-Weierstrass theorem to a class of entire functions (which he called E-functions) satisfying linear differential equations. Some restrictions in the hypotheses of Siegel's results were relaxed by A. B. Shidlovskii in 1953, and during the last thirty years many publications have been devoted to this subject (A. B. Shidlovskii, S. Lang, V. A. Oleinikov, T. V. Pershikova, I. I. Belogrivov, K. Mahler, A. I. Galochkin, V. G. Sprindzuck, Ju. V. Nesterenko, A. A. Shmelev, M. S. Nurmagomedov, K. Väänänen, V. H. Salihov, W. D. Brownawell, ...).[1])

---

[1]) cf. A. B. Shidlovskii. *Diophantine approximations and transcendental numbers* (in russian); Izd. Mosk. Univ., МГУ 1982.

3   Also in 1929, K. Mahler studied transcendental functions satisfying certain functional equations, and succeeded to prove in certain cases the algebraic independende of their values. In the first issue of the Journal of Number Theory, in 1969, Mahler remarked that these old papers had been forgotten. Starting in 1976, new progress has been achieved with this method (K. K. Kubota, J. H. Loxton and A. J. van der Poorten, Y. Z. Flicker, D. W. Masser).

4   The fourth method, created by A. O. Gel'fond in 1949, will be the subject of this survey.
    Here is one classical example of a problem on which this method gives information. It was raised by Gel'fond (p. 127 of the english translation of his book) and by Schneider (7th problem of his book).

Problem of Gel'fond-Schneider. Let α be a non-zero algebraic number, $\log \alpha$ a non-zero determination of its logarithm, and β an algebraic number of degree $d = [\mathbb{Q}(\beta) : \mathbb{Q}]$ with $d \geq 2$. Then the $d-1$ numbers

$$\alpha^{\beta}, \ \alpha^{\beta^2}, \ \ldots, \ \alpha^{\beta^{d-1}}$$

(*)

are algebraically independent.

An equivalent formulation of this problem is to ask for the algebraic independence of $\alpha^{\beta_1}, \ldots, \alpha^{\beta_k}$ for algebraic α with $\alpha \neq 0$, $\log \alpha \neq 0$, and for algebraic $\beta_1, \ldots, \beta_k$ with $1, \beta_1, \ldots, \beta_k$ linearly independent over $\mathbb{Q}$.

Hilbert's seventh problem stated that each number in (*) is transcendental, and this was proved by Gel'fond and Schneider in 1934. One main step achieved by Gel'fond in 1948 was to prove that for $d \geq 3$, at least two of the numbers in (*) are algebraically independent. In particular for $d = 2$ and $d = 3$ the problem is solved.

Until 1970, Gel'fond's method was restricted to the proof of algebraic independence of two numbers, among certain values of the exponential function (see I §1 below). In the 70's the first works arise which yield the algebraic independence of at least three numbers for the exponential function (I §2), and at the same time the first results appear connected with elliptic functions (I §3).

The next step is to produce fields of large transcendence degree generated by values of the exponential function, or elliptic functions (and more generally by numbers connected with the exponential of an algebraic group). We know essentially three ways of moving in this direction. The first one originates from a preprint of Chudnovsky, Kiev in 1974 (II §1). The second one, which was published very recently, has been developed by Masser and Wüstholz, starting from their joint work on zero estimates on group varieties (II §2). The third one has been initiated by P. Philippon in connection with the elliptic analog of the Gel'fond-Schneider problem, and used by Philippon and Wüstholz for the elliptic analog of the Lindemann-Weierstrass theorem (II §3).

One important tool in most of the previously mentioned works is a transcendence criterion. The first one appears already in the work of Gel'fond in

1949, and has been improved later (III §1). Next we will see a criterion of algebraic independence of Chudnovsky-Reyssat (III §2), and finally another criterion due to Philippon (III §3).

It should be clear, at least at the end of this survey, that the theory is just at the beginning of its life. We will finish by listing a few conjectures in this field (IV).

I   Small transcendence degree

In this section, our aim is to produce some fields generated by values of exponential (§1 and §2) or elliptic (§3) functions, for which we know that the transcendence degree is at least two or at least three.

§ 1 Two algebraically independent values of the exponential function

The first result of algebraic independence produced by A. O. Gel'fond [1948] was the independence of the two numbers α^β, α^{β^2} for α algebraic, $\alpha \neq 0$, $\log \alpha \neq 0$, and β cubic. In [1949] and in his book [1], he gave more general results which were extended or refined later by A. A. Shmelev [1967], [1968a], [1968b], R. Tijdeman [1970a], W. D. Brownawell [1969], [1971b], [1971g], [1971i] and others [1971a], [1971f], [1971j], [3] Chap. 7, [4] Chap. 12, [6] Chap. 9 (see also the surveys [1966b], [1974c], and [1979e]).

The following result is now well-known.

Theorem 1. Let $x_1, \ldots, x_m$ be complex numbers which are linearly independent over $\mathbb{Q}$, and $y_1, \ldots, y_l$ be also complex numbers which are linearly independent over $\mathbb{Q}$.

a) *Assume $lm \geq 2\,(l+m)$. Then two at least of the lm numbers*

$$e^{x_i y_j}, \qquad (1 \leq i \leq m,\ 1 \leq j \leq l)$$

are algebraically independent.

b) *Assume $lm \geq l + 2m$. Then two at least of the $lm + l$ numbers*

$$y_j, e^{x_i y_j}, \qquad (1 \leq i \leq m,\ 1 \leq j \leq l)$$

are algebraically independent.

c) *Assume $lm > l + m$. Then two at least of the $lm + l + m$ numbers*

$$x_i, y_j, e^{x_i y_j}, \qquad (1 \leq i \leq m,\ 1 \leq j \leq l)$$

are algebraically independent.

d) *Assume $l = m = 2$, and assume also that $e^{x_1 y_1}$ and $e^{x_1 y_2}$ are algebraic. Then two at least of the six numbers*

$$x_1, \ x_2, \ y_1, \ y_2, \ e^{x_2 y_1}, \ e^{x_2 y_2}$$

are algebraically independent.

Further results, using Baker's method, have been given in [1972c], and by A. A. Shmelev [1973d] and G. V. Chudnovsky [1975c].

A measure of algebraic independence of α^β and α^{β^2} (for β cubic) was derived already in 1950 by Gel'fond and Fel'dman [1950], and improved by W. D. Brownawell [1975i], [1977a].

In [1977a], W. D. Brownawell constructs Liouville numbers a such that for β of degree 3, the three numbers a, a^β, a^{β^2} are algebraically independent (see also [1975b], [1975e] and [1977b]).

Further approximation estimates are due to A. A. Shmelev [1970c], [1973b], [1975g], [1978i], [1983a] p. 166—167.

Related works for the p-adic case have been done by W. W. Adams [1964] and T. N. Shorey [1972b] (see also [1972a]). It is worth it to emphasize here the fact that the p-adic results are still sometimes weaker than their complex analog. For example in the p-adic version of theorem 1 one needs a strict inequality in the assumptions a) and b). In particular for algebraic α and β in $\mathbb{C}_p$ with α neither zero nor root of 1, $|\alpha - 1|_p < 1$ and $[\mathbb{Q}(\beta) : \mathbb{Q}] = 2$, with $|\beta^j \log \alpha|_p < p^{-1/(p-1)}$ for $j = 1$ and $j = 2$, it is not yet proved that α^β and α^{β^2} are algebraically independent (see [3] Appendix).

Existing methods could give the feeling that the problem of algebraic independence of π and e^π should be easier than the problem of the algebraic independence of π and e. Anyway, both problems are still open.

§ 2 *Three algebraically independent values of the exponential function*

The easiest way to give a lower bound for the transcendence degree of some fields has been proposed by S. Lang [1966a], [2] Chap. 5. More or less, the main difficulty which is at the heart of most proofs of algebraic independence is now put in the hypotheses, under the name of "transcendence type". But checking this assumption involves the same kind of difficulties as proving a result of algebraic independence.

One can compare the level of difficulty for giving a measure of algebraic independence of s numbers and that for producing a transcendence degree $\geq s + 1$. For instance, transcendence measures are derived in [1949], [1], [1968b], using the method of algebraic independence of two numbers.

On the other hand, if one combines a measure of algebraic independence of s numbers with a method which yields a transcendence degree $\geq t$, one can hope to produce a transcendence degree $\geq t + s$.

For instance, using a transcendence measure, due to A. O. Gel'fond [1] Chap. III § 4, of α^β, W. D. Brownawell [1971b], [1974a] and A. A. Shmelev [1971d] proved the algebraic independence of three at least of the numbers

$$\alpha^\beta, \alpha^{\beta^2}, \ldots, \alpha^{\beta^{d-1}}$$

in the problem of Gel'fond-Schneider, provided that $d = [\mathbb{Q}(\beta):\mathbb{Q}]$ is sufficiently large: $d \geqq 19$ for [1971d], $d \geqq 15$ for [1974a].

After a first attempt by A. A. Shmelev [1971e], G. V. Chudnovsky succeeded to remove this assumption to $d \geqq 7$, by avoiding the use of a transcendence type [1973a]. This method is explained in [1975a], [1976d], and [1979e], and used in [1975b], [1975e], [1975i], [1977a] and [1977b].

From a very general claim by Chudnovsky in [1978g] th. 2 p. 347 (which should be considered as a conjecture) one deduces that the above result should hold also for $d \geqq 5$, and this seems to follow from the arguments given by Chudnovky in [1978h].

It is very likely that one can get the algebraic independence of three numbers with the hypotheses of theorem 1, provided that one replaces the assumptions by:

$$lm \geqq 3(l+m) \quad \text{for a),}$$

$$lm \geqq 2l+3m \quad \text{for b),}$$

and

$$lm > 2(l+m) \quad \text{for c).}$$

However, for technical reasons, one needs measures of linear independence of $x_1, \ldots, x_m$ and of $y_1, \ldots, y_l$.

A corresponding generalization of part d) of theorem 1 is also possible (see [1978j] for a first step in this direction).

§ 3 Small transcendence degree and elliptic functions

The possibility of producing fields of transcendence degree $\geqq 2$ generated by values of elliptic functions was suggested by S. Lang in [1966a] and [2] Chap. V § 3 (see also A. Altman [1970b] for abelian functions), and a kind of axiomatization was proposed in [1972a], but the first concrete examples were given by W. D. Brownawell and K. K. Kubota [1975d]. Further results are due to A. A. Shmelev [1975f], [1979d], but in all these cases the proven results are very far from the conjectural ones.

The first paper with sharp estimates is Chudnovsky's [1975h], which actually yields the algebraic independence of two numbers: if $\wp$ is an elliptic function with algebraic invariants g_2, g_3, and with complex multiplications in an imaginary quadratic field, each non-zero period ω of $\wp$ is algebraically independent of π. A consequence is the algebraic independence of $\Gamma(1/4)$ and π (the transcendence of $\Gamma(1/4)$ was not yet known!), and also of $\Gamma(1/3)$ and π (see [1975j] and [1976d]). Further results were announced by Chudnovsky at the Oberwolfach conference in 1977 (reference [6] of [1978g]; the lecture was delivered by W. D. Brownawell), and the following result can be proved essentially by the same arguments as [1975h] (see [1978g] th. 4 and [5] Lecture 8).

Theorem 2. Let $\wp$ be a Weierstrass elliptic function with algebraic invariants g_2, g_3, let ω be a non-zero period of $\wp$, η be the associated quasi-period of the Weierstrass zeta function:

$$\zeta(z + \omega) = \zeta(z) + \eta,$$

and u a complex number, which is not a pole of $\wp$, with u and ω linearly independent over $\mathbb{Q}$, and $\wp(u)$ algebraic. Then the two numbers

$$\zeta(u) - \frac{\eta}{\omega} u, \ \frac{\eta}{\omega}$$

are algebraically independent.

Several consequences of this result on the modular function and its derivatives have been given by D. Bertrand in [1978c], together with a *p*-adic analog. Bertrand pursued these investigations in [1978d] and [1978e] where he gave results of algebraic independence of the values of modular functions. (See also Ianchenko, [1983a] p. 174—175.)

At Helsinki [1978g], G. V. Chudnovsky announced further results (esp. th. 5 and 6 p. 344, and th. 4 p. 347), and at the same time he produced lists of similar statements (e. g. reference [7] of [1978g]; here we decided not to include unpublished manuscripts in our bibliography). The proofs of these statements are not published, but it is now a routine matter to check these results (and it has been done for the most interesting of them).

Concerning this subject we mention two papers by N. I. Fel'dman [1979c], [1980b], (cf. also [1983a] p. 148) and a paper by P. Philippon [1979f] where he introduces his "fausses variables" which will play an important role in 1982.

In the above mentioned meeting of Oberwolfach in March 1977 (which is referred to also p. 270 of [1979a]), G. V. Chudnovsky explained how Gel'fond's method for the algebraic independence of two numbers can be used to yield the case $n = 2$ of the Lindemann-Weierstrass theorem. From [1979a] § 3, using the arguments of [1978h], the same is true for $n = 3$. Then, in [1979a], he extends his method to the elliptic case (thanks to a subtle but technical device known as the Baker-Coates-Anderson lemma):

if $\wp$ has algebraic invariants g_2, g_3, and if α_1, ..., α_5 are algebraic numbers linearly independent over $\mathbb{Q}$, then among $\wp(\alpha_1)$, ..., $\wp(\alpha_5)$, there are at least three algebraically independent numbers.

In the case of complex multiplication, this yields the algebraic independence of $\wp(\alpha_1)$, $\wp(\alpha_2)$, $\wp(\alpha_3)$ when α_1, α_2, α_3 are linearly independent over the ring of endomorphisms.

Better results are announced in:

[1979b], th. 2.3 and 2.4: algebraic independence of four numbers

[1978g], th. 1 p. 348, and [1979a] §6: algebraic independence of six numbers

[1980d], th. 10.4 p. 71: algebraic independence of n numbers (with a measure!), but proofs are not given (see II §3 below).

The main difficulty in the proofs for the small transcendence degrees is to exhibit a non-zero value of the auxiliary function. Fundamental progress on this matter have been achieved in the last ten years, first by D. W. Masser, then by W. D. Brownawell and D. W. Masser, and later by D. W. Masser and G. Wüstholz (see the address [1983c] of D. W. Masser at the international congress of Warsaw).

Using the method of zero-estimates of Brownawell-Masser, G. Wüstholz [1979g] gave a general statement (in the style of the Schneider-Lang theorem) which yields the algebraic independence of two numbers among values of functions satisfying certain differential equations. In concrete examples this result gives rather crude estimates, but these can be improved as shown by E. Reyssat [1980a], [1980f]. One motivation for these works is the unsolved problem of the algebraic independence of the three numbers π, e^π, $\Gamma(1/4)$; [1980f] contains the best known partial results in this direction, involving elliptic integrals of second and third kind, and transcendence degree at least two or three.

The elliptic analog of theorem 1 has been obtained in [1980c] by Masser and Wüstholz as a direct consequence of their works on zero estimates on algebraic groups. For instance when $\wp$ has algebraic g_2, g_3 and complex multiplication by an imaginary quadratic field k, when u is a non-zero complex number such that $\wp(u)$ is defined and algebraic, and when β is algebraic of degree 3 over k, then the two numbers $\wp(\beta u)$ and $\wp(\beta^2 u)$ are algebraically independent. In [1982i], R. Tubbs gives a measure of algebraic independence for these two numbers.

Finally general results can now be obtained involving n-parameter subgroups of algebraic groups [1981d] §5, [1982h].

II Large transcendence degree

We begin with a method of Chudnovsky (§1), which works only for exponential functions, and yields rather weak estimates. Then we mention a method of Masser and Wüstholz (§2), which gave the first results for elliptic functions, with estimates of essentially the same quality. Fortunately, the third method (§ 3) gives sometimes best possible results.

§ 1 A method of Chudnovsky and related works

During his "Tagung über diophantische Approximationen" at Oberwolfach in July 1974, Th. Schneider received from G. V. Choodnovsky the canonical sheet (see the copy on the next page).

The first theorem was also stated in [1973a] (p. 398 in the english translation), [1974b] (with measures of algebraic independence in th. 4), and [1974d] th. 1.1. A sketch of the proof can be found in [1976d] and [1979e].

The only place where Chudnovsky provides a proof for large transcendence degree is the Kiev preprints [1974e]. It is difficult to answer the question of whether or not the proof given there is complete. The arguments are quite convoluted, and attempts to write down all the details involved non-trivial problems. The questions (especially from Warkentin) raised to Chudnovsky at Oberwolfach in May—June 1979 were not answered.

On the other hand there is no doubt that the method of [1974e] leads to fields of large transcendence degree generated by values of the exponential function. This has been checked by P. Warkentin in [1978f]. The proof of Philippon [1981a] and Reyssat [1981b] has been inspired by Chudnovsky's one, but the details are comparatively extremely easy. Further works on this subject, due to R. Endell [1981c], have been completed by W. D. Brownawell.

A rather different approach, due to Ju. V. Nesterenko [1982f], yields the same result.

Theorem 3. Let α and β be algebraic numbers, $\alpha \neq 0$, $\log \alpha \neq 0$, and let d $= [\mathbb{Q}(\beta):\mathbb{Q}]$ satisfy $d \geq 2$. Then the transcendence degree over $\mathbb{Q}$ of the field generated by

$$\alpha^\beta, \alpha^{\beta^2}, \ldots, \alpha^{\beta^{d-1}}$$

is at least $\left[\dfrac{\log(d+1)}{\log 2}\right]$.

(The brackets denote the integral part.)

In the p-adic case, the transcendence degree is at least $\left[\dfrac{\log d}{\log 2}\right]$ as shown by Philippon in [1981a].

A quantitative refinement of theorem 3 is given in [1982f] th. 2 (compare with th. 4 of [1974b]). For the proof of this estimate, Nesterenko develops the methods of commutative algebra which he introduced in the subject for estimating the multiplicity of zeros. (See also [1983a] p. 102—104.)

Mathematisches Forschungsinstitut
Oberwolfach

Tagung:

Vortragender: G.V. Choodnovsky (Kiev, USSR)

Thema des Vortrages: Several questions in the theory of transcendental and algebraically independent numbers

Vortragsdauer:

Kurze Zusammenfassung: (höchstens 15 Zeilen)

This work is devoted to the basic investigation of the arithmetic nature of numbers, connected with the values of exponent.
Let's denote by M, N and n any natural numbers (≥ 1) and by $\alpha_1, \ldots, \alpha_N$
$\beta_1, \ldots, \beta_M$ linearly independent over Q sequences of complex num-
bers. We choose three sets of numbers: $\overline{S}_1 = \{e^{\alpha_i \beta_j}\}$, $\overline{S}_2 = \{\beta_j,$
$e^{\alpha_i \beta_j}\}$, $\overline{S}_3 = \{\alpha_i, \beta_j, e^{\alpha_i \beta_j}\}$, $1 \leq i \leq N$, $1 \leq j \leq M$ and
such constants $\mathscr{R}_1 = MN/(M+N)$, $\mathscr{R}_2 = M(N+1)/(M+N)$, $\mathscr{R}_3 = \mathscr{R}_1 + 1$.

Theorem 1. If $\mathscr{R}_i \geq 2^n$ ($\mathscr{R}_3 > 2^n$ for $i = 3$), then in
the set $\overline{S}_i$ there exist $n+1$ algebraically independent numbers.

We obtain a good estimation of measure of transcendence in
the case of bounded height:

Theorem 3. For $P(x) \in Z[x]$, $P(x) \not\equiv 0$ of degree $\leq d$ and
height $\leq H$ we have

$$|P(e^\pi)| > exp(-c_1 \, d \ln H \ln^2 (d \ln H)),$$

$$|P(e)| > exp(-c_2 \, d^2 \ln(d H)).$$

The conclusion of theorem 3 can be written $2^t > (d+1)/2$, where t is the transcendence degree. The right hand side $(d+1)/2$ is, to a certain extent, the limit in the present stage of Gel'fond's method. The left hand side 2^t arises from the criterion of algebraic independence or the induction procedure, and one should be able to replace it by $t+1$. The conclusion would be $t \geq [(d+1)/2]$, which is "half" of the Gel'fond-Schneider problem. This result was claimed first by Gel'fond in [1948] p. 280, then by Chudnovsky (th. 2 p. 347 of his Helsinki adress [1978g]; see also p. 288—290 of [1976d] which is taken from a letter of Chudnovsky dated May 1976). The only information given by Chudnovsky on his proof is that it uses the resolution of singularities ([1978h] p. 1—2, [1979a] p. 267, [1980d] p. 50).

He tried also to generalize his method to elliptic functions. In [1979a] p. 268 he agrees that he did not succeed to solve the technical difficulties, but in [1980d] (p. 15 and p. 70—71) he states general results (even with a measure of algebraic independence). Overly optimistic statements can be found in [1980e] (see Masser's comments in Zbl. 456.10016).

Apart from his Kiev preprints [1974e], all the proofs provided by Chudnovsky so far deal only with small transcendence degrees.

§ 2 A method of Masser and Wüstholz

In their joint paper [1982d], Masser and Wüstholz develop a new method for obtaining results of algebraic independence. Here is the concrete case they work out.

Theorem 4. Let $\wp$ be an elliptic function with algebraic invariants g_2, g_3, and without complex multiplication. Let $x_1, \ldots, x_d, y_1, \ldots, y_l$ be complex numbers. Assume that $\varkappa > 0$ is such that, for sufficiently large A, B, and for all integers $(a_1, \ldots, a_d, b_1, \ldots, b_l)$ with $\max |a_i| = A$ and $\max |b_j| = B$, we have

$$|a_1 x_1 + \ldots + a_d x_d| \geq \exp(-A^\varkappa)$$

and

$$|b_1 y_1 + \ldots + b_l y_l| \geq \exp(-B^\varkappa).$$

Let K be the field generated by the values of $\wp$ at the points $x_i y_j$, $(1 \leq i \leq d, 1 \leq j \leq l)$ where $\wp$ is defined, and let t be the transcendence degree of K over $\mathbb{Q}$. Then

$$2^{t+2}(t+8) + 4\varkappa > ld/(l+2d).$$

This is the first result giving many algebraically independent values of elliptic functions.

The starting point is an improved version of their zero estimate. The second main tool is an effective version of the Hilbert Nullstellensatz. Next they need an explicit description of all algebraic subgroups of a power of an elliptic curve, and for that they have to improve a result of Kolchin.

This method involves a rather heavy machinery, and it is not clear whether it can be extended to a general commutative algebraic group (defined over a finitely generated extension of $\mathbb{Q}$).

The final lower bounds are the logarithms of that one expects. However this is due only to the weak estimate in the effective Nullstellensatz. An important problem is to improve this estimate.

§3 A method of Philippon and related works

This method arised in connection with the elliptic analog of Gel'fond-Schneider's problem. One main idea [1982a] is to use the action of certain endomorphisms on an algebraic group (in a different situation, Bertrand and Masser had used similar actions to deduce results of linear independence from the theorem of Schneider-Lang). Then Philippon uses a transcendence criterion of his own (see III §3 below). The following statement of his thesis [1983b] improves his earlier results of [1982a] and [1982b].

Theorem 5. Let $\wp$ be an elliptic function with algebraic g_2, g_3, k be the field of endomorphisms of the elliptic curve, u a non-zero complex number, and β an algebraic number of degree $d \geq 2$ over k. Assume that $u, \beta u, \ldots, \beta^{d-1} u$ are not poles of $\wp$. Then the transcendence degree t over $\mathbb{Q}$ of the field generated by

$$\wp(u), \ \wp(\beta u), \ \ldots, \ \wp(\beta^{d-1} u)$$

satisfies

$$t \geq [(d-1)/3] \quad if \quad k = \mathbb{Q}$$

and

$$t \geq [(d-1)/2] \quad if \quad k \neq \mathbb{Q}.$$

One expects $t \geq d-1$ in both cases. It is remarquable that the elliptic result (th. 5) is stronger than its exponential analog (th. 3).

The results of [1982a], [1982b] and [1983b] are more general, dealing with abelian varieties, and the action of certain endomorphisms.

The same kind of action occurs also in the papers [1982c], [1982g] of Wüstholz, and [1982e] of Philippon, connected with Chudnovsky's method of [1979a]. We do not quote the general statements, but only the following striking consequence:

Theorem 6. Let $\wp$ be an elliptic function with algebraic invariants g_2, g_3, and with complex multiplications in an imaginary quadratic field k. Let $\alpha_1, \ldots, \alpha_n$ be algebraic numbers which are linearly independent over k. Then the n numbers

$$\wp(\alpha_1), \ldots, \wp(\alpha_n)$$

are algebraically independent.

The method is extended in [1982h] and [1983b] to n-parameter subgroups of an abelian variety.

A difficult problem here is to extend the method for the study of non-complete group varieties. The difficulty arises from the use of Philippon's criterion or the elimination procedure.

Another limitation of the method is due to the hypotheses on the endomorphisms which do not enable one to study general situations like in th. 1 and th. 4.

III   Criteria of algebraic independence

One of the basic tools in Gel'fond's method is a criterion which replaces the fundamental Liouville estimate (size inequality) of transcendence proofs. We consider first the criteria in one or two variables (§ 1), then an extension to t variables (§ 2). The method, by induction, yields an exponent 2^t in general. A different approach, based on Nesterenko's work on commutative algebra, enables Philippon to get another criterion with the right exponent $t + 1$.

§ 1 Gel'fond's criterion, and first generalizations

a) *One variable over* $\mathbb{Q}$

Gel'fond's criterion appears first in [1949] and in his book [1] (Chap. III § 4 lemma 7). Refinements are given in [1965] § 6, [1970a] lemmas 6 and 6', [1971a] § 3, [1971h], [1976c] critère 2.4 et remarque 2.10. See also [3] Chap. 5, [4] Chap. 12 § 4, [5] § 8.2, [6] lemma 3.9, and the expositions of Brownawell [1979e] and Chudnovsky [1979h].

For $P \in \mathbb{C}[X]$, $P \neq 0$, we write $\deg P$ for its degree, and $H(P)$ for the maximal absolute value of its coefficients.

Theorem 7. Let θ be a complex number, $(t_N)_{N \geq 0}$ and $(d_N)_{N \geq 0}$ two sequences of numbers ≥ 1, and $(P_N)_{N \geq 1}$ a sequence of non-zero elements of $\mathbb{Z}[X]$. We define

$$T_N = 2(d_{N-1} + d_N + d_{N+1})(t_{N-1} + t_N + t_{N+1}), \quad (N \geq 1),$$

and we assume $\lim T_N = +\infty$ as $N \to +\infty$. We assume further that for all $N \geq 1$,

$$\deg P_N \leqq d_N, \ \deg P_N + \log H(P_N) \leqq t_N,$$

and

$$|P_N(\theta)| \leqq e^{-T_N}.$$

Then θ is algebraic and $P_N(\theta) = 0$ for all $N \geq 1$.
    Here is a sketch of the proof. Since $|P_N(\theta)|$ is small, θ is close to a root α_N of P_N. Let s_N be its multiplicity:

$$|\theta - \alpha_N|^{s_N} \leqq e^{-T'_N}, \quad \text{where} \quad T'_N = T_N - 2 d_N t_N.$$

Define $d'_N = d_N/s_N$, $t'_N = t_N/s_N$. It is readily verified that

$$|\alpha_N - \alpha_{N+1}| \leqq 2 \exp\left(-d'_N d'_{N+1} - t'_N d'_{N+1} - t'_{N+1} d'_N\right).$$

On the other hand Liouville's inequality shows that this estimate cannot hold unless $\alpha_N = \alpha_{N+1}$. The theorem follows at once.

    b) *One variable with finite transcendence type*
    The preceeding proof works as well if one replaces the rational field by a finitely generated extension of $\mathbb{Q}$ on which one assumes some type of Liouville inequality (Lang's transcendence type [2] Chap. V). This was worked out by W. D. Brownawell [1969], [1974a] and A. A. Shmelev [1973c]. A further result was obtained by Chudnovsky [1976a] using a rather subtle argument (coloured sequences; see also [1975a], [1976c], [1978h] Coroll. 5.4, [1979e], [1979h]).

    c) *Two variables*
    From a well-known example due to Cassels (see for instance [1971c] p. 670), the obvious generalization of Gel'fond's criterion to several variables ([1965] p. 191—192) does not hold without further assumptions.
    The first generalization of this type was given already for t variables (see §2 below). However it is worthwhile mentionning that better results are available for two variables. The two papers [1978h] and [1980d] by Chudnovsky contain description of methods which lead to sharp results for the algebraic independence of three numbers, and his paper [1979h] contains explicitly criteria of algebraic independence in two variables ([1979h], th. 4.1 p. 339—340; see also [1980e], Prop. 1 and 13).

§ 2 *A criterion of Chudnovsky-Reyssat*

    The Kiev preprints [1974e] contain a description of Chudnovsky's method for algebraic independence of t numbers, with some details. They do not contain an explicit criterion (the quotation in [1979h] p. 326 is misleading). Such a criterion is stated without proof in [1974d] lemma 1.2 p. 23 (probably the conclusion should be algebraically dependent, rather than algebraically independent, but anyway one can show [1983d] that the hypotheses are never satisfied—thus the result is true!).

The idea of Chudnovsky is to assume not only an upper bound for the values of the polynomials, but also a lower bound. In the transcendence proof this new hypothesis is checked by using a "small value lemma" for exponential polynomials, due to R. Tijdeman.

A more general criterion has been stated—and proved— by E. Reyssat in [1981b]. The proof uses several ideas of Chudnovsky in [1974e], and especially the semi-resultant (see also [1976b] and [1979e]). The induction performed by Reyssat in [1981b] is rather more tricky than it appears. See also [1983d].

Unfortunately the induction procedure leads to an exponent 2^t where one expects $t+1$. There are claims by Chudnovsky of a criterion with the right exponent (e. g. [1974d] p. 30, [1979h] Prop. 5.1 p. 357—358). However they are not supported by proofs, so far, and Philippon noticed that Cassel's construction leads to counter-examples to the statements at the end of § 5 of [1979h] p. 359.

§ 3 A criterion of Philippon

A generalization of Gel'fond's criterion to several variables was proposed by R. Dvornicich in [1978a]. He replaces the sequence of polynomials by a sequence of ideals in a polynomial ring. However for the proof he had to assume that the ideals are prime, and this hypothesis is too strong for applications.

Using methods of commutative algebra (see also [1982f]), Philippon succeeded in proving a very good criterion ([1982a] Prop. 1, [1982b] Th. 1.4, and [1983b] Chap. 1), which he used for his proof of theorem 5 (see II § 3), and which will certainly have other applications later.

For simplicity we quote only a special case of a criterion in [1983a].

Theorem 8. For an integer $n \geq 1$, let $\theta \in \mathbb{P}_n(\mathbb{C})$, let $(\theta_0, \ldots, \theta_n) \in \mathbb{C}^{n+1}$ be projective coordinates of θ, and $\mathscr{E}$ a prime homogeneous ideal of the ring $A = \mathbb{Q}[X_0, \ldots, X_n]$, of codimension $n - t$, such that $e(\theta) = 0$ for all $e \in \mathscr{E}$. Let a be a real number, $a > 1$.

There exists a constant $C > 0$ with the following property. Let $(t_N)_{N \geq 1}$, $(d_N)_{N \geq 1}$ be two sequences of numbers ≥ 1, with

$$d_N \leq d_{N+1} \leq a d_N, \quad t_N \leq t_{N+1} \leq a t_N \quad \text{for all} \quad N \geq 1,$$

and

$$\lim d_N = +\infty \quad as \quad N \rightarrow +\infty.$$

Let $(I_N)_{N \geq 1}$ be a sequence of homogeneous ideals of A of codimension $\geq n$; assume that for each $N \geq 1$ there exist an integer $m(N) \geq 1$ and homogeneous polynomials $Q_1^{(N)}, \ldots, Q_{m(N)}^{(N)}$ in $\mathbb{Z}[X_0, \ldots, X_n]$ with

$$I_N = (\mathscr{E}, Q_1^{(N)}, \ldots, Q_{m(N)}^{(N)}),$$
$$\deg Q_j^{(N)} + \log H(Q_j^{(N)}) \leq t_N, \ \deg Q_j^{(N)} \leq d_N,$$

and

$$|Q_j^{(N)}(\theta_0, \ldots, \theta_n)| \leq \exp(-C t_N d_N^t),$$

for $1 \leq j \leq m(N)$. Then for all sufficiently large N, θ is a zero of I_N. In particular $\theta \in \mathbb{P}_n(\overline{\mathbb{Q}})$, where $\overline{\mathbb{Q}}$ is the field of algebraic numbers.

When infinitely many of the ideals I_N are of codimension $n + 1$, Philippon uses an effective elimination procedure (see e. g. [1982b] th. 1.3). For this special case (which is sufficient in the proof of theorem 6, II § 3 above), Wüstholz uses the *U*-resultant of Macaulay [1982g].

The method of Masser-Wüstholz in [1982d] (see II § 2) does not use a transcendence criterion explicitly, but the elimination is performed through the use of Hilbert's Nullstellensatz.

IV   Further conjectures

We consider first the usual exponential function, then elliptic functions, and finally algebraic groups.

§ 1 Exponential function

A very general conjecture concerning the transcendence and algebraic independence properties of the exponential function has been made by S. Schanuel [2] p. 30.

Schanuel's conjecture. Let $x_1, \ldots, x_n$ be complex numbers which are linearly independent over $\mathbb{Q}$. Then the transcendence degree of the field

$$\mathbb{Q}(x_1, \ldots, x_n, e^{x_1}, \ldots, e^{x_n})$$

over $\mathbb{Q}$ is at least n.

Most statements (either proved or conjectural ones) concerning algebraic independence of numbers connected with exponential or logarithms are consequences of this conjecture (see [3] §7.5 + Exercices 7.5 a, ..., e). This is the case for instance of the Gel'fond-Schneider problem stated in the introduction, and of the problem of the algebraic independence of $\log \alpha_1, \ldots, \log \alpha_n$, when $\alpha_1, \ldots, \alpha_n$ are non-zero algebraic numbers and $\log \alpha_1, \ldots, \log \alpha_n$ are $\mathbb{Q}$-linearly independent ([1] p. 127 and p. 177, [2] p. 31, [3] conj. 7.5.3, [4] p. 119—120).

The p-adic analog of Schanuel's conjecture is stated in [1981a]. An open problem is to prove a p-adic version of the Lindemann-Weierstrass theorem (see [3] p. A 9).

Finally, let us try a first generalization of Schanuel's conjecture in the realm of diophantine approximations.

Let $x_1, \ldots, x_n$ be complex numbers which are linearly independent over $\mathbb{Q}$. Let d be a positive integer. Then there exists a positive number $C = C(x_1, \ldots, x_n, d)$ with the following property: for all $P_1, \ldots, P_{n+1}$ in $\mathbb{Z}[X_1, \ldots, X_n, Y_1, \ldots, Y_n]$ of degrees $\leq d$ and heights $\leq H_1, \ldots, H_{n+1}$, which generate an ideal of $\mathbb{Q}[X_1, \ldots, X_n, Y_1, \ldots, Y_n]$ of rank $n+1$, we have

$$\sum_{j=1}^{n+1} |P_j(x_1, \ldots, x_n, e^{x_1}, \ldots, e^{x_n})| \cdot H_j^C \geq 1/C.$$

§ 2 Elliptic functions

Let $\wp$ be an elliptic function of Weierstrass. We denote by g_2, g_3 the invariants, ω_1, ω_2 a pair of fundamental periods, and η_1, η_2 the associated quasiperiods. According to Chudnovsky [1978g] [4] p. 343, the transcendence degree of the field $\mathbb{Q}(g_2, g_3, \omega_1, \omega_2, \eta_1, \eta_2)$ over $\mathbb{Q}$ is at least 2. It can be as small as 2 (when $\wp$ has complex multiplications). It would be interesting to decide whether this bound can be improved when there is no complex multiplication: in this case, for g_2 and g_3 algebraic, is it true that the four numbers $\omega_1, \omega_2, \eta_1, \eta_2$ are algebraically independent?

Another problem is to list the algebraic relations between the periods and quasi-periods of several $\wp$ functions. For instance one would like to know that the three numbers $\pi, \Gamma(1/4), \Gamma(1/3)$ are algebraically independent.

The analog of the Lindemann-Weierstrass theorem for elliptic functions without complex multiplication is not yet known. One would like to prove at least "half" of it: the transcendence degree over $\mathbb{Q}$ of $\mathbb{Q}(\wp(\alpha_1), \ldots, \wp(\alpha_n))$ is at least $n/2$ (for $\alpha_1, \ldots, \alpha_n$ algebraic numbers $\mathbb{Q}$-linearly independent). Philippon [1983b] proves this result under the additional assumption $\mathbb{Q}(\alpha_1, \ldots, \alpha_n) = \mathbb{Q}\alpha_1 + \ldots + \mathbb{Q}\alpha_n$.

§ 3 Algebraic groups

A proof of the transcendence of $\Gamma(1/5)$ (and other values of the gamma function) would follow from a complete description of the algebraic relations between the coordinates of the periods of abelian functions (see [2] Chap. IV, historical note, [1971c], and the papers by Deligne, Ribet and Shimura in "Fonctions Abéliennes et Nombres Transcendants", Mémoire S. M. F. n° 2 [1980]).

The following problem is proposed by Philippon [1983b] as a generalization of the Lindemann-Weierstrass theorem: *let G be an algebraic group which is defined over the field $\bar{\mathbb{Q}}$ of algebraic numbers, $\alpha \in T_G(\bar{\mathbb{Q}})$, X the Zariski closure of $\exp_G \alpha$ over $\bar{\mathbb{Q}}$, and H the smallest algebraic subgroup of G containing $\exp_G \alpha$. Assuming $\mathrm{Hom}_{\bar{\mathbb{Q}}}(H, \mathbb{G}_a) = 0$, is it true that X is a connected component of H?*

As a consequence, if A is a simple abelian variety of dimension d (without assumptions on its endomorphisms), for $\alpha \in T_A(\bar{\mathbb{Q}})$, $\alpha \neq 0$, the point $\exp_G \alpha$ has a transcendence degree $\geq d$ (see [2] p. 42).

References

We first quote several biographies of Alexandre Ossipovich Gel'fond, which contain complete lists of his published work:
Uspekhi Mat. Nauk, **11** n° 5 (1956), 239—248; **22** n° 3 (1967), 247—256; **24** n° 3 (1969), 219—220; **25** n° 1 (1970), 201—202; Acta Arith., **17** (1970/71), 315—336.
The second and third are translated in:
Russian Math. Surveys, **22** n° 3 (1967), 234—242, and **24** n° 3 (1969), 177—178.
Selected works of A. O. Gel'fond have been published in Russian (Izd. Nauk Mosc. 1973).

Bibliography

[1]   Gel'fond, A. O., *Transcendental and algebraic numbers*, GITTL Moscow 1952. English transl.: Dover, New-York, 1960.
[2]   Lang, S., *Introduction to transcendental numbers*, Addison Wesley 1966.
[3]   Waldschmidt, M., *Nombres transcendants*, Lecture Notes in Math., **402** (1974), Springer Verlag.
[4]   Baker, A., *Transcendental number theory*, Cambridge Univ. Press, 1975; 2nd Ed.: 1979.
[5]   Waldschmidt, M., *Transcendence methods*, Queen's papers in pure and applied Math., **52** (1979), Kingston (Ont., Canada).
[6]   Fel'dman, N. I., *Hilbert's seventh problem*, Izdat. Mosk. Univ., 1982.

Papers

[1948] Gel'fond, A. O., *On the algebraic independence of algebraic powers of algebraic numbers*, Dokl. Akad. Nauk SSSR, **64** (1949), 277—280.

[1949] Gel'fond, A. O., *On the algebraic independence of transcendental numbers of certain classes*, Dokl. Akad. Nauk SSSR, **67** (1949), 13—14; Usp. Mat. Nauk **4** fasc. 5 (1949), 14—48. Engl. transl.: Amer. Math. Soc. Transl., (1) 2, **66** (1952), 125—169.

[1950] Gel'fond, A. O. and N. I. Fel'dman, *On the measure of mutual transcendence of certain numbers*, Izv. Akad. Nauk SSSR Ser. Mat., **14** (1950), 493—500.

[1964] Adams, W. W., *Transcendental numbers in the p-adic domain*, Amer. J. Math., **88** (1966), 279—308.

[1965] Lang, S., *Report on diophantine approximations*, Bull. Soc. Math. France, **93** (1965), 177—192.

[1966a] Lang, S., *Nombres transcendants*, Sém. Bourbaki, 18è année (1965/66), n° 305, 8 p.

[1966b] Fel'dman, N. I. and A. B. Shidlovskii, *The development and present state of the theory of transcendental numbers*, Usp. Mat. Nauk SSSR, **22** n° 3 (1967), 3—81. Engl. transl.: Russian Math. Surveys, **22** n° 3 (1967), 1—79.

[1967] Shmelev, A. A., *Concerning algebraic independence of some transcendental numbers*, Mat. Zam., **3** (1968), 51—58. Engl. Transl.: Math. Notes, **3** (1968), 31—35.

[1968a] Shmelev, A. A., *On algebraic independence of some numbers*, Mat. Zam., **4** (1968), 525—532. Engl. transl.: Math. Notes, **4** (1968), 805—809.

[1968b] Shmelev, A. A., *A. O. Gel'fond's method in the theory of transcendental numbers*, Mat. Zam., **10** (1971), 415—426. Engl. transl.: Math. Notes, **10** (1971), 672—678.

[1969] Brownawell, W. D., Ph. D., Cornell Univ., 1969.

[1970a] Tijdeman, R., *On the algebraic independence of certain numbers*, Proc. Nederl. Akad. Wet. Ser. A, **74** (= Indag. Math. **33**), (1971), 146—162.

[1970b] Altman, A., *The size function on abelian varieties*, Trans. Amer. Math. Soc., **164** (1972), 153—161.

[1970c] Shmelev, A. A., *On joint approximations of values of an exponential function at non-algebraic points*, (in russian, with engl. summ.), Vestn. Mosk. Univ. Ser. Mat. Mec., **27** n° 4 (1972), 25—33.

[1971a] Waldschmidt, M., *Indépendance algébrique des valeurs de la fonction exponentielle*, Bull. Soc. Math. France, **99** (1971), 285—304.

[1971b] Brownawell, W. D., *Some transcendence results for the exponential function*, Norske Vid. Selsk. Sk., **11** (1972), 2 p.

[1971c] Lang, S., *Transcendental numbers and diophantine approximations*, Bull. Amer. Math. Soc., **77** (1971), 635—677.

[1971d] Shmelev, A. A., *On the question of the algebraic independence of algebraic powers of algebraic numbers*, Mat. Zam., **11** (1972), 635—644. Engl. transl.: Math. Notes, **11** (1972), 387—392.

[1971e] Shmelev, A. A., *The arithmetic properties of the values of the exponential function at non algebraic points*, Izv. Visš. Uč. Zav. Mat., **10** (137), (1973), 90—99.

[1971f] Waldschmidt, M., *Solution du huitième problème de Schneider*, J. Number Theory, **5** (1973), 191—202.

[1971g] Brownawell, W. D., *The algebraic independence of certain values of the exponential function*, Norske Vid. Selsk. Sk., **23** (1972), 5 p.

[1971h] Brownawell, W. D., *Sequences of diophantine approximations*, J. Number Theory, **6** (1974), 11—21.

[1971i] Brownawell, W. D., *The algebraic independence of certain numbers related by the exponential function*, J. Number Theory, **6** (1974), 22—31.

[1971j] Wallisser, R., *Habilitation Schrift*, Freiburg 1971 (cf. Zbl. 234, 10021).

[1972a] Waldschmidt, M., *Propriétés arithmétiques des valeurs de fonctions méromorphes algébriquement indépendantes*, Acta Arith., **23** (1973), 19—88.

[1972b] Shorey, T. N., *Algebraic independence of certain numbers in the p-adic domain*, Proc. Nederl. Akad. Wet. Ser. A, **75** (= Indag. Math., **34**), (1972), 423—435.

[1972c] Waldschmidt, M., *Utilisation de la méthode de Baker dans des problèmes d'indépendance algébrique*, C. R. Acad. Sci. Paris Sér. A, **275** (1972), 1215—1217.

[1973a] Chudnovskii, G. V., *Algebraic independence of some values of the exponential function*, Mat. Zam., **15** (1974), 661—672. Engl. transl.: Math. Notes, **15** (1974), 391—398.

[1973b] Shmelev, A. A., *Simultaneous approximations of exponential functions by polynomials in a given transcendental number*, Ukr. Mat. Z., **27** (1975), 555—563. Engl. transl.: Ukr. Math. J., **27** (1975), 459—466.

[1973c] Shmelev, A. A., *A criterion for algebraic dependence of transcendental numbers*, Mat. Zam., **16** (1974), 553—562. Engl. transl.: Math. Notes, **16** (1974), 921—926.

[1973d] Shmelev, A. A., *Algebraic independence of exponent*, Mat. Zam., **17** (1975), 407—418. Engl. transl.: Math. Notes, **17** (1975), 236—243.

[1974a] Brownawell, W. D., *Gel'fond's method for algebraic independence*, Trans. Amer. Math. Soc., **210** (1975), 1—26.

[1974b] Čudnovskiĭ, G. V., *A mutual transcendence measure for some classes of numbers*, Dokl. Akad. Nauk SSSR, **218** (1974), n° 4, 771—774. Engl. transl.: Soviet Math. Dokl., **15** (1974), 1424—1428.

[1974c] Tijdeman, R., *On the Gel'fond-Baker method and its applications*, in: *Mathematical developments arising from Hilbert problems*, Proc. Symp. Pure Math., **28** (1976), 241—268.

[1974d] Čudnovskiĭ, G. V., *The Gel'fond-Baker method in problems of diopthantine approximation*, Coll. Math. Soc. Janos Bolyai, **13** Topics in Number Theory, Debrecen 1974, North Holland (1974), 19—30.

[1974e] Čudnovskiĭ, G. V., *Some analytic methods in the theory of transcendental numbers*, Inst. of Math., Ukr. SSR Acad. Sci., Preprint IM 74—8 (48 p.) and 74—9 (52 p.), Kiev 1974.

[1975a] Waldschmidt, M., *Indépendance algébrique par la méthode de G. V. Čudnovskij*, Sém. Delange-Pisot-Poitou (Groupe d'Etude de théorie des nombres), 16è année (1974/75), G 8, 18 p.

[1975b] Mignotte, M., *Indépendance algébrique de certains nombres de la forme α^β et α^{β^2}* (*d'après W. Dale Brownawell et Michel Waldschmidt*), Sém. Delange-Pisot-Poitou (Groupe d'Etude de théorie des nombres), 16è année (1974/75), G9, 5 p.

[1975c] Chudnovskii, G. V., *Baker's method in the theory of transcendental numbers*, Usp. Mat. Nauk, **31** n° 4 (190), (1976), 281—282.

[1975d] Brownawell, W. D. and K. K. Kubota, *The algebraic independence of Weierstrass functions and some related numbers*, Acta Arith., **33** (1977), 111—149.

[1975e] Brownawell, W. D. and M. Waldschmidt, *The algebraic independence of certain numbers to algebraic powers*, Acta Arith., **32** (1977), 63—71.

[1975f] Shmelev, A. A., *Algebraic independence of values of exponential and elliptic functions*, Mat. Zam., **20** (1976), 195—202. Engl. transl., Math. Notes, **20** (1976), 669—673.

[1975g] Shmelev, A. A., *Simultaneous approximations of exponent by transcendental numbers of certain classes*, Mat. Zam., **20** (1976), 305—314. Engl. transl.: Math. Notes, **20** (1976), 731—736.

[1975h] Čudnovskiĭ, G. V., *Algebraic independence of constants connected with exponential and elliptic functions*, Dokl. Ukr. SSR, Ser A, n° 8 (1976), 698—701.

[1975i] Brownawell, W. D., *Pairs of polynomials small at a number to certain algebraic powers*, Sém. Delange-Pisot-Poitou (Théorie des Nombres), 17è année (1975/76), n° 11, 12 p.

[1975j] Bertrand, D., *Transcendance de valeurs de la fonction gamma (d'après G. V. Čudnovskij)*, Sém. Delange-Pisot-Poitou (Groupe d'Etude de théorie des nombres), 17è année (1975/76), G 8, 5 p.

[1976a] Čudnovskiĭ, G. V., *Towards the Schanuel hypothesis. Algebraic curves near the point*. I: *general theory of coloured sequences*. II: *fields of finite type of transcendence and coloured sequences. Resultants*; (in russian); Studia Sci. Math. Hungar., **12** (1977), 125—144 and 145—157.

[1976b] Brownawell, W. D., *Some remarks on semi-resultants*, Chap. 14 (p. 205—210) of: *Transcendence theory: advances and applications*, ed. A. Baker and D. W. Masser, Proc. Conf. Cambridge, 1976, Academic Press 1977.

[1976c] Waldschmidt, M., *Suites colorées (d'après G. V. Čudnovskij)*, Sém. Delange-Pisot-Poitou (Groupe d'Etude de théorie des nombres), 17è année (1975/76), G 21, 11 p.

[1976d] Waldschmidt, M., *Les travaux de G. V. Čudnovskiĭ sur les nombres transcendants*, Sém. Bourbaki, 28è année (1975/76), n° 488; Lecture Notes in Math., **567** (1977), 274—292.

[1977a] Brownawell, W. D., *On the Gel'fond-Fel'dman measure of algebraic independence*, Comp. Math., **38** (1979), 355—368.

[1977b] Laurent, M., *Indépendance algébrique de nombres de Liouville élevés à des puissances algébriques*, C. R. Acad. Sci. Paris, Sér. A, **286** (1978), 131—133. Thèse 3è cycle, Univ. Paris VI, 1977.

[1978a] Dvornicich, R., *A criterion for the algebraic independence of two complex numbers*, Boll. UMI, (5) **15** A (1978), 678—687.

[1978b] Brownawell, W. D., *Methods for algebraic independence*, Sém. Théorie des Nombres (Bordeaux), 1977/78, n° 21, 7 p.

[1978c] Bertrand, D., *Fonctions modulaires, courbes de Tate et indépendance algébrique*, Sém. Delange-Pisot-Poitou (Théorie des Nombres), 19è année (1977/78), n° 36, 11 p.

[1978d] Bertrand, D., *Modular functions and algebraic independence*, Proc. Conf. "*p*-adic analysis", Nijmegen 1978; Kath. Univ. Report n° 7806.

[1978e] Bertrand, D., *Fonctions modulaires et indépendance algébrique II*, Journées Arithmétiques Luminy, Soc. Math. France Astérisque, **61** (1979), 29—34.

[1978f] Warkentin, P., *Algebraische Unabhängigkeit gewisser p-adischer Zahlen*, Diplomarbeit, Freiburg, 1978.

[1978g] Chudnovsky, G. V., *Algebraic independence of values of exponential and elliptic functions*, Proc. Intern. Cong. Math., Helsinki 1978, 339—350 (cf. M. R. 81j: 10051).

[1978h] Chudnovsky, G. V., *Algebraic grounds for the proof of algebraic independence. How to obtain measure of algebraic independence using elementary methods. Part I: elementary algebra*, Comm. Pure Appl. Math., **34** (1981), 1—28. [*Elementary approach II. Intersection of two curves*. Preprint, Bures sur Yvette, 1979].

[1978i] Shmelev, A. A., *Simultaneous approximations of exponentials by elements of a field* $\mathbb{Q}_1$, Mat. Zam., **30** (1981), 3—12. Engl. transl.: Math. Notes, **30** (1981), 487—492.

[1978j] Shmelev, A. A., *Analog of the Brownawell-Waldschmidt theorem on transcendental numbers*, Mat. Zam., **32** (1982), 765—775. Engl. transl.: Math. Notes, **32** (1982), 868—874.

[1979a] Chudnovsky, G. V., *Algebraic independence of the values of elliptic functions at algebraic points; elliptic analogue of the Lindemann-Weierstrass theorem*, Invent. Math., **61** (1980), 267—290.

[1979b] Chudnovsky, G. V., *Indépendance algébrique des valeurs d'une fonction elliptique en des points algébriques. Formulation des résultats*, C. R. Acad. Sci. Paris Sér. A, **288** (1979), 439—440.

[1979c] Fel'dman, N. I., *The algebraic independence of certain numbers*, Vestn. Mosk. Un.-ta., Ser. 1 Mat. Mec., fasc. 4 (1980), 46—50.

[1979d] Shmelev, A. A., *Algebraic independence of certain numbers connected with the exponential and elliptic functions*, Ukr. Mat. Z., **33** (1981), 277—282. Engl. transl.: Ukr. Math. J., **33** (1981), 216—220.

[1979e] Brownawell, W. D., *On the development of Gel'fond's method*, Proc. Number Theory Carbondale 1979, Lecture Notes in Math., **751** (1979), 16—44.

[1979f] Philippon, P., *Indépendance algébrique de valeurs de fonctions elliptiques p-adiques*, Proc. Queen's Number Theory Conf. 1979 (Ed. P. Ribenboim), Queen's Papers in pure and applied Math., **54** (1980), 223—235.

[1979g] Wüstholz, G., *Algebraische Unabhängigkeit von Werten von Funktionen, die gewissen Differentialgleichungen genügen*, J. reine angew. Math. (Crelle), **317** (1980), 102—119.

[1979h] Chudnovsky, G. V., *Criteria of algebraic independence of several numbers*, in: *The Riemann problem, complete integrability and arithmetic applications*, Proc. IHES and Columbia Univ., 1979/80, ed. D. Chudnovsky and G. Chudnovsky, Lecture Notes in Math., **925** (1982), 323—368.

[1980a] Reyssat, E., *Fonctions de Weierstrass et indépendance algébrique*, C. R. Acad. Sci. Paris Sér. A, **290** (1980), 439—441.

[1980b] Fel'dman, N. I., *Algebraic independence of some numbers II*, Ann. Univ. Sci. Budapest Sec. Math., **25** (1982), 109—123.

[1980c] Masser, D. W. and G. Wüstholz, *Algebraic independence properties of values of elliptic functions*, in: *Journées Arithmétiques 1980*, Ed. J. V. Armitage, London Math. Soc. Lect. Note Ser., **56** (1982), 360—363, Cambridge Univ. Press.

[1980d] Chudnovsky, G. V., *Measures of irrationality, transcendence and algebraic independence*, in: *Journées Arithmétiques 1980*, Ed. J. V. Armitage, London Math. Soc. Lect. Note Ser., **56** (1982), 11—82, Cambridge Univ. Press.

[1980e] Chudnovsky, G. V., *Indépendance algébrique dans la méthode de Gel'fond-Schneider*, C. R. Acad. Sci. Paris Sér. A, **291** (1980), 365—368 (cf. Zbl., **456**.10016).

[1980f] Reyssat, E., *Propriétés d'indépendance algébrique de nombres liés aux fonctions de Weierstrass*, Acta Arith., **41** (1982), 291—310.

[1981a] Philippon, P., *Indépendance algébrique de valeurs de fonctions exponentielles p-adiques*, J. reine angew. Math. (Crelle), **329** (1981), 42—51.

[1981b] Reyssat, E., *Un critère d'indépendance algébrique*, J. reine angew. Math. (Crelle), **329** (1981), 66—81.

[1981c] Endell, R., *Zur algebraischen Unabhängigkeit gewisser Werte der Exponentialfunktion (nach Chudnovsky)*, Diplomarbeit, Düsseldorf, 1981.

[1981d] Waldschmidt, M., *Sous-groupes analytiques de groupes algébriques*, Ann. of Math., **117** (1983), 627—657.

[1982a] Philippon, P., *Indépendance algébrique et variétés abéliennes*, C. R. Acad. Sci. Paris Sér. I, **294** (1982), 257—259.

[1982b] Philippon, P., *Variétés abéliennes et indépendance algébrique I*, Invent. Math., **70** (1983), 289—318.

[1982c] Wüstholz, G., *Sur l'analogue abélien du théorème de Lindemann*, C. R. Acad. Sci. Paris Sér. I, **295** (1982), 35—37.

[1982d] Masser, D. W. and G. Wüstholz, *Fields of large transcendence degree generated by values of elliptic functions*, Invent. Math., **72** (1983), 407—464.

[1982e] Philippon, P., *Variétés abéliennes et indépendance algébrique II: un analogue abélien du théorème de Lindemann-Weierstrass*, Invent. Math., **72** (1983), 389—405.

[1982f] Nesterenko, Ju. V., *On the algebraical independence of algebraic numbers to algebraic powers*, in: *Approximations diophantiennes et nombres transcendants*, Luminy 1982, Progress in Math., **31**, Birkhäuser (1983), 199—220.

[1982g] Wüstholz, G., *Über das abelsche Analogon des Lindemannschen Satzes I*, Invent. Math., **72** (1983), 363—388.

[1982h] Philippon, P., *Sous-groupes à n paramètres et indépendance algébrique*, in: *Approximations diophantiennes et nombres transcendants*, Luminy 1982, Progress in Math., **31**, Birkhäuser (1983), 221—234.

[1982i] Tubbs, R., *A transcendence measure for some special values of elliptic functions*, Proc. Amer. Math. Soc., **88** (1983), 189—196.

[1983a] *Transcendental number theory and its applications*, Proc. Conf. Moscov Univ. 2—4 Feb. 1983, Izd. Mosk. Univ. 1983.

[1983b] Philippon, P., *Pour une théorie de l'indépendance algébrique*, Thèse, Orsay, 1983.

[1983c] Masser, D. W., *Some recent results in transcendence theory*, Proc. Intern. Cong. Math., Warsaw, 1983.

[1983d] Waldschmidt, M. et Zhu Yao Chen, *Une généralisation en plusieurs variables d'un critère de transcendance de Gel'fond*, C. R. Acad. Sc. Paris, Sér. I, **297** (1983), 229—232.

Perspectives in Mathematics
Anniversary of Oberwolfach 1984
© Birkhäuser Verlag, Basel

Fixed Points of Symplectic Maps and a Classical Variational Principle for Forced Oscillations

EDUARD ZEHNDER

Abteilung für Mathematik, Ruhr-Universität Bochum,
Universitätsstraße 150, D-4630 Bochum 1 (FRG)

Introduction

Many fixed point results originate in the search for periodic solutions in celestial mechanics. By following, for example, every point x along its orbit over a fixed interval of time, a map $\phi^T(x)$ is defined whose fixed points are the initial conditions for periodic solutions having period T. Poincaré constructed other maps whose fixed points give rise to periodic solutions. In the special case of a Hamiltonian equation the maps considered belong to the restricted class of symplectic maps, which in dimension two, are the area preserving maps. For such maps the usual topological methods as for example degree theory and Lefschetz fixed point theory are in general not adequate to establish fixed points. In the following we shall describe a special fixed point problem belonging to the circle of old questions in celestial mechanics which was suggested by V. I. Arnold. By returning to tools well known in mechanics the fixed points will be established using methods from functional analysis and from topology. The result states that on the two dimensional torus every area preserving diffeomorphism homologous to the identity map possesses as many fixed points as a smooth function possesses critical points on the torus, provided the map leaves the center of gravity invariant.

In order to find fixed points for symplectic maps the so called generating function technique is commonly used. It goes back to Poincaré and has so far lead to many results, which are, however, perturbation results only. We proceed differently. Instead of looking for fixed points of maps we look for periodic solutions of a related Hamiltonian equation. These periodic solutions are the critical points of a classical variational principle on the infinite dimensional loopspace of the manifold. The functional is bounded neither from above nor from below. But as described in section 2, the required critical points can quite easily be found in different ways, for example by means of the classical Ljusternik-Schnirelman technique and the Morse theory or, alternatively, by means of a direct topological approach for continuous flows.

Our goal in this paper is not to give technical details, which can be found in [8], but rather to explain to nonspecialists the underlying ideas of the proof which was already presented last year during the Oberwolfach-meeting.

1   Fixed points for symplectic maps on tori and a conjecture by V. Arnold

Every continuous map f of S^2 homotopic to the identity map possesses at least one fixed point. This is of course well known and follows for example from the Lefschetz fixed point theory since the Lefschetz-number is

$$L(f) = L(id) = \chi(S^2) = 2 \neq 0 .$$

The map f may have only one fixed point. For example, the translation $z \to z + 1$ on $\mathbb{C}$ induces a map on the Riemann sphere whose only fixed point is the north pole. It is, however, a very striking fact, that f possesses at least 2 fixed points if it is in addition area preserving.

Theorem 1. *A homeomorphism f of S^2 homotopic to the identity, which preserves a regular measure μ has at least 2 fixed points. In particular every diffeomorphism of S^2, which leaves an area form invariant, $f^* \omega = \omega$, possesses at least 2 fixed points.*

Proof. The second statement is a consequence of the first one. In fact the map has degree 1 and hence is homotopic to the identity by Hopf's theorem. To prove the first statement assume p^* is the fixed point of f guaranteed by the Lefschetz theory. Then the map $g = f | S^2 \setminus p^*$ can be identified with a homeomorphism of the plane $\mathbb{R}^2$. If g had no fixed point, then by Brouwer's translation theorem there would be an open set U with the property that $g^j(U) = f^j(U)$ are mutually disjoint for all $j > 0$. Consequently

$$\sum_{j=0}^{n} \mu(f^j(U)) = (n+1)\, \mu(U) \leq \mu(S^2)$$

for every n and hence $\mu(U) = 0$, contradicting the assumption that $\mu(U) \neq 0$ for an open set U.   ●

The same argument, which was used by C. Loewner in the sixties in his Stanford lectures also shows that a measure preserving homeomorphism of the open disc possesses a fixed point. The proofs by C. Simon [17], and N. Nikishin [14] of the above theorem are different and use the fact that the index $j(p)$ of a fixed point 0 for an area preserving diffeomorphism in two dimensions is always ≤ 1, so that the fixed point formula $\chi(S^2) = \sum j(p)$ requires at least 2 fixed points. As an aside we observe that this theorem is strictly twodimensional — no higher dimensional analogue is known. The underlying phenomenon however, that mappings allow more fixed points, if they are restricted to the class of symplectic mappings, is not restricted to the dimension two. Similarly, many bifurcation results demonstrate that vectorfields admit more periodic solutions if they are restricted to be Hamiltonian vectorfields. In this connection it should be recalled that examples show that not every vectorfield on an odd dimensional sphere allows a periodic orbit in contrast to a conjecture by Seifert. It is conceivable, however, that in case of a Hamiltonian vectorfield a compact energysurface diffeomorphic to a sphere carries a periodic solution.

In contrast to S^2 the Lefschetz fixed point theory is not adequate to find fixed points of maps on the torus $T^2 = \mathbb{R}^2/\mathbb{Z}^2$ whose Euler-characteristic vanishes. If $f: T^2 \to T^2$ is a map on the torus with induced map f^* in $H^1(T^2)$ one finds for the Lefschetz number $L(f) = \det(1 - f^*)$. In particular $L(f) = 0$ if f is homologous to the identity map, so that one cannot expect a fixed point. There are indeed many maps on the torus without fixed points. Take for example a map g_1 on S^1 having an irrational rotation number, if then g_2 is any map of S^1, the map $f = g_1 \times g_2$ has no fixed point on T^2. Also, on the covering space $\mathbb{R}^2$ of the torus, the areapreserving translation map

$$X = x + c_1$$
$$Y = y + c_2$$

has clearly no fixed point on T^2, if $c = (c_1, c_2) \notin \mathbb{Z}^2$. The class of mappings on T^2 has therefore to be restricted, if they necessarily should possess fixed points.

In the following we consider measure preserving diffeomorphisms ψ of T^2, which are homologous to the identity, hence are, on $\mathbb{R}^2$ given by

$$\psi: \begin{array}{l} X = x + p(x, y) \\ Y = y + q(x, y) \end{array}$$

with two periodic functions p and q and require that ψ preserves the center of mass, hence excluding in particular the above translation map. Summarizing we require that ψ is a diffeomorphism of T^2 satisfying:

(i) ψ is homologous to id
(ii) $dX \wedge dY = dx \wedge dy$
(iii) $[p] = [q] = 0$,

where $[\ ]$ stands for the meanvalue of a function over the torus T^2. Due to (ii) the condition (iii) is indeed equivalent to

$$\int_{T^2} X\, dX \wedge dY = \int_{T^2} x\, dx \wedge dy$$

$$\int_{T^2} Y\, dX \wedge dY = \int_{T^2} y\, dx \wedge dy,$$

as one readily verifies.

One could try to find fixed points of ψ by means of the socalled generating function technique recalling that there is a relation between the fixed points of a symplectic map and the critical points of a function on the corresponding manifold. Indeed, following H. Poincaré in "Les méthodes nouvelles de la mécanique céleste" (Gauthiers Villars, Paris 1899, tome III, p. 214), we consider the oneform on $\mathbb{R}^2$:

$$(X-x)\,(\mathrm{d}\,Y+\mathrm{d}\,y)-(Y-y)\,(\mathrm{d}\,X+\mathrm{d}\,x)=\mathrm{d}\,S(x,y).$$

Due to the assumption (ii) this form is closed, hence, on $\mathbb{R}^2$, exact. Since $\mathrm{d}S$ is periodic, it has the form $S(x,y)=cx+dy+s(x,y)$, the function s being periodic. One checks readily that $c=d=0$ if and only if $[p]=[q]=0$. Therefore, the function S is periodic, hence a function on T^2. Obviously the fixed points of ψ are critical points of S, and conversely a critical point p^* of S is a fixed point of the map, if the forms $(\mathrm{d}\,Y+\mathrm{d}\,y)$ and $(\mathrm{d}\,X+\mathrm{d}\,x)$ are linearly independent at p^*. This is the case if and only if (-1) is not an eigenvalue of the Jacobian matrix $\mathrm{d}\psi(p^*)$, as one easily verifies. Since, by Ljusternik-Schnirelman, a function on T^2 possesses at least 3 critical points, we have verified the following

Statement: A diffeomorphism ψ of T^2 satisfying (i)−(iii) possesses at least 3 fixed points, *provided* (-1) is not an eigenvalue for $\mathrm{d}\psi(p)$, for all $p \in T^2$. (e. g. provided $|\psi - id|_{C^1}$ is sufficiently small).

The idea of relating fixed points of symplectic maps to critical points of a related function is being used quite frequently in order to establish existence results. For example, A. Weinstein [19] uses it in order to show that a symplectic diffeomorphism of a compact and simply connected manifold M possesses at least as many fixed points as a function on M has critical points, provided the map is sufficiently C^1-close to the identity map on M. For more general results and references we refer to J. Moser [13]. So far, however, the method has lead to perturbation results only and one may ask for more global results.

V. Arnold conjectured in [1] and in [2] that the above statement for a diffeomorphism of T^2 holds true globally, that is without the proviso. This is in fact true, but was proved only recently in [8]:

Theorem 2. *Every diffeomorphism ψ on T^2 satisfying* (i)−(iii) *possesses at least 3 fixed points. Moreover, if all the fixed points are nondegenerate, then ψ possesses at least four of them.*

A fixed point p is called nondegenerate, if 1 is not an eigenvalue of $\mathrm{d}\psi(p)$.
The proof of this statement which will be sketched later on uses quite a different idea. The fixed points will also be found as critical points of a function, which, however, is not defined on T^2 but on a high dimensional manifold $T^2 \times \mathbb{R}^{2N}$ and which originates in a variational problem for forced oscillations of a Hamiltionian equation on T^2. This has the advantage that one avoids the difficulty of the eigenvalues -1.

We point out that there are maps even close to the identity having 3 fixed points only. To see this we take a smooth function G on T^2 with precisely 3 critical points. An example of such a function is

$$G(x, y) = \sin \pi x \cdot \sin \pi y \cdot \sin \pi (x+y),$$

whose level lines look as follows:

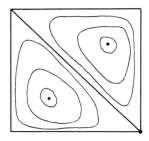

G takes positive values in the lower triangle and negative values in the upper triangle. Its critical points are the maximum, the minimum and a monkey saddle. Define now the map $\psi : (x, y) \to (X, Y)$ of T^2 implicitly by

$$X = x + \varepsilon \, \frac{\partial G}{\partial Y} \, (x, Y)$$

$$y = Y + \varepsilon \, \frac{\partial G}{\partial x} \, (x, Y).$$

If $\varepsilon > 0$ is sufficiently small, the map ψ is close to the identity, satiesfies (i) $-$ (iii) and has 3 fixed points, namely the critical points of G on T^2.

Historically the interest in theorem 2 comes from a dynamical problem, more precisely from the Poincaré-Birkhoff fixed point theorem, which has applications in celestial mechanics. In his search for periodic solutions in the restricted three body problem of celestial mechanics, H. Poincaré constructed a symplectic section map of an annulus A on the energy surface. This annulus is bounded by the socalled direct and retrograde periodic orbits. It lead him in 1912 to the formulation of the following theorem, see [15].

Every area preserving homeomorphism of an annulus $A = S^1 \times [a, b]$ rotating the two boundaries in opposite direction possesses at least 2 fixed points in the interior.

The strength of this theorem is that it provides at once infinitely many periodic points by applying to the iterates which leads to infinitely many periodic solutions in the application, see [6]. G. Birkhoff succeeded in 1913 to prove this theorem in an ingenious way, [4] and [5]. As pointed out by V. Arnold the

Poincaré-Birkhoff theorem could also be derived from theorem 2, at least in the differentiable case. Indeed take two copies of the annulus A and glue the boundaries together to get a torus T^2. Extending the diffeomorphism ψ on T^2 without adding any new fixed point and such that the conditions (i) – (iii) are met, one deduces from theorem 2 at least 2 fixed points on each "side" A, see [1].

Instead of looking for fixed points of the mapping ψ we rather search for periodic orbits of a dynamical system. For this purpose we have to construct a vectorfield with a flow ϕ^t such that $\phi^T = \psi$ for some T, for example $T = 1$. In fact it can be shown that the assumptions (i) – (iii) of the map ψ on T^2 are equivalent to $\psi = \phi^1$, where ϕ^t is the flow of a timedependent Hamiltonian vectorfield on T^2, i. e. satisfies:

$$\frac{d}{dt} \phi^t(x) = J \nabla h(t, \phi^t(x)) \quad \text{and} \quad \phi^0(x) = x,$$

where $x \in \mathbb{R}^2$ and where $h = h(t, x)$ is periodic in all its variables of period 1. This equivalence is not quite obvious and we refer to [8] for a proof, which is strictly two dimensional. As usual, the matrix J stands for the symplectic structure:

$$J = \begin{pmatrix} 0 & 1 \\ -1 & 0 \end{pmatrix}.$$

Clearly a periodic solution of the Hamiltonian equation on T^2 having period 1 gives rise to a fixed point of ψ. This way the problem of finding fixed points is reduced to the problem of finding periodic solutions of a Hamiltonian vectorfield. More generally, we can now consider Hamiltonian vectorfields on $T^{2n} = \mathbb{R}^{2n}/\mathbb{Z}^{2n}$ for any $n \geq 1$:

$$\dot{x} = J \nabla h(t, x), \ (t, x) \in \mathbb{R} \times \mathbb{R}^{2n}, \tag{1}$$

with $h \in C^2(\mathbb{R} \times \mathbb{R}^{2n})$ being periodic in all its variables of period 1.

Theorem 3 [8]. *Every Hamiltonian vectorfield* (1) *on* T^{2n} *possesses at least* $(2n + 1)$ *periodic solutions of period* 1. *Moreover, if all the* 1-*periodic solutions are nondegenerate, then there are at least* 2^{2n} *of them.*

Here a periodic solution is called nondegenerate, if none of its Floquet multipliers is equal to 1. This condition requires effectively, that the forced oscillations are isolated among such solutions. It will turn out that the periodic solutions found by the theorem belong to the same homotopy class of loops on T^{2n}, in fact they are all contractible. Other periodic solutions may not exist. This is in contrast to the closed geodesics on T^{2n}, where one finds very easily a closed geodesic in every homotopy class.

Theorem 2 is a consequence of Theorem 3 in the special case $n = 1$ as previously discussed. Also in higher dimensions a fixed point theorem for a restricted class of symplectic maps on T^{2n} follows, which is somewhat weaker than in the special case $n=1$. Let G be the identity component of the group $\text{Diff}^\infty (T^{2n}, \omega)$ of symplectic diffeomorphism of T^{2n} with respect to a symplectic structure ω. It can be shown to be the identity component by smooth arcs in $\text{Diff}^\infty (T^{2n}, \omega)$. For a symplectic diffeomorphism ψ the following statements are equivalent:

(1) ψ is the time 1 map of the flow of an exact Hamiltonian vectorfield on T^{2n}.

(2) ψ belongs to the commutator subgroup of G.

(3) The Calabi-invariant of ψ vanishes.

In fact this equivalence holds true for any compact symplectic manifold as was proved by A. Banyaga [3]. The following statement now follows immediately from Theorem 3.

Theorem 4. Let ψ be a symplectic diffeomorphism of T^{2n} with respect to the standard symplectic structure, which meets one of the equivalent conditions (1) $-(3)$. Then ψ possesses at least $2n+1$ fixed points. If, moreover, all the fixed points of ψ are nondegenerate, then there are at least 2^{2n} of them.

In the special case $n=1$ the statement of the theorem holds true for every symplectic structure. This follows in fact from J. Moser's theorem [11] that two volume forms are equivalent if they have the same total volume. For $n>1$ the classification of symplectic forms according to equivalence is not understood. Nevertheless, the statement of theorem 3 is probably true for symplectic structures which are not equivalent to the standard symplectic structure J considered. More generally, on every compact symplectic manifold M one might hope to find at least as many forced oscillations for a Hamiltonian vectorfield depending periodically on time, as a function on M has critical points. Under an additional smallness condition requiring the Hamiltonian vectorfield to be sufficiently C^1-small this is not difficult to prove. By means of the ideas of the proof of theorem 3 it is actually sufficient to assume a smallnes condition in the C^0 topology only. This has been observed by A. Weinstein [18].

2   The variational principle for forced oscillations

The forced oscillations claimed in theorem 3 will be found by means of a classical variational principle for which the periodic orbits are the critical points. In general the flow of a Hamiltonian vectorfield is very complicated since solutions of quite different behaviour over an infinite interval of time are mixed. There is, however, a variational principle which picks out precisely the periodic solutions among all the solutions, thereby avoiding the complexity of the flow. Consider the contractible loops on T^{2n} which are, on the covering space $\mathbb{R}^{2n}$

described by the periodic functions $t \to x(t) \in \mathbb{R}^{2n}$ with $x(0) = x(1)$. The action functional on the periodic functions is then defined to be

$$f(x) = \int\limits_0^1 \left\{ \frac{1}{2} \langle \dot{x}, Jx \rangle - h(t, x(t)) \right\} dt.$$

We claim that the critical points of f are the required periodic solutions of the Hamiltonian equation having period 1. Indeed for the derivative one finds:

$$f'(x)y = \frac{d}{d\varepsilon} f(x + \varepsilon y)|_{\varepsilon = 0} = \int\limits_0^1 \langle -J\dot{x} - \nabla h(t, x(t)), y \rangle dt$$

$$=: (\nabla f(x), y),$$

and one sees that $f'(x) = 0$ precisely if x satisfies $\nabla f(x) = -J\dot{x} + \nabla h(t, x) = 0$, hence, with $J^2 = -1$, if x satiesfies the equation (1) as claimed.

It remains to find critical points of f. Here one is confronted with the difficulty that f is bounded neither from below nor from above so that standard variational methods do not apply directly. All the critical points for example are saddle points having infinite dimensional stable und unstable invariant manifolds, which can be seen from the Hessian of f at a critical point x:

$$f''(x)(y_1, y_2) = \int\limits_0^1 \langle -J\dot{y}_1 - h_{xx}(t, x(t)) y_1, y_2 \rangle dt.$$

This is in sharp contrast to the energy functional for closed geodesics on a Riemannian manifold, which is bounded from below. There are of course other variational principles for periodic solutions of Hamiltonian systems which are coercive, but they require conditions on the Hamiltonian, as for example convexity, not satisfied in our case; we refer to I. Ekeland [10] and the references therein. That in the indefinite case the variational approach can be used effectively for existence proofs was first demonstrated by P. Rabinowitz [16] and subsequently used by many authors. It turns out that in our special case the analytical difficulties are only minor due to the fact, that the function h is periodic.

In view of the periodicity of h the function f satisfies $f(x + j) = f(x)$ for every constant $j \in \mathbb{Z}^{2n}$ and for every loop x. Splitting $x = [x] + \xi$ into its mean-value $[x] \in \mathbb{R}^{2n}$ over a period and the remainder we can identify the meanvalues with points on the torus $T^{2n} = \mathbb{R}^{2n}/\mathbb{Z}^{2n}$ and can view the function f as a function on $T^{2n} \times E$:

$$f: T^{2n} \times E \to R,$$

where E is the linear space of periodic functions having meanvalue zero. From a geometrical point of view it can be advantageous in order to find critical points of a functional f to study the rather artifical gradient flow $\frac{d}{ds} x = -\nabla f(x)$, whose restpoints are the critical points of f. This is an ordinary differential equation on an infinite dimensional space but of very special structure so that the bounded solutions tend in forward and backward time to the set of rest points.

For example, if in our action functional $h = 0$ then the gradient flow $\frac{d}{ds} x = J\dot{x}$ is linear and one sees that the torus T^{2n} is a hyperbolic invariant manifold consisting of the rest points. Since h and its derivatives are bounded, the dominant term in f is in fact its quadratic part. This can be used in order to reduce the problem of finding critical points of f on the infinite dimensional space $T^{2n} \times E$ to the problem of finding the critical points of a related function $g \in C^2(T^{2n} \times \mathbb{R}^{2N}, \mathbb{R})$ which is defined on a finite dimensional submanifold $T^{2n} \times \mathbb{R}^{2N}$. Here N is a large integer depending on the C^2-size of h.

The approximation by finite dimensional problems is a technical device often used in variational problems. One is reminded of the approximation by broken geodesics; also in his existence proof of normal modes in a neighborhood of an equilibrium point J. Moser [12] reduces the action functional to a finite dimensional problem by means of an averageing procedure. In our case the reduction is a standard but global Lyapunov-Schmidt reduction. We do not carry it out but refer to [8].

The critical points of g will now be found as the rest points of the gradient flow

$$\frac{d}{ds} z = -\nabla g(z), \quad z \in T^{2n} \times \mathbb{R}^{2N}, \tag{2}$$

which are easily localized. Indeed if we set $z = ([z], \xi) \in T^{2n} \times \mathbb{R}^{2N}$, the flow is, more explicitly, of the following form

$$\frac{d}{ds} [z] = v_0(z)$$

$$\frac{d}{ds} \xi = A\xi + v_1(z),$$

with a differentiable and uniformly bounded vectorfield $v = (v_0, v_1)$ which vanishes with h. Moreover, there is an invariant splitting $\xi = (\xi_+, \xi_-)$ for the linear map A such that $A\xi = (A_+\xi_+, A_-\xi_-)$ satisfies $(A_+\xi_+, \xi_+) \geq 2\pi|\xi_+|^2$ and $(A_-\xi_-, \xi_-) \leq -2\pi|\xi_-|^2$. From this one finds immediately the existence of a constant $K > 0$ satisfying

$$\frac{d}{ds}|\xi_+|^2 \geq 1 \qquad \text{if} \quad |\xi_+| \geq K$$

$$\frac{d}{ds}|\xi_-|^2 \leq -1 \quad \text{if} \quad |\xi_-| \geq K.$$

In order to interpret these estimates geometrically we define the compact sets

$$\begin{aligned}
B &= T^{2n} \times D_1 \times D_2 \\
B^- &= T^{2n} \times \partial D_1 \times D_2 \\
B^+ &= T^{2n} \times D_1 \times \partial D_2,
\end{aligned} \qquad (3)$$

with the discs $D_1 = \{\xi_+ \in \mathbb{R}^N \mid |\xi_+| \leq K\}$ and $D_2 = \{\xi_- \in \mathbb{R}^N \mid |\xi_-| \leq K\}$. One sees that the flow leaves the set B through the boundary part B^- of ∂B in forward time and through B^+ in backward time. Hence B^- is called the strict exit set of B, similarly B^+ is called the strict entrance set of B. The set B is an example of an isolating block in the sense of C. Conley [9]. Moreover, it is clear from the estimates that the set of bounded solutions, hence in particular the rest points, are contained in the interior of B.

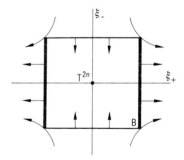

There are now several ways to find the rest points in B. They can be found for example by applying the Ljusternik-Schnirelman minimax theory and the Morse-theory to the gradient flow in (B, B^-), recalling that B^- is the exit set. If for example $[h] \in H_*(B, B^-)$ is a nonzero singular homology class, then

$$c([h]) = \inf_{h \in [h]} \sup_{z \in |h|} g(z)$$

is a critical value of g, where $|h|$ denotes the support of the singular chain $h \in [h]$.

That one finds sufficiently many classes follows from $H^*(B, B^-) \cong H^*(T^{2n})$ $\otimes H^*(S^N, *)$, by the Künneth formula, so that

$$H^j(B, B^-) \cong H^{j-N}(T^{2n}). \tag{4}$$

From this one can deduce a socalled chain of subordinated homology classes having length $2n+1$ and therefore $(2n+1)$ critical points of g. These correspond to the required $(2n+1)$ forced oscillations.

If there are only finitely many rest points, z_j, $1 \leq j \leq k$ in B, then the algebraic invariants of (B, B^-), represented by the Poincaré-polynomial $p(t, (B, B^-))$, are related to the local algebraic invariants $p(t, z_j)$ of the rest points by the Morse inequalities:

$$\sum_{j=1}^{k} p(t, z_j) - p(t, (B, B^-)) = (1+t)\, q(t),$$

q being a polynomial having nonnegative integer coefficients. In case the critical points are nondegenerate, $p(t, z_j) = t^{d_j}$, with d_j being the Morse index of the critical point so that in view of (4) the Morse-inequalities become

$$\sum_{j=1}^{k} t^{d_j} - \sum_{j=0}^{2n} \binom{2n}{j} t^{j+N} = (1+t)\, q(t).$$

We find in particular for the number of critical points $k \geq 2^{2n}$, which is equal to the sum of the Betti-numbers of T^{2n}. Since it can be shown that a critical point z_j of g is nondegenerate precisely if the corresponding 1-periodic solution has no Floquet-multiplier equal to 1, the statement of theorem 3 follows.

Alternatively we shall next describe a direct and more flexible approach to determine the structure of the set of bounded solutions of the gradient flow in B. The method is in particular applicable to topological flows which are not necessarily gradient like. We first claim that in fact every topological flow defined in a neighborhood of B contains an invariant set S in B whose topology is inherited from that of the surrounding B, provided only it behaves at the boundary like the gradient flow as in the above picture. More precisely the following statement is proved in [8] by using properties of a flow and duality theorems in algebraic topology.

Theorem 5. *Every topological flow near B having B^- as strict exit set and B^+ as strict entrance set has an invariant set S in B, $S = \{\gamma \in B \mid \gamma \cdot t \in B$ for all $t \in \mathbb{R}\}$, for which*

$$l(S) \geq l(B) = l(T^{2n}) = 2n+1.$$

Here $l(X)$ stands for the cup long of a compact space X. If $H^*(X)$ is the Alexander cohomology it is defined as

$$l(X) = 1 + \sup \{k \in \mathbb{N} \mid \text{there are classes } \alpha_1, \ldots, \alpha_k \in H^*(X) \setminus 1 \text{ with } \alpha_1 \cup \ldots \cup \alpha_k \neq 0\},$$

and $l(X) = 1$ if no such class exists.

It should be pointed out, that it is, in general not clear how the cohomology of an invariant set S of a flow is related to that of an isolating block around it. The information, for example, given by the Morse-index as generalized to isolated invariant sets in [9] is not sufficient; one can make a degenerate critical point having the same index as the set S under considerations. In the above theorem considerably finer information is used about the way B^+ and B^- sit in B. Of course in the quite different case of a positively invariant block where the exit set is empty, in which case S is an attractor, the cohomology of S and of B is the same. This fact also helps to clarify a difference between the variational problem of closed geodesics on a Riemannian manifold and that for periodic solutions of general Hamiltonian systems. The first case, the functional being bounded from below, is an attractor situation and each cohomology class in the loop space is represented in the index of some critical point. In the free Hamiltonian case this is obviously not true, as in the torus example only the cohomology of the manifold itself has to be represented. An elementary example should illustrate this. The following gradient flow in the neighborhood of the annulus A has only one critical point P. It is hyperbolic and it is the invariant set S contained in A. Here $l(A) = 2 > l(S) = 1$. In case, however, the annulus is positively invariant one concludes from the existence of an attracting critical point P a second critical point Q as illustrated by the following picture:

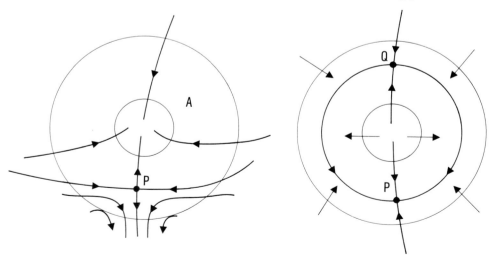

Going back to theorem 5 it remains to count the rest points of the flow in S. We first recall the useful concept of a Morse-decomposition which in this generality is found in [9].

Definition: Let S be a compact invariant set of a continuous flow. A Morse-decomposition of S is a finite collection $\{M_p\}_{p \in P}$ of disjoint compact and invariant subsets of S which can be ordered, say $\{M_1, M_2, \ldots, M_k\}$ so that the following property holds true. If

$$\gamma \in S \setminus \bigcup_{p \in P} M_p,$$

then there is a pair of indices $i < j$ such that the positive $(t \to +\infty)$ and the negative $(t \to -\infty)$ limit sets $\omega(\gamma)$ and $\omega^*(\gamma)$ of γ satisfy

$$\omega(\gamma) \subset M_i \quad \text{and} \quad \omega^*(\gamma) \subset M_j.$$

We illustrate this concept by the special but familiar example of the gradient flow $\dot{x} = -\nabla f(x)$ of a function f on a Riemannian manifold M. If the flow has only finitely many restpoints, which are of course the critical points of f, then every orbit tends in forward and in backward time to a restpoint due to the gradient structure of the flow. These restpoints, therefore, constitute a Morse decomposition of the invariant set S which is here the manifold M itself. An ordering of the critical points is simply an ordering of their critical values.

From the definition of a Morse-decomposition one deduces quite easily the following estimate, see [8]:

$$l(S) \le \sum_{p \in P} l(M_p).$$

In the previous example of the gradient flow on M the sets M_p are points and hence have cup long equal to 1 and we deduce for the number of critical points of f on M the familiar estimate

$$l(M) \le \sum_P 1 = \# \{\text{critical points of } f\}.$$

The same argument applies to every compact invariant set S of a continuous flow, which is on S gradientlike, i.e. for which there is a continuous real valued function on S which is strictly decreasing on nonconstant orbits. Indeed if there are only finitely many rest points in S, they constitute a Morse-decomposition of S, since the flow is gradientlike. As above one sees that $l(S)$ is a lower bound of the number of restpoints.

If now S is, in particular, the set of bounded solutions of the gradientflow

$$\frac{d}{ds} z = -\nabla g(z) \text{ on } T^{2n} \times \mathbb{R}^{2N},$$ then S is a compact, invariant and gradientlike set satisfying $l(S) \geqq 2n+1$ with regard to theorem 5. We therefore find in a different way that g has at least $2n+1$ critical points.

Concluding Remarks

The proofs of theorem 3 outlined above make heavy use of a global coordinate system on which moreover the symplectic structure is constant. For this reason the quadratic part of the action functional dominates the behaviour of the gradient flow so that an isolating block for the set of bounded solutions is easily found for every Hamiltonian function. On an arbitrary symplectic and compact manifold M such coordinates do exist only locally, but the actionfunctional for $h = 0$ still dominates in a neighborhood of the constant loops if the Hamiltonian vectorfield is sufficiently C^0 small. This way A. Weinstein [18] succeeded to establish for C^0-small Hamiltonian vectorfields the existence of at least cup long (M) forced oscillations.

Finally it should be pointed out that the above approach which turns a problem of symplectic geometry into a variational problem on an infinite dimensional space is applicable to global intersection problems of Lagrangian manifolds. We do not go into this but refer to M. Chaperon [7].

I would like to thank J. Moser and C. Conley for valuable discussions and the Stiftung Volkswagenwerk for its support.

References

[1]   Arnold, V. I., *Mathematical methods of classical mechanics.* (Appendix 9) Berlin— Heidelberg—New York: Springer 1978.

[2]   Arnold, V. I., *Proceedings of symposia in pure mathematics.* Vol. XXVIII A. M. S., p. 66, 1976.

[3]   Banyagá, A., *Sur la structure du groupe des difféomorphismes qui préservent une forme symplectique.* Comment. Math. Helvetici **53** (1978), 174—227.

[4]   Birkhoff, G. D., *Proof of Poincaré's Geometric Theorem.* Trans. Amer. Math. Soc. **14** (1913), 14—22.

[5]   Birkhoff, G. D., *An Extension of Poincaré's Last Geometric theorem.* Acta Math. **47** (1925).

[6]   Birkhoff, G. D., *The restricted problem of three bodies.* Rend. Circolo Mat. Palermo **39** (1915), 265—334.

[7]   Chaperon, M., *Quelques Questions de Géométrie symplectique.* Séminaire Bourbaki 1982/83, n° 610, to appear in Astérisque.

[8]   Conley, C., and Zehnder, E., *The Birkhoff-Lewis Fixed point theorem and a Conjecture of V. I. Arnold.* Invent. math. **73** (1983), 33—49.

[9]   Conley, C., *Isolated invariant sets and the Morse index.* CBMS, Regional Conf. Series in Math. Vol. **38** (1978).

[10]   Ekeland, I., *Une théorie de Morse pour les systèmes hamiltoniens convexes*. Ann. Inst. Henri Poincaré, Analyse non linéaires, Vol. **1** (1984), 19—78.

[11]   Moser, J., *On the volume elements on a manifold*. Transactions Amer. Math. Soc. **120** (1965), 286—294.

[12]   Moser, J., *Periodic orbits near an equilibrium and a theorem by Alan Weinstein*. Comm. Pure Appl. Math. **29** (1976), 727—747.

[13]   Moser, J., *A fixed point theorem in symplectic geometry*. Acta Math. **141** (1978), 17—34.

[14]   Nikishin, N., *Fixed points of diffeomorphisms on the twosphere that preserve area*. Funkcional Anal. i Prelozen **8** (1974), 84—85.

[15]   Poincaré, H., *Sur un théorème de Géométrie*. Rend. Circolo Mat. Palermo **33** (1912), 375—407.

[16]   Rabinowitz, P., *Periodic solutions of Hamiltonian systems*. Comm. Pure Appl. Math. **31** (1978), 157—184.

[17]   Simon, C. P., *A bound for the Fixed point Index of an Areapreserving map with Applications to Mechanics*. Invent. math. **26** (1974), 187—200.

[18]   Weinstein, A., C^0 *perturbation theorems for symplectic fixed points and Lagrangian intersections*. Lecture Notes of the AMS summer Institute on nonlinear Functional Analysis and Applications, Berkeley 1983.

[19]   Weinstein, A., *Lectures on symplectic manifolds*. CBMS Regional Conference series in math. Vol. **29** (1977).